21ˢᵗ Century Nanoscience –

21st Century Nanoscience – A Handbook

Bioinspired Systems and Methods (Volume Seven)

Edited by

Klaus D. Sattler

CRC Press
Taylor & Francis Group
Boca Raton London New York

CRC Press is an imprint of the
Taylor & Francis Group, an **informa** business

CRC Press
Taylor & Francis Group
6000 Broken Sound Parkway NW, Suite 300
Boca Raton, FL 33487-2742

First issued in paperback 2022

© 2020 by Taylor & Francis Group, LLC

CRC Press is an imprint of Taylor & Francis Group, an Informa business

No claim to original U.S. Government works

ISBN-13: 978-0-815-35703-2 (hbk)
ISBN-13: 978-1-03-233650-3 (pbk)
DOI: 10.1201/9780429351525

Library of Congress Cataloging-in-Publication Data

Names: Sattler, Klaus D., editor.
Title: 21st century nanoscience : a handbook / edited by Klaus D. Sattler.
Description: Boca Raton, Florida : CRC Press, [2020] | Includes bibliographical references and index. | Contents: volume 1. Nanophysics sourcebook—volume 2. Design strategies for synthesis and fabrication—volume 3. Advanced analytic methods and instrumentation—volume 5. Exotic nanostructures and quantum systems—volume 6. Nanophotonics, nanoelectronics, and nanoplasmonics—volume 7. Bioinspired systems and methods. | Summary: "This 21st Century Nanoscience Handbook will be the most comprehensive, up-to-date large reference work for the field of nanoscience. Handbook of Nanophysics, by the same editor, published in the fall of 2010, was embraced as the first comprehensive reference to consider both fundamental and applied aspects of nanophysics. This follow-up project has been conceived as a necessary expansion and full update that considers the significant advances made in the field since 2010. It goes well beyond the physics as warranted by recent developments in the field"—Provided by publisher.
Identifiers: LCCN 2019024160 (print) | LCCN 2019024161 (ebook) | ISBN 9780815384434 (v. 1 ; hardback) | ISBN 9780815392330 (v. 2 ; hardback) | ISBN 9780815384731 (v. 3 ; hardback) | ISBN 9780815355281 (v. 4 ; hardback) | ISBN 9780815356264 (v. 5 ; hardback) | ISBN 9780815356417 (v. 6 ; hardback) | ISBN 9780815357032 (v. 7 ; hardback) | ISBN 9780815357070 (v. 8 ; hardback) | ISBN 9780815357087 (v. 9 ; hardback) | ISBN 9780815357094 (v. 10 ; hardback) | ISBN 9780367333003 (v. 1 ; ebook) | ISBN 9780367341558 (v. 2 ; ebook) | ISBN 9780429340420 (v. 3 ; ebook) | ISBN 9780429347290 (v. 4 ; ebook) | ISBN 9780429347313 (v. 5 ; ebook) | ISBN 9780429351617 (v. 6 ; ebook) | ISBN 9780429351525 (v. 7 ; ebook) | ISBN 9780429351587 (v. 8 ; ebook) | ISBN 9780429351594 (v. 9 ; ebook) | ISBN 9780429351631 (v. 10 ; ebook)
Subjects: LCSH: Nanoscience—Handbooks, manuals, etc.
Classification: LCC QC176.8.N35 A22 2020 (print) | LCC QC176.8.N35 (ebook) | DDC 500—dc23
LC record available at https://lccn.loc.gov/2019024160
LC ebook record available at https://lccn.loc.gov/2019024161

Visit the Taylor & Francis Web site at
http://www.taylorandfrancis.com

and the CRC Press Web site at
http://www.crcpress.com

Contents

Editor

Klaus D. Sattler pursued his undergraduate and master's courses at the University of Karlsruhe in Germany. He earned his PhD under the guidance of Professors G. Busch and H.C. Siegmann at the Swiss Federal Institute of Technology (ETH) in Zurich. For three years he was a Heisenberg fellow at the University of California, Berkeley, where he initiated the first studies with a scanning tunneling microscope of atomic clusters on surfaces. Dr. Sattler accepted a position as professor of physics at the University of Hawaii, Honolulu, in 1988. In 1994, his group produced the first carbon nanocones. His current work focuses on novel nanomaterials and solar photocatalysis with nanoparticles for the purification of water. He is the editor of the sister references, *Carbon Nanomaterials Sourcebook* (2016) and *Silicon Nanomaterials Sourcebook* (2017), as well as *Fundamentals of Picoscience* (2014). Among his many other accomplishments, Dr. Sattler was awarded the prestigious Walter Schottky Prize from the German Physical Society in 1983. At the University of Hawaii, he teaches courses in general physics, solid state physics, and quantum mechanics.

Contributors

Anja Aarva
Department of Electrical Engineering
    and Automation
School of Electrical Engineering
Aalto University
Espoo, Finland

Christopher F. Adams
School of Life Sciences
Faculty of Natural Sciences
Keele University
Staffordshire, United Kingdom

Mohammad Sadegh Amiri
Department of Biology
Payame Noor University
Tehran, Iran

Sohrab Behnia
Department of Physics
Urmia University of Technology
Urmia, Iran

Raghvendra A. Bohara
Centre for Interdisciplinary Research
D. Y. Patil University
Kolhapur, Maharashtra, India

Miguel Caro
Department of Electrical Engineering
    and Automation
School of Electrical Engineering
Aalto University
Espoo, Finland
and
QTF Centre of Excellence, Quantum
    Technology Finland
Department of Applied Physics
Aalto University
Espoo, Finland

Bridget Crawford
Fitzpatrick Institute for Photonics
Duke University
Durham, North Carolina
and
Department of Biomedical
    Engineering
Duke University
Durham, North Carolina

Apurba K. Das
Department of Chemistry
Indian Institute of Technology Indore
Indore, Madhya Pradesh, India

Majid Darroudi
Nuclear Medicine Research Center
Mashhad University of Medical
    Sciences
Mashhad, Iran
and
Department of Modern Sciences and
    Technologies
School of Medicine
Mashhad University of Medical
    Sciences
Mashhad, Iran

Mohamed S. Draz
Division of Engineering in Medicine
Department of Medicine
Brigham and Women's Hospital
Harvard Medical School
Boston, Massachusetts
and
Faculty of Science
Tanta University
Tanta, Egypt

Kelong Fan
Key Laboratory of Protein and
    Peptide Pharmaceuticals
CAS-University of Tokyo Joint
    Laboratory of Structural Virology
    and Immunology
Institute of Biophysics
Chinese Academy of Sciences
Beijing, China

Samira Fathizadeh
Department of Physics
Urmia University of Technology
Urmia, Iran

Alexey Ferapontov
Biophysical Immunology Laboratory
Department of Biomedicine
Aarhus University & Interdisciplinary
    Nanoscience Center (iNANO)
Aarhus University
Aarhus, Denmark

Ilia Gelfat
Wyss Institute for Biologically
    Inspired Engineering
Harvard University
Boston, Massachusetts
and
School of Engineering and Applied
    Sciences
Harvard University
Cambridge, Massachusetts

Tapas Ghosh
Department of Chemistry
Indian Institute of Technology Indore
Indore, Madhya Pradesh, India

Ali Golchin
Department of Tissue Engineering
    and Applied Cell Sciences
School of Advanced Technologies in
    Medicine
Urmia University of Medical Sciences
Tehran, Iran

Cristiana Gonçalves
3B's Research Group, I3Bs – Research
    Institute on Biomaterials,
    Biodegradables and Biomimetics
University of Minho
Guimarães, Portugal
and
ICVS/3B's–PT Government Associate
    Laboratory
Braga/Guimarães, Portugal
and
The Discoveries Centre for
    Regenerative and Precision
    Medicine
University of Minho
Guimarães, Portugal

Victoria S. Haritos
Department of Chemical Engineering
Monash University
Clayton, Victoria, Australia

Lizhong He
Department of Chemical Engineering
Monash University
Clayton, Victoria, Australia

Simzar Hosseinzadeh
Department of Tissue Engineering
    and Applied Cell Sciences
School of Advanced Technologies in
    Medicine
Shahid Beheshti University of Medical
    Sciences
Tehran, Iran

Stuart I. Jenkins
School of Medicine, Institute of
    Science and Technology in
    Medicine
Keele University
Staffordshire, United Kingdom

Neel S. Joshi
Wyss Institute for Biologically
    Inspired Engineering
Harvard University
Boston, Massachusetts
and
School of Engineering and Applied
    Sciences
Harvard University
Cambridge, Massachusetts

Kristian Juul-Madsen
Biophysical Immunology Laboratory
Department of Biomedicine
Aarhus University & Interdisciplinary
    Nanoscience Center (iNANO)
Aarhus University
Aarhus, Denmark

Parisa Kangari
Department of Tissue Engineering
    and Applied Cell Sciences
School of Advanced Technologies in
    Medicine
Shiraz University of Medical Sciences
Shiraz, Iran

Tomi Laurila
Department of Electrical Engineering
    and Automation
School of Electrical Engineering
Aalto University
Espoo, Finland

Youjin Lee
Department of Chemistry
The University of Chicago
Chicago, Illinois

Wuled Lenggoro
Department of Chemical Engineering
Tokyo University of Agriculture and
    Technology
Tokyo, Japan

Qian Liang
Key Laboratory of Protein and
    Peptide Pharmaceuticals
CAS-University of Tokyo Joint
    Laboratory of Structural Virology
    and Immunology
Institute of Biophysics
Chinese Academy of Sciences
Beijing, China

Juewen Liu
Department of Chemistry
Waterloo Institute for Nanotechnology
University of Waterloo
Waterloo, Ontario, Canada

Anand Lopez
Department of Chemistry
Waterloo Institute for Nanotechnology
University of Waterloo
Waterloo, Ontario, Canada

Avinash Manjula-Basavanna
Wyss Institute for Biologically
    Inspired Engineering
Harvard University
Boston, Massachusetts
and
School of Engineering and Applied
    Sciences
Harvard University
Cambridge, Massachusetts

Faeza Moradi
Department of Tissue Engineering
Faculty of Medical Sciences
Tarbiat Modares University
Tehran, Iran

Sepideh Mousazadehe
Department of Tissue Engineering &
    Regenerative Medicine
Faculty of Advanced Technologies in
    Medicine
Iran University of Medical Sciences
Tehran, Iran

Nayeem A. Mulla
Centre for Interdisciplinary Research
D. Y. Patil University
Kolhapur, Maharashtra, India

Peter Q. Nguyen
Wyss Institute for Biologically
    Inspired Engineering
Harvard University
Boston, Massachusetts
and
School of Engineering and Applied
    Sciences
Harvard University
Cambridge, Massachusetts

Shinji Nozaki
Graduate School of Informatics and
    Engineering
The University of
    Electro-Communications
Tokyo, Japan

Isabel M. Oliveira
3B's Research Group, I3Bs – Research
    Institute on Biomaterials,
    Biodegradables and Biomimetics
University of Minho
Guimarães, Portugal
and
ICVS/3B's–PT Government Associate
    Laboratory
Braga/Guimarães, Portugal

Joaquim M. Oliveira
3B's Research Group, I3Bs – Research
    Institute on Biomaterials,
    Biodegradables and Biomimetics
University of Minho
Guimarães, Portugal
and
ICVS/3B's–PT Government Associate
    Laboratory
Braga/Guimarães, Portugal
and
The Discoveries Centre for
    Regenerative and Precision
    Medicine
University of Minho
Guimarães, Portugal

Geoffrey A. Ozin
Department of Chemistry
University of Toronto
Toronto, Ontario, Canada

Shivaji H. Pawar
Centre for Interdisciplinary Research
D. Y. Patil University
Kolhapur, Maharashtra, India
and
Centre for Innovative and Applied
    Research
Anekant Education Society
Baramati, Maharashtra, India

Andrew Phillips
Department of Chemistry
The University of Chicago
Chicago, Illinois

Pichet Praveschotinunt
Wyss Institute for Biologically
    Inspired Engineering
Harvard University
Boston, Massachusetts
and
School of Engineering and Applied
    Sciences
Harvard University
Cambridge, Massachusetts

Chenxi Qian
Division of Chemistry and Chemical
    Engineering
California Institute of Technology
Pasadena, California

Rui L. Reis
3B's Research Group, I3Bs – Research
    Institute on Biomaterials,
    Biodegradables and Biomimetics
University of Minho
Guimarães, Portugal
and
ICVS/3B's–PT Government Associate
    Laboratory
Braga/Guimarães, Portugal
and
The Discoveries Centre for
    Regenerative and Precision
    Medicine

University of Minho
Guimarães, Portugal

Max Rose
Deutsches Elektronen-Synchrotron
    DESY
Hamburg, Germany

Suryani Saallah
Biotechnology Research Institute
Universiti Malaysia Sabah
Kota Kinabalu, Malaysia

Sachindra Nath Sarangi
Institute of Physics
P.O.: Sainik School
Bhubaneswar, Odisha, India

Bhuvana K. Shanbhag
Department of Chemical Engineering
Monash University
Clayton, Victoria, Australia

Yiwei Tang
Faculty of Food Science and
    Technology
Agricultural University of Hebei
Baoding, China

Bozhi Tian
Department of Chemistry
The University of Chicago
Chicago, Illinois

Ivan A. Vartanyants
Deutsches Elektronen-Synchrotron
    DESY
Hamburg, Germany
and
National Research Nuclear University
    MEPhI (Moscow Engineering
    Physics Institute)
Moscow, Russia

Tuan Vo-Dinh
Fitzpatrick Institute for Photonics
Duke University
Durham, North Carolina
and

Department of Biomedical
    Engineering
Duke University
Durham, North Carolina
and
Department of Chemistry
Duke University
Durham, North Carolina

Thomas Vorup-Jensen
Biophysical Immunology Laboratory
Department of Biomedicine
Aarhus University & Interdisciplinary
    Nanoscience Center (iNANO)
Aarhus University
Aarhus, Denmark

Xiyun Yan
Key Laboratory of Protein and
    Peptide Pharmaceuticals
CAS-University of Tokyo Joint
    Laboratory of Structural Virology
    and Immunology
Institute of Biophysics
Chinese Academy of Sciences
Beijing, China

**Mohammad Ehsan Taghavizadeh
Yazdi**
Neurogenic Inflammation Research
    Center
Mashhad University of Medical
    Sciences
Mashhad, Iran

Pengfei Zhang
Tongji University School of Medicine
Shanghai Skin Disease Hospital
Shanghai

Ruofei Zhang
Key Laboratory of Protein and
    Peptide Pharmaceuticals
CAS-University of Tokyo Joint
    Laboratory of Structural Virology
    and Immunology
Institute of Biophysics
Chinese Academy of Sciences
Beijing, China

Plant-Mediated Biosynthesis of Nanoparticles

Majid Darroudi and Mohammad
Ehsan Taghavizadeh Yazdi
Mashhad University of Medical Sciences

Mohammad Sadegh Amiri
Payame Noor University

1.1 Introduction

The major solicitous idea that nanotechnology has to deal with has been shown to be the synthesis of nanoparticles (NPs) that are of varied sizes and shapes, have different chemical compositions and controlled dispersity, and are ultimately capable of being utilized for human benefit. Considering the already-existing chemical and physical techniques capable of successfully producing pure and well-defined NPs, they are quite costly and can have potentially hazardous effects on the environment. Lately, progress in the use of efficient green chemistry procedures in the synthesis of metal NPs has turned into the major focal point of researchers. Researches and enquiries have been done in order to discover an eco-friendly method for assessing the composition of well-characterized NPs. The arrangement of metal NPs through the use of organisms is one of the most noted procedures. Plants, being one such organism, have captured the title of most acceptable for the large-scale biosynthesis of NPs. When compared with the NPs originated from microorganisms, those originated from plants have proven to be more balanced and firm, along with possessing much quicker synthesis times. Additionally, these particular NPs have been observed to be more diverse in their shape and size when compared with the ones that are produced by other organisms. The existing benefits and superiority of utilizing plant and plant-derived substances in the biosynthesis of metal NPs have tempted many researchers to examine the mechanisms of metal ions uptake and bioreduction by plants, as well as to better comprehend the feasible mechanism of metal NP genesis in plants.

1.2 Importance of Plants in the Biosynthesis of NPs

As one of the most promising technologies in the world, nanotechnology is mainly associated with biological and nonbiological (physical and chemical) approaches, since it leads to the discovery and emergence of novel therapeutic nano-sized materials for the purpose of biomedical and pharmaceutical implementations [1]. Due to the inducement of environmental pollution as a result of utilizing heavy metals, which affect living organisms with their high toxicity, different nonbiological (physiochemical) methods for the synthesis of NPs have been limited. Nowadays, the world is facing an expanding requirement for the discovery of reliable, sustainable, and environmental-friendly methods for the purpose of manufacturing a wide range of metal and metal oxide NPs. Therefore, considering its advantages such as nontoxicity, reproducibility in production, ease of scale-up, and well-defined morphology, synthesis through biological resources has turned into a new approach in the fabrication of NPs. More specifically, it has been observed that microorganisms and plants can stand in as new resources for this process due to their exceptional capabilities suitable for the synthesis of NPs. Microorganisms have been shown to possess astonishing potential as vital nanofactories, including being eco-friendly and cost-effective tools, having the ability to avoid toxic and harsh chemicals, and capable of fulfilling the demand of high energy required for physiochemical synthesis. In addition, microorganisms contain different reductase enzymes that can cause metal salts to reduce down to metal NPs with a narrow size distribution; this quality has enabled microorganisms to accumulate and detoxify heavy metals, thus resulting in less polydispersity. Nevertheless, most of the microorganism-based synthesis procedures for NPs have been observed to be quiet time consuming with low efficiency, and the recovery of NPs is in need of further downstream processing. Also, there are some complex steps, including microbial sampling, isolation, culturing, and maintenance, which are the other obstacles that are associated with microorganism-based synthesis methods [2]. Recently, plant-mediated biosynthesis of NPs has become one of the popular alternatives over the conventional methods. It has several advantages of green synthesis over other procedures and provides new avenues for the synthesis of NPs because of its unique properties such as the environmentally friendly nature, high biocompatibility, scalability, simplicity, rapidity, stability, and cost-effectiveness [3]. Furthermore, the plant-mediated biosynthesis of NPs, usually called as the "green synthesis", is safe and a one-step protocol for the synthesis of natural NPs. Plants contain a broad spectrum of biologically active compounds and secondary metabolites (SMs) that enable these plants to act as biological factories for the manufacture of metal and metal oxide NPs [4]. Acquired from different parts of plants such as stems, roots, leaves, fibers, flowers, callus, barks, and seeds, the mentioned natural products are able to take on the roles of reducing and stabilizing agents in the bioreduction reaction of synthesized novel NPs [1]. The rich biodiversity and easy availability of plants along with simplistic manner of NP synthesis using them provides a plausible explanation for increase in the number of reports in the scientific literature on this topic originating from different countries around the world. Some of the plant-mediated NP biosyntheses and their pharmacological applications are presented in Table 1.1. Therefore, the plant-derived resources appear to be the best natural

candidates for the production of nanostructures and NPs. The plant-based NPs have multifunctional applications in various fields including nutrition, cosmetics, agriculture, biomedicine, and energy science.

1.3 Plant Metabolites and Their Properties

Two types of metabolites can be obtained when using a plant, which are as follows: primary metabolites that are directly included in the flourishing and metabolism (carbohydrates, lipids, and proteins), and then the SMs, which are the end products of primary metabolism since they are not used through metabolic activity [16]. One of the significant characteristics of plants is their competence in synthesizing and accumulating a boundless diversity of low molecular weight compounds, which are commonly called the secondary metabolites (SMs), also known as phytochemicals or natural products. To shield and protect the plant from herbivores and microbes as well as to maintain its health and fitness, the biosynthesis of SMs generally transpires in higher plants while demonstrating a very inflated structural assortment as it is also counted as a signal compound in enticing pollinators and fruit dispersers [17]. In accordance with their biosynthetic pathways and arrangements, the structurally diverse SMs can be categorized into two large groups, the kind that contains SMs with nitrogen in their structure and the kind without nitrogen (Table 1.2). The number of basic biosynthetic pathways that are in agreement with the vast diversity of SMs seems to be confined and discrete. An excessive majority of the mentioned metabolites are initiated from five particular precursor pathways: acetyl coenzyme A (polyketides such as anthraquinones and flavonoids), active isoprene (various terpenoids), shikimic acid (aromatic amino acids, cinnamic acids, tannins, indole, and isoquinoline alkaloids), glycolysis (sugars and gallic acid), and the citric acid cycle (alkaloids) [18]. Since the structures of SMs have been constructed and optimized over the course of more than 500 million years of evolution, many of them are capable of wielding amazing biological and pharmacological qualities, thus earning the label of profitable candidates for

TABLE 1.1   List of Plant-Mediated Biosyntheses of NPs and Phytochemicals and Their Pharmacological Applications

Parts of Plant	Scientific Name of Plant	NPs	Type of Phytochemicals	Medical Applications	Ref.
Leaves	*Lawsonia inermis* L.	Fe	Phenolic compounds	Antibacterial	[1]
	Melia azedarach L.	Ag	Tannic acid & polyphenols	Cytotoxicity	[2]
	Mentha piperita L.	Ag, Au	Menthol	Antibacterial	[3]
Stem	*Cassia fistula* L.	Au	Hydroxyl group	Anti-hypoglycemic	[4]
Shoots	*Rheum turkestanicum* Janisch.	Ag	Flavonoids, phenols, peptides	Antibacterial	[5]
Root	*Panax ginseng* C.A.Mey.	Ag, Au	Terpenes and phenolic compounds	Antibacterial	[6]
Flower	*Mirabilis jalapa* L.	Au	Polysaccharides	Antimicrobial	[7]
	Nyctanthes arbor-tristis L.	Ag	Phenolic compounds	Antibacterial and cytotoxic	[8]
Fruit	*Carica papaya* L.	Ag	Hydroxyl flavones, catechins	Antimicrobial	[9]
	Tribulus terrestris L.	Ag	Alkaloids, terpenes, and phenolic compounds	Antimicrobial	[10]
	Vitis vinifera L.	Pd	Polyphenol	Antibacterial	[11]
Seed	*Trigonella foenum-graecum* L.	Au	Flavonoids	Catalytic	[12]
Peel	*Citrus sinensis* (L.) Osbeck	Ag	Water-soluble compounds	Antibacterial	[13]
Gum	*Astragalus verus* Olivier	CeO$_2$	Polysaccharides	Cytotoxicity	[14]
	Boswellia serrata Roxb. ex Colebr.	Ag	Protein, enzyme	Antibacterial	[15]

TABLE 1.2 Approximate Numbers of Known SMs from Higher Plants [19]

Group		Type of SMs	Approximate No. of Structures
SMs containing nitrogen		Alkaloids	27,000
		Nonprotein amino acids (NPAA)	700
		Cyanogenic Glucosides/HCN	60
		Mustard oils (Glucosinolates)	150
		Amines	100
		Lectins, peptides, AMPs	2,000
SMs without nitrogen	Terpenes	Monoterpenes (C10) (including iridoids)	3,000
		Sesquiterpenes (C15)	5,000
		Diterpenes (C20)	2,500
		Triterpenes (C30), steroids, and saponins (including cardiac glycosides)	5,000
		Tetraterpenes (C40)	500
	Phenols	Phenylpropanoids, coumarins, and lignans	2,000
		Flavonoids, anthocyanins, tannins	4,000
		Polyketides (anthraquinones)	800

medicinal applications or biorational pesticides [17]. From the standpoint of evolutionary pharmacology, besides illustrating a great arrangement of biological and pharmacological features, SMs have been proven to contain a captivating library of preselected bioactive compounds. Several SMs have been observed to be determined and fixed for one or a restricted number of molecular targets (e.g., alkaloids, cardiac glycosides), whereas most of the SMs that exist in the utilized extracts for herbal medicine (various phenols and terpenoids) are multi-target agents that serve the purpose of regulating the activity of proteins, nucleic acids, and biomembranes in a less targeted procedure. In addition, some specific SMs of different extracts can interact in a synergistic manner in order to increase the power of their bioactivities. This topic is very engrossing and fascinating, and it deserves much more attention from pharmacologists [19]. From the nanotechnology point of view, the primary and secondary metabolites are continuously taken into account while assessing the redox reaction toward eco-friendly synthesized NPs [20]. As has been clearly suggested, proteins, amino acids, organic acid, and vitamins, and SMs such as flavonoids, alkaloids, polyphenols, terpenoids, heterocyclic compounds, and polysaccharides play notable roles in the synthesis and stabilization of metal and metal oxide NPs that are constructed with the desired sizes and shapes [21,22]. Various taxa in the plant kingdom, particularly dicot plants species, possess many SMs that may be ideal for the preparation of NPs. Literature review surveys have indicated that different plant species contain diverse mechanisms to achieve synthesized NPs. Peculiar ingredients such as emodin stand as purgative resins with quinone compounds that exist in xerophyte plants (plants that have acclimatized to survive in deserts or environments that contain scarce amounts of water), which can be used for the synthesis of silver NPs; cyperoquinone, dietchequinone, and remirin in mesophytic plants (terrestrial plants that have accommodated to a neither particularly dry nor wet environment) seem to have practical applications in metal NP synthesis. Eugenol, the dominant terpenoid of *Cinnamomum zeylanisum* Nees, has been found to play a significant role in the synthesis of gold and silver NPs [23]. In addition, phytochemicals that carry antioxidant or reducing qualities are normally useful for the synthesis of NPs [24]. Many plant species contain antioxidative compounds

(reductant) of different structure types like flavonoids, phenylpropanoids, coumarins, xanthones, anthraquinones, and terpenoids, which are excellent candidates for production of natural NPs [25]. Nowadays, there is remarkable interest in the food and pharmaceutical industries for development of antioxidants from natural sources. In accordance with the reports available to date, only 20%–30% of the discovered 350,000 plant species have been studied from a phytochemical perspective in a bit of detail, and thus, the actual number of SMs that present in the plant kingdom most definitely exceeds 200,000 compounds [18]. Consequently, there is immense potential of novel chemical structures and new types of action of the compounds from plants resources, which will then play outstanding roles in new drug discovery for treating various diseases in future.

1.4 Biosynthesis of Metal NPs

The biosynthesis of metal NPs through utilization of plants is a quite complex process. The biological synthesis of metal NPs (to be specific, gold and silver NPs) from plants (inactivated plant tissue, plant extracts, and living plant) has attracted the interest of many as a genuine candidate for chemical and physical techniques. Synthesis of metal NPs is an economical and respected method for production of NPs on a large scale, because using plants is very cost-effective and easy. Plants extracts have the ability to function as stabilizing agents, as well as capping agents, in the synthesis of NPs, while biomolecules are responsible for the bioreduction of metal NPs in plant extracts (e.g., enzymes, proteins, phenols, polysaccharides, and organic acids).

1.4.1 Silver Nanoparticles (Ag-NPs)

Silver (Ag) is a white and shiny element that is found at the 47th position of the periodic table and is indexed as Ag, prehensed from Argentum. Pure silver has the highest electrical and heat conductivity compared with other metals. The medicinal qualities of silver have been discovered and acknowledged for over 2,000 years. Since the 19th century, silver-based compounds have been applied in many different antimicrobial implementations. Although for the purpose of producing antimicrobial results, silver is habitually utilized in its nitrate form, when Ag-NPs are used, there is a significant increase in the accessible surface area for the microbe to be revealed to. The biosynthesis of Ag-NPs via plants is probably the most uncomplicated and effortless technique that encompasses the bioreduction of metal salt through the application of biocomponents. Transmission electron microscopy (TEM) and UV–visible (UV–vis) absorption spectroscopy have been performed to distinguish the achieved NPs in order to assess their quality. The applications of silver varies, ranging from kitchen tools to antimicrobial materials [26], sensor pages [27], etc. The antiseptic properties of silver are also acceptable and widely used for treatment of wounds and burns because of its ability to destroy all kinds of pathogenic organisms [28,29]. Silver can be found in various forms such as silver nitrate ($AgNO_3$), silver zeolite, silver chloride (AgCl), and silver cadmium powder.

NPs are normally produced through different chemical and physical procedures that are known to require big budgets while also being quite perilous to the environment due to the involvement of toxic and unsafe chemicals that are usually held accountable for biological risks. The growth of exploratory exercises that have been inspired through biological means for the synthesis of NPs is advancing into a salient field of nanotechnology. In general, two different approaches can be taken for synthesizing Ag-NPs: a "top-to-bottom" method or a "bottom-to-top" procedure. In the bottom-to-top approach, the synthesis of NPs is performed using chemical and biological techniques for the self-assembly of atoms to new nuclei that germinate into a particle of nano-scale; on the other hand, in the top-to-bottom approach, the requisite bulky ingredients are broken down into fine particles using different lithographic methods, e.g., grinding, milling, sputtering, and thermal/laser ablation. The expanding development of green NP synthesis is proceeding as a principal branch of nanotechnology, since the employment of biological entities, including plant extract or plant biomass, for the synthesis of NPs can stand as a substitute for chemical and physical techniques in an eco-friendly manner [5,30,31]. Numerous reports have been published by many researchers for the biosynthesis of metal NPs through plant extracts and their possible implementations. Some of them are summarized in Table 1.3. As it is noted as a fact, silver ions and silver-based compounds can cause extreme toxic effects on the leading species of bacteria and other microorganisms; therefore, in various public areas, Ag-NPs are utilized as antimicrobial agents [32–34]. The

precise mechanism by which Ag-NPs bring about the antimicrobial effect is still a mystery and a topic of debate [35]. Nevertheless, different theories have been presented regarding the course of actions that Ag-NPs take for causing microbicidal effects on microbes [36]. Ag-NPs are capable of anchoring and eventually puncturing the bacterial cell wall, which results in alterations in the permeability of cell membrane structure, thus leading to cell destruction. One can observe the emergence of "pits" on the cell surface, along with the NPs gathering in the same area [37]. The genesis of free radicals through the use of Ag-NPs could be noted as another mechanism that results in the destruction of cells. Moreover, it has been suggested that NPs seem to be capable of unbinding silver ions [38], and these particular ions have the ability to interact with the thiol groups of several essential enzymes and disable their functionality [39]. Silver is known as a soft acid, and since a natural tendency exists in the group of acids to respond to bases, a similarly soft acid is required to react with a soft base in this particular case [40]. Acknowledged as soft bases, sulfur and phosphorus are the principal elements that build up a cell. The NPs' course of operation on a cell can result in the inducement of a reaction and gradually end up in cell annihilation. Yet another fact, DNA is mainly constructed from sulfur and phosphorus; thus while NPs are reacting with these soft bases, DNA damage occurs, which would obviously result in the death of cell [41]. In addition, it has been discovered that NPs have the ability to regulate signal transduction in bacteria. Dephosphorylation has only been mentioned in the tyrosine residues of Gram-negative bacteria, and the phosphotyrosine profile of bacterial peptides can be changed through the effects of NPs. As has been perceived, NPs can dephosphorylate the peptide substrates that belong to the tyrosine residues, resulting in obstruction of signal transduction, which then terminates the growth process. Nonetheless, in order to validate the present claim, further investigation and assessment on the topic is definitely required [42]. Due to its numerous advantages such as being rapid, environmentally friendly, nonpathogenic, economical, and capable of being done as a single-step procedure, many researchers have focused their attention on utilizing plants in the production and assembly of Ag-NPs. The reduction and stabilization of silver ions, the amalgamation of biomolecules including proteins, amino acids, enzymes, polysaccharides, alkaloids, tannins, phenolics, saponins, terpenoids, and vitamins have been put into usage; although they are already habitual in the plant extracts, while containing medicinal values and being environmental benign, yet they seem to have complex chemical structures. An enormous number of plants have been observed to be of use for easing and enabling the Ag-NPs synthesis process, and these have been mentioned (Table 1.3) and will be discussed shortly in the current chapter. For the purpose of providing stability to Ag-NPs, comparatively high levels of steroids, sapogenins, carbohydrates, and flavonoids have been employed to function as the reducing agents, while phyto-constituents operated as the capping agents.

TABLE 1.3 Plants Used for Spherical Ag-NPs

No.	Plant	Part	Size (nm)	Ref.
1	Rheum turkestanicum	Shoot	26	[5]
2	Jatropha curcas	Latex	10–20	[43]
3	Pelargonium hortorum	Leaf	5–150	[44]
4	Ficus religiosa	Leaf	21	[45]
5	Tecomella undulata	Flower	12.5	[46]
6	Bauhinia purpurea	Flower	20	[47]
7	Ephedra intermedia	Stem	24	[48]
8	Lycium barbarum	Fruit	3–15	[49]
9	Stereospermum suaveolens	Root	49	[50]
10	Prosopis fracta	Leaf	25	[51]

Despite the abovementioned facts, some reports and studies have revealed the unfavorable side effects of Ag-NPs that can allegedly harm humans and the environment. Many tons of silver produced by industrial waste is carelessly released into the environment, which is dangerous considering the toxicity mainly caused by the free silver ions in the aqueous phase. The world first took note of Ag-NPs at the beginning of the 21st century, and its popularity has been increasing rapidly, judging how it is being employed in almost every related field, particularly in the medical field. It is a must to mention that NPs have a tendency to agglomerate; therefore, while the primary particle size was observed to be less than 100 nm, the agglomerating effects that were greater than 100 nm have also been considered and documented. Insecure constructions of primary particles that are clasped together by means of adhesion are known as agglomerates, and their shape and size can be altered in varied situations, and hence they should be differentiated from aggregation to prevent confusion. Aggregation is the procedure in which the primary particles start to form a defined crystalline structure while the total surface area is calculated to be less than the sum of the surface areas of the primary particles [52]. Although they are known as the Ag-NPs' hallmark of toxicity, the increased reactive oxygen species (ROS) can create positive growth outcomes in varying species of plants by heightening their activity [53]. Although the toxicity of Ag-NPs to model bacteria has been reported, there is evidence that Ag-NPs are capable of altering the appearances of key genes and proteins that are included in denitrification, as well as restraining the activities of nitrite and nitrate reductase [54,55]. Also, there is documented evidence that shows the toxic effects of Ag-NPs on human sperm [56], but other investigations show that it induces hormesis in humans [57]. Nowadays, the most commonly used NPs in the industry are known to be Ag-NPs because of their peculiar biocidal qualities and excessive advantages for our society from their employment in industrial and therapeutic implementations. Nevertheless, there are obstacles that limit the utilization of these particular NPs, such as lack of standardization based on their size and shape, and the absence of dose-dependent toxicity elucidation. Apart from these hindrances, there is a shortage of assessment in measuring the toxicity of these NPs after their interaction with organic matter and the specifics of the intracellular mechanisms that lead to their activity. Therefore, this topic requires further investigations in Ag-NPs' unique features to potentiate its usage.

1.4.2 Gold Nanoparticles (Au-NPs)

Labeled with the symbol Au (Latin: *aurum*) and having the atomic number of 79, gold is a chemical element that can resist most of the discovered acids, although it has been observed to dissolve in aqua regia, a combination of nitric acid and hydrochloric acid, which results in the formation of a soluble tetrachloroaurate ion. Despite the fact that nitric acid can dissolve silver and base metals, gold stands insoluble in this particular acid. Medicinal applications of gold and its complexes date back to ancient times. In order to medicate rheumatoid arthritis, various gold composites have been employed. Considering how they stand as achievable anti-cancer drugs, both Au (I) and Au (III) compounds have been thoroughly researched. The topic of colloidal gold has been perceived for research implementations in medicine, biology, and materials science. Recently, biomedical applications using Au-NPs have turned into a popular and active research field [58]. A vast array of feasible biomedical implementations including drug and gene delivery [59,60], cancer therapy [61], protein and pathogen detection [62,63], fluorescent probes [64], and tissue engineering [65] have been surveyed and investigated. A significant amount of research on this topic has been dedicated to the synthesis, stabilization, and functionalization of Au-NPs, which is mainly due to their relatively uncomplicated synthesis process, satisfactory control over sizes and shapes, optical characteristics, and fine biocompatibility. Repeated administrations of Au-NPs to the body can result in its accumulation to toxic levels [66]. For this reason, most of the researches regarding the utilization of Au-NPs are still at preclinical stages [67]. It is more suitable to employ nontoxic reagents for the purpose of upgrading the biocompatibility of Au-NPs. All of the available techniques regarding the composition of Au-NPs are established on the reduction of gold ions, most preferably as solutions of $HAuCl_4$. Although different reducing agents have been mentioned in the literature, in addition to those, Au-NPs synthesis needs a protecting agent in order to adsorb onto the surface of the newly arranged NPs, while hindering its germination and particle agglomeration. Therefore, for the purpose of controlling the size and shape of Au-NPs, suitable reduction and agitation procedures must be applied while genuine kinds of protection agents and specific concentrations must be chosen to enable practical synthesis conditions, i.e., correct temperature and pH. For example, the size and shape of triangular Au-NPs was managed by altering the concentration of the existing plant extract in the reaction medium [68,69]. As an outcome of the mentioned considerations, many publications have proposed new methods for the synthesis of Au-NPs in which green reduction and protection agents have been employed [70]. As per the data abridged in Table 1.4, most of the available reducing and stabilizing agents have been acquired from plants. Spherical Au-NPs can display an array of different colors (e.g., brown, orange, red, and purple) when the core size is increased from 1 to 100 nm in aqueous solution, while demonstrating a size-relative absorption peak from 500 to 550 nm [71]. Due to their general applications in medical fields, toxicological facets stand in need of certain consideration and evaluation in the case of Au-NPs. Without any notably apparent ion release, Au becomes comparatively motionless under physiological conditions as a base metal [72]. Moreover, as one can observe, there is less ROS genesis due to the fact that Au-NPs do not possess the band gap arrangement that commonly exists

TABLE 1.4   Plants Used for Biosynthesis of Au-NPs

No.	Plant	Part	Shape	Size (nm)	Ref.
1	*Rosa rugosa*	Leaf	Triangular and HEXAGONAL	11	[85]
2	*Morinda citrifolia*	Root	Spherical	12–38	[86]
3	*Salvia officinalis*	Leaf	Truncated triangular	29	[67]
4	*Stevia rebaudiana*	Leaf	Spherical,	5–20	[87]
5	*Pogestemon benghalensis*	Leaf	Spherical and triangular	10–50	[88]
6	*Plumeria alba*	Flower	Spherical	15–28	[89]
7	*Citrus maxima*	Fruit	Spherical	25.7	[90]
8	*Nepenthes khasiana*	Leaf	Triangular and spherical	50–80	[91]
9	*Abelmoschus esculentus*	Pulp	Spherical	14	[92]
10	*Cucurbita pepo* L	Leaf	Polygonal	10–15	[93]

in semiconductor metal oxide NPs [73]; however, there are reports that have announced the harmful effects of Au-NPs [74]. Despite the abovementioned facts, there are various studies and reports that indicate the lack of toxicity both in vivo and in vitro [75,76]. Owing to the divergent observations, it is very difficult and almost impossible to conclude definitively regarding the safety of Au-NPs. The existence of toxicity stands as a possibility for similar variables that ameliorate the therapeutic features of Au-NPs. Examples of such variables would be (i) physical parameters (shape and size), (ii) dose, (iii) biological system tested, (iv) respective route, and (v) time of exposure employed for perceiving the feasible dangerous outcomes. Currently, it is impractical to come up with a final conclusion regarding the toxicity of Au-NPs based on their size range. As a whole, an Au-NP that has a diameter of 1–2 nm is considered as toxic due to the inducement of quick cell deaths by necrosis, while Au-NPs that are larger than 15 nm seem to be nontoxic regardless of the cell type that has been examined [77–79]. However, bare Au-NPs of sizes between 17 and 37 nm have sometimes been observed to cause anorexia, weight loss, change in fur color, and liver damage [80]. Owing to the increased vulnerability to nanomaterials, systematic exposure investigations by environmental and health agencies have been advanced [81–83]. Regarding therapeutic methods, most of the researchers have focused on assessing the spherical Au-NPs, which is quite interesting. Since spherical Au-NPs have shown greater levels of toxicity, evaluating various shapes of Au-NPs is more prominent when toxicity assessment is the main focus [84]. Therefore, for the purpose of evaluating the function of shape in therapeutic effectiveness, researchers must consider other nanoshapes in their future investigations on Au-NPs.

1.4.3   Other Metal NPs (Cu, Co, and Se)

The useful optical and thermal qualities of copper nanoparticles (Cu-NPs) have attracted much more attention compared with other metal NPs [94,95], owing to their use as sensors [96] and photo-catalysts [97] and its antimicrobial effect [98]. Although Cu is highly toxic to microorganisms, it has no effects on animal cells, and various human uses are in food packaging and water treatment [99,100] owing to its acknowledged role as an effective bactericidal metal [101]. Compared with other noble metals such as Ag,

Au, and Pt, Cu-NPs are much less expensive and have caught the interest of many researchers; for this reason, they have been potentially employed in the fields of catalysis, cooling fluids, and conductive links. Due to surface plasmon resonance (SPR), Cu-NPs have displayed improved nonlinear optical features, which has led to its implementation in optical devices and nonlinear optical materials including optical switches or photochromic glasses [102]. Cu-NPs have been successfully synthesized through *gamma*-radiolysis [103], laser irradiation [104], thermal decomposition [105], polyol [106], and chemical reduction [107] methods. Nonetheless, these procedures are all afflicted with drawbacks such as insecure reaction condition, utilization of high quantities of chemicals and costly instruments, and longer reaction time. Several green methods have been reported for the purpose of overcoming the aforementioned obstacles that stand in the way of synthesizing Cu-NPs, including the use of plant extracts from such varieties as Abutilon indicum [108], *Eclipta prostrate* [109], *Citrus medica* Linn [110], *Ginkgo biloba* [111], and *Euphorbia esula* L [112]. Therefore, there is still opportunity for the development of new approaches for the synthesis of Cu-NPs.

Cobalt (Co) is one of the most favorable elements among the metallic NPs, and it has captured the attention of many researchers from a wide and diverse range of fields for various applications. In most cases, Co-NPs have been proved to possess outstanding magnetic, electrical, and catalytic features, suitable for applications such as recording media, magnetic sensors, magnetic memories, magnetic composites, which has attracted the attention of many from different scientific and technological fields [113–115]. Specific characteristics of Co-NPs such as maximum saturation magnetization and massive anisotropy field have led to them being labeled as suitable candidates for high-density information storage and permanent magnets [116]. In recent years, Co-NPs have been observed to be of practical use in electromagnetic wave absorption implementations, including the expansion of wireless communications and high-frequency circuit devices [117]. Aside from these benefits, Co-NPs have proven to be more functional and contain an inherent advantage in biomedical fields such as drug delivery and magnetic resonance imaging [118,119]; all of the abovementioned applications require the agent to have high quality and purity in order to prevent alteration in magnetization or response stability, which can be fulfilled through the use of Co-NPs [120]. The amount of achievable success and functionality of the magnetic NPs is conditional on the microstructure, size, and morphology of the particle. Cobalt has been assessed and examined thoroughly from the aspect of its microstructure since it has been observed to contain variable crystal structures, such as hexagonal close-packed, face-centered cubic (fcc), and ε-phases [121]. In addition, various morphologies of Co have been reported to accommodate outstanding catalytic features; for this very reason, different experiments have been performed to achieve diverse morphologies and shapes such as nanosheets, spheres, snowflakes or cauliflower-like particles, leaf-shaped flakes, and microchains [122].

To date, selenium (Se), although being identified as an essential trace element for plants and mammals, has been acknowledged as a notorious ingredient. Se is commonly found as selenate (SeO_4^{2-}) and selenite oxyanions. In order to separate soluble Se^{4+} and Se^{6+} from contaminated soil, water, and drainage, reducing these elements into insoluble nontoxic elemental Se, with the help of microbes, would be an efficient method [123]. Se stands as one of the chalcogens that transpires as selenate, selenite(SeO^2-_3), and selenide (Se^{2-}), while it has the capability of being reduced to an atomic state by a precursor that carries a suitable reducing agent. Biogenic synthesis of Se-NPs can be regularly accomplished through the reduction of selenate/selenite along with the plant extracts that hold phenols, flavonoids amines, alcohols, proteins, and aldehydes. Over 40 illnesses of man have been acknowledged to be related to the deficiency of Se [124,125]; at low doses, it can invigorate the growth of a plant but cause damage if high dosages are used [126,127]. Also, it has been discovered that Se can act against cancer [128,129]. These particular composites are metabolized in the biological system in the form of selenocysteine and selenomethionine [130,131]. Although for the purpose of synthesizing Se-NPs a variety of plants extracts have been used, available data on this topic are not vast enough and must be further developed [132,133]. Se is particularly used in rectifiers, solar cells, photocopiers, and semiconductors. Moreover, this element has displayed biological functionality due to its interaction with proteins and other biomolecules, which are apparent in the bacterial cells and plant extracts, that accommodate functional groups such as =NH, C=O, COO^-, and C-N [134]. Despite the fact that Se prevails in various crystalline and amorphous forms, the shape, size, and structure of the NPs are conditional on the concentration, temperature, nature of biomolecules, and pH of the reaction mixture. The main feature of this element's NPs is its diversity with regard to size and shape. As an example, Se nanospheres contain high biological functionality and low toxicity, while Se nanowires have been observed to possess high photoconductivity [135]. These kinds of techniques are usually associated with toxic chemicals or high temperature and high pressure, resulting in further contamination of the environment. Typically, the biogenic protocol is altered for the purpose of circumventing the outcomes of toxic chemicals in the manufacturing procedure of NPs

[136]. Innocuous techniques have been discovered by various scientists for the production of NPs through the utilization of plant extracts. A clear line exists between the optimum limit/or deficiency and the excess of Se that remain in a living system, which can be the cause of toxicity. As has been assessed and proved, the Se-NPs fabricated through biological materials are much less toxic than the bulk of Se. The apparent biomolecules existing in the extract are capable of functioning as reducing and stabilizing agents of Se-NPs [137].

1.5   Biosynthesis of Metal Oxide NPs

The existence of metal oxides is quite crucial in various fields of chemistry, physics, and materials science [138–140]. Besides being capable of arranging a large variation of oxide composites, these metal elements can affect a large number of structural geometries with an electronic construction that is able to display metallic, semiconductor, or insulator characteristics. Throughout technological implementations, oxides are employed in the manufacturing of microelectronic circuits, sensors, piezoelectric devices, fuel cells, and catalysts, as well as coatings for the passivation of surfaces in opposition to corrosion. One of the aims of nanotechnology, which is becoming an apparent field, is to construct nanostructures or nanoarrays with exceptional qualities in regard to those of bulk or single-particle species [141–143]. Metal oxide NPs contain the ability to unravel the unique physical and chemical features, considering their limited size and the high density of corner or edge surface sites. Three vital groups of fundamental qualities in any material are thought to be affected by particle size. The inducement of a decrease in the average size of an oxide particle can alter the magnitude of the band gap [144,145] by means of strongly affecting the conductivity and chemical reactivity [146]. As their significance in chemistry is noted, surface qualities are considered a rather particular group in this study. Generally, it is feasible to enclose the solid–gas or solid–liquid chemical reactions to the surface and/or subsurface regions that belong to the solid.

1.5.1   Zinc Oxide Nanoparticles (ZnO-NPs)

Zinc (Zn) is the single discovered metal in all of the six enzyme classes, i.e., oxidoreductases, lyases, isomerases, transferases, hydrolases, and ligases [147]. Being an essential micronutrient, this particular element has a vital functionality in many different integral metabolic procedures [148]. Zn has the ability to heighten the biosynthesis of chlorophylls and carotenoids, as well as intensify the photosynthetic apparatus of the plant [149]. A massive potential can be provided by the important optoelectrical, physical, and antimicrobial properties of zinc oxide nanoparticles (ZnO-NPs) for the purpose of improving agricultural productivity [150]. Several investigations have demonstrated the existence powerful absorption abilities in a series of organic compounds, heavy metals, ZnO, and ZnO-NPs [151]. In comparison to the bulk material, the large surface area-to-volume ratio of ZnO-NPs can cause a noticeable increase in the efficacy of blocking the UV radiation [152]. Different studies have assessed the effects of ZnO-NPs on crop plants, with the results indicating that nano-dimensional ZnO particle can improve germination, pigments, and sugar and protein contents by extending the activities of antioxidant enzymes in several vegetable crops [153]. ZnO-NPs possess antimicrobial activity in opposition to pathogenic organisms [154]. The use of various plant extracts or microorganisms in the biogenic synthesizing procedure of ZnO-NPs has shown an advantage over traditional chemical and physical techniques.

As a critical factor, in order to calculate the effect of NPs on plant metabolism, they should be transported through the environment into the plants. As a significant component of the environment, plants play a crucial role as a pathway for the uptake, transportation, and accumulation of NPs into the food chain [155]. Uptake, translocation, and accumulation of NPs are conditional on the situation and properties of the plant NPs, such as species, size, chemical composition, stability, and concentration.

1.5.2 Cerium Oxide Nanoparticles (CeO_2-NPs)

Cerium oxide nanoparticles (CeO_2-NPs) have captured the interest of many researchers in nanotechnology since they have many practical applications such as catalysts [156], sustainable pollutant removers [157], antimicrobial agents [158], antioxidants [159], neuronal protectors [160], biosensors [161], and environment [162]. Normally, cerium has two oxidation states, i.e., Ce^{3+} and Ce^{4+}, and it can be taken from this fact that cerium dioxide can have two diverse oxide forms, i.e., CeO_2 (Ce^{4+}) or Ce_2O_3 (Ce^{3+}), in bulk material [163]. The cerium oxide lattice accommodates a cubic fluorite structure, with both forms (Ce^{3+} and Ce^{4+}) capable of coexisting on its surface. Thus, it can be stated that CeO_2-NPs contain upgraded redox features in comparison to the bulk materials. In addition, one can observe the crucial function of mixed valance state in hunting the reactive oxygen and nitrogen species. CeO_2-NPs have been shown to be effective in opposing pathologies that are related to chronic oxidative stress and inflammation. Also, in accordance with various reports, CeO_2-NPs seem to contain multiple enzymes, such as superoxide oxidase, catalase, and oxidase, and mimetic qualities; for this reason, they have been of interest in biological fields, including bioanalysis [164,165], biomedicine [166], and drug delivery [167]. Various routes and synthesis procedures have been assessed for the purpose of fabricating CeO_2-NPs, such as solution precipitation [168], sono-chemical method [169], hydrothermal process [170], spray pyrolysis [171], and sol–gel methods [172]. Nonetheless, the use of these techniques is rife with different drawbacks, such as employment of toxic solvents and reagents, requisition of high temperature and

pressure, and the demand for external additives such as stabilizing or capping agents during the reaction. Although similar core elements apply for every type of CeO_2-NPs, the same biological impacts are not exhibited by all of them. Due to the utilization of different NPs and the inducement of various physiochemical parameters, some reports have announced the pro-oxidant toxicity of these particular particles in some specific cases, while other studies have suggested their antioxidant protective outcomes in different situations. The production of CeO_2-NPs with diverse physicochemical qualities is dependent on the synthesizing procedure, type of the employed stabilizing agent, and the Ce^{3+}/Ce^{4+} surface ratio [173,174]. Particles as small as <10 nm have been achieved through some green synthesizing procedures of CeO_2-NPs. As has been suggested by some reports, the plant-based synthesis of CeO_2-NPs can furnish larger NPs than those from biopolymer and nutrient-based techniques, although they possess antibacterial properties that have displayed high levels of cytotoxicity to bacterial cells [175,176].

1.5.3 Iron Oxide Nanoparticles (Fe_3O_4 and Fe_2O_3)

Iron oxides (Fe_3O_4 and Fe_2O_3) can be used in various applications such as nanosorbents for heavy metal [177] and organic contaminants [178], biomedicine [179], and photocatalysts [180]. The most common forms of iron oxides are magnetite (Fe_3O_4), maghemite (γ-Fe_2O_3), and hematite (α-Fe_2O_3) [181]. As a result of containing novel features and qualities such as being sized in the nano-range, having high surface area-to-volume ratios, and possessing superparamagnetism, the synthesizing procedure and applications of iron oxide nanomaterials have been extensively investigated [182–184]. In particular, the simplicity of synthesis, coating, or modification and the ability to direct or handle the matter at an atomic scale results in unparalleled versatility [185,186]. Moreover, the low toxicity, chemical inertness, and biocompatibility have given iron oxide nanomaterials the capacity to produce amazing results in the field of biotechnology [179]. The unique qualities of these nanomaterials are the reason behind its implementations, as well as the fact that has made them different from iron oxide bulk materials [187,188]. As can be observed from various reports, in the process of ascertaining the size distribution, morphology, magnetic features, and surface chemistry of nanomaterials, the fabrication method plays a crucial role [189,190]. The enhancement of chemical and physical procedures for the synthesis of metal NPs has been at the center of many research investigations and, as a result in recent years, a diversity of synthesis approaches for the purpose of constructing high-quality NPs have emerged [191,192]. In order to achieve precise control over active surface sites, synthesizing methods are required that allow for the production of monodisperse and shape-controlled iron oxide nanomaterials [193,194]. Evolution in this regard is necessary in the various emerging techniques such as plant-mediated biological method. For the purpose of creating new systematic and particular magnetic nanomaterials, future assessments need to focus on and aim at addressing divergent challenges. Furthermore, evolving the iron oxide nanomaterials into a field scale can open the possibility of an inventive area of research, even as it is necessary to perform more investigations into the application potential of these novel nanomaterials. Through the process of surface adjustment with acceptable functional groups, such as phosphonic acids, carboxylic acids, and amines, the stability of iron oxide colloid suspensions can be noticeably supplemented [185]. Considering how the feasible implementation is conditional on the type of altered medium, it is vital to examine this aspect with different mediums [195]. It must be mentioned that the application of iron oxide nanomaterials is tightly associated with their intrinsic features, which is conditional on the fabricating method and adapted mediums [196].

1.5.4   Other Metal Oxide Nanoparticles

Aluminum Oxide (Alumina, Al_2O_3)

Considering its significance as a catalyst component or absorbent and ceramic material in a large number of industrial procedures, Al_2O_3 stoichiometry has been the center of focus among all Al-O systems. Novel nanostructured alumina is, at the present time, employed as the active phases in the area of catalysis, as well as being used with other substances, including yttrium aluminum garnet (YAG) or nano-Ni/W, for the purpose of fabricating materials with unprecedented mechanical qualities, which seem to be associated with the strong resistance to distortion at moderates temperatures (e.g., YAG) or retain their hardness above 30 GPa (e.g., Ni,W) [139]. Seven types of Al_2O_3polymorphs have been acknowledged, although only the α, δ, θ, and γ are normally implied in most of the industrial operations [197]. Theoretical assessments of $(Al_2O_3)_n$ ($n \leq 15$) small clusters have shown specific structural, electronic, and chemical (behavior against adsorbates) resemblances with several α-Al_2O_3 surfaces [198]. After γ-Al_2O_3, which is the nanostructured phase that is usually achieved through most of the synthetic procedures, there is also the α-Al_2O_3polymorph that can be synthesized with high surface area [199]. Although the corundum (Aluminium oxide), α-Al_2O_3 structure, is known as the bulk thermodynamically stable phase, yet the calorimetry work of McHale et al. [199] has provided incontrovertible proof that γ-Al_2O_3 contains a lower surface energy and can turn into energetically stable at a size of below a point close to 10 nm (surface BET area ca. 75 m^2/g). They have also illustrated the significance of surface hydroxyls or water molecules regarding the energetics of surfaces.

Magnesium Oxide (MgO)

MgO stands as a crucial element in the chemical industry for uses such as being a scrubber for air pollutant gases (CO_2, NO_x, SO_x) and catalyst support [200]. It has been discovered to contain a rock salt structure similar to the oxides of other alkaline earth metals. The nonpolar (100) face is known as the most solid surface [201], and particles of MgO typically have a cubic shape. Although it is quite rare to come across a method for the green synthesis of other metal oxides using plants, the results of these projects can be seen only in the future. As an example, MgO nanoparticles (MgO-NPs) have been synthesized through a chemical precipitation procedure, and thus have two varying concentrations (10% and 20%) of acacia gum in order to obtain flower-shaped MgO-NPs; also, the synthesized samples proved to be well organized during the discarding process of all the selected divalent metallic species [202].

Titanium Oxide (Titania, TiO_2)

The administration of TiO_2 structures is a substantial field of research and demands a separate part. The objectives behind the utilization of surface and bulk doping include stabilizing the anatase or rutile phases, influencing the temperature of the anatase rutile phase transformation, and adjusting the optical band gap or changing the ionic/electrical conductivity through the appearance of intrinsic vacancies. The acquired qualities of the combined oxide are primarily conditional on the nature of the doping operation. Substitutional mixed oxides have been observed to be manufactured in the cases of Ca, Sr, and Ba [203]. In addition, TiO_2-NPs seem to carry appealing optical, dielectric, antimicrobial, and antibacterial properties, as well as satisfactory chemical stability and catalytic features that have resulted in their being applied for various industrial purposes including pigment, fillers, catalyst supports, and photocatalysts [204–206]. Some of the typically employed synthetic techniques include nonsputtering, solvo-thermal, reduction, and sol-gel technique as well as the electrochemical procedure [207,208]. However, these methods are expensive, toxic, possess high pressure, demand high energy, show difficulty in separation, and are potentially dangerous [24]. Consequently, the lack of a dependable, biosynthetic, and environmentally friendly method is deeply felt with the growing need of producing eco-friendly products, ensuring biocompatibility and economic viability in the long run, as well as preventing unfavorable outcomes that can be induced during their application, especially in the medical field.

1.6   Benefits of Plant-Mediated Biosynthesis of NPs and Future Perspectives

Green chemistry opens up new methods for implementation of plants in the role of chemical substances in synthesis of NPs, thus reducing the hazards that can harm the health and environment. The benefits and advantages of the new approaches include economical and energy efficiency, lower cost of production and regulation, reduction of waste, fewer number of accidents, safer products, competitive advantages, and healthier workplaces and communities, thus leading to protection of human health and environment while being compatible for pharmaceutical and other biomedical applications.

1.7 Conclusions

For the construction of metal and metal oxides NPs, many chemical, physical, and biological synthetic methods have been examined and tested. While most of these approaches still remain in the progressing stage, various obstacles stand in their path such as the stability and aggregation of NPs, control of crystal growth, morphology, size, and size distribution. Another significant factor would be the process of separating the produced NPs to be utilized in further applications. In contrast to the other constructed products, the metal and metal oxide NPs that are fabricated using plants have proved to be more stable. Plant extracts seem to be capable of reducing metal ions at a faster speed than fungi or bacteria. Also, when it comes to utilizing a simple and secured green procedure in scale-up and industrial fabrication of well-dispersed metal NPs, plant extracts have demonstrated more satisfactory results compared with plant biomass or living plants. Most of the observations and focus of researchers have been aimed at comprehending the biological operation and enzymatic procedures of NP biosynthesis, along with perceiving and distinguishing the biomolecules that are incorporated in the synthesis of metallic NPs. Proteins/enzymes, amino acids, polysaccharides, alkaloids, alcoholic compounds, and vitamins are the biomolecules of plants that might be related to the processes of bioreduction, emergence, and stabilization of metal NPs, reduction being conditional on the existence of polyphenols, enzymes, and other chelating agents that are present in plants. The reduction potential of ions and the reducing capacity seem to impart significant effects on the amount of NPs that is manufactured. It is necessary to point out the fact that future assessments might be required to proceed toward the optimization of reaction conditions and for the engineering the recombinant organisms for the purpose of fabricating massive amounts of proteins, enzymes, and biomolecules that are incorporated in the biosynthesis and stabilization processes of NPs. In order to further the construction of NPs, it is crucial to comprehend the biochemical processes/pathways that are included in plant heavy metal detoxification, accumulation, and resistance. Future perspectives for increasing the efficiency of the mentioned organisms in the synthesizing procedure of NPs would be the genetic qualification of plants that contain advanced metal tolerance and amassing potential.

References

1. Naseem, T. and M.A. Farrukh, Antibacterial activity of green synthesis of iron nanoparticles using Lawsonia inermis and Gardenia jasminoides leaves extract. *Journal of Chemistry*, 2015. **2015**: pp. 1–7.

2. Sukirtha, R. et al., Cytotoxic effect of Green synthesized silver nanoparticles using Melia azedarach against in vitro HeLa cell lines and lymphoma mice model. *Process Biochemistry*, 2012. **47**(2): pp. 273–279.

3. MubarakAli, D. et al., Plant extract mediated synthesis of silver and gold nanoparticles and its antibacterial activity against clinically isolated pathogens. *Colloids and Surfaces B: Biointerfaces*, 2011. **85**(2): pp. 360–365.

4. Daisy, P. and K. Saipriya, Biochemical analysis of Cassia fistula aqueous extract and phytochemically synthesized gold nanoparticles as hypoglycemic treatment for diabetes mellitus. *International Journal of Nanomedicine*, 2012. **7**: p. 1189.

5. Yazdi, M.E.T. et al., Biosynthesis, characterization, and antibacterial activity of silver nanoparticles using Rheum turkestanicum shoots extract. *Research on Chemical Intermediates*, 2018. **44**(2): pp. 1325–1334.

6. Singh, P., Y.J. Kim, and D.C. Yang, A strategic approach for rapid synthesis of gold and silver nanoparticles by Panax ginseng leaves. *Artificial Cells, Nanomedicine, and Biotechnology*, 2016. **44**(8): pp. 1949–1957.

7. Vankar, P.S. and D. Bajpai, Preparation of gold nanoparticles from Mirabilis jalapa flowers. *Indian J Biochem Biophys*, 2010. **47**(3):pp. 157–160.

8. Gogoi, N. et al., Green synthesis and characterization of silver nanoparticles using alcoholic flower extract of Nyctanthes arbortristis and in vitro investigation of their antibacterial and cytotoxic activities. *Materials Science and Engineering: C*, 2015. **46**: pp. 463–469.

9. Jain, D. et al., Synthesis of plant-mediated silver nanoparticles using papaya fruit extract and evaluation of their anti microbial activities. *Digest Journal of Nanomaterials and Biostructures*, 2009. **4**(3): pp. 557–563.

10. Gopinath, V. et al., Biosynthesis of silver nanoparticles from Tribulus terrestris and its antimicrobial activity: a novel biological approach. *Colloids and Surfaces B: Biointerfaces*, 2012. **96**: pp. 69–74.

11. Amarnath, K. et al., Synthesis and characterization of chitosan and grape polyphenols stabilized palladium nanoparticles and their antibacterial activity. *Colloids and Surfaces. B, Biointerfaces*, 2012. **92**: pp. 254–261.

12. Aromal, S.A. and D. Philip, Green synthesis of gold nanoparticles using Trigonella foenum-graecum and its size-dependent catalytic activity. *Spectrochimica Acta Part A: Molecular and Biomolecular Spectroscopy*, 2012. **97**: pp. 1–5.

13. Kaviya, S. et al., Biosynthesis of silver nanoparticles using Citrus sinensis peel extract and its antibacterial activity. *Spectrochimica Acta Part A: Molecular*

and *Biomolecular Spectroscopy*, 2011. **79**(3): pp. 594–598.

14. Darroudi, M. et al., Nanoceria: gum mediated synthesis and in vitro viability assay. *Ceramics International*, 2014. **40**(2): pp. 2863–2868.

15. Kora, A.J., R. Sashidhar, and J. Arunachalam, Aqueous extract of gum olibanum (Boswellia serrata): a reductant and stabilizer for the biosynthesis of antibacterial silver nanoparticles. *Process Biochemistry*, 2012. **47**(10): pp. 1516–1520.

16. Kumar, A. et al., Metabolites in plants and its classification. *World Journal of Pharmaceutical Sciences*, 2015. **4**(1): pp. 287–305.

17. Wink, M., *Annual Plant Reviews, Biochemistry of Plant Secondary Metabolism*. Vol. 40. 2011, John Wiley & Sons, Singapore.

18. Wink, M., Evolution of secondary plant metabolism, in *eLS*. 2016, John Wiley & Sons, Ltd, Chichester. doi:10.1002/9780470015902.a0001922.pub3.

19. Wink, M., Modes of action of herbal medicines and plant secondary metabolites. *Medicines*, 2015. **2**(3): pp. 251–286.

20. Kim, J.S. et al., Antimicrobial effects of silver nanoparticles. *Nanomedicine: Nanotechnology, Biology and Medicine*, 2007. **3**(1): pp. 95–101.

21. Duan, H., D. Wang, and Y. Li, Green chemistry for nanoparticle synthesis. *Chemical Society Reviews*, 2015. **44**(16): pp. 5778–5792.

22. Mohammadinejad, R. et al., Plant-derived nanostructures: types and applications. *Green Chemistry*, 2016. **18**(1): pp. 20–52.

23. Makarov, V. et al., "Green" nanotechnologies: synthesis of metal nanoparticles using plants. *Acta Naturae (англоязычная версия)*, 2014. **6**(1): pp. 35–44.

24. Sundrarajan, M. and S. Gowri, Green synthesis of titanium dioxide nanoparticles by Nyctanthes arbortristis leaves extract. *Chalcogenide Letters*, 2011. **8**(8): pp. 447–451.

25. Rai, M. and C. Posten, *Green Biosynthesis of Nanoparticles: Mechanisms and Applications*. 2013, CABI, Boston.

26. Hoge, W., G. Winch, and D. Yearsley, *Anti-microbial utility and kitchen wipe utilizing metallic silver as an oligodynamic agent*. 2003, Google Patents.

27. Zhao, B. et al., Silver microspheres for application as hydrogen peroxide sensor. *Electrochemistry Communications*, 2009. **11**(8): pp. 1707–1710.

28. Bowler, P. et al., Microbicidal properties of a silver-containing Hydrofiber® dressing against a variety of burn wound pathogens. *Journal of Burn Care & Rehabilitation*, 2004. **25**(2): pp. 192–196.

29. Lansdown, A.B., Silver in health care: antimicrobial effects and safety in use, in Hipler, U.-C. and Elsner, P. (eds), *Biofunctional Textiles and the Skin*. Current Problems in Dermatology, vol 33. 2006, Karger Publishers, Basel. pp. 17–34.

30. Korbekandi, H. et al., Green biosynthesis of silver nanoparticles using Althaea officinalis radix hydroalcoholic extract. *Artificial Cells, Nanomedicine, and Biotechnology*, 2016. **44**(1): pp. 209–215.

31. Iravani, S., Green synthesis of metal nanoparticles using plants. *Green Chemistry*, 2011. **13**(10): pp. 2638–2650.

32. Kumar, S.S.D. et al., Cellular imaging and bactericidal mechanism of green-synthesized silver nanoparticles against human pathogenic bacteria. *Journal of Photochemistry and Photobiology B: Biology*, 2018. **178**: pp. 259–269.

33. Rai, M., A. Yadav, and A. Gade, Silver nanoparticles as a new generation of antimicrobials. *Biotechnology Advances*, 2009. **27**(1): pp. 76–83.

34. Fayaz, A.M. et al., Biogenic synthesis of silver nanoparticles and their synergistic effect with antibiotics: a study against gram-positive and gram-negative bacteria. *Nanomedicine: Nanotechnology, Biology and Medicine*, 2010. **6**(1): pp. 103–109.

35. You, C. et al., The progress of silver nanoparticles in the antibacterial mechanism, clinical application and cytotoxicity. *Molecular Biology Reports*, 2012. **39**(9): pp. 9193–9201.

36. Prabhu, S. and E.K. Poulose, Silver nanoparticles: mechanism of antimicrobial action, synthesis, medical applications, and toxicity effects. *International Nano Letters*, 2012. **2**(1): p. 32.

37. Sondi, I. and B. Salopek-Sondi, Silver nanoparticles as antimicrobial agent: a case study on E. coli as a model for Gram-negative bacteria. *Journal of Colloid and Interface Science*, 2004. **275**(1): pp. 177–182.

38. Danilczuk, M. et al., Conduction electron spin resonance of small silver particles. *Spectrochimica Acta Part A: Molecular and Biomolecular Spectroscopy*, 2006. **63**(1): pp. 189–191.

39. Matsumura, Y. et al., Mode of bactericidal action of silver zeolite and its comparison with that of silver nitrate. *Applied and Environmental Microbiology*, 2003. **69**(7): pp. 4278–4281.

40. Morones, J.R. et al., The bactericidal effect of silver nanoparticles. *Nanotechnology*, 2005. **16**(10): p. 2346.

41. Hatchett, D.W. and H.S. White, Electrochemistry of sulfur adlayers on the low-index faces of silver. *The Journal of Physical Chemistry*, 1996. **100**(23): pp. 9854–9859.

42. Shrivastava, S. et al., Characterization of enhanced antibacterial effects of novel silver nanoparticles. *Nanotechnology*, 2007. **18**(22): p. 225103.

43. Bar, H. et al., Green synthesis of silver nanoparticles using latex of Jatropha curcas. *Colloids and Surfaces A: Physicochemical and Engineering Aspects*, 2009. **339**(1–3): pp. 134–139.

44. Rivera-Rangel, R.D. et al., Green synthesis of silver nanoparticles in oil-in-water microemulsion and nano-emulsion using geranium leaf aqueous extract as a reducing agent. *Colloids and Surfaces A: Physicochemical and Engineering Aspects*, 2018. **536**: pp. 60–67.

45. Nakkala, J.R., R. Mata, and S.R. Sadras, Green synthesized nano silver: synthesis, physicochemical profiling, antibacterial, anticancer activities and biological in vivo toxicity. *Journal of Colloid and Interface Science*, 2017. **499**: pp. 33–45.

46. Chaudhuri, S.K., S. Chandela, and L. Malodia, Plant mediated green synthesis of silver nanoparticles using tecomella undulata leaf extract and their characterization. *Nano Biomedicine and Engineering*, 2016. **8**(1): pp. 1–8.

47. Chinnappan, S. et al., Biomimetic synthesis of silver nanoparticles using flower extract of Bauhinia purpurea and its antibacterial activity against clinical pathogens. *Environmental Science and Pollution Research*, 2018. **25**(1): pp. 963–969.

48. Ebrahiminezhad, A. et al., Green synthesis of silver nanoparticles capped with natural carbohydrates using ephedra intermedia. *Nanoscience and Nanotechnology-Asia*, 2017. **7**(1): pp. 104–112.

49. Dong, C. et al., Wolfberry fruit (Lycium barbarum) extract mediated novel route for the green synthesis of silver nanoparticles. *Optik-International Journal for Light and Electron Optics*, 2017. **130**: pp. 162–170.

50. Francis, S., E.P. Koshy, and B. Mathew, Green synthesis of Stereospermum suaveolens capped silver and gold nanoparticles and assessment of their innate antioxidant, antimicrobial and antiproliferative activities. *Bioprocess and Biosystems Engineering*, 2018. **41**: pp. 1–13.

51. Miri, A. et al., Plant-mediated biosynthesis of silver nanoparticles using Prosopis farcta extract and its antibacterial properties. *Spectrochimica Acta Part A: Molecular and Biomolecular Spectroscopy*, 2015. **141**: pp. 287–291.

52. Walter, D., Primary particles–agglomerates–aggregates. in Deutsche Forschungsgemeinschaft (DFG) (ed), *Nanomaterials*. 2013, Wiley-VCH Verlag GmbH & Co. KGaA, Bonn. pp. 9–24.

53. Cox, A. et al., Reprint of: silver and titanium dioxide nanoparticle toxicity in plants: a review of current research. *Plant Physiology and Biochemistry*, 2017. **110**: pp. 33–49.

54. Zheng, X. et al., Comprehensive analysis of transcriptional and proteomic profiling reveals silver nanoparticles-induced toxicity to bacterial denitrification. *Journal of Hazardous Materials*, 2018. **344**: pp. 291–298.

55. Bollyn, J. et al., Transformation-dissolution reactions partially explain adverse effects of metallic silver nanoparticles to soil nitrification in different soils. *Environmental Toxicology and Chemistry*, 2018. **37**(8): pp. 2123–2131.

56. Wang, E. et al., Silver nanoparticle induced toxicity to human sperm by increasing ROS (reactive oxygen species) production and DNA damage. *Environmental Toxicology and Pharmacology*, 2017. **52**: pp. 193–199.

57. Sthijns, M.M. et al., Silver nanoparticles induce hormesis in A549 human epithelial cells. *Toxicology in Vitro*, 2017. **40**: pp. 223–233.

58. Mieszawska, A.J. et al., Multifunctional gold nanoparticles for diagnosis and therapy of disease. *Molecular Pharmaceutics*, 2013. **10**(3): pp. 831–847.

59. Kumar, A., X. Zhang, and X.-J. Liang, Gold nanoparticles: emerging paradigm for targeted drug delivery system. *Biotechnology Advances*, 2013. **31**(5): pp. 593–606.

60. Niu, J. et al., Transdermal gene delivery by functional peptide-conjugated cationic gold nanoparticle reverses the progression and metastasis of cutaneous melanoma. *ACS Applied Materials & Interfaces*, 2017. **9**(11): pp. 9388–9401.

61. Satriano, C. et al., Angiogenin-mimetic peptide functionalised gold nanoparticles for cancer therapy applications. *Microchemical Journal*, 2018. **136**: pp. 157–163.

62. Lai, T.-S., T.-C. Chang, and S.-C. Wang, Gold nanoparticle-based colorimetric methods to determine protein contents in artificial urine using membrane micro-concentrators and mobile phone camera. *Sensors and Actuators B: Chemical*, 2017. **239**: pp. 9–16.

63. Zhan, L. et al., Magnetic bead-based sandwich immunoassay for viral pathogen detection by employing gold nanoparticle as carrier. *Journal of Analysis and Testing*, 2017. **1**(4): pp. 298–305.

64. Ma, H. et al., A gold nanoparticle based fluorescent probe for simultaneous recognition of single-stranded DNA and double-stranded DNA. *Microchimica Acta*, 2018. **185**(2): p. 93.

65. Vial, S., R.L. Reis, and J.M. Oliveira, Recent advances using gold nanoparticles as a promising multimodal tool for tissue engineering and regenerative medicine. *Current Opinion in Solid State and Materials Science*, 2017. **21**(2): pp. 92–112.

66. Hartung, G. and G. Mansoori, In vivo general trends, filtration and toxicity of nanoparticles. *Journal of Nanomaterials and Molecular Nanotechnology*, 2013. **2**(3): pp. 17–22.

67. Elia, P. et al., Green synthesis of gold nanoparticles using plant extracts as reducing agents. *International Journal of Nanomedicine*, 2014. **9**: p. 4007.

68. Ikram, S., Synthesis of gold nanoparticles using plant extract: an overview. *Archivos De Medicina*, 2015. **1**(1): p. 5.

69. Grzelczak, M. et al., Shape control in gold nanoparticle synthesis. *Chemical Society Reviews*, 2008. **37**(9): pp. 1783–1791.

70. Shukla, D. and P.S. Vankar, Synthesis of plant parts mediated gold nanoparticles. *International Journal of Green Nanotechnology*, 2012. **4**(3): pp. 277–288.

71. Sharma, P. et al., Green synthesis of silver nanoparticle capped with Allium cepa and their catalytic reduction of textile dyes: an ecofriendly approach. *Journal of Polymers and the Environment*, 2018. **26**(5): pp. 1795–1803.

72. Dykman, L. and N. Khlebtsov, Gold nanoparticles in biomedical applications: recent advances and perspectives. *Chemical Society Reviews*, 2012. **41**(6): pp. 2256–2282.

73. Zhang, H. et al., Use of metal oxide nanoparticle band gap to develop a predictive paradigm for oxidative stress and acute pulmonary inflammation. *ACS Nano*, 2012. **6**(5): pp. 4349–4368.

74. Kim, K.-T. et al., Gold nanoparticles disrupt zebrafish eye development and pigmentation. *Toxicological Sciences*, 2013. **133**(2): pp. 275–288.

75. Leonavičiené, L. et al., Effect of gold nanoparticles in the treatment of established collagen arthritis in rats. *Medicina (Kaunas)*, 2012. **48**(2): pp. 91–101.

76. Tiedemann, D. et al., Reprotoxicity of gold, silver, and gold–silver alloy nanoparticles on mammalian gametes. *Analyst*, 2014. **139**(5): pp. 931–942.

77. Coradeghini, R. et al., Size-dependent toxicity and cell interaction mechanisms of gold nanoparticles on mouse fibroblasts. *Toxicology Letters*, 2013. **217**(3): pp. 205–216.

78. Selim, M.E., Y.M. Abd-Elhakim, and L.Y. Al-Ayadhi, Pancreatic response to gold nanoparticles includes decrease of oxidative stress and inflammation in autistic diabetic model. *Cellular Physiology and Biochemistry*, 2015. **35**(2): pp. 586–600.

79. Truong, L. et al., Persistent adult zebrafish behavioral deficits results from acute embryonic exposure to gold nanoparticles. *Comparative Biochemistry and Physiology Part C: Toxicology & Pharmacology*, 2012. **155**(2): pp. 269–274.

80. Chen, Y.-S. et al., Assessment of the in vivo toxicity of gold nanoparticles. *Nanoscale Research Letters*, 2009. **4**(8): p. 858.

81. Bondarenko, O. et al., Toxicity of Ag, CuO and ZnO nanoparticles to selected environmentally relevant test organisms and mammalian cells in vitro: a critical review. *Archives of Toxicology*, 2013. **87**(7): pp. 1181–1200.

82. Hadrup, N. and H.R. Lam, Oral toxicity of silver ions, silver nanoparticles and colloidal silver–a review. *Regulatory Toxicology and Pharmacology*, 2014. **68**(1): pp. 1–7.

83. Kermanizadeh, A. et al., A multilaboratory toxicological assessment of a panel of 10 engineered nanomaterials to human health—ENPRA project—the

highlights, limitations, and current and future challenges. *Journal of Toxicology and Environmental Health, Part B*, 2016. **19**(1): pp. 1–28.

84. Hornos Carneiro, M.F. and F. Barbosa Jr, Gold nanoparticles: a critical review of therapeutic applications and toxicological aspects. *Journal of Toxicology and Environmental Health, Part B*, 2016. **19**(3–4): pp. 129–148.

85. Dubey, S.P., M. Lahtinen, and M. Sillanpää, Green synthesis and characterizations of silver and gold nanoparticles using leaf extract of Rosa rugosa. *Colloids and Surfaces A: Physicochemical and Engineering Aspects*, 2010. **364**(1–3): pp. 34–41.

86. Suman, T. et al., The green synthesis of gold nanoparticles using an aqueous root extract of Morinda citrifolia L. *Spectrochimica Acta Part A: Molecular and Biomolecular Spectroscopy*, 2014. **118**: pp. 11–16.

87. Sadeghi, B., M. Mohammadzadeh, and B. Babakhani, Green synthesis of gold nanoparticles using Stevia rebaudiana leaf extracts: characterization and their stability. *Journal of Photochemistry and Photobiology B: Biology*, 2015. **148**: pp. 101–106.

88. Paul, B. et al., Green synthesis of gold nanoparticles using Pogestemon benghalensis (B) O. Ktz. leaf extract and studies of their photocatalytic activity in degradation of methylene blue. *Materials Letters*, 2015. **148**: pp. 37–40.

89. Mata, R., A. Bhaskaran, and S.R. Sadras, Green-synthesized gold nanoparticles from Plumeria alba flower extract to augment catalytic degradation of organic dyes and inhibit bacterial growth. *Particuology*, 2016. **24**: pp. 78–86.

90. Yu, J. et al., Facile one-step green synthesis of gold nanoparticles using Citrus maxima aqueous extracts and its catalytic activity. *Materials Letters*, 2016. **166**: pp. 110–112.

91. Bhau, B. et al., Green synthesis of gold nanoparticles from the leaf extract of Nepenthes khasiana and antimicrobial assay. *Advanced Materials Letters*, 2015. **6**(1): pp. 55–58.

92. Mollick, M.M.R. et al., Anticancer (in vitro) and antimicrobial effect of gold nanoparticles synthesized using Abelmoschus esculentus (L.) pulp extract via a green route. *RSC Advances*, 2014. **4**(71): pp. 37838–37848.

93. Gonnelli, C. et al., Green synthesis of gold nanoparticles from extracts of cucurbita pepo L. Leaves: insights on the role of plant ageing, in Piotto, S., Rossi, F., Concilio, S., Reverchon, E. and Cattaneo G. (eds), *Advances in Bionanomaterials*. Lecture Notes in Bioengineering. 2018, Springer, Cham. pp. 155–164.

94. Bae, Y.-H. et al., Synergistic effects of segregated network by polymethylmethacrylate beads and sintering of copper nanoparticles on thermal and electrical properties of epoxy composites.

Composites Science and Technology, 2018. **155**: pp. 144–150.

95. Fan, D. et al., Synthesis, thermal conductivity and anti-oxidation properties of copper nanoparticles encapsulated within few-layer h-BN. *Ceramics International*, 2018. **44**(1): pp. 1205–1208.

96. Bhamore, J.R. et al., One-step eco-friendly approach for the fabrication of synergistically engineered fluorescent copper nanoclusters: sensing of Hg 2+ ion and cellular uptake and bioimaging properties. *New Journal of Chemistry*, 2018. **42**: pp. 1510–1520.

97. Mosleh, S. et al., Sonochemical-assisted synthesis of CuO/Cu₂O/Cu nanoparticles as efficient photocatalyst for simultaneous degradation of pollutant dyes in rotating packed bed reactor: LED illumination and central composite design optimization. *Ultrasonics Sonochemistry*, 2018. **40**: pp. 601–610.

98. Sivaranjana, P. et al., Formulation and characterization of in situ generated copper nanoparticles reinforced cellulose composite films for potential antimicrobial applications. *Journal of Macromolecular Science, Part A*, 2018. **55**(1): pp. 58–65.

99. Subhankari, I. and P. Nayak, Synthesis of copper nanoparticles using Syzygium aromaticum (Cloves) aqueous extract by using green chemistry. *World Journal of Nano Science and Technology*, 2013. **2**(1): pp. 14–17.

100. Saranyaadevi, K. et al., Synthesis and characterization of copper nanoparticle using Capparis zeylanica leaf extract. *International Journal of ChemTech Research*, 2014. **6**(10): pp. 4533–4541.

101. Kolekar, R. et al., Biosynthesis of copper nanoparticles using aqueous extract of Eucalyptus sp. plant leaves. *Current Science*, 2015. **109**: p. 255.

102. Soomro, R.A. et al., Synthesis of air stable copper nanoparticles and their use in catalysis. *Advanced Materials Letters*, 2014. **5**(4): pp. 191–198.

103. Joshi, S. et al., Radiation induced synthesis and characterization of copper nanoparticles. *Nanostructured Materials*, 1998. **10**(7): pp. 1135–1144.

104. Yeh, M.-S. et al., Formation and characteristics of Cu colloids from CuO powder by laser irradiation in 2-propanol. *The Journal of Physical Chemistry B*, 1999. **103**(33): pp. 6851–6857.

105. Kim, Y.H. et al., Synthesis of oleate capped Cu nanoparticles by thermal decomposition. *Colloids and Surfaces A: Physicochemical and Engineering Aspects*, 2006. **284**: pp. 364–368.

106. Park, B.K. et al., Synthesis and size control of monodisperse copper nanoparticles by polyol method. *Journal of Colloid and Interface Science*, 2007. **311**(2): pp. 417–424.

107. Dang, T.M.D. et al., Synthesis and optical properties of copper nanoparticles prepared by a chemical reduction method. *Advances in Natural Sciences: Nanoscience and Nanotechnology*, 2011. **2**(1): p. 015009.

108. Ijaz, F. et al., Green synthesis of copper oxide nanoparticles using Abutilon indicum leaf extract: antimicrobial, antioxidant and photocatalytic dye degradation activities. *Tropical Journal of Pharmaceutical Research*, 2017. **16**(4): pp. 743–753.

109. Chung, I.M. et al., Green synthesis of copper nanoparticles using Eclipta prostrata leaves extract and their antioxidant and cytotoxic activities. *Experimental and Therapeutic Medicine*, 2017. **14**(1): pp. 18–24.

110. Shende, S. et al., Green synthesis of copper nanoparticles by Citrus medica Linn.(Idilimbu) juice and its antimicrobial activity. *World Journal of Microbiology and Biotechnology*, 2015. **31**(6): pp. 865–873.

111. Nasrollahzadeh, M. and S.M. Sajadi, Green synthesis of copper nanoparticles using Ginkgo biloba L. leaf extract and their catalytic activity for the Huisgen [3+ 2] cycloaddition of azides and alkynes at room temperature. *Journal of Colloid and Interface Science*, 2015. **457**: pp. 141–147.

112. Nasrollahzadeh, M., S.M. Sajadi, and M. Khalaj, Green synthesis of copper nanoparticles using aqueous extract of the leaves of Euphorbia esula L and their catalytic activity for ligand-free Ullmann-coupling reaction and reduction of 4-nitrophenol. *RSC Advances*, 2014. **4**(88): pp. 47313–47318.

113. Thanh, N.T. and L.A. Green, Functionalisation of nanoparticles for biomedical applications. *Nano Today*, 2010. **5**(3): pp. 213–230.

114. Lim, J. and S.A. Majetich, Composite magnetic–plasmonic nanoparticles for biomedicine: manipulation and imaging. *Nano Today*, 2013. **8**(1): pp. 98–113.

115. Puntes, V.F., K.M. Krishnan, and A.P. Alivisatos, Colloidal nanocrystal shape and size control: the case of cobalt. *Science*, 2001. **291**(5511): pp. 2115–2117.

116. Dumestre, F. et al., Shape control of thermodynamically stable cobalt nanorods through organometallic chemistry. *Angewandte Chemie*, 2002. **114**(22): pp. 4462–4465.

117. He, C. et al., Facile synthesis of hollow porous cobalt spheres and their enhanced electromagnetic properties. *Journal of Materials Chemistry*, 2012. **22**(41): pp. 22160–22166.

118. Pankhurst, Q. et al., Progress in applications of magnetic nanoparticles in biomedicine. *Journal of Physics D: Applied Physics*, 2009. **42**(22): p. 224001.

119. Dey, C. et al., Improvement of drug delivery by hyperthermia treatment using magnetic cubic cobalt ferrite nanoparticles. *Journal of Magnetism and Magnetic Materials*, 2017. **427**: pp. 168–174.

120. Montiel, M.G. et al., Synthesis and thermal behavior of metallic cobalt micro and nanostructures. *Nano-Micro Letters*, 2011. **3**(1): pp. 12–19.

121. Zhang, Y.-J. et al., Solvothermal synthesis of magnetic chains self-assembled by flowerlike cobalt submicrospheres. *Crystal Growth and Design*, 2008. **8**(9): pp. 3206–3212.

122. Yan, D. et al., Shape-controlled synthesis of cobalt particles by a surfactant-free solvothermal method and their catalytic application to the thermal decomposition of ammonium perchlorate. *CrystEngComm*, 2015. **17**(47): pp. 9062–9069.

123. Dungan, R.S. and W. Frankenberger, Microbial transformations of selenium and the bioremediation of seleniferous environments. *Bioremediation Journal*, 1999. **3**(3): pp. 171–188.

124. Tapiero, H., D. Townsend, and K. Tew, The antioxidant role of selenium and seleno-compounds. *Biomedicine & Pharmacotherapy*, 2003. **57**(3–4): pp. 134–144.

125. Cox, D.N. and K. Bastiaans, Understanding Australian consumers' perceptions of selenium and motivations to consume selenium enriched foods. *Food Quality and Preference*, 2007. **18**(1): pp. 66–76.

126. Turakainen, M., H. Hartikainen, and M.M. Seppänen, Effects of selenium treatments on potato (Solanum tuberosum L.) growth and concentrations of soluble sugars and starch. *Journal of Agricultural and Food Chemistry*, 2004. **52**(17): pp. 5378–5382.

127. Lyons, G.H. et al., Selenium increases seed production in Brassica. *Plant and Soil*, 2009. **318**(1–2): pp. 73–80.

128. Vinceti, M. et al., *Selenium for preventing cancer*. The Cochrane Library, 2018.

129. Khandelwal, S. et al., Cytotoxicity of selenium trastuzumab and bevacizumab immunoconjugates against triple negative breast cancer cells. *International Journal of Molecular Sciences*, 2018. **19**(11): pp. 3352.

130. Ip, C. et al., In vitro and in vivo studies of methylseleninic acid: evidence that a monomethylated selenium metabolite is critical for cancer chemoprevention. *Cancer Research*, 2000. **60**(11): pp. 2882–2886.

131. Miller, S. et al., Selenite protects human endothelial cells from oxidative damage and induces thioredoxin reductase. *Clinical Science*, 2001. **100**(5): pp. 543–550.

132. Sharma, G. et al., Biomolecule-mediated synthesis of selenium nanoparticles using dried Vitis vinifera (raisin) extract. *Molecules*, 2014. **19**(3): pp. 2761–2770.

133. Kong, H. et al., Synthesis and antioxidant properties of gum arabic-stabilized selenium nanoparticles. *International Journal of Biological Macromolecules*, 2014. **65**: pp. 155–162.

134. Zhang, S.-Y. et al., Synthesis of selenium nanoparticles in the presence of polysaccharides. *Materials Letters*, 2004. **58**(21): pp. 2590–2594.

135. Husen, A. and K.S. Siddiqi, Plants and microbes assisted selenium nanoparticles: characterization and application. *Journal of Nanobiotechnology*, 2014. **12**(1): p. 28.

136. Husen, A. and K.S. Siddiqi, Phytosynthesis of nanoparticles: concept, controversy and application. *Nanoscale Research Letters*, 2014. **9**(1): p. 229.

137. Ramamurthy, C. et al., Green synthesis and characterization of selenium nanoparticles and its augmented cytotoxicity with doxorubicin on cancer cells. *Bioprocess and Biosystems Engineering*, 2013. **36**(8): pp. 1131–1139.

138. Fernandez-Garcia, M. et al., Nanostructured oxides in chemistry: characterization and properties. *Chemical Reviews*, 2004. **104**(9): pp. 4063–4104.

139. Rodríguez, J.A. and M. Fernández-García, *Synthesis, Properties, and Applications of Oxide Nanomaterials*. 2007, John Wiley & Sons, New Jersey.

140. Blesa, M.A., *Chemical Dissolution of Metal Oxides*. 2018, CRC Press, Boca Raton.

141. Bäumer, M. and H.-J. Freund, Metal deposits on well-ordered oxide films. *Progress in Surface Science*, 1999. **61**(7–8): pp. 127–198.

142. Rodriguez, J. et al., Activation of gold on titania: Adsorption and reaction of SO_2 on $Au/TiO_2(110)$. *Journal of the American Chemical Society*, 2002. **124**(18): p. 5242–5250.

143. Charbgoo, F., M. Ramezani, and M. Darroudi, Biosensing applications of cerium oxide nanoparticles: advantages and disadvantages. *Biosensors and Bioelectronics*, 2017. **96**: pp. 33–43.

144. Fernández-García, M., J.C. Conesa, and F. Illas, Effect of the Madelung potential value and symmetry on the adsorption properties of adsorbate/oxide systems. *Surface Science*, 1996. **349**(2): pp. 207–215.

145. Rodriguez, J.A. et al., Reaction of H_2S and S_2 with metal/oxide surfaces: band-gap size and chemical reactivity. *The Journal of Physical Chemistry B*, 1998. **102**(28): pp. 5511–5519.

146. Fernández-García, M. and J.A. Rodriguez, Metal oxide nanoparticles, in Scott, R. A. (ed), *Encyclopedia of Inorganic and Bioinorganic Chemistry*, 2011, John Wiley and Sons, Inc, New Jersey. pp. 1–22.

147. Auld, D.S., Zinc coordination sphere in biochemical zinc sites, in Maret, W. (ed.), *Zinc Biochemistry, Physiology, and Homeostasis*. 2001, Springer, Netherlands. pp. 85–127.

148. Rout, G.R. and P. Das, Effect of metal toxicity on plant growth and metabolism: I. Zinc, in Lichtfouse, E., Navarrete, M., Debaeke, P., Véronique, S. and Alberola, C. (eds), *Sustainable Agriculture*. 2009, Springer, Netherlands. pp. 873–884.

149. Aravind, P. and M. Prasad, Zinc protects chloroplasts and associated photochemical functions in cadmium exposed Ceratophyllum demersum L., a freshwater macrophyte. *Plant Science*, 2004. **166**(5): pp. 1321–1327.

150. Hussain, I. et al., Green synthesis of nanoparticles and its potential application. *Biotechnology Letters*, 2016. **38**(4): pp. 545–560.

151. Singh, A. et al., Green synthesis of nano zinc oxide and evaluation of its impact on germination and metabolic activity of Solanum lycopersicum. *Journal of Biotechnology*, 2016. **233**: pp. 84–94.

152. Yadav, A. et al., Functional finishing in cotton fabrics using zinc oxide nanoparticles. *Bulletin of Materials Science*, 2006. **29**(6): pp. 641–645.

153. Singh, N. et al., Zinc oxide nanoparticles as fertilizer for the germination, growth and metabolism of vegetable crops. *Journal of Nanoengineering and Nanomanufacturing*, 2013. **3**(4): pp. 353–364.

154. Raja, A. et al., Eco-friendly preparation of zinc oxide nanoparticles using Tabernaemontana divaricata and its photocatalytic and antimicrobial activity. *Journal of Photochemistry and Photobiology B: Biology*, 2018. **181**: pp. 53–58.

155. Wang, P. et al., Fate of ZnO nanoparticles in soils and cowpea (Vigna unguiculata). *Environmental Science & Technology*, 2013. **47**(23): pp. 13822–13830.

156. Song, H. et al., Fabrication of CeO_2 nanoparticles decorated three-dimensional flower-like BiOI composites to build pn heterojunction with highly enhanced visible-light photocatalytic performance. *Journal of Colloid and Interface Science*, 2018. **512**: pp. 325–334.

157. Tang, J. et al., Sustainable pollutant removal by periphytic biofilm via microbial composition shifts induced by uneven distribution of CeO_2 nanoparticles. *Bioresource Technology*, 2018. **248**: pp. 75–81.

158. Kumar, K.M. et al., Green synthesis of Ce3+ rich CeO_2 nanoparticles and its antimicrobial studies. *Materials Letters*, 2018. **214**: pp. 15–19.

159. Eriksson, P. et al., Cerium oxide nanoparticles with antioxidant capabilities and gadolinium integration for MRI contrast enhancement. *Scientific Reports*, 2018. **8**(1): p. 6999.

160. Self, W.T. et al., *Neuronal protection by cerium oxide nanoparticles*. 2018, Google Patents.

161. Nguyet, N.T. et al., Highly sensitive DNA sensors based on cerium oxide nanorods. *Journal of Physics and Chemistry of Solids*, 2018. **115**: pp. 18–25.

162. Mishra, P.K. et al., Surfactant-free one-pot synthesis of low-density cerium oxide nanoparticles for adsorptive removal of arsenic species. *Environmental Progress & Sustainable Energy*, 2018. **37**(1): pp. 221–231.

163. Charbgoo, F., M.B. Ahmad, and M. Darroudi, Cerium oxide nanoparticles: green synthesis and biological applications. *International Journal of Nanomedicine*, 2017. **12**: p. 1401.

164. Kaittanis, C. et al., A cerium oxide nanoparticle-based device for the detection of chronic inflammation via optical and magnetic resonance imaging. *Nanoscale*, 2012. **4**(6): pp. 2117–2123.

165. Perez, J.M. et al., *Cerium oxide nanoparticle-based device for the detection of reactive oxygen species and monitoring of chronic inflammation*. 2014, Google Patents.

166. Celardo, I. et al., Pharmacological potential of cerium oxide nanoparticles. *Nanoscale*, 2011. **3**(4): pp. 1411–1420.

167. Li, M. et al., Cerium oxide caged metal chelator: anti-aggregation and anti-oxidation integrated H 2 O 2-responsive controlled drug release for potential Alzheimer's disease treatment. *Chemical Science*, 2013. **4**(6): pp. 2536–2542.

168. Chen, H.-I. and H.-Y. Chang, Synthesis of nanocrystalline cerium oxide particles by the precipitation method. *Ceramics International*, 2005. **31**(6): pp. 795–802.

169. Jimmy, C.Y., L. Zhang, and J. Lin, Direct sonochemical preparation of high-surface-area nanoporous ceria and ceria–zirconia solid solutions. *Journal of Colloid and Interface Science*, 2003. **260**(1): pp. 240–243.

170. Yan, Z. et al., Hydrothermal synthesis of CeO_2 nanoparticles on activated carbon with enhanced desulfurization activity. *Energy & Fuels*, 2012. **26**(9): pp. 5879–5886.

171. Feng, X. et al., Converting ceria polyhedral nanoparticles into single-crystal nanospheres. *Science*, 2006. **312**(5779): pp. 1504–1508.

172. Darroudi, M. et al., Green synthesis and evaluation of metabolic activity of starch mediated nanoceria. *Ceramics International*, 2014. **40**(1): pp. 2041–2045.

173. Karakoti, A. et al., Redox-active radical scavenging nanomaterials. *Chemical Society Reviews*, 2010. **39**(11): pp. 4422–4432.

174. Alili, L. et al., Downregulation of tumor growth and invasion by redox-active nanoparticles. *Antioxidants & Redox Signaling*, 2013. **19**(8): pp. 765–778.

175. Priya, G.S. et al., Biosynthesis of Cerium oxide nanoparticles using Aloe barbadensis miller gel. *International Journal of Scientific Research Publications*, 2014. **4**(6): pp. 199–224.

176. Munusamy, S. et al., Synthesis and characterization of cerium oxide nanoparticles using Curvularia lunata and their antibacterial properties. *International Journal of Innovative Research in Science and Engineering*, 2014. **2**(1): pp. 318–323.

177. Pang, Y. et al., Preparation and application of stability enhanced magnetic nanoparticles for rapid removal of Cr (VI). *Chemical Engineering Journal*, 2011. **175**: pp. 222–227.

178. Luo, L.H. et al., Fe$_3$O$_4$/Rectorite composite: preparation, characterization and absorption properties from contaminant contained in aqueous solution. *Advanced Materials Research*, 2011. **287–290**: pp. 592–598.

179. Gupta, A.K. and M. Gupta, Synthesis and surface engineering of iron oxide nanoparticles for biomedical applications. *Biomaterials*, 2005. **26**(18): pp. 3995–4021.

180. Pal, B. and M. Sharon, Preparation of iron oxide thin film by metal organic deposition from Fe (III)-acetylacetonate: a study of photocatalytic properties. *Thin Solid Films*, 2000. **379**(1–2): pp. 83–88.

181. Xu, P. et al., Use of iron oxide nanomaterials in wastewater treatment: a review. *Science of the Total Environment*, 2012. **424**: pp. 1–10.

182. Unni, M. et al., Thermal decomposition synthesis of iron oxide nanoparticles with diminished magnetic dead layer by controlled addition of oxygen. *ACS Nano*, 2017. **11**(2): pp. 2284–2303.

183. Wang, L. et al., Exerting enhanced permeability and retention effect driven delivery by ultrafine iron oxide nanoparticles with T 1–T 2 switchable magnetic resonance imaging contrast. *ACS Nano*, 2017. **11**(5): pp. 4582–4592.

184. Shah, A. and M.A. Dobrovolskaia, Immunological effects of iron oxide nanoparticles and iron-based complex drug formulations: therapeutic benefits, toxicity, mechanistic insights, and translational considerations. *Nanomedicine: Nanotechnology, Biology and Medicine*, 2018. **14**(3): pp. 977–990.

185. Boyer, C. et al., The design and utility of polymer-stabilized iron-oxide nanoparticles for nanomedicine applications. *NPG Asia Materials*, 2010. **2**(1): p. 23.

186. Dias, A. et al., A biotechnological perspective on the application of iron oxide magnetic colloids modified with polysaccharides. *Biotechnology Advances*, 2011. **29**(1): pp. 142–155.

187. Bystrzejewski, M. et al., Carbon-encapsulated magnetic nanoparticles as separable and mobile sorbents of heavy metal ions from aqueous solutions. *Carbon*, 2009. **47**(4): pp. 1201–1204.

188. Selvan, S.T. et al., Functional and multifunctional nanoparticles for bioimaging and biosensing. *Langmuir*, 2009. **26**(14): pp. 11631–11641.

189. Jeong, U. et al., Superparamagnetic colloids: controlled synthesis and niche applications. *Advanced Materials*, 2007. **19**(1): pp. 33–60.

190. Machala, L., R. Zboril, and A. Gedanken, Amorphous iron (III) oxide a review. *The Journal of Physical Chemistry B*, 2007. **111**(16): pp. 4003–4018.

191. Hassanjani-Roshan, A. et al., Synthesis of iron oxide nanoparticles via sonochemical method and their characterization. *Particuology*, 2011. **9**(1): pp. 95–99.

192. Gotić, M., G. Dražić, and S. Musić, Hydrothermal synthesis of α-Fe$_2$O$_3$ nanorings with the help of divalent metal cations, Mn2+, Cu2+, Zn2+ and Ni2+. *Journal of Molecular Structure*, 2011. **993**(1–3): pp. 167–176.

193. Bautista, M.C. et al., Surface characterisation of dextran-coated iron oxide nanoparticles prepared by laser pyrolysis and coprecipitation. *Journal of Magnetism and Magnetic Materials*, 2005. **293**(1): pp. 20–27.

194. Li, Y. and G.A. Somorjai, Nanoscale advances in catalysis and energy applications. *Nano Letters*, 2010. **10**(7): pp. 2289–2295.

195. Mohanraj, V. and Y. Chen, Nanoparticles-a review. *Tropical Journal of Pharmaceutical Research*, 2006. **5**(1): pp. 561–573.

196. Girginova, P.I. et al., Silica coated magnetite particles for magnetic removal of Hg2+ from water. *Journal of Colloid and Interface Science*, 2010. **345**(2): pp. 234–240.

197. Stumpf, H.C. et al., Thermal transformations of aluminas and alumina hydrates-reaction with 44% technical acid. *Industrial & Engineering Chemistry*, 1950. **42**(7): pp. 1398–1403.

198. Fernandez, E. et al., Adsorption and dissociation of water on relaxed alumina clusters: a first principles study. *Physica Status Solidi (b)*, 2005. **242**(4): pp. 807–809.

199. McHale, J. et al., Surface energies and thermodynamic phase stability in nanocrystalline aluminas. *Science*, 1997. **277**(5327): pp. 788–791.

200. Föller, A., *Magnesium Oxide and Its Applications*. 1978, Vollhardt, Berlin, Germany.

201. Stener, M., G. Fronzoni, and R. De Francesco, Core excitations in MgO: a DFT study with cluster models. *Chemical Physics*, 2005. **309**(1): pp. 49–58.

202. Srivastava, V., Y. Sharma, and M. Sillanpää, Green synthesis of magnesium oxide nanoflower and its application for the removal of divalent metallic species from synthetic wastewater. *Ceramics International*, 2015. **41**(5): pp. 6702–6709.

203. Al-Salim, N.I. et al., Characterisation and activity of sol–gel-prepared TiO$_2$ photocatalysts modified with Ca, Sr or Ba ion additives. *Journal of Materials Chemistry*, 2000. **10**(10): pp. 2358–2363.

204. Barbe, C.J. et al., Nanocrystalline titanium oxide electrodes for photovoltaic applications. *Journal of the American Ceramic Society*, 1997. **80**(12): pp. 3157–3171.

205. Carp, O., C.L. Huisman, and A. Reller, Photoinduced reactivity of titanium dioxide. *Progress in Solid State Chemistry*, 2004. **32**(1–2): pp. 33–177.

206. Ruiz, A.M. et al., Microstructure control of thermally stable TiO$_2$ obtained by hydrothermal process for

gas sensors. *Sensors and Actuators B: Chemical*, 2004. **103**(1–2): pp. 312–317.

207. Balantrapu, K. and D.V. Goia, Silver nanoparticles for printable electronics and biological applications. *Journal of Materials Research*, 2009. **24**(9): pp. 2828–2836.

208. Saxena, A., R. Tripathi, and R. Singh, Biological synthesis of silver nanoparticles by using onion (Allium cepa) extract and their antibacterial activity. *Digest Journal of Nanomaterials and Biostructures*, 2010. **5**(2): pp. 427–432.

Self-assembled Peptide Nanostructures and Their Applications

Apurba K. Das and
Tapas Ghosh
Indian Institute of Technology Indore

2.1 Introduction

Self-assembly is a process in which molecules assemble themselves to form an organized stable structure. Life emerged through different chemical reactions driven via self-assembly of biomolecules with different scales of complexity. This is a fully spontaneous process occurring with the help of several noncovalent interactions including van der Waals forces, hydrophobic effect, hydrogen bonding, and π–π stacking interactions (Tu and Tirrell 2004, Shi and Xu 2015). Self-assembly may occur at both the macroscale as well as the molecular level. One example of molecular-level-self-assembly is the "**bottom–up approach**" (Feynman 1959) in the imitation of nano-dimensional structure, where molecules self-assemble in the sub-nano range. In nature, there are two types of self-assembling processes: **intramolecular** and **intermolecular** self-assembly processes. The intramolecular self-assembly is generally seen in the "folding" of biomolecules like peptides, proteins, and DNA. The intermolecular self-assembly leads to nanostructure formation. Again, based on the dissipation of energy, the self-assembly processes are classified into **dynamic** and **static** (Whitesides and Grzybowski 2002). In a dynamic process, the self-assembled organized structure dissipates energy while in static category there is no dissipation of energy. Nowadays, various efforts for the development of new techniques based on the self-assembly of small bioactive and biocompatible molecules are showing huge opportunity in the field of molecular design bearing a broad range of applications.

Supramolecular chemistry, chemical biology, and materials engineering are now smartly interconnected to create self-assembling materials that achieve mechanical action, mimicking biological microstructures, and generation of molecular electronics or sensor devices (Hirst et al. 2008).

In nature, many biological scaffolds like sugars, amino acids, and nucleic acids offer the widest variety of functionality and cell signaling capacity with rapid and easy synthesis of complex molecules (Andrews and Tabor 1999). So, amino acid-based peptides are the most important scaffolds for designing a self-assembled structure. For the construction of self-assembled peptides, both natural and unnatural amino acids are used (Kerr et al. 1993, Hodgson and Sanderson 2004). By simply changing or manipulating the structure of amino acids, various self-assembled peptide-based nanostructures can be designed. In most of the cases, thermodynamically stable structures are formed through enthalpic and entropic effects present in both assembling subunits and the surroundings solvent molecules (Stephanopoulos et al. 2013). The α-helices and β-sheet structures in proteins are formed through self-assembly processes (Zhang et al. 2008, Yu et al. 2016). Peptide-based self-assembled nanostructures have significant advantages in the field of biomedical applications (Sun et al. 2017), cell culture (Sun et al. 2017, Zhang 1995), molecular electronics applications (Dinca 2007), and nanotechnological applications (Rosenman et al. 2011, Zhang 2003b) (Figure 2.1). Now objects and devices have been designed through the "top–down" and "bottom–up" approaches (Figure 2.2).

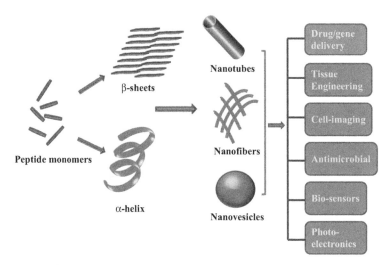

FIGURE 2.1 Schematic diagram of self-assembled peptides to different nanostructures and their various applications.

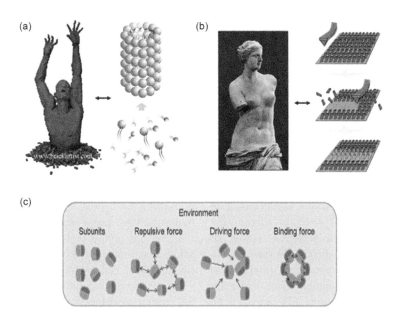

FIGURE 2.2 Bottom–up and top–down fabrication: (a) self-assembly of a micellar structure and the building of a sculpture using Lego pieces and (b) a nano-grafting patterned substrate and the crafting of a stone sculpture. (c) Schematic diagram of five key parts of a self-assembling system: the subunits; the repulsive, driving and binding forces; and the environment. (Reprinted with permission from Smith, K. H. et al. *Chem. Soc. Rev.*, 2011, 40, 4563–4577.)

2.1.1 Top–Down Approach

The top–down approach, a synonym of decomposition, is defined as the making of prearranged structures by breaking down or manipulating a system or components in a specific location (Smith et al. 2011). In the simplest form, the top–down approach starts with big elements, which are then broken down into smaller portions. The "top–down" technique was invented in 1978 when Alois Asnefeler proposed the method of lithography. Now, this strategy has started a new era in materials and biological sciences (Groves and Boxer 2002). Top–down techniques are used to prepare functional materials based on peptides and proteins. So, the combination of self-assembling materials that can be produced with this top–down approach may

have various uses in micro- or nanofabricated devices and bioelectronics.

2.1.2 Bottom–Up Approach

A famous quotation by Richard Feynman during a lecture entitled "there's plenty of room at the bottom" generated an idea for "bottom–up" approach for the fabrication of higher ordered structures via self-assembly process using individual atoms and molecules as building blocks (Feynman 1959). In supramolecular chemistry, the construction of molecules with well-defined design has potential applications in various fields of chemistry and material sciences. Generally, the bottom–up approach is the arrangement of smaller components into a more complex assembly (Zhang 2003a).

These techniques are often similar to a "seed" model, in which the starting material is small but eventually grows in complexity and completeness. Normally, the molecular assembly process is based on the bottom–up technique that helps in developing and constructing the higher ordered functional materials. Natural bioactive macromolecules and bioarchitectures like amino acids, sugars, nucleic acids or lipids are the building blocks for the development of higher ordered self-assembled functional materials by using bottom–up strategies. Now, the design of self-assembled supramolecular structures of peptides, proteins, and other amphiphiles includes tapes (Anilkumar and Jayakannan 2009), belts (Cui et al. 2009), fibers (Komatsu et al. 2009), tubes (Childers et al. 2010), and vesicles (Koley et al. 2011). The bottom–up approach is the most useful technique to construct different nanostructures.

2.2 Noncovalent Interactions Responsible for Molecular Self-assembly

The self-assembly of peptide molecules is driven by several noncovalent interactions. Noncovalent interactions like H-bonding interaction, electrostatic interaction, π–π stacking interaction, and van der Waals and hydrophobic interactions (Tokosz et al. 2010) are the key contributors toward self-assembly process. These interactions vary from one amino acid to the other due to the changes in their side chain. Nonpolar side chain-containing amino acids self-assemble through hydrophobic interactions, whereas polar side chain-containing amino acids self-assemble using H-bonding and electrostatic interactions. The aromatic side chain-containing amino acids self-assemble via π–π stacking interaction (Figure 2.3). Aside from an individual part, the whole peptide backbone also contributes to the self-assembly process by noncovalent interactions. These noncovalent interactions are weak in energy, but the assembly of binding residues over multiple self-assembling units may provide the foundation for stable assemblies. In general, self-assembling systems are composed of five important parts: the subunits, a driving force, a repulsive force, a binding force, and environments (Figure 2.2) (Smith et al. 2011). Subunits such as peptides or proteins are the building blocks of the assembled material, and the driving force allows them to interact by the random movement of the subunits. Repulsive forces like electrostatic or hydrophobic/hydrophilic repulsions are very essential for the reversibility of the system. The binding forces are the noncovalent interactions, including hydrogen bonding, van der Waals, ionic, dipole-dipole, and hydrophobic interactions, which provide stability to the assembled structures. The reversibility in the system is mainly generated due to comparable magnitudes of binding and repulsive forces. Finally, in the environment, the subunits, forces, and assembled structures interact with each other in order to form stable complexes.

Hydrogen bonding interaction is one type of electrostatic interaction that occurs between H and more electronegative atoms like O, N, and F. H-bonding plays important role in the formation of secondary structures of peptides and proteins. The multiple peptide backbones are stabilized by the H-bonding interaction formed between the amide and carbonyl groups in the backbone. Therefore, they can self-assemble to form parallel or antiparallel β-sheet-type structures depending upon the directions of peptide sequences. By utilizing the H-bonding pattern formed in α-helices, β-sheets, and coiled coils, one can design various peptide sequences which can self-assemble to form various nanostructures.

Hydrophobic interaction is likely one of the most important factors in the self-assembly process, mostly for the nonpolar side chain-containing amino acids. In most of the peptide amphiphiles (PA), the main driving force is hydrophobic interaction for the formation of various nanostructures. In PAs, both hydrophobic and hydrophilic parts are present. When these amphiphatic molecules are introduced into water, the hydrophobic parts tend to assemble together in order to decrease the surface area in contact with water, leaving the other hydrophilic parts exposed in water (Brack and Orgel 1975). This hydrophobic interaction is stabilized by favorable entropy rather than favorable enthalpy (Wang et al. 2016b).

Electrostatic interactions involving both attractive and repulsive forces have important effects on the self-assembly of charged amino acids-containing peptides. The positively charged peptides aggregate because of attractive electrostatic interactions with negatively charged peptides or even drugs. This phenomenon helps to form nanostructures that could be used in drug delivery applications (Kataoka et al. 2001). Various cell-penetrating peptides have been designed to form self-assembled nanostructures through electrostatic interaction (Chen et al. 2015a). In aromatic peptides, the main driving force for self-assembly is π–π stacking interaction. In pure organic solvents such as toluene and TFA, the π–π stacking interaction has a more efficient driving force (Zhu et al. 2010). The interactions for π–π stacking can promote directional growth, and they are strong in water due to their low solubility of molecules containing aromatic groups. The self-assembly of simple dipeptide FF (F = phenylalanine) in various nanostructures is the key example of the contribution of π–π stacking interaction in the self-assembly process (Guo et al. 2012).

Apart from the abovementioned interactions, van der Waals interactions also play a crucial role in the self-assembly process (Tahara et al. 2010). The strength of van der Waals interactions is weaker than that of a hydrogen bond. This interaction, which is associated with the interaction between the aliphatic side chains in peptides, provides important contribution toward various noncovalent interactions seen commonly in the self-assembly process. This interaction is generated as a result of the fluctuations of electron distribution of two closely spaced molecules. So, van der Waals force can be considered as one type of instantaneous electrostatic interaction.

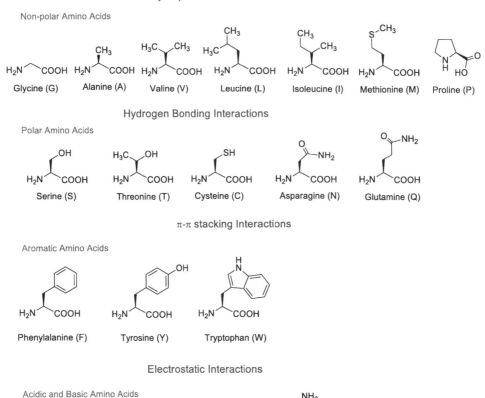

FIGURE 2.3 Category-wise representation of 20 coded amino acids.

2.3 Tuning of Self-assembly to Different Nanostructures

Molecular self-assembly is a spontaneous process in which molecules in a nonaggregated state transform to a well-defined ordered organized state. Though self-assembly is a spontaneous process, the morphology can be controlled by tuning external parameters. Several external stimuli like solvent, temperature, concentration, pH, ionic strength, ultrasound, shaking, and stirring are responsible for tuning the nanostructures (Lowik et al. 2010, Mart et al. 2006). Liu et al. have reviewed the control of molecular design and external stimuli to tune the morphologies of the nanostructures in self-assembled supramolecular gels (Zhang et al. 2014). Thus, here, we have discussed the tuning of self-assembled peptide nanostructures by controlling the external parameters (Figure 2.4).

The self-assembly process is driven by noncovalent interactions, and these interactions can be varied by temperature. An increase in temperature could break the intermolecular H-bonding interactions; on the other hand, however, it could increase hydrophobic interactions.

The change of the self-assembled structure of amphiphilic peptide RADA 16-I with increase in temperature is an example of tuning of peptide self-assembled nanostructure by temperature. RADA 16-I can be transformed from a well-defined 3D nanofiber to become aggregated/separated globular structures as the temperature increases from 25°C to 80°C (Figure 2.5) (Ye et al. 2008).

Switching of pH is one of the most facile techniques for controlling and directing the self-assembly of peptides molecules. Due to the structural diversity of amino acids, several peptides are extremely sensitive to pH change, thus leading to various structural transformations. Cote et al. described the mechanism of pH-controlled self-assembly of PAs into nanofibers (Cote et al. 2014). Again in case of Tyr-Aib-Ala (YUA), the tripeptide showed different self-assembled structures at different pH. In acidic pH (4.3–5.5), a hollow nanotubular structure was seen while at pH 6.4, YUA was seen as both nanotubes and nanovesicles. However, at higher pH (7.0–9.2), only nanovesicles were observed. These pH-sensitive nanovesicles were used for entrapment and slow release of several dyes (Bose et al. 2007). In this case, self-assembled sheet-like structures were formed. Further, the

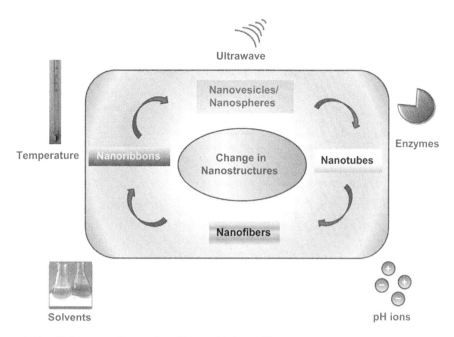

FIGURE 2.4 External stimuli that may change the self-assembled peptide nanostructures.

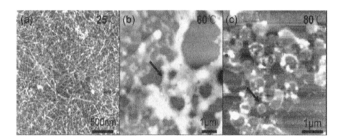

FIGURE 2.5 AFM images of RAD16-I peptide at (a) 25°C, (b) 60°C, and (c) 80°C temperature. (Reprinted with permission from Ye, Z. et al. *J. Pept. Sci.*, 2008, 14, 152–162. Copyright 2008, John Wiley and Sons.)

formation of nanotubes and nanovesicles was governed by the folding of layered β-sheet structures.

PAs also self-assemble to form various nanostructures like nanotubes, nanofibers, and nanovesicles (Huang et al. 2011). The amphiphiles are mainly composed of hydrophilic and hydrophobic parts. So the interactions of these hydrophilic and hydrophobic segments play a key role in the self-assembly process. Meng et al. reported the change in effect of the self-assembly of PAs by changing the number and type of amino acids in their hydrophilic and hydrophobic region (Meng et al. 2012). They changed the nature of the hydrophobic segment by altering the amino acids from alanine to valine to leucine and the number of amino acids in hydrophilic part from 1 to 5 in the peptide sequence X_6K_n (X = alanine, valine, leucine; n = 1–5). So, the self-assembly is controlled by altering the interactions between the hydrophobic region and the repulsive forces between the charged part. They observed that the self-assembled nanostructures changed from vesicles to tubes to ribbons with an increase in the hydrophobicity of the peptides. The

change in nanostructures occurred due to the change in their critical micellar concentration (CMC) upon increase in the hydrophobicity (Figure 2.6). The PAs A_6K, V_6K_2, and L_6K_3 formed nanotubes; A_6K_2, V_6K_3, and L_6K_4 formed vesicles; and A_6K_3, V_6K_4, and L_6K_5 formed irregular aggregates. Thus, peptides having identical hydrophobic segments but greater number of hydrophilic amino acids have higher CMC values and therefore self-assemble to form vesicle-like structures rather than nanotubes. So, the self-assembled nanostructures are based on the composition of peptides, indicating to an effective method for tuning the peptide self-assembly (Meng et al. 2012).

Ultrasonication is one of the techniques used in supramolecular chemistry to cross the energy barrier and disturb the intermolecular interactions. Therefore, ultrasonication is commonly used to control the self-assembly and gelation processes. Sonication is mainly used for dissolution and dispersion of molecules via disruption of weak noncovalent interactions among molecules. So the sonication method helps the peptide molecules aggregate and also modifies the morphology by providing suitable energy. Maity et al. reported the formation of sonication-induced stable, self-supporting, transparent hydrogels formed by peptide-appended bolaamphiphiles. The bolaamphiphiles formed hydrogels with different self-assembled structures just after sonication at room temperature. The H-bonding and π–π stacking interactions are responsible for this self-assembly process (Maity et al. 2012). The self-assembly pattern of peptides [D]FFD (D-phenylalanine-L-phenylalanine-L-aspartic acid) and [D]FFI (D-phenylalanine-L-phenylalanine-L-isoleucine) were tuned by ultrasonication (Pappas et al. 2015). The peptides formed well-defined tubular nanostructure with an average diameter of 15 nm from a disordered fibrillar structures when they were

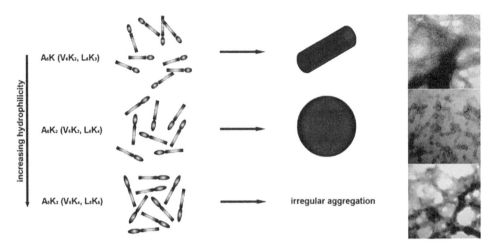

FIGURE 2.6 Schematic representation of PAs in different morphologies upon increasing hydrophobicity. (Reprinted with permission from Meng, Q. et al., *Langmuir*, 2012, 28, 5017–5022. Copyright 2012 American Chemical Society.)

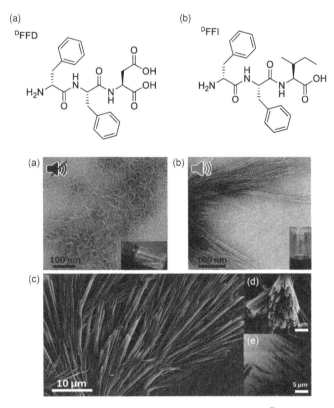

FIGURE 2.7 Molecular structure of peptides (^{D}FFD and ^{D}FFI), TEM images (a) before sonication, (b) after sonication. Inset pics for (a) solution state before sonication (b) gel formed after sonication. (c and d) SEM images (e) two-photon microscopy image of ultrasonicated hydrogel. (Reproduced with permission from Pappas, C.G. et al. *Chem. Commun.*, 2015, 51, 8465–8468.)

treated with ultrasound followed by cooling (Figure 2.7). By increasing the hydrophobicity of peptides from aspartic acid to isoleucine, gel formation was observed in aqueous medium. These results suggested that ultrasonication could induce self-assembly and also tune the formation of highly ordered nanostructures.

The solvent is one of the important parameters in the self-assembly process as it affects supramolecular assembly by forming solvent-bridged H-bonding network. The molecular assembly may be changed upon changing the solvent composition. Solvents with different polarity and H-bonding ability may influence the self-assembly process. The FF dipeptide undergoes self-assembly in polar solvents to form flower-like microcrystal. The dipeptide also forms long nanofibers in nonpolar solvents like toluene. The reason behind this anomaly is that the nonpolar solvent (toluene) breaks the relationship between hydrophobic and hydrogen bonding interactions and induces the aromatic π–π stacking interactions, which then determine the self-assembly pathway (Zhu et al. 2010). Another recent study revealed that trace amounts of polar solvents in DCM played a key role in nanofibril formation (Wang et al. 2016a). A perylene bisimide-(di)glycine-tyrosine based (PBI-[GY]$_2$) bolaamphiphile formed nanofibers in water. However, in the presence of trace amounts of THF in water followed by solvent evaporation, the bolaamphiphile formed a nanosphere structure (Figure 2.8). A trace amount of THF as a proton acceptor solvent may change the H-bonding network, leading to tuning of the morphology from nanofibers to spherical assembly (Bai et al. 2014, Wang et al. 2017).

The use of enzymes is an effective way to control and direct molecular self-assembly through redefining molecular interactions. Nature produces enzymes to perform many essential functions in biological systems. Enzyme-catalyzed self-assembly of peptide derivatives has been well known due to its potential applications in the biomedical field (Williams et al. 2010). For example, Das et al. reported a lipase-catalyzed self-assembly of a thixotropic peptide bolaamphiphile to a nanofibrillar hydrogel that could be used as a scaffold for human umbilical cord stem-cell proliferation (Das et al. 2015). Peptide systems may also undergo enzyme-mediated disassembly and enzyme-triggered self-assembly (Williams et al. 2010). Dehsorkhi et al. reported the tuning of self-assembled nanostructures of PAs through enzymatic

(a)
PBI-[GY]₂ in water

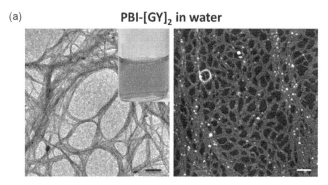

(b)
PBI-[GY]₂/trace THF in water

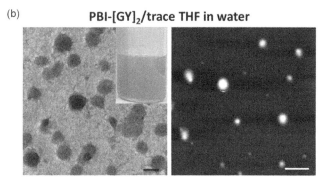

FIGURE 2.8 TEM images of peptide bolaamphiphile (a) in water, and (b) with trace amount of THF in water showed two different morphologies. (Reprinted with permission from Wang, A. et al. *ACS Appl. Mater. Interfaces*, 2017, 9, 21390–21396. Copyright 2017 American Chemical Society.)

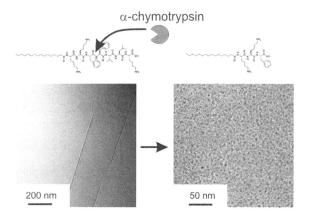

FIGURE 2.9 TEM images of PAs before and after enzyme cleavage. (Reprinted with permission from Dehsorkhi, A. et al. *Langmuir*, 2013, 29, 6665–6672. Copyright 2013 American Chemical Society.)

degradation. An enzyme α-chymotrypsin was used to cleave the PA C_{16}-KKFFVLK at two distinct sites, leading to the formation of products C_{16}-KKF with FVLK and C_{16}-KKFF with VLK (Figure 2.9). The two cleaved PAs (C_{16}-KKF and C_{16}-KKFF) self-assembled into spherical micelles rather than the nanotubes seen for PA C_{16}-KKFFVLK (Dehsorkhi et al. 2013). So, these reports suggest the ability of enzymes to regulate the self-assembly of PAs and also the ability to tune the nanostructures through the use of enzymatic

reactions. A pentapeptide, Nap-FFGEY, self-assembled into nanofibers, but this nanostructure was disrupted in the presence of kinase enzymes, leading to the formation of a gel-sol transition state (Yang et al. 2006). On the other hand, the phosphatase enzyme again restored the self-assembly of this peptide and led to the formation of supramolecular hydrogels *in vivo*.

2.4 Self-assembled Peptide Nanostructures

2.4.1 Nanotubes

Peptides self-assemble to form nanometer-size tube-like structures called nanotubes. These elongated nanostructures have a defined inner cavity. Recently, self-assembled peptide nanotubes have received considerable attention due to their potential applications in biology, chemistry, and physics (Gazit 2007). Noncovalent interactions play an important role in the formation of nanotubes. It has significant advantages due to easy synthesis, self-aggregation, and controllable size. The simplest example is a dipeptide diphenylalanine (L-Phe-L-Phe, FF) that forms self-assembled nanotube (Reches and Gazit 2006a). The length of that nanotube is approximately 100 μm. In case of FF, the noncovalent interactions like H-bonding and π–π stacking of aromatic moieties are involved in the self-assembly process. There are several types of peptides which form tubular-like nanostructures.

Cyclic Peptides

Cyclic peptides are those in which the polypeptide chain is connected through a circular sequence of bonds. The self-assembled tubular structures formed by cyclic-like peptides were first reported by Ghadiri et al. (1993). They reported the cyclic peptide with an even number of alternating D- and L- amino acids that formed β-sheet-like tubular nanostructures. The first cyclic peptide synthesized by this group was cyclo-(L-Gln-D-Ala-L-Glu-D-Ala)₂ (Figure 2.10) This self-assembly is mediated in an acidic medium upon H-bonding interaction between the backbone amide groups, which are oriented perpendicular to the side chains, and the plane of the ring. The individual cyclic peptide monomers are stacked in a ring-shaped conformation to form a hollow structure while the side chains of the amino acid lie on the outside surface of the nanotubes. At alkaline pH, the more intermolecular repulsion between the negatively charged carboxylate side chains of aspartic acid would discourage the ring stacking and at the same time promote the dissolution of the peptide subunits in aqueous medium. However at acidic pH, upon protonation, the repulsive parameter is no longer present and so the attractive side-chain/side-chain H-bonding should occur, which provides the principal driving force in the self-assembly process. So, the controlled acidification of peptide solution promotes the spontaneous self-assembly of peptide subunits into the form of a tubular

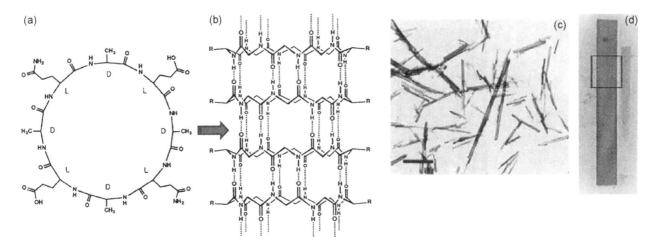

FIGURE 2.10   (a) 2D representation of peptide subunits alternating D- and L- amino acids. (b) Tubular representation of peptide subunits and their extensive antiparallel stacking and H-bonding interactions. Electron microscopy: (c) low-magnification image of nanotube, (d) low-dose image of a frozen, hydrated single nanotube. (Particle measurement $\approx 86 \times 1180$ nm). (Reprinted with permission from Ghadiri, M.R. et al. *Nature*, 1993, 366, 324–327. Copyright 1993, Springer Nature.)

nanostructure. This tubular structure has a length of 100 nm and internal diameter of 7–8 Å. Electron microscopy images also support the tubular structure (Figure 2.10) (Ghadiri et al. 1993, Hartgerink et al. 1996).

Lanreotide octapeptide is another example of a cyclic peptide that self-assembles to form tubular structure. The octapeptide self-assembles into nanotubes with a tube diameter of 244 Å and wall thickness of 18 Å. Self-assembly of this peptide into nanotubes is mediated through the association of β-sheets driven by the amphiphilicity and a systematic aromatic/aliphatic side chain with an alternating pattern (Valery et al. 2003).

Amphiphilic Peptide Nanotubes

PAs also form self-assembled tubular-like nanostructures. PAs have hydrophilic amino acid sequences attached to a hydrophobic alkyl tail. Various noncovalent interactions like H-bonding, van der Waals, and electrostatic interactions hold the PA chains together and promote the self-assembly process. Zhang et al. reported the glycine-rich PAs (G_nD_2, $n = 4, 6, 8, 10$) that could form nanotubes upon self-assembly in water at neutral pH. The dynamic light scattering (DLS) measurement revealed that this nanostructure has a dimension of the order of 40–80 nm and the dimension reached around 100–200 nm with increase in the tail length (glycine residues). The transmission electron microscopy (TEM) image showed that G_4D_2 formed a nanotubular structure with a diameter of 40 nm; G6D2 formed vesiclular structures along with nanotubes, whereas G_8D_2 and $G_{10}D_2$ formed entangled nanotubes (Figure 2.11) (Santoso et al. 2002). These observations indicate that the closely packed monomers inhibit the formation of vesicles and tubes as the membrane becomes prominent. The nonpolar glycine units remain away from water, and the polar aspartic acids are exposed to water. Another set of designed PAs with 7–8 amino acid residues (containing

aspartic acid as hydrophilic head group and valine, alanine as hydrophobic tail) also show tubular morphologies under the transmission electron microscope. The valine, alanine-rich hydrophobic tail induced a hydrophobic interaction with the charged groups outside, which helps to form the nanotube wall (Vauthey et al. 2002). Therefore, the formations of self-assembled nanostructures from short peptides depend on the amphiphilicity of a peptide. Meng et al. designed several PAs by changing the number and type of hydrophobic and hydrophilic amino acids for tuning the self-assembly behavior of PAs. They have synthesized a series of self-assembled peptides X_6K_n, where X changes from alanine to valine to leucine and the number of lysines varies (n) from 1 to 5. The peptides A_6K, V_6K_2, and L_6K_3 formed uniform nanotubes with a diameter of 5 nm upon self-assembly in aqueous solution. However, upon increasing the lysine unit, the nanotube morphology was changed into nanovesicles (Meng et al. 2012).

Another class of PAs is called peptide bolaamphiphile, in which the hydrophobic core is linked with two hydrophilic parts at both ends of the chain. Peptide bolaamphiphiles also formed nanotubes upon self-assembly. The self-assembly behavior of glutamic acid-based peptide bolaamphiphile into nanotubes was also described. The designed EFL$_4$FE octapeptide consisted of four hydrophobic leucine units and hydrophilic glutamic acids-based polar head groups. EFL$_4$FE undergoes self-assembly into nanotubes in alkaline medium by the disruption of β-sheet structure. The electrostatic interaction between the carboxylate anion and Na^+ ions play the driving force leading to nanotube formation (Rodrigo da Silva et al. 2015).

Apart from PAs, hybrid peptides possessing both natural and unnatural amino acids also showed tubular-like morphologies upon self-assembly in various solvents. Hybrid peptides Boc-Gbn-Aib-Phe-Aib-OMe (Gbn: gababutin, Aib: α-amino isobutyric acid, Phe: phenylalanine) and

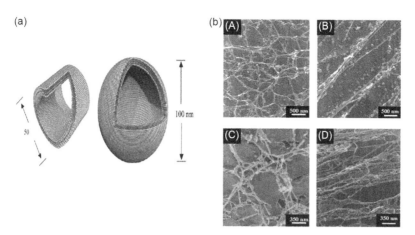

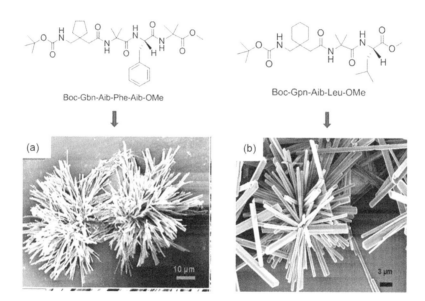

FIGURE 2.11 (a) Molecular modeling of cut-away structure of PA. Left side pictures depict the nanotube and right side the nanovesicle. The outer part of the vesicle contains the polar aspartic acids region and the inner part of vesicle contains the non-polar glycine tails. (b) TEM images of PA (A) G_4D_2, (B) G_6D_2, (C) G_8D_2, and (D) $G_{10}D_2$. (Reprinted with permission from Santoso, S. et al. *Nano Letters.*, 2012, 2, 687–691. Copyright 2012 American Chemical Society.

FIGURE 2.12 Molecular structure and FE-SEM image of (a) Boc-Gbn-Aib-Phe-Aib-OMe tetrapeptide. (Reprinted with permission from Konda, M. et al. *Org. Biomol Chem.,* 2018, 16, 1728–1735.) (b) Boc-Gpn-Aib-Leu-OMe tripeptide. (Reprinted with permission from Konda, M. et al. *ChemistrySelect*, 2016, 1, 2586–2593. Copyright 2016, John Wiley and Sons.)

Boc-Gpn-Aib-Leu-OMe (Gpn: gabapentin, Leu: leucine) exhibited self-assembled microrod- and nanorod-like architectures in methanol/water and THF/water solvents, respectively (Figure 2.12) (Konda et al. 2016, 2018).

2.4.2 Nanofibers

Peptides also self-assemble into fiber-like nanostructures, having diameters less than 100 nm, called nanofibers. Taking inspiration from natural nanofibers, self-assembly of the peptides into nanofibers also have a wide range of applications (Adams and Topham 2010). Various noncovalent interactions such as hydrophobic interactions, hydrogen bonding, π–π stacking interactions, and van der Waals interactions play important role in the self-assembly process, but other

parameters like pH, ionic strength, and assembly are also responsible for this process (Chen et al. 2014, Li et al. 2015, Korevarr 2014). TEM and atomic force microscopy (AFM) are useful techniques for studying the self-assembly of the formation of peptide nanofibers.

Amphiphilic Peptide Nanofibers

As mentioned earlier regarding the self-assembly character of PAs into nanotubes, PAs can also form nanofibers upon self-assembly in water at specific solution conditions depending upon the charge, shape, and environment (Israelachvili et al. 1977). The chemical structures of PAs are generally composed of four different segments that simultaneously play important roles in the formation of nanofibrillar structures (Figure 2.13). Region 1 contains a long chain fatty acid (hydrophobic) part that accounts for the amphiphilicity. Region 2 consists of small peptide sequences that enable the formation of secondary structures through noncovalent interactions. Region 3 contains charged amino acids which enhance the water solubility and formation of pH- and salt-responsive nanostructures. Region 4 consists of some bioactive peptide sequences that demonstrate bioactive signals. In PAs, the self-assembly process is mainly governed by the hydrophobic interactions (region 1), intermolecular hydrogen bonding interactions (region 2), and the electrostatic interactions (region 3). For example, the hydrophobic alkyl tail containing palmitic acid in the case of IKVAV-PA is linked with the AAAAGGG peptides sequence, charged glutamic acids, and bioactive IKVAV epitope. These PAs are self-assembled in an aqueous medium to form nanofibers that result in the formation of 3D networks and gel-like semi-solid materials. The nanofibers have a diameter of 5–8 nm with lengths of hundreds of nanometers to micrometers. (Silva et al. 2004, Niece et al. 2003) (Figure 2.13) Hartgerink et al. developed 12 different types of PAs by changing the amino acids in the head region and by modification of the alkyl chain at the tail region (Hartgerink et al. 2002). The alkyl chain provides hydrophobic interaction in self-assembly. The PAs form hydrogels upon changing the pH, and then the PA fibers achieve enhanced stability by reversible polymerization. These PAs then self-assemble into one-dimensional fibrous motifs. However, upon mixing these PAs with hydrophobic oligo(*p*-phenylene ethylene) core, their one-dimensional fiber structures were restricted. AFM and TEM data also supported the change in supramolecular

aggregation of PAs from nanofibers to small monodispersive nanostructures upon mixing (Bull et al. 2008). Further nanofibrous morphology was observed upon comixing of two oppositely charged PAs at physiological pH. Coassembled nanofibers were formed due to the electrostatic interactions between the two oppositely charged PAs.

The shape and size of the self-assembled nanostructures formed by PAs can be controlled through the modification of amino acid sequences at the middle peptide segment. It has been reported that upon introducing hydrophilic amino acid sequences in that segment, the self-assembly mechanism is changed. Due to the loss of all interfacial curvature required for nanofiber formation, a fat nanobelt-like morphology was observed. AFM data showed that the giant fat nanobelt assemblies exhibited lengths of over 10 mm and widths of 150 nm. The alternative hydrophilic and hydrophobic amino acid sequences in VEVE peptides promoted the formation of β-sheet structures and therefore played key roles in nanobelt formation. The proposed mechanism suggests that the hydrophobic and hydrophilic amino acid residues align with the opposite site on the peptide backbone during the formation of β-sheet conformation. In aqueous medium, the hydrophilic valine residues aggregate to each other in order to minimize their exposure to the solvent, resulting in dimeric lipid-like structures which have a flat bilayer packing geometry upon assembly (Cui 2009) (Figure 2.14).

Maity et al. designed two peptide-based bolaamphiphiles that formed strong and rigid self-supporting hydrogels at physiological conditions after only sonication. In the gel state, these bolaamphiphiles adopted nanofiber and nanoribbon structures through self-assembly by π–π

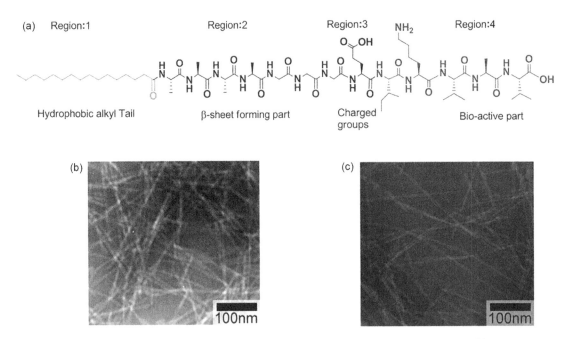

FIGURE 2.13 (a) Molecular structure of a PA with four functional regions. (b) TEM image of IKAV PA (c) TEM image upon mixing of two oppositely charged PAs. (Reprinted with permission from Niece, K.L. et al. *J. Am. Chem. Soc.*, 2003, 125, 7146–7147. Copyright 2003 American Chemical Society.)

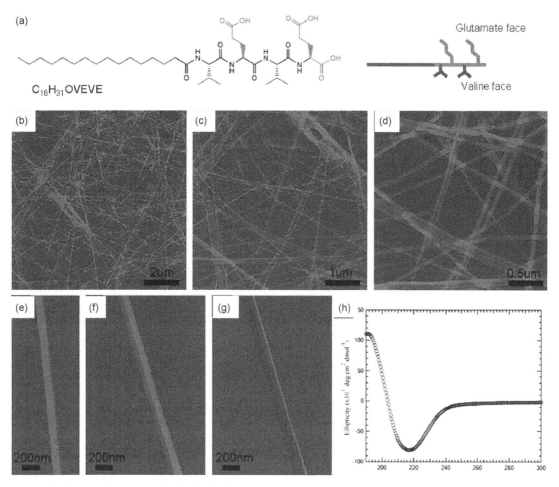

FIGURE 2.14 (a) Chemical structure of PA. (b–d) AFM images of PAs nanobelts at different scanning sizes. (e–f) AFM images of single- and double-layer nanobelt morphology (g) AFM amplitude image of (f). (h) CD spectrum of PA at a concentration of 0.05 wt% shows the β-sheet characteristics in the supramolecular assemblies. (Reprinted with permission from Cui, H. et al. *Nano Lett.*, 2009, *9*, 945–951. Copyright 2009 American Chemical Society.)

stacking and H-bonding interactions. The nanofibers could serve as a template for the in situ generation of Pt nanoparticles. Further, the synthesized nanofiber-decorated Pt nanoparticles were used as catalyst for the hydrogenation reaction (Maity et al. 2012).

Other Peptide Nanofibers

Various short and long peptides are also found to show fiber-like nanostructures upon self-assembly. EAK-16-II (AEAEAKAKAEAEAKAK) (Hong et al. 2003) was the first member of the self-assembling peptide family. The scanning electron microscope (SEM) techniques revealed that EAK-16-II peptide formed well-defined nanofibers. Another set of peptides, KFE-8, KLD-12, undergoes self-assembly to form a left-handed nanofiber with a diameter of 7 nm as a single fiber and becomes thicker when bundled with other fibers. It was observed that the length of the nanofibers increases after 2 h (Marini et al. 2002). It is well observed that the formation of the peptide scaffolds and their mechanical properties are influenced by the level of hydrophobicity (Caplan 2002).

A short pentapeptide sequence (KLVFF) from β-amyloid peptides could also be self-assembled into a nanofibrous structure, which led to gel formation in PBS buffer solution (Krysmann et al. 2008). Experimental data suggested that KLVFF transformed its structure form β-sheet to nanofibrillar hydrogel structure by electrostatic interactions.

The use of aromatic moieties at the N-terminal site is another approach to induce the self-assembly by π–π stacking interactions. One of the common capped aromatic moieties is Fmoc (9-fluorenylmethoxycarbonyl) group. Various peptide combinations have been capped with Fmoc group. Cheng et al. designed two Fmoc-capped tripeptides, one of which formed highly anisotropic fibrils (Fmoc-VLK-Boc) and the other (Fmoc-K(Boc)LV) formed highly branched fibrils. Both the peptides formed hydrogels in borate buffer solution. The interchange of the position of K and V in tripeptide sequences leads to the formation of different self-assembled structures, indicating an important approach to control hydrogel behavior (Cheng et al. 2010). Fmoc-FG, Fmoc-FRGD, Fmoc-RGDF, and Fmoc-FF peptides also showed self-assembled nanofibrous structures in their hydrogel state (Figure 2.15) (Orbach et al. 2009).

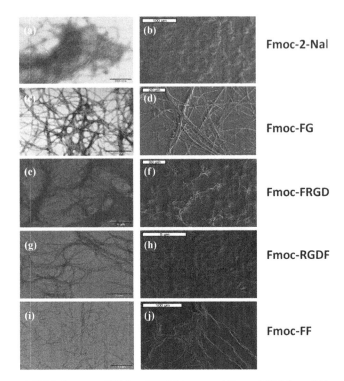

FIGURE 2.15 SEM and TEM images of Fmoc-2-Nal (a and b), Fmoc-FG (c and d), Fmoc-FRGD (e and f), Fmoc-RGDF (g and h) and Fmoc-FF (i and j) showed nanofibers morphology. (Reprinted with permission from Orbach, R. et al. *Biomacromolecules*, 2009, 10, 2646–2651. Copyright 2009 American Chemical Society.)

Rasale et al. described the biocatalytic evolution of naphthalene-2-methoxycarbonyl (Nmoc)-capped dynamic combinatorial peptide libraries in hydrogel state. The Nmoc-F/FF and Nmoc-L/LL peptides showed nanofiber-like morphology in the hydrogel state (Rasale et al. 2015).

Napthalenediimide (NDI, another aromatic core)-appended peptides showed well-defined nanofiber network in gels in various organic solvents. The NDI aromatic core induced π–π stacking interactions, association with hydrogen bonding, and other noncovalent interactions of peptides, which thus promoted the self-assembly process (Basak et al. 2013).

2.4.3 Nanovesicles

Peptide-based materials have great potential for use as drug delivery systems. Peptides self-assemble to form nanovesicles, which have considerable applications in drug delivery. In general, lipid-based carriers such as liposomes and micelles are useful for delivering bioactive compounds into living systems (Allen and Cullis 2004). Several amphiphilic oligopeptides also self-assemble to form vesicular structures in aqueous medium at neutral pH. An aromatic diphenylalanine-based dipeptide self-assembled into nanotubes and nanorod-like structures. Analogues of FF dipeptide like cationic FF, Fmoc-FF, phenylglycine self-assemble into nanovesicles and nanospheres.

The phenylglycine peptide suppressed the rotation of the C–C bond and increased the steric hindrance compared to the Phe residue and helped in the formation of nanovesicles from Phe-Phe dipeptide. Again the FF dipeptide conjugated with cysteine residue at N-terminus also showed vesicle-like morphology. The introduction of thiol groups in FF dipeptide promotes the emergence of nanosphere structure via the disulfide bridge formation. (Reches and Gazit 2004).

Ghosh et al. have reported a short tetrapeptide PWWP (derived from an anti-microbial peptide indolicdin), which upon self-assembly formed vesicular structure (Figure 2.16) (Ghosh et al. 2007b). The same group also designed a tripodal dipeptide derivative, in which tris(2-aminoethyl)amine(tren) was used as the scaffold for the conjugation of three ditryptophan dipeptide units. The synthetic triskelion ditryptophan conjugate showed rapid self-organization into spherical nanostructures upon incubation in methanol/water mixture at 37°C for 7 days (Ghosh et al. 2007a). The molecular modeling structures suggested that the indole aromatic ring-induced π–π stacking played a crucial role in the self-assembly process.

In recent research, the self-assembling polymeric vesicles have gained considerable interest in replacement of liposomes and other lipid-based delivery systems for targeting cells or tissues. Zhang et al. first reported 7–8 residue-based amphiphilic peptides which were able to self-assemble into nanovesicles (Vauthey et al. 2002). Van Hell et al. reported the self-assembly of oligopeptide (SA2) into nanovesicles (van Hell et al. 2007). These vesicles are more stable than lipid and polysaccharide vesicles and also more biocompatible and biodegradable than other synthetic polymeric vesicles. Another example showed a set of branched

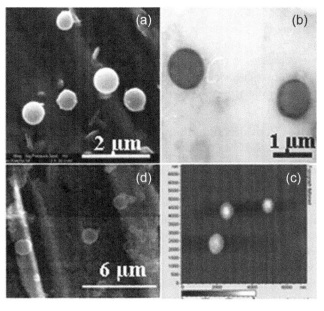

FIGURE 2.16 (a) SEM image after peptide solution after 5 min of incubation, (b) TEM image, (c) AFM image, and (d) E-SEM image of PWWP tetrapeptide. (Reprinted with permission from Ghosh. S. et al. *Chem. Commun.*, 2007b, 2296–2298.)

amphiphilic peptides self-assembled to form solvent-filled, bilayer-delimited spheres with 50–200 nm diameter, which were confirmed by TEM, STEM, and DLS methods (Gudlur et al. 2012). The hydrophobic and hydrogen bonding interactions stabilized the assembly process. The peptide vesicles could also entrap fluorescent dye molecules inside their interior part.

The self-assembly nature of copolypeptides constituting by poly(L-lysine)-X-poly(L-leucine), poly(L-glutamic acid)-X-poly(L-leucine), and poly(L-arginine)-X-poly(L-leucine) polypeptides (X represents block) was reported. The oppositely charged block copolypeptides formed unilamellar vesicle-like structure upon self-assembly in aqueous medium. The reason for this self-aggregation was due to the α-helical hydrophobic rod formation of poly(L-leucine) moiety. The arginine-containing block copolypeptides showed potential applications in drug delivery (Bellomo et al. 2004, Holowka et al. 2005, 2007).

2.4.4 Amyloid-like Nanofibrils

The formation of amyloid fibril is due to the aggregation of proteins in biological systems. Amyloid fibrils are commonly seen in several diseases like Alzheimer's disease and Parkinson's diseases (Harper and Lansbury 1997, Wickner et al. 2000, Sunde and Blake 1998, Gazit 2002b, Reches et al. 2002). In different diseases, many proteins without having any structural and functional homology also form amyloid-like fibrils with similar properties. The large, highly ordered, organized amyloid fibrils formed by proteins exhibited a diameter of 7–10 nm and an X-ray diffraction pattern with 4.6–4.8 Å on the meridian (Gilead and Gazit 2005). In general, the formation of amyloid fibrils occurred in long chain polypeptides having greater than 30 amino acids residues. These fibrils are formed via β-sheet conformation. Not only the long chain polypeptides, short tetra-, penta- and hexapeptides also form similar amyloid-like fibrillar nanostructures with similar biophysical and structural properties like the nanofibrils that are formed by larger polypeptides. Tenidis et al. reported fibrillar-like structures with short pentapeptide FGAIL and hexapeptide NFGAIL that are very similar to the fibrillar structures formed by islet amyloid polypeptide (IAPP) (Tenidis et al. 2000). This IAPP is a 37 amino acid containing polypeptide hormone that is responsible for type II diabetes.

Again, further studies have revealed that a pentapeptide (DFNKF) and a tetrapeptide (DFNK) also formed amyloid-like fibrils. These short sequences were derived from human calcitonin (hCT) polypeptide hormone, which is found in medullary carcinoma of the thyroid (Reches et al. 2002).

Balbach et al. reported a heptapeptide sequence (KLVFFAE), which is the main constituent of amyloid plaques in the brain of Alzheimer's disease patients. KLVFFAE self-assembled to form amyloid-like fibrils upon incubation in an aqueous solution, and the fibril formation was confirmed by X-ray powder diffraction and optical birefringence measurements (Balbach et al. 2000). Later,

Gazit et al. reported that the aromatic amino acid-based dipeptide FF plays a crucial role in the process of fibril generation. After comparing the self-assembly nature of all aromatic residues, they concluded that the diphenylalanine is the core recognition motif of Alzheimer's β-amyloid polypeptide. The π–π stacking interaction provides the main contribution in self-assembled amyloid formation (Gazit 2002a).

A water-soluble tripeptide (VIA) which was identical to the C-terminal portion of Alzheimer Aβ-peptides self-assembled to form straight unbranched nanofibrils exhibiting amyloid-like behavior. The self-assembly behavior of VIA was driven by the association of intermolecular hydrogen-bonded supramolecular β-sheet structure. However, another peptide AVI formed branched nanofibrillar structure but did not exhibit amyloid-like behavior (Figure 2.17). The nonamyloid characteristic of AVI peptide was supported by proof from data of TEM images and a Congo-red binding study. The amyloid-specific dye Congo red and thioflavin T showed enhanced fluorescence upon binding with peptide VIA, but no significance change was observed in case of tripeptide AVI (Ray et al. 2006).

2.5 Application of Self-assembled Peptide Nanostructures

The demand for self-assembling peptides is increasing in the field of regenerative medicine. Well-defined peptide-based nanomaterials are highly attractive for several biomedical applications in the field of drug delivery, vaccination, tissue engineering, gene delivery, antibacterial agent preparations, and nanosensors (Sun et al. 2017, Zhang 1995). However, by changing molecular length, charge, and functional groups, different bioactive PAs were designed with tunable mechanical properties which can be used for the preparation of hydrogels, biomimetic materials, extracellular matrix, and so on. Here, the applications of self-assembling peptide-based nanomaterials in various fields are discussed.

2.5.1 Drug Delivery

Drug delivery systems are often characterized by their compact size, the conjugation of drug molecules to the carrier of the nanoparticles by removal of the unorganized drugs, followed by conjugation of the targeting ligands to obtain cell specificity. This laborious procedure suffers from several disadvantages including poor oral availability, poor water solubility, quick biodegradation, nonspecific delivery, and other serious effects. The advantages of self-assembled peptide nanostructures over other delivery agents are attributed to their biocompatibility, chemical diversity, and high loading capacity toward both hydrophobic and hydrophilic drugs and their ability to target recognition sites (Habibi et al. 2016, Pawar et al. 2004). Most importantly, self-assembled peptide nanostructures provide a very promising and effective method for gene-drug delivery

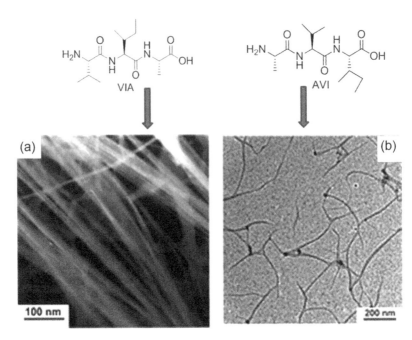

FIGURE 2.17 Molecular structure TEM image of (a) VIA tripeptide with amyloid-like fibrillar structure, but (b) AVI peptide shows branched fiber without any amyloid behavior. (Reprinted with permission Ray, S. et al. *Chem. Commun.*, 2006, 4230–4232.)

due to their intrinsic properties and precisely controllable fabrication approaches. Peptide self-assembled nanotubes played an important role in gene-drug delivery through the transformation of nanotublar structures into nanovesicles in the endocytosis process (Yan et al. 2007). Therefore, self-assembled peptide nanostructures like nanospheres, nanotubes, nanovesicles and nanofibers could be used for various drug delivery applications such as anticancer drug delivery, gene-drug delivery, targeted drug delivery, and stimuli-responsive drug delivery systems (Habibi et al. 2016, Fan et al. 2017).

The *in situ* injectable hydrogels based on KLD peptide motif could be useful to control the delivery of a conventional cytotoxic drug doxorubicin (DOX) or Smac-derived proapototic peptide for inhibiting tumor cell growth *in vitro*. The slow release of drug molecules from hydrogels helps more prolonged and sustained cytotoxic action than conventional chemotherapeutic agents, therefore enhancing therapeutic efficacy (Yishay-Safranchik et al. 2004).

The synthesized Nap-GFFYGRGD peptide forms nanofibers upon self-assembly but could not form hydrogel due to weak interfiber interaction. Thus, the peptide nanofibers could be utilized as a vehicle for anticancer drugs. The DOX formed nanospheres at the surface of peptide nanofibers. The electrostatic interactions between the positively charged DOX-nanospheres and negatively charged nanofibers enhanced the weak interfiber interaction, which lead to the formation of stable 3D nanofiber networks and hydrogels. The DOX-peptide-based hydrogels showed sustained release of drugs and also showed comparable cytotoxicity toward cancerous cells. Furthermore, about 95% of DOX was released from the hydrogel over a time period of 72 h (Figure 2.18) (Xue et al. 2015). Again a

thixotropic, injectable nanofibrous hydrogel network of N-terminal protected long chain amino acid containing tripeptide Boc-AUDA-Phe-Phe-OH showed entrapment and sustained release of an antibiotic and vitamin B_{12} at physiological pH and temperature. The three dimensional fibrous network structures inside the hydrogel provide a cage-like environment, which suggest the template for entrapment and release of several drug molecules (Figure 2.18) (Baral et al. 2014).

Other self-assembling peptides like EAK16 II and RAD16 II could serve as a potential carrier for hydrophobic drugs as well as the anticancer drug ellipticine through encapsulation method. The self-assembly and drug delivery applications were analyzed by two methods: (i) UV–vis and (ii) fluorescence approaches. In vitro experiments also suggest that the encapsulated anticancer drug in the EAK peptide nanofiber in the protonated stage is much more efficient than in the crystalline stage for cancer therapy (Lu et al. 2012, Keyes-Baig et al. 2004, Li et al. 2009).

An enzyme-triggered self-assembled PA could be used as a nanocarrier for anticancer drugs and inhibition of the tumor growth. Kalafatovic et al. reported a PA that can be triggered by matrix metalloproteinase (MMP) enzyme. MMP-9 catalyzed the hydrolysis of PAs to form a nanofiber structure, which provides a depot for prolonged drug delivery of DOX in tumor tissues (Kalafatovic et al. 2016).

Though we have discussed several peptide-based drug delivery systems, carriers for efficient delivery of genes still need major improvements. The ideal gene carriers must show higher cellular uptake efficiency, nontoxicity, and efficient loading and should overcome limitations such as immunogenic response and short-lived transgene expression. Self-assembled peptide nanostructures also illustrate

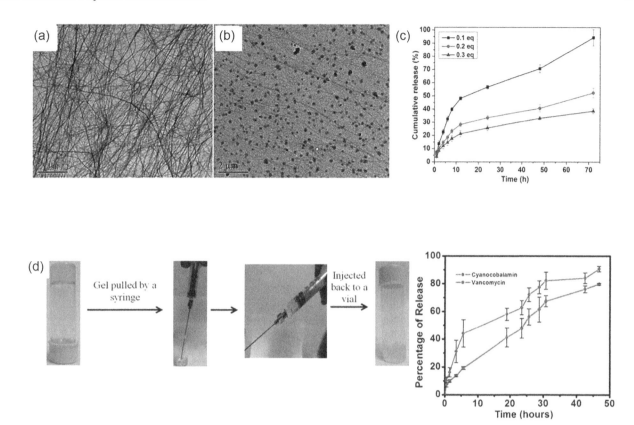

FIGURE 2.18 TEM image of (a) Nap-GFFYGRGD peptide in PBS solution at a concentration of 0.5 wt%, (b) the peptide hydrogel formed after adding 0.1 equiv. of DOX. (Reprinted with permission from Xue, Q. et al. *Sci. Rep.,* 2015. 5, 8764. Copyright 2015, Springer Nature.) (c) The release profile of DOX from hydrogel. (d) Injectable behavior of Vitamin B12-loaded peptide hydrogel (left) and release profile of drugs (cyanocobalamin and vitamin B12) from hydrogel (right). (Reprinted with permission from Baral, A. et al. *Langmuir,* 2014, 30, 929–936. Copyright 2014 American Chemical Society.)

a novel approach for efficient gene delivery. In recent studies, a combination of viral genomes and self-assembled peptides is used for the development of bioinspired gene delivery vehicles. Cell-penetrating peptides (CPPs) are useful as gene delivery agents as they are able to transfer genes by means of endosomal escape. Combrez et al. developed a CPP, named as CADY, a 20 amino acid residue containing amphiphatic peptide with a sequence of Ac-GLWRALWRLLRSLWRLLWRA-cysteamide, which formed helical structure and was able to deliver siRNA into the cells. CADY also helped to lower the expression of GAPDH at both mRNA and protein levels. The helical conformation of CADY plays an important role in the process of cell penetration and interactions with the cell membrane (Crombez 2009).

2.5.2 Tissue Engineering

The advantage of using peptide-based building blocks is that it can be readily extended to include the biologically active sequence to facilitate cell growth. In the recent past, several self-assembled peptide-based hydrogels have been used as scaffolds for 3D cell growth and tissue engineering purposes (Kyle et al. 2009). Advances in the field

of cell research and technical capabilities make possible to regrow damaged tissue using the patient's own cells. The cell signaling mechanism by different ligand interactions in ECM can be encapsulated by the motifs and bioactive factor which modulate the generation of specific tissues (Lutolf and Hubbell 2005). There are several reports on self-assembling peptide enabling *in vitro* tissue regeneration (Holmes 2002, Zhang and Webster 2009). Self-assembling peptides can be employed to regenerate several tissues in vivo. Self-assembled nanostructured (nanofiber, nanotubular) peptides could mimic the ECM of many tissues and help to release the drug at the desired site. Stile et al. designed thermo-responsive peptide-modified hydrogels for tissue regeneration (Stile and Healy 2001).

Tissue engineering methods can target restoration of the function of injured cells, tissues, and organs. A peptide RADA-16-I (sequence of arginine, alanine, aspartate, and alanine) was used to support the growth of PC12 cells, which help the formation of functional synapses in vitro upon using rat primary hippocampal neurons. RADA-16-I peptide formed self-assembling nanofibers, which supported wide neuronal growth and development by using both in vitro and in vivo cell culture systems (Holmes et al. 2000, Ellis-Behnke et al. 2006).

Zhang et al. reported a temperature-responsive self-assembled PA ($V_3A_3E_3$) which formed nanofiber structure in the hydrogel state upon addition of calcium chloride at 80°C followed by cooling at room temperature for 30 min. The string-like $V_3A_3E_3$ peptide hydrogel could direct the orientation of cells in a 3D environment. Upon incubation of human mesenchymal stem cells in PA hydrogel, elongation was observed and both cell bodies and filopodia grew toward the aligned direction of the hydrogel. So, these results indicate the initiation of the development of tissue engineering process, which required directed cell growth and directed cell migration (Figure 2.19) (Zhang et al. 2010).

Lee et al. designed biomimetic amphiphilic peptide nanofibers that could bind strongly with heparin sulfate chains and could therefore be used as a delivery system for bone morphogenetic protein (BMP-2) to promote bone regeneration. The heparin-binding peptide amphiphile scaffold enhanced the BMP-2 retention, leading to a bone graft with adequate biodegradation rate as new bone forms. Histological evaluation showed the presence of more matured bone in the new stiffened tissues when a low dose of BMP-2 was delivered using the biomimetic amphiphilic system (Figure 2.20) (Lee et al. 2013).

Miller et al. reported a self-assembling peptide sequence Ac-$(KLDL)_3$-$CONH_2$, which formed a hydrogel. The hydrogel could be used for cartilage tissue repair by encapsulation of chondrocytes and use of insulin-like cofactor-1.

The Ac-$(KLDL)_3$-$CONH_2$ peptide hydrogel could directly be injected at the injured site for cartilage tissue regeneration upon mixing with growth factors and the patient's chondrocytes (Miller et al. 2011).

Self-assembling peptide amphiphilic nanofibers have also been reported to act as a scaffold for neural progenitor cells and dental stem cells (Galler et al. 2008) and also have been used for cell entrapment and simulation of angiogenesis (Rajangam et al. 2006). The IKAV-containing PA could form nanofibers and also induce rapid differentiation of cells into neurons (Silva et al. 2004).

Panda et al. reported 3D growth of mammalian cells on a FΔF dipeptide hydrogel which was functionalized with a pentapeptide containing Arg-Gly-Asp (RGD) motif. This functionalized noncytotoxic hydrogel promoted 3D cell growth and proliferation of cells for almost 2 weeks, maintaining cell viability and other metabolic activities (Panda et al. 2010). These results provided an excellent example of the use of simple peptide-based hydrogel in 3D cell growth, tissue engineering, and cell biology.

2.5.3 Antimicrobial Peptides

The emergence of the field of bioactive peptides is gaining massive importance due to their various antimicrobial activities. Antimicrobial peptides (AMPs) also named as host defense peptides target and kill Gram-positive and

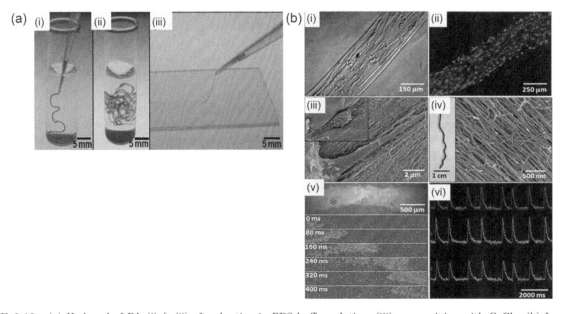

FIGURE 2.19 (a) Hydrogel of PA (i) & (ii) after heating in PBS buffer solution, (iii) upon mixing with $CaCl_2$. (b) Images of (i) Mesenchymal stem cells growing and differentiated at along the direction aligned with the hydrogel string formed from the peptides. (ii) Fluorescence image of calcein-labeled human mesenchymal stem cells in the peptide hydrogel. (iii) SEM images at different magnifications. (inset is the zoomed out view, with arrow indicating alignment direction) (iv) A conductive black string formed by dispersing carbon nanotubes in PA solutions before heating. The SEM micrograph on the right shows aligned nanofiber bundles along the black string. (v) (Top) Calcium fluorescence image of HL-1 cardiomyocytes encapsulated in a noodle-like string. (Below) Successive spatial maps of calcium fluorescence intensity traveling at 80 ms intervals, showing the propagation of an electrical signal throughout the entire string and demonstrating a functional cardiac syncytium (vi) Calcium fluorescence intensity signal in time at three points in the string. (Reprinted with permission from Zhang, S. et al. *Nat. Mater.*, 2010, 9, 594–601. Copyright 2010, Springer Nature.)

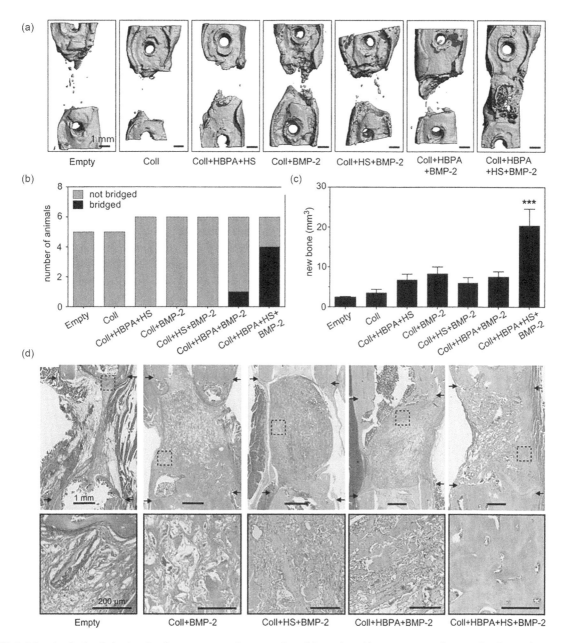

FIGURE 2.20 Analysis of the in vivo bone regeneration capacity of heparin sulfate-presenting heparin-binding PA nanofibers. (a) Representative femur reconstructions from micro-computed tomography are shown for the various treatment groups. (b) The number of animals used per condition and the number of animals with a bridged femur after treatment. (c) Quantitative analysis of new bone volume. (d) Representative histological longitudinal sections of demineralized femora stained with Goldner's Trichrome. In the upper box, arrows indicate the proximal (top) and distal (bottom) edges of the defect (scale bars = 1 mm). The lower box shows a higher magnification image of the inset indicated above (scale bars, 200 μm). (Reprinted with permission from Lee, S.S. et al. *Biomaterials,* 2013, 34, 452–459.)

Gram-negative bacteria, enveloped viruses, fungi, and even cancerous cells (Seo et al. 2012). These potential behaviors of AMPs lead them to be used as novel therapeutic agents. The action of AMPs in the microbial killing process can be determined by several methods like microscopic techniques, atomic emission spectroscopy, fluorescent dyes, ion channel formation, circular dichroism solid-state NMR spectroscopy, neutron and X-ray diffraction, and dual polarization interferometry. AMPs are classified into four different categories (α-helical, β-sheet, extended, and loop peptides) based on their structures (Zasloff 2002, Hancock and Sahl 2006, Nguyen et al. 2011, Hancock and Lehrer 1998). The α-helical AMPs may disrupt the bacterial membrane by forming barrel-like bundles, carpet-like clusters, or toroidal pores (Seo et al. 2012, Zasloff 2002, Hancock and Sahl 2006, Nguyen et al. 2011, Hancock and Lehrer 1998, van't Hof et al. 2001, Teixeira et al. 2012). The β-sheet peptides also show antimicrobial activity through disrupting the bacterial membrane by forming toroidal pores upon perpendicular insertion into the lipid membrane.

The hydrophilic parts of the peptides are associated with the polar head group of the membrane. The extended AMPs mainly consist of proline, arginine, histidine, and tryptophan-rich amino acids. AMPs show antimicrobial activity by penetrating across the membrane and interacting with the bacterial protein inside (Nguyen et al. 2011). A tryptophan-rich extended AMP Indolicidine formed a poly-L-II helical-like structure in presence of liposomes and the high tryptophan residues induced the interaction with lipid membranes (Falla et al. 1996). The AMPs adopt a loop formation with one disulfide bridge during activity.

The peptide GRRRRSVQWCA showed antimicrobial activity against both Gram-positive and Gram-negative bacteria and various fungi. GRRRRSVQWCA peptide showed activity against methicillin-resistant *Staphylococcus aureus*, multidrug-resistant *Acinetobacter baumannii*, and

fluconazole-resistant *Candida albicans* (Brouwer et al. 2011, Dijkshoorn et al. 2004).

Chen et al. reported a photoluminescent antimicrobial gold nanodot (NDs) conjugated with an antimicrobial peptide (surfactin, SFT) and 1-dodecanethiol, on gold nanoparticles. SFT is a cyclic peptide with the sequence of Glu-Leu-D-Leu-Val-Asp-D-Leu-Leu connected by a lactone bond with a β-hydroxy fatty acid chain. The hybrid nanoparticles were formed through self-assembly of peptides on the DT-Au surface by the hydrophobic interaction between the alkyl chain of the peptide and the DT molecules. The photoluminescent property and antimicrobial activity of the combined nanoparticles depend upon the density of SFT on Au surface. The SFT/DT-Au showed antimicrobial activity to both non-multidrug- and multidrug-resistant bacteria due to the synergetic effect of SFT and DT-Au on the disruption of bacterial membrane. In vitro studies revealed that the

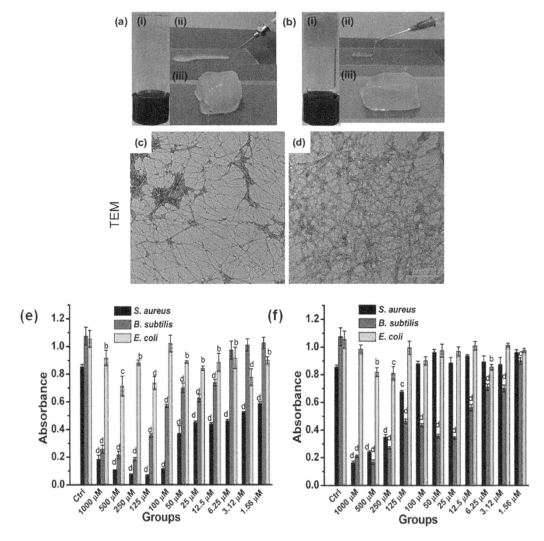

FIGURE 2.21 Optical photographs (a and b): (i) Vial inverted image, (ii) syringe injection, (iii) 3D self-supported pattern of hydrogels formed by peptides Amoc-FL and Amoc-FY. TEM images (c) Amoc-FL hydrogel, (d) Amoc-FY hydrogel. Antibacterial study of (e) Amoc-FL hydrogel and (f) Amoc-FY hydrogel at various concentrations on two Gram-positive and one Gram-negative bacteria. (Reprinted with permission from Gavel, P.K. et al. *ACS Appl. Mater. Interfaces*, 2018, 10, 10729–10740. Copyright 2018 American Chemical Society.)

SFT/DT-Au NDs were more biocompatible than the free SFT, and therefore NDs could be useful for wound healing treatment (Chen et al. 2015b).

The self-assembly process can be driven by adding aromatic moieties at N-terminal sites by providing amphiphilicity to the peptide backbone. Anthracenemethoxycarbonyl (Amoc)-capped dipeptides Amoc-FL-OH and Amoc-FY-OH self-assembled into a cross-linked nanofibrillar structure to form biocompatible, injectable, and shape-supported hydrogels. These hydrogels exhibit inherent antibacterial property against both Gram-positive and Gram-negative bacteria. These hydrogels effectively decrease the oxidative stress on human red blood cells and therefore provide cellular stability against oxidative stress (Figure 2.21) (Gavel et al. 2018).

2.5.4 Others Applications

Tubular nanostructures have gained a wide range of applications in nanotechnology. Reches et al. used peptide nanotubes to generate 20 nm silver nanowires by using D-phenylalanine containing dipeptide nanotubes as degradable casting gold. The peptide nanotubes were used as a template in the reduction of silver ions to metallic silver, and the template was erased by enzymatic degradation. These results indicated the applications in molecular electronics as small nanowires could not be fabricated by conventional lithography (Reches and Gazit 2003).

Supramolecular aromatic organic molecular systems are used as active materials in photoelectronic devices, particularly in photovoltaics. Self-assembled-Nmoc capped PAs were hybridized by upon electrochemical deposition of inorganic $Zn(OH)_2$. The aromatic π–π stacking interaction among the aromatic Nmoc moieties and the H-bonding interactions of the PAs led to the formation of lamellar nanostructures. The peptide-ZnO-based hybrid semiconductor nanomaterials showed optoelectronic behavior, which was supported by UV–vis and photoluminescence spectroscopy. The results suggested the importance of using peptide-based hybrid optoelectronic materials for photovoltaics (Manna et al. 2015).

A pyrene-labeled histidine-rich PA (HG12) with linear or branched hydrophobic chain can undergo self-assembly to form nanofibrils in aqueous medium. The PAs could be used to detect metal ions. The hydrophilic histidine part of this PA had strong binding ability toward Cu^{2+} ions, and therefore the fluorescence could be blocked ("light off"), but the PAs with branched alkyl chain revealed the fluorescence "light up" response by PET inhibition upon Ag^+ binding. The PAs nanofibers could serve as a template for metal ion

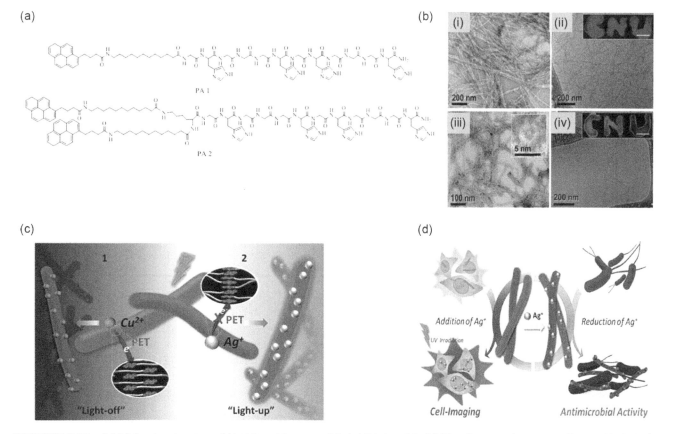

FIGURE 2.22 (a) Molecular structure of histidine-rich pyrene-labeled PA 1 and 2. (b) Visualization of aqueous self-assembled of PAs: (i) & (iii) negative strained TEM images and (ii) & (iv) cyro-TEM images of PA 1 and 2, respectively. (c) Pictorial representation of fluorescent light off and light up detection to Cu and Ag ions. (d) Schematic representation of PA nanofiber application in antimicrobial effect and cell imaging. (Reprinted with permission from Kim. I. et al. *J. Mater. Chem. B.*, 2014, 2, 6478–6486.)

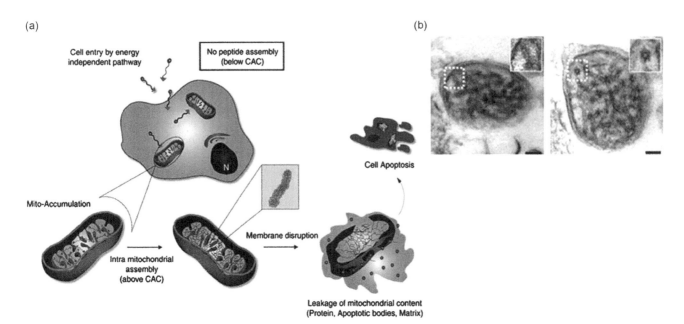

FIGURE 2.23 (a) Schematic representation of molecular assembly of mitochondria-targeted peptide inside the mitochondria leading to the cellular dysfunction. (b) TEM images show the mitochondria at different cross-sections. (Reprinted with permission from Jeena, M.T. et al. *Nature Communications, 2017*, 8, 1–10. Copyright 2015, Springer Nature.)

TABLE 2.1 Examples of Peptides, Their Self-Assembled Structures, and Applications in Various Fields

Peptides	Self-assembled Nanostructures	Applications	References
FF	Nanotubes, β-sheets, nanovesicles, nanofibrils	Gene delivery, anticancer drug delivery, biosensors	Scanlon et al. *Nano Today*, 2008; Hendler et al. *Adv. Mater.*, 2007
LS, IV, VA, AV	Nanopores	Storage of many gases (CO_2, H_2, CH_4)	Soldatov et.al *Angew. Chem. Int. Ed.*, 2004; Comotti et al. *Chem. Commun.*, 2009
Aromatics homodipeptides	Nanospheres, nanoplates, nanofibrils	Biosensing, tissue engineering, drug delivery	Reches and Gazit *Phys. Biol.*, 2006b
KLVFF	Nanofibers	Drug delivery	Krysmann. et al. *Biochemistry*, 2008
C_{16}-$V_2A_2E_2$	Nanofibers	Anti-inflammatory, drug delivery	Webber et al. *Biomaterials*, 2012
$(C_{16})_2$-Glu-PEO-GRGDSP	Vesicles	Promote fibroblast cell adhesion and growth	Stroumpoulis et al. *Langmuir*, 2007
C_{16}-$V_3A_3F_3$	Nanofibers	Cell encapsulation	Zhang et al. *Nature Mater.*, 2010
(Trp-D-Leu)$_4$-Gln-D-Leu	Cyclic nanotubes	Delivery of anticancer drug (5-Fu)	Liu et al. *Mol. Pharmaceutics.*, 2010
Ac-A_6K-$CONH_2$, KA$_6$-$CONH_2$	Nanovesicles	Delivery of diagnostic and therapeutic agents	Fatouros et al. *ACS Appl. Mater. Interfaces*, 2014
YGAAKKAAKAAKKAAKAA	Nanofibers	Antimicrobial activity	Chu-Kung et al. *J. Colloid Interface Sci.*, 2010
KK(K)K-LL-CCC-K-C_{16} KGRGDS(K)K-LLL-AAA-K-C_{16}	Spherical and tubular structures	Diagnostic imaging	Bull et al. *Nano. Lett.*, 2005
C16-A4G3(KLAKLAK)$_2$	Cylindrical nanofiber	Anticancer activity	Standley et al. *Cancer Res.*, 2010
NTFR-PAs	Nanofibers	Selective drug delivery to fractalkine.	Kokkoli et al. *Biomacromolecules*, 2005
$C_{18}GR_7RGDS$	Spherical structures	Inhibit selectively bone cancer cells	Chang et al. *Int. J. Nanomedicine*, 2015

sensing as well as for detection of Ag nanoparticles via cell-imaging (Figure 2.22) (Kim et al. 2014).

In situ self-assembly of peptide building blocks inside the cells and their functions with the cellular components are an important strategy to control the cellular fate. The diphenylalanine-based mitochondria-targeted tripeptide (FFK) formed self-assembled fibrillar nanostructures inside the mitochondria. The peptides achieved the critical aggregation concentration inside the organelle, which induced the fibril formation. The formation of nanofibrils resulted in the disruption of mitochondrial membrane, thereby activating the intrinsic apoptotic pathway against

cancer cells. As mitochondrial disruption initiates cellular death, the organelle localization-induced supramolecular self-assembly will provide a new platform for cancer therapy (Figure 2.23) (Jeena et al. 2017).

Table 2.1 shows other examples of self-assembled peptide nanostructures and their applications in various fields.

2.6   Conclusion

Owing to the several noncovalent interactions, peptides self-assemble to form various nanostructures including nanotubes, nanofibers, nanovesicles, and amyloid-like nanofibrils. Peptide nanostructures have gained considerable importance in various fields including drug/gene delivery, tissue engineering, antimicrobial activity, cell imaging, biosensors, and materials sciences. The peptide nanostructures can be tuned by changing the size and positions of amino acids and also by controlling the effects of temperature, pH, sonication, and change of solvent. The investigations on self-assembled peptides will generate a new pathway for the development of new functional peptide-based architectures. The self-assembled peptides not only help the study of biological phenomena but also help fight against the diseases and improve human health. Therefore, in the 21st century, scientists and researchers have given importance on fabrication of novel materials with new properties by using supramolecular architectures. Further studies are required to develop alternative systems for the prediction of the structure of the self-assembled peptides on their sequence and other physiochemical properties. Nevertheless, the applications of peptide self-assembled nanostructures in the field of gene/ oligonucelotide delivery have not yet been developed extensively.

Acknowledgment

A.K.D. sincerely acknowledges the Department of Science & Technology, NanoMission, New Delhi, India (Project SR/NM/NS-1458/2014), for financial support. T.G. acknowledges to the Department of Science & Technology for his doctoral fellowship.

References

Adams, D.J., and P.D. Topham. 2010. Peptide conjugate hydrogelators. *Soft Matter* 6(16):3707–3721.

Allen, T.M., and P.R. Cullis. 2004. Drug delivery systems: Entering the mainstream. *Science* 303(5665):1818–1822.

Andrews, M.J.I., and A.B. Tabor. 1999. Forming stable helical peptides using natural and artificial amino acids. *Tetrahedron* 55(40):11711–11743.

Anilkumar, P., and M. Jayakannan. 2009. Self-assembled cylindrical and vesicular molecular templates for polyaniline nanofibers and nanotapes. *J. Phys. Chem. B* 113(34):11614–11624.

Bai, S., S. Debnath, N. Javid, P.W.J.M Frederix, S. Fleming, C. Pappas, and R.V. Ulijn. 2014. Differential self-assembly and tunable emission of aromatic peptide bola-amphiphiles containing perylene bisimide in polar solvents including water. *Langmuir* 30(25):7576–7584.

Balbach, J.J., Y. Ishii, O.N. Antzutkin et al. 2000. Amyloid fibril formation by $A\beta_{16-22}$, a seven-residue fragment of the Alzheimer's β-amyloid peptide, and structural characterization by solid state NMR. *Biochemistry* 39(45):13748–13759.

Baral, A., S. Roy, A. Dehsorkhi et al. 2014. Assembly of an injectable noncytotoxic peptide-based hydrogelator for sustained release of drugs. *Langmuir* 30(3):929–936.

Basak, S., J. Nanda, and A. Banerjee. 2013. Assembly of naphthalenediimide conjugated peptides: Aggregation induced changes in fluorescence. *Chem. Commun.* 49(61):6891–6893.

Bellomo, E.G., M.D. Wyrsta, L. Pakstis, D.J. Pochan, and T.J. Deming. 2004. Stimuli-responsive polypeptide vesicles by conformation-specific assembly. *Nat. Mater.* 3:244–248.

Bose, P.P., A.K. Das, R.P. Hegde, N. Shamala, and A. Banerjee. 2007. pH-sensitive nanostructural transformation of a synthetic self-assembling water-soluble tripeptide: Nanotube to nanovesicle. *Chem. Mater.* 19(25):6150–6157.

Brack, A., and L.E. Orgel. 1975. β-structures of alternating polypeptides and their possible prebiotic significance. *Nature* 256:383–387.

Brouwer, C.P., M. Rahman, and M.M. Welling. 2011. Discovery and development of a synthetic peptide derived from lactoferrin for clinical use. *Peptides* 32(9):1953–1963.

Bull, S.R., L.C. Palmer, N.J. Fry et al. 2008. A templating approach for monodisperse self-assembled organic nanostructures. *J. Am. Chem. Soc.* 130(9):2742–2743.

Bull, S.R., M.O. Guler, R.E. Bras, T.J. Meade, and S.I. Stupp. 2005. Self-assembled peptide amphiphile nanofibers conjugated to MRI contrast agents. *Nano Lett.* 5(1):1–4.

Caplan, M.R., E.M. Schwartzfarb, S. Zhang, R.D. Kamm, and D.A. Lauffenburger. 2002. Effects of systematic variation of amino acid sequence on the mechanical properties of a self-assembling, oligopeptide biomaterial. *J. Biomater. Sci. Polym. Ed.* 13(2):225–236.

Chang, R., L. Sun, and T.J. Webster. 2015. Selective inhibition of MG-63 osteosarcoma cell proliferation induced by curcumin-loaded self-assembled arginine-rich-RGD nanospheres. *Int. J. Nanomed.* 10(1):3351–3365.

Chen, B., X.-Y. He, X.-Q. Yi, R.-X. Zhou, and S.-X. Cheng. 2015a. Dual peptide-functionalized albumin-based nanoparticles with pH dependent self-assembly behaviour for drug delivery. *ACS Appl. Mater. Interfaces* 7(28):15148–15153.

Chen, C., Y. Gu, and L. Deng et al. 2014. Tuning gelation kinetics and mechanical rigidity of β-hairpin peptide hydrogels via hydrophobic amino acid substitutions. *ACS Appl. Mater. Interfaces* 6(16):14360–14368.

Chen, W.-Y., H.-S. Chang, J.-K. Lu et al. 2015b. Self-assembly of antimicrobial peptides on gold nanodots: Against multidrug-resistant bacteria and wound-healing application. *Adv. Funct. Mater.* 25(46):7189–7199.

Cheng, G., V. Castelletto, C.M. Moulton, G.E. Newby, and I.W. Hamley. 2010. Hydrogelation and self-assembly of Fmoc-tripeptides: unexpected influence of sequence on self-assembled fibril structure, and hydrogel modulus and anisotropy. *Langmuir* 26(7):4990–4998.

Childers, W.S., A.K. Mehta, R. Ni, J.V. Taylor, and D.G. Lynn. 2010. Peptides organized as bilayer membranes. *Angew. Chem. Int. Ed.* 49(24):4104–4107.

Chu-Kung, A.F., R. Nguyen, K.N. Bozzelli, and M. Tirrell. 2010. Chain length dependence of antimicrobial peptide–fatty acid conjugate activity. *J. Colloid Interface Sci.* 345(2):160–167.

Comotti, A., S. Bracco, G. Distefano, and P. Sozzani. 2009. Methane, carbon dioxide and hydrogen storage in nanoporous dipeptide-based materials. *Chem. Commun.* (3):284–286.

Cote, Y., I.W. Fu, E.T. Dobson, J.E. Goldberger, H.D. Nguyen, and J.K. Shen. 2014. Mechanism of the pH controlled self-assembly of nanofibers from peptide amphiphiles. *J. Phys. Chem. C* 118(29):16272–16278.

Crombez, L., M.C. Morris, S. Dufort et al. 2009. Targeting cyclin B1 through peptide-based delivery of siRNA prevents tumour growth. *Nucl. Acids Res.* 37(14):4559–4569.

Cui, H., T. Muraoka, A.G. Cheetham, and S.I. Stupp. 2009. Self-assembly of giant peptide nanobelts. *Nano Lett.* 9(3):945–951.

Das, A.K., I. Maity, H.S. Parmer, T.O. Mcdonald, and M. Konda. 2015. Lipase-catalyzed dissipative self-assembly of a thixotropic peptide bolaamphiphile hydrogel for human umbilical cord stem-cell proliferation. *Biomacromolecules* 16(4):1157–1168.

Dehsorkhi, A., I.W. Hamley, J. Seitsonen, and J. Ruokolainen. 2013. Tuning self-assembled nanostructures through enzymatic degradation of a peptide amphiphile. *Langmuir* 29(22):6665–6672.

Dijkshoorn, L., C.P. Brouwer, S.J. Bogaards, A. Nemec, P.J. van den Broek, and P.H. Nibbering. 2004. The synthetic N-terminal peptide of human lactoferrin, hLF(1–11), is highly effective against experimental infection caused by multidrug-resistant acinetobacter baumannii. *Antimicrob. Agents Chemother.* 48(12):4919–4921.

Dinca, V. 2007. Directed three-dimensional patterning of self-assembled peptide fibrils. *Nano Lett.* 8(2): 538–543.

Ellis-Behnke, R.G., Y.X. Liang, S.-W. You et al. 2006. Nano neuro knitting: Peptide nanofiber scaffold for brain repair and axon regeneration with functional return of vision. *Proc. Nat. Acad. Sci. U. S. A.* 103(13):5054–5059.

Falla, T.J., D.N. Karunaratne, and R.E. Hancock. 1996. Mode of action of the antimicrobial peptide indolicidin. *J. Biol. Chem.* 271:19298–19303.

Fan, T., X. Yu, B. Shen, and L. Sun. 2017. Peptide self-assembled nanostructures for drug delivery applications. *J. Nanomat.* 2017:1–17.

Fatouros, D.G., D.A. Lamprou, A.J. Urquhart et al. 2014. Lipid-like self-assembling peptide nanovesicles for drug delivery. *ACS Appl. Mater. Interfaces* 6(11): 8184–8189.

Feynman, R.P. 1959. There's plenty of room at the bottom. *Caltech. Eng. Sci.* 23(5):22–36.

Galler, K.M., A. Cavender, V. Yuwono et al. 2008. Self-assembling peptide amphiphile nanofibers as a scaffold for dental stem cells. *Tissue Eng. Part A* 14(12):2051–2058.

Gavel, P.K., D. Dev, H.S. Parmer, S. Basin, and A.K. Das. 2018. Investigations of peptide-based biocompatible injectable shape-memory hydrogels: Differential biological effects on bacterial and human blood cells. *ACS Appl. Mater. Interfaces* 10(13):10729–10740.

Gazit, E. 2002a. A possible role for π-stacking in the self-assembly of amyloid fibrils. *FASEB J* 16(1):77–83.

Gazit, E. 2002b. The "correctly folded" state of proteins: Is it a metastable state? *Angew. Chem. Int. Ed.* 41(2):257–259.

Gazit, E. 2007. Self-assembled peptide nanostructures: The design of molecular building blocks and their technological utilization. *Chem. Soc. Rev.* 36(8):1263–1269.

Ghadiri, M.R., J.R. Granja, R.A. Milligan, D.E. McRee, and N. Khazanovich. 1993. Self-assembling organic nanotubes based on a cyclic peptide architecture. *Nature* 366(6453):324–327.

Ghosh, S., M. Reches, E. Gazit, and S. Verma. 2007a. Bioinspired design of nanocages by self-assembling triskelion peptide elements. *Angew. Chem. Int. Ed.* 46(12):2002–2004.

Ghosh, S., S.K. Singh, and S. Verma. 2007b. Self-assembly and potassium ion triggered disruption of peptide-based soft structures. *Chem. Commun.* 14(22):2296–2298.

Gilead, S. and E. Gazit. 2005. Self-organization of short peptide fragments: From amyloid fibrils to nanoscale supramolecular assemblies. *Supramol. Chem.* 17(1–2): 87–92.

Groves, J.T. and S.G. Boxer. 2002. Micropattern formation in supported lipid membranes. *Acc. Chem. Res.* 35(3):149–157.

Gudlur, S., P. Sukthankar, J. Gao et al. 2012. Peptide nanovesicles formed by the self-assembly of branched amphiphilic peptides. *PLoS One* 7(9):e45374.

Guo, C., Y. Luo, R. Zhou, and G. Wei. 2012. Probing the self-assembly mechanism of diphenylalanine-based peptide nanovesicles and nanotubes. *ACS Nano* 6(5):3907–3918.

Habibi, N., N. Kamaly, A. Memic, and H. Shafiee. 2016. Self-assembled peptide-based nanostructures: Smart nanomaterials toward targeted drug delivery. *Nano Today* 11(1):41–60.

Hancock, R.E., and H.G. Sahl. 2006. Antimicrobial and host-defense peptides as new anti-infective therapeutic strategies. *Nat. Biotechnol.* 24(12):1551–1557.

Hancock, R.E., and R. Lehrer. 1998. Cationic peptides: A new source of antibiotics. *Trends Biotechnol.* 16(2):82–88.

Harper, J.D., and P.T. Jr. Lansbury. 1997. Models of amyloid seeding in Alzheimer's disease and scrapie: Mechanistic truths and physiological consequences of the time-dependent solubility of amyloid proteins. *Annu. Rev. Biochem.* 66:385–407.

Hartgerink, J.D., E. Beniash, and S.I. Stupp. 2002. Peptide-amphiphile nanofibers: A versatile scaffold for the preparation of self-assembling materials. *Proc. Nat. Acad. Sci. U. S. A.* 99(8):5133–5138.

Hartgerink, J.D., J.R. Granja, R.A. Milligan, and M.R. Ghadiri. 1996. Self-assembling peptide nanotubes. *J. Am. Chem. Soc.* 118(1):43–50.

Hendler, N., N. Sidelman, M. Reches, E. Gazit, Y. Rosenberg, and S. Richter. 2007. Formation of well-organized self-assembled films from peptide nanotubes. *Adv. Mater.* 19(11):1485–1488.

Hirst, A.R., B. Escuder, J.F. Miravet, and D.K. Smith. 2008. High-tech applications of self-assembling supramolecular nanostructured gel-phase materials: From regenerative medicine to electronic devices. *Angew. Chem. Int. Ed.* 47(42):8002–8018.

Hodgson, D.R.W., and J.M. Sanderson. 2004. The synthesis of peptides and proteins containing non-natural amino acids. *Chem. Soc. Rev.* 33(7):422–430.

Holmes, T.C. 2002. Novel peptide-based biomaterial scaffolds for tissue engineering. *Trends Biotechnol.* 20(1):16–21.

Holmes, T.C., S.D. Lacalle, X. Su, G. Liu, A. Rich, and S. Zhang. 2000. Extensive neurite outgrowth and active synapse formation on self-assembling peptide scaffolds. *Proc. Nat. Acad. Sci. U. S. A* 97(12):6728–6733.

Holowka, E.P., D.J. Pochan, and T.J. Deming. 2005. Charged polypeptide vesicles with controllable diameter. *J. Am. Chem. Soc.* 127(35):12423–12428.

Holowka, E.P., V.Z. Sun, D.T. Kamei, and T.J. Deming. 2007. Polyarginine segments in block copolypeptides drive both vesicular assembly and intracellular delivery. *Nat. Mater.* 6(1):52–57.

Hong, Y., R.L. Legge, S. Zhang, and P. Chen. 2003. Effect of amino acid sequence and pH on nanofiber formation of self-assembling peptides EAK16-II and EAK16-IV. *Biomacromolecules* 4(5):1433–1442.

Huang, R.L., W. Qi, R.X. Su, J. Zhao, and Z.M. He. 2011. Solvent and surface controlled self-assembly of diphenylalanine peptide: From microtubes to nanofibers. *Soft Matter* 7(14):6418–6421.

Israelachvili, J.N., D.J. Mitchell, and B.W. Ninham. 1977. Theory of self-assembly of lipid bilayers and vesicles. *Biochim. Biophys. Acta* 470(2):185–201.

Jeena, M.T., L. Palanikumar, E.M. Go et al. 2017. Mitochondria localization induced self-assembly of peptide amphiphiles for cellular dysfunction. *Nat. Commun.* 8(1):1–10.

Kalafatovic, D., M. Nobis, J. Son, K.I. Anderson, and R.V. Ulijn. 2016. MMP-9 triggered self-assembly of doxorubicin nanofiber depots halts tumor growth. *Biomaterials* 98:192–202.

Kataoka, K., A. Harada, and Y. Nagasaki. 2001. Block copolymer micelles for drug delivery: Design, characterization and biological significance. *Adv. Drug Delivery. Rev.* 47(1):113–131.

Kerr, J.M., S.C. Banville, and R.N. Zuckermann. 1993. Encoded combinatorial peptide libraries containing non-natural amino acids. *J. Am. Chem. Soc.* 115(6):2529–2531.

Keyes-Baig, C., J. Duhamel, S.Y. Fung, J. Bezaire, and P. Chen. 2004. Self-assembling peptide as a potential carrier of hydrophobic compounds. *J. Am. Chem. Soc.* 126(24):7522–7532.

Kim, I., H.-H. Jeong, Y.-J. Kim et al. 2014. A "Light-up" 1D supramolecular nanoprobe for silver ions based on assembly of pyrene-labeled peptide amphiphiles: Cell-imaging and antimicrobial activity. *J. Mater. Chem. B* 2(38):6478–6486.

Kokkoli, E., R.W. Kasinskas, A. Mardilovich, and A. Garg. 2005. Fractalkine targeting with a receptor-mimicking peptide-amphiphile. *Biomacromolecules* 6(3):1272–1279.

Koley, P., M.G.B. Drew, and A. Pramanik. 2011. Salts responsive nanovesicles through π-stacking induced self-assembly of backbone modified tripeptides. *J. Nanosci. Nanotechnol.* 11(8):6747–6756.

Komatsu, H., S. Matsumoto, S.-I. Tamaru, K. Kaneko, M. Ikeda, and I. Hamachi. 2009. Supramolecular hydrogel exhibiting four basic logic gate functions to fine-tune substance release. *J. Am. Chem. Soc.* 131(15):5580–5585.

Konda, M., R.G. Jadhav, S. Maiti, S.M. Mobin, B. Kauffmann, and A.K. Das. 2018. Understanding the conformational analysis of gababutin based hybrid peptides. *Org. Biomol. Chem.* 16(10):1728–1735.

Konda, M., S. Bhowmik, S.M. Mobin, S. Biswas, and A.K. Das. 2016. Modulating hydrogen bonded self-assembled patterns and morphological features by a change in side chain of third amino acid of synthetic γ-amino acid based tripeptides. *ChemistrySelect* 1(11):2586–2593.

Korevaar, P.A., C.J. Newcomb, E.W. Meijer, and S.I. Stupp. 2014. Pathway selection in peptide amphiphile assembly. *J. Am. Chem. Soc.* 136(24):8540–8543.

Krysmann, M.J., V. Castelletto, A. Kelarakis, I.W. Hamley, R.A. Hule, and D.J. Pochan. 2008. Self-assembly and hydrogelation of an amyloid peptide fragment. *Biochemistry* 47(16):4597–4605.

Kyle, S., A. Aggeli, E. Ingham, and M.J. McPherson. 2009. Production of self-assembling biomaterials for tissue engineering. *Trends Biotechnol.* 27(7):423–433.

Lee, S.S., B.J. Huang, S.R. Kaltz et al. 2013. Bone regeneration with low dose BMP-2 amplified by biomimetic supramolecular nanofibers within collagen scaffolds. *Biomaterials* 34(2):452–459.

Li, F., J. Wang, F. Tang et al. 2009. Fluorescence studies on a designed self-assembling peptide of RAD16-II as a potential carrier for hydrophobic drug. *J. Nanosci. Nanotechnol.* 9:1611–1624.

Li, R., C.C. Horgan, B. Long et al. 2015. Tuning the mechanical and morphological properties of self-assembled peptide hydrogels via control over the gelation mechanism through regulation of ionic strength and the rate of pH change. *RSC Adv.* 5(1):301–307.

Liu, H., J. Chen, Q. Shen, W. Fu, and W. Wu. 2010. Molecular insights on the cyclic peptide nanotube-mediated transportation of antitumor drug 5-fluorouracil. *Mol. Pharmaceutics* 7(6):1985–1994.

Lowik, D.W., E. Leunissen, M. Van den Heuvel, M. Hansen, and J.C. Van Hest. 2010. Stimulus responsive peptide based materials. *Chem. Soc. Rev.* 39(9):3394–3412.

Lu, S., H. Wang, Y. Sheng, M. Liu, and P. Chen. 2012. Molecular binding of self-assembling peptide EAK16-II with anticancer agent EPT and its implication in cancer cell inhibition. *J. Controlled Release* 160(1):33–40.

Lutolf, M.P., and J.A. Hubbell. 2005. Synthetic biomaterials as instructive extracellular microenvironments for morphogenesis in tissue engineering. *Nat. Biotechnol.* 23:47–55.

Maity, I., D.B. Rasale, and A.K. Das. 2012. Sonication induced peptide-appended bolaamphiphile hydrogels for *In situ* generation and catalytic activity of Pt nanoparticles. *Soft Matter* 8(19):5301–5308.

Manna, M.K., S.K. Pandey, I. Maity, S. Mukherjee, and A.K. Das. 2015. Electrodeposited lamellar photoconductor nanohybrids driven by peptide self-assembly. *Chem. Plus Chem.* 80(3):583–590.

Marini, D.M., W. Hwang, D.A. Lauffenburger, S. Zhang, and R.D. Kamm. 2002. Left-handed helical ribbon intermediates in the self-assembly of a β-sheet peptide. *Nano Lett.* 2(4):295–299.

Mart, R.J., R.D. Osborne, M.M. Stevens, and R.V. Ulijn. 2006. Peptide-based stimuli-responsive biomaterials. *Soft Matter* 2(10):822–835.

Meng, Q., Y. Kou, X. Ma et al. 2012. Tunable self-assembled peptide amphiphile nanostructures. *Langmuir* 28(11):5017–5022.

Miller, R.E., P.W Kopesky, and A.J. Grodzinsky. 2011. Growth factor delivery through self-assembling peptide scaffolds. *Clin. Orthop. Relat. Res.* 469(10):2716–2724.

Nguyen, L.T., E.F. Haney, and H.J. Vogel. 2011. The expanding scope of antimicrobial peptide structures and their modes of action. *Trends Biotechnol.* 29(9):464–472.

Niece, K.L., J.D. Hartgerink, J.J.J.M. Donners, and S.I. Stupp. 2003. Self-assembly combining two bioactive peptide-amphiphile molecules into nanofibers by electrostatic attraction. *J. Am. Chem. Soc.* 125(24):7146–7147.

Orbach, R., L. Alder-Abramovic, S. Zigerson, I. Mironi-Harpaz, D. Seliktar, and E. Gazit. 2009. Self-assembled Fmoc peptides as a platform for the formation of nanostructures and hydrogels. *Biomacromolecules* 10(9):2646–2651.

Panda, J.J., R. Dua, A. Mishra, B. Mittra, and V.S. Chauhan. 2010. 3D cell growth and proliferation on a RGD functionalized nanofibrillar hydrogel based on a conformationally restricted residue containing dipeptide. *ACS Appl. Mater. Interfaces* 2(10):2839–2848.

Pappas, C.G., P.W.J.M. Frederix, T. Mutasa et al. 2015. Alignment of nanostructured tripeptide gels by directional ultrasonication. *Chem. Commun.* 51(40):8465–8468.

Pawar, R., A. Ben-Ari, and A.J. Domb. 2004. Protein and peptide parenteral controlled delivery. *Expert Opin. Biol. Ther.* 4(8):1203–1212.

Rajangam, K., H.A. Behanna, M.J. Hui et al. 2006. Heparin binding nanostructures to promote growth of blood vessels. *Nano Lett.* 6(9):2086–2090.

Rasale, D.B, S. Biswas, M. Konda, and A.K. Das. 2015. Exploring thermodynamically downhill nanostructured peptide libraries: From structural to morphological insight. *RSC Adv.* 5(2):1529–1537.

Ray, S., A.K. Das, M.G.B. Drew, and A. Banerjee. 2006. A short water soluble self-assembly peptide forms amyloid-like fibrils. *Chem. Commun.* (40):4230–4232.

Reches, M., and E. Gazit. 2003. Casting metal nanowires within discrete self-assembled peptide nanotubes. *Science* 300(5619):625–627.

Reches, M., and E. Gazit. 2004. Formation of closed-cage nanostructures by self-assembly of aromatic dipeptides. *Nano Lett.* 4(4):581–585.

Reches, M., and E. Gazit. 2006a. Controlled patterning of aligned self-assembled peptide nanotubes. *Nat. Nanotechnol.* 1(3):195–200.

Reches, M., and E. Gazit. 2006b. Designed aromatic homo-dipeptides: Formation of ordered nanostructures and potential nanotechnological applications. *Phys. Biol.* 3(1):S10–S19.

Reches, M., Y. Porat, and E. Gazit. 2002. Amyloid fibril formation by pentapeptide and tetrapeptide fragments of human calcitonin. *J. Biol. Chem.* 277(38):34575–35480.

Rodrigo da Silva, E., A.W. Andrade, V. Castelletto et al. 2015. Self-assembly pathway of peptide nanotubes formed by a glutamatic acid-based bolaamphiphile. *Chem. Commun.* 51(59):11634–11637.

Rosenman, G., P. Beker, I. Koren et al. 2011. Bioinspired peptide nanotubes: Deposition technology, basic physics and nanotechnology applications. *J. Pept. Sci.* 17(2):75–87.

Santoso, S., W. Hwang, H. Hartman, and S. Zhang. 2002. Self-assembly of surfactant-like peptides with variable glycine tails to form nanotubes and nanovesicles. *Nano Lett.* 2(7):687–691.

Scanlon, S. and A. Aggeli. 2008. Self-assembling peptide nanotubes. *Nano Today* 3(3):22–30.

Seo, M.-D., H.-S. Won, J.-H. Kim, T. Mishig-Ochir, and B.-J. Lee. 2012. Antimicrobial peptides for therapeutic applications: A review. *Molecules* 17(10):12276–12286.

Shi, J.F., and B. Xu. 2015. Nanoscale assemblies of small molecules control the fate of cells. *Nano Today* 10(5):615–630.

Silva, G.A., C. Czeisler, K.L. Niece et al. 2004. Selective differentiation of neural progenitor cells by high-epitope density nanofibers. *Science* 303(5662):1352–1355.

Smith, K.H., E. Tejeda-Montes, M. Poch, and A. Mata. 2011. Integrating top-down and self-assembly in the fabrication of peptide and protein-based biomedical materials. *Chem. Soc. Rev* 40(9):4563–4577.

Soldatov, D.V., I.L. Moudrakovski, and J.A. Ripmeester. 2004. Dipeptides as microporous materials. *Angew. Chem. Int. Ed.* 43(46):6308–6311.

Standley, S.M., D.J. Toft, H. Cheng et al. 2010. Induction of cancer cell death by self-assembling nanostructures incorporating a cytotoxic peptide. *Cancer Res.* 70(8):3020–3026.

Stephanopoulos, N., J.H. Ortony, and S.I. Stupp. 2013. Self-assembly for the synthesis of functional biomaterials. *Acta Mater.* 61(3):912–930.

Stile, R.A., and K.E. Healy. 2001. Thermo-responsive peptide-modified hydrogels for tissue regeneration. *Biomacromolecules* 2(1):185–194.

Stroumpoulis, D., H. Zhang, L. Rubalcava, J. Gliem, and M. Tirrell. 2007. Cell adhesion and growth to peptide-patterned supported lipid membranes. *Langmuir* 23(7):3849–3856.

Sun, L., C. Zheng, and T.J. Webster. 2017. Self-assembled peptide nanomaterials for biomedical applications: Promises and pitfalls. *Int. J. Nanomed.* 12:73–86.

Sunde, M., and C.C.F. Blake. 1998. From the globular to the fibrous state: Protein structure and structural conversion in amyloid formation. *Q. Rev. Biophys.* 31(1): 1–39.

Tahara, K., S. Lei, J. Adisoejoso, S.D. Feyter, and Y. Tobe. 2010. Supramolecular surface-confined architectures created by self-assembly of triangular phenylene–ethynylene macrocycles via van der Waals interaction. *Chem. Commun.* 46(45):8507–8525.

Teixeira, V., M.J. Feio, and M. Bastos. 2012. Role of lipids in the interaction of antimicrobial peptides with membranes. *Prog. Lipid Res* 51(2):149–177.

Tenidis, K., M. Waldner, J. Bernhagen et al. 2000. Identification of a penta- and hexapeptide of islet amyloid polypeptide (IAPP) with amyloidogenic and cytotoxic properties. *J. Mol. Biol.* 295(4):1055–1071.

Tokosz, S., H. Acar, and M.O. Guler. 2010. Self-assembled one dimensional soft nanostructures. *Soft Matter* 6(23):5839–5849.

Tu, R.S., and M. Tirrell. 2004. Bottom-up design of biomimetic assemblies. *Adv. Drug Delivery Rev.* 56(11):1537–1563.

Valery, C., M. Paternostre, B. Robert et al. 2003. Biomimetic organization: Octapeptide self-assembly into nanotubes of viral capsid-like dimension. *Proc. Nat. Acad. Sci. U. S. A* 100(18):10258–10262.

van Hell, A.J., C.I.C.A. Costa, F.M. Flesch et al. 2007. Self-assembly of recombinant amphiphilic oligopeptides into vesicles. *Biomacromolecules* 8(9):2753–2761.

van't Hof, W., E.C. Veerman, E.J. Helmerhorst, and A.V. Amerongen. 2001. Antimicrobial peptides: Properties and applicability. *Biol. Chem.* 382(4):597–619.

Vauthey, S., S. Santoso, H. Gong, N. Watson, and S. Zhang. 2002. Molecular self-assembly of surfactant-like peptides to form nanotubes and nanovesicles. *Proc. Nat. Acad. Sci. U. S. A* 99(8):5355–5360.

Wang, A., L. Cui, S. Debnath et al. 2017. Tuning supramolecular structure and functions of peptide bolaamphiphile by solvent evaporation: Dissolution. *ACS Appl. Mater. Interfaces* 9(25):21390–21396.

Wang, J., K. Liu, L. Yan, A. Wang, S. Bai, and X. Yan. 2016a. Trace solvent as predominant factor to tune dipeptide self-assembly. *ACS Nano* 10(2):2138–2143.

Wang, J., K. Liu, R. Xinga, and X. Yan. 2016b. Peptide self-assembly: Thermodynamics and kinetics. *Chem. Soc. Rev.* 45(20):5589–5604.

Webber, M.J., J.B. Matson, V.K. Tamboli, and S.I. Stupp. 2012. Controlled release of dexamethasone from peptide nanofiber gels to modulate inflammatory response. *Biomaterials* 33(28):6823–6832.

Whitesides, G.M., and B. Grzybowski. 2002. Self-assembly at all scale. *Science* 295(5564):2418–2421.

Wickner, R.B., K.L. Taylor, H.K. Edskes, M.L. Maddelein, H. Moriyama, and B.T. Roberts. 2000. Prions of yeast as heritable amyloidoses. *J. Struct. Biol.* 130:310–322.

Williams, R.J., R.J. Mart, and R.V. Ulijn. 2010. Exploiting biocatalysis in peptide self-assembly. *Biopolymers* 94(1):107–117.

Xue, Q., H. Ren, C. Xu et al. 2015. Nanospheres of doxorubicin as cross-linkers for a supramolecular hydrogelation. *Sci. Rep.* 5:8764.

Yan, X., Q. He, K. Wang, L. Duan, Y. Cui, and J. Li. 2007. Transition of cationic dipeptide nanotubes into vesicles and oligonucleotide delivery. *Angew. Chem. Int. Ed.* 46(14):2431–2434.

Yang, Z., G. Liang, L. Wang, and B. Xu. 2006. Using a kinase/phosphatase switch to regulate a supramolecular hydrogel and forming the supramolecular hydrogel in vivo. *J. Am. Chem. Soc.* 128(9):3038–3043.

Ye, Z., H. Zhang, H. Luo et al. 2008. Temperature and pH effects on biophysical and morphological properties of self-assembling peptide RADA16-I. *J. Pept. Sci.* 14(2):152–162.

Yishay-Safranchik, E., M. Golan, and A. David. 2014. Controlled release of doxorubicin and Smac-derived Pro-apoptotic Peptide from Self-assembled KLD-based Peptide Hydrogels. *Polym. Adv. Technol.* 25(5): 539–544.

Yu, Z., Z. Cai, and Q. Chen. 2016. Engineering β-sheet peptide assemblies for biomedical applications. *Biomater. Sci.* 4(3):365–374.

Zasloff, M. 2002. Antimicrobial peptides of multicellular organisms. *Nature* 415(6870):389–395.

Zhang, L., and T.J. Webster. 2009. Nanotechnology and nanomaterials: Promises for improved tissue regeneration. *Nano Today* 4(1):66–80.

Zhang, L., J. Zhong, L. Huang, L. Wang, Y. Hong, and Y. Sha. 2008. Parallel-oriented fibrogenesis of a β-sheet forming peptide on supported lipid bilayers. *J. Phys. Chem. B* 112(30):8950–8954.

Zhang, L., X. Wang, T. Wang, and M. Liu. 2014. Tuning soft nanostructures in self-assembled supramolecular gels: From morphology control to morphology dependent functions. *Small* 11(9–10):1–14.

Zhang, S. 1995. Self-complementary oligopeptide matrices support mammalian cell attachment. *Biomaterials* 16(18):1385–1393.

Zhang, S. 2003a. Building from the bottom up. *Materials Today* 6:20–27.

Zhang, S. 2003b. Fabrication of novel biomaterials through molecular self-assembly. *Nat. Biotechnol.* 21(10):1171–1178.

Zhang, S., M.A. Greenfield, A. Mata et al. 2010. A self-assembly pathway to aligned monodomain gels. *Nat. Mater* 9(7):594–601.

Zhu, P., X. Yan, Y. Su, Y. Yang, and J. Li. 2010. Solvent-induced structural transition of self-assembled dipeptide: From organogels to microcrystals. *Chem. Eur. J.* 16(10):3176–3183.

Bacterial Detection with Magnetic Nanoparticles

Nayeem A. Mulla and
Raghvendra A. Bohara
D. Y. Patil University

Shivaji H. Pawar
D. Y. Patil University
Anekant Education Society

3.1 Introduction to Bacteria

Bacteria are microscopic single-celled organisms that are all around us. They come in many different sizes and shapes, and a common way to classify them is by their morphology, or shape and appearance [1]. The three basic shapes of bacteria are spherical, rod shaped, and spiral. Spiral-shaped bacteria can be further categorized depending in part on how much spiraling they show. Not all bacteria are capable of causing disease, but each morphology-based group has at least some disease-causing representatives [2].

Bacteria are the most abundant and diverse life forms on the earth. A bacterium belongs to the prokaryote group, which means they are unicellular organism without nucleus. A typical bacteria is only several micrometers in size, and because they are such small organisms light microscopes only demonstrate their morphology. The reproduction rate of bacteria is the fastest among all living organisms [3]. They have a rapid adaptive nature not only morphologically and metabolically but also genetically. Therefore bacterial populations can adapt to new environments very quickly, and they evolve to meet the demands of any new localized stressors. They can change so that they can use other substances as their food and can develop resistance to antibiotics [4,5].

Certain bacteria, like actinomycetes, produce antibiotics such as streptomycin and nocardicin; others live symbiotically in the gut of other animals (including humans) or elsewhere in their bodies, or in the roots of certain plants, converting nitrogen into a usable form [6]. Bacteria put the tang in yogurt and the sour in sourdough bread; bacteria help to break down dead organic matter; bacteria make up the base of the food web in many environments. Bacteria are of such immense importance because of their extreme flexibility, capacity for rapid growth and reproduction, and their age—the oldest fossils known, nearly 3.5 billion years old, are fossils of bacteria-like organisms [7].

A bacterium is a single-celled, unicellular microorganism that does not have a nucleus or any other membrane-bound organelles. Bacteria are sometimes called "prokaryotes". In Greek, "prokaryote" literally means "before the nut". Bacteria adapt to become well suited to their environments, and therefore come in many shapes and forms [8]. However, they all have a few features in common (Figure 3.1).

1. **Capsule**: A protective, often slimy, coating mainly made up of sugars that help to protect the bacterium. It also makes bacteria virulent. This means the bacteria are more likely to cause disease, since it aids the cell survive against immunological attack. For example, the bacteria may survive an attack from the human body's immune system.

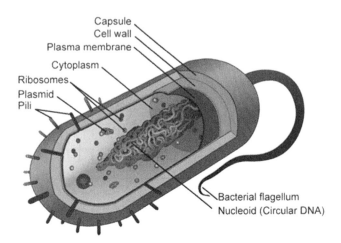

FIGURE 3.1   Anatomy of a typical bacterium.

2. **Cell wall**: In bacteria, the cell wall is usually made of peptidoglycan, a protein and sugar compound. This structure gives the cell some rigidity and protection.

3. **Plasma membrane**: As in most cells, the bacterium's plasma membrane acts to coordinate the passage of molecules into and out of the cell.

4. **Cytoplasm**: Again, as in many cells, the cytoplasm serves as a medium through which molecules are transported as well as a system to maintain homeostatic conditions (like temperature and pH) that are best for the cell.

5. **Ribosomes**: These are the main site for the bacterium's protein synthesis. They are scattered throughout the cell cytoplasm.

6. **Plasmid**: It is an extrachromosomal, small, circular double-stranded DNA. It replicates itself independently of the bacterial genome and many times contains important genes.

7. **Nucleoid**: This is the region where the bacterium's DNA is located. It is not the same as a nucleus in eukaryotic cells because it is not surrounded by a membrane.

8. **Flagellum**: In many bacteria, a flagellum is present, and it is the means by which the cell moves around.

9. **Pili**: Many species of the bacterium contains hair-like structure on the cell surface called pili. They are an important organ in the bacterial reproduction process.

3.2   Bacteria and Their Detection

Fundamentals of morphology, molecular chemistry, and surface physiochemical are important considerations while detecting bacterial species. With respect to conventional sense of biology, bacteria are known as microscopic organisms, ranging a few micrometers in length and of various different kinds. They are unicellular, prokaryotic in nature with no internal organization, and multiply by undergoing fission. They have different shapes like rod, spherical, and cuboidal. They can be found either singly or in pairs or even as chains or clusters. They have a single chromosome with a closed circle of double-stranded DNA [9]. Sometimes, they possess characteristic appendages like flagella. The cell wall is rigid and made up of phospholipid bilayers. Bacteria can be categorized on different bases such as by staining (e.g., Gram-positive, Gram-negative), culturing requirements (aerobic, anaerobic), etc. Mostly, they are recognized based on Gram staining. Gram-positive bacteria possess a thick cell wall containing many layers of peptidoglycan and teichoic acid. In contrast, Gram-negative bacteria have a relatively thin cell wall consisting few layers of peptidoglycan surrounded by a second lipid layer containing lipopolysaccharides and lipoproteins [10].

3.3   Biomagnetism

Magnetoreception is a sense that allows an organism to detect a magnetic field to perceive direction, altitude, or location. This sensory modality is used by a range of animals for orientation and navigation [11] and as a method for animals to develop regional maps. For the purpose of navigation, magnetoreception deals with the detection of the earth's magnetic field. Magnetoreception is present in bacteria, arthropods, molluscs, and members of all major taxonomic groups of vertebrates. Humans are not thought to have a magnetic sense, but there is a protein (a cryptochrome) in the eye, which could serve this function. Unequivocal demonstration of the use of magnetic fields for orientation within an organism is seen in a class of bacteria known as magnetotactic bacteria. These bacteria demonstrate a behavioral phenomenon known as magnetotaxis, in which the bacterium orients itself and migrates in the direction that follows the earth's magnetic field lines. The bacteria contain magnetosomes, which are nanometer-sized particles of magnetite or iron sulfide enclosed within the bacterial cells [12]. The magnetosomes are surrounded by a membrane composed of phospholipids and fatty acids and contain at least 20 different proteins. They form as chains where the magnetic moments of each magnetosome align in parallel, causing each singular bacterium to act as a magnetic dipole, giving the bacteria their permanent-magnet characteristics. In animals, the mechanism for magnetoreception is unknown, but there are two main hypotheses to explain this phenomenon [13]. According to one model, magnetoreception is possible via the radical-pair mechanism. The radical-pair mechanism is well established in spin chemistry and was speculated to be applicable to magnetoreception [14]. In year 2000, cryptochrome was proposed as the "magnetic molecule", so to speak, that could harbor magnetically sensitive radical-pairs. Cryptochrome, a flavoprotein found in the eyes of European robins and other animal species, is the only protein known to form photo induced radical-pairs in animals. The function of cryptochrome is diverse across species; however, the photo induction of radical-pairs occurs by exposure to blue light, which excites an electron in a chromophore. The Earth's magnetic field is only 0.5 Gauss, so it is difficult to conceive of a mechanism other than phase shift, by which such a field could lead to any chemical changes other than those affecting the weak magnetic fields between radical pairs [14]. Cryptochromes are therefore thought to be essential for the light-dependent ability of the fruit fly *Drosophila melanogaster* to sense magnetic fields. The second proposed model for magnetoreception relies on Fe_3O_4, also referred to as iron (II, III) oxide or magnetite, a natural oxide with strong magnetism. Iron (II, III) oxide remains permanently magnetized when its length is larger than 50 nm and becomes magnetized when exposed to a magnetic field if its length is less than 50 nm. In both of these situations, the earth's magnetic field leads to a transducible signal via a physical effect on this magnetically sensitive oxide. The movement of charged particles causes magnetism. To understand the magnetic properties of a substance, one would need to look at the motion of electrons within the material [15]. Without the force of magnetism, or the knowledge of it, we would not be able to navigate without the sun or stars. In addition, we would not be able to run most electronics, from a loudspeaker to a car or a plane without magnetism. The medical field would not be advanced enough to diagnose diseases within hours, or to detect cancerous tumors. Without magnetism, stores and libraries would not be able to have anti-theft security systems. Similarly, detector systems such as those in airports would not be able to scan people for weapons with metal detectors. Migrating animals use magnetism to find their proper habitats; without it, entire species could die. In fact, without the Earth's magnetic field, the entire planet would erode away. Magnetism is clearly an unseen force that our world depends on [16].

3.4 Magnetic Nanoparticles

Magnetic nanoparticles (MNPs) have shown great promise in many fields. MNPs are nanoparticles which can be manipulated by using magnetic field gradients. When the size of the nanoparticles becomes smaller, from a few nanometers to a couple of tenths of nanometers, depending on the material, superparamagnetism appears. Superparamagnetic nanoparticles are single-domain particles with all their magnetic moments aligned in the same direction, and with a short relaxation time. These particles have unique properties such as nearly instantaneous change of magnetization in the applied magnetic field, which allows them to be directed toward a target using an external magnetic field and heating in alternating magnetic fields [17]. Owing to their unique properties, superparamagnetic nanoparticles have been intensively developed and have found numerous applications in biomedical, optical and electronic fields such as biosensing [18], targeted drug delivery [19], destruction of cancer tissues through hyperthermia, magnetic resonance

imaging, jet printing, cell separation, DNA separation [20,21], pathogen detection [22] immunoassay, and tissue repair [23].

The two main forms of nanoparticles with superparamagnetic properties are magnetite (Fe_3O_4) and its oxidized form maghemite (c-Fe_2O_3). As the superparamagnetic behavior of iron oxide nanoparticles strongly depends upon the dimension of the nanoparticles, control of uniform size distribution is very important in synthesis. The size and shape of the nanoparticles, however, depend on many factors such as pH, ionic strength, temperature, nature of the salts, and the Fe(II)/Fe(III) concentration ratio [24].

Various procedures have been described to synthesize MNPs, and among these methods, chemical coprecipitation of Fe^{2+} and Fe^{3+} ions by an alkali such as NH_4OH or ammonia, in an aqueous solution, is the most commonly used solution-phase procedure. Nucleation and growth of nanocrystals are the two main steps that are involved in this bottom–up method [25]. The advantage of the precipitation method is that comparatively large quantities of nanoparticles can be prepared. Nevertheless, there are several drawbacks for this method such as difficulty in controlling the size distribution and the formation of impurities such as goethite and maghemite [26]. The main factors affecting the formation of impurities during coprecipitation are the initial and final pH of the solution and the reaction temperature [27]. It is reported that iron oxide nanoparticles with sizes below 20 nm cannot be labeled as "magnetite". When the particle size is less than 10 nm, increase in surface area/volume ratio results in a large number of surface atoms, which would oxidize readily to Fe^{3+}, thereby leading to the formation of maghemite on the surface of the magnetic particle. Longer storage periods (6 months) and exposure to high temperatures ($>180°C$) would also cause the Fe^{2+} ions to oxidize to Fe^{3+} ions leading to the formation of maghemite. The disadvantage is that the maghemite has slightly less saturation magnetization than magnetite. However, when it comes to the technological application of the nanoparticles, both magnetite and maghemite are suitable, as both of them are ferromagnetic in nature with superparamagnetism. In contrast, goethite is antiferromagnetic in nature. The formation of goethite can increase with an increase in pH above 4.7 and, thus in co-precipitation, the ideal ratio of Fe^{2+}/Fe^{3+} is about 1:2 [28].

3.5 Role of MNPs in Microbiology and Biotechnology

MNPs possess certain remarkable physicochemical properties such as superparamagnetic behavior, the ability to become heated under alternating magnetic field, high surface/volume ratio, and versatility in synthesis [29]. These particles are used to kill/inhibit growth of bacteria or to act as carriers of antibiotic and other antibacterial nanomaterials to help in the prevention, diagnosis, and treatment of infectious disease. At the same time, these

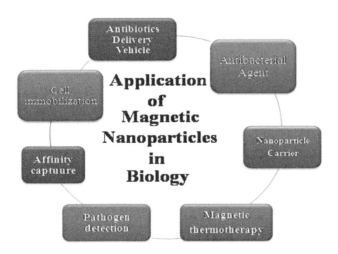

FIGURE 3.2 Role of MNPs in biological applications.

characteristics can also be used in microbial biotechnology to make or modify products or processes using microbial systems. The bioapplications of MNPs are shown in Figure 3.2.

3.6 Methods to Generate Bioconjugated MNPs

Many strategies have been developed to immobilize specific molecules with high affinity for bacteria onto the surface of nanoparticle substrates. These methods include both covalent and noncovalent attachment strategies. A detailed discussion is provided (*vide infra*). A popular coupling method to bind molecules or materials to biomolecules is the use of the *N*-hydroxysuccinimide ester (NHS-ester) functionalization group. The NHS-ester can readily react with primary amines, resulting in an amide bond covalent linkage [30,31]. For example, NHS-ester functionalized MNPs can be conjugated to the amino groups of biomolecules (see Figure 3.3a). Another approach is the use of a dual NHS ester reagent, bis-(*N*-hydroxysuccinimide ester) (DSS), which can link amino groups of biomolecules to amino-functionalized MNPs (see Figure 3.3b). Indeed, biomolecules with amide bond linkages to nanoparticles can also be created by first converting the amino group of amino-functionalized MNPs to a carboxylic acid by reaction with succinic anhydride. The amino group of the biomolecule can then react with the carbonyl group forming an amide bond linkage (Figure 3.3c) [32,33]. A click chemistry approach to connect biomolecules and MNPs can be realized by employingtrans-cyclooctene (TCO) and tetrazine (Tz) functional groups (Figure 3.3d). Another common method to tether molecules to proteins utilizes the streptavidin–biotin interaction. The streptavidin–biotin interaction is a well-known recognition event with extremely high binding affinity (Figure 3.3e). The hydroxyl group on the surface of the Fe_4O_3 can be converted into an epoxide by silanization in the presence of 3-glycidyloxypropyltrimethoxysilan (GOPTS) and ethanol.

The amino group of the biomolecule can then react with the epoxy group to form bioconjugated MNPs (Figure 3.3f) [33].

3.7 Rapid and Real-Time Detection of Bacteria Using MNP

Pathogenic bacteria are major concerns when it comes to human health, food industry, and water facilities. Accurate and definitive microorganism identification, including bacterial identification and detection, is essential for correct disease diagnosis, treatment of infection, and trace-back of disease outbreaks associated with microbial infections. For this, innovative, rapid, sophisticated, and highly sensitive detection methods are necessary. Successful attempts have been made in the development of molecular analyzing techniques like ELISA, PCR, ribotyping, microarray, etc. Aside from high sensitivity and reliability, these techniques suffer from high cost of performance, sample pretreatment, and lower limit of detection [17,18]. The drawbacks of conventional and current molecular diagnostics can be overcome with the help of nanoscience. Nanotechnology is a multidisciplinary branch of science which that deals with technology relating to nanosized materials. It has a huge significance in biomedical, pharmaceutical, agricultural, environmental, and many more branches of science. Nanobiotechnology is the branch of science that deals with the fabrication and use of nanomaterials for biological and biochemical applications. It demonstrates all facets of research of biology assisted with nanotechnology [19]. As regards the properties of sub 100-nm materials and devices, their surface modification has contributed significantly to biomedical fields such as cellular repair, drug delivery, therapeutic applications, and diagnostic aids [20]. Knowledge and application of nanomaterials enable in-depth understanding of all bimolecular processes easy. Although many techniques are still in the nascent stage of development, some are actually being employed in daily practices [21]. Use of nanobiotechnology extend the limits of current molecular diagnostics, allows point-of-care diagnostics, and integrates diagnostics with therapeutics. It enables diagnosis at the single-cell and molecular level [22]. MNPs possessing nanoscale size ranges are examples of bionanomaterials that mimic the size of molecules in nature and possess favorable characteristics making them multifunctional for bio-nanoapplications. The use of MNP's high surface area and superparamagnetic property provides a promising and sophisticated platform for detection techniques so that conventional and molecular diagnostics become much easier and more accurate [23] (Figure 3.4).

3.8 Role of MNPs in Bacterial Detection

Nanoparticles are the key focus of research for a variety of novel applications, not only because of their wonderful properties but also due to their nanosize, smaller compared

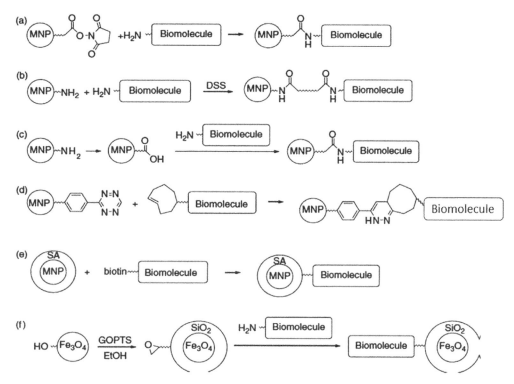

FIGURE 3.3 A summary of methods to synthesize bioconjugated MNPs. (a) The amino group of a biomolecule reacts with the NHS-ester coated nanoparticles forming an amide linkage. (b) Using the dual NHS-ester reagent (DSS), amino-functionalized nanoparticles are linked to biomolecules. (c) By conversion of amino-functionalized nanoparticles to carboxyl-modified particles, the amide bond linkage can be achieved by reaction with the amino group of the biomolecule. (d) Teatrazine-modified nanoparticles are attached to trans-cyclooctene modified biomolecules through a cycloaddition reaction. (e) Streptavidin-coated nanoparticles can bind to biomolecules via the biotin–streptavidin interaction. (f) Fe_3O_4 nanoparticles are converted to an epoxy-functionalized nanoparticle in the presence of 3-glycidyoxypropyltrimethoxysilan and ethanol. The antibody amino groups can then react with the epoxy on the nanoparticles forming antibody-modified nanoparticles.

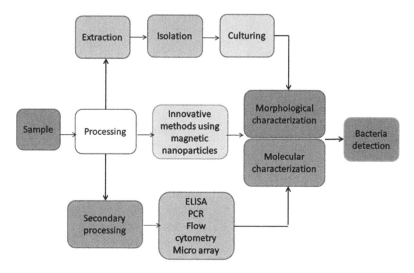

FIGURE 3.4 Rapid and real-time detection of bacteria by using MNPs.

with their bulk counterparts. Nanoparticles are intermediate between atomic- and bulk-level particles. At the nano level, the properties greatly change as the size of the particles change owing to their large surface to volume ratio. Owing to their widespread applications, much research has been carried on the synthesis various nanostructures. For decades,

MNPs have been in focus as they show high potential in a variety of different application fields, ranging from chemistry, biology, and medicine to physics. Magnetic nanoparticles have a wide range of applications ranging from magnetic fluids recording, catalysis [34], and biomedicine [35] to material sciences, photocatalysis [36,37], and so on. Nowadays, many novel binary and ternary magnetic nanocomposites have also been synthesized with various core–shell structures including grapheme [38], carbon nanotube [39], conducting polymer [40], metal oxide, and other inorganic materials. Owing to its unique and creative applications in every field of life, researchers are highly focused in developing a number of synthetic ways to synthesize MNPs with different sizes, morphologies, and compositions, but the successful application of MNPs for the abovementioned applications is highly dependent on the stability of the particles under a plethora of different conditions. The significance of size for controlling the various properties is obvious, because in most of the cases, the properties of the magnetic nanoparticles are dependent on their dimension and morphology. Therefore, the synthesis of magnetic nanoparticles with controlled size and exposed facets is of core importance. The main problem associated with magnetic particles is their agglomeration, which tends to reduce the energy associated with the high surface area to volume ratio of the nanosized particles. In addition, the magnetic nanoparticles are highly chemically active. Therefore, it is of utmost importance to protect these magnetic nanoparticles against oxidation, which may involve functionalization and coating with certain protective layers to form a core–shell structure that completely modifies the MNPs. Various methods have been used for the synthesis of different kinds of magnetic nanostructures including iron oxide and different metal alloys. MNPs of core–shell nature and composites structures have also been synthesized for various applications with suitable modifications. Throughout the last decades, synthesis methods for MNPs using different techniques have been evolved and specialized for fine-tuning of the nanostructures [41].

3.9 Synthesis of MNPS for Bacterial Detection

Here, we will give a short description of only those methods that offer excellent size and shape control (Figure 3.5).

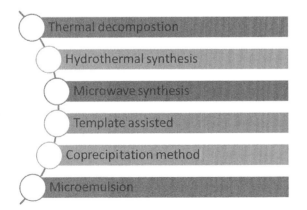

FIGURE 3.5 Different synthesis methods opted for development of MNPs for bacterial detection.

3.9.1 Thermal Decomposition

This method of synthesis based on the chemical decomposition of the substance at elevated temperature involves breaking of the chemical bonds. It mostly uses organometallic compounds such as acetylacetonates in organic solvents with surfactants. In this method, the composition of various precursors that are involved in the reaction determine the final size and morphology of the magnetic nanostructures. Using this method, nanocrystals with very narrow-sized distribution (4–45 nm) could be synthesized along with excellent control of morphology [42].

3.9.2 Hydrothermal Synthesis

Another important chemical synthesis technique that involves the use of liquid–solid–solution (LSS) reaction and gives excellent control over the size and shape of the magnetic nanoparticles is hydrothermal synthesis. This method involves the synthesis of magnetic nanoparticles from high boiling point aqueous solution at high vapor pressure. It is a unique approach for the fabrication of metal, metal oxide, rare earth transition metal magnetic nanocrystals, semi-conducting, dielectric, rare-earth fluorescent, and polymeric nanomaterials [43]. This synthetic technique involve the fabrication of magnetic metallic nanocrystals under different reactions conditions. The reaction strategy is based upon phase separation which occurs at the interface of the solid–liquid–solution phases present in the reaction. For example, the fabrication of monodisperse (6, 10, and 12 nm) Fe_3O_4 and MFe_2O_4 nanocrystals has been demonstrated by Sun et al [44]. The synthesized magnetic nanoparticles were used for photocatalytic degradation of organic dyes, and it was observed that truncated nanocubes possess much higher photocatalytic degradation activity as compared to oblique nanocubes [45].

3.9.3 Microwave-Assisted Synthesis

The microwave-assisted method is a chemical method that use microwave radiation for heating materials containing electrical charges, for instance polar molecule in the solvent or charged ions in the solid. As compared to the other heating methods, microwave-assisted solution fabrication methods get more research attention because of rapid processing, high reaction rate, reduced reaction time and high yield of product [46].

3.9.4 Template-Assisted Fabrication

Another fabrication method used for the synthesis of MNPs is the template-assisted fabrication [47]. Active template-based synthesis involves the growth of the nuclei in the holes and defects of the template. Subsequently, the growth of the nuclei at the preformed template yields the desired morphology of the nanostructures. So, through proper selection of a base template, the size and shape of the magnetic nanoparticles can be controlled. This technique has two important advantages over the chemical routes:

- Template use in the fabrication process determines the final size and morphology of the nanostructures.
- Complex nanostructures such as nano-barcode (segmented nanorods); nanoprism; nanocube – hexagon, and -octahedron MNPs can be fabricated in an easy manner, with full control of size and morphology.

However, this method has also some drawbacks. It is a multi-step process that first of all requires the fabrication of base templates followed by the subsequent deposition of magnetic material within the template [48].

3.9.5 Coprecipitation

Coprecipitation is a facile and convenient way to synthesize of MNPs from aqueous salt solutions by the addition of a base under inert atmosphere at room temperature or at elevated temperature. The size, shape, and composition of the MNPs very much depends on the type of salts used, the salt ratio, the reaction temperature, the pH value, and ionic strength of the media. With this synthesis, once the synthetic conditions are fixed, the quality of the magnetite nanoparticles is fully reproducible [49].

3.9.6 Microemulsion

A microemulsion is a thermodynamically stable isotropic dispersion of two immiscible liquids, where the microdomain of either or both liquids is stabilized by an interfacial film of surfactant molecules. In water-in-oil microemulsions, the aqueous phase is dispersed as microdroplets (typically 1–50 nm in diameter) surrounded by a monolayer of surfactant molecules in the continuous hydrocarbon phase. The size of the reverse micelle is determined by the molar ratio of water to surfactant ratio [50]. By mixing two identical water-in-oil microemulsions containing the desired reactants, the microdroplets will continuously collide, coalesce, and break again, and finally a precipitate forms in the micelles [51]. By the addition of a solvent, such as acetone or ethanol, to the microemulsions, the precipitate can be extracted by filtering or centrifuging the mixture. In this case, a microemulsion can be used as reactor for the formation of nanoparticles. Using the microemulsion technique, nanoparticles can be prepared as spheroids [52]. Although many types of MNPs have been synthesized in a controlled manner using the microemulsion method, the particle size and shapes usually vary over a relative wide range. Moreover, the working window for synthesis of microemulsions is usually quite narrow, and the yield of nanoparticles is low compared to other methods, such as thermal decomposition and coprecipitation. Large amounts of solvent are necessary to synthesize appreciable amounts of material. It is thus not a very efficient process and also rather difficult to scale up [53].

3.10 Innovative Techniques for Bacterial Detection by Using MNPs

The integration of bioconjugate MNPs with different analytical methods has paved a new path for bacteria, protein, and cancer cell sensing, purification, and quantitative analysis. The scope of superparamagnetic nanoparticles in many technological applications like magnetic storage media, biosensing applications, and medical applications caused this field to develop intensively [54]. In the absence of an external magnetic field, the overall magnetization value of superparamagnetic nanoparticles is randomized to zero. Such fluctuations in magnetization direction result in minimization of the magnetic interactions between any two NPs in the dispersion, making the dispersion stable in physiological solutions and facilitating NP coupling with biological agents [55]. When exposed to an external magnetic field, these MNPs align along the direction of magnetic field, achieving magnetic saturation at a magnitude that far exceeds any of the known biological entities. Due to this unique property of MNPs, detection of the MNP-containing biological samples is enhanced on manipulation of these biological samples with an external magnetic field [56].

3.10.1 Recognition Moieties Used for Enrichment of Bacteria

Surface modification of MNPs with recognition moieties such as antibodies, antibiotics (vancomycin, daptomycin, etc.), and carbohydrate enables its use for bacterial detection. These recognition moieties help to detect the bacteria selectively and at low concentration. Different approaches have been used to isolate bacteria using MNPs, such as those described in the following (Figure 3.6).

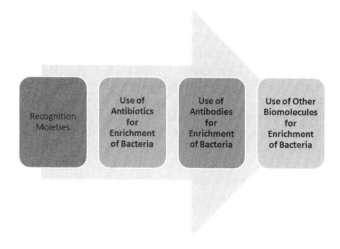

FIGURE 3.6 Strategies to develop MNPs for efficient capture and detection of bacteria.

3.10.2 Use of Antibiotics for Enrichment of Bacteria

Vancomycin belongs to the glycopeptide group of antibiotics, which is known to interact strongly with a broad range of Gram-positive bacteria. Vancomycin kills bacteria by inhibiting bacterial cell wall synthesis. This interaction is mediated via five hydrogen bond motifs between the heptapeptide backbone of vancomycin and the D-alanyl-D-alanine dipeptide from the cell wall [57]. As a result, vancomycin-functionalized MNPs are capable of recognizing the cell surfaces of different bacteria. It has been demonstrated that vancomycin offers less specificity when compared with a monoclonal antibody but can bind to different Gram-positive bacteria such as *Enterococcus faecalis, Streptococcus pneumoniae, and Staphylococcus aureus* but not against Gram-negative bacteria. Gu et al. reported a strategy to use vancomycin-conjugated FePt MNPs of around 4 nm that are water soluble in nature to detect Gram-negative as well as Gram-positive bacteria at low concentrations [58]. As a control experiment, they have used FePt nanoparticles capped with an amine group (FePt–NH2), which failed to capture the bacteria because of the lack of specific molecular recognition. Lin et al. reported vancomycin-immobilized iron oxide nanoparticles that can be used to trap Gram-positive bacteria such as *Staphylococcus saprophyticus, S. aureus, and E. faecalis* selectively from urine samples, followed by detection with matrix-assisted laser desorption/ionization mass spectrometry (MALDI-MS). Their result suggests that this method is capable of rapidly identifying trace pathogens in urine samples [59]. Kell et al. have reported a series of vancomycin-modified Fe_3O_4 MNPs used in magnetic confinement assay to isolate different Gram-positive and Gram-negative bacteria at a low concentration. Their results demonstrate that small moieties are an excellent alternative to antibody-mediated detection of bacteria, where more precaution is required as compared to small moieties like vancomycin [60]. In 2011, Chung et al. reported that the bio-orthogonal modification of vancomycin and daptomycin, which is lipopeptide in nature and binds to the cell wall of Gram-positive bacteria via its hydrophobic tail, resulted in the depolarization of the bacterial cell membrane. Primarily, they have synthesized trans-cyclooctene derivatives of these antibiotics, which are attached to tetrazine-decorated Fe_3O_4 fluorescent MNPs. Their result shows that using a two-step labeling procedure, their assay is superior to using direct antibiotic–nanoparticle conjugates [61]. Recently, Chen et al. synthesized fluorescent MNPs with a core–shell structure followed by conjugation of gentamycin, which is a FDA-approved thermal-resistant antibiotic belonging to the aminoglycoside group and used for the treatment of infection caused by Gram-negative bacteria. Their results demonstrate that gentamicin-bioconjugated fluorescent MNPs can capture Gram-negative bacteria, i.e., *Escherichia coli* (1×10^7 CFU/mL) within 20 min from 10 mL of solution. In addition to this, these gentamycin-modified MNPs are also able to detect diluted *E. coli* cells at a concentration as low as 1×10^3 CFU/mL [62]. Several such approaches are reported [63].

3.10.3 Use of Antibodies for Enrichment of Bacteria

Antibody-conjugated MNPs can selectively capture target bacteria from the given biological sample. Here, application of a magnetic field separates the particle–bacteria complexes from the solution, thereby enriching the concentration of the bacteria and enabling the detection of target bacteria without a culturing process. The use of this approach seems to be more specific in nature. Recently, Tran et al. reported the use of protein A-conjugated chitosan-modified Fe_3O_4 MNPs for separation *Vibrio cholerae* at low concentration. In their study, they have prepared a conjugation of chitosan-coated MNPs and protein A. This conjugate was incubated with specific IgG antibodies against *V. cholera* that could be detected by a conventional diagnostic method as well as immune chromatographic strip test. This method serves as a convenient stage for enrichment and separation of various pathogens from different liquid samples, after incubation with specific IgG antibodies [32]. Several such immunomagnetic approaches have been developed for the enrichment of MNPs with bacteria and used for detection at a low concentration [64,65].

3.10.4 Use of Other Biomolecules for Enrichment of Bacteria

Biomolecules such as carbohydrate, protein, and nucleic acid are used for enrichment of MNPs. It is known that many bacteria use mammalian cell surface carbohydrates as anchors for attachment, which subsequently results in infection [66]. The unique combination of MNPs and carbohydrate group helps to enrich MNPs with bacteria, which can thus be detected [67]. Pigeon ovalbumin (POA), a phosphoprotein, contains high levels of terminal Gal $\alpha(1/4)$ Gal units. Thus, MNPs with immobilized POA can be used as affinity potential probes for bacteria enrichment [66]. Lee et al. have recently developed a magneto-DNA nanoparticle system for rapid detection of bacteria. In their work, they have used oligonucleotide probes to detect specifically targeted nucleic acids, particularly 16S rRNAs, from the pathogen. Furthermore, the assay is rapid in nature and able to simultaneously detect 13 bacterial specimens within 2 h [68]. A study conducted by Huang et al. has reported the use of amine-functionalized MNPs for capturing bacteria from water, food, and urine samples. This developed method does not require the use of affinity molecules on the surface and is able to detect different Gram-positive and Gram-negative bacteria. The detection is based upon the positive charge present on the surface of the MNPs and negatively charged bacterial cell, which promotes a strong electrostatic interaction that results in

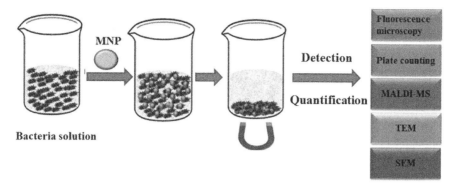

FIGURE 3.7 Schematic representation showing the surface-modified MNPs for detection of pathogens.

efficient adsorptive ability [69]. Schematic representation for the detection and isolation of pathogenic bacteria is presented in Figure 3.7.

3.11 Conclusion

MNPs gained much interest due to their ability to be manipulated upon application of a magnetic field. MNPs are now being used extensively for multiple functions that are much needed for biological applications. Factors such as biocompatibility, toxicity, in vivo and in vitro targeting efficiency, and long-term stability of MNPs should always be addressed with respect to their use in bioapplications. The increasing demands and versatile requirements imposed on MNPs intended for bioapplications require close monitoring of the size, structure, and surface properties of MNPs. Although functionalized MNPs have great potential applications for bacterial detection, it can be difficult to detect organisms in real-life samples when the bacterial concentration is low. In addition, attempts need to be made to detect the microbes in the presence of other microbes of different other genus and in the presence of contaminants. Optimum surface modification of nanoparticles is hard to control. Hence, strategies that are more consistent have to be developed to enable precise composition and uniform surface modification with reproducible functionalization. It is also seen that much work is being carried on Fe_3O_4 MNPs. Other ferrites should be explored in this direction. Future research should focus more on sensitivity, reproducibility, and improving the ability of the technique, so that it can be used for normal samples.

References

1. Love, T.E. and Jones, B. (2008). Introduction to pathogenic bacteria. In: Zourob, M., Elwary, S., Turner, A. (eds) *Principles of Bacterial Detection: Biosensors, Recognition Receptors and Microsystems.* New York: Springer.
2. de Pedro, M.A. and Cava, F. (2015). Structural constraints and dynamics of bacterial cell wall architecture. *Frontiers in Microbiology*, 6, 449.
3. Prakash, S., Tomaro-Duchesneau, C., Saha, S., and Cantor, A. (2010). The gut microbiota and human health with an emphasis on the use of microencapsulated bacterial cells. *Journal of Biomedicine and Biotechnology*, 2011, 981214.
4. van Teeseling, M.C.F., de Pedro, M.A., and Cava, F. (2017). Determinants of bacterial morphology: From fundamentals to possibilities for antimicrobial targeting. *Frontiers in Microbiology*, 8, 1264.
5. Yang, D.C., Blair, K.M., and Salama, N.R. (2016). Staying in shape: The impact of cell shape on bacterial survival in diverse environments. *Microbiology and Molecular Biology*, 80(1), 187–203.
6. Cava, F., de Pedro, M.A., Lam, H., Davis, B.M., and Waldor, M.K. (2011). Distinct pathways for modification of the bacterial cell wall by non-canonical D-aminoacids. *The EMBO Journal*, 30, 3442–3453.
7. Clark, J.A. and Coopersmith, C.M. (2007). Intestinal crosstalk: A new paradigm for understanding the gut as the "motor" of critical illness. *Shock*, 28, 384–393.
8. Honda, K. and Takeda, K. (2009). Regulatory mechanisms of immune responses to intestinal bacteria. *Mucosal Immunology*, 2, 187–196.
9. Jain, K.K. (2007). Applications of nanobiotechnology in clinical diagnostics. *Clinical Chemistry*, 53(11), 2002–2009.
10. Heo, J. and Hua, S.Z. (2009). An overview of recent strategies in pathogen sensing. *Sensors (Switzerland)*, 9(6), 4483–4502.
11. Rezende, L. (2006). *Chronology of Science*. New York: Facts on File.
12. Roberts, G.W. (2009). Magnetism and chronometers: The research of Reverend George Fisher. *British Journal for the History of Science*, 42(1), 57–72.
13. Semenivk, I. (2009). Can magnetism save a vaporizing planet? *Sky & Telescope*, 118(6), 16.
14. Schulten, K., Swenberg, C.E., and Weller, A. (1978). A biomagnetic sensory mechanism based on magnetic field modulated coherent electron spin motion. *Zeitschrift für Physikalische Chemie*, 111, 1–5.
15. Wysession, M., Frank, D., and Yancopoulos, S. (2005). *Physical Science Concepts in Action: Teacher's*

Edition for North Carolina. Upper Saddle River, NJ: Pearson Education.

16. Yong, E. (2010). Masters of magnetism. *New Scientist*, 208(2788), 2.

17. Newman, E.B. (1994). General microbiology. *Research in Microbiology*, 2508(94), 90009–90014.

18. Gracias, K.S. and McKillip, J.L. (2004). A review of conventional detection and enumeration methods for pathogenic bacteria in food. *Canadian Journal of Microbiology*, 50(11), 883–890.

19. Jain, K.K. (2005). Nanotechnology in clinical laboratory diagnostics. *Clinica Chimica Acta*, 358, 37–54.

20. Muthukumar, A., Zitterkopf, N.L., and Payne, D. (2008). Molecular tools for the detection and characterization of bacterial infections: A review. *Laboratory Medicine*, 39(7), 430–436.

21. Sapsford, K.E., Algar, W.R., Berti, L., Gemmill, K.B., Casey, B.J., Oh, E., and Medintz, I.L. (2013). Functionalizing nanoparticles with biological molecules: Developing chemistries that facilitate nanotechnology. *Chemical Reviews*, 113(3), 1904–2074.

22. Subbiah, R., Veerapandian, M., and Yun, K.S. (2010). Nanoparticles: Functionalization and multifunctional applications in biomedical sciences. *Current Medicinal Chemistry*, 17, 4559–4577.

23. Sapsford, K.E., Tyner, K.M., Dair, B.J., Deschamps, J.R., and Medintz, I.L. (2011). Analyzing nanomaterial bioconjugates: A review of current and emerging purification and characterization techniques. *Analytical Chemistry*, 83, 4453–4488.

24. Brand, Mike, Sharon Neaves, and Emily Smith (1995). *"Lodestone". Museum of Electricity and Magnetism.* Tallahassee, FL: Mag Lab U. US National High Magnetic Field Laboratory.

25. Castelvecchi, D. (2012). The compass within. *Scientific American*, 306(1), 48.

26. Did Olmecs have first compass? (1975). *Science News*, 108(10), 148. www.jstor.org/stable/3960320?seq=1#page·scan·tab·contents.

27. Gunderson, P.E. (2005). Magnetism, electromagnetism, and electronics. In *Handy Physics Answer Book*. Canton, MI: Visible Ink Press.

28. Hamzelou, J. (2011). Magnets cut diagnosis time for infections by days. *New Scientist*, 210(2810), 9.

29. Highfield, R. (2010). Electromagnetism. *New Scientist*, 207(2777), 34.

30. Chu, Y.W., Engebretson, D.A., and Carey, J.R. (2013). Bioconjugated magnetic nanoparticles for the detection of bacteria. *Journal of Biomedical Nanotechnology*, 9, 1951–1961.

31. Ray, P.C., Khan, S.A., Singh, A.K., Senapati, D., and Fan, Z. (2012). Nanomaterials for targeted detection and photothermal killing of bacteria. *Chemical Society Reviews*, 41, 3193.

32. Tran, Q.H., Pham, V.C., Nguyen, T.T., Blanco Andujar, C., and Thanh, N.T.K. (2014). Protein a conjugated iron oxide nanoparticles for separation of Vibrio cholerae from water samples. *Faraday Discussions*, 175, 73–82.

33. Chu, Y.W., Engebretson, D.A., and Carey, J.R. (2013). Bioconjugated magnetic nanoparticles for the detection of bacteria. *Journal of Biomedical Nanotechnology*, 9, 1951–1961.

34. Bayer, M.E. and Sloyer, J.L. (1990). The electrophoretic mobility of gram-negative and gram-positive bacteria: An electrokinetic analysis. *Journal of General Microbiology*, 136, 867–874.

35. Nazzaro, F., Fratianni, F., De Martino, L., Coppola, R., and De Feo, V. (2013). Effect of essential oils on pathogenic bacteria. *Pharmaceuticals (Basel, Switzerland)*, 6, 1451–1474.

36. Zourob, M., Elwary, S., and Turner, A. (2008). *Principles of Bacterial Detection: Biosensors Recognition Receptors and Microsystems.* New York: Springer.

37. Niemirowicz, K., Markiewicz, K.H., Wilczewska, A.Z., and Car, H. (2012). Magnetic nanoparticles as new diagnostic tools in medicine. *Advances in Medical Sciences*, 57(2), 196–207.

38. Salunkhe, A.B., Khot, V.M., and Pawar, S.H. (2014). Magnetic hyperthermia with magnetic nanoparticles: A status review. *Current Topics in Medicinal Chemistry*, 14(5), 572–594.

39. Tartaj, P., Morales, M.A.D.P., Veintemillas-Verdaguer, S., Gonzlez-Carreo, T., and Serna, C.J. (2003). The preparation of magnetic nanoparticles for applications in biomedicine. *Journal of Physics D: Applied Physics*, 36(13), R182–R197.

40. Du, P.F., Song, L.X., Xiong, J., Xi, Z.Q., Chen, J.J., Gaoand, L.H., and Wang, N.Y. (2011). High-efficiency photocatalytic degradation of methylene blue using electrospun ZnO nanofibers as catalyst. *Journal of Nanoscience and Nanotechnology*, 11, 7723–7728.

41. Bin Na, H., Lee, I.S., Seo, H., Il Park, Y., Lee, J.H., Kimand, S.-W., and Hyeon, T. (2007). Versatile PEG-derivatized phosphine oxide ligands for water-dispersible metal oxide nanocrystals. *Chemical Communications*, 8, 5167–5169.

42. Cheng, K., Peng, S., Xu, C., and Sun, S. (2009). Porous hollow Fe_3O_4 nanoparticles for targeted delivery and controlled release of cisplatin. *Journal of the American Chemical Society*, 131, 10637–10644.

43. McNeil, S.E. (2009). Nanoparticle therapeutics: A personal perspective. *Wiley Interdisciplinary Reviews: Nanomedicine and Nanobiotechnology*, 1, 264–271.

44. Sun, S., Zeng, H., David, B.R., Raoux, S., Rice, P.M., Wang, S.X., and Li, G. (2004). Monodisperse MFe_2O_4 (M=Fe, Co, Mn) nanoparticles. *Journal of the American Chemical Society*, 126, 273–279.

45. Wu, W., Hao, R., Liu, F., Su, X., and Hou, Y. (2013). Single-crystalline α-Fe_2O_3 nanostructures: controlled synthesis and high-index plane-enhanced

photodegradation by visible light. *Journal of Materials Chemistry A*, 1, 6888–6894.

46. Chu, X., Yu, J., and Hou, Y.-L. (2015). Surface modification of magnetic nanoparticles in biomedicine. *Chinese Physics B*, 24, 014704.

47. Patil, R.M., Thorat, N.D., Shete, P.B., Otari, S.V., Tiwaleand, B.M., and Pawar, S.H. In vitro hyperthermia with improved colloidal stability and enhanced SAR of magnetic core/shell nanostructures. *Material Science and Engineering C*, 2016, 59, 702–709.

48. Sahoo, Y., Goodarzi, A., Swihart, M.T., Ohulchanskyy, T.Y., Kaur, N., Furlani, E.P., and Prasad, P.N. (2005). Aqueous ferrofluid of magnetite nanoparticles: Fluorescence labeling and magnetophoretic control. *The Journal of Physical Chemistry B*, 109, 3879–3885.

49. Laurent, S., Forge, D., Port, M., Roch, A., Robic, C., VanderElst, L., and Muller, R.N. (2008). Magnetic iron oxide nanoparticles: Synthesis, stabilization, vectorization, physicochemical characterizations, and biological applications. *Chemical Reviews*, 108, 2064–2110.

50. Portet, D., Denizot, B., Rump, E., Hindre, F., Le Jeune, J.-J., and Jallet, P. (2001). Comparative biodistribution of thin-coated iron oxide nanoparticles TCION: Effect of different bisphosphonate coatings. *Drug Development Research*, 54, 173–181.

51. Kreller, D.I., Gibson, G., Novak, W., Van Loon, G.W., and Horton, J.H. (2003). Competitive adsorption of phosphate and carboxylate with natural organic matter on hydrous iron oxide as investigated by chemical force microscopy. *Colloids and Surfaces A Physicochemical and Engineering Aspects*, 212, 249–264.

52. Czakler, M., Artner, C., and Schubert, U. (2014). Acetic acid mediated synthesis of phosphonate-substituted titanium oxo clusters. *European Journal of Inorganic Chemistry*, 2014, 2038–2045.

53. Yallapu, M.M., Othman, S.F., Curtis, E.T., Gupta, B.K., Jaggi, M., and Chauhan, S.C. (2011). Multifunctional magnetic nanoparticles for magnetic resonance imaging and cancer therapy. *Biomaterials*, 32, 1890–1905.

54. Gupta, A.K. and Curtis, A.S. (2004). Surface modified superparamagnetic nanoparticles for drug delivery: Interaction studies with human fibroblasts in culture. *Journal of Materials Science Materials in Medicine*, 15, 493–496.

55. Nikam, D.S., Jadhav, S.V., Khot, V.M., Phadatare, M.R., and Pawar, S.H. (2014). Study of AC magnetic heating characteristics of $Co0.5Zn 0.5Fe_2O_4$ nanoparticles for magnetic hyperthermia therapy. *Journal of Magnetism and Magnetic Materials*, 349, 208–213.

56. Patil, R.M., Shete, P.B., Thorat, N.D., Otari, S.V., Barick, K.C., Prasad, A., Ningyhojam, R.S.,

Tiwale, B.M., and Pawar, S.H. (2014). Non-aqueous to aqueous phase transfer of oleic acid coated iron oxide nanoparticles for hyperthermia application. *RSC Advances*, 4(9), 4515.

57. Cai, W. and Wan, J. (2007). Facile synthesis of superparamagnetic magnetite nanoparticles in liquid polyols. *Journal of Colloid and Interface Science*, 305(2), 366–370.

58. Gu, H., Ho, P.L., Tsang, K.T., Wang, L., and Xu, B. (2003). Using biofunctional magnetic nanoparticles to capture vancomycin-resistant enterococci and other gram-positive bacteria at ultralow concentration. *Journal of the American Chemical Society*, 125, 15702–15703.

59. Lin, J., Zhou, W., Kumbhar, A., Wiemann, J., Fang, J., Carpenter, E. E., and O'Connor, C. J. (2001). Gold-coated iron (Fe@Au) nanoparticles: synthesis, characterization, and magnetic field-induced self-assembly. *Journal of Solid State Chemistry*, 159, 26–31.

60. Kell, A.J., Stewart, G., Ryan, S., Peytavi, R., Boissinot, M., Huletsky, A., Bergeron, M.G., Simard, B. (2008). Vancomycin-modified nanoparticles for efficient targeting and preconcentration of Gram-positive and Gram-negative bacteria. *ACS Nano*, 2, 1777–1788.

61. Xu, C. and Sun, S. (2013). New forms of superparamagnetic nanoparticles for biomedical applications. *Advanced Drug Delivery Reviews*, 65(5), 732–743.

62. Gao, J., Li, L., Ho, P.-L., Mak, G.C., Gu, H., and Xu, B. (2006). Combining fluorescent probes and biofunctional magnetic nanoparticles for rapid detection of bacteria in human blood. *Advanced Materials*, 18(23), 3145–3148.

63. Wan, Y., Zhang, D., and Hou, B. (2010). Determination of sulphate-reducing bacteria based on vancomycin functionalized magnetic nanoparticles using a modification-free quartz crystal microbalance. *Biosensors and Bioelectronics*, 25, 1847–1850.

64. Chen, L. (2012). Bioconjugated magnetic nanoparticles for rapid capture of gram-positive bacteria. *Journal of Biosensors and Bioelectronics*. doi:10.4172/2155-6210.S11-005. S:11.

65. Tran, Q.H., Pham, V.C., Nguyen, T.T., Blanco Andujar, C., and Thanh, N.T.K. (2014). Protein a conjugated iron oxide nanoparticles for separation of Vibrio cholerae from water samples. *Faraday Discussions*, 175, 73–82.

66. Kumar, V., Nath, G., Kotnala, R.K., Saxena, P.S., and Srivastava, A. (2013). Biofunctional magnetic nanotubeprobe for recognition and separation of specific bacteria from a mixed culture. *RSC Advances*, 3(34), 14634.

67. Varshney, M., Yang, L., Su, X.-L., and Li, Y. (2005). Magnetic nanoparticle-antibody conjugates for the separation of Escherichia coli O157:H7 in ground beef. *Journal of Food Protection*, 68, 1804–1811.

68. Lee, W., Kwon, D., Chung, B., Jung, G.Y., Au, A., Folch, A., and Jeon, S. (2014). Ultrarapid detection of pathogenic bacteria using a 3D immuno-magnetic flow assay. *Analytical Chemistry*, 86, 6683–6688.

69. Huang, Y.-F., Wang, Y.-F., and Yan, X.-P. (2010). Amine-functionalized magnetic nanoparticles for rapid capture and removal of bacterial pathogens. *Environmental Science & Technology*, 44, 7908–7913.

4

Nanoparticles Carrying Biological Molecules

Suryani Saallah
Universiti Malaysia Sabah

Wuled Lenggoro
Tokyo University of Agriculture and Technology

4.1 Introduction

The rapid growth of nanotechnology has paved the way for the establishment of bionanotechnology, a field that is an interdisciplinary branch of nanotechnology and biology, biotechnology, and medicine (Reisner et al., 2017). This emerging field is generally characterized as having two somewhat opposite goals: (i) exploiting the intrinsic features of biomolecules such as specific recognition and catalytic properties to develop new composite materials or devices; (ii) utilizing the unique features of nanomaterials (NMs) within a biological setting (Sapsford et al., 2013). For example, nanoparticles (NPs) can be used to image biological processes *in vivo*.

NPs are materials with dimensions that are typically less than 100 nm. In many cases, NPs are prepared from inorganic materials such as metals and their oxides, quantum dots, polymers, and silica. NPs also can be derived from biological materials which include viral particles, liposomes, or biopolymers such as polysaccharides (Algar et al., 2011). The interest in using NPs in various applications arises from the unique size-dependent physical, optical, electronic, and chemical properties that they can contribute to the resulting conjugate (Sapsford et al., 2013).

While NPs have their own intrinsic properties, it is generally necessary in biological applications to impart additional properties or functions through physical or chemical coupling between an NP and one or more molecules (Algar et al., 2011). The conjugation of NPs and biomolecules has found applications in bioimaging, biocatalysis, drug delivery, and biosensors. Recently, the applications of NP–biomolecule conjugates were shown in various fields beyond biological applications such as material science, physics and energy, nanosensors, and biofuel cells, and even in new

tools for assembly, to name a few (Aubin-Tam and Hamad-Schifferli, 2008).

The nanoparticle-carrying biological molecules or, in short, NP–biomolecule conjugate is characterized by the association of one or more biologically relevant molecules at the interface of an NP (Algar et al., 2011; Sapsford et al., 2013). Biomolecules of interest may include all forms of proteins and peptides, antibodies, enzymes, nucleic acids, or oligonucleotides such as aptamers, carbohydrates, and lipids. Biologically active small molecules such as contrast agents and reporters are also included.

Research on the preparation of NP–biomolecule conjugates has grown considerably over the past decade, but the definite understanding of the interaction between NPs and biomolecules is far from mature (Stephanopoulos and Francis, 2011). Furthermore, numerous conjugation strategies have been developed that mainly depend on the classical chemistries associated with biomolecule labeling. While these strategies are adequate for proof-of-concept studies, the development of NP–biomolecule conjugates for real applications necessitate optimization and much greater control than these chemistries can offer (Algar et al., 2011). In a simplified view, the selected conjugation strategies should be able to tackle important issues associated with NP–biomolecule interactions including minimizing nonspecific interactions, eliminating undesirable side effects, and improving reproducibility.

In this chapter, the interactions of NPs and biomolecules are put into perspective by providing an overview on various relevant NP–biomolecule interactions. Before discussing the strategies to develop the chemistry between NPs and biomolecules, factors affecting such interactions are highlighted. Then, state of the art on the development of NP–biomolecule conjugates is presented.

4.2 NP–Biomolecule Interactions

4.2.1 Overview

Depending on the nature of the biomolecules and the properties of NP, interaction between NPs and biomolecules can possibly occur through various configurations and involve varying levels of complexity, which was best described by Sapsford et al. (2011) as multilayered structures. Figure 4.1 highlights various possible structural iterations of NP–biomolecule conjugates. Generally, biomolecules make up the outer layer of the conjugate. In the simplest configuration, biomolecules can be attached directly to the surface of the NP core, for instance the nonspecific attachment of fluorescent dye (Stephanopoulos and Francis, 2011), or on the other hand, the biomolecules decorate the or entrapped within the NP.

Since most NPs are intrinsically hydrophobic, surface-bound stabilizing ligands are often required to mediate their

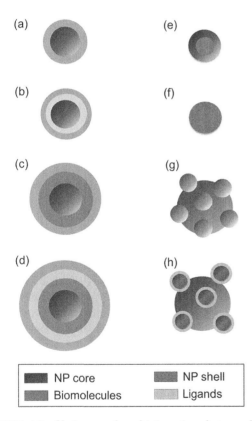

FIGURE 4.1 Various modes of interactions between NP and biomolecules. (a) Direct interaction of biomolecule with NP core. (b) Interaction of biomolecule with an NP core through intermediate ligands. (c) Interaction of biomolecule with NP core surrounded by NP shell layer. (d) Interaction of biomolecule with NP shell layer through intermediate ligands. (e) Biomolecule entrapped in porous NP core. (f) Biomolecule entrapped in porous NP core surrounded by NP shell layer. (g) Interaction of NP with the much larger biomolecule. (h) Interaction of NP with the much larger biomolecule attached via intermediate ligands. (Adapted from Sapsford et al., 2011.)

hydrophilicity and biocompatibility. These ligands would also serve as chemical "handles" for attaching biomolecules of interest, either directly or using chemical cross-linkers and other mediators. The conjugation of a targeting ligand to chemically modified NPs will aid in biorecognition or selective delivery of the desired NP therapeutics, or it can even imbue the NP–biomolecule conjugate with a bioderived activity such as enzyme catalysis (Sapsford et al., 2013).

Many NPs also consist of a core–shell structure in which the outer shell protects and insulates the core as well as provides solubility and linkage area to biomolecules. Preparation of more complex NP–biomolecule conjugate structures can take place through various arrangements and iterations of the core/shell/ligand/biomolecule multilayered structure (Sapsford et al., 2013), for example in the development of well-defined antibody conjugates with highly functionalized drugs (Stephanopoulos and Francis, 2011).

Interaction of NP and biomolecule within a conjugate is complex and dynamic. The interaction involves the interplay between the physicochemical properties of both the NPs and biomolecules, and their interfacial chemistry may affect the binding and induce dynamic changes during and after the interaction (Figure 4.2).

4.2.2 Effect of NP Properties on the NP–biomolecule Interaction

Nanoparticles can be synthesized, assembled into various geometries and configurations, and decorated with targeting agents, to yield novel NP properties for the desired applications. Theoretically, such modularity offers an infinite matrix of nanoparticles with different properties to interact with biomolecule of interest (Chou, Ming and Chan, 2011). Overall, the interaction is influenced by a variety of factors with respect to the NP properties itself that can be classified into NP composition, physical properties, surface chemistry, and the type of surface ligand attached to the NP (Figure 4.3).

NP Composition

NP composition can affect the structure and function of biomolecules in the conjugate. For instance, some proteins may have strong affinity for specific NP materials as observed in the different adsorption behavior of plasma proteins to single-walled carbon nanotubes (SWCNT) and silica NP where serum albumin shows better affinity toward the SWCNT than the silica NP. Investigation on adsorption of various types of plasma proteins such as alipoprotein, immunoglobin, fibrinogen, transferrin, and albumin on NP ZnO, SiO_2, and TiO_2 NP by gel-electrophoresis and mass spectrometry indicate that, although these NPs having similar surface charge, they have different binding affinity on different plasma proteins (Phogat, Kohl and Uddin, 2018).

Moreover, depending on the composition, nanoparticles can have diverse properties such as high electron density and strong optical absorption (e.g., metals, particularly gold),

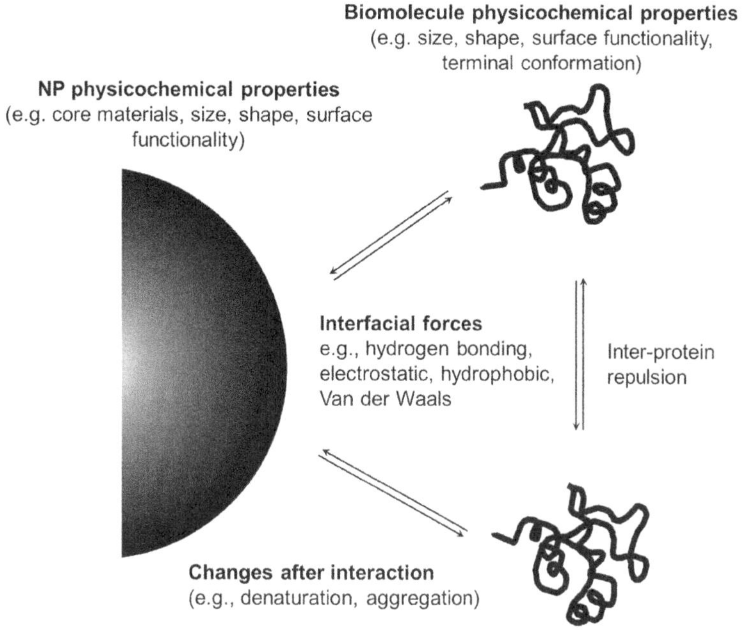

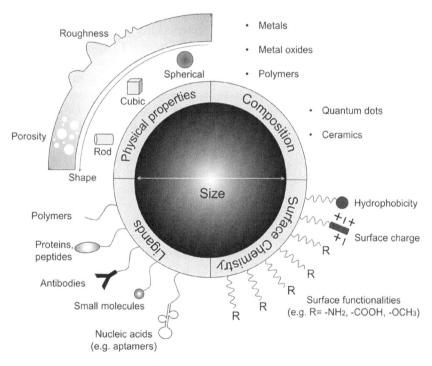

FIGURE 4.2 Schematic illustration of NP–biomolecule interaction and factors affecting the interaction. (Adapted from Huang and Lau 2016 and Sapsford et al. 2011.)

FIGURE 4.3 Important factors that affecting the NP design for biomolecule conjugations. (Adapted from Shen, Nieh and Li 2016.)

photoluminescence (quantum dots such as CdSe or CdTe) or phosphorescence (doped oxide materials such as Y_2O_3), or magnetization (magnetic nanoparticles such as iron oxide) (Sperling and Parak, 2010).

NP Physical Properties

Another important NP-dependent property is the physical characteristics, such as size, shape, and surface curvature.

siRNA–AuNP conjugates of different core sizes and shapes (13-nm spherical, 50-nm spherical, and 40-nm star) have been shown to have different effects on cellular uptake, with the larger NPs showing higher potential as carriers for the delivery of siRNA while the particle shape influenced the localization and distribution of the conjugated NP in the cell (Jun et al., 2017). In the past decade, silica NPs with sizes of 4, 20, and 100 nm have been used to investigate the effect of NP size on structure and biological function of the adsorbed lysozyme having dimension equivalent to those of the 4 nm NP (Vertegel, Siegel and Dordick, 2004). Stronger protein–NP interaction, more protein unfolding, and reduction of enzyme activity were observed in the case of the larger NPs (Figure 4.4). Similarly, adsorption of cytochrome c onto larger silica NPs results in remarkable structural disruption of the cytochrome c, which eventually affect the protein stability and caused more significant changes of the local heme microenvironment compared to the smaller size silica NPs (Shang et al., 2009). Generally, for larger NPs, the higher effective biomolecule surface area accessible by the NPs may increase the chances of denaturation. Moreover, the larger NPs tend to sterically block the biomolecule, preventing the substrate from reaching the binding or active sites, thus lowering the biomolecule activity. For the smaller NPs, on the other hand, due to their higher surface curvature, only a small number of NP ligands can interact with the biomolecules, thus lowering the likelihood of structural distortion.

Existing findings suggest that adsorption on highly curved surfaces causes fewer changes in biomolecules conformation compared with that on relatively flat surfaces. Basically, these results can be explained by assuming that biomolecules adsorbed on a flat surface come into contact with it more extensively, creating stronger interactions and experiencing greater changes in their native conformation, hence resulting in loss of biological activity. This holds true in most cases, although quite a few exceptions have been reported (Avvakumova et al., 2014). It is also possible that as NPs reached a critical size, the effects of surface curvature will no longer be as significant, since there is no more spatial restriction on the maximum interaction between the NP and biomolecule (Huang and Lau, 2016).

Surface Chemistry

Most naturally occurring biomolecules present functional groups that are useful for conjugation or as chemical handles, while most currently available NPs have rather limited diversity of functional groups (Sapsford et al., 2013). Various types of reactions have been established reliant on the functional groups presented on the biomolecules. Chemical targets that account for most of conjugation techniques include primary amines (–NH$_2$) present at the N-terminus of each polypeptide chain and lysine residues side chain, carboxyl groups (–COOH) existing at the C-terminus of each polypeptide chain and in the side chains of aspartic acid and glutamic acid, cysteine thiols (–SH), carbonyl groups (–CHO), and carbohydrates, which usually offer hydroxyl groups or aldehydes as alternative reaction sites (Liebana and Drago, 2016).

Synthetic biomolecules, particularly peptides and nucleic acids, offer more versatility in terms of functional groups. The uniquely reactive chemical functional groups can be introduced into the biomolecule sequence in the early synthesis stage or through subsequent modification. The advancement of a library of chemistries for producing and modifying nucleic acids through genomic revolution allows a variety of functional groups such as amines, thiols, carboxyls, biotin, azides, and alkynes to be site-specifically incorporated into their structures (Sapsford et al., 2013).

NP surface charge is another important property that affects the NP–biomolecule interaction. The degree of protein adsorption on NP is strongly dependent on the surface charge. To investigate the effect of NP surface charge on serum protein adsorption and cellular uptake, sulfated (dPGS) and non-sulfated (dPGOH) dendritic polyglycerol were prepared and synthesized on a gold core. Higher protein affinity was observed for the sulfated dPGS compared with the nonsulfated dPGOH, confirming the influence of surface charge on protein interactions (Bewersdorff et al., 2017). Another study on the interaction of ubiquitin and fibrinogen with AuNPs revealed that both serum proteins can adsorb onto the NPs at pH above and below the IEPs of the proteins, primarily driven by electrostatic interaction through the protein's positive patches (e.g., lysine or arginine). Surface charge modulation of the protein that occurred through protonation or deprotonation

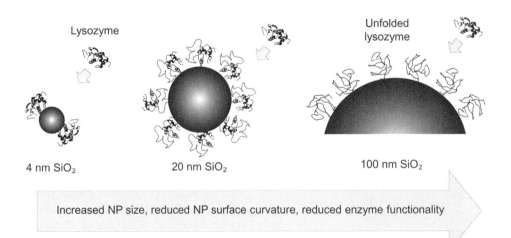

FIGURE 4.4 Effect of NP size and surface curvature on adsorption of lysozyme. (Adapted from Vertegel, Siegel and Dordick, 2004.)

(due to pH changes) determines the structural stability of the adsorbed proteins. In some cases, the NP–biomolecule conjugate remained stable, while in other cases aggregation following structural changes occurred (Huang et al., 2013).

The surface hydrophobicity of NPs also plays a significant role in the development of NP–biomolecule conjugate. The enhanced adsorption of proteins on hydrophobic surface in comparison with the hydrophilic counterpart increases the rate of opsonization of hydrophobic NPs (Lynch and Dawson, 2008). For serum albumin, a relatively hydrophobic protein, the binding stoichiometry and affinity increase with NP surface hydrophobicity. For NPs having similar charge property, hydrophobicity may enhance the affinity of the protein to the NP surface.

Choice of NP Ligands

The type of ligand attached to the NP can significantly influence the structure of both non-covalently and covalently attached biomolecules. Proper selection of ligand is crucial as the ligand is in close contact with the biomolecule and can interact with adjacent residues. Typically, NP ligands include polymers such as PEG, antibodies, proteins, peptides, nucleic acids (aptamers), small molecules, or other compounds (vitamins and carbohydrates). For the covalently bound biomolecule–NP conjugates, charged ligands are likely to denature the biomolecule through electrostatic effects. Moreover, charged ligands can either inhibit or improve the linking chemistry through electrostatic repulsive or attractive forces and may induce a dramatic effect on non-specific adsorption (Aubin-Tam and Hamad-Schifferli, 2008). PEG has been widely used as a ligand to prevent such an effect due to its floppy chains and neutral charge. A study on the effect of AuNPs with positive, negative, and neutral ligands attached on cytochrome c revealed that the protein structure is not affected with neutral ligands (PEG) but denatures in the existence of charged species. Increasing the salt concentration for the negatively and positively charged ligands has been found to reduce the extent of denaturation, which confirms that the interaction between the protein and the NP that promotes denaturation is primarily electrostatic. Furthermore, the findings also suggest that neutral ligands are best for preserving the protein structure in the conjugate.

The NP ligand also affects the structure of non-covalently bound biomolecule–NP conjugates. Exploiting the ability of the surface-coating ligand of the NP to interact with biomolecules in an explicit way can allow specific control over the biomolecule structure. For example, functionalization of NP surfaces with peptides is increasingly being used to control the interaction of NPs with proteins.

4.2.3 Principle Criteria in Developing NP–biomolecule Chemistries

Beyond the distinct properties of NPs and biomolecules, several criteria should be considered for bioconjugation of NPs and biomolecules as they illuminate the potential impact on the conjugate's ultimate function. The "ideal" criteria anticipated from virtually all NP–biomolecule conjugation chemistries should involve control over the following properties (as illustrated in Figure 4.5):

i. Valence or ratio of biomolecule per NP.
   Depending on the intended applications, valency could affect the binding effectiveness of biomolecules on the NP. Lower valency is required especially for identification of single binding events, while higher valency is usually associated with enhanced binding interactions and avidity. For example, the efficient uptake of cell-penetrating-peptide (CPP) into cells requires a high ratio of peptide on the NP surface, particularly when small concentrations of NPs are used. In some cases, however, high packing density of biomolecules will lead to overconjugation, increased inter-biomolecule interaction, and reorientation that might impair the binding interactions, leading to a loss of activity (Chou, Ming and Chan, 2011; Sapsford et al., 2013).

ii. Orientation of biomolecules on the NP.
   The way in which biomolecules arrange themselves on the NP surface is dependent upon their binding sites and may affect their bioactivity.

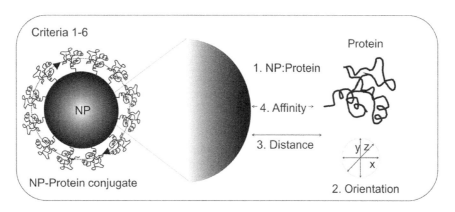

FIGURE 4.5 Schematic of "ideal" criteria for conjugation of biomolecules (e.g., protein) to NPs.

For proteins involved in molecular recognition, such as immunoglobulin, if the recognition site was hidden, sterically blocked, or attached in a heterogeneous manner due to the orientation of the protein relative to the NP surface, the recognition process will be compromised (Algar et al., 2011). Site-specific and gold-thiol surface chemistry approaches were shown to be effective in controlling the orientation of proteins conjugated to AuNPs (Rodriguez-Quijada et al., 2018).

iii. Relative separation distance of the biomolecule from the NP.

Some applications require specific control over the separation distance for optimal function. Increasing proximity of a recombinant protein with the NP has been shown to have a negative effect on the structure and function of the conjugate. For sensing applications such as the preparation of QD probes for protease activity, the Förster resonance energy transfer (FRET) created between the probe terminals is sensitive toward changes in the separation distance between the QD and the dye. In principle, the peptide connected to the QD terminus generates the proximity required for FRET, and this proximity is lost due to the protease activity, which then generates an optical signal for transduction (Algar et al., 2011; Sapsford et al., 2013).

iv. NP–biomolecule binding affinity.

Information on the binding affinities for different combinations of biomolecules and nanoparticles is of relevance to design the conjugate for the intended application. For drug delivery within body fluids or inside cellular compartments, a tightly associated protein may follow the nanoparticle if it endocytoses from the extracellular fluid into an intracellular location, while a weakly bound protein will be replaced by an intracellular protein during or after such transfer. A good example is the modulation of glutathione S-transferase (GST) affinity toward AuNPs, which allows the hierarchical assembly of an extra protein layer to prevent rapid binding of human serum proteins. No serum protein was detected in the AuNPs protected with GST, while the unprotected AuNPs showed extensive binding of proteins (Lee et al., 2018; Ma et al., 2018).

v. Optimal function and activity of both the NP and the biomolecule.

Although the conjugation of biomolecules and NP potentially yields new materials with a diverse range of properties and applications, it is possible that the conjugation somehow compromises their functionalities, leading to such issues as loss of structural and functional properties of protein (see Figure 4.6) and affecting NP stability and

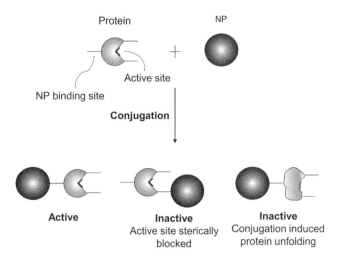

FIGURE 4.6 Conjugation of NP can lead to protein deactivation via structural denaturation or blocking the active site.

optoelectronic properties. Therefore, research on NP–biomolecule conjugation progressively emphasizes optimizing the activity and functionality of the conjugated molecule. This is particularly important in nanomedicine as NPs used in drug delivery often fail to fully manifest their clinical potential due to limitations associated with their targeting ability and subsequent biodistribution issues. Clearly, to create an NP–biomolecule conjugate that properly functions biologically, the interface must be optimized in such a manner to preserve the biological function of the biomolecule as well as the intrinsic properties of the NP.

vi. A corollary of the previous criteria is that the display of one type of biomolecule is reproducible and can be replicated with other type of biomolecules, between experiments and different types or batches of NP (Algar et al., 2011; Sapsford et al., 2013).

4.2.4 NP–biomolecule Linking Chemistries

Selection of the appropriate conjugation strategy is primarily depending on synergy of the aforementioned NP and biomolecules properties. Basically, the conjugation of biomolecule to the NP can occur through covalent chemistries or non-covalent interactions with the latter covering the conjugation strategies based on physical adsorption and electrostatic interaction.

Noncovalent Interactions

Noncovalent conjugation strategies include those based on physical adsorption driven by hydrogen bonding, van der Waals forces, and hydrophobic interaction along with those based on electrostatic and affinity interactions. The nature of the forces involved in physical adsorption is reversible, depending on the governing factors that influence the interaction (pH, ionic strength, temperature, solvent polarity). The underlying principle of electrostatic interaction is the attractive forces between oppositely charged species: the biomolecule and the NP. Most nucleic acids such as DNA have a strong negative charge associated with the phosphate backbone that could attract the positively charged NP materials (Sapsford et al., 2013). Although electrostatic interaction is relatively simple and spontaneous, it is inherently non-specific; thus, the biomolecule can adopt several orientations on the NP. Therefore, attention must be paid to the modulation and control of the pH and ionic strength of the medium (Algar et al., 2011; Avvakumova et al., 2014). In certain circumstances, regio-specific interactions can be established through strategic modification of the NP or biomolecule surface chemistry (Aubin-Tam and Hamad-Schifferli, 2008). For example, charge and hydrophobicity of the NP ligand can be tuned to improve the selectivity of protein orientation and enable conjugation at the specific site.

Generally, physical conjugation offers a simple and rapid route to bind biomolecule to the NP and requires minimal modification steps (Saallah and Lenggoro, 2018). Thus, the biomolecules' functionalities can be preserved. Nevertheless, this method suffers from weak binding, random orientation, high likelihood of desorption, and poor reproducibility (Liebana and Drago, 2016). Such drawbacks can be overcome by introducing specific functional groups or targeting ligands to the nanoparticles through affinity interactions. The most well-known example in the last several decades is the avidin–biotin system. Avidin comprises four identical subunits that provide four binding pockets which specifically recognize and bind to biotin, resulting in a strong and stable interaction with a dissociation constant, K_D, of the order of 10^{-15} M. The combination of basic pI and carbohydrate content, however, results in nonspecific binding as observed in several applications (Sapsford et al., 2013). Alternatively, streptavidin, a non-glycosylated homologous tetrameric protein displaying similar affinity to biotin, can be used as avidin analog (Saallah and Lenggoro, 2018). The mechanism of the most widely applied non-covalent interactions is illustrated in Figure 4.7.

Classical Covalent Chemistries

The traditional covalent chemistries take advantage of the existence of reactive functional groups such as amino–NH$_2$ (lysine), carboxylic acid–COOH (aspartic, glutamic), hydroxyl–OH (serine, tyrosine), and –SH (cysteine) on biomolecules which can be achieved by means of thiol chemistry or using a bifunctional linker (Marco et al., 2010). Thiol chemistry typically utilizes cysteine residues present in the polypeptide sequence or introduced at specific positions by site-directed mutagenesis. Disulfide bond formation and Michael addition are among the well-known reactions involved (Avvakumova et al., 2014) (see Figure 4.8). Gold NPs are highly reactive toward thiol

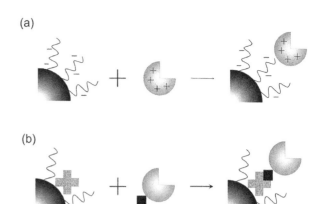

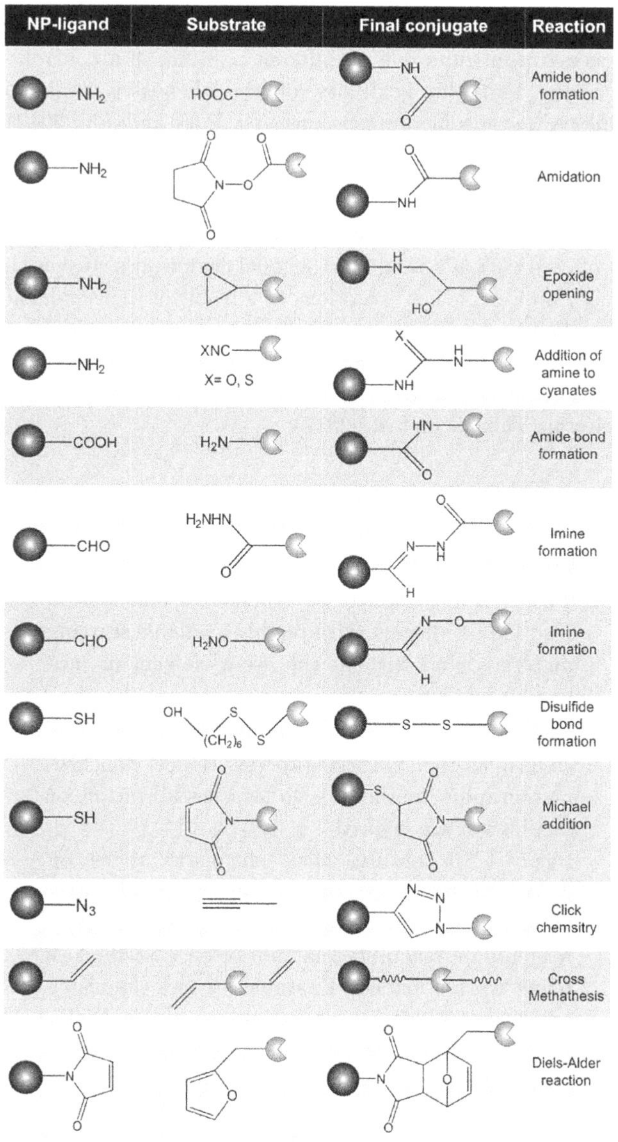

FIGURE 4.7 Illustration of non-covalent interactions of biomolecules and NPs through electrostatic and avidin–biotin affinity bindings. (a) Electrostatic Interaction. (b) Affinity interaction.

NP-ligand	Substrate	Final conjugate	Reaction
–NH$_2$	HOOC–		Amide bond formation
–NH$_2$			Amidation
–NH$_2$			Epoxide opening
–NH$_2$	XNC– X= O, S		Addition of amine to cyanates
–COOH	H$_2$N–		Amide bond formation
–CHO	H$_2$NHN–		Imine formation
–CHO	H$_2$NO–		Imine formation
–SH	OH (CH$_2$)$_6$ S–S		Disulfide bond formation
–SH			Michael addition
–N$_3$			Click chemistry
			Cross Methathesis
			Diels-Alder reaction

FIGURE 4.8 Classical covalent chemistries for NP–biomolecule conjugation.

groups, enable formation of strong Au-S bond, and can be directly conjugated with various thiol-functionalized biomolecules (Saallah and Lenggoro, 2018).

Biomolecules are often covalently linked to NPs using bifunctional linker molecules which have anchor groups such as thiols, disulfides, and phosphine ligands containing terminal carboxy, amino, or maleimide groups via carbodiimide-mediated amidation and esterification. Most proteins have several primary amines in the side chain of lysine residues and the N-terminus of each polypeptide that can be offered as targets for N-hydroxysuccinimide-ester and carbodiimide reagents (Marco et al., 2010). NP–biomolecule conjugates produced through this method are prone to structural alterations and interference from multiple cysteine residues at various positions. Such a limitation can be partially overcome through native chemical ligation which involves modification of the target moieties with cysteine residue at the N-terminal end followed by reaction with a thioester at the C terminus of a short peptide attached on the NP. This strategy enabled successive introduction of different peptides in a controlled manner, and the flexibility of the chemistry could be tailored for specific applications (Yu, Park and Jon, 2012).

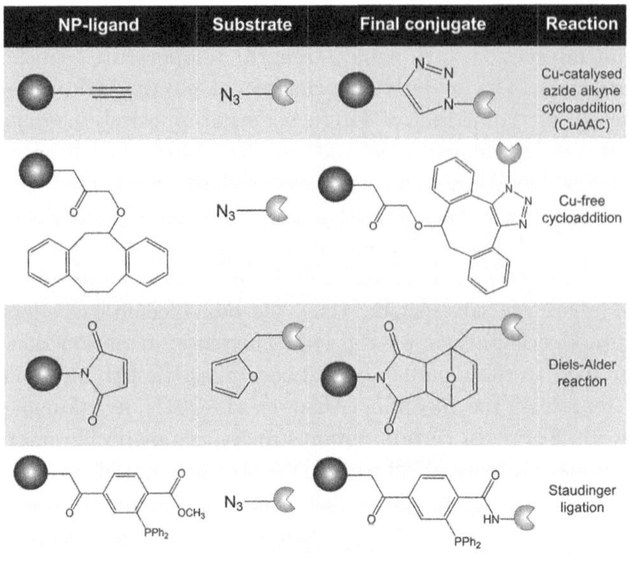

"Click" Chemistry

Recent research on NP–biomolecule conjugation has focused profoundly on adapting biorthogonal chemistries such as the superfamily of "click" reactions as a result of their exquisite specificity, thus bringing significant advances related to improving reaction efficiency and reducing toxicity. Reactions were classified as "click" if they met the following criteria (Tăbăcaru et al., 2017):

i. The reaction was modular, stereospecific, wide of scope, producing high yields, and only generated safe by-products.

ii. The reaction had to be conducted under mild reaction conditions, with readily available starting materials, and without the use of solvent or in a benign solvent.

iii. Purification of the reaction products should be as easy as the synthesis process. Hence, nonchromatographic methods such as crystallization or distillation are favored.

Inspired by nature, click chemistry relies on carbon-heteroatom bond formation. This irreversible reaction counts on highly energetic reagents or reactants that are often described as being "spring-loaded". Examples of click chemistry reactions include cycloaddition reactions, Diels–Alder reactions, nucleophilic ring-opening reactions (e.g., epoxides), and Stauginder ligation (Nwe and Brechbiel, 2009) (Figure 4.9).

Copper-catalyzed azide-alkyne cycloaddition is one of the most well-documented click reactions and results in the formation of a strong triazole

FIGURE 4.9 "Click" chemistry reactions.

linkage, which is water-soluble and biocompatible. The reaction occurs under mild conditions and is highly specific, ensuring conjugation of biomolecules at the target location. This reaction has turned out to be a valuable ligation method for the preparation of bioconjugates, which has found numerous applications in drug synthesis, sensing, protein labeling, and activity-based protein profiling (Algar et al., 2011). The use of Cu(I) catalyst to drive the reaction however can cause toxicity which trigger diseases such as hepatitis, neurological disorders, or Alzheimer's *in vivo*. The Diels−Alder reaction, characterized by a [4 + 2]-cycloaddition between a diene and a dienophile, provides fast and chemoselective strategy for biofunctinalization of NP materials and have been used mainly for molecular imaging at the cellular level. In the case of Staudinger ligation, it involves the formation of an amide bond between an arylphosphine and an azido group. This reaction does not require the use of catalysts but is generally slower and in most cases does not run to full conversion.

4.3 State of the Art on the Development of NP–biomolecule Conjugate

Maximizing the efficiency and functionality of the NP–biomolecule conjugate requires the formation of robust, irreversible binding to reduce the system free energy, elimination of undesirable side reactions, and controlled spatial orientation of the biomolecule. Although a plethora of chemistries have been developed to enable pairing of NPs and biomolecules by adapting standard biomolecule conjugation strategies for NP functionalization, control of the spatial arrangement of biomolecules on the NP remains a major challenge. Recently, novel strategies specifically aimed to tackle this issue and improving the overall NP–biomolecule conjugation efficiency have been developed.

4.3.1 Site-selective Methods for Biomolecules Conjugation

Site-selective strategies are important, particularly for conjugating valuable synthetic molecules such as fluorescent probes and drugs to proteins in a well-defined manner (Lee et al., 2018). Selectively reactive sequence can be achieved by means of modification of natural amino acids and the introduction of synthetic amino acids via genetic expansion to install bioorthogonal chemical handles. Low natural abundancy and high nucleophilicity make cysteine a main target for conjugation of proteins. Lack of protein residues in some proteins can be overcome by site-directed mutagenesis, which adapted the conventional reactions for protein engineering to introduce a single cysteine at a user-defined site within the protein sequence (Avvakumova et al., 2014; Lee et al., 2018). For example, engineering of an anti-HER2 scFv variant enabled the introduction of two orthogonal conjugation functionalities, a cysteine residue and a histidine tag, for the immobilization of the targeting ligand on NP surface. The different localizations of cysteine residue and a histidine tag forced the molecule to assume a different orientation on the NP, thus affecting cell labelling efficiency. This site-selective strategy enables good control of the position and orientation of ligands in NP–biomolecule conjugates which can then play an important role especially in biorecognition.

4.3.2 Enzyme-mediated Site-Specific Bioconjugation

Enzyme-mediated ligation exploits the ability of enzymes (e.g., Sortase A, transglutaminase, phosphopantetheinyl transferase) to promote the covalent binding of specific molecules in a site-selective manner and under mild conditions. Conjugation mediated by Sortase A relies on the specificity of the transpeptidase for short peptide sequences (LPXTG and GGG). Sortase A binds the LPXTG (X is any amino acid) substrate, hydrolyzes the backbone between threonine and glycine, and generates an acyl-enzyme intermediate. The reaction creates a new amide bond between a C-terminal sorting motif LPXTG and an N-terminal oligoglycine which acts as nucleophile. One particularly exciting application of sortase-mediated ligations is the generation of antibody–drug conjugates (Chen et al., 2016; Falck, 2018). Phosphopantetheinyl transferase enzyme was being used for site-specific covalent immobilization of enzymes bearing a small 12-mer "ybbR" tag on polystryrene NP functionalized with coenzyme. This one-step immobilization strategy was conducted under mild reaction conditions results in a homogeneous population of enzymes that are covalently and site-specifically linked to the NP surface. The immobilized enzymes could retain their activity and showed high operational stability (Wong, Okrasa and MicKlefield, 2010).

4.3.3 Modular Assembly of Proteins on NPs via Catcher/Tag Technology

The Catcher/Tag technology is a covalent protein ligation strategy consisting of a peptide called as Tag and a reactive domain called as Catcher. The ability of SpyCatcher, an engineered protein, to interact specifically with a SpyTag peptide enables the construction of a self-catalyzed self-assembly system by formation of a covalent isopeptide bond between the amino and carboxylic groups on the lysine and aspartate residue side chains. Interestingly, covalent conjugation can be achieved without the need for chemical cross-linking agents which enable the development of multi-protein mega-molecules from individual building blocks. Very recently, hierarchical assembly of proteins has been made possible by the modification of AuNPs with GST via Au-S cross-linking followed by covalent isopeptide bond formation between SpyCatcher and the SpyTag peptide. This modular approach to conjugation of recombinant proteins to NPs requires no optimization of every specific protein–NP pair, thus providing a universal platform for conjugating functional proteins on AuNPs. This technology has also been applied to construct dual plug-and-display assembly based on GST-SpyCatcher fusion protein to accelerate the development of a vaccine for malaria (Ma et al., 2018). A dually addressable synthetic NPs was designed by engineering the multimerizing coiled-coil IMX313 and two orthogonally reactive split proteins. SpyCatcher/SpyTag and Snoop-Catcher/SnoopTag isopeptide bonds between protein and peptide were formed through amidation and transamidation, respectively. A modular platform was offered by SpyCatcher–IMX–SnoopCatcher, while the SpyTag–antigen and SnoopTag–antigen can be simply multimerized on opposite faces of the NP upon mixing (Figure 4.10).

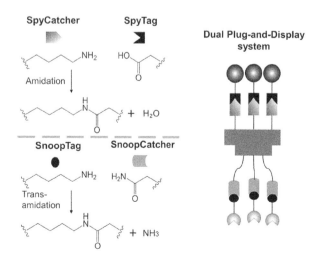

FIGURE 4.10 Schematic of the development of dual plug-and-display system for vaccination. (Adapted from Brune et al., 2017.)

The assembly boosted the antibody response to two malarial transmission-blocking antigens by almost 100-fold and enabled a strong antibody response after only a single immunization. For this reason, it is proposed that the developed dual plug-and-display assembly can be applied to other advanced synthetic biology techniques such as biosensors and enzyme scaffolding (Brune et al., 2017).

4.4   Conclusion

Conjugation of biomolecules to NPs has shown unprecedented opportunities for exploiting biological functions of biomolecules and expanded potential applications in areas such as targeting and delivery of valuable therapeutics, development of biosensors and bioanalytical devices, and industrial biocatalysis. Elucidation of the interaction between NPs and biomolecules is important as there are numerous challenges to be met to design truly efficient NP–biomolecule conjugates that can manifest their full potential for the intended applications. The interaction of NPs and biomolecules is complex and dynamic and involves the interplay between the physicochemical properties of both the NPs and biomolecules and their interfacial chemistry. While the interaction could potentially yield novel NP–biomolecule conjugates with a diverse range of properties and applications, it is possible that the interactions somehow affect the structural and functional properties of the biomolecules and reduce the NP stability and optoelectronic properties. For this reason, researchers have identified several principle criteria for a universal "toolset" that would allow attachment of biomolecules to NPs or any surface in a controlled manner as presented in Section 4.2.3. These principle criteria serve as a "guideline" for selection of suitable conjugation chemistries. Although various efforts have been devoted to improving NP–biomolecule interactions and functions through various conjugation strategies including the non-covalent interactions, classical covalent chemistries and "click" chemistries, controlling the spatial orientation of the biomolecule on the NPs, remain a daunting task. This issue has paved the way for the development of advanced conjugation strategies which primarily focused on enhancing the specificity and selectivity of NP–biomolecule interactions. For this endeavor to be successful, future works must shift from proof-of-concept studies to implementation of the concept in "real" applications.

References

Algar, W. R. et al. (2011) 'The controlled display of biomolecules on nanoparticles: A challenge suited to bioorthogonal chemistry', *Bioconjugate Chemistry*, 22(5), pp. 825–858.

Aubin-Tam, M. E. and Hamad-Schifferli, K. (2008) 'Structure and function of nanoparticle-protein conjugates', *Biomedical Materials*, 3(3), pp. 034001.

Avvakumova, S. et al. (2014) 'Biotechnological approaches toward nanoparticle biofunctionalization', *Trends in Biotechnology*, 32(1), pp. 11–20.

Bewersdorff, T. et al. (2017) 'The influence of surface charge on serum protein interaction and cellular uptake: studies with dendritic polyglycerols and dendritic polyglycerol-coated gold nanoparticles', *International Journal of Nanomedicine*, 12, pp. 2001–2019.

Brune, K. D. et al. (2017) 'Dual plug-and-display synthetic assembly using orthogonal reactive proteins for twin antigen immunization', *Bioconjugate Chemistry*, 28(5), pp. 1544–1551.

Chen, L. et al. (2016) 'Improved variants of SrtA for site-specific conjugation on antibodies and proteins with high efficiency', *Scientific Reports*, 6, pp. 1–12.

Chou, L. Y. T., Ming, K. and Chan, W. C. W. (2011) 'Strategies for the intracellular delivery of nanoparticles', *Chemical Society Reviews*, 40(1), pp. 233–245.

Falck, G. (2018) 'Enzyme-based labeling strategies for antibody–drug conjugates and antibody mimetics', *Antibodies*, 7(1), p. 4.

Huang, R. and Lau, B. L. T. (2016) 'Biomolecule-nanoparticle interactions: Elucidation of the thermodynamics by isothermal titration calorimetry', *Biochimica et Biophysica Acta - General Subjects*, 1860(5), pp. 945–956.

Huang, R. et al. (2013) 'Protein–nanoparticle interactions: the effects of surface compositional and structural heterogeneity are scale dependent', *Nanoscale*, 5(15), p. 6928.

Jun, Y. et al. (2017) 'Gold nanoparticle size and shape effects on cellular uptake and intracellular distribution of siRNA nanoconstructs', *Bioconjugate Chemistry*, 28(6), pp. 1791–1800.

Lee, B. et al. (2018) 'Site-selective installation of an electrophilic handle on proteins for bioconjugation', *Bioorganic and Medicinal Chemistry*, 26(11), pp. 3060–3064.

Liebana, S. and Drago, G. A. (2016) 'Bioconjugation and stabilisation of biomolecules in biosensors', *Essays In Biochemistry*, 60(1), pp. 59–68.

Lynch, I. and Dawson, K. A. (2008) 'Protein-nanoparticle interactions', *Nano Today*, 3(1–2), pp. 40–47.

Ma, W. et al. (2018) 'Modular assembly of proteins on nanoparticles', *Nature Communications*, 9(1), pp. 1–9.

Di Marco, M. et al. (2010) 'Overview of the main methods used to combine proteins with nanosystems: absorption, bioconjugation, and encapsulation', *International Journal of Nanomedicine*, 5, pp. 37–49.

Nwe, K. and Brechbiel, M. W. (2009) 'Growing applications of "click chemistry" for bioconjugation in contemporary biomedical research', *Cancer Biotherapy & Radiopharmaceuticals*, 24(3), pp. 289–302.

Phogat, N., Kohl, M. and Uddin, I. (2018) *Interaction of Nanoparticles with Biomolecules, Protein, Enzymes, and Its Applications, Precision Medicine: Tools and Quantitative Approaches*. Elsevier Inc., London, pp. 253–276.

Reisner, D. E. et al. (2017) *Bionanotechnology, Handbook of Research on Biomedical Engineering Education and Advanced Bioengineering Learning: Interdisciplinary Concepts.* IGI Global, Hershey, pp. 436–489.

Rodriguez-Quijada, C. et al. (2018) 'Physical properties of biomolecules at the nanomaterial interface', *Journal of Physical Chemistry B*, 122(11), pp. 2827–2840.

Saallah, S. and Lenggoro, I. W. (2018) 'Nanoparticles carrying biological molecules: Recent advances and applications', *KONA Powder and Particle Journal*, 2018(35), pp. 89–111.

Sapsford, K. E. et al. (2011) 'AuNP130-Analyzing nanomaterial bioconjugates: A review of current and emerging purification and characterization techniques', *Analytical Chemistry*, 83(12), pp. 4453–4488.

Sapsford, K. E. et al. (2013) 'Functionalizing nanoparticles with biological molecules: Developing chemistries that facilitate nanotechnology', *Chemical Reviews*, 113(3), pp. 1904–2074.

Shang, W. et al. (2009) 'Cytochrome c on silica nanoparticles: Influence of nanoparticle size on protein structure, stability, and activity', *Small*, 5(4), pp. 470–476.

Shen, Z., Nieh, M.-P. and Li, Y. (2016) 'Decorating nanoparticle surface for targeted drug delivery: opportunities and challenges', *Polymers*, 8(3), p. 83.

Sperling, R. A. and Parak, W. J. (2010) 'Surface modification, functionalization and bioconjugation of colloidal Inorganic nanoparticles', *Philosophical Transactions of the Royal Society A: Mathematical, Physical and Engineering Sciences*, 368(1915), pp. 1333–1383.

Stephanopoulos, N. and Francis, M. B. (2011) 'Choosing an effective protein bioconjugation strategy', *Nature Chemical Biology*, 7(12), pp. 876–884.

Tăbăcaru, A. et al. (2017) 'Recent advances in click chemistry reactions mediated by transition metal based systems', *Inorganica Chimica Acta*, 455, pp. 329–349.

Vertegel, A. A., Siegel, R. W. and Dordick, J. S. (2004) 'Silica nanoparticle size influences the structure and enzymatic activity of adsorbed lysozyme', *Langmuir*, 20(16), pp. 6800–6807.

Wong, L. S., Okrasa, K. and MicKlefield, J. (2010) 'Site-selective immobilisation of functional enzymes on to polystyrene nanoparticles', *Organic and Biomolecular Chemistry*, 8(4), pp. 782–787.

Yu, M. K., Park, J. and Jon, S. (2012) 'Targeting strategies for multifunctional nanoparticles in cancer imaging and therapy', *Theranostics*, 2(1), 3–44.

Silicon-Based Nanoscale Probes for Biological Cells

Youjin Lee*, Andrew Phillips*, and Bozhi Tian
The University of Chicago

5.1 Introduction

5.1.1 Biological Processes and Traditional Methods

Sensing Electrical Signal

A primary goal of cellular biology has been to understand how cells process and use electrical, chemical, and biological signals both within cells (intracellular signaling) and between cells (intercellular signaling). In particular, electrical signaling is ubiquitous in biological cells. Cells that can generate electrical signals, specifically action potentials, are termed excitable cells and include muscle fibers that initiate contraction, and nerve cells, i.e. neurons, that transmit electrical signals to communicate. All cells, including the ones that are not excitable, have a membrane potential, which is established by a concentration difference of sodium and potassium ions between the cell membrane and extracellular matrix. For example, the resting potential of excitable cells is about 70 mV.

The patch-clamp technique is a traditional electrophysiology tool used to record intracellular electrical currents. In this technique, a thin glass micropipette with a recording microelectrode inside is filled with an electrolyte and inserted into a cell. A reference electrode is inserted into the solution surrounding the cell and ionic currents are recorded between the electrodes. Even though the technique can yield high spatial resolution—up to single ion channel recording—with high signal-to-noise ratio, it has nontrivial setbacks (Sakmann and Neher 1984). Decreasing the size of the micropipette is crucial for increasing the spatial resolution. However, small micropipettes will give high impedance between the micropipette and the cell interior, thus decreasing the temporal resolution as well as the signal-to-noise ratio (Prohaska et al. 1986). Another set of techniques used for probing cellular electrical activities include voltage- and calcium-sensitive dyes, which display high temporal and spatial resolution (Grinvald and Hildesheim 2004; Rochefort, Jia, and Konnerth 2008). The dye-based techniques however, suffer from challenges such as photo-bleaching, cytotoxicity from the dye, and differential dye loading efficiency.

In this chapter, a silicon-based nanomaterial that can sense and even register electrical signals from both individual cells and groups of cells will be introduced.

Sensing Chemical and Biological Signals

Modulating biochemical and biophysical processes naturally requires a method to sense changes to these natural mechanisms. The biological effect should be considered, particularly at the level where cellular regulation it occurs. For instance, if an induced change to cellular physiology is caused by an alteration of gene expression, this result could likely be assayed at the level of gene expression by quantifying changes in messenger RNA by methods such as real-time quantitative polymerase chain reaction (RT-PCR) or RNA sequencing (RNA-Seq). Alternatively, if the modulated gene encodes for a protein, this could be used to quantify the effect induced on the cell via protein quantification methods like Western blotting, or more sophisticated

---

*These authors are contributed equally to this book chapter

proteomics methods like stable isotope labeling by/with amino acids in cell culture (SILAC) and isobaric tags for relative and absolute quantitation (iTRAQ) (Bantscheff et al. 2007). Proteomics methods and Western blotting are typically destructive to the cell. Noninvasive methods for protein detection and quantification include using fluorescently labeled antibodies specific for the protein of interest. Permeabilizing and staining the cells can allow for protein quantification while preserving the cell. Nucleic acids can similarly be quantified using complementary nucleic acid hairpin structures containing a fluorescent dye and a quencher. These probes exhibit turn-on fluorescence when the complementary nucleic acid sequence binds the probe and subsequently spatially separate the fluorophore and quencher, turning off Förster resonance energy transfer from the dye to the quencher, restoring fluorescence to the dye. A variety of small molecule fluorophores that have a specific interaction with an analyte, resulting in a spectral change, can also be used for live-cell imaging. Calcium imaging using Fura-2 is a popular method to visualize calcium fluxes in live cells, which can be correlated with electrical activity (Williams et al. 1985).

As cellular perturbations could affect only existing populations of biomolecules in the cell during the perturbation event, it is important to consider which populations a given assay actually measures and which is the relevant population to interrogate. Many extracellular signals—for instance, soluble molecules, insoluble biopolymers, mechanical forces, temperature, and ion concentrations—are mediated by signal transduction pathways, connecting exterior cues to physiological changes inside the cell. When considering effects of cellular stimulation mediated by signal transduction pathways, the levels of the secondary messengers should be quantified using appropriate methods as a control.

5.1.2 Silicon Materials and Rational Design

Synthetic Flexibility of Silicon Materials

Silicon nanowires (SiNWs) are grown in a chemical-vapor-deposition (CVD) system via vapor-liquid-solid (VLS) mechanism. First, gold nanoparticles (AuNPs) are deposited on a silicon (Si) wafer, which is placed in a CVD chamber at high temperature and low pressure. A silicon precursor gas, silane, and dopant precursor gas are introduced, which break down at high temperature and dissolve in the AuNPs, forming a Au-Si alloy. Once the Si concentration of the alloy reaches its maximum solubility, solid Si precipitates out in the form of a SiNW. This allows more Si to be dissolved in the AuNPs and leads to further growth of the SiNWs. In this process, numerous parameters such as temperature, gas flow rate, gas composition—including dopant concentration—and pressure can be controlled, giving flexibility in the synthesis. To synthesize thicker NWs, larger AuNPs are used. The length of the NWs is often controlled by the growth time. To introduce a kink, the vacuum pump on the CVD chamber is switched on and off to introduce drastic pressure change during the growth (Tian et al. 2009). This abrupt decrease in pressure introduces a 120° angle in a NW because during the purge the reactant concentration drops and elongation ceases, then heterogeneous nucleation occurs upon reintroduction of reactant. The particular angle is due to the fast growth direction switching from <112> to <110> then back to <112>, which can be explained by thermodynamic stability. By throttling the pressure twice within a short growth interval, the NWs will develop two 120° kinks, effectively a 60° angle joint.

After the synthesis, NWs can be selectively etched to yield various morphologies along their long axis. Kim et al. have modulated dopant level during the synthesis and selectively etched the doped region with hydrochloric acid (Kim et al. 2017) (See Figure 5.1). Luo et al. has embedded a periodic pattern of diffused Au along SiNWs by periodically modulating silane pressure during the synthesis (Luo et al. 2015). Subsequent potassium hydroxide etching selectively etched the non-Au diffused regions of the NW.

Just like core–shell nanocrystals, shells can be grown on SiNW cores. Tian et al. has synthesized p-i-n coaxial SiNWs: p-type core/intrinsic inner shell/n-type outer shell (Tian et al. 2007). Here, p-type SiNWs are grown by VLS mechanism. Then, the shells are deposited at a higher temperature to prevent further axial growth of the NW. This one-dimensional p-i-n nanostructure has shown effective

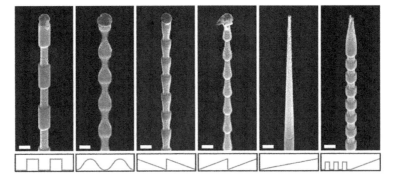

FIGURE 5.1 Various morphologies of SiNWs. Upper panels: SEM images of NWs of various morphologies. P-doped regions of the NWs were etched by buffered hydrofluoric acid. Scale bars: 200 nm, Lower panels: pressure profiles of p dopant precursor, phosphine gas, flow during the final 1 μm growth of above NW.

charge separation upon light absorption, making it suitable for photovoltaic material. The application of using this p-i-n SiNW to photo-stimulate cells will be discussed in Section 5.3.2.

Apart from SiNWs, a different SiNW-based mesostructure, nanoporous Si, was developed by Jiang et al. This nanoporous Si matches the mechanical properties of soft biological cells and tissues and exhibits a Young's modulus of 1.84 GPa, approximately two orders of magnitude lower than bulk Si with a Young's modulus of 160 GPa. The material can be made softer by immersing it in phosphate-buffered saline solution, which results in a Young's modulus of 0.41 GPa, comparable to hydrated collagen fibers (Jiang et al. 2016). Such mechanical property matching is critical for biological application as (i) it is a better representation of the cellular environment when used as an in vitro support to grow cells and (ii) is less likely to cause an autoimmune response when integrated in biological systems.

5.2 Sensing Probes

5.2.1 Electrical Signal and Field Effect Transistor

Learning Point 1: Basics of Transistor and Field Effect Transistor

Transistors are semiconductor devices that can gate electronic signals dependent on input signals. First practically implemented by John Bardeen, Walter Brattain, and William Shockley in 1947, they have become an integral part of circuits found in almost all modern electronic devices ("The Nobel Prize in Physics 1956" n.d.). Transistors usually have at least three terminals and are made of Si. Transistors are categorized into two groups: current-controlled bipolar junction transistors (BJT) and voltage-controlled field-effect transistors (FET). BJT uses both holes and electrons for conductance, while FET utilizes either solely electrons or solely holes. For this chapter, we will focus on the FET, which is more common in practice.

FET consists of three terminals: the source, the gate, and the drain. Charge carriers—either electrons or holes—flow through a channel from the source to the drain. The carrier density, or the conductivity, of the channel is modulated by the electric field generated by the voltage across the gate and the source (see Figure 5.2).

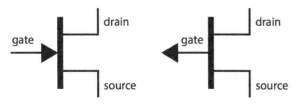

FIGURE 5.2 Schematic electric symbol diagram of FET with n-channel (left) and FET with p-channel (right) comprising three terminals: drain, source, and gate.

FETs are sensitive to small voltage differences across the gate and source, and so can significantly amplify signals in the current readout. With the same device principal, sensing of electrical/chemical/biological signal can be done on a much smaller scale. For such applications, SiNW has been an excellent material due to its one-dimensional structure, its nanoscale size comparable to biological systems, and its synthetic tunability. Compared to a bulk or planar material, one-dimensional SiNW has high surface-to-volume ratio. The one-dimensional nature of NWs makes them intrinsically sensitive since any signal imposed on the NW surface will translate to the entirety of its volume as opposed to being attenuated by the large volume in a planar FET channel. For example, a heavily doped SiNW was used as a channel in a planar nanowire FET (NW-FET) (see Figure 5.3).

To quantitatively explain the properties of NW-FET, we must understand the relationship between the change in conductance (ΔG) and the change in electrostatic potential ($\Delta\varphi_{Si}$). A comprehensive derivation is provided in Gao et al (Gao et al. 2010).

$$\Delta G = q\mu \int_0^R 2\pi r \Delta p \, dr$$

Here, q is the elementary charge, μ, the carrier mobility, and r, the distance from the NW center. R is the NW radius and p is the majority carrier density.

In one limiting case, either when the NW radius is much less than the Debye length (defined in Section 1.2.2) or when the dopant concentration of the NW is very low, the conductance change varies exponentially with surface charge change.

$$\Delta G = q\mu\pi R^2 p \left[e^{\frac{-q\Delta\varphi_{Si}}{K_B T}} - 1 \right]$$

In the other limiting case, when the NW is heavily doped or thick, where the NW radius is much greater than the Debye length, the conductance change depends linearly on the change in surface charge.

$$\Delta G = 2\pi R p \lambda_D \frac{q^2 \Delta\varphi_{Si}}{K_B T}$$

Because the conductance change responds to change in surface charge differently based on the NW dopant level, NW dopant modulation can further localize the detection area. For example, the p–n junction of an axially doped kinked NW was used as a point-like FET detector (Jiang et al. 2012).

FET Principal Applied to Sensing Mechanism

This dopant level-dependent response property along with established surface modification techniques make SiNWs an excellent material to be used as a high-sensitivity nano-biosensor in FET-like devices. The native silicon oxide layer on the NW surface can be modified to transform NW-based FET devices to biosensors that can detect chemical and biomolecules. For example, n-doped SiNWs functionalized

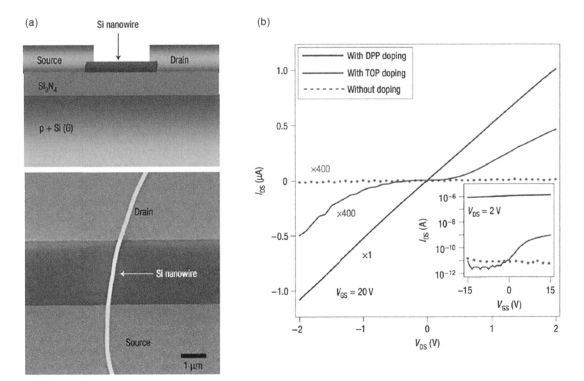

FIGURE 5.3 A back-gated SiNW-based FET. (a) Schematics diagram of the device, where a SiNW is used as a channel, silicon nitride as an insulating layer, and a silicon wafer as a back-gating material (top). Falsely colored SEM image of the device (bottom). (b) Source–drain current versus source–drain voltage measured with FET devices with an undoped and doped SiNW. Different precursors were used for monolayer doping: diethyl 1-propylphosphonate (DPP) and trioctylphosphine oxide (TOP).

with amine and oxide species were used in a nanosensor to detect the environment's pH as well as protein binding (Cui et al. 2001). In this example, 3-aminopropyltriethoxysilane (APTES) was used to modify the surface charge of the amine group of APTES since the molecule's protonation state depends on the pH. Therefore, changes in pH regulate the surface charges of APTES-modified SiNWs, which enables it to act as a chemical gate. In summary, pH change is read out as a change in the current flow through the NW nanosensor. Surface modification that enables biomolecule detection will be further explored in Section 5.2.2.

Extracellular Signal Recording

SiNW-FETs can be used for extracellular field potential sensing. Unlike its bulk counterpart, nanometer-thick SiNWs are flexible and can be made into a flexible device to interface with biological systems' that typically have low Young's modulus. Cohen-Karni et al. have fabricated a multiplexed SiNW-FET array that can resolve a propagation of extracellular potential spikes in a cardiomyocyte monolayer (Cohen-Karni et al. 2009). The recording was possible due to its flexible and thus robust interface between the NW-FET device and cells as well as the comparable length scale of the NW and the biological system. To achieve a high signal-to-noise ratio, a strong interface between the device and biological sample is crucial. Further discussion about FET–biological sample interfacing will be provided in section 1.3.1.

Intracellular Signal Recording

Compared to extracellular potential, intracellular signal is harder to measure, since the probe has to be inserted into a cell. Extracellular sensors cannot detect intracellular signal as the lipid membrane has electrical resistance and capacitance, thus acting as an electrical barrier. Intracellular electric signal recording can contain rich information and can help reveal subcellular information flow and cell physiology.

A traditional method, such as the patch clamp technique, is mechanically invasive and its operation and setup are not trivial. NW-FET probes, on the other hand, present a less invasive alternative for such measurements. Tian et al. has integrated a multiply kinked NW into a metal-polymer based support to form a flexible NWFET device (Tian et al. 2010) (See Figure 5.4). The probing tip has two 120° kinks that are only about 160 nm apart. This pointy morphology facilitates the insertion of the probe into the cell and is less invasive. In addition, the probe was coated with a phospholipid bilayer that was critical for successfully inserting the NW through the membrane to access the intracellular region. The kinked NW has varying dopant concentration: a lightly doped region is sandwiched between heavily doped arms. In this case, the heavily doped arms act as source and drain terminals while the lightly doped region in between them acts as a gate. The point-like detection area is advantageous as it increases both the signal-to-noise ratio of the FET device and the spatial resolution. Localization was

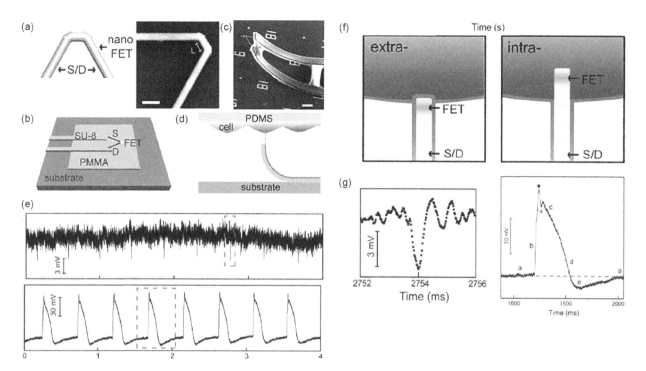

FIGURE 5.4 3D kinked SiNW probes and electrical recording of cardiomyocytes. (a) Schematic diagram of 60° doubly kinked nanowire (left) and the SEM image of a cis-doubly kinked nanowire (right). L is the distance between the two kinks. Scale bar: 200 nm, (b) a schematic diagram of the fabricated device. Poly(methyl methacrylate) (PMMA) was used as a sacrificial layer and SU-8 microribbons as a flexible support. (c) SEM image of the device. The arrow points to the NW-FET and the pink star is drawn on the SU-8 support. (d) Schematic diagram setups for recording beating cardiomyocytes on polydimethylsiloxane (PDMS) (left) (e) electrical beating readout for extracellular recording (top) and intracellular recording (bottom), (f) schematic diagram of NW-FET device and cell interface for extracellular recording (left) and intracellular recording (right). The purple outline marks the cell membrane and the lipid bilayer coating of the FET device, (g) zoomed-in recording of the red-dashed boxes in extracellular recording (left) and intracellular recording (right).

possible because the surface charge-dependent FET conductance change is sensitive to the local dopant level. The device has read intracellular potential while minimally destructing the cell, both chemically and mechanically.

5.2.2 Biomolecular Signal and Field Effect Transistor (BioFET)

SiNWs have been designed to detect a variety of biomolecules, including DNA (Hahm and Lieber 2004), viruses (Patolsky et al. 2004), proteins (Zheng et al. 2005; Cui et al. 2001), and biologically relevant ions (Anand et al. 2017). Key to any sensor is its selectivity—a measure of how well the sensor can discriminate the desired analytes versus other interfering species (Vessman et al. 2001). Analytes serve as the gate in SiNW FETs by modulating the nanowire surface charge and thus conductivity—the output signal. This means that analyte selectivity can easily be designed into SiNW FETs by taking advantage of the plethora of known biomolecular interactions. Examples of these include antibodies, nucleic acids, and aptamers, all of which can have extremely low dissociation constant (K_D) values, engendering highly specific sensing without the need to design new sensory mechanisms, given the inherent self-signaling mechanism of FETs.

This strategy involves surface modification of the SiNW with a receptor specific for the desired analyte that has to be detected. In this case, when the analyte meets the complementary receptor bound to the nanowire, the surface charge of the nanowire would be altered, transducing this biochemical interaction into an electrical output signal, in real time and in a label-free fashion. Additionally, given that SiNW FETs can be fabricated in an individually addressable manner, multiplexed biosensing is possible (Zheng and Lieber 2005).

Silicon and silicon dioxide surfaces have been the subject of much study over the past century. Alkoxysilanes and chlorosilanes, in particular, are commonly used reagents that condense on silica to form Si-O-Si bonds, imparting whatever functional group is present on the original silane. Many different silanes are commercially available, allowing for facile introduction of functional groups like amines, carboxylic acids, aldehydes, and thiols, enabling easy conjugation with biomolecules serving as the receptor in the SiNW FET. Figure 5.5 details a common scheme used by Patolsky et al. for surface-modifying SiNW FET devices with an antibody to introduce specificity to a complementary analyte. First, the prepared FET devices are subjected to an oxygen plasma to remove any organic contaminants on the silicon dioxide surface and to expose

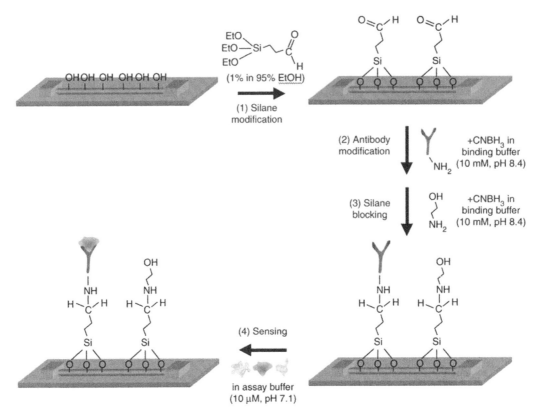

FIGURE 5.5 Surface functionalization of SiNW FET with an antibody to make a BioFET. SiNW are first treated with oxygen plasma to form surface silanol groups. (1) SiNW are immersed in a solution of a functionalized silane. (2) An antibody with a functional group complementary to the functional silane is added. (3) Silanes that did not react with the antibody are blocked. (4) The antibody binds its complementary analyte during operation of the BioFET as a sensor. (Schematic modified from Patolsky et al. 2006.)

free silanols. Next, a functional alkoxysilane, in this case 3-(trimethoxysilyl)propyl aldehyde, is used to introduce an aldehyde functional group on the SiNW surface. This enables antibody attachment to the SiNW via a reductive amination mechanism, wherein an imine is initially formed from the surface aldehyde with a native amine on the antibody, followed by reduction with sodium cyanoborohydride to afford a stable amine linkage. Unreacted silanes were similarly blocked using ethanolamine via reductive amination (Patolsky et al. 2006). Similar approaches to introduce specificity elements have been accomplished using carbodiimide chemistry (Lee et al. 2009).

Learning Point 2: Debye Screening and Detection Limit

When placed in an electrolyte, the charged surface of a nanowire will attract counter-ions, forming a double ion layer. This phenomenon, known as Debye screening, limits the detection range of the NW since charged species that are farther than the Debye screening length from the NW will not gate the NW-FET.

The double ion layer electrical potential (ψ) exponentially decays as a function of distance from the surface (y) over the Debye length (λ_D).

$$\psi(y) \sim e^{\left(\frac{-y}{\lambda_D}\right)}$$

The Debye length measures how well charges, often ions, in solution screen electric fields resulting from other charged species. It is a function of the dielectric constant (ε_r), the permittivity of free space (ε_0), the Boltzmann constant (k_B), and temperature in Kelvin (T). N_A is Avogadro's number, q is the elementary charge, and C is the ionic concentration of the electrolyte.

$$\lambda_D = \sqrt{\frac{\varepsilon_r \varepsilon_0 k_B T}{2 N_A q^2 C}}$$

The electric potential will decrease by $1/e$ for every Debye length. Thus, electrolyte solutions with higher ionic concentration have shorter Debye lengths. As a point of reference, in physiological environments (1x PBS), the Debye length is 0.7 nm.

This short Debye length present in physiological environments presents challenges to FET-based biosensors since the charged biomolecules that are 0.7 nm away or further from the device will be effectively screened and not be able to modulate the device conductance. One obvious way to overcome this challenge is to decrease the salt content of the solution. This can be accomplished by using microfluidic chambers to purify samples or by using the dialysis method to desalinate the sample. Altering the ionic strength of the solution can however affect the behavior of bimolecular species. Therefore, other methods are being explored. For example, alternating current (AC) signals were used in drain–source voltage to extend Debye length by breaking down the electric-double-layer around the FET channel surface (Munje et al. 2015). In 2017, Chu et al. used antibody-immobilized chemical pair in a high electron mobility transistor to overcome the Debye screening and to detect proteins in 1x PBS (Chu et al. 2017).

The Debye screening can, on the other hand, screen out bulk of the species in solution that are outside the Debye sphere and increase the signal-to-noise ratio in some cases. For instance, in the case of monitoring binding and unbinding of biomolecular species to the NW, charge screening of other charged species further away from the NW will decrease the background noise.

5.2.3 Mechanical Signal and Free-Standing Nanowire

While the roles of biochemical, and electrical, cues in biological cells have been studied in many aspects in depth, that of mechanical cues has only begun to be understood. Mechanical cues, however, are ubiquitous in cells and play a significant role such as in regulating cells' death, proliferation, differentiation, wound healing, and, even in B cells, antigen formation (Lafaurie-Janvore et al. 2013; Moore, Roca-Cusachs, and Sheetz 2010; Natkanski et al. 2013). There are studies on force-sensitive proteins, but there have not been many probes that can directly measure the force exerted on cells by the environment (Yusko and Asbury 2014).

Zimmerman et al. in 2015 utilized kinked SiNW to probe inter- and intracellular force dynamics (see Figure 5.6). With a kink that anchors on a cell, the NW bends as a

neighboring cell exerts force. When the kinked SiNW is internalized in a cell, contraction or deformation of the cell bends the NW. The bending of the NW is quantified by modeling the kinked SiNW as a simply supported beam using Euler–Bernoulli beam theory. SiNW and cells can be imaged concurrently under optical microscope in a scatter-enhanced phase contrast mode. Internal lamp was used in bright field with a phase contrast filter to image the cells while ring LED lens was used to image the SiNW. Silicon material scatters light very well, and so it can be easily visualized in both bright and dark field.

This particular study has focused on investigating individual cells that had not necessarily reached confluency. With kinked SiNW as a tool, further studies on systems that mimic biological environment, for example where cells are under constant shear force such as by blood flow, will be interesting.

5.2.4 Multiplexing and Tissue Level Devices

Lieber and coworkers extended the concept of recording from cells cultured on planar devices by integrating stimulation with recording. In a seminal paper, they fabricated arrays of SiNW FETs and patterned regions of polylysine, to promote desired cell growth orientations. To stimulate the cells, they used conventional glass microelectrodes to inject current. With their device, they were able to record intracellular potentials using extracellularly placed FETs and observed action potentials (Patolsky et al. 2006).

In later work, Lieber and coworkers developed tissue-scaffold-mimicking 3D arrays of FETs with

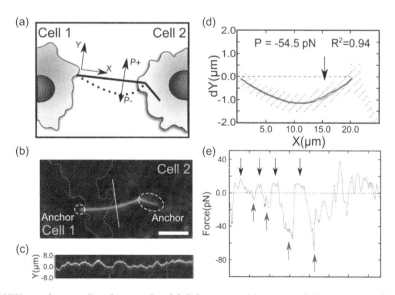

FIGURE 5.6 Kinked SiNW as a freestanding force probe. (a) Schematics of human umbilical vein cells (HUVECs) and kinked SiNW, (b) optical micrograph image of the HUVECs and a bent kinked SiNW. The cyan dashed line indicates the outline of two cell membranes, (c) A kymograph of the NW positions from the yellow line in (b), (d) An example force fitting data of bent kinked SiNW. The dots indicate the detected region of the NW and the thick curve indicates the fitted centerline of the detected region. The arrow indicates the calculated load position, (e) Calculated graph of bending force over time. The black and blue arrows indicate ratcheting peaks in the force plot.

stimulating electrodes (Tian et al. 2012; Dai et al. 2016). The material was designed to be mimetic in terms of the dimensions and mechanical properties to synthetic materials used as cardiac tissue engineering scaffolds, such as electrospun biodegradable poly(lactic-co-glycolic acid) (PLGA) fibers. They initially fabricated a 2D freestanding mesh, and folded this material on top of itself to form a 3D structure. The material can have up to 64 individually addressable SiNW FETs and four stimulation electrodes constructed from platinum/palladium (Dai et al. 2016). To this 3D folded material, they seeded neonatal rat ventricular cells and cultured them inside the scaffold. This platform was then utilized to map and manipulated the electrophysiology of 3D tissue in real time. The nano-electronic scaffold enabled the mapping of conduction pathways in developing cardiac tissue with sub-milisecond resolution, in 3D. Further, using the four electrodes, they demonstrate stimulation of the culture (Dai et al. 2016).

5.3 Bio-nano Interfacing for Cell Modulation

5.3.1 Cell Membrane–Material Interactions

Bianxiao Cui's and Yi Cui's groups examined the cell-nanopillar interface in detail, showing that nanopillar arrays can pin embryonic cortical neurons from rats. They show an intimate interaction between the nanopillars and the neurons, causing the pillars to deflect and the pinned cells to have vastly reduced migration. Their studies have shown asymmetric membrane response toward nanostructures, in that cell membranes easily deform inward to wrap around nanoscale objects and rarely deform outward (Xie et al. 2010; Santoro et al. 2017). Similar work was carried out by Yang and coworkers, culturing human hepatic stellate (LX-2) and a liver hepatocellular carcinoma cell line (Hep G2) on arrays of vertically aligned SiNW arrays, demonstrating that cell spreading was restricted, as compared to cells grown on silicon wafers. The showed that cells cultured on SiNW arrays interface the substrate through filopodia and qualitatively demonstrated through centrifugation that cells more strongly adhered to nanowire arrays than to flat silicon wafers. Using Western immunoblotting, they also showed that collagen I and α-actin were downregulated on SiNW arrays, relative to flat wafers, potentially consistent with the reduced spreading on SiNW arrays. Integrin and focal adhesion kinase were seen to be upregulated on SiNW arrays, compared to flat wafers, consistent with the stronger adhesion of the cells to the SiNW arrays relative to flat wafers (Qi et al. 2009).

It is perhaps instructive to consider interactions of cells with highly anisotropic topographical features as a form of mechanical stimulation. Bianxiao Cui's group and others have looked at the effect of such topography on intracellular signaling. Cells naturally contain mechanisms to sense and alter membrane curvature. These processes are relevant for cellular processes involving vesicle formation and fusion, like endocytosis and exocystosis. As Cui's group describes what they call the curvature hypothesis, the same proteins involved in endogenous curvature-recognizing mechanisms are activated by artificial sources of curvature; namely, inward deformations of the cell membrane around anisotropic materials. Further, mechanical stimulation of the nucleus by anisotropic structures can also affect gene expression (Lou et al. 2018).

Besides nanoscale features, ones at the mesoscale can also affect the biointerface. Luo et al. has synthesized a skeleton-like SiNW by modulating Au diffusion during synthesis followed by silicon etching (Luo et al. 2015). The anisotropic structure makes the retraction of the NW from collagen hydrogels harder compared to initial insertion of the NW, similar to the case of a bee-stinger. The AFM cantilever approach/retract experiment has shown the potential of the NW to form a tight and semi-irreversible interface with soft biological samples such as biological tissues. Unlike other isotropic SiNWs or metal pillars, the anisotropic interaction between the mesh and skeleton-like SiNW mesostructure will be an important characteristic in clinical applications like in vivo chronic recording.

5.3.2 Wired Probe Designs

Silicon has long been used as the working material in traditional electrodes for recording and stimulating electrically excitable tissues, such as neuronal tissue. The most common of these are planar microelectrode arrays, such as the Michigan and Utah arrays, which are still frequently used in the neural interface community and have been used in highly advanced brain–computer interfaces in humans. Multielectrode arrays, however, have diameters on the order of microns and length on the order of millimeters, meaning that there is a limit to their spatial precision—that is, the volume of tissue stimulated or recorded from is rather large, leading to spatial undersampling, and single-cell resolution is challenging with traditional electrode arrays (Tsai et al. 2017). Their bulky size and need to be wired to computers for signal processing also means that they are highly invasive. In addition to the short-term immune response resulting from surgical implantation, the mechanical mismatch of silicon electrodes and host brain tissue means that there is chronic inflammation due to micro-motions of the brain, leading to glial scarring at the implantation site (Prodanov and Delbeke 2016).

Figure 5.7 shows the BrainGate sensor, a multielectrode array of 100 electrodes used to enable the user control an external system by recording their thoughts, decoding them, and transducing this into an action. Figure 5.7a and b shows the active sensor portion in comparison to a US penny, showing its relatively large size. Figure 5.7d shows the external wiring required to attach the sensor to the computers for signal processing and decoding (Hochberg et al. 2006).

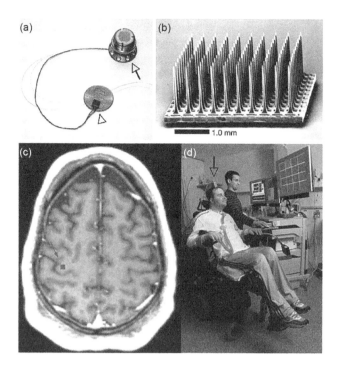

FIGURE 5.7 The BrainGate sensor, depicted by the arrowhead, is shown in relation to a US penny. (a) It is connected to a titanium pedestal, depicted by the arrow, affixed to the patients' skull. (b) A scanning electron micrograph of the sensor portion of the BrainGate, showing the 10 × 10 array of 100 microneedle electrodes, each 1 mm long with a pitch of 400 μm. (c) Preoperative magnetic resonance imaging scan (T1-weighted), showing the implantation site in the precentral gyrus of Participant 1's brain. Participant 1 in the BrainGate trial (MN). The arrow shows the location of the connection of the sensor, through the titanium pedestal, to the amplifier and signal conditioning hardware (d) (Hochberg et al. 2006).

As the fundamental unit of the brain is increasingly understood to be neural circuits, ideal tools to probe these structures should be on similar length scales (Purves et al. 2001). To assess the activity and function of neural circuits, probes should be able to sense and stimulate with spatial resolution on the order of the individual components of neural circuits: neurons. The requirement of single-cell resolution naturally suggests the nano- or meso-length scale, given that cells are on the order of microns. Advances in nanoscale fabrication of semiconductors, developed for the electronics industry, have enabled the creation of arrays of nano- and meso-scale probes for biological systems, and are uniquely suited to probing large circuits of cells, with single-cell resolution.

5.3.3 Freestanding Probes

Photothermal Mechanism

Freestanding probes not attached to substrates are attractive as they are more versatile tools than their substrate-bound counterparts. Free particles could

conceivably be delivered in a drug-like fashion in ways that surfaces cannot, meaning that these freestanding probes could have therapeutic merit, whereas surface-bound devices are likely better suited as tools for fundamental cellular and tissue studies. Further, particles can also have substantially different mechanical, electrical, and degradation properties as compared to their bulk counterparts. Jiang et al. fabricated nanoporous silicon particles as photothermal stimulating agents to depolarize dorsal root ganglia neurons. They used a nanocasting approach wherein an ordered nanoporous silica material was used to template the decomposition of silane into amorphous silicon, followed by etching of the silica template, affording an all-silicon material. The material was drop-cast on cell membranes and was shown to depolarize neurons by heating the cell and changing the membrane capacitance (Jiang et al. 2016). The photothermal mechanism of the nanoporous silicon particles, as well as bulk silicon, is due to the indirect bandgap of silicon, wherein optical excitations thermally decay, radiating heat (Roder et al. 2014). Further, given the nanoporous nature of the particles, they have degradation kinetics on the order of days, meaning they could be used for transient biological modulation (Jiang et al. 2016).

Photoelectrochemical Mechanism

Parameswaran et al. recently reported a freestanding photoelectrochemical modulation platform using SiNWs. Drawing inspiration from the solar cell literature, they synthesized core–shell SiNW fabricated into nanoscale diodes—p-type core/intrinsic shell/n-type shell (PIN). The PIN SiNWs were synthesized using the VLS growth mechanism in a CVD reactor. They used PIN SiNWs to wirelessly stimulate primary rat dorsal root ganglion neurons, showing that action potentials could be produced in a manner indistinguishable from those elicited by traditional patch clamp methods (Figure 5.8). As the material was dropcast on cells without any pretreatment, it is suitable for use in a drug-like fashion. Shining laser light on the PIN SiNW interfaced with the cell membrane resulted in a photocathodic current that locally depolarized neurons. The mechanism was reported to be light-induced production of electrons and holes in the material, with the built-in electric field of the diode enhancing charge separation. Photoexcited electrons move to the n-type shell and participate in a cathodic process at the semiconductor–electrolyte interface. The authors also examined the role of atomic gold on the nanowires. Between the core and shell growths, they intentionally diffused the gold catalyst down the sidewalls of the nanowire at low pressure, resulting in atomic gold-induced surface states. Despite most considerations of gold as a source of deep traps in silicon, the authors showed that atomic gold was necessary to elicit action potentials. The effect of the gold was attributed to alteration of silicon surface states such that the impedance of the nanowires in aqueous solutions was reduced, which would enhance the faradic currents that depolarized the cells. The measured photoelectrochemical currents were unipolar in nature. This interesting

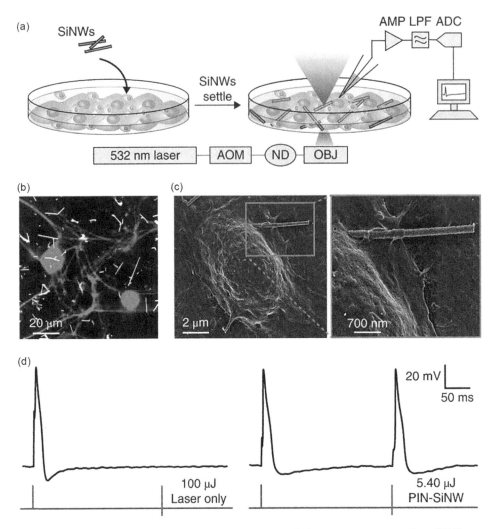

FIGURE 5.8 Schematic depicting the setup used to record action potentials from primary neonatal rat DRG neurons interfaced with SiNW. SiNWs were dropcast onto DRG neurons in culture to form an intimate interface. A 532 nm laser was used to stimulate individual neuron–SiNW interfaces. The setup used a typical patch-clamp electrophysiology setup in current-clamp mode, including an amplifier (AMP), low-pass filter (LPF), and an analog-to-digital converter (ADC). The 532 nm laser added onto this system was controlled by an acoustic modulator (AOM), with neutral density (ND) filters to attenuate the laser power. (a)The laser was aligned to the central optical axis of the objective lens (OBJ) of an inverted microscope. (b) Confocal micrograph of DRG neurons cocultured with PIN-SiNWs (white) stained with anti-β-tubulin. (c) Scanning electron micrograph of a SiNW–DRG neuron interface. (d) Action potential in a DRG neuron induced by injected current (left pulse) or a laser pulse (right pulse) recorded in current-clamp mode of patch-clamp electrophysiology, showing membrane voltage (Parameswaran et al. 2018).

phenomenon was suggested to be due to the larger exposed area of the n-type shell relative to the p-type core exposed at the ends of the nanowires. The photoanodic reaction would be expected to occur at the p-type core–electrolyte interface due to the built-in electric field of the diode accelerating holes to this region. The authors also hypothesized that the nonequilibrium current production was in part due to the surface states altered by the atomic gold (Parameswaran et al. 2018).

Parameswaran et al. also demonstrated a unique method for estimating the photoelectrochemical behavior of SiNWs species in an interconnect-free manner by growing nanowires inside quartz capillary tubes. These capillary tubes were pulled into patch pipettes and attached to a patch-clamp

rig. By shining a laser onto a single nanowire inside the capillary and performing the measurement in voltage-clamp mode, with the voltage held at zero as a virtual ground, they were able to measure the photocurrent of single nanowires. Importantly, the measurement was done in such a way to minimize the effect from photothermal heating (Parameswaran et al. 2018).

5.3.4 Future Direction

The biggest challenges for nanoscale FET-based devices are in increasing the signal-to-noise ratio and overcoming the Debye screening phenomena. Additionally, there is a need for nanoscale techniques that can accurately measure

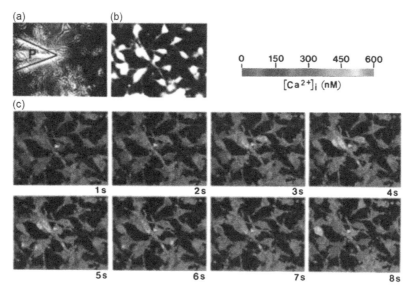

FIGURE 5.9 (a) Phase-contrast micrograph of 7-day-old mixed glial culture. Single-cell mechanical stimulation is achieved with a micropipette (P). (b) Ca^{2+} imaging of the same field of cells in (a) using Fura-2. (c) Sequential time points of the same field of cells in after mechanical stimulation of one cell, showing Ca^{2+} wave propagation to other cells, in all directions. All images are scaled to 155 μm vertically by 210 μm horizontally (Charles et al. 1991).

mechanical stress at the intracellular and subcellular levels. At the extracellular level, mechanical stimulation is only recently being appreciated as an important biological process, despite having been known for several decades (Figure 5.9). We predict that intracellular and extracellular mechanical stimulation of cells will therefore become important cellular modulation tools. We envision freestanding probes capable of mechanically modulating cells, both in extracellular and intracellular conditions, as being an important next step in the direction of biological modulation.

References

Anand, A.; Liu, C.-R.; Chou, A.-C. et al. 2017. Detection of K+ efflux from stimulated cortical neurons by an Aptamer-modified silicon nanowire field-effect transistor. *ACS Sens.* 2: 69–79.

Bantscheff, M.; Schirle, M.; Sweetman, G. et al. 2007. Quantitative mass spectrometry in proteomics: A critical review. *Anal. Bioanal. Chem.* 389: 1017–1031.

Charles, A.C.; Merrill, J.E; Dirksen, E.R. et al. 1991. Intercellular signaling in glial cells: calcium waves and oscillations in response to mechanical stimulation and glutamate. *Neuron* 6: 983–992.

Chu, C.-H.; Sarangadharan, I.; Regmi, A. et al. 2017. Beyond the Debye length in high ionic strength solution: Direct protein detection with field-effect transistors in human serum. *Sci. Rep.* 7: 5256.

Cohen-Karni, T.; Timko, B. P.; Weiss, L. E.; Lieber, C. M. 2009. Flexible electrical recording from cells using nanowire transistor arrays. *Proc. Natl. Acad. Sci. U.S.A.* 106: 7309–7313.

Cui, Y.; Wei, Q.; Park, H. et al. 2001. Nanowire nanosensors for highly sensitive and selective detection of biological and chemical species. *Science* 293: 1289–1292.

Dai, X.; Zhou, W.; Gao, T. et al. 2016. Three-dimensional mapping and regulation of action potential propagation in nanoelectronics innervated tissues. *Nat. Nanotechnol.* 11: 776–782.

Gao, X. P. A.; Zheng, G; Lieber, C. M. 2010. Subthreshold regime has the optimal sensitivity for nanowire FET biosensors. *Nano Lett.* 10: 547–552.

Grinvald, A.; Hildesheim, R. 2004. VSDI: A new era in functional imaging of cortical dynamics. *Nat. Rev. Neurosci.* 5: 874–885.

Hahm, J.-I.; Lieber, C. M. 2004. Direct ultrasensitive electrical detection of DNA and DNA sequence variations using nanowire nanosensors. *Nano Lett.* 4: 51–54.

Hochberg, L. R.; Serruya, M. D.; Friehs, G. M. et al. 2006. Neuronal ensemble control of prosthetic devices by a human with tetraplegia. *Nature* 442: 164–171.

Jiang, Y.; Carvalho-de-Souza, J. L.; Wong, R. C. S. et al. 2016. Heterogeneous silicon mesostructures for lipid-supported bioelectric interfaces. *Nat. Mater.* 15: 1023–1030.

Jiang, Z.; Qing, Q.; Xie, P.; Gao, R.; Lieber, C. M. 2012. Kinked P–n junction nanowire probes for high spatial resolution sensing and intracellular recording. *Nano Lett.* 12: 1711–1716.

Kim, S.; Hill, D. J.; Pinion, C. Q.; Christesen, J. D.; McBride, J. R.; Cahoon, J. F. 2017. Designing morphology in epitaxial silicon nanowires: The role of gold, surface chemistry, and phosphorus doping. *ACS Nano* 11: 4453–4462.

Lafaurie-Janvore, J.; Maiuri, P.; Wang, I. et al. M. 2013. ESCRT-III assembly and cytokinetic abscission are induced by tension release in the intercellular bridge. *Science* 339: 1625–1629.

Lee, H.-S.; Kim, K. S.; Kim, C.-J. et al. 2009. Electrical detection of VEGFs for cancer diagnoses using anti-vascular endotherial growth factor aptamer-modified Si nanowire FETs. *Biosens. Bioelectron.* 24: 1801–1805.

Lou, H.-Y.; Zhao, W.; Zeng, Y. et al. 2018. The role of membrane curvature in nanoscale topography-induced intracellular signaling. *Acc. Chem. Res.* doi:10.1021/acs.accounts.7b00594.

Luo, Z.; Jiang, Y.; Myers, B. D. et al. 2015. Atomic gold–enabled three-dimensional lithography for silicon mesostructures. *Science* 348: 1451–1455.

Moore, S. W.; Roca-Cusachs, P.; Sheetz, M. P. 2010. Stretchy proteins on stretchy substrates: The important elements of integrin-mediated rigidity sensing. *Dev. Cell* 19: 194–206.

Munje, R. D.; Muthukumar, S.; Selvam, A. P.; Prasad, S. 2015. Flexible nanoporous tunable electrical double layer biosensors for sweat diagnostics. *Sci. Rep.* 5: 14586.

Natkanski, E.; Lee, W.-Y.; Mistry, B.; Casal, A.; Molloy; J. E.; Tolar, P. 2013. B cells use mechanical energy to discriminate antigen affinities. *Science* 340: 1587–1590.

Parameswaran, R.; Carvalho-de-Souza, J. L.; Jiang, Y. 2018. Photoelectrochemical modulation of neuronal activity with free-standing coaxial silicon nanowires. *Nat. Nanotechnol.* 13: 260–266.

Patolsky, F.; Timko, B. P.; Yu, G. et al. 2006. Detection, stimulation, and inhibition of neuronal signals with high-density nanowire transistor arrays. *Science* 313: 1100–1104.

Patolsky, F.; Zheng, G.; Hayden, O. et al. 2004. Electrical detection of single viruses. *Proc. Natl. Acad. Sci. U.S.A.* 101: 14017–14022.

Patolsky, F.; Zheng, G.; Lieber, C. M. 2006. Fabrication of silicon nanowire devices for ultrasensitive, label-free, real-time detection of biological and chemical species. *Nat. Protoc.* 1: 1711–1724.

Prodanov, D.; Delbeke, J. 2016. Mechanical and biological interactions of implants with the brain and their impact on implant design. *Front. Neurosci.* 10: doi:10.3389/fnins.2016.00011.

Prohaska, O. J.; Olcaytug, F.; Pfundner, P.; Dragaun, H. 1986. Thin-film multiple electrode probes: Possibilities and limitations. *IEEE Trans. Biomed. Eng.* 33: 223–229.

Purves, D.; Augustine, G. J.; Fitzpatrick, D.; Katz, L. C.; LaMantia, A.-S.; McNamara, J. O.; Williams, S. M. 2001. *Neuroscience.* Sunderland, MA: Sinauer Associates.

Qi, S.; Yi, C.; Ji, S. et al. 2009. Cell adhesion and spreading behavior on vertically aligned silicon nanowire arrays. *ACS Appl. Mater. Interfaces* 1: 30–34.

Rochefort, N. L.; Jia, H.; Konnerth, A. 2008. Calcium imaging in the living brain: Prospects for molecular medicine. *Trends Mol. Med.* 14: 389–399.

Roder, P. B.; Smith, B. E.; Davis, E. J. 2014. Photothermal heating of nanowires. *J. Phys. Chem. C* 113: 1407–1416.

Sakmann, B.; Neher, E. 1984. Patch clamp techniques for studying ionic channels in excitable membranes. *Ann. Rev. Physiol.* 46, no. 1: 455–472.

Santoro, F.; Zhao, W.; Joubert, L.-M. et al. 2017. Revealing the cell–material interface with nanometer resolution by focused ion beam/scanning electron microscopy. *ACS Nano* 11: 8320–8328.

"The Nobel Prize in Physics 1956." n.d. Accessed April 28, 2018.

Tian, B.; Cohen-Karni, T.; Qing, Q.; Duan, X.; Xie, P.; Lieber, C. M. 2010. Three-dimensional, flexible nanoscale field-effect transistors as localized bioprobes. *Science* 329: 830–834.

Tian, B.; Liu, J.; Dvir, T. et al. 2012. Macroporous nanowire nanoelectronic scaffolds for synthetic tissues. *Nat. Mater.* 11: 986–994.

Tian, B.; Xie, P.; Kempa, T. J.; Bell, D. C.; Lieber, C. M. 2009. Single-crystalline kinked semiconductor nanowire superstructures. *Nat. Nanotechnol.* 4: 824–829.

Tian, B; Zheng, X.; Kempa, T. J. et al. 2007. Coaxial silicon nanowires as solar cells and nanoelectronic power sources. *Nature* 449: 885–889.

Tsai, D.; Sawyer, D.; Bradd, A. et al. 2017. A very large-scale microelectrode array for cellular-resolution electrophysiology. *Nat. Commun.* 8: doi:10.1038/s41467-017-02009-x.

Vessman, J.; Stefan, R. I.; Van Staden, J. F. et al. 2001. Selectivity in analytical chemistry. *Pure Appl. Chem.* 73: 1381–1386.

Williams, D. A.; Fogarty, K. E.; Tsien, R. Y. et al. 1985. Calcium gradients in single smooth muscle cells revealed by the digital imaging microscope using Fura-2. *Nature* 318: 558–561.

Xie, C.; Hanson, L.; Xie, W. et al. 2010. Noninvasive neuron pinning with nanopillar arrays. *Nano Lett.* 10: 4020–4024.

Yusko, E. C.; Asbury, C. L. 2014. Force is a signal that cells cannot ignore. *Molecular Biology of the Cell* 25: 3717–3725.

Zheng, G.; Patolsky, F.; Cui, Y. et al. 2005. Multiplexed electrical detection of cancer markers with nanowire sensor arrays. *Nat. Biotechnol.* 10: 1294–1301.

Zimmerman, J. F.; Murray, G. F.; Wang, Y. et al. 2015. Free-standing kinked silicon nanowires for probing inter- and intracellular force dynamics. *Nano Lett.* 15: 5492–5498.

Ptychographic Imaging of Biological Samples with Soft X-Ray Radiation

Max Rose
Deutsches Elektronen-Synchrotron DESY

Ivan A. Vartanyants
Deutsches Elektronen-Synchrotron DESY
National Research Nuclear University NRNU

X-RAY Ptychography is a lensless and coherent X-ray diffractive imaging method that is superior to conventional and lens-based X-ray imaging in various respects. Lenses are replaced by computer algorithms that have the advantage of less dose required to image at ultimate resolution.

With its short wavelength in the nanometer range, soft X-rays especially are used to image specimen, with nanometer resolution. From structural and chemical analysis channels provided by imaging, biologists may infer on the function of sub-cellular parts in cells and material scientists can study new functional materials used in advanced energy storage technology.

6.1 Introduction

A dream for biologists is to obtain high-resolution images of the building blocks of life, i.e. cells and skeleton, to understand complex structures on the nanoscale [5,25].

However, the spatial resolution of a microscope is typically limited by its lens and ultimately limited by its wavelength. The transmission of X-rays through objects enables the fascinating techniques of tomography without sectioning the sample. The most abundant elements in biological objects are carbon, nitrogen, oxygen and sulfur. They are located at the beginning of the periodic table with low number of electrons. Hard X-rays have wavelengths well below 1 nm but the interaction with matter is weak, possibly too weak to detect contrast in biological objects with sufficient accuracy. The scattering cross-section, which is used to characterize the interaction strength, for X-rays depends on the number of electrons associated with an atom [1]. With X-rays of a few nanometer wavelength, nanometer resolution may be achievable [24]. A solution to obtain enough scattered signal and sufficient resolution is to use soft X-rays. As it turned out, the water window, a special part of the soft X-ray spectrum, is a sweet spot for high-contrast and high-resolution imaging of biological objects like cells [53]. The price to pay for this unique spectral range comes in the form of experimental challenges of performing experiments under vacuum. This complicates sample handling significantly. A restriction is that only samples of a few tens of micrometer in thickness are transmissive and can be imaged.

The history of X-ray microscopy can be divided in roughly two main development branches [28]: The first one, adapting concepts from visible light microscopy, uses optical elements to form full images on a detector (conventional full-field imaging). Optical elements are notoriously difficult to manufacture with sufficient accuracy, and roughness and structure sizes smaller than the wavelength of the X-rays are required. High-quality X-ray mirror and lens development was enabled by nano-fabrication technology. As an alternative concept, so-called lensless approaches became popular, using coherent X-rays and eliminating optical elements altogether from the image forming process. Coherence is a required ingredient to facilitate image reconstruction from measured diffraction patterns. Here, large computing facilities and corresponding algorithm development gave a tremendous boost, and the progress is still ongoing. Meanwhile, both lensless and conventional techniques are advanced and both coexist with increasing number of applications.

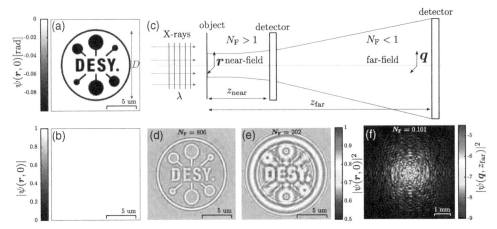

FIGURE 6.1   Near- and far-field diffraction of the DESY logo as a function of the Fresnel number. (a) Phase profile and (b) amplitude profile of the DESY logo. This test sample is a pure phase object which develops amplitude contrast images by propagation. (c) Diffraction geometry where X-rays impinge on the object and the resulting exit wave behind the object plane travels distance z to a detector. (d) The near-field image closely behind the object and (e) the near-field image at a distance z_{near} behind the object (both with linear color scale; gray in print). (f) The far-field diffraction pattern (logarithmic color scale; gray in print) after a large propagation distance z_{far}. Note that the near- and far-field regions depend also on wavelength λ and the object size D which are expressed by the Fresnel number $N_{\mathrm{F}} = \frac{D^2}{\lambda z}$.

A conceptual key advantage of lensless techniques is that no absorption losses occur at optical elements and eventually the undisturbed diffraction signal can be measured. With this, lensless techniques are very dose efficient. Typically, one wants to expose a sensitive biological object only to a minimum dose [23,26]. Another advantage of computational image reconstruction in lensless coherent diffractive X-ray imaging (CXDI) is the full complex valued object that gives access to two contrast mechanisms of phase and absorption. Among a variety of CXDI methods, ptychography aims to increase the limited field of view and the reliability of CXDI reconstructions.

6.1.1   Lensless X-Ray Diffractive Imaging

The demand for lensless imaging can be attributed to many problems arising with lenses such as limited resolution by the practical numerical aperture and aberrations. Most important are the encountered transmission losses and technological fabrication limitations of sub-nanometer precision for X-ray lenses.

In the general picture of free space propagation of a wave field diffracted from an object, one finds different types of intensity distributions. With a measurement of one of these intensity distributions, the object can be reconstructed by applying the mathematical inverse process of the propagation. Here the illumination probes an object that is described by a complex function

$$O(\boldsymbol{r}) = T(\boldsymbol{r}) \exp[\mathrm{i}\varphi(\boldsymbol{r})] \ . \tag{6.1}$$

Equation (6.1) contains the optical properties of transmission $T(\boldsymbol{r})$ and phase $\varphi(\boldsymbol{r})$ of the specimen.

The example in Figure 6.1 illustrates how to gain amplitude contrast by simple means of wave propagation from a pure phase object[1]. Here, the wave field propagation of a pure phase object $O(\boldsymbol{r}) = \exp[\mathrm{i}\varphi(\boldsymbol{r})]$ without amplitude variations (i.e. its transmission everywhere is $T(\boldsymbol{r}) = 1$) illuminated with a coherent plane wave probe function (i.e. $P(\boldsymbol{r}) = $ constant) is shown. The projection approximation in the description of X-ray interaction with sufficiently thin objects is used here [44]. As a result, the exit wave field can be formulated by multiplication of the probe with the object function

$$\psi(\boldsymbol{r}) = O(\boldsymbol{r})P(\boldsymbol{r}) \ . \tag{6.2}$$

In the object plane at $z = 0$, the resulting exit wave is equal to the phase object function $\psi(\boldsymbol{r}, z = 0) = \exp[\mathrm{i}\varphi(\boldsymbol{r})]$. The measurable quantity of a detector is the optical intensity $I(\boldsymbol{r}) = |\psi(\boldsymbol{r}, z_{\mathrm{near}})|^2$ which is depicted at two propagation distances in the near-field (see Figure 6.1d and e). Note, in the near-field the measured intensity has the same spatial coordinates as in the object plane, in the far-field; however, the spatial coordinates are scaled with respect to the propagation distance and the measurable far-field intensity is expressed by $I(\boldsymbol{q}) = |\psi(\boldsymbol{q}, z_{\mathrm{far}})|^2$. Here a coordinate transformation from the real space ($\boldsymbol{r}$) to the reciprocal space ($\boldsymbol{q}$) is implied by the reciprocity relation

$$\Delta q = \frac{2\pi}{D} \ , \tag{6.3}$$

where Δq is the smallest change in reciprocal space and D is the size of the object, i.e. the maximum coordinate in the object plane. The interpretation of the reciprocity relation is that in the far-field picture small objects (small D) will

---

[1]This is to a very good approximation fulfilled with hard X-rays as a microscopic object may be essentially transparent and free from absorption. For soft X-rays an upper boundary of sample thickness should be ensured.

cause diffraction signal with large range of Δq. The far-field is in many experiments the preferred region to measure the diffraction signal as it provides magnification of small structures. A simple Fourier transform relationship connects the exit wave and the far-field diffraction [16]

$$\psi(\boldsymbol{q}) = \mathcal{F}\{\psi(\boldsymbol{r})\} \ . \tag{6.4}$$

The factor 2π in (6.3) depends on the mathematical definition of the Fourier transform.

Interestingly, although the amplitude transmitted by the object was uniform and unity in Figure 6.1b, the propagated wave field shows intensity modulations as a consequence of interference. Interference however can only encode unique phase information into defined intensity modulations when the illumination is coherent. This was implied by a constant and non-fluctuating illumiantion function as mentioned above.

It is important to recognize that the far-field intensity contains only the squared magnitude of the complex valued far-field $|\psi(\boldsymbol{q})|^2 = I(\boldsymbol{q})$, where the full complex far-field is

$$\psi(\boldsymbol{q}) = |\psi(\boldsymbol{q})| \exp\left[\mathrm{i}\Phi(\boldsymbol{q})\right] \tag{6.5}$$

The phase exponential $\exp\left[\mathrm{i}\Phi(\boldsymbol{q})\right]$ associated with the magnitude is not directy accessible in the measurement. However, it is encoded as interference into the diffraction intensity. Obtaining the object function (assuming the illumination $P(\boldsymbol{r})$ is known) contained in Eq. (6.2) by inverse Fourier transform of only the magnitude $|\psi(\boldsymbol{q})| = \sqrt{I(\boldsymbol{q})}$ is not possible without the corresponding phase. This is called the phase problem. Efficient numerical implementations of the Fast Fourier Transform (FFT) algorithm can be used to retrieve the phases of the far-field diffraction in order to obtain the complex object function by inverse Fourier transform. Iterative phase retrieval algorithms have been developed to reconstruct the object and probe functions from far-field diffraction intensities [35,38].

A necessary requirement for phase retrieval is that the diffraction intensity is sufficiently sampled,[2] as stated by the Shannon sampling theorem [52]. This demands at least two samples Δq per reciprocal size $1/D$ of the illuminated object which is expressed by the reciprocity relation. Equation (6.3) can be understood as sampling criterion and if it is fulfilled, namely the object can be treated as isolated, i.e. surrounded by vacuum or another uniform medium, the unknown phases of the diffraction measurement can be reconstructed iteratively.

As described above, images can be obtained without lenses by reconstruction from coherent diffraction intensities when the interference can be resolved as stated by the sampling criterion. That the illumination is required to be known is an important conceptual weak point in many

experiments. This has practical implications that can be overcome by the scanning variant of CXDI called ptychography.

6.1.2 X-Ray Ptychography

X-ray ptychography is a novel lensless imaging technique that uses the principles of CXDI but acquires multiple diffraction patterns from overlapping illumination regions on the specimen. Ptychography originated in the electron microscopy community and was soon adopted for X-rays to overcome the specimen size limitations of CXDI and to facilitate a faster and more reliable[3] phase retrieval process [46,47]. A ptychographic data set consists of a stack of diffraction patterns obtained from translations $\boldsymbol{R}_j$ of the object with respect to the X-ray beam [12,45]

$$\psi_j(\boldsymbol{q}) = \mathcal{F}\{O_j(\boldsymbol{r})P(\boldsymbol{r})\} \ . \tag{6.6}$$

Here different parts of the object are illuminated which is reflected by $O_j(\boldsymbol{r}) = O_j(\boldsymbol{r} + \boldsymbol{R}_j)$, where j is the index of position. Important is that the translations are smaller than the size D of the probe in order to create sufficient overlap

$$o = 1 - \frac{s}{2D} \tag{6.7}$$

in the illuminations of the object. In this definition, s is the step size from one position to the next. A minimum of $o = 60\%$ was suggested as the optimal choice [3]. With this, the diffraction data contains redundant information to accelerate the phase retrieval process. X-ray ptychography has gained much attention, and many method advances have been reported in connection with X-ray experiments. The advantage is especially scanning specimens of arbitrary lateral size. In addition, more sophisticated algorithms allow for retrieval of the object and the probe at the same time [33,59,61] and make ptychography more versatile and robust than CXDI.

6.1.3 Soft X-Rays and the Water Window

In a classical description, the response of matter on incident X-rays depends on the refractive index

$$n_\omega(\boldsymbol{r}) = 1 - \delta_\omega(\boldsymbol{r}) + \mathrm{i}\beta_\omega(\boldsymbol{r}) \ . \tag{6.8}$$

Here, $\delta_\omega(\boldsymbol{r})$ represents the refraction, responsible for phase shift, and $\beta_\omega(\boldsymbol{r})$ the absorption of X-rays. Both are dependent on photon energy, where in the hard X-ray range $\beta_\omega/\delta_\omega \ll 1$. This means that image contrast due to phase shift dominates and absorption may be negligible. Also, when using photon energies above 12.4 keV (hard X-rays) the corresponding wavelength of 0.1 nm is on the same scale as inter-atomic distances and leads eventually to atomic

---

[2]Here sampling is performed by the pixels (typically 10–300 μm) of 2D X-ray image detectors.

[3]An estimate of the illumination function is retrieved during phase retrieval.

resolution information in diffraction experiments. While hard X-rays are favorable for ultimate resolution, the interaction strength with the specimen may be reduced dramatically, and thus the image contrast as well. Especially biological specimen consist mainly of light atoms and the interaction needs to be strong enough to resolve small changes in atomic compositions.

In the soft X-ray range, the refractive index is larger, and thus stronger interaction between X-rays and biological specimen will result in more reliable image contrast both in phase and absorption. For the research of biological specimens a rather strong X-ray–matter interaction is desired to observe contrast between organic compounds of often very similar refractive indices. A second demand is the maintenance of naturally aqueous sample environments that require the presence of water or solution in the specimen. In the so-called water window, defined by the K-absorption edges of carbon at 284 eV and oxygen at 532 eV [29], carbon-rich biological specimens such as proteins have a much stronger absorption than water, as shown in Figure 6.2. Here, for example, at 500 eV, the difference in absorption between water and protein is an order of magnitude and results in a valuable contrast between these two materials. Thus, soft X-rays in this region will offer a rather undisturbed view into a specimen through water or ice [54].

The wavelength in the water window (2.3–4.4 nm) is much larger than atomic distances. The expected resolution in scattering experiments is not on the atomic scale, but still up to a factor of 100 better compared to the visible light domain. The resolution combined with the opportunity of phase and absorption imaging of unstained biological samples in a natural environment constitutes a major attraction for biologists [22].

Despite the success of Fresnel zone plate–based X-ray microscopy [10,29,37], the radiation damage by X-rays due to ionization sets limitations [8]. A logical consequence of radiation damage is to search for dose- efficient microscopy

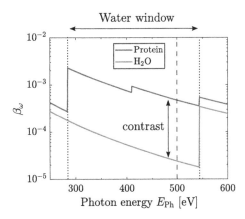

FIGURE 6.2 Absorption characteristics in the water window energy range. Here water (lower curve) is by a factor of ten more transparent than protein (upper curve). For the protein, a composition of $C_{94}H_{139}N_{24}O_{31}S$ with a density of $2.2 \, \text{g/cm}^3$ was assumed and data from [20] used to calculate its absorption.

techniques. From this point of view, lensless techniques are indispensable, because they eliminate any optical elements between the specimen and the detector [23]. The lensless CXDI technique was first demonstrated on a test object [39] and later on a biological cell [55]. A key advantage of CXDI and ptychography is the retrieval of absorption and phase information of the specimen. It has been shown that this leads to more reliable chemical specificity [56].

Ptychographic imaging using the water window energy range has been tested several times on fabricated and natural test objects [13,51]. Here, the relatively strong interaction was exploited to obtain high contrast images of fossil diatom samples that show similar phase characteristics compared to unstained biological cells. Recent research in the soft X-ray region demonstrated chemical sensitivity on fuel cell cathodes [67]. Here especially the capability of retrieved high-resolution absorption and phase images led to more precise results of the important cathode structure. Other recent ptychography work obtained reliable chemical information inside a frozen hydrated biological cell by additional hard X-ray fluorescence [7]. Here, a high resolution ptychography phase image was combined with the fluorescence signals of the elements S, P, K and Ca. Using hard X-rays, it also becomes possible to study crystalline materials by Bragg ptychography [15]. Recent works shown here retrieved strain information in nanowires [9]. The strain is crucial to the nanowire function as potential successors of solid state-light emitting devices.

Iterative phase retrieval algorithms are required to obtain images from ptychographic data sets. In the following, the basic concept of phase retrieval for ptychography is reviewed. Note that ptychography can be viewed as a generalization of CXDI for a laterally extended specimen.

6.2 Ptychography and Algorithms

Since the first X-ray ptychography experiments, a variety of iterative algorithms have been proposed to solve the phase problem, among them the extended ptychographic iterative engine (ePIE) [33], difference map (DM) [59,61] and maximum-likelihood refinement [42,60]. Common to all algorithms is that they apply a modulus constraint, also called the modulus projection

$$\pi_M \{\psi_j(\boldsymbol{q})\} = \sqrt{I_j(\boldsymbol{q})} \exp\left[\mathrm{i}\varphi_j(\boldsymbol{q})\right], \qquad (6.9)$$

where the measured diffraction intensity replaces the iteratively calculated modulus and only the calculated phases $\exp\left[\mathrm{i}\varphi_j(\boldsymbol{q})\right]$ are kept. In ptychography, this is done at each illumination position j on the object. Since an estimate of the complex valued far-field diffraction is given by Eq. (6.5), one can now use the Fourier transform relationship of Eq. (6.4) to switch between reciprocal space ($\boldsymbol{q}$ coordinates in the detector plane) and real space ($\boldsymbol{r}$ coordinates in the object plane).

Since we are interested in the real space reconstruction of the object, the definition of the modulus projection from the initial exit wave guess $\psi_j(\boldsymbol{r})$ to an updated exit wave $\widetilde{\psi}_j(\boldsymbol{r})$ is convenient

$$\widetilde{\psi}_j(\boldsymbol{r}) = \pi_{\mathrm{M}}^{\mathcal{F}}\{\psi_j(\boldsymbol{r})\} = \mathcal{F}^{-1}\{\pi_{\mathrm{M}}\{\mathcal{F}\{\psi(\boldsymbol{r})\}\}\} \quad (6.10)$$

where $\mathcal{F}$ and $\mathcal{F}^{-1}$ denote forward and inverse Fourier transform, respectively. After obtaining the updated exit wave, we want to separate object and probe function and perform an update on both of them separately. This task now is different among the variety of algorithms mentioned above. In the following, the ePIE algorithm is used to illustrate the iterative phase retrieval procedure. Two cases will be considered: Fully coherent illumination and partial coherent illumination. This results in different update rules for object and probe.

6.2.1 Fully Coherent Illumination

In the case of fully coherent illumination, a single probe function needs to be updated at each position on the object. The ePIE probe update rule is [33]

$$\widetilde{P}(\boldsymbol{r}) = P(\boldsymbol{r}) + \beta_{\mathrm{P}}\frac{O_j^*(\boldsymbol{r})}{(|O_j(\boldsymbol{r})|^2)_{\max}}\Delta\psi_j(\boldsymbol{r}) . \quad (6.11)$$

The update rule contains the previous estimate of probe $P(\boldsymbol{r})$ and a correction term that is added to form the updated quantity weighted by a feedback factor β_{P} (a value between 0 and 1). The complex conjugated object $O_j^*(\boldsymbol{r})$ is normalized by its maximum absolute square value $(|O_j(\boldsymbol{r})|^2)_{\max}$ and thus separates the object from the probe

function update. The most important part of the correction term is the exit wave difference $\Delta\psi_j(\boldsymbol{r})$ from the previous and updated exit wave

$$\Delta\psi_j(\boldsymbol{r}) = \widetilde{\psi}_j(\boldsymbol{r}) - \psi_j(\boldsymbol{r}) . \quad (6.12)$$

Analogous to the probe update, one can formulate the object update as [33]

$$\widetilde{O}_j(\boldsymbol{r}) = O_j(\boldsymbol{r}) + \beta_{\mathrm{O}}\frac{\widetilde{P}^*(\boldsymbol{r})}{\left(|\widetilde{P}(\boldsymbol{r})|^2\right)_{\max}}\Delta\psi_j(\boldsymbol{r}) . \quad (6.13)$$

Note that here the feedback factor β_{O} (a value between 0 and 1) can be adjusted independently from the probe update. Also note that the probe is updated first, and then $\widetilde{P}(\boldsymbol{r})$ will be used in the object update rule (Figure 6.3).

6.2.2 Partially Coherent Illumination

In spite of their name, coherent X-ray diffractive imaging and ptychography rely on highly coherent illumination. However, partial coherence needs to be accounted for. This leads to relaxed experimental conditions. Partial coherence can be efficiently described in the frame-work of coherent modes. By a superposition of coherent modes, a partially coherent wave field can be composed. In the context of ptychography, advanced algorithms have been developed that use such coherent modes approach [62].

The following sections are of instructive character and intended to facilitate an implementation of coherent modes into the ptychographic algorithm. While the experienced

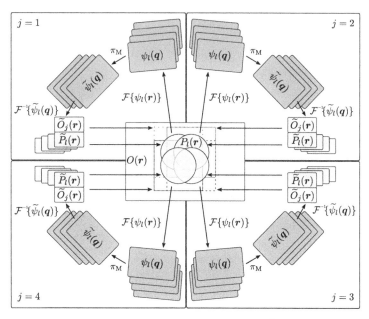

FIGURE 6.3 Block diagram of ptychographic phase retrieval algorithm for full ($l = 1$) and partial coherence ($l > 1$). Real space ($\boldsymbol{r}$) of probe P_l and object O are denoted by light gray boxes. Reciprocal space is denoted by dark gray boxes. With forward ($\mathcal{F}$) and inverse Fourier transform ($\mathcal{F}^{-1}$) the transition between real and reciprocal space is achieved. The modulus constraint (π_{M}) is applied in reciprocal space. The coherent modes (l) are visualized by stacks of boxes which are required in real and reciprocal space. At each scan position j, the same cycle of Fourier transforms, modulus constraint and factorization into probe and object is used to produce updated quantities denoted by the $\widetilde{}$ symbol.

reader may appreciate this section, for a more broad overview of ptychographic imaging results, the reader may skip to Section 6.3.

Coherent Modes

In the coherent mode description, the measured intensity at position j from a ptychographic experiment is the incoherent sum of the individual modes (indexed with l)

$$I_j(\boldsymbol{q}) = \sum_l \beta_l |\mathcal{F}\{\psi_{j,l}(\boldsymbol{r})\}|^2 \;, \qquad (6.14)$$

with the weights β_l of the modes and the exit wave field depending on the l-th mode of the probe

$$\psi_{j,l}(\boldsymbol{r}) = P_l(\boldsymbol{r})O_j(\boldsymbol{r}) \;. \qquad (6.15)$$

A numerically efficient method for decomposing a partially coherent probe wave field into coherent modes and obtaining their weights is the singular value decomposition (SVD). For practical reasons, the SVD used for mode decomposition is truncated to the number of modes desired. This is very fast from a computational point of view and gives the largest singular values corresponding to the probe modes of most importance.

Coherent Mode Decomposition with SVD

The mutual intensity (assuming quasi-monochromatic radiation) of the incident illumination relates to the mutual intensity by

$$J(\boldsymbol{r}_1, \boldsymbol{r}_2) = \sum_l P_l(\boldsymbol{r}_1)P_l^*(\boldsymbol{r}_2) \;. \qquad (6.16)$$

Assuming quasi-monochromatic radiation leads to a significant simplification of the cross-spectral density (CSD). In this case, the CSD $W(\boldsymbol{r}_1, \boldsymbol{r}_2)$ and the mutual intensity $J(\boldsymbol{r}_1, \boldsymbol{r}_2)$ are equal. Then the mutual intensity can be expressed as the sum of the coherent modes via

$$J(\boldsymbol{r}_1, \boldsymbol{r}_2) = W(\boldsymbol{r}_1, \boldsymbol{r}_2) \qquad (6.17)$$

$$= \sum_l \beta_l(\omega)P_l^\perp(\boldsymbol{r}_1)P_l^{\perp*}(\boldsymbol{r}_2) \;, \qquad (6.18)$$

where $P_l^\perp(\boldsymbol{r}_1)$ and $P_l^\perp(\boldsymbol{r}_2)$ are orthonormal modes of the probe. Each probe mode expressed as 2D matrix contains a total number of $\mathcal{N} = N^2$ elements representing a 2D complex valued amplitude distribution on a plane of the radiation. In matrix notation, Eqs. (6.16) and (6.17) can be rewritten as [43]

$$\mathcal{P}\mathcal{P}^* = \mathcal{P}^\perp \mathcal{B} \mathcal{P}^{\perp*} \;, \qquad (6.19)$$

where $\mathcal{P}$ contains all modes reshaped into columns so that

$$\mathcal{P} = \begin{bmatrix} P_1(r_1) & \cdots & P_L(r_1) \\ \vdots & \ddots & \vdots \\ P_1(r_\mathcal{N}) & \cdots & P_L(r_\mathcal{N}) \end{bmatrix} \qquad (6.20)$$

with the dimensions $\mathcal{N} \times L$, where L is the total number of modes. The matrix $\mathcal{P}^\perp$ contains columns of orthonormal

modes of the probe and $\mathcal{P}^{\perp*}$ is the conjugate transpose of $\mathcal{P}^\perp$. All mode weights β_l are represented in the diagonal, positive and real valued matrix

$$\mathcal{B} = \begin{bmatrix} \beta_1 & 0 & 0 \\ 0 & \ddots & 0 \\ 0 & 0 & \beta_L \end{bmatrix} \;. \qquad (6.21)$$

Solving Eq. (6.19) is computationally expensive but usually not required as only the most dominant modes significantly contribute to the partial coherent probe in the case of synchrotron and XFEL radiation [63].

The SVD method decomposes a given matrix into its orthonormal components by

$$\mathcal{P} = \mathcal{U}\mathcal{S}\mathcal{V}^* \;. \qquad (6.22)$$

Here, $\mathcal{U}$ is a unitary and orthogonal complex valued matrix of dimensions $\mathcal{N} \times L$, $\mathcal{V}$ is a unitary and orthogonal complex valued matrix of dimensions $L \times L$, and

$$\mathcal{S} = \begin{bmatrix} \sigma_1^{\mathrm{SVD}} & 0 & 0 \\ 0 & \ddots & 0 \\ 0 & 0 & \sigma_L^{\mathrm{SVD}} \end{bmatrix} \qquad (6.23)$$

is a diagonal matrix of dimensions $L \times L$ containing so-called singular values which are positive and real. The above shown definition corresponds to a compact SVD that decomposes into a low rank of L. This is less computationally expensive compared to a full rank SVD and matches the demands of retrieving only the largest singular values.

In order to obtain a set of orthogonal and thus coherent modes, a multiplication of the orthonormal matrix with the singular values has to be used

$$\mathcal{P}^\perp = \mathcal{U}\mathcal{S} \;. \qquad (6.24)$$

As it turns out, one can reproduce the mutual intensity with

$$\mathcal{P}\mathcal{P}^* = \mathcal{U}\mathcal{S}^2\mathcal{U}^* \;, \qquad (6.25)$$

where now the singular values and the coherent mode weights are related by

$$\mathcal{B} = \mathcal{S}^2 \qquad (6.26)$$

with their elements

$$\beta_l = \left(\sigma_l^{\mathrm{SVD}}\right)^2 \qquad (6.27)$$

equal to the square of the singular values of the SVD of the probe function. Now given a partially coherent illumination, it can be replaced by a number of coherent modes which can be updated independently.

Partial Coherent ePIE Update Rules

In case of partially spatial coherent illumination, the update rules for a partially coherent probe have to be adjusted according to [62]. The probe is then represented by an ensemble of probe modes $P_l(\boldsymbol{r})$ with the update for each mode according to

$$\widetilde{P}_l(\boldsymbol{r}) = P_l(\boldsymbol{r}) + \beta_P \frac{\sum_l O_j^*(\boldsymbol{r}) \Delta \psi_{j,l}(\boldsymbol{r})}{(|O_j(\boldsymbol{r})|^2)_{\max}} . \quad (6.28)$$

Analogously, the object update is defined with

$$\widetilde{O}_j(\boldsymbol{r}) = O_j(\boldsymbol{r}) + \beta_O \frac{\sum_l P_l^*(\boldsymbol{r}) \Delta \psi_{j,l}(\boldsymbol{r})}{\left(\sum_l |P_l(\boldsymbol{r})|^2\right)_{\max}} . \quad (6.29)$$

In the object update a sum over all modes l is used to create a single object function. In both update functions Eq. (6.28) and Eq. (6.29), an exit wave difference for each coherent mode is used

$$\Delta \psi_{j,l}(\boldsymbol{r}) = \widetilde{\psi}_{j,l}(\boldsymbol{r}) - \psi_{j,l}(\boldsymbol{r}) , \quad (6.30)$$

where the updated wave field is

$$\widetilde{\psi}_{j,l}(\boldsymbol{r}) = \pi_M^{\mathcal{F}} \{ \psi_{j,l}(\boldsymbol{r}) \} . \quad (6.31)$$

A partially coherent probe adds complexity to the algorithms in the sense of a stack of probe modes. This inevitably demands more computational memory. The probe update for each mode has to be performed. This can be done in parallel. The computational memory demand thus scales with the number of modes.

The description of ptychographic algorithms thus far was focused on the well-studied ePIE algorithm. The reason for this is that a number of systematic investigations on the performance exist for this algorithm as well as many scientific works are based on it [32,33,45]. There are a number of other algorithms [18,36,59] that share a global solution search (gradient descent approach), which is different from the local solution search (incremental gradient descent approach) of ePIE.

Due to the wide usage of ePIE in ptychography experiments, there are also a number of cases where the algorithms fail to retrieve plausible reconstruction results. This was investigated and modifications were suggested to make the algorithms of the PIE family more robust, more rapidly converging to a solution or simply capable of reconstructing some pathological cases where ePIE fails completely [32].

The extension to partial spatial coherence leaves additional degrees of freedom for the algorithm. One could argue that, the redundancy of ptychographic data is in fact limited, and as such cannot solve for all cases of partial coherence. However, in cases of synchrotron undulator X-ray experiments the degree of coherence is by definition rather high. This suggests that one can effectively use the data instead of using only the object reconstruction of the most coherent part of the data. This is reflected by using only the most dominant coherent mode in the object update

$$\widetilde{O}_j(\boldsymbol{r}) = O_j(\boldsymbol{r}) + \beta_O \frac{P_{l=1}^*(\boldsymbol{r}) \Delta \psi_{j,l=1}(\boldsymbol{r})}{(|P_{l=1}(\boldsymbol{r})|^2)_{\max}} . \quad (6.32)$$

Temporal coherence in ptychographic phase retrieval with coherent modes was discussed in [11].

Partial Coherent Illumination Reconstruction

Using a partial coherent ptychography algorithm helped to mitigate the impact of partial coherence on the image reconstruction results of a Fibroblast cell [50]. As a result, the probe can be decomposed into coherent modes. This can be done by the SVD orthogonalization method described in Section 6.2.2. This resulted in the following probe mode weights (sum normalized to unity): $\beta_1 = 0.741$, $\beta_2 = 0.132$, $\beta_3 = 0.067$, $\beta_4 = 0.033$ and $\beta_5 = 0.028$. The reconstructed five complex valued probe modes $P_l(\boldsymbol{r})$ are shown in Figure 6.4. A dominant mode is observed (Figure 6.4a) and the two following modes (Figure 6.4b and c) are clearly orthogonal to each other but much weaker with contributions of 13.2 % and 6.7 %, respectively. The last two modes have contributions of less than 4 % (Figure 6.4d and e). They contain fringes that are resolved nearly to the pixel level. The number of modes is limited to five since the energy contribution of further modes is less than one per cent. Assuming all modes originate from the partial coherent nature of the incoming X-ray beam, we can calculate the degree of transverse coherence from the mode weights by [63]

$$\zeta = \frac{\sum_l \beta_l^2}{(\sum_l \beta_l)^2} . \quad (6.33)$$

With the above weights, the degree of coherence of the X-ray beam passing through a 5 µm pinhole was 57 % [50].

6.2.3 Algorithm Implementations

There is a demand to use ptychography more routinely, especially in X-ray facilities, and some efforts can be observed to deliver software packages for users. Some documented implementations exist that focus on user interface, a reduced number of tuning parameters and of course reconstruction speed by parallelization [36,40]. At the same time, ptychographic algorithms are themselves the subject of ongoing research and a large number of undocumented implementations in a variety of programming languages may exist to satisfy the demand for flexibility of new options or systematic studies [11,34]. As the data sets increase in size with ever larger areas and volumes scanned and at the same time a

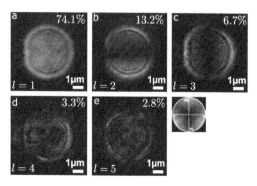

FIGURE 6.4 Probe modes and corresponding weights. The global degree of coherence is 57 %. (Figure adapted from *Optics Express* [50].)

higher resolution is sought, this demands for clever programming schemes to cope with the computational complexity [41] or even data compression strategies that may become relevant [31].

6.3 X-Ray Ptychography Experiments

6.3.1 Imaging of Biological Samples with Ptychography

Imaging of unstained cells and biological tissues with X-rays within the water window spectrum offers quantitative analysis routes for biology. X-ray microscopy provides higher spatial resolution than conventional visible light microscopy due to the comparably short wavelength of X-rays. Opposed to high-resolution electron microscopy, the longer absorption length of X-rays allows imaging the interior of comparably thick samples in the micrometer range.

The soft X-ray range offers unique imaging opportunities for high-contrast bio-imaging. It provides a maximum contrast between biological and aqueous components. Remarkably high contrast can be obtained without any staining for additional contrast enhancement [30,54,55]. A number of technical challenges had and still have to be solved in soft X-ray ptychographic imaging for biological samples. Among them is the limited dynamic range of the pixelated image detectors. In order to capture the large dynamic range of the diffraction signal, high dynamic range modalities are important as described below. The coherent mode decomposition relaxes the experimental constraints with respect to the X-ray source. This is illustrated by a coherence analysis of the X-ray illumination presented along with the reconstruction of a biological sample. That the ptychography method retrieves quantitative results is illustrated in several examples. It stems from the fact that the full optical properties of absorption and phase of an object can be retrieved and recalculated into mass and electron density or even into chemical maps.

Water window ptychography, which will be described in the following sections, is among the more challenging experiments because many additional experimental conditions not related to the X-ray source have to be met, i.e. a vacuum chamber with stable nano-positioning system, large area detector with small pixels and high dynamic range and eventually also a cryogenic sample holder. This effort is rewarded with high contrast images that is difficult to obtain by using hard X-rays.

High-Resolution Soft X-Ray Diffraction

The method of multiple exposures and beam stops (BS) is often used to extend the limited detector dynamic range (DR). This is necessary to record the full diffraction pattern intensities that typically span over several orders of magnitude. Two exposures may then be merged into a single high-dynamic range (HDR) diffraction pattern by using the

rescaled region of the exposure without beam and place these values into the exposure with beam stop [49]. The extension in DR is given by the exposure time ratio

$$DR_{ext} = \frac{t_{BS}}{t_{noBS}} . \tag{6.34}$$

The DR extension based on multiple exposures will inevitably increase the incident dose on the sample. However, the dose scales with the exposure time and as such the long exposure (with BS) crucially determines the dose. For a dynamic range extension of $DR_{ext} = 100$, one finds that the short exposure adds only 1 % of the dose applied by the long exposure. With this argument, the dose is not significantly increased by the multiple exposure scheme compared to a single long exposure.

The double exposure scheme to enhance the dynamic range (DR) of the detector was used in the measurements of a fibroblast cell [50]. In total, two scans were acquired with a fixed exposure time for each scan. The first scan consisted of short exposures (0.4 s) without beam stop (Figure 6.5a). With this scan, a spatial frequency of $3.1\,\mu m^{-1}$ was obtained. The second scan consisted of longer exposures (3 s) with beam stop and slightly increased beamline slits for a ten times higher photon flux ($\Phi_{ph}^{Pinhole} = 10^9$ ph/s) (Figure 6.5b). With beam stop, the maximum spatial frequency was $11\,\mu m^{-1}$. The double exposure HDR data acquisition including the two scans took about 25 min. The significant overhead of 10 s per exposure is imposed by the detector readout rate and added about 2.5 h to the full data acquisition time. The stability for long exposures and large ptychographic scans is explicitly addressed by the mechanical design of the HORST chamber and its sample holder [17]. A circular, semitransparent golden beam stop was used with a diameter of 1 mm and a thickness of 200 nm, sputtered on a 500 nm thick SiN membrane that is glued on all sides to a motorized metal frame.

After background subtraction, both scans are merged into a HDR data set. The DR of the single exposure data set without beam stop was 2.3×10^4 and the enhanced DR of the acquired HDR data set was 1.7×10^6 as shown in Figure 6.5. The combination of two exposures per position extended the effective dynamic range of our detector by roughly two orders of magnitude. As described, a drawback may be the overall longer time to acquire data. The benefit of using the beam stop is, however, that high-resolution HDR signal up to the detector edges can be measured.

Diatom Skeleton

Diatoms are organisms found abundantly in oceans. Their skeleton consists of silicon dioxide and presents itself as an ideal biological test object for ptychography. The fossil diatom dataset (single exposure) consisted of 119 diffraction patterns (17 horizontal × 7 vertical) covered an area of $14.4\,\mu m \times 6.4\,\mu m$ [51]. The reconstructed amplitude of the fossil diatom is shown in Figure 6.6a. The reconstructed and unwrapped phase of the sample is shown in Figure 6.6b. Here the color scheme displays the relative phase shift map

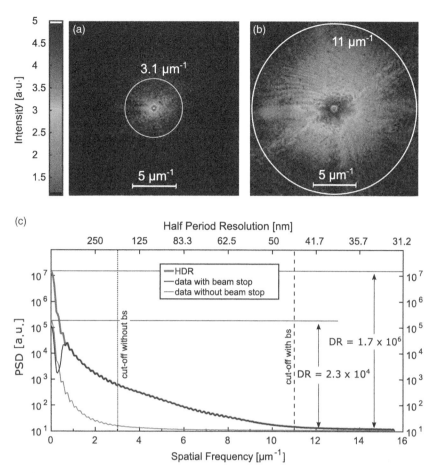

FIGURE 6.5 Ptychographic diffraction data from fibroblast cell specimen. (a) Diffraction pattern of a single scan position without beam stop. (b) Diffraction pattern with semi transparent beam stop. (c) Angular average of both exposures indicating the DR for the single exposure and the merged HDR data. (Figure adapted from *Optics Express* [50].)

$\Delta\varphi(x,y)$ between the substrate (shown in white) and the diatom. Assuming uniform density of SiO_2 ($\rho = 2.2\,\mathrm{g/cm^3}$), the refraction coefficient δ at 500 eV photon energy was determined with $\delta = 1.39 \cdot 10^{-3}$ [20]. This converted the relative phase shift $\Delta\varphi(x,y)$ between the substrate and the diatom to the projected material thickness by applying the relation $\Delta z(x,y) = \Delta\varphi(x,y)/(k\delta)$. Diatoms of the *Fragilariopsis cylindrus* species have a cylindrically curved shape [27]. From the two-dimensional projection image, the integrated SiO_2 mass along the depth of the diatom with $m(x,y) = \rho \cdot \Delta z(x,y)$ was deduced. The (x,y) coordinates denote the discrete pixels of the reconstruction with a pixel size of 53 nm × 53 nm. In Figure 6.6c, the surface plot of the integrated SiO_2 mass is shown up to a threshold of 3.5 fg (femto gram) to preserve the visibility of the fine structure.

Ten equidistantly spaced ribs are visible with a period of 1 μm. The length and width of each rib was estimated with 3 μm and 250 nm, respectively. The fine structure that appears in the form of a perforation and which is well pronounced in the amplitude image in Figure 6.6a had a period of 200 nm in the vertical direction. All the discovered features of the fossil diatom are specific for the species *Fragilariopsis cylindrus* and are comparable with earlier studies [2,27].

From line scans, which are extracted from the phase reconstruction, the FWHM values of the fitted error functions were determined (see Figure 6.6d). Using the FWHM values, the half-period resolution of the ptychographic diatom reconstruction was just below 90 nm and, consequently, 30 % better than in a previous experiment [13].

Fibroblast Cell

Despite the success of test samples like the diatom, the image of a more relevant biological fibroblast cell based is described in the following. Here the double exposure beam stop method is mandatory to obtain high resolution (see Figure 6.5). A visible light microscopy image of the fibroblast cell [50] under investigation is shown in Figure 6.7a. The white square indicates the ptychographically scanned region. Phase wrapping is a known problem for phase shifts larger than 2π and occurs also in the ptychographic reconstruction result. Here an automatic phase unwrapping algorithm was used [21]. The unwrapped high-resolution phase image is shown in Figure 6.7b. An enlarged region that shows the details of fine features can be found in Figure 6.7b. The large oval nucleus (marked as **N**) is visible at the top of the fibroblast image. Inside the nucleus, evidence of a protein- and RNA-rich nucleolus (a ribosome synthesis site, marked

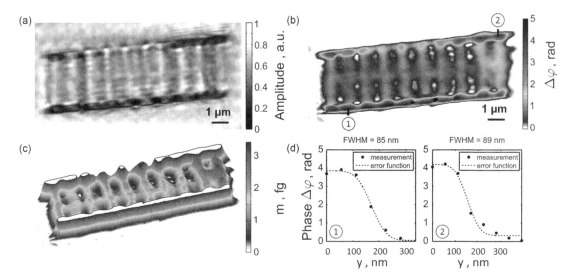

FIGURE 6.6 Ptychographic reconstruction of fossil diatom skeleton structure (a) and the phase (b) of the fossil diatom. (c) Integrated SiO_2 mass along the depth of the diatom (see main text for further details) and (d) FWHM values of two error function fits along the black lines indicated in the phase reconstruction in (b). Figure adapted from *Journal of Synchrotron Radiation* [51].

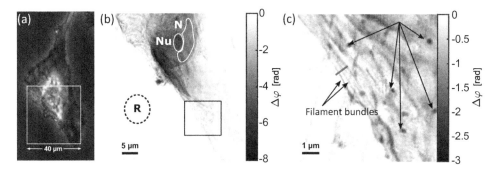

FIGURE 6.7 Fibroblast images obtained from ptychographic reconstruction. (a) Visible light microscopy image for comparison. (b) Image with large FOV $40\,\mu m \times 40\,\mu m$ of the fibroblast cell ptychographic reconstruction. Here, nucleus (N) and nucleolus (Nu) are encircled in white. An empty reference area (R) is indicated for calculating the absorbed dose of the cell. (c) Enlarged view and of the lower right part of the cell showing fine details of the cellular structures. The black line indicates the profile taken for contrast analysis of periodic ultrastructure of the cell. Figure adapted from *Optics Express* [50].

as Nu) with a size about $2\,\mu m$ by $2\,\mu m$ was found. Individual proteins or RNA inside of the nucleolus were not resolved in the ptychographic reconstruction.

In the enlarged view of the reconstructed cell image seen in Figure 6.7c, one finds irregular distributed dots with diameters between $225\,nm$ and $360\,nm$, which suggested the presence of shedding vesicles (they play an important role in exocytosis [58]). In addition, parts of the fibrous cytoskeleton are visible. As typical for eucaryotic cells, those fibrous elements consist mainly of actin and vimentin, intermediate and micro-filament proteins. It is well known that individual filaments of a typical size of $10\,nm$ form aligned bundles with a size between $50\,nm$ and $100\,nm$ [19]. Although single filaments were not resolved, the analysis of aligned bundles of these filaments was feasible. The visibility or contrast C of image features with a periodic pattern can be described by

$$C = \frac{\varphi_{\max} - \varphi_{\min}}{\varphi_{\max} + \varphi_{\min}}. \qquad (6.35)$$

Here, $\varphi_{\max}$ and $\varphi_{\min}$ denote the average maximum and minimum phase values along the black line profile from Figure 6.7c. A contrast of $C = 0.53$ is found at a spatial frequency of $7.1\,\mu m^{-1}$ and resolved filament bundles with an average width of $71\,nm$ [50].

To quantify the obtained resolution, an evaluation using the *phase retrieval transfer function* (PRTF) [6] is helpful. The PRTF is the ratio of the reconstructed and measured modulus as a function of spatial frequency that can be adopted for ptychography [65]. In order to take into account the decreasing signal to noise ratio for increasing spatial frequency, the Wiener-filtered PRTF (wPRTF) was used [57]. This resulted in a resolution of $250\,nm$ and $54\,nm$ for the normal and HDR data, respectively [50].

The general limits in resolution, however, are imposed by structural changes during the exposure as a consequence of radiation damage [8,22]. Ptychography and other coherent imaging techniques are, by principle, dose efficient because no lenses are used between the sample and detector [23].

In the described experiment, the aim was to use the minimum dose in agreement with the resolution demand. In the experiment, the dose was estimated by adding up the doses with and without beam stop for both exposures to obtain the total dose per illumination position of $D_{\text{Nu}}^{\text{total}} = 9.5 \times 10^4$ Gy applied in the region of the cell nucleus. This dose value was far from the destructive regime [22]. No significant changes in the diffraction patterns during the exposure were observed. This leads to the conclusion that no radiation damage effects are induced. The absorbed dose on the sample in the experiment is similar to the reported dose values for the resolution range between 70 nm and 90 nm [22]. With these arguments, the resolution estimate from the contrast analysis (71 nm) of the periodic image features of the ptychographic fibroblast cell reconstruction are validated. The configuration of the experiment shows that the resolution limitation is mainly given by the detector size, the detector dynamic range and the radiation dose. Better resolution down to the radiation damage limit at 10 nm may be expected by technological advances in soft X-ray detectors.

Tomographic Ptychography Reconstruction of a Yeast Cell

Ptychographic reconstructions show projection images, and more significant results about biological samples can be expected from tomographic reconstructions using many projections [48].

The step from projected electron density images towards electron density distribution has been demonstrated using hard X-ray ptychographic images from several rotations of a cryo-preserved yeast cell [14]. As an experimental simplification, hard X-rays (7.9 keV) instead of soft X-rays were used. Although this implies a weaker interaction with the sample and less contrast of the cell to the amorphous ice matrix, the sample has not necessarily to be mounted and kept under vacuum. Hard X-rays are also mandatory when samples are mounted on flat-surface holders that result in increasing effective background thickness upon large rotation angles. In order to keep the sample frozen during the experiment, a cryogenic gas jet was used. The cryo-preservation ensured a radiation-protected environment. The tomographic series required an overall dose of 1.1×10^8 Gy that was much larger compared to the average dose of a single projection image of 1.8×10^6 Gy.

The tomographic electron density distribution in Figure 6.8 was reconstructed from a series of ptychographic projections of the yeast cell. The most significant detail is the internally visible low density site inside the yeast cell that was difficult to identify from two-dimensional projections only. Despite the extension to three-dimensional renderings, which help identifying internal subunits inside the cell, the conservative resolution estimate using Fourier shell correlation of the tomographic reconstruction with 234 nm is rather low. In order to reach better resolution, a straight-forward solution using higher dose was suggested.

Combined Ptychography and X-Ray Fluorescence Reconstruction of a Chlamydomonas Alga

Besides tomographic approaches, the combination of ptychographic imaging with X-ray fluorescence (XRF) complements projection images with relevant biological information of important subunits in cells [64].

Hard X-ray (5.2 keV) ptychography and simultaneous XRF mapping of a frozen hydrated Chlamydomonas alga serves as excellent example [7]. The frozen hydrated state of the alga cell preserves the structural integrity and adds stability against radiation damage. Due to the simultaneous acquisition of diffraction patterns and the XRF signal, the elemental information does not need additional alignment procedures in order to correlate the images from both techniques as shown in Figure 6.9.

In this example four XRF signals, for sulfur (S), phosphor (P), potassium (K) and calcium (Ca) were obtained. While the XRF signal resolution corresponds directly to the X-ray illumination size (100 nm), the ptychographic image resolution was much better with 17.8 nm. Ptychography also retrieves the X-ray probe function and deconvolution techniques were applied in order to enhance the contrast in the XRF images. The correlative overlay of all image channels yields a comprehensive image with elemental specificity that contains significant biological information in a single projection image.

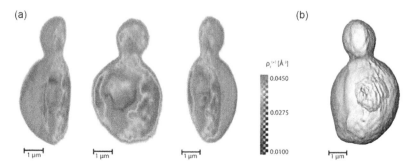

FIGURE 6.8 Tomographic reconstruction from X-ray ptychography projection images. (a) Shows the electron density mean in the range of $0.01\,\text{Å}^{-3}$ to $0.045\,\text{Å}^{-3}$ and (b) an isosurface at $0.015\,\text{Å}^{-3}$. (Figure adapted from *BioPhysical Journal* [14].)

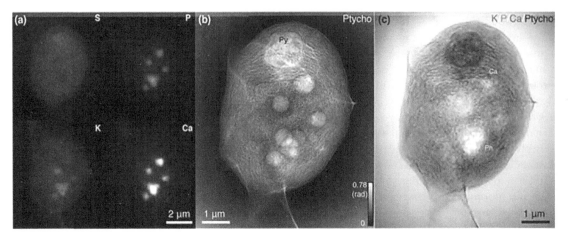

FIGURE 6.9 Combined X-ray fluorescence and ptychography of chlamydomonas alga. (a) XRF images corresponding to the four elements S, P, K and Ca. (b) High-resolution ptychography projection image and (c) co-aligned XRF and ptychography image. (Figure adapted from *Scientific Reports* [7].)

6.3.2 Imaging of Functional Materials with Ptychography

New functional materials like battery electrodes and catalysis are less radiation-sensitive specimens. Using soft X-rays, absorption edges of metals can be probed and also used as complementary chemical information beside the structural image.

Quantitative Reconstructions of a Catalyst Particle

Catalysts play an important role in life science, and soft X-ray ptychographic imaging provides sufficient resolution to investigate the degradation of catalyst particles. In addition to high resolution also, spectroscopic channels can be used to trace iron in such materials [66].

Here spectroscopic information was gained by using multiple photon energies between 704 eV and 710 eV to trace iron (Fe) at the L_3 absorption edge and 830 eV and 834.5 eV to trace lanthanum (La) at the M_5 absorption edge. With multiple photon energies in combination with ptychography, elemental maps at resolution between 12.2 nm to 14.2 nm were achieved. Quantitative spectroscopic results can be obtained by analysis of the per pixel NEXAFS data.

The ptychographic experiment yielded direct visual evidence for the catalyst particle deactivation process, as seen from Figure 6.10. It was found that Fe clogs the pores of the catalyst particle which shows reduced catalysis activity. This example demonstrates the unique high-resolution capabilities of soft X-ray ptychography with elemental specificity.

Quantitative Reconstructions of Battery Reactions

While spectroscopic projection images obtained through ptychography help obtain element specific maps, their information content is limited. Much more precise conclusions can be drawn from three-dimensional chemical distributions. This was tested by using spectroscopic ptychography with

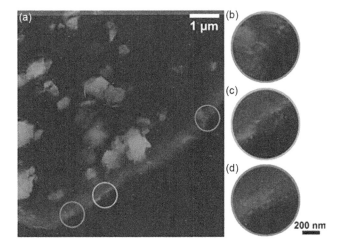

FIGURE 6.10 Ptychographic catalyst particle projection image with element specific color for La (light gray) and Fe (dark gray). (a) Shows the edge of the particle and (b-d) enlarged sites of Fe contamination. (Figure adapted from *ACS Catalysis* [66].)

many rotation angles in order to produce a tomographic data set.

Tomographic X-ray spectro-ptychography (708.2 eV and 710.2 eV) was demonstrated in an experiment that observed battery reactions and the corresponding chemical phases that determine the material properties [68]. Due to the comparably large probed volume with a resolution of 11 nm, this technique provides great opportunities for the study of ensemble statistics of the active structures in the battery electrodes as seen in Figure 6.11. In this example, particles of lithium iron phosphate ($LiFePO_4$) and their oxidation states were investigated.

The results suggested that the detected internal heterogeneous chemically state structure of the particles, as shown in Figure 6.11, plays an important role in battery function. The class of targeted particle sizes smaller than 100 nm was claimed to play a more important role in the chemical reaction than larger particles that were investigated previously

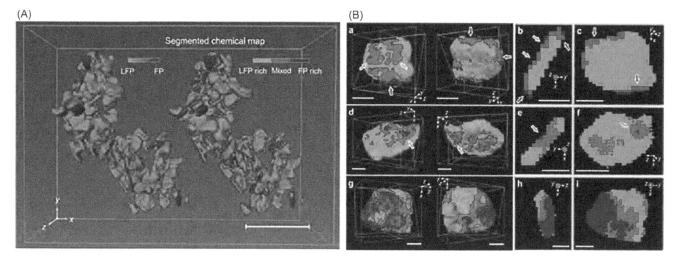

FIGURE 6.11 Chemical state mapping of battery electrode particles using tomographic reconstruction from multiple ptychography images. (A) Ensemble of particles with color encoded chemical phase. (B) Chemical segmentation map for a single particle of 100 nm in size. All scale bars indicate 50 nm. (Figure adapted from *Nature Communications* [68].)

by other methods. Such small particles are difficult to study with sufficient spatial resolution and chemical information at the same resolution. Here, spectroscopic soft X-ray ptychography offers quantitative information on structure and chemical state.

6.4   Summary

Ptychography is a variant of coherent imaging techniques that uses multiple far-field diffraction patterns from overlapping illuminations on an object. The far-field offers a reciprocal relationship of sizes where objects and their microscopic features produce diffraction signal at large angles, which is essentially the magnification that is a demand in microscopy. Using pixelated image detectors, the coherent diffraction intensity pattern can be used to retrieve an estimate of the phase that is required to reconstruct the complex valued specimen by simple Fourier transform.

A necessary condition for the phase retrieval process is sufficient sampling of the diffraction pattern. The sampling is reciprocal to the size of the sample to be imaged. This binds the size of an object to a given pixel size of a pixel detector. In ptychography, this limitation is relaxed because a sufficiently small X-ray illumination can be used to scan an arbitrarily large area. Many diffraction patterns from overlapping positions are measured where the sampling criterion has to be fulfilled only for each illumination area and not for the whole scanned area of interest. Important advances in ptychography include partial coherence effects that require modifications of the reconstructions algorithms. An efficient scheme for the coherent mode decomposition based on SVD was reviewed.

Water window cryo-microscopy was identified early as a promising technique for high-resolution projection imaging tool [54]. Ptychography in the water window promises to be more dose efficient because it is a lensless diffraction

technique. Thus this can be regarded as a suitable tool microscopy. However, water window diffraction experiments are challenging due to their strict technical demands of a vacuum sample environment and on the dynamic range of pixel detectors. This technical effort may be rewarded with high contrast image results that are dose efficiently measured. Ptychography is inherently quantitative as it retrieves the complex valued wave field together with the complex valued object function. With ptychographic results the index of refraction of the material found the object reconstruction can be analyzed and related to quantitative values of the projected mass density as described in one of the presented examples.

Furthermore, the important conceptual advance of coherent probe modes was demonstrated with biological specimens measured in the water window. Here also, the partial coherence was quantified using the ptychographic reconstructions. It was shown that the dynamic range directly influences the resolution in CXDI and ptychography. Especially in the soft X-ray range, where small pixels are required to obtain sufficient sampling, the full well capacity of each pixel limits the DR. The experiment presented an improvement in resolution from 250 nm down to 50 nm for a biological cell specimen using HDR diffraction patterns.

A substantial improvement for biological impact of the soft X-ray ptychography method would be a tomographic mode. This, however, requires adequate cryogenic rotation sample holder under vacuum conditions. Then a major step in relevance of soft X-ray ptychography for biology may be expected.

While dose may become a major issue in tomographic reconstructions, lensless projection imaging techniques are ideally suited to ensure minimal dose on sensitive biological specimen. However, clear evidence on the expected advantage in dose efficiency in comparison to Fresnel zone plate soft X-ray tomography (SXT) [4] is yet to be demonstrated.

Relaxed complexity in the experiments was achieved using hard X-rays. The described example shows that ptychographic tomography yields quantitative electron density of a yeast cell. Better resolution would be required for substantial biological impact of such tomographic reconstructions.

An important topic of recent research is the combination of ptychographic imaging with other techniques. XRF is an important example as it allows bio-imaging with hard X-rays with additional elemental specificity based on fluorescence. The XRF image has a lower resolution compared to the ptychographic image. However, due to the simultaneous acquisition of diffraction patterns and the XRF signal the elemental information does not need additional alignment procedures in order to correlate the images from both techniques. Thus, the enrichment of a projection image by XRF occurs with less experimental and data processing cost.

X-ray ptychography with multiple photon energies results in better resolution of the elemental maps compared to XRF. Chemical states can be investigated by accessing spectroscopic information in each pixel/voxel. Spectroscopic soft X-ray ptychography yields impressive resolution that is, however, reserved for functional materials in catalysis and battery technology that contain metals.

The presented experiments with synchrotron radiation showed recent development in the methodology of scanning lensless ptychographic coherent diffractive microscopy technique. The benefits over conventional CXDI are the capability of scanning laterally extended samples. Especially in biological imaging applications, specimens are measured without additional staining, and quantitative phase analysis can be performed to yield projected optical material properties and electron density. When this work was finished the following book [69] overviewing the field of X-ray microscopy came out.

6.5   Glossary

XFEL:	X-ray free-electron laser.
CXDI:	Coherent X-ray diffractive imaging.
PRTF:	Phase retrieval transfer function.
SXT:	Soft X-ray tomography.
FSC:	Fourier shell correlation.
FFT:	Fast Fourier transform.
XRF:	X-ray fluorescence.
DR:	Dynamic range.
HDR:	High-dynamic range.
NEXAFS:	Near edge X-ray absorption fine spectrum.
SVD:	Singular value decomposition.
RNA:	Ribonucleic acid.
FWHM:	Full width at half maximum.
BS:	Beam stop.
ePIE:	Extended ptychographic iterative engine.
CSD:	Cross spectral density.
HORST:	Holographic Roentgen scattering chamber.

Bibliography

1. D. Attwood and A. Sakdinawat. *X-rays and Extreme Ultraviolet Radiation.* Cambridge University Press, New York, 2nd edition, 2016.

2. M. Bertilson, O. von Hofsten, U. Vogt, A. Holmberg, and H.M. Hertz. High-resolution computed tomography with a compact soft x-ray microscope. *Optics Express*, 17(13):11057, 2009.

3. O. Bunk, M. Dierolf, S. Kynde, I. Johnson, O. Marti, and F. Pfeiffer. Influence of the overlap parameter on the convergence of the ptychographical iterative engine. *Ultramicroscopy*, 108(5):481–487, 2008.

4. R. Carzaniga, M.-C. Domart, L.M. Collinson, and E. Duke. Cryo-soft x-ray tomography: A journey into the world of the native-state cell. *Protoplasma*, 251(2):449–458, 2013.

5. H.N. Chapman. Microscopy: A new phase for x-ray imaging. *Nature*, 467(7314):409–410, 2010.

6. H.N. Chapman, A. Barty, S. Marchesini, A. Noy, S.P. Hau-Riege, C. Cui, M.R. Howells, R. Rosen, H. He, J.C.H. Spence, U. Weierstall, T. Beetz, C. Jacobsen, and D. Shapiro. High-resolution ab initio three-dimensional x-ray diffraction microscopy. *Journal of the Optical Society of America A*, 23(5):1179–1200, 2006.

7. J. Deng, D.J. Vine, S. Chen, Q. Jin, Y.S.G. Nashed, T. Peterka, S. Vogt, and C. Jacobsen. X-ray ptychographic and fluorescence microscopy of frozen-hydrated cells using continuous scanning. *Scientific Reports*, 7:445, 2017.

8. M. Du and C. Jacobsen. Relative merits and limiting factors for x-ray and electron microscopy of thick, hydrated organic materials. *Ultramicroscopy*, 184:293–309, 2018.

9. D. Dzhigaev, T. Stankevič, Z. Bi, S. Lazarev, M. Rose, A. Shabalin, J. Reinhardt, A. Mikkelsen, L. Samuelson, G. Falkenberg, R. Feidenhans'l, and I.A. Vartanyants. X-ray bragg ptychography on a single InGaN/GaN core–shell nanowire. *ACS Nano*, 11(7):6605–6611, 2017.

10. A.A. Ekman, J.-H. Chen, J. Guo, G. McDermott, M.A. Le Gros, and C.A. Larabell. Mesoscale imaging with cryo-light and x-rays: Larger than molecular machines, smaller than a cell. *Biology of the Cell*, 109(1):24–38, 2016.

11. B. Enders, M. Dierolf, P. Cloetens, M. Stockmar, F. Pfeiffer, and P. Thibault. Ptychography with broad-bandwidth radiation. *Applied Physics Letters*, 104(17):171104, 2014.

12. H.M.L. Faulkner and J.M. Rodenburg. Movable aperture lensless transmission microscopy: A novel phase retrieval algorithm. *Physical Review Letters*, 93:023903, 2004.

13. K. Giewekemeyer, M. Beckers, T. Gorniak, M. Grunze, T. Salditt, and A. Rosenhahn. Ptychographic

coherent x-ray diffractive imaging in the water window. *Optics Express*, 19(2):1037–1050, 2011.

14. K. Giewekemeyer, C. Hackenberg, A. Aquila, R.N. Wilke, M.R. Groves, R. Jordanova, V.S. Lamzin, G. Borchers, K. Saksl, A.V. Zozulya et al. Tomography of a cryo-immobilized yeast cell using ptychographic coherent x-ray diffractive imaging. *Biophysical Journal*, 109(9):1986–1995, 2015.

15. P. Godard, G. Carbone, M. Allain, F. Mastropietro, G. Chen, L. Capello, A. Diaz, T.H. Metzger, J. Stangl, and V. Chamard. Three-dimensional high-resolution quantitative microscopy of extended crystals. *Nature Communications*, 2:568, 2011.

16. J.W. Goodman. *Introduction to Fourier Optics*. W. H. Freeman and Company, New York, 4th edition, 2017.

17. T. Gorniak and A. Rosenhahn. Ptychographic x-ray microscopy with the vacuum imaging apparatus horst. *Zeitschrift für Physikalische Chemie*, 228(10–12):1089, 2014.

18. M. Guizar-Sicairos and J. R. Fienup. Phase retrieval with transverse translation diversity: A nonlinearoptimization approach. *Optics Express*, 16(10):7264–7278, 2008.

19. C.Y.J. Hémonnot, J. Reinhardt, O. Saldanha, J. Patommel, R. Graceffa, B. Weinhausen, M. Burghammer, C.G. Schroer, and S. Köster. X-rays reveal the internal structure of keratin bundles in whole cells. *ACS Nano*, 10(3):3553–3561, 2016.

20. B.L. Henke, E.M. Gullikson, and J.C. Davis. X-ray interactions: Photoabsorption, scattering, transmission, and reflection at e = 50–30,000 ev, z = 1-92. *Atomic Data and Nuclear Data Tables*, 54(2):181–342, 1993.

21. M.A. Herráez, D.R. Burton, M.J. Lalor, and M.A. Gdeisat. Fast two-dimensional phase-unwrapping algorithm based on sorting by reliability following a noncontinuous path. *Applied Optics*, 41(35):7437–7444, 2002.

22. M.R. Howells, T. Beetz, H.N. Chapman, C. Cui, J.M. Holton, C.J. Jacobsen, J. Kirz, E. Lima, S. Marchesini, H. Miao, D. Sayre, D.A. Shapiro, J.C.H. Spence, and D. Starodub. An assessment of the resolution limitation due to radiation-damage in x-ray diffraction microscopy. *Journal of Electron Spectroscopy and Related Phenomena*, 170(1–3):4–12, 2009.

23. X. Huang, H. Miao, J. Steinbrener, J. Nelson, D. Shapiro, A. Stewart, J. Turner, and C. Jacobsen. Signal-to-noise and radiation exposure considerations in conventional and diffraction x-ray microscopy. *Optics Express*, 17(16):13541–13553, 2009.

24. C. Jacobsen, J. Kirz, and S. Williams. Resolution in soft x-ray microscopes. *Ultramicroscopy*, 47(1–3):55–79, 1992.

25. C. Jacobsen. Soft x-ray microscopy. *Trends in Cell Biology*, 9(2):44–47, 1999.

26. C. Jacobsen. Future challenges for x-ray microscopy. *AIP Conference Proceedings*, 1696(1):020035, 2016.

27. S.-H. Kang and G.A. Fryxell. Fragilariopsis cylindrus (grunow) krieger: The most abundant diatom in water column assemblages of antarctic marginal ice-edge zones. *Polar Biology*, 12(6-7):609–627, 1992.

28. J. Kirz and C. Jacobsen. The history and future of x-ray microscopy. *Journal of Physics: Conference Series*, 186(1):012001, 2009.

29. J. Kirz, C. Jacobsen, and M. Howells. Soft x-ray microscopes and their biological applications. *Quarterly Reviews of Biophysics*, 28(01):33–130, 1995.

30. C.A. Larabell and K.A. Nugent. Imaging cellular architecture with x-rays. *Current Opinion in Structural Biology*, 20(5):623–631, 2010.

31. L. Loetgering, M. Rose, D. Treffer, I.A. Vartanyants, A. Rosenhahn, and T. Wilhein. Data compression strategies for ptychographic diffraction imaging. *Advanced Optical Technologies*, 6(6):475–483, 2017.

32. A. Maiden, D. Johnson, and P. Li. Further improvements to the ptychographical iterative engine. *Optica*, 4(7):736, 2017.

33. A.M. Maiden and J.M. Rodenburg. An improved ptychographical phase retrieval algorithm for diffractive imaging. *Ultramicroscopy*, 109(10):1256–1262, 2009.

34. O. Mandula, M.E. Aizarna, J. Eymery, M. Burghammer, and V. Favre-Nicolin. PyNX.ptycho: A computing library for x-ray coherent diffraction imaging of nanostructures. *Journal of Applied Crystallography*, 49(5):1842–1848, 2016.

35. S. Marchesini, H. He, H.N. Chapman, S.P. Hau-Riege, A. Noy, M.R. Howells, U. Weierstall, and J.C.H. Spence. X-ray image reconstruction from a diffraction pattern alone. *Physical Review B*, 68(14):140101, 2003.

36. S. Marchesini, H. Krishnan, B.J. Daurer, D.A. Shapiro, T. Perciano, J.A. Sethian, and F.R.N.C. Maia. SHARP: A distributed GPU-based ptychographic solver. *Journal of Applied Crystallography*, 49(4):1245–1252, 2016.

37. G. McDermott, M.A.L. Gros, C.G. Knoechel, M. Uchida, and C.A. Larabell. Soft x-ray tomography and cryogenic light microscopy: The cool combination in cellular imaging. *Trends in Cell Biology*, 19(11):587–595, 2009.

38. J. Miao, D. Sayre, and H.N. Chapman. Phase retrieval from the magnitude of the Fourier transforms of nonperiodic objects. *Journal of the Optical Society of America A*, 15(6):1662, 1998.

39. J. Miao, P. Charalambous, J. Kirz, and D. Sayre. Extending the methodology of x-ray crystallography to allow imaging of micrometre-sized noncrystalline specimens. *Nature*, 400(6742):342–344, 1999.

40. Y.S.G. Nashed, D.J. Vine, T. Peterka, J. Deng, R. Ross, and C. Jacobsen. Parallel ptychographic reconstruction. *Optics Express*, 22(26):32082, 2014.

41. Y.S.G. Nashed, T. Peterka, J. Deng, and C. Jacobsen. Distributed automatic differentiation for ptychography. *Procedia Computer Science*, 108:404–414, 2017.

42. M. Odstrčil, A. Menzel, and M. Guizar-Sicairos. Iterative least-squares solver for generalized maximum-likelihood ptychography. *Optics Express*, 26(3):3108, 2018.

43. H.M. Ozaktas, S. Yüksel, and M.A. Kutay. Linear algebraic theory of partial coherence: Discrete fields and measures of partial coherence. *Journal of the Optical Society of America A*, 19(8):1563, 2002.

44. David Paganin. *Coherent X-Ray Optics*. Oxford University Press, Oxford, 2006.

45. J.M. Rodenburg and H.M.L. Faulkner. A phase retrieval algorithm for shifting illumination. *Applied Physics Letters*, 85(20):4795–4797, 2004.

46. J.M. Rodenburg, A.C. Hurst, A.G. Cullis, B.R. Dobson, F. Pfeiffer, O. Bunk, C. David, K. Jefimovs, and I. Johnson. Hard-x-ray lensless imaging of extended objects. *Physical Review Letters*, 98:034801, 2007.

47. J.M. Rodenburg, A.C. Hurst, and A.G. Cullis. Transmission microscopy without lenses for objects of unlimited size. *Ultramicroscopy*, 107(2-3):227–231, 2007.

48. J.A. Rodriguez, R. Xu, C.-C. Chen, Z. Huang, H. Jiang, A.L. Chen, K.S. Raines, A. Pryor Jr., D. Nam, L. Wiegart et al. Three-dimensional coherent x-ray diffractive imaging of whole frozen-hydrated cells. *IUCrJ*, 2(5):575–583, 2015.

49. M. Rose, D. Dzhigaev, T. Senkbeil, A.R. von Gundlach, S. Stuhr, C. Rumancev, I. Besedin, P. Skopintsev, J. Viefhaus, A. Rosenhahn, and I.A. Vartanyants. High-dynamic-range water window ptychography. *Journal of Physics: Conference Series*, 849(1): 012027, 2017.

50. M. Rose, T. Senkbeil, A.R. von Gundlach, S. Stuhr, C. Rumancev, D. Dzhigaev, I. Besedin, P. Skopintsev, L. Loetgering, J. Viefhaus, Axel Rosenhahn, and Ivan A. Vartanyants. Quantitative ptychographic bio-imaging in the water window. *Optics Express*, 26(2):1237, 2018.

51. M. Rose, P. Skopintsev, D. Dzhigaev, O. Gorobtsov, T. Senkbeil, A. von Gundlach, T. Gorniak, A. Shabalin, J. Viefhaus, A. Rosenhahn, and I. Vartanyants. Water window ptychographic imaging with characterized coherent X-rays. *Journal of Synchrotron Radiation*, 22(3):819–827, 2015.

52. D. Sayre. Some implications of a theorem due to Shannon. *Acta Crystallographica*, 5(6):843–843, 1952.

53. G. Schmahl. X-ray microscopy. *Nuclear Instruments and Methods in Physics Research*, 208(1-3): 361–365, 1983.

54. G. Schneider. Cryo x-ray microscopy with high spatial resolution in amplitude and phase contrast. *Ultramicroscopy*, 75(2):85–104, 1998.

55. D. Shapiro, P. Thibault, T. Beetz, V. Elser, M. Howells, C. Jacobsen, J. Kirz, E. Lima, H. Miao, A. M. Neiman, and D. Sayre. Biological imaging by soft x-ray diffraction microscopy. *Proceedings of the National Academy of Sciences*, 102(43): 15343–15346, 2005.

56. D.A. Shapiro, Y.-S. Yu, T. Tyliszczak, J. Cabana, R. Celestre, W. Chao, K. Kaznatcheev, A.L.D. Kilcoyne, F. Maia, S. Marchesini, Y.S. Meng, T. Warwick, L. L. Yang, and H.A. Padmore. Chemical composition mapping with nanometre resolution by soft x-ray microscopy. *Nat Photon*, 8(10):765–769, 2014.

57. J. Steinbrener, J. Nelson, X. Huang, S. Marchesini, D. Shapiro, J.J. Turner, and C. Jacobsen. Data preparation and evaluation techniques for x-ray diffraction microscopy. *Optics Express*, 18(18):18598–18614, 2010.

58. S. Taverna, G. Ghersi, A. Ginestra, S. Rigogliuso, S. Pecorella, G. Alaimo, F. Saladino, V. Dolo, P. Dell'Era, A. Pavan, G. Pizzolanti, P. Mignatti, M. Presta, and M.L. Vittorelli. Shedding of membrane vesicles mediates fibroblast growth factor-2 release from cells. *Journal of Biological Chemistry*, 278(51):51911–51919, 2003.

59. P. Thibault, M. Dierolf, A. Menzel, O. Bunk, C. David, and F. Pfeiffer. High-resolution scanning x-ray diffraction microscopy. *Science*, 321(5887):379–382, 2008.

60. P. Thibault and M. Guizar-Sicairos. Maximum-likelihood refinement for coherent diffractive imaging. *New Journal of Physics*, 14(6):063004, 2012.

61. P. Thibault, M. Dierolf, O. Bunk, A. Menzel, and F. Pfeiffer. Probe retrieval in ptychographic coherent diffractive imaging. *Ultramicroscopy*, 109(4): 338–343, 2009.

62. P. Thibault and A. Menzel. Reconstructing state mixtures from diffraction measurements. *Nature*, 494(7435):68–71, 2013.

63. I.A. Vartanyants and A. Singer. Coherence properties of hard x-ray synchrotron sources and x-ray free-electron lasers. *New Journal of Physics*, 12(3):035004, 2010.

64. D.J. Vine, D. Pelliccia, C. Holzner, S.B. Baines, A. Berry, I. McNulty, S. Vogt, A.G. Peele, and K.A. Nugent. Simultaneous x-ray fluorescence and ptychographic microscopy of cyclotella meneghiniana. *Optics Express*, 20(16):18287–18296, 2012.

65. R.N. Wilke, M. Priebe, M. Bartels, K. Giewekemeyer, A. Diaz, P. Karvinen, and T. Salditt. Hard

x-ray imaging of bacterial cells: nano-diffraction and ptychographic reconstruction. *Optics Express*, 20(17):19232–19254, 2012.

66. A.M. Wise, J.N. Weker, S. Kalirai, M. Farmand, D.A. Shapiro, F. Meirer, and B.M. Weckhuysen. Nanoscale chemical imaging of an individual catalyst particle with soft x-ray ptychography. *ACS Catalysis*, 6(4):2178–2181, 2016.

67. J. Wu, X. Zhu, M.M. West, T. Tyliszczak, H.-W. Shiu, D. Shapiro, V. Berejnov, D. Susac, J. Stumper, and A.P. Hitchcock. High-resolution imaging of polymer electrolyte membrane fuel cell cathode layers by soft x-ray spectro-ptychography. *The Journal of Physical Chemistry C*, 122(22):11709–11719, 2018.

68. Y.-S. Yu, M. Farmand, C. Kim, Y. Liu, C.P. Grey, F.C. Strobridge, T. Tyliszczak, R. Celestre, P. Denes, J. Joseph, H. Krishnan, F.R. N.C. Maia, A.L. David Kilcoyne, S. Marchesini, T.P.C. Leite, T. Warwick, H. Padmore, J. Cabana, and D.A. Shapiro. Three-dimensional localization of nanoscale battery reactions using soft x-ray tomography. *Nature Communications*, 9(1):921, 2018.

69. Chris Jacobsen. *X-ray Microscopy*, Cambridge University Press, Cambridge, 2019.

7

Bioapplication of Inorganic Nanomaterials

Sachindra Nath Sarangi
Institute of Physics

Shinji Nozaki
The University of Electro-Communications

7.1 Introduction

Interdisciplinary nanoscale science has attracted researchers/engineers in the fields of materials science, chemistry and bioscience/engineering. Nano-biodevices may be widely used for disease diagnosis and gene therapy in the near future at hospitals. The interaction of biomaterials, such as deoxyribonucleic acid (DNA) and glucose, with low-dimensional semiconductors (Sarangi 2015 et al., Sarangi 2018 et al., Sarangi et al. 2007) and metals (Hussain et al. 2011) leads to a variety of applications in the areas of biosensing and DNA labeling. There are two main approaches to the bioapplications of inorganic nanomaterials. One is to fabricate nano-biosensors using inorganic nanomaterials (Wang et al. 2005). Compared with bulk materials, nanomaterials, in general, respond more sensitively to biological targets, such as glucose and DNA, due to the increased surface-to-volume ratio, which is more beneficial to optical sensing than electrochemical sensing (Sarangi et al. 2015). The other is to synthesize inorganic nanomaterials using biotemplates such as DNAs. The growth of nanostructures was significantly affected by the DNA and its conjugate, as were their optical and electrical properties. Therefore, the growth of nanostructures with a DNA template will lead to a label-free DNA detection (Peng et al. 2009).

Although nano-biosensors can be made of materials other than semiconductors, a semiconductor is more attractive because of its unique property, namely photoluminescence (PL) (Sarangi et al. 2015, Sarangi et al. 2007). When a semiconductor is illuminated by light with photon energies higher than its bandgap energy, electron–hole pairs are photo-generated. The electrons in the conduction band then recombine with the holes in the valence band. This recombination may result from the direct transition of the electrons from the conduction band to the valence band or to the valence band via a defect or impurity level within the bandgap. The latter often exhibits a luminescence with a photon energy less than the bandgap energy. In general, the PL spectra of semiconducting nanomaterials are significantly influenced by the materials, surface defects, strain, shapes, sizes, etc. Since they can be easily modified by interactions with biomolecules, such as glucose and DNA, non-electrical biosensors, which do not require any electrodes and are simple to make, may be developed using a PL spectrum as a probe (Sarangi et al. 2015, Sarangi et al. 2007). The carrier confinement in the low-dimensional semiconductor makes significant PL observable at room temperature. Nano-biosensors have been made of II–VI semiconductors such as CdS (Sarangi et al. 2018), CdSe (Sarangi et al. 2007), HgTe (Rath et al. 2007) and ZnO (Sarangi et al. 2015, Sarangi et al. 2018). In the next section, our recent successful fabrication of a glucose sensor made of ZnO nanorods are extensively reviewed.

In particular, the glucose biosensor measures the glucose concentration in blood and is critical for the diagnosis of diabetes mellitus, which affects about 150 million people around the world. Thirty-two million diabetic patients need to have their blood daily tested for glucose levels. Since Clark and Lyons proposed the original concept of glucose enzyme electrodes in 1962 (Clark et al. 1962), impressive progress has been made in the development of glucose biosensors, which now account for about 85% of the entire biosensor market (Wang et al. 2008, Heller et al. 2008).

In the first part of the next section, conventional electrochemical glucose sensors are reviewed and compared to non-electrical glucose sensors.

We recently made a glucose sensor based on ZnO nanorods grown on GaN substrates by the hydrothermal technique (Sarangi et al. 2015). The glucose on the surface of the ZnO nanorods was oxidized without the enzyme under UV illumination, and the UV irradiation of the glucose-treated ZnO nanorods decomposed the glucose into hydrogen peroxide (H_2O_2) and gluconic acid by UV oxidation. The ZnO nanorods played the role of a catalyst similar to the oxidase used in the enzymatic glucose sensors. The PL intensity of the near-band edge emission of the ZnO nanorods linearly decreased with the increase in concentration of H_2O_2. The non-enzymatic ZnO-nanorod sensor was demonstrated using human serum samples from both normal persons and diabetic patients. There was a good agreement between the glucose concentrations measured by the PL quenching and standard clinical methods.

In the second section of this chapter, the growth of nanomaterials using biotemplates, such as DNAs, is discussed. A template is used as a base for the growth, and it could be a substrate for epitaxial growth. In a broad sense, a template is defined as a material that controls and affects the chemical synthesis of inorganic materials. Apart from various inorganic templates, we discuss only the templates that are organic and biological in nature and refer to them as biotemplates, i.e., made of DNA and other biomolecules (Knez et al. 2003). The biotemplated growth of nanorods and nanowires with well-controlled sizes, morphologies, structures and compositions is facilitated by the epitaxial growth of a preferential orientation. The interaction of a biomolecule and inorganic material often suppresses growth in a particular direction. In the last several years, numerous efforts have been made to develop methods to synthesize nanowires and nanorods with good crystallinity and fabricate nanodevices in which the electrical and optical properties of the nanomaterials are utilized for recognition and labeling of biomolecules (Sarangi et al. 2018, Hussain et al. 2011).

We recently successfully synthesized CdS nanowires using a chemical route with a template made of two conjugated single-stranded (ss) DNA molecules, poly G (30) and poly C (30) (Sarangi et al. 2018). During the early stage of synthesis with the DNA molecules, the Cd^{2+} interacts with Poly G and Poly C and produces Cd^{2+}–poly GC complex. As growth proceeds, it results in nanowires. In contrast to the DNA-templated synthesis, the sample made without the DNA template was nanoparticles. The CdS nanoparticles exhibit a PL associated with the surface-dangling bond, while the DNA-templated CdS nanowires do not show such a PL. The absence of the surface-related peak suggests good passivation of the surface-dangling bonds of the nanowires by the DNA molecules. The quenching can be used to detect and label the DNAs (Fang et al. 2003, Chen et al. 2018).

In this chapter, the biosensors made of inorganic nanostructures and the biotemplated synthesis of inorganic nanostrucutures are extensively reviewed along with our recent related studies. In particular, DNA, the molecular origin of life, has taken the center stage in bioscience research during the past few decades (Gullu et al. 2008). The elucidation of the biomolecule's structure 50 years ago and the unraveling of the genetic code revolutionized research in bioscience and technology. Among the molecular-scale systems, DNA is one of the most influential materials in biosceince because the diseases and clinical syndromes were well associated with the DNAs. Further development of the techniques to identify the structure and label the DNA will lead to disease diagnoses and gene therapy in the near future. Since the molecular size of the DNAs is on a nanometer scale, the interaction of the inorganic nanomaterials becomes extremely significant and is key to bioscience and technology.

7.2   Glucose Sensors Made of Nanomaterials

Diabetes is a disease in which the glucose concentration in the blood cannot be regulated. It sometimes leads to death. In addition to measurement of the blood glucose concentration for the diagnosis of diabetes, routine measurement of the glucose concentration is required for diabetic patients to live healthy lives. Various glucose measurement technologies have been developed for more than three decades. A glucose sensor, which is one of the most popular biosensors, has been extensively investigated due to its important clinical application (Ahmad et al. 2010). Although many sensor technologies are being pursued to develop a novel glucose sensor, a simpler, cheaper and more highly sensitive sensor for faster glucose measurement is yet to be developed.

7.2.1   Electrochemical Glucose Sensors

In recent years, an increasing number of researchers have made efforts to develop glucose sensors using nanomaterials (Ahmad et al. 2010, Chaiyo et al. 2018, Rahman et al. 2010). Because of a larger surface-to-volume ratio of the nanomaterial, a higher sensitivity of the biological sensor can be expected. Among the various biosensor materials, oxide semiconductors, such as zinc oxide (ZnO), are the best due to their chemical stability and non-toxicity. ZnO, in particular, has a high isoelectric point (IEP) which helps with the adsorption of biomolecules such as proteins and glucose (Ahmad et al. 2017). One can find many reports on bio- and glucose sensors made of ZnO.

Most commercially available glucose sensors are enzymatic electrochemical sensors in which the oxidant enzyme, such as glucose oxidase (GO_x), oxidizes glucose as a catalyst and produces gluconic acid as shown in Eq. (7.1):

$$D - Glucose + O_2 + H_2O \xrightarrow{GO_x} D - Gluconic\ Acid + H_2O_2$$

$$(7.1)$$

In the electrochemical sensors, an electrode measures the glucose concentration based on monitoring H_2O_2, which is decomposed into oxygen and a positive hydrogen ion. The decomposition is accompanied by the release of electrons. These electrons are monitored as an electrochemical current. The enzymatic glucose sensors, however, suffer from drawbacks such as insufficient stability and loss of their enzyme activity before attaching to the glucose (Harris et al. 2013). A recently developed non-enzymatic electrochemical sensor was made with a glassy carbon electrode modified with platinum nanostructures on graphite oxide (Wua et al. 2013). The sensor demonstrated an improved response to glucose over a wide concentration range. These electrochemical sensors, however, require three electrodes, i.e., working, reference and counter-electrodes, and are too complex in their structure for mass production. Therefore, although there has been great progress in the development of non-enzymatic electrochemical sensors using nanomaterials, efforts must be shifted to developing non-enzymatic and non-electrochemical glucose sensors.

7.2.2 Non-Electrochemical Glucose Sensors

There are commercially available clinical sensors based on optical absorption and reflection to measure the oxygen levels in human blood. As seen in Figure 7.1, when light is irradiated on human blood, it is reflected, scattered and absorbed. The intensity of each reflected, scattered and absorbed light depends on the content of the blood and wavelength of the incident light. Therefore, the contents can be identified and their amounts can be measured by monitoring the reflection, scattering and absorption. In addition to the reflection, scattering and absorption, if the blood is anchored on the luminescent material, the PL, the intensity of which could be affected by interaction with the blood, can be monitored. In particular, oxidation significantly affects the PL intensity of a luminescing material. The enzymatic optical glucose sensor is made of ZnO nanocrystals (Kim et al. 2012). In this glucose sensor, the PL intensity decreased with the glucose concentration because of PL quenching caused by H_2O_2 resulting from oxidation of the glucose with GO_x. The PL quenching occurred when the radiative transition of the excited electrons was suppressed by H_2O_2 as a quencher. The energies of the valence and conduction band edges increased by the quantum confinement that facilitated the electron transfer

from ZnO to H_2O_2. Since the amount of PL quenching correlates with the amount of H_2O_2, which also correlates with the glucose concentration, the PL intensity can be used as a measure of the glucose concentration.

Although the developed optical glucose sensor demonstrated a high sensitivity, mercaptoundecanoic was needed to cap the ZnO for immobilization of GO_x. The structure will become much simpler without the enzyme. Furthermore, the luminescence intensity will be much higher if nanorods are used instead of nanocrystals because of their improved crystallinity. We successfully made a non-enzymatic and non-electrochemical glucose sensor made of ZnO nanorods by making the best use of their material properties of photo-oxidation and PL quenching.

7.2.3 Hydrothermal Growth of ZnO Nanorods

Before reviewing our ZnO glucose sensor, let us briefly describe a simple chemical synthesis method of nanorods. Till now, various fabrication techniques have been established for the synthesis of ZnO nanorods. Simple chemical synthesis methods, such as electrochemical and non-electrochemical deposition, have recently been developed to grow ZnO nanorods. Non-electrochemical methods are preferred because the nanorods can be deposited both on conducting and non-conducting substrates. Among the non-electrochemical methods, the hydrothermal growth method has been widely used to grow a variety of nanorods. The diameter and length are well controlled by the concentration of the chemicals.

The ZnO nanorods were grown on various substrates, and their effects on the optical properties and structure were analyzed in detail (Nayak et al. 2008). The cleaned substrate was dipped in a mixture of 10 mM zinc nitrate hydrate ($Zn\ (NO_3)_2 \cdot 6\ H_2O$) and hexamethylenetetramine HMT ($(CH_2)_6N_4$) contained in a screw-capped stainless steel bottle, whose bottom was partially submerged in an oil bath. The substrate was suspended in the solution for 3 h, and the temperature of the oil bath was maintained at 90°C. The substrate was subsequently removed from the bottle and rinsed in DI water.

Figure 7.2 shows a schematic of the hydrothermal growth of the ZnO nanorods. The stainless bottle in the oil bath is heated from the bottom. There is a temperature difference in the solutions at the bottom and top. The solution at the bottom is at a higher temperature and under saturation. The solution moves up by convection and is cooled. The oversaturated solution precipitates ZnO on the substrate. The solution then moves down, and solution movement by convection continues. The precipitates result in ZnO nanorods. As a sealed system, high pressure prevents the solution from evaporating. The chemical reaction is shown in Figure 7.3. The Zn^{2+} from $Zn(NO_3)_2$ and OH^- from NH_3 forms $Zn(OH)_2$ then ZnO. HMT not only produces NH_3 but also suppresses the lateral growth along the diameter and plays an important role in ZnO nanorod growth.

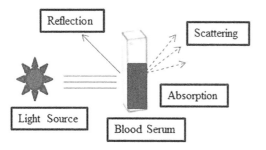

FIGURE 7.1 Possible optical probes for contents in blood.

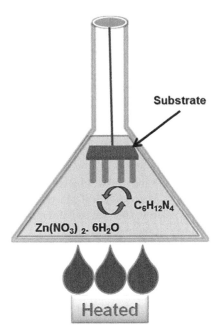

FIGURE 7.2 Schematic of hydrothermal growth of ZnO nanorods.

$$Zn(NO_3)_2 \cdot 6H_2O \rightarrow \underline{Zn^{2+}} + 2NO_3^-$$

$$C_6H_{12}N_4 + 6H_2O \rightarrow 6HCHO + \underline{4NH_3}$$

$$\underleftarrow{NH_3} + H_2O \rightarrow NH_4^+ + \underline{OH^-}$$

$$2OH^- + Zn^{2+} \rightarrow Zn(OH)_2$$

$$Zn(OH)_2 \rightarrow ZnO + H_2O$$

FIGURE 7.3 Reaction process for ZnO synthesis.

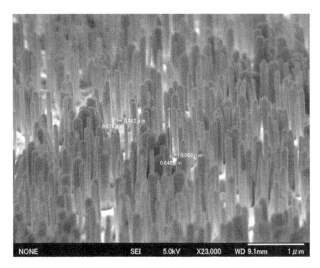

FIGURE 7.4 SEM image of the ZnO nanorods grown on GaN.

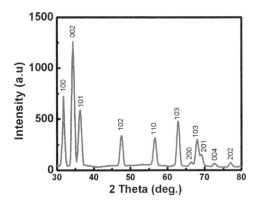

FIGURE 7.5 X-ray diffraction pattern of the ZnO nanorods grown on GaN.

Arrays of vertically standing, single crystalline nanorods were directly grown on GaN substrates due to the good lattice match between GaN and ZnO, as seen in Figure 7.4. In the scanning electron microscopy (SEM) image, the ZnO nanorods are densely packed and fully cover the GaN surface. Their diameters and lengths are 70–80 nm and 0.8–0.9 μm, respectively. Figure 7.5 shows the X-ray diffraction pattern of the ZnO nanorods grown on GaN. The dominant reflection peak of (002) suggests that these ZnO nanorods are highly crystallized and grew along the c-axis.

7.2.4 PL Quenching and Photo-Oxidation

As mentioned in Section 7.2.2, PL quenching is a good probe for measurement of the glucose concentration and may also be employed to develop an optical glucose sensor made of ZnO nanorods without the GO_x enzyme. The solutions of various glucose concentrations were dropped on ZnO nanorods, and their PL spectra were collected at room temperature. In Figure 7.6a, the PL spectra are normalized to the PL peak intensity of the ZnO nanorods before treatment. The PL peak at 377 nm is associated with the near-band edge emission. As seen in the figure, the PL peak intensity decreases with increased glucose concentration.

The PL quenching is confirmed even without GO_x. The calibration curve shown in Figure 7.6b exhibits a good linearity in the glucose concentration range of 0.5–30 mM. The sensitivity estimated from the calibration curve is on the order of 1.4 %/mM. Compared to the linearity ranges of optical glucose sensors using silver nanoparticles and carbon nanotubes (Wu et al. 2010, Yum et al. 2012), the linearity range of the ZnO nanorods on GaN is broader and is considered to be more suitable for glucose sensors.

If the observed PL quenching is analogous to that with GO_x, there must be oxidation taking place without an oxidizing agent. Figure 7.7 compares the Fourier Transform Infrared (FTIR) spectra of the glucose-treated ZnO nanorods before and after intentional exposure to UV light. In Figure 7.7, the arrows marked with # and * indicate the locations of the characteristic peaks of H_2O_2 and gluconic acid, respectively. The characteristic peaks of gluconic acid and H_2O_2 become more pronounced after UV exposure, and thus the oxidation of glucose by UV exposure is confirmed. Sarangi et al. prepared the H_2O_2 solutions at various concentrations and drop-cast them on ZnO nanorods. A linear decrease in the PL peak intensity with the increased H_2O_2 confirmed that the PL quenching was due to H_2O_2 resulting from the oxidation of glucose

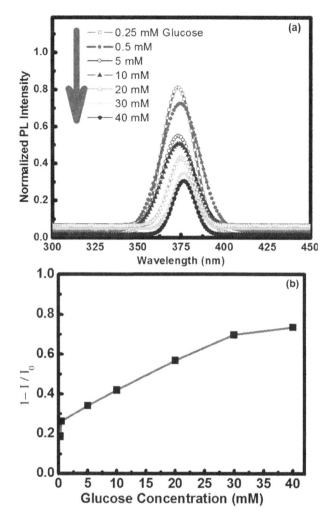

FIGURE 7.6 (a) PL spectra of the glucose-treated ZnO nanorods. The intensity is normalized to the PL peak intensity of the as-grown ZnO nanorods. (b) Change in the PL peak intensity relative to that of the as-grown ZnO nanorods.

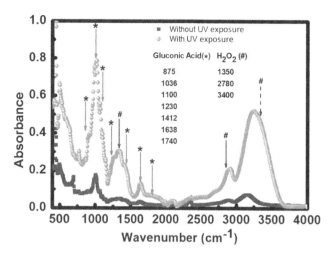

FIGURE 7.7 FTIR spectra of the ZnO nanorods treated with glucose before and after intentional exposure to UV light.

even without GO_x. ZnO nanorods play the role of a photocatalyst for the oxidation of glucose. The photo-generated electrons in the ZnO nanorods transfer to H_2O_2, which acts as a PL quencher of the ZnO nanorods.

A similar oxidation was observed in the Ge nanostructures (Sato et al. 1997). Such a photo-oxidation was well explained by the modified electron-active oxidation model for semiconducting nanomaterials. In the electron-active model proposed for the photo-oxidation of Si, the rate-determining step of the oxidation of Si is to break up the molecular oxygen into its smaller atomic components with the help of electrons injected from Si into the SiO_2: (Young 1988)

$$O_2 + e^- \rightarrow O + O^- \qquad (7.2)$$

During the photo-oxidation of Si, the photo-generated electrons are excited into the conduction band of SiO_2 at the Si surface only if the electron energy is high enough to overcome the conduction band offset at the SiO_2/Si interface. On the other hand, the generated holes

left in the Si weaken the Si–Si bonds. In the case of semiconducting nanomaterials, the increased bandgap due to the quantum confinement lowers the band offset at the oxide/semiconductor and facilitates the transfer of photo-generated electrons into the oxide. In the system of glucose/ZnO nanorods, the photo-generated electrons in the ZnO transfer to the glucose and decompose the O_2 into oxygen radicals. The photo-oxidation of glucose occurs during the PL measurement because the incident UV light from the He-Cd laser is absorbed by ZnO, in which the electron–hole pairs are generated and facilitate the decomposition of O_2. Analogous to Eq. (7.1),

$$\text{Glucose} + O_2 + H_2O \xrightarrow[\text{UV light}]{\text{ZnO nanorods}} D - \text{gluconic acid}$$
$$+ H_2O_2 \qquad (7.3)$$

ZnO nanorods play the role of GO_x and act as a photocatalyst.

7.2.5 ZnO Nanorods as a Non-Enzymatic and Non-Electrochemical Glucose Sensor

The high sensitivity and wide range of linearity in the calibration curve of ZnO nanorods grown on GaN shown in Figure 7.6b ensure their potential application as a non-enzymatic and non-electrochemical glucose sensor. In particular, the sensing range includes the normal glucose concentrations in human serum of 4.4–6.6 mM (Xu et al. 2011), and even higher ranges for diabetics. For further confirmation of their feasibility as a glucose sensor, interference from other substances possibly found in human serum, such as ascorbic acid (AA), uric acid (UA) and bovine serum albumin (BSA), must be studied. Sarangi et al. treated the ZnO naorods with AA, UA and BSA to study the interference (Sarangi et al., 2015). Each concentration of the substances was chosen to be the same as that in human serum (Safavi et al. 2009, Bourdon et al. 1999). No effect of the treatment of the ZnO nanorods with AA or UA on the PL spectrum confirms no interference from these substances. In contrast to AA and UA, BSA itself exhibits a significant PL peak at 340 nm, not far from the near-band edge emission peak of the ZnO nanorods. The shoulder of the BSA's peak on the longer wavelength side affects the shoulder of the ZnO peak on the shorter wavelength side and not only blue-shifts the peak wavelength, but also increases the peak intensity of the ZnO nanorods. The effect is more significant for a lower PL peak intensity of the ZnO nanorods, and the glucose concentration in the human serum could be underestimated. Fortunately, the PL spectrum of BSA is not affected by immobilization of the glucose on the ZnO nanorods, and the measured PL spectra of the human serum are capable of being corrected by subtracting the contribution from the BSA. However, Sarangi et al. found that the contribution from the BSA in human serum even with the highest glucose concentrations among the blood samples is not significant enough to affect the glucose concentration determined by the calibration curve in Figure 7.6b (Sarangi et al., 2015). Therefore, it can be concluded that there is no interference from the major substances other than glucose in the determination of the glucose concentrations in human serum.

Finally, the feasibility of ZnO nanorods grown on GaN as a glucose sensor was tested with blood samples, S1–S7 (Sarangi et al., 2015). The samples were collected from healthy persons and diabetic patients and provided to Sarangi with clinically measured data by a hospital. Figure 7.8a shows the PL spectra of these samples. The PL spectrum of the as-grown ZnO nanorods is also shown as a reference. As previously mentioned, the PL peak of BSA is seen in all the blood samples. Since the presence of BSA does not significantly affect the PL quenching due to glucose, the peak intensity of the ZnO nanorods is not corrected to estimate the glucose concentration from the calibration curve in Figure 7.6b. The glucose concentrations estimated from the PL quenching of the ZnO nanorods on GaN are well correlated to those provided by the hospital, as seen in Figure 7.8b. A good correlation in the wide glucose concentration has proven the potential of ZnO nanorods as a non-enzymatic and non-electrochemical glucose sensor. It should be noted that the PL peak intensity of the as-grown ZnO nanorods varies among the samples, and a fresh sample must be used to measure the amount of the PL quenching for each blood specimen. Nevertheless, glucose sensors made of ZnO nanorods by the hydrothermal growth method are easily prepared and can be widely used as disposable clinical biosensors in the future.

7.3 Growth of Nanomaterials Using Biotemplates

Biotemplates are attractive for fabricating nanostructured materials and have uses in diverse areas of applications such as catalysis, drug delivery, biomedicine, composites,

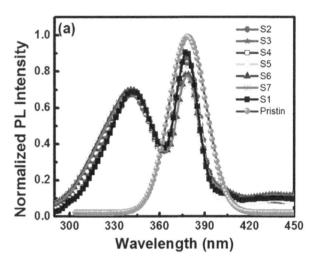

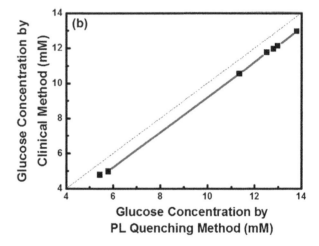

FIGURE 7.8 (a) PL spectra of the as-grown ZnO nanorods and ZnO nanorods treated with different blood specimens, S1–S7. The intensity is normalized to the peak intensity of the as-grown ZnO nanorods. (b) Correlation of the glucose concentrations determined by the PL quenching method and by the standard clinical method.

etc. In particular, the use of DNA, protein and glucose as a biotemplate has attracted increasing attention in nano-bioscience. Understanding of the interaction between the inorganic and biomolecular materials is important for the controlled growth by the self-assembly of inorganic nanomaterials. The chalcogenide-based nanotechnology for biomedical applications is a growing field and is to be developed to tailor their unique functional and structural properties (Zhou et al., 2014). The nanomaterials synthesized using different biomolecules as templates exhibit a variety of shapes and structures and can be produced in a large quantity using simple chemical and electrochemical growth techniques (Sarangi et al. 2015, Hussain et al. 2011). Among the biomolecular templates, DNAs have been mostly used to grow inorganic nanomaterials because of their physicochemical stability and unique structure (Barun et al. 1998, Liu et al. 2008, Yao et al. 2008, Dwivedi et al. 2010). The shape and size of the nanomaterials can be controlled by selecting a particular DNA as well as the properties of the grown nanomaterials. Therefore, the structure of the DNA can be well understood by studying the properties of the grown nanomaterials. In other words, the biotemplated growth of nanomaterials can be used for biomolecular recognition and DNA labeling. In this section, the DNA-templated growth of chalcogenide nanomaterials is reviewed and the templated growth process for self-assembly is modeled in detail.

7.3.1 Self-Assembly Using Biotemplates

There has been significant interest in recent years to develop the technology for the self-assembly of nanomaterials. Self-assembly can be defined as the spontaneous and reversible organization of molecular units into ordered structures by physical or chemical interactions. The interactions responsible for the formation of the self-assembled system act on a strictly local level; in other words, the nanostructure builds itself. The self-assembled nanostructure is an object that appears as a result of ordering and aggregation of individual nanoscale objects guided by some physical or chemical principle. The self-assembly of a nanostructure becomes an easy and inexpensive way to fabricate them if the physical or chemical reactions with the DNA templates favor self-assembly. In addition to the good control of shapes and sizes, one major advantage of using self-assembly is that it does not suffer from the limitation of the size reduction unlike lithography. Since self-assembly is a solution to the construction of nanodevices, it is extremely important to explore the physical or chemical reactions favoring the self-assembly of inorganic nanostructures.

Many self-assembly strategies rely on electrostatic interactions, groove binding or intercalation for the absorption of metal or semiconductor ions directly into the DNA templates. After absorption, reduction of these ions facilitates the formation of nanoparticles along the DNA templates. Single-stranded DNA (ssDNA) is a long polymer made from repeating units called nucleotides. The nucleotide repeats contain both the segment of the backbone of the molecule, which holds the chain together, and a base. The backbone of the DNA strand is made from alternating phosphate and sugar residues. The four bases found in DNA are adenine (abbreviated A), cytosine (C), guanine (G) and thymine (T). Each type of base on one strand overwhelmingly prefers a bond with just one type of base on the other strand. This is called complementary base pairing; A bonds preferentially to T, and C bonds preferentially to G. This arrangement of two nucleotides binding together across the double helix is called base pairing. In living organisms, DNA does not usually exist as a single molecule, but instead, as a pair of molecules called the double-stranded DNA, in which a pair of ssDNAs are held tightly together via a reaction known as DNA hybridization. The ssDNA is flexible and has a small persistence length when compared to the double-stranded DNA. The unique properties of DNA—hybridization and base-pairing—are major factors influencing not only the DNAs themselves but also other materials. The single and double-stranded DNAs

should make a difference in the self-assembly of inorganic nanostructures as templates. There are many reports on the growth of chalcogenide nanomaterials using DNA templates. The chalcogenide nanomaterials are of significant interest for biosensor/biomolecular applications (Sarangi et al. 2007, Sarangi et al. 2018), and among them, DNA-assisted CdS nanomaterials have received the most attention. The early work on CdS nanomaterials mainly focused on the use of DNA to grow CdS nanoparticles for quantum confinement (Coffer et al. 1992). The cadmium chalcogenide quantum dots (QDs) grown using DNA templates were studied for the biomimetic and bioinspired synthesis of DNA-linked nanomaterials and their applications (Nithyaja et al. 2012, Coffer et al. 1996, Kulkarni et al. 2005, Godman et al. 2015, Wang et al. 2016, and Zan et al.2016).

Next, the growth of chalcogenide nanostructures using single, double and two ssDNAs as templates will be discussed in detail, and the base-pairing and hybridization are addressed in the model of the growth.

7.3.2 Growth of Cadmium Selenide Nanomaterials Using DNA templates

Sarangi et al. grew cadmium selenide (CdSe) nanomaterials using DNA templates by electrodeposition (Sarangi et al. 2010). Commercially available ssDNAs, poly G (30) and poly C (30), were used as templates. The solution of $CdSO_4$ and SeO_2 was prepared, and poly G and poly C DNAs were added to the solution. Three sets of samples were made: one without DNAs, another with only poly G DNA and the other with both the poly G and poly C DNAs. A constant current was applied to the solution to start deposition on the indium tin oxide-coated glass substrates. The detailed experimental procedure can be found elsewhere (Sarangi et al. 2010).

Figure 7.9 shows the transmission electron microscopy (TEM) and transmission electron diffraction (TED) images of these three samples. The samples without the DNAs and with the poly G DNA are in the form of nanoparticles, while that with both the poly G and poly C DNAs is in the form of nanowires. The sizes of the nanoparticles without the DNAs and with the poly G DNA are 15 and 3 nm in the high-resolution TEM (HRTEM) images, respectively. In addition to the one in the HRTEM image of the nanoparticles without DNAs, there are also smaller nanoparticles. The nanoparticles in the sample with poly G are tangled with the DNA strings and much smaller than those in the sample without the DNAs. The DNA seems to further suppress growth of the CdSe nanoparticles by tangling them. All the CdSe nanostructures are single crystalline. The grazing incident angle x-ray diffraction (GXRD) patterns in Figure 7.10 confirm a zinc blende CdSe structure. There are three reflection peaks of (111), (220) and (312) in all the samples. However, the nanostructures with the DNAs show a suppressed peak of (312) in the XRD pattern. Although the nanoparticles without the DNAs are larger, the peak intensities are weak. This suggests that the

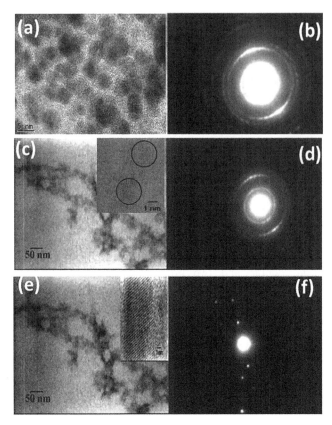

FIGURE 7.9 TEM images of the CdSe samples: (a) without DNAs, (c) with poly G and (e) with both poly G and poly C DNAs. The insets in (c) and (e) show high-resolution TEM images. TED pattern of the CdSe samples: (b) without DNAs, (d) with poly G and (f) with poly G and poly C.

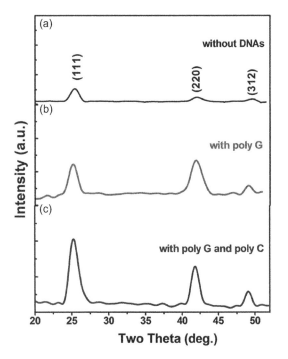

FIGURE 7.10 GXRD patterns of the CdSe samples: (a) without DNAs, (b) with poly G and (c) with both poly G and poly C.

crystal quality is lower than those with the DNAs. The crystallinity seems to be further improved by forming nanowires as seen in the higher intensity of the (111) reflection peak in the sample with both the poly G and poly C DNAs. Because of its preferential growth direction, a nanowire, in general, has a better crystallinity than a nanocrystal. The 3D growth tends to introduce stacking faults. The lattice constants of the samples, which were determined from the (111) reflection peaks, without the DNAs, with the poly G DNA and with both the poly G and poly C DNAs are 0.60, 0.61 and 0.61 nm, respectively. Compared with the bulk lattice constant of 0.605 nm, the sample without the DNAs have a compressive strain, while the rest have a tensile strain. A compressive strain is more common in a material deposited by electrodeposition (Kongstein et al. 2005). The lattice constant of the bare DNA is 3 nm (Evans et al. 2003) and much larger than that of CdSe. As a result, a tensile strain will be introduced in the CdSe nanostructures.

It is interesting to note that the nanostructures with both the poly G and poly C DNAs are in the form of nanowires. Their average diameter of the nanowires is 4 nm, and their average length is found to be much longer than that of a ssDNA (10.2 nm) used as template. It implies that there are a few more DNA molecules along the nanowire length. Since the end points of the nanowires are more active than the surface (Zhang et al. 1999), successive interdiffusion at the end points of the nanowires has resulted in such lengthy nanowires compared to the DNA length. More details about nanowire synthesis will be discussed later.

Although the DNAs may not be clearly observed in the TEM image, tangling of the CdSe nanoparticles and nanowires with the DNAs was confirmed by the FTIR. The peaks at 968 and 1051 cm^{-1} associated with the bare DNAs (G–C duplex) were merged together and shifted to a higher wavenumber. This shift was attributed to the interaction between the cations and the phosphate groups of the poly G DNA in the sample with the poly G DNA. The tangling of the CdSe with the DNAs was further confirmed by the shift in the peaks associated with the C = N$_7$ and C$_6$ = O bonds of the DNA.

The PL spectra of the nanoparticles without the DNAs and nanowires grown with the poly G and poly C DNAs are significantly different, as seen in Figure 7.11. The PL peak wavelengths of the CdSe nanoparticles and nanowires are 578 and 544 nm, respectively. Since the bandgap of the bulk CdSe is 1.75 eV (700 nm), the blueshift of the PL peak in both nanostructures suggests quantum confinement. Although the size of a nanoparticle in the inset of Figure 7.11 is too large for quantum confinement, some smaller nanoparticles contribute to the PL at 578 nm. A further blueshift and broadening of the PL peak observed for the CdSe nanowires are attributed to the increased quantum confinement. The band gap of a CdSe nanoparticle with a size of about 2.27 nm is 2.96 eV (546 nm) (Hegazy and El-Hameed 2014). The average diameter of the CdSe nanowires is 4 nm, while the diameters of some nanowires may be smaller. The decreased intensity of the PL peak associated with the

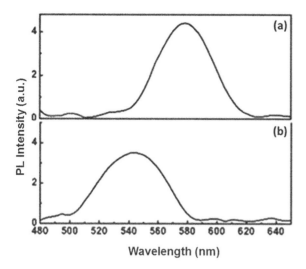

FIGURE 7.11 Room-temperature PL spectra of the CdSe samples (a) without DNAs and (b) with poly G and poly C.

band edge emission in the CdSe nanowires with the poly G and poly C DNAs is caused by charge transfer to the DNAs [72].

7.3.3 Growth of Cadmium Sulfide Nanomaterials Using DNA templates

Sarangi et al. grew cadmium sulfide (CdS) nanomaterials by a simple chemical technique using DNA templates (Sarangi et al., 2018). Commercially available ssDNAs, poly G (30) and poly C (30), were used as the templates. Unlike the growth of the CdSe nanomaterials, the starting solution contains only Cd^{2+} from an ammonia-mixed $CdSO_4$ solution. The S^{2-} was provided by a thiourea solution. For the growth of the CdSe nanomaterials, the starting solution contained both Cd^{2+} and Se^{2-} as well as the DNAs. The growth was initiated by imposing a current in the case of electrodeposition. However, for a pure chemical synthesis, the sequential addition of the constituents, Cd^{2+} and S^{2-}, of CdS and the poly G and poly C DNAs is needed to avoid the uncontrolled formation of the CdS adducts from Cd^{2+} and S^{2-} and double-stranded DNA from the poly G and poly C DNAs. The detailed sequential process is discussed in the next subsection. Two sets of samples were made: one without the DNAs and the other with both the poly G and poly C DNAs.

Figure 7.12 shows the TEM images and TED pattern of the grown CdS nanomaterials. The nanoparticles are observed in the samples without the DNAs, while a network of nanowires is observed with the poly G and poly C DNAs. The sizes of the nanoparticles vary from 4 to 12 nm in the TEM image. The HRTEM image of the network confirms overlapped nanowires, each of which has a diameter of 4 nm. Some lengths are longer than 20 nm, much longer than those of the DNAs. Like the DNA-tangled CdSe nanowires, the DNA strings seem to be connected at the end and form lengthy strings. The TED patterns and HRTEM image

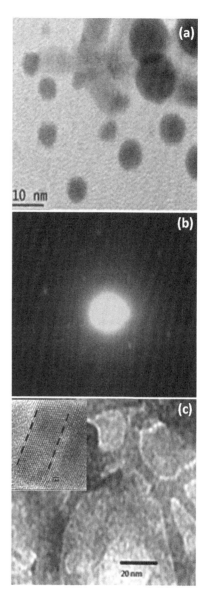

FIGURE 7.12 (a) Low-magnification TEM image and (b) TED pattern of the CdS sample without DNAs. (c) Low-magnification TEM image of the CdS sample with poly G and poly C. The inset shows high-resolution TEM image.

suggest that both samples are single-crystal and have the zinc blende CdS structure. The GXRD patterns show three peaks associated with (111), (200) and (220), as seen in Figure 7.13, and confirm that both samples are the zinc blende CdS.

The room-temperature PL spectra of the CdS nanoparticles and nanowires show a significant difference in Figure 7.14. The PL spectrum of the CdS nanoparticles without the DNAs exhibits two significant peaks at 425 and 472 nm, while that of the CdS nanowires tangled with the DNAs show only one significant peak at 485 nm. The wavelength of 425 nm is similar to the excitonic peak wavelength in the absorption spectrum and is associated with the band edge emission. The size estimated from the 3D confinement for the PL peak wavelength is 4 nm (Sarangi and Sahu 2004).

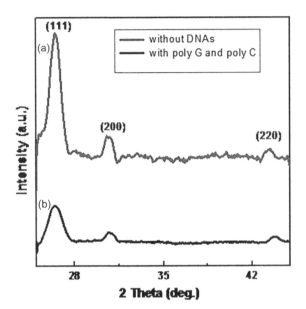

FIGURE 7.13 GXRD patterns of the CdSe samples (a) without DNA and (b) with poly G and poly C.

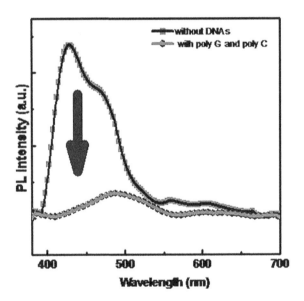

FIGURE 7.14 Room-temperature PL spectra of the CdS nanoparticles and nanowires.

As observed in the TEM image, there are nanoparticles with a size of 4 nm in the sample grown without the DNAs. Since the excitonic peak wavelength of the nanowires is not much different from that of the nanoparticles in the absorption spectra, the red shift of the PL peak cannot be explained by the size effect, but by a change in the effective dielectric constant of the medium resulting from the hybridization of the DNAs and CdS nanowires (Heller et al. 2006).

The PL peak at 472 nm in the nanoparticles without the DNAs is associated with radiative recombination via the surface states. The absence of the corresponding surface-state associated emission in the nanowires is attributed to the well-passivated surface by the hybridization of the DNAs and CdS nanowires. Hybridization enhances the charge

transfer from the nanowires to the DNAs and results in the quenching of the band edge PL (Wu et al. 2012, Zhang et al. 2011). It was reported that the quenching phenomenon in the PL of the CdS nanoparticle was an indirect approach to detect the DNA structure and depended on the surface charge and functionality of the nanoparticles (Murphy et al. 2001).

7.3.4 Model of Growth of Chalcogenide Nanomaterials Using Conjugated DNA Templates

During the growth of the CdSe and CdS nanomaterials using the poly G and poly C DNAs, there are several critical steps to form nanomaterials. These steps involve the interactions between (i) Cd^{2+} and ssDNA, (ii) Cd^{2+} and Se^{2-}/S^{2-} and (iii) Se^{2-}/S^{2-} and ssDNA. It should be noted that the poly G and poly C DNAs are conjugated DNAs, because guanine (G) and cytosine (C) form a base pair, as mentioned earlier. This is a natural phenomenon whereby a ssDNA can recognize its complementary strand by the highly predictable Watson–Crick base pairing (Tan et al. 2011). The nanoparticles are formed only using poly G and the nanowires using the conjugated DNAs. One of the best approaches for the self-assembly of nanostructures employs Watson–Crick base-pairing interactions between the conjugated DNAs. In general, the nanoparticles tend to introduce more crystal defects during 3D growth than the nanowires during 1D growth. The size of the nanoparticles must be small enough to suppress the formation of crystal defects such as stacking faults and dislocations. Such smaller nanomaterials may not be suitable for commercial applications in highly sensitive biosensors. Therefore, we prefer the growth of nanowires using the DNA templates, such as the conjugated DNAs, and would like to model the growth process of the CdSe and CdS nanowires using the poly G and poly C DNAs. The sequential process discussed for the growth of the CdS nanowires is preferred to the process using the solution containing both anions and cations and conjugated DNAs together for the growth of the CdSe nanowires. However, the growth process even using such a solution becomes synergized by applying an electric current, which helps in ordering of the anions and cations and suppressing the uncontrolled formation of chalcogenides. As a result, the CdSe nanowires are formed even using the mixed solution. Although the CdSe nanowires were formed by applying an electric current to the solution containing all materials, they can also be formed by the sequential process without an electric current. Hence, the sequential growth process is important to be modeled for self-assembly and will be described in a step-by-step manner next.

There are 5 steps in the growth process shown in Figure 7.15. In Step I, the starting solution was prepared by mixing poly G DNA and $CdSO_4$. In a few minutes, the electrostatic interaction between the negatively charged PO_2^{4-} backbone of the poly G and Cd^{2+} leads to the self-assembly of Cd^{2+} on the poly G by the charge condensation process,

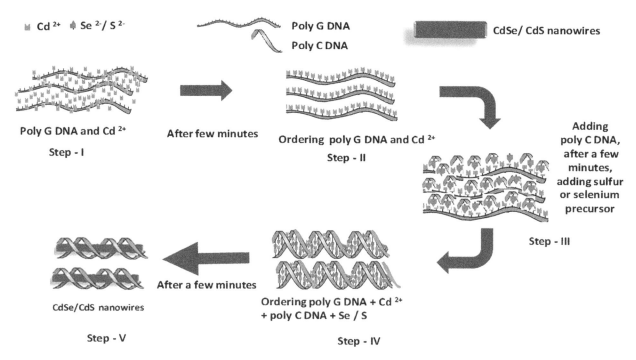

FIGURE 7.15   Schematic of growth process of DNA-templated CdSe and CdS nanowires.

as shown in Step II. In Step III, poly C DNA is added. Poly C interacts with the complex of Cd^{2+}–poly G, which possesses a net positive charge resulting from the charge reversal process (Sarangi et al. 2007). The base-to-base interaction between poly G and poly C leads to 30 base pairs of G and C. As a result, poly G and poly C form a stable double-stranded DNA, which forms a complex with Cd^{2+}. After a few minutes, the precursor for Se^{2-} or S^{2-}, SeO_2 or thiourea (CH_4N_2S) is added to the solution in Step III. In Step IV, Se^{2-} or S^{2-} hybridizes the complex of Cd^{2+} and the double-stranded DNA made of poly G and poly C. The hybridization proceeds with time to achieve a minimum energy by ordering. In Step V, the nanowires are formed by surface and end point diffusion in a self-assembly manner.

One may ask why the double-stranded DNAs cannot be used instead of two conjugated ssDNAs for the growth of the nanowires by self-assembly. In Step III of the growth process, a double-stranded DNA is formed from the poly G and poly C. A similar growth process could be expected even using double-stranded DNAs. However, CdS nanoparticles, not nanowires, were formed with double-stranded salmon sperm DNA (Kulkarni et al. 2005). The salmon sperm DNA contains all four bases A, G, T and C. Watson–Crick base pairing of A–T and G–C takes place in a double-stranded DNA. The self-assembly of Cd^{2+} may not occur in a row on one of the two strings of the DNAs but results in clustering of the Cd^{2+}. By adding a precursor for S^{2-}, Cd^{2+} reacts with S^{2-} to form nanoparticles instead of lengthy nanowires. The DNA caps the CdS nanoparticles and limits the size of the nanoparticles. Therefore, the ssDNA should react with the Cd^{2+} first, then the conjugated ssDNA hybridizes a complex of Cd^{2+} and the ssDNA for the controlled growth of the chalcogenide nanowires.

7.4   Conclusion

In conclusion, two approaches for the bioapplication of inorganic nanomaterials have been reviewed in detail. One is to fabricate nano-biosensors using inorganic nanomaterials. Compared to the bulk materials, nanomaterials, in general, respond more sensitively to biological targets, such as glucose and DNA, due to the increased surface-to-volume ratio, which is more beneficial for optical sensing than electrochemical sensing. In particular, non-enzymatic and non-electrochemical glucose sensors are highly awaited by over 1 million diabetic patients around the world because of their simplicity and stability. The glucose sensor made of ZnO nanorods demonstrated feasibility in terms of clinical applications. The developed glucose sensor makes the best use of photo-oxidation and PL quenching, which are unique to ZnO nanorods. ZnO plays the role of photo-catalyst and enhances the oxidation of the immobilized glucose without glucose oxidase, which is usually required for most commercial glucose sensors and often degrades the stability of the sensor performance. PL quenching is caused by H_2O_2 molecules, which are produced by the oxidation of glucose and enhance the carrier transfer from the ZnO nanorods as a luminescence quencher. The H_2O_2 molecules also affect the resistance and current in the electrochemical sensor. A change in the current is also monitored in the enzyme electrochemical glucose sensors. However, the electrochemical sensor requires electrodes and is more complex in terms of device structure. ZnO nanorods function as a non-enzyme and non-electrochemical glucose sensor. In addition, high-quality ZnO nanorods can be grown on GaN substrates by a simple chemical technique called the hydrothermal growth technique. Therefore, the glucose sensor made of

ZnO nanorods has a high potential as a disposable commercial glucose sensor.

The other approach for the bioapplication of inorganic nanomaterials is to use DNAs as templates for the growth of nanomaterials, which is significantly affected by the DNA templates as well as the structure and optical and electrical properties of the grown nanomaterials. Since there are four bases in DNA, i.e., A, C, G and C, their combination and arrangement in the DNA should affect the DNA-templated growth of the nanomaterials. Therefore, the study of self-assembly growth using DNA templates will ultimately lead to a label-free DNA detection. As the first step, conjugated ssDNAs, poly G and poly C, were selected for the growth of the chalcogenide in our study. In contrast to the growth of CdS nanoparticles using the double-stranded DNA, the CdS nanowires were grown using the poly G and poly C DNAs as a template. The difference in the structure was attributed to the ordered formation of the $Cd^{2+}-PO_2^{4-}$ complex before the Watson–Crick base pairing of poly G and poly C occurs. The electrostatic interaction, charge condensation, charge reverse and base pairing are well controlled for self-assembly by the sequential chemical synthesis process.

There has been an increase in interest in nanomaterials. Inorganic nanomaterials possess unique electrical and optical properties, which improve the performance of conventional electronic and optical devices and even lead to novel functional devices. However, the application of nanomaterials in commercial devices is always hindered by factors such as control of the size, uniformity and positioning. It is always a challenge to develop nanotechnology for the controlled synthesis of nanomaterials. Self-assembly using DNA templates may be a good solution to the controlled growth. The DNA molecules cap the nanoparticles and inhibit further growth. As a result, the size distribution can be narrowed. The positioning can be well controlled by DNA scaffolds. The DNA itself exhibits self-assembly. The well-organized DNA scaffolds can be used as a base for the growth of inorganic nanomaterials.

More research is needed on the topics of DNA self-assembly and DNA-assisted self-assembly, which are highly interdisciplinary and use techniques from multiple disciplines such as biochemistry, physics, chemistry, material science, computer science and mathematics. A breakthrough in nano-bioscience and technology leading to commercial applications is only possible by a team effort of researchers and engineers from different fields.

Acknowledgement

The authors are grateful to the late Professor S. N. Sahu then at Institute of Physics for his dedicated contribution to glucose sensors and the DNA-templated growth of chalcogenide nanomaterials reviewed in the chapter. We are certain that we would not have been able to achieve great success in the field of nano-bioscience without the dedication of Professor Sahu. He is sorely missed by his students and colleagues in the nanoscience community. Some of the projects reviewed in the chapter were supported in part by the Indo-JSPS project (DST-JSPS Project No. DST/JAP/P68/09). One of the authors (S. N. Sarangi) gratefully acknowledges the JSPS Postdoctral Research Fellowship for Foreign Researchers.

References

Ahmad, M.; Pan, C.; Luo, Z.; Zhu, J. 2010. A single ZnO nanofiber-based highly sensitive amperometric glucose biosensor. *J. Phys. Chem. C* 114: 9308–9313.

Ahmad, R.; Tripathy, N.; Ahn, M.-S.; Hahn, Y.-B. 2017. Solution process synthesis of high aspect ratio ZnO nanorods on electrode surface for sensitive electrochemical detection of uric acid. *Sci. Rep.* 7. Article number: 46475.

Bourdon, E.; Loreau, N.; Blanche, D. 1999. Glucose and free radicals impair the antioxidant properties of serum albumin. *FASEB J.* 13: 233.

Braun, E.; Eichen, Y.; Sivan, U.; Ben-Yoseph, G. 1998. DNA-templated assembly and electrode attachment of a conducting silver wire. *Nature* 391: 775–778.

Chaiyo, S.; Mehmeti, E.; Siangproh, W.; Hoang, T. L.; Nguyen, H. P.; Chailapakul, O.; Kalcher, K. 2018. Non-enzymatic electrochemical detection of glucose with a disposable paper-based sensor using a cobalt phthalocyanine–ionic liquid–graphene composite. *Biosens. Bioelectron.* 102: 113–120.

Chen, J.; Luo, Z.; Wang, Y.; Huang, Z.; Li, Y.; Duan, Y. 2018. DNA specificity detection with high discrimination performance in silver nanoparticle coupled directional fluorescence spectrometry. *Sens. Actuators B Chem.* 255: 2306–2313.

Clark, L. C.; Lyons, C. 1962. Electrode systems for continuous monitoring in cardiovascular surgery. *Ann. N. Y. Acad. Sci.* 102: 29–45.

Coffer, J. L.; Bigham, S. R.; Pinizzotto, R. F.; Yang, H. 1992. Characterization of quantum-confined CdS nanocrystallites stabilized by deoxyribonucleic acid (DNA). *Nanotechnology* 3: 69–76.

Coffer, J. L.; Bigham, S. R.; Li, X.; Pinizzotto, R. F.; Rho, Y.; Young, G.; Pirtle, R. M.; Pirtle, I. L. 1996. Dictation of the shape of mesoscale semiconductor nanoparticle assemblies by plasmid DNA. *Appl. Phys. Lett.* 69: 3851–3854.

Dwivedi, D. K.; Dayashankara; Dubey, M. 2010. Synthesis, structural and optical characterization of CdS nanoparticles. *J. Ovonic Res.* 6: 57–62.

Evans, H.M.; Ahmad, A.; Ewert, K.; Pfohl, T.; Martin-Herranz, A.; Briunsma, R.F.; Safnya, C.R. 2003. Structural polymorphism of DNA-dendrimer complexes. *Phys. Rev. Lett.* 91: 075501–075504.

Fang, Bi-Y.; Li, C.; An, J.; Zhao, S. D.; Zhuang, Z. Y.; Zhao, Y. D.; Zhang, Y. X. 2018. HIV-related DNA detection through switching on hybridized quenched fluorescent DNA-Ag nanoclusters. *Nanoscale* 10: 5532–5538.

Goodman, S. M.; Noh, H.; Singh, V.; Cha, J. N. 2015. Charge transport through exciton shelves in cadmium chalcogenide quantum dot-DNA nano-bioelectronic thin films. *Appl. Phys. Lett.* 106: 83109–83114.

Güllü, Ö.; Çankaya, M.; Bariş, Ö.; Türüt, A. 2008. DNA-modified indium phosphide Schottky device. *Appl. Phys. Lett.* 92: 212106.

Harris, J. M.; Reyes, C.; Lopez, G. P. 2013. Common causes of glucose oxidase instability in in vivo biosensing: A brief review. *J Diabetes Sci. Technol.* 7(4): 1030–1038.

Hegazy, M. A.; El-Hameed, A. M. A. 2014. Characterization of CdSe-nanocrystals used in semiconductors for aerospace applications: Production and optical properties. *NRIAG J. Astron. Geophys.* 3: 82–87.

Heller, A.; Feldman, B. 2008. Electrochemical glucose sensors and their applications in diabetes management. *Chem. Rev.* 108: 2482–2505.

Heller, D.A.; Jeng, E.S.; Yeung, T.K.; Martinez, B.M.; Moll, A.E.; Gastala, J.B.; Strano, M.S. 2006. Optical detection of DNA conformational polymorphism on single-walled carbon nanotubes. *Science* 311: 508–511.

Hussain, A. M. P.; Sarangi, S. N.; Kesarwani, J. A.; Sahu, S. N. 2011. Au-nanocluster emission based glucose sensing. *Biosens. Bioelectron.* 29: 60–65.

Kim, K.; Kim, T. G.; Sung, Y. 2012. Enzyme-conjugated ZnO nanocrystals for collisional quenching-based glucose sensing. *CrystEngComm* 14: 2859.

Knez, M, Bittner, A. M.; Boes, F.; Wege, C., Jeske, H.; Mai, E.; Kern, K. 2003. Biotemplate Synthesis of 3-nm nickel and cobalt nanowires. *Nano Lett.* 3: 179–182.

Kongstein, O. E.; Bertocci, U.; Stafford, G. R. 2005. In situ stress measurements during copper electrodeposition on (111)-textured Au. *J. Electrochem. Soc.* 152: C116–C122.

Kulkarni, S. K.; Athiraj, A. S.; Kharrzi, S; Deobagkar D. N.; Deobagkar D. D. 2005. Synthesis and spectral properties of DNA capped CdS nanoparticles in aqueous and non-aqueous media. *Biosens. Bioelectron.* 21: 95–102.

Liu, Z.; Zu, Y.; Fu, Y.; Zhang, Y.; Liang, H. 2008. Growth of the oxidized nickel nanoparticles on a DNA template in aqueous solution. *Mater. Lett.* 62: 2315.

Murphy, C. J.; Mahtab, R.; Caswell, K.; Gearheart, L.; Jana, N. R. 2001. Inorganic nanoparticles as optical sensors of DNA. *Proceedings of SPIE* Vol. 4258, *The International Symposium on Biomedical Optics*, San Jose, CA.

Nayak, J.; Sahu, S. N.; Kasuya, J.; Nozaki, S. 2008. Effect of substrate on the structure and optical properties of ZnO nanorods. *J. Phys. D Appl. Phys.* 41: 115303–115308.

Nithyaja, B.; Vishnu, K.; Mathew, S.; Radhakrishnan P.; Nampoori, V. P. N. 2012. Studies on CdS nanoparticles prepared in DNA and bovine serum albumin based biotemplates. *J. Appl. Phys.* 112: 064704.

Peng, H. I.; Strohsah, C. M.; Leach, K. E.; Krauss, T. D.; Miller, B. L. 2009. Label-free DNA detection on nanostructured Ag surfaces. *ACS Nano* 3: 2265–2273.

Rahman, Md. M.; Ahammad, A. J. S.; Jin, J.-H.; Ahn, S. J.; Lee, J.-J. 2010. A comprehensive review of glucose biosensors based on nanostructured metal-oxides. *Sensors* 10: 4855–4886.

Rath, S.; Sarangi, S.N.; Sahu, S. N. 2007. Biocatalytic growth of semiconductor nanowires. *J. Appl. Phys.* 101 (7): 074306.

Safavi, A.; Maleki, N.; Farjami, E. 2009. Fabrication of a glucose sensor based on a novel nanocomposite electrode. *Biosens. Bioelecron.* 24: 1655–1660.

Sarangi, S. N.; Sahu, S.N. 2004. CdSe nanocrystalline thin films: composition, structure and optical properties. *Physica E Low Dimens. Syst. Nanostruct.* 23: 159–167.

Sarangi, S. N.; Goswami, K.; Sahu, S. N. 2007. Biomolecular recognition in DNA tagged CdSe nanowires. *Biosens. Bioelectron.* 22: 3086–3091.

Sarangi, S. N.; Rath, R.; Goswami, K.; Nozaki, S.; Sahu, S. N. 2010. DNA template driven CdSe nanowires and nanoparticles: Structure and optical properties. *Physica E Low Dimens. Syst. Nanostruct.* 42: 1670–1674.

Sarangi, S. N.; Nozaki, S.; Sahu, S. N. 2015. ZnO nanorod based non-enzymatic optical glucose sensing. *J. Biomed. Nanotechnol.* 11: 988–996.

Sarangi, S. N.; Sahu, S. N.; Nozaki, S. 2018. CdS nanowires formed by chemical synthesis using conjugated single-stranded DNA molecules. *Physica E Low Dimens. Syst. Nanostruct.* 97: 64–68.

Sato, S.; Nozaki, S.; Morisaki, H. 1997. Photo-oxidation of germanium nanostructures deposited by the cluster-beam evaporation technique. *J. Appl. Phys.* 81: 1518.

Tan, S. J.; Campolongo, M. J.; Luo, D.; Cheng, W. 2011. Building plasmonic nanostructures with DNA. *Nat. Nanotechnol.* 6: 268–276.

Wang, J. 2008. Electrochemical glucose biosensors. *Chem. Rev.* 108: 814–825.

Wang, L.; Sun, Y.; Li, Z.; Wu A.; Wei, G. 2016. Bottom-Up synthesis and sensor applications of biomimetic nanostructures. *Materials* 9: 53 (1–28).

Wang, Y; Tang, Z; Kotov, N. A. 2005. Bio-application of nano-semiconductors. *Nanotoday* 8: 20–31.

Wu, G.; Song, X.; Wua, Y.; Chen, X.; Luo, F.; Chen, X. 2013. Non-enzymatic electrochemical glucose sensor based on platinum nanoflowers supported on grapheme oxide. *Talanta* 105: 379–385.

Wu, W.; Mitra, N.; Yan, E. C. Y.; Zhou, S. 2010. Multifunctional hybrid nanogel for integration of optical glucose sensing and self-regulated insulin release at physiological pH. *ACS Nano* 4: 4831.

Wu, Y.; Eisele, K.; Doroshenko, M.; Algara-Siller, G.; Kaiser, U.; Koynov, K.; Weil, T. 2012. A quantum dot photoswitch for dna detection, gene transfection, and live-cell imaging. *Small* 8: 3465–3475.

Xu, S.; Sheng, Z.; Wang, L. 2011. One-dimensional ZnO nanostructures: Solution growth and functional properties. *Nano Res.* 4: 1013 (86 page).

Yao, Y.; Song, Y.; Wang, L. 2008. Synthesis of CdS nanoparticles based on DNA network templates. *Nanotechnology* 19: 405601.

Young, E. M. 1988. Electron-active silicon oxidation. *Appl. Phys. A* 47: 259–269.

Yum, K.; Ahn, J. H.; McNicholas, T. P.; Barone, P. W.; Mu, B.; Kim, J. H.; Jain, R. M.; Strano, M. S. 2012. Boronic acid library for selective, reversible near-infrared fluorescence quenching of surfactant suspended single-walled carbon nanotubes in response to glucose. *ACS Nano* 6: 819–830.

Zan, G.; Wu, Q. 2016. Biomimetic and bioinspired synthesis of nanomaterials/nanostructures. *Adv. Mater.* 28: 2099–2147.

Zhang, B.; Zhang, Y.; Mallapragada, S. K.; Clapp, A. R. 2011. Sensing polymer/DNA polyplex dissociation using quantum dot fluorophores. *ACS Nano*, 5: 129–138.

Zhang, Y.; Ichaihashi, T.; Landree, E.; Nihey, F.; Iijima, S. 1999. Heterostructures of single-walled carbon nanotubes and carbide nanorods. *Science* 285: 1719–1722.

Zhou, Z. Y.; Bedwell, G. J.; Li, R.; Prevelige, P. E.; Gupta, A. 2014. Formation mechanism of chalcogenide nanocrystals confined inside genetically engineered virus-like particles. *Sci. Rep.* 4: 3832.

8

Engineering Living Materials: Designing Biological Cells as Nanomaterials Factories

Peter Q. Nguyen,
Pichet Praveschotinunt,
Avinash Manjula-Basavanna,
Ilia Gelfat, and Neel S. Joshi
Harvard University

8.1 Introduction

The field of synthetic biology has integrated decades of molecular biology, biochemistry, and systems biology knowledge toward applied engineering efforts for modular and forward engineering of biological systems. In parallel, biomaterials research has focused on the engineering of wholly biological or biologically compatible large-scale materials with defined micro- or nanoscale properties for applications in organ and tissue engineering, drug delivery, and a myriad of other applications. Where these efforts begin to overlap, there is a rich area for exploring the design and engineering of materials with novel properties combining elements of living systems with advanced materials. In particular, we will focus on materials that derive their characteristics from nanoscale and molecular structures or patterns.

At the core of the ELM concept is the realization that there are many large-scale materials created and maintained by living organisms that materials scientists struggle to replicate (Figure 8.1a). Perhaps there is no better example than wood, a ubiquitous construction material that humans have used since prehistoric times. The mechanical properties of wood are a result of its multi-scale hierarchical architecture and compositional complexity (Burgert 2007; Fratzl and Weinkamer 2007), resulting in a versatile material that can be strong, flexible and light enough to construct 70-story buildings, exquisite violins, or even mammoth cargo airplanes. But consider that wood is stitched together, atom by atom, from nothing more than air and water by plant cells following an evolved genetic construction algorithm. Other wonderfully complex materials created from natural biological systems include mechanically robust bone in animals, tough and durable nacre from abalone (Tang et al. 2003), ultrastrong teeth made of magnetite in chitons and limpets (Barber, Lu, and Pugno 2015; Gordon and

Joester 2011), nano- and micro-structures for absorbing, reflecting, harvesting, or tuning of light in various organisms (Jordan, Partridge, and Roberts 2012; McCoy et al. 2018; Zhang and Chen 2015), tough protein-based spider silk polymers (Agnarsson, Kuntner, and Blackledge 2010), linear arrays of magnetic nanoparticle "compasses" (Bazylinski and Frankel 2004), and naturally antimicrobial cicada and dragonfly wings (Bhadra et al. 2015; Pogodin et al. 2013), among many, many others. Given our advances in synthetic biology engineering, can we begin to create similar synthetic advanced materials in which living cells are considered a part of the material itself? To delineate ELMs from the wide range of engineered hybrid materials that may contain cells in close association with other synthetic materials, we define ELMs as synthetic materials in which the living component of the ELM creates, organizes, or modifies the material on the nano- to macroscale in a programmable manner (Nguyen 2017; Nguyen et al. 2018). It should be noted, however, that the resulting materials could be processed to remove the living cells after fabrication to yield an abiotic material.

One central theme of biological systems, and by extension, ELMs, is that the materials are formed through the principles of self-assembly. This powerful design principle utilizes intermolecular interactions that occur on the atomic scale to drive the bottom–up organization of molecules to yield hierarchically complex materials (Kushner 1969). Although this has its obvious advantages, integrating self-assembly into a rational engineering design paradigm is challenging due to the complexity of these spatiotemporal interactions. How do we control the assembly process in space and time to create desired structures on the nano-, micro-, and macroscale? If we can build them, how can we predict the cooperative interactions of these structures with each other on length scales from the atomic to macroscopic levels? These questions of

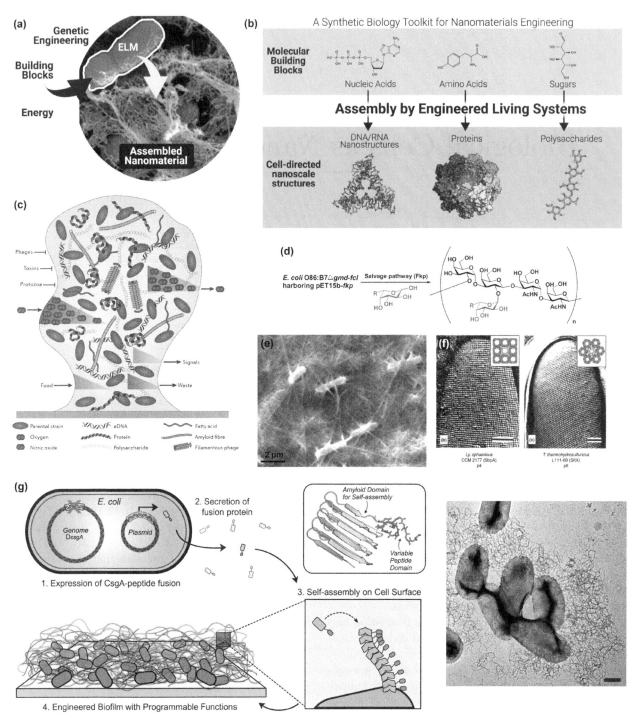

FIGURE 8.1   (a) ELMs use cells to fabricate, organize and modulate self-assembling materials. (b) Fundamental classes of biomolecules that can be designed on the molecular level to create user-defined nanoscale structures. (c) Biofilms are composed of a community of bacterial cells that create their own polymeric environment, known as the extracellular matrix, which is composed largely of polysaccharides, proteins and extracellular DNA. (Reproduced with permission from McDougald et al. 2011.) (d) By altering the microbial pathways for polysaccharide assembly, various chemical functional groups (shown in gray) can be integrated into polysaccharides. (Reproduced with permission from Yi et al. 2009.) (e) Scanning electron micrographs of *K. rhaeticus* iGEM encased in bacterial cellulose that they are engineered to secrete. Images are taken after 8 days of growth at 6,000× magnification. (Reproduced with permission from Florea et al. 2016.) (f) Examples of S-layer proteins that self-assemble on the surface of some bacteria to form nanostructured crystal lattices. Scale bars are 100 nm. (Reproduced with permission from Ilk, Egelseer, and Sleytr 2011.) (g) Programmed functionalization of protein nanofibers by genetic engineering of the curli amyloid system of *E. coli*. Left, schematic of the pathway. Right, transmission electron micrograph of engineered *E. coli* cells secreting curli nanofibers. Scale bar, 500 nm. (Reproduced with permission from Nguyen et al. 2014.)

process and production control and hierarchically organized self-assembly are key challenges in ELM development.

All living cells consist of structural elements that can be broadly classified into nucleic acids (DNA and RNA), proteins, polysaccharides, and lipids. Of these, nucleic acids and proteins are most amenable to direct programming, through genetic engineering (Figure 8.1b). Recently, there have been exciting advances in the design of self-assembling DNA and RNA assemblies to create complex user-defined nanoscale structures, including drug-delivering robots, and arbitrary two- and three-dimensional shapes (Han et al. 2017; Ong et al. 2017; Amir et al. 2014; Douglas et al. 2009). However, much of this research uses in vitro reactions of nucleic acids to assemble the structures, although there is a recent attempt to use programmed DNA in cells for assembling nanoscale structures (Elbaz, Yin, and Voigt 2016). Proteins, on the other hand, can perform a diverse array of functions, such as chemical catalysis by enzymes, forming structural elements, information processing, and templating of inorganic molecules into useful materials or nanostructures. Largely considered the "workhorses" of the cell, assemblies of proteins work in concert to synthesize all other components of the cells, including nucleic acids, polysaccharides, lipids, and even proteins themselves. As these biological polymers are genetically encoded, they can be extensively reengineered and altered rationally by programming the DNA sequence. Given their central importance in living systems, the functional and structural diversity, ability to direct the assembly inorganic elements, and the high programmability of proteins, they are likely to be the material platform of choice for developing ELMs. Indeed, much of the ELM examples presented in this chapter exclusively use proteins as the fundamental material for direct engineering. The ability of proteins to form nanoparticles, nanofibers, and crystalline assemblies along with their evolved ability to draw from environmental inorganic molecules to form highly complex materials also enables the design of materials. Protein engineering can also enable the design of synthetic polysaccharides, by either random incorporation of unnatural monosaccharides or directing the efficient assembly of polysaccharides with novel properties, such as nanocellulose. Viruses that infect bacteria, known as bacteriophages, will also be covered as a potential platform for ELM development.

8.2 ELMs for Constructing Bionanomaterials

In this section, current and past efforts toward building engineered living materials that produce biological materials that can be genetically programmed with desired nanoscale properties will be discussed. With the goal of creating the living materials that are capable of autonomous growth, self-healing, and environmental responsiveness, scientists need to use genetic tools, chemical modifications or spatial patterning to direct the organisms to produce the materials in a programmable manner (Nguyen et al. 2018). Among organisms that can be suitably engineered by such techniques, bacteria are the simplest system that can secrete extracellular matrix components including polysaccharides, proteins, and DNA during the formation of biofilms (Figure 8.1c), which reflect the most elementary form of living materials (i.e., cells that generate an extracellular matrix) (Flemming and Wingender 2010). These biological materials together with living bacteria can be controlled and modified, to different extents, to have functional properties that are of interest to nanomaterial engineers. Therefore, much effort has recently been focused into developing engineered bacteria as stepping stones for developing advanced ELMs.

As we previously discussed, two major classes of biological materials that are suitable for ELM development are polysaccharides and proteins. Within the polysaccharide category, scientists have attempted to make structural modifications to or alter synthesis pathways of cell surface polysaccharides, exopolysaccharides and bacteria celluloses within engineered microbes. For the cell surface polysaccharides, the Wang group has demonstrated a novel method to incorporate monosaccharide analogs into bacterial polysaccharides by altering the bacteria's sugar nucleotide (GDP-Fucose) biosynthetic pathway. These novel sugar analogs (Figure 8.1d) contain functional groups such as azide, alkyne, ketone, and amine, which are extremely useful for selective and bioorthogonal chemical reactions (Yi et al. 2009).

Hyaluronic acid, which is a polysaccharide found in extracellular matrices of all vertebrates, plays important roles in joints and connective tissues of animals. It is another important target for nanoscale engineering because of its applications in cosmetic and medicinal fields. Researchers used metabolic engineering to enhance the ability of *Streptococcus zooepidemicus* and *Bacillus subtilis* to produce hyaluronic acid for commercial use (Blank Lars, McLaughlin Richard, and Nielsen Lars 2005; Chong et al. 2005). Moreover, microbes such as *Pseudomonas aeruginosa* and *Azotobacter vinelandii* can produce alginate, which is widely used for tissue engineering scaffolds and is usually extracted from the brown seaweed in the industrial scale. Scientists have attempted to increase the alginate production and alter alginate's molecular structure and properties using metabolic engineering in these bacteria as well (Ruffing and Chen 2006). If researchers are successful in engineering these two polysaccharides as living materials, such materials will be extremely useful toward medical usage in the future.

Another important exopolysaccharide known to be produced by bacteria is bacterial cellulose. Due to its vast variety of applications, bacterial cellulose has been one of the main focuses in developing ELMs. *Gluconacetobacter xylinum*, the first discovered and the most well-studied cellulose-producing bacteria, has been engineered to alter its cellulose producing properties in various ways, ranging from enhancing the cellulose production (Chandrasekaran, Bari, and Sinha 2017) to incorporating other exopolysaccharides, such as curdlan, via genetic engineering to enhance mechanical properties of cellulose (Fang et al. 2015). Another strain of cellulose-producing bacteria, *Komagataeibacter rhaeticus*, has also gained attention because it can produce cellulose at high yields, grow in low-nitrogen conditions, and resist toxic chemicals. Researchers have created genetic tools to manipulate and rationally reprogram *K. rhaeticus*. They were able to have control over cellulose production, and to create patterned and functionalized cellulose with living bacteria (Figure 8.1e) (Florea et al. 2016). Overall, the engineering of polysaccharides for complex living materials will continue to advance over time. However, to modify polysaccharides, researchers will have to alter their complex biosynthesis pathways, which has been one of the major hurdles in the field. Next, we will discuss the engineering of protein scaffolds for living materials, which could be more convenient from a genetic engineering point of view.

Researchers focus on manipulating proteins as structural and functional components of ELMs for multiple reasons. Protein engineering offers precise control of peptide sequences based on genetic codes introduced into the organisms, enabling intricate features, such as programmed self-assembly, specific binding to different materials, and the display of modular functional proteinaceous domains. Early efforts of engineering programmable ELM-based scaffolds from bacteria involve bacterial cell surface display, which enables bacterial cells to interact with specified targets (Daugherty 2007). Along a similar trajectory, scientists have also engineered S-layer proteins, which are parts of cell envelopes found in archaea and many bacteria and play roles in self-defense, adhesion, and barrier functions by adopting two-dimensional crystalline patterns (Figure 8.1f). S-layer proteins can be fused to heterologous, functional domains such as enzymes, ligands, antibodies, or antigens to create novel nanobiomaterials beneficial for creating biocompatible surfaces, mucosal vaccines, bioremediation and biomineralization platforms, affinity matrices, and microcarriers (Ilk, Egelseer, and Sleytr 2011; Schuster and Sleytr 2014).

Functional amyloids, especially bacteria associated curli nanofibers, are bacterial extracellular matrix proteins that have been an intense focus for development of engineered living material due to its modularity, robustness, and self-assembly properties. Curli fibers, found in most Enterobacteriaceae, consist of monomeric, β-sheet-rich CsgA protein monomers that can self-polymerize extracellularly and remain associated with the bacterial cells (Blanco et al. 2012). Each CsgA monomer can be genetically mutated or fused with heterologous domains to create fully assembled, engineered curli fibers with desired properties (Figure 8.1g). The Lu Group has explored the possibility of producing hybrid curli fibers *in situ* using two groups of *Escherichia coli* harboring different engineered CsgA plasmids with separate inducible systems. There were able to create hybrid fibers with tunable, co-block polymer-like features (Chen et al. 2014). The group also recently explored the use of engineered curli fibers for gold nanoparticle

templating to create electrically conductive fibers (Seker et al. 2017). Concurrently, the Joshi group has developed the biofilm-integrated nanofiber display (BIND) platform which initially surveyed the possibility of displaying various functional domains at different lengths and structures to test the limit of the system and demonstrate the functional properties of displayed domains on the engineered fibers, such as affinity binding and specific nanoparticle templating (Nguyen et al. 2014). The Joshi group has expanded upon multiple possibilities of this platform as living materials. For example, curli fibers as catalytic surfaces were made by displaying SpyTag peptides on CsgA monomers, which in the presence of its cognate protein partner, SpyCatcher, is able to form specific biorthogonal covalent linkages. The SpyCatcher proteins heterologously fused with various enzymes of interest were used to create monolithic living catalytic surfaces (Botyanszki et al. 2015; Nussbaumer et al. 2017). The group has also engineered curli fibers to bind contaminated mercury in seawater for bioremediation applications with the use of synthetic biology to control expression of curli based on the presence of mercury (Tay, Nguyen, and Joshi 2017). In addition, the group has explored the use of BIND platform to interact with biological systems by displaying therapeutic factors on curli fibers, which implies the development of a living therapeutic platform inside the gut (Duraj-Thatte et al. 2018). Engineering of extracellular matrix proteins such as curli fibers is a significant step toward building truly advanced ELMs as the strategy involves secreting engineerable, self-assembling proteins which remain associated with live cells that continuously produce more of the materials. With the tweaking of components in the curli biogenesis pathway, and further development of the system, researchers may be able to create more complex, hierarchical living materials in the future.

These efforts mark the initial stages of developing engineered living materials with the long-term goal of having the ability to engineer structural and functional properties directly into the genomes of the organisms without the need of temporary scaffolds, or external and environmental cues to direct the formation of the living materials. Further advances in scientific knowledge and technology toward manipulating the development of cells and their interactions in three dimensions will be required to achieve the future goals of cutting-edge ELM-based materials (DARPA 2016).

8.3 ELMs for Guiding the Formation of Inorganic Nanomaterials

Integration of the living cellular milieu with inorganic components gives rise to distinct mechanical, electrical, optical, magnetic, and catalytic properties (Nguyen et al. 2018; Kehoe and Kay 2005; Wen and Steinmetz 2016). In order to develop such functional materials, living cells and/or its components have been strategically employed to template inorganic nanomaterials. Herein, we present exclusive examples of ELMs for the growth and formation of functional inorganic nanomaterials.

The engineered *E. coli* curli nanofibers ($\sim$4 nm) described in the last chapter were genetically engineered to nucleate gold nanoparticles, and in turn form gold nanowires (Seker et al. 2017). The gold nanoparticle size was found to vary depending on the binding affinity of the peptide tag displayed on CsgA. The curli nanofibers that are inherently non-conducting became electrically conducting ($\sim$1 nanosiemens conductance) upon coating with gold nanoparticles. By a tight regulation on the expression of CsgA via an inducer anhydrotetracycline, an environmentally responsive biofilm-based electrical switch was developed (Figure 8.2a). Curli nanofibrils were also utilized to template quantum dots such as CdTe, CdS and ZnS (Chen et al. 2014). When gold nanoparticles were co-assembled with CdTe/CdS on curli nanofibers, the fluorescence lifetimes and intensities of the quantum dot was significantly altered and such plasmon–exciton interactions of gold–quantum dot heteroarchitectures might allow effective modulation of photon-emission properties (Figure 8.2b).

Besides bacteria, viruses such as the filamentous bacteriophage M13 are also employed to bind to inorganic nanomaterials (Kehoe and Kay 2005; Wen and Steinmetz 2016). Wild-type M13 is 6.5 nm in diameter and 930 nm in length. A typical M13 virion is covered with 2700 copies of the major coat protein (pVIII) and 5 copies of minor coat proteins (pIII, pVI, pVII and PIX). pIII and pVI form the one end of the virion, while pVII and pIX are at the other end. Peptides/proteins having affinity to a specific material are typically selected from a phage display library. In one particular example, a 15-mer peptide library displayed on the filamentous phage was used to investigate the crystallization of calcium carbonate (Li, Botsaris, and Kaplan 2002). The template effect of the phage-displayed peptides resulted in hollow spheres of calcium carbonate nanoparticles (Figure 8.2c). Interestingly, the bacteriophage was found to slow down the phase transformation from vaterite to calcite.

In another report, zinc sulfide-binding peptides were displayed on pIII minor coat proteins of the M13 bacteriophage (Lee et al. 2002). When ZnS solution precursors were coupled with the bacteriophage, a self-supporting hybrid film having nanoscale and microscale ordering extending over centimetre length scales was obtained. Moreover, the bacteriophage–ZnS hybrid films showed liquid crystalline behaviour, while their crystalline phases could be modulated by solvent concentration and by the use of an external magnetic field. On the other hand, when tetra-glutamate residues were genetically fused to the N-terminus of major coat pVIII protein, bacteriophages were enabled to bind cobalt ions, which upon reduction and spontaneous oxidation resulted in crystalline cobalt oxide (Co_3O_4) nanowires (Nam et al. 2006). Further, a bifunctional virus template was designed that simultaneously expressed cobalt- and gold-binding peptides on its major coat proteins. Incubation of gold nanoparticles and cobalt chloride with

a bifunctional-virus template yielded gold nanoparticles interspersed within the Co_3O_4 wires (Figure 8.2d). The virus-templated Au-Co_3O_4 hybrid electrode showed remarkable improvement in electrochemical performance, and the specific capacity of the hybrid nanomaterial was 30% greater than that of Co_3O_4 nanowires.

By engineering two genes of M13, amorphous iron phosphate and single-walled carbon nanotube (SWNT)-binding peptide tags were displayed on pVIII and pIII coat proteins, respectively (Lee et al. 2009). The amorphous iron phosphate assembled on the major coat protein enabled fabrication of lithium-ion batteries with good power performance. Interestingly, with the inclusion of highly conducting SWNTs, the resulting hybrid electrode of M13-templated amorphous iron phosphate showed excellent capacity retention upon recycling at 1°C, and the power performance was comparable to that of crystalline lithium iron phosphate. Similarly, M13-templated manganese oxide nanowires having spherulitic surface morphology were obtained by displaying the peptide tags on pVIII (Oh et al. 2013). The hierarchical morphology of manganese oxide nanowire engenders large catalytic area and storage space for discharge products. To enhance the oxygen reduction reaction activity and surface conductivity, catalytic metal nanoparticles like gold and palladium were incorporated onto the surface of the M13-templated manganese oxide nanowires. To facilitate homogeneous distribution of metal nanoparticles, polyacrylic acid was introduced on to the manganese oxide nanowires before incorporating nanoparticles. The resulting hybrid cathode not only outperformed the material made by mere mechanical mixing but also improved the capacity and life cycle of lithium–oxygen batteries. M13 bacteriophage was also genetically engineered to bind monodisperse iron oxide nanoparticles along its coat and in addition, by displaying the SPARC glycoprotein (over-expressed in various cancers) on its distal end, allowing for efficient magnetic resonance imaging of cancer cells and tumours in mice (Ghosh et al. 2012).

Recently, blue-light-responsive *E. coli* was developed to bind to a wide variety of quantum dots having distinct fluorescence emission colours such as red CdSeS@ZnS, green CdZnSeS@ZnS, and blue CdZnS@ZnS (Wang et al. 2018). Here, the CsgA having Histidine tag (CsgA$_{His}$) bound to nitrilotriacetic acid-decorated quantum dots via metal coordination chemistry. The nitrilotriacetic acid-decorated quantum dots were co-incubated with the designed cells, which upon illumination with 470 nm blue-light promotes the expression, secretion and self-assembly of CsgA$_{His}$ into curli fibers that in turn templates the coordination with the quantum dots. By using masks having specific hole patterns with pore diameters as small as 50 μm, macro-/micro-scale as well as hierarchical patterned assembly of red, green, and blue quantum dots on *E. coli* biofilms was achieved.

It is thus strongly believed that the cellular-guided organization of heterogeneous inorganic nanomaterials would not only integrate the properties of living and non-living entities but also enable the fabrication of advanced ELM materials

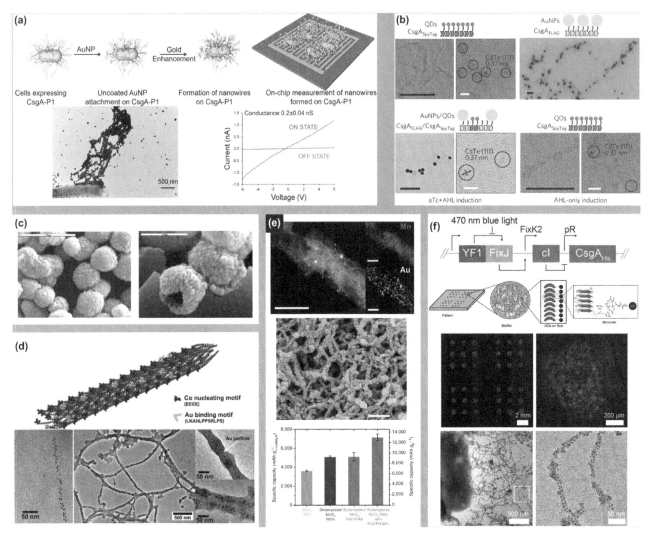

FIGURE 8.2 (a) Conductive gold nanowires templated on engineered curli nanofibers by selective attachment of gold nanoparticles followed by gold enhancement. Top, schematic of process. Bottom left, TEM image of formed templated nanowires. Bottom right, conductance measurement; "ON" state is when the genetic circuit is activated. "OFF" state uses uninduced cells. (Reproduced with permission from Seker et al. 2017.) (b) Curli nanofibers can be used to template gold–quantum dot nanoparticle assemblies. Images shown are representative transmission electron micrographs. Black scale bars, 200 nm. White scale bars, 5 nm. (Reproduced with permission from Chen et al. 2014.) (c) Bacteriophage with displayed peptides that selectively bind to CaCO3 were used to template hollow nanoscale vaterite crystalline spheres when incubated in a solution of calcium and carbonate. Left, after 2 hour incubation, scale bar, 5 microns. Right, 24 hour incubation, scale bar 2 microns. (Reproduced with permission from Li, Botsaris, and Kaplan 2002.) (d) Hybrid Au/Co3O4 nanowires for improved lithium battery electrodes, generated from bacteriophage engineered to co-display two different types of templating peptides. Various transmission electron microscopy images are shown. (Reproduced with permission from Nam et al. 2006.) (e) Au/MnOx nanowires for improved lithium–oxygen battery electrodes from bacteriophage display. Top: scanning transmission electron micrograph of the templated nanowires with the corresponding energy dispersive X-ray elemental mapping images of manganese and gold shown on the right, scale bars are 50 nm. Middle: SEM images of the final biotemplated electrode. Scale bar, 500 nm. Bottom: Specific capacity measurements of various tested electrode materials. (Reproduced with permission from Oh et al. 2013.) (f) Photopatterned biofilm displayed quantum dots (QDs). Top: genetic circuit diagram for light-controlled expression of engineered curli and the general scheme of the process. Middle: patterned QDs imaged by fluorescence microscopy. Bottom: transmission electron micrograph of the QDs immobilized on curli nanofibers. (Reproduced with permission from Wang et al. 2018.)

having unprecedented complexity. Most importantly, biological systems offer tremendous opportunities to synthesize materials under ambient conditions, and thereby environmentally benign materials can be economically realised.

8.4 Toward Large-scale and Hierarchically Designed Materials

Although at the time of this publication most efforts on the development of ELMs for the production of nanomaterials are at the proof-of-concept stages, it would be prudent to briefly consider future hurdles in scaling up this technology for practical usage. Laboratory methods used for traditional nanomaterial synthesis (e.g., vapor deposition, sol-gel, or hydrothermal synthesis) differs from that which is used for ELM production, which is based on traditional bioreactors or fermentation processes, due to the living cells. This distinction may initially seem advantageous, in that various industrial-scale bioproduction processes – for example, beer production, food processing, or recombinant protein production – are very mature technologies. However, this assumption may be misleading, in the same way that traditional chemical synthesis on the industrial scale cannot be directly adapted to the scalable production of many nanomaterials. This disconnect between small laboratory synthesis methods and the need for novel large-scale fabrication methods has led to a severe bottleneck in moving nanomaterials out of the lab and into practical use (A matter of scale 2016).

Similarly, the nuances of designed ELM nanomaterials make it likely that novel fabrication processes and/or substantial modifications of current biofermentation methods will be necessary for scalable production. Much like traditional nanomaterials production, these scaled-up methods should be expected to be individually tailored depending on the peculiarities of the ELM system. In cases where templating onto ELM structures requires the addition of inorganic components, there may be a multistep process combining biofermentation and chemical reactors. Specialty industries, such as pulp manufacturing or wastewater treatment, with their high volume processing and turnover of biomass, may be particularly suited to offer insights when designing such production schemes. A medium-scale manufacturing method for the spatial patterning of ELMs that has been explored recently by a number of different groups is 3D printing. This includes the photopatterning biofilm technology reviewed in the previous section (Wang et al. 2018), as well as additive techniques using hydrogels laden with the engineered cells (Connell et al. 2013; Schaffner et al. 2017; Schmieden et al. 2018). The ability to precisely arrange ELMs allows for the development of complex architectures and compositions not possible by other manufacturing methods. It remains to be seen, however, if these 3D printing technologies can be scaled up for industrial production. One exciting prospect for nanomaterials production that is uniquely suited to ELMs is replacing the model of centralized manufacturing with in situ production, where the desired material is generated on-site. The living cellular nanomaterial factories could draw upon environmental or embedded resources, all packaged together as a freeze-dried inoculant mixture. However, such an in situ production method would be highly dependent on the particular application in which the material would not need to be purified, can form under ambient conditions, and where the release of the genetically modified cells can be controlled.

8.5 Applications of Engineered Living Materials

In the previous sections, we outlined a definition of ELMs, discussed the design principles, highlighted some of the early efforts in the field, and considered enabling technologies necessary for their further development. Here, we present several applications and industries in which ELMs would be particularly relevant.

In biomedical research and the pharmaceutical industry, there is growing recognition of the potential of genetically engineered cells and organisms as therapeutics. Notable examples, which have reached clinical trials, include the use of engineered lymphocytes for cancer immunotherapy (Mellman, Coukos, and Dranoff 2011) and the treatment of Crohn's disease using lactic acid bacteria modified to secrete interleukin-10 (Braat et al. 2006). Bacteria are of particular interest for ELM development, as many commensal and probiotic species are genetically tractable and are known to produce biofilm components that could be engineered. Early efforts in the field include the fusion of trefoil factors – a family of anti-inflammatory cytokines – to curli fibers produced by the probiotic strain *E. coli* Nissle 1917 (Figure 8.3a) (Duraj-Thatte et al. 2018). In addition to treating inflammation, ELMs have also been proposed for *in situ* patching of defects and lesions in the mucus layer of various epithelial surfaces (Axpe et al. 2018). Conceptually, ELMs are particularly well suited for this application, since it requires the design of a material with desirable nanoscale properties (network morphology and surface properties facilitating both nutrient transport and pathogen capture) as well as the continuous production and maintenance of this material by a living organism. For similar reasons, an additional application could be found in sequestration of undesirable or excessive chemical species from the gastrointestinal tract. Currently, large bolus doses of cross-linked polymers are administered to patients to sequester excessive amounts of inorganic ions (mainly potassium, phosphate, or iron) or bile acids (Connor, Lees, and Maclean 2017). The continuous production of a living matrix with tailored nanoscale binding properties could provide an alternative to traditional polymer sequestrants, potentially circumventing many of their side effects and limitations.

The potential of biofilms for wastewater treatment has long been recognized, owing to their high biomass concentration, large surface area, and self-renewal abilities

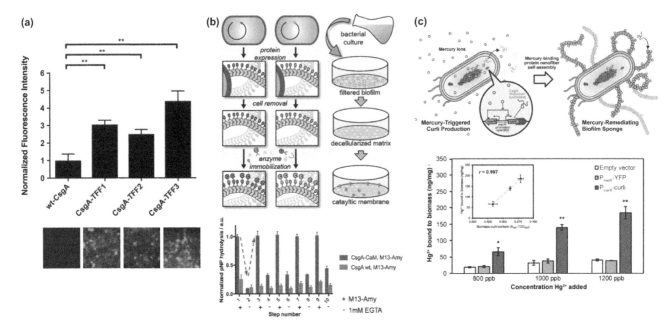

FIGURE 8.3 Curli nanofibers engineered with the mucin-interacting trefoil factor domains can bind to mucus. Shown are fluorescence images using fluorescent anti-mucin 2 antibodies. (Reproduced with permission from Duraj-Thatte et al. 2018.) (b) Top: Schematic for engineered curli nanofibers used to reversibly immobilize enzymes for catalytic applications is presented. Bottom: The enzymes can be alternatively immobilized and removed over ten cycles onto the same curli matrix. Dark gray is the engineered curli biofilms, light gray is the unengineered control biofilm. (Reproduced with permission from Nussbaumer et al. 2017.) (c) Amyloid nanofibers are engineered to express in the presence of mercury and absorb toxic mercury ions. Top: schematic of biofilm sequestration concept. Bottom: quantitative mercury removal by adsorption onto curli nanofibers. Inset shows correlation between total mercury removed and the amount of curli fibers that are present. (Reproduced with permission from Tay, Nguyen, and Joshi 2017.)

(Van Loosdrecht and Heijnen 1993; Qureshi et al. 2005). By rational engineering of microorganisms to incorporate enzymes into the biofilm matrix, they can be tailored to bind and process specific contaminants. The same strategy can be applied to catalysis and manufacturing (Rosche et al. 2009; Halan, Buehler, and Schmid 2012), with some existing work demonstrating the formation of catalytically active engineered biofilms (Figure 8.3b) (Botyanszki et al. 2015; Nussbaumer et al. 2017). ELMs have also been proposed as sequestrants of environmental pollutants, owing to their unique combination of sensing capabilities and material properties. In a recent study by Tay et al. (Tay, Nguyen, and Joshi 2017), the feasibility of this approach was demonstrated by producing biofilm proteins in response to the presence of mercury in the surrounding medium and preferentially binding mercury ions (Figure 8.3c).

The environmentally responsive nature of ELMs can be applied to their use in the construction industry as "smart" building materials, sealants, or coatings. These materials, potentially with self-templated nanomaterials, would be able to sense conditions such as temperature, light, mechanical force or pollutants in the air, and adjust their properties, composition or rate of production accordingly. In a series of studies by Jonkers et al., this concept was implemented for the fabrication of self-healing concrete (de Koster et al. 2015; Jonkers 2011). Spores of various *Bacillus* species were incorporated into blocks of concrete. Upon crack formation and

exposure to water, the spores germinate, and the revived bacteria produce enzymes that catalyze calcium carbonate precipitation. Notably, the incorporation of coated bacterial spores into the concrete altered its mechanical performance, highlighting the need for better control over material properties.

Extrapolating this further, one could imagine the bottom–up formation of macroscopic architectural elements through ELMs, going from nanoscale biomaterials to furniture, public infrastructure, or even entire buildings. Interestingly, a similar fabrication philosophy has already been implemented – if on a small scale – in tree shaping (Ludwig, Schwertfreger, and Storz 2012), though the control over macrostructure is accomplished through physical force and not genetic manipulation.

Finally, the application of ELMs for the electronics industry has been discussed in several publications, striving for the development of bacterial fuel cells and sensors. Significant effort has been made toward harnessing the naturally occurring electrical conductivity of *Geobacter sulfurreducens* and *Shewanella oneidensis* pili (Malvankar et al. 2011; El-Naggar et al. 2010). Advances have also been made in the metabolic engineering of bacteria to facilitate the conversion between cellular metabolism and electrical energy through the use of self-assembling nanostructures (Lovley 2008). Successful practical implementation of ELMs as electronic components, based on some of the exciting demonstrations presented above, would be of particular importance to the field, as it would bridge the interface between ELMs and the vast array of modern electronics and digital processing tools.

Engineered living materials constitute a relatively young field of research, found at the intersection of self-assembling materials and synthetic biology. As our control and understanding of both cells and nanoscale materials expands, so does our ability to utilize ELMs in a myriad of exciting new ways. The applications highlighted here represent a sampling of prominent ideas from various industries but should by no means limit our imagination regarding the possible benefits ELMs can offer.

8.6 Challenges and Future Directions

The past few years have seen many significant advances in the development of foundational ELM technologies and a conceptual exploration of this fruitful intersection of disciplines (Chen, Zhong, and Lu 2015; Nguyen 2017; Nguyen et al. 2018). Although bacterial biofilms have been extensively explored as nanomaterial ELM platforms due to the relative ease of microbial engineering, the large-scale nature of the biofilm material, and self-assembling characteristics, expanding ELMs to use more complex eukaryotic organisms will greatly increase the complexity of the nanomaterials that can be generated. In addition, multi-species consortia that can coordinate the assembly of such nanomaterials would leverage the specific robustness of each species for adapting to the environment and maintaining the material. This would emulate naturally evolved symbiotic communities such as lichen, which can be considered an example of an evolved large-scale living material. A number of outstanding challenges can be seen from the current perspective of ELMs. One is how to rationally design and link self-assembling components in a hierarchical manner. Much of the effort in this area is being undertaken in a separate field, known as synthetic morphogenesis, which is focused on design principles for bottom–up self-organization and pattern formation in living systems (Teague, Guye, and Weiss 2016). We anticipate the future integration of ELM technology with synthetic morphogenesis advances to yield a new generation of living materials that can organize nanoscale elements with greater precision, complexity, and robustness. As ELMs are further developed, novel production methods appropriate for the desired production scale will likely be required to translate these efforts for practical deployment. And all of these efforts will be driven by the demand for these novel living functional materials for specific applications. Thus, it is critical that the application space for these unique materials be thoroughly explored, to identify key unmet needs that cannot be addressed using conventional materials that are purely biological or synthetic. ELMs represent a novel attempt to integrate the properties of self-replicating biological systems with

that of nanoscale advanced materials, hinting at a different bioinspired strategy for the development of nanotechnology replicators capable of programmable fabrication of designer materials (Drexler 1990). As more and more nanomaterial researchers embrace and expand ELM concepts, we look to an emerging future in which instead of being physically or chemically assembled, nanotechnology is instead grown.

References

Agnarsson, Ingi, Matjaž Kuntner, and Todd A. Blackledge. 2010. Bioprospecting finds the toughest biological material: Extraordinary silk from a giant riverine Orb Spider. *PLOS ONE* 5 (9):e11234.

Amir, Yaniv, Eldad Ben-Ishay, Daniel Levner, Shmulik Ittah, Almogit Abu-Horowitz, and Ido Bachelet. 2014. Universal computing by DNA origami robots in a living animal. *Nature Nanotechnology* 9:353.

Axpe, Eneko, Anna Duraj-Thatte, Yin Chang et al. 2018. Fabrication of amyloid curli fibers–alginate nanocomposite hydrogels with enhanced stiffness. *ACS Biomaterials Science & Engineering* 4:2100–2105.

Barber, Asa H., Dun Lu, and Nicola M. Pugno. 2015. Extreme strength observed in limpet teeth. *The Royal Society Interface* 12: 20141326.

Bazylinski, Dennis A., and Richard B. Frankel. 2004. Magnetosome formation in prokaryotes. *Nature Reviews Microbiology* 2:217.

Bhadra, Chris M., Vi Khanh Truong, Vy T. H. Pham et al. 2015. Antibacterial titanium nano-patterned arrays inspired by dragonfly wings. *Scientific Reports* 5:16817.

Blanco, Luz P., Margery L. Evans, Daniel R. Smith, Matthew P. Badtke, and Matthew R. Chapman. 2012. Diversity, biogenesis and function of microbial amyloids. *Trends in Microbiology* 20 (2):66–73.

Blank Lars, M., L. McLaughlin Richard, and K. Nielsen Lars. 2005. Stable production of hyaluronic acid in Streptococcus zooepidemicus chemostats operated at high dilution rate. *Biotechnology and Bioengineering* 90 (6):685–693.

Botyanszki, Zsofia, Pei Kun R. Tay, Peter Q. Nguyen, Martin G. Nussbaumer, and Neel S. Joshi. 2015. Engineered catalytic biofilms: Site-specific enzyme immobilization onto E. coli curli nanofibers. *Biotechnology and Bioengineering* 112 (10):2016–2024.

Braat, Henri, Pieter Rottiers, Daniel W. Hommes et al. 2006. A phase I trial with transgenic bacteria expressing interleukin-10 in Crohn's disease. *Clinical Gastroenterology and Hepatology* 4 (6):754–759.

Burgert, Ingo. 2007. Exploring the mechancial design of wood at the micro-and nanoscale. In: Gdoutos E.E. (ed) *Experimental Analysis of Nano and Engineering Materials and Structures*. Springer, Dordrecht.

Chandrasekaran, Prathna T., Naimat Kalim Bari, and Sharmistha Sinha. 2017. Enhanced bacterial cellulose production from Gluconobacter xylinus using super optimal broth. *Cellulose* 24 (10):4367–4381.

Chen, Allen Y., Zhengtao Deng, Amanda N. Billings et al. 2014. Synthesis and patterning of tunable multiscale materials with engineered cells. *Nature Materials* 13:515.

Chen, Allen Y., Chao Zhong, and Timothy K. Lu. 2015. Engineering living functional materials. *ACS Synthetic Biology* 4 (1):8–11.

Chong, Barrie Fong, Lars M. Blank, Richard McLaughlin, and Lars K. Nielsen. 2005. Microbial hyaluronic acid production. *Applied Microbiology and Biotechnology* 66 (4):341–351.

Connell, Jodi L., Eric T. Ritschdorff, Marvin Whiteley, and Jason B. Shear. 2013. 3D printing of microscopic bacterial communities. *Proceedings of the National Academy of Sciences* 110 (46):18380–18385.

Connor, Eric F., Inez Lees, and Derek Maclean. 2017. Polymers as drugs—Advances in therapeutic applications of polymer binding agents. *Journal of Polymer Science Part A: Polymer Chemistry* 55 (18):3146–3157.

DARPA. *Living Structural Materials Could Open New Horizons for Engineers and Architects*. Defense Advanced Research Projects Agency 2016. Available from www.darpa.mil/news-events/2016-08-05.

Daugherty, Patrick S. 2007. Protein engineering with bacterial display. *Current Opinion in Structural Biology* 17 (4):474–480.

de Koster, Stephan A.L., Renée M. Mors, Henk W. Nugteren, Henk M. Jonkers, Gabrie M.H. Meesters, and J. Ruud van Ommen. 2015. Geopolymer coating of bacteria-containing granules for use in self-healing concrete. *Procedia Engineering* 102:475–484.

Douglas, Shawn M., Hendrik Dietz, Tim Liedl, Björn Högberg, Franziska Graf, and William M. Shih. 2009. Self-assembly of DNA into nanoscale three-dimensional shapes. *Nature* 459:414.

Drexler, K. Eric. 1990. *Engines of Creation, The Anchor Library of Science*. New York: Anchor Books.

Duraj-Thatte, Anna M., Pichet Praveschotinunt, Trevor R. Nash, Frederick R. Ward, and Neel S. Joshi. 2018. Modulating bacterial and gut mucosal interactions with engineered biofilm matrix proteins. *Scientific Reports* 8 (1):3475.

El-Naggar, Mohamed Y., Greg Wanger, Kar Man Leung et al. 2010. Electrical transport along bacterial nanowires from Shewanella oneidensis MR-1. *Proceedings of the National Academy of Sciences* 107 (42):18127.

Elbaz, Johann, Peng Yin, and Christopher A. Voigt. 2016. Genetic encoding of DNA nanostructures and their self-assembly in living bacteria. *Nature Communications* 7:11179.

Fang, Ju, Shin Kawano, Kenji Tajima, and Tetsuo Kondo. 2015. In vivo curdlan/cellulose bionanocomposite synthesis by genetically modified gluconacetobacter xylinus. *Biomacromolecules* 16 (10):3154–3160.

Flemming, Hans-Curt, and Jost Wingender. 2010. The biofilm matrix. *Nature Reviews Microbiology* 8:623.

Florea, Michael, Henrik Hagemann, Gabriella Santosa et al. 2016. Engineering control of bacterial cellulose production using a genetic toolkit and a new cellulose-producing strain. *Proceedings of the National Academy of Sciences* 113 (24):E3431.

Fratzl, Peter, and Richard Weinkamer. 2007. Nature's hierarchical materials. *Progress in Materials Science* 52 (8):1263–1334.

Ghosh, Debadyuti, Youjin Lee, Stephanie Thomas et al. 2012. M13-templated magnetic nanoparticles for targeted in vivo imaging of prostate cancer. *Nature Nanotechnology* 7:677.

Gordon, Lyle M., and Derk Joester. 2011. Nanoscale chemical tomography of buried organic–inorganic interfaces in the chiton tooth. *Nature* 469:194.

Halan, Babu, Katja Buehler, and Andreas Schmid. 2012. Biofilms as living catalysts in continuous chemical syntheses. *Trends in Biotechnology* 30 (9):453–465.

Han, Dongran, Xiaodong Qi, Cameron Myhrvold et al. 2017. Single-stranded DNA and RNA origami. *Science* 358 (6369):eaao2648.

Ilk, Nicola, Eva M. Egelseer, and Uwe B. Sleytr. 2011. S-layer fusion proteins—Construction principles and applications. *Current Opinion in Biotechnology* 22 (6):824–831.

Jonkers, Henk Marius. 2011. Bacteria-based self-healing concrete. *Heron* 56 (1/2).

Jordan, T. M., J. C. Partridge, and N. W. Roberts. 2012. Non-polarizing broadband multilayer reflectors in fish. *Nature Photonics* 6:759.

Kehoe, John W., and Brian K. Kay. 2005. Filamentous phage display in the new millennium. *Chemical Reviews* 105 (11):4056–4072.

Kushner, D. J. 1969. Self-assembly of biological structures. *Bacteriol Rev* 33 (2):302–345.

Lee, Seung-Wuk, Chuanbin Mao, Christine E. Flynn, and Angela M. Belcher. 2002. Ordering of quantum dots using genetically engineered viruses. *Science* 296 (5569):892–895.

Lee, Yun Jung, Hyunjung Yi, Woo-Jae Kim et al. 2009. Fabricating genetically engineered high-power lithium-ion batteries using multiple virus genes. *Science* 324 (5930):1051–1055.

Li, Chunmei, Gregory D. Botsaris, and David L. Kaplan. 2002. Selective in vitro effect of peptides on calcium carbonate crystallization. *Crystal Growth & Design* 2 (5):387–393.

Lovley, Derek R. 2008. The microbe electric: Conversion of organic matter to electricity. *Current Opinion in Biotechnology* 19 (6):564–571.

Ludwig, Ferdinand, Hannes Schwertfreger, and Oliver Storz. 2012. Living systems: Designing growth in baubotanik. *Architectural Design* 82 (2):82–87.

Malvankar, Nikhil S., Madeline Vargas, Kelly P. Nevin et al. 2011. Tunable metallic-like conductivity in microbial nanowire networks. *Nature Nanotechnology* 6:573.

A matter of scale. 2016. *Nature Nanotechnology* 11 (9):733.

McCoy, Dakota E., Teresa Feo, Todd Alan Harvey, and Richard O. Prum. 2018. Structural absorption by barbule microstructures of super black bird of paradise feathers. *Nature Communications* 9 (1):1.

McDougald, Diane, Scott A. Rice, Nicolas Barraud, Peter D. Steinberg, and Staffan Kjelleberg. 2011. Should we stay or should we go: Mechanisms and ecological consequences for biofilm dispersal. *Nature Reviews Microbiology* 10:39.

Mellman, Ira, George Coukos, and Glenn Dranoff. 2011. Cancer immunotherapy comes of age. *Nature* 480:480.

Nam, Ki Tae, Dong-Wan Kim, Pil J. Yoo et al. 2006. Virus-enabled synthesis and assembly of nanowires for lithium ion battery electrodes. *Science* 312 (5775):885–888.

Nguyen, Peter Q., Noémie-Manuelle Dorval Courchesne, Anna Duraj-Thatte, Pichet Praveschotinunt, and Neel S. Joshi. 2018. Engineered living materials: Prospects and challenges for using biological systems to direct the assembly of smart materials. *Advanced Materials* 30 (19):e1704847.

Nguyen, Peter Q. 2017. Synthetic biology engineering of biofilms as nanomaterials factories. *Biochemical Society Transactions* 45 (3):585.

Nguyen, Peter Q., Zsofia Botyanszki, Pei Kun R. Tay, and Neel S. Joshi. 2014. Programmable biofilm-based materials from engineered curli nanofibres. *Nature Communications* 5:4945.

Nussbaumer, Martin G., Peter Q. Nguyen, Pei Kun R. Tay et al. 2017. Bootstrapped biocatalysis: Biofilm-derived materials as reversibly functionalizable multienzyme surfaces. *ChemCatChem* 9 (23):4328–4333.

Oh, Dahyun, Jifa Qi, Yi-Chun Lu, Yong Zhang, Yang Shao-Horn, and Angela M. Belcher. 2013. Biologically enhanced cathode design for improved capacity and cycle life for lithium-oxygen batteries. *Nature Communications* 4:2756.

Ong, Luvena L., Nikita Hanikel, Omar K. Yaghi et al. 2017. Programmable self-assembly of three-dimensional nanostructures from 10,000 unique components. *Nature* 552:72.

Pogodin, Sergey, Jafar Hasan, Vladimir A Baulin et al. 2013. Biophysical model of bacterial cell interactions with nanopatterned cicada wing surfaces. *Biophysical Journal* 104 (4):835–840.

Qureshi, Nasib, Bassam A. Annous, Thaddeus C. Ezeji et al. 2005. Biofilm reactors for industrial bioconversion processes: Employing potential of enhanced reaction rates. *Microbial Cell Factories* 4 (1):24–24.

Rosche, Bettina, Xuan Zhong Li, Bernhard Hauer, Andreas Schmid, and Katja Buehler. 2009. Microbial biofilms: A concept for industrial catalysis? *Trends in Biotechnology* 27 (11):636–643.

Ruffing, Anne, and Rachel Ruizhen Chen. 2006. Metabolic engineering of microbes for oligosaccharide and polysaccharide synthesis. *Microbial Cell Factories* 5 (1):25.

Schaffner, Manuel, Patrick A. Rühs, Fergal Coulter, Samuel Kilcher, and André R. Studart. 2017. 3D printing of bacteria into functional complex materials. *Science Advances* 3 (12): eaao6804.

Schmieden, Dominik T., Samantha J. Basalo Vázquez, Héctor Sangüesa, Marit van der Does, Timon Idema, and Anne S. Meyer. 2018. Printing of patterned, engineered E. coli biofilms with a low-cost 3D printer. *ACS Synthetic Biology* 7:1328–1337.

Schuster, Bernhard, and Uwe B. Sleytr. 2014. Biomimetic interfaces based on S-layer proteins, lipid membranes and functional biomolecules. *Journal of The Royal Society Interface* 11 (96):20140232.

Seker, Urartu Ozgur Safak, Allen Y. Chen, Robert J. Citorik, and Timothy K. Lu. 2017. Synthetic biogenesis of bacterial amyloid nanomaterials with tunable inorganic–organic interfaces and electrical conductivity. *ACS Synthetic Biology* 6 (2):266–275.

Tang, Zhiyong, Nicholas A. Kotov, Sergei Magonov, and Birol Ozturk. 2003. Nanostructured artificial nacre. *Nature Materials* 2:413.

Tay, Pei Kun R., Peter Q. Nguyen, and Neel S. Joshi. 2017. A synthetic circuit for mercury bioremediation using self-assembling functional amyloids. *ACS Synthetic Biology* 6 (10):1841–1850.

Teague, Brian P., Patrick Guye, and Ron Weiss. 2016. Synthetic morphogenesis. *Cold Spring Harbor Perspectives in Biology* 8:a023929.

Van Loosdrecht, Mark C. M., and Sef J. Heijnen. 1993. Biofilm bioreactors for waste-water treatment. *Trends in Biotechnology* 11 (4):117–121.

Wang, Xinyu, Jiahua Pu, Bolin An et al. 2018. Programming cells for dynamic assembly of inorganic nano-objects with spatiotemporal control. *Advanced Materials* 30 (16):e1705968.

Wen, Amy M., and Nicole F. Steinmetz. 2016. Design of virus-based nanomaterials for medicine, biotechnology, and energy. *Chemical Society Reviews* 45 (15):4074–4126.

Yi, Wen, Xianwei Liu, Yanhong Li et al. 2009. Remodeling bacterial polysaccharides by metabolic pathway engineering. *Proceedings of the National Academy of Sciences* 106 (11):4207.

Zhang, Sichao, and Yifang Chen. 2015. Nanofabrication and coloration study of artificial Morpho butterfly wings with aligned lamellae layers. *Scientific Reports* 5:16637.

Nanoparticles for Bone Tissue Engineering

Cristiana Gonçalves
University of Minho
ICVS/3B's–PT Government Associate
Laboratory
The Discoveries Centre for Regenerative and
Precision Medicine

Isabel M. Oliveira
University of Minho
ICVS/3B's–PT Government Associate
Laboratory

Rui L. Reis and Joaquim M. Oliveira
University of Minho
ICVS/3B's–PT Government Associate
Laboratory
The Discoveries Centre for Regenerative and
Precision Medicine

9.1 Introduction

The Nobel Prize-winning and quantum physicist Richard Feynman is known as the father of nanotechnology. He described for the first time a process wherein scientists would be able to manipulate and control individual atoms and molecules, in the classic lecture *There's Plenty of Room at the Bottom: An Invitation to Enter a New Field of Physics*, in the 1959 annual meeting of the American Physical Society (Caltech) (Pitkethly, 2008; Hochella Jr, 2002; Voss, 1994). Later in 1974, Professor Norio Taniguchi introduced the term "nanotechnology". This term coined by Taniguchi in 1974 is nowadays widely used, and the given definition is considered the basic principle of nanotechnology, till the present day: the processing of separation, consolidation, and deformation of materials by one atom or one molecule (Wolf, 2015).

Besides that trigger on the nanotechnology revolution, the era of modern nanotechnology had not actually begun (Pitkethly, 2008), since the right tools were not available back then. And it was not, until 1981 with the development of the scanning tunneling microscope (STM) that it did begin. In fact, its inventors, Gerd Binnig and Heinrich Rohrer (IBM Zürich) won the Nobel Prize in Physics in 1986 for this innovation (Coleman, 2014). Likewise, the creators of the atomic force microscope (AFM), developed in the beginning of the 80s, were also awarded a Nobel Prize in Physics in the year 1986. At this stage, it is important

to clarify that nanoscience and nanotechnology work with matter at dimensions around 1–100 nm, a nanometer being a billionth of a meter, i.e., 10^{-9} of a meter. Moreover, to give a clear idea of the scale used in nanotechnology, relatively to our everyday objects, it is just like to comparing a marble with our planet. That is why it was only possible to work at this scale with the development of the correct tools, such as STM and AFM, for the manipulation of matter on an atomic, molecular, and supramolecular scale (Ganji and Kachapi, 2015). The production and use of nanoparticles (NPs) is not a contemporary invention; plenty of examples of the earlier use of nanomaterials to produce materials with enhanced properties exist (Krukemeyer et al., 2015). These objects were typically obtained by processing them using *high*-heated glass or ceramic with gold or gold chloride, silver, copper or other metallic NPs or metal oxides; for instance in decoration objects. But even carbon nanotubes were present in saber blades. Modern nanotechnology is an interdisciplinary science concerning the NPs' exceptional chemical, physical and mechanical properties embracing materials science, mechanical engineering, physics, life sciences, chemistry, biology, electrical engineering, computer science and information technology (Krukemeyer et al., 2015; Jurvetson and Jurvetson, 2007).

Nanoscience and nanotechnology are now well-established areas of knowledge, and also in deep and growing expansion in terms of data and scope and variety of applications. In fact, NPs are now in a commercial exploration period.

It is conceivable to differentiate nanoscience from nanotechnology. In nanoscience, the objects of study are materials smaller than 100 nm, at least in one dimension, while nanotechnology creates and uses valuable and functional materials, devices, and systems at the nanoscale (Dahman, 2017), atomic and/or molecular scale, using different techniques, taking advantage of their enhanced properties (strength, lightness, electrical and thermal conductance, and also reactivity), uniting boundaries of the knowledge areas of engineering, material chemistry and physics.

Thus, nanomaterials are defined as materials containing particles or constituents of nanoscale dimensions or produced by nanotechnology (Nagarajan, 2008; Abdullaeva, 2017). In the same way, NPs are described as particles with size in the range of 1–100 nm, at least in one of the three possible dimensions (Nagarajan, 2008; Rotello, 2004). These are considered the "building blocks" of nanotechnology and exhibit distinct characteristics from the bulk material (Virlan et al., 2016). They can be composed by a wide range of materials of diverse chemical nature. The most common materials in the NPs composition are metals, metal oxides, silicates, non-oxide ceramics, polymers, organics, carbon and biomolecules (Nagarajan, 2008).

NPs can be from natural or anthropogenic sources. There are countless naturally occurring examples. Self-assembly in biology plays several important roles and triggers the formation of a wide variety of complex biological structures (Mendes et al., 2013). Making use of this, nature has shown an extraordinary aptitude to generate nanoscale elements (Rogers, 2016). Naturally occurring NPs can be found everywhere in nature. Soils are the most fertile place for nanomaterials in the planet, while oceans make available the largest collective reservoir of these materials (Hochella Jr et al., 2012). NPs are also present in the atmosphere and include volcanic eruption ash, forest fires, ocean spray, ultrafine sand grains of mineral origin and dust from cosmic sources (Strambeanu et al., 2015; Raab et al.). Moreover, it is possible to find several further nanostructured materials, such as minerals (e.g., clays), natural colloids (e.g., milk, blood, fog and gelatine), mineralized natural materials (shells, corals and bones), paper and cotton, insect wings and opals, silk, gecko feet and even biological matter (e.g., viruses, bacteria and fungi) (Sadik, 2013).

Anthropogenic NPs are man-made and, similar to the naturally occurring ones, are very wide-ranging, being created through well-designed fabrication processes. Manmade NPs can be divided in two main groups: (i) a group of those with no predetermined size, which may exhibit undefined chemistry, and (ii) another of those, also known as engineered nanoparticles (ENPs), which exhibit specific size ranging from 1 to 100 nm. Examples of the first group are combustion particulates, diesel exhaust, welding fumes and coal fly ash; those of the second are carbon nanotubes, dendrimers, quantum dots (QDots), TiO_2, and gold and silver NPs. NPs can also be divided into classes regarding their physical and chemical characteristics: (i) Carbon-based NPs, (ii) Metal NPs, (iii) Ceramic NPs, (iv) Semiconductor

NPs, (v) Polymeric NPs and (vi) Lipid-based NPs (Khan et al., 2019; Choi and Frangioni, 2010). In the nanometer range, it is possible to find a variety of NPs and NP systems ranging n size from few nanometers to hundreds of nanometers (Figure 9.1) (Choi and Frangioni, 2010). NPs have size-dependent characteristics that make these materials superior and indispensable for many applications.

Anthropogenic NPs can be classified into four subtypes: carbon-based materials, metal-based materials, dendrimers and composites (Azmi and Shad, 2017). NPs can be produced in numerous different morphologies, such as spheres, cylinders, platelets and tubes, among others, being designed with surface modifications for specific applications (Nagarajan, 2008).

NPs have huge range of applications (Raab et al.) and can be used in diverse technology and industry fields, but mainly in medicine, manufacturing and materials, environmental science and also in energy and electronics (information technology and homeland technology) (Bonzani et al., 2006). Specifically, in the biological and medical field, NPs may be used in fluorescent biological labels, drug and gene delivery, biodetection of pathogens, detection of proteins,

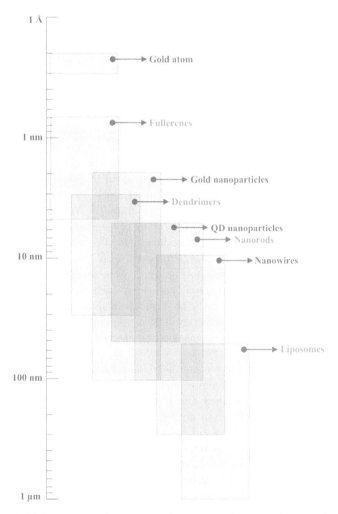

FIGURE 9.1   Relative sizes of some NPs, based on (Choi and Frangioni, 2010).

probing of DNA structure, tumor destruction via heating (hyperthermia), separation and purification of biological molecules and cells, MRI contrast enhancement, photokinetic studies and tissue engineering (TE) (Salata, 2004). The first developed product in nanomedicine was in the 1970s, with drug-loaded liposomes used to deliver the drugs to diseased cells; nowadays, this is considered a vital science.

TE and Regenerative Medicine (RM), also known together as TERM, are multidisciplinary and rapidly evolving fields that embrace the three main areas of engineering, biology and medicine, aiming to repair, replace and regenerate diseased tissues with an engineered construct equivalent to the damaged one (Vial et al., 2017; Vieira et al., 2017). Thus, TERM intends to restore, maintain or even improve the function of the lost or injured tissue, by using scaffolds and/or other materials in combination with cells or growth factors. In fact, it has a huge scope for application since millions of people have tissue damage and organ failure annually (Ahmed et al., 2018). TE is a subfield of RM, with thinner scope and is strictly defined as the engineering of body parts *ex vivo*, by seeding sells on and/or into a scaffold, not requiring cellular regeneration. RM, on the other hand, is a broader and more generalized field that aims to replace or regenerate human cells, tissues, or organs (*in vivo* or *ex vivo*) in order to restore or establish normal function, which may involve cells, natural or artificial scaffolding materials, growth factors and/or gene manipulation (Katari et al., 2015).

The incidence of bone disorders and other conditions around the globe has been growing, mostly due to aging together with obesity and poor physical activity (Amini et al., 2012). Bones are vascularized living tissues, known by their role in providing the body with shape and support, recognized by its complexity and for being in a constant process of renewal (due to the osteoclasts that form part of its composition). Bones can be classified in four main groups: (i) long bones (e.g., humerus and femur), (ii) short bones (e.g., carpals and tarsals), (iii) flat bones (e.g., skull and ribs), and (iv) irregular bones (e.g., facial and vertebrae bones) (Clarke, 2008). A small portion of the bones' composition is organic matter (mostly collagen type I and some non-collagenous proteins and lipids) while a major part (60%–70%) is mineral/inorganic matter (mostly hydroxyapatite [HAp]), and even a part of water (10%–20%) (Boskey, 2013; Feng, 2009). Depending on the kind of bone and other factors, each bone could have different proportions of those building blocks. Flat bones are made by membranous bone and long bones by a blend of endochondral and membranous bone (Clarke, 2008). The long bones are structured by epiphysis (trabecular bone bounded by a thin shell of dense cortical bone, the round ended regions) and diaphysis (dense cortical bone, the long central shaft) and are mainly made of cancellous/spongy bone, compact bone, marrow cavity, endosteum and periosteum (Clarke, 2008).

Bone diseases encompass all the diseases and injuries affecting the skeletal system of the human body and can be due to several causes, such as metabolic origin, inflammation, bone resorption and others. These defects may negatively affect the structural stability and the biomechanical function of bones. Bone regeneration has been an important challenge in clinical surgery, due to conditions such as trauma, tumor and diseases (e.g., osteomyelitis and osteitis), with the existing therapeutic and research approaches not sufficiently effective due to several limitations (Torgbo and Sukyai, 2018). In fact, one of the main challenges of TERM is precisely bone tissue regeneration (Vieira et al., 2017). Moreover, it was three decades ago that the field of bone tissue engineering (BTE) had started to form, showing since then an increasingly fast pace of development (Amini et al., 2012; Henkel et al., 2013).

In fact, the particular nature of the bones and the mechanical stress to which they are subjected makes it almost impossible to produce structures of the same size and characteristics of the naturally occurring bone; but recent advances in nanotechnology have made this achievable, opening a new era for TERM/BTE (Vieira et al., 2017).

9.2 NPs

NPs could be grouped in several ways, but we herein divided them in two main groups: i) organic NPs, and ii) inorganic NPs, which can be from natural origin or anthropogenic in nature (Figure 9.2). Natural NPs are the oldest ones that existed and also possess huge diversity, but the man-made ones also come in various forms. NPs could be prepared with organic polymers or inorganic elements, but could also have a hybrid structure. Organic and inorganic compounds are the basis of chemistry. A concise description of each group and their main NPs is further given.

9.2.1 Organic NPs

Organic NPs, derived from organic compounds, can be found in nature but also in many industrial products. Examples of organic NPs are micelles, liposomes, nanogels, ferritin and dendrimers (Rajabi and Mousa, 2016).

The IUPAC definition for micelles says that these are "particles of colloidal dimensions that exist in equilibrium with the molecules or ions in solution from which it is formed", and polymeric micelles are defined as those "organized auto-assembly formed in a liquid and composed of amphiphilic macromolecules, in general amphiphilic di-or tri-block copolymers made of solvophilic and solvophobic block". Thus, micelles are colloidal dispersions with nanometric dimensions (5–100 nm), prepared from amphiphilic molecules, with a hydrophobic tail and a hydrophilic head (Mukherjee et al., 2016). Micelles may have cores that are liquid-like, glassy or crystalline, inside their spherical structures of surfactant covers (Allen et al., 1992; Rapp, 2016). Polymeric micelles are self-assembling nanoconstructs of amphiphilic copolymers (Movassaghian et al., 2015).

Critical micelle concentration (CMC) or critical association concentration (CAC) is the minimum concentration of polymer needed to form micelles. Thus, above

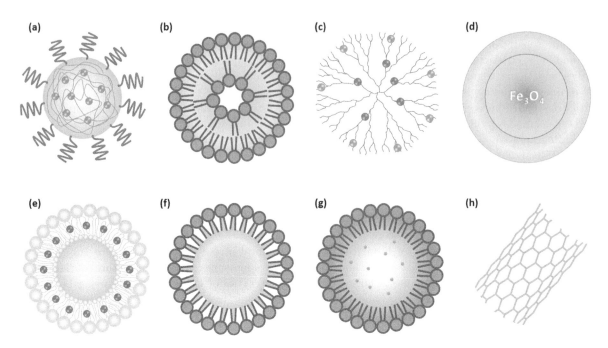

FIGURE 9.2 Schematic of the NP systems. (a) Polymeric NPs, (b) liposomes, (c) dendrimers, (d) magnetic NPs, (e) metallic NPs, (f) polymeric micelles, (g) solid lipid NPs, and (h) nanotubes. (Reprinted by permission from Springer Nature: Springer, *Nano Research*, Oliveira et al., 2018.)

that concentration is formed the core of the micelles (hydrophobic polymer segments spontaneously aggregate), the surfactants assemble in curved layers that eventually lead to the formation of a sphere (Rapp, 2016), but below the CMC value the core breakdowns and if a drug is loaded it will be released. This is one of the foremost concerns associated with the use of micelles for drug delivery (Overstreet et al., 2011; Zhang et al., 2016). The surfactant molecules will be oriented toward the inside or outside depending on the nature of the surroundings, with the head groups oriented toward the inside in a polar environment but toward the outside in in non-polar environments, thus able to enclose molecules of corresponding polarity (Rapp, 2016). These micelle formation characteristics mimic a normal physiological process, namely the bile acids that are secreted into the duodenum, which play a vital function in fat absorption by the intestine (Hill, 2012).

Several chemical forms of micelles have been developed for pharmaceutical use. Some advantages related with the micelles include easy surface manipulation and easy encapsulation of drug, as it is a good nanocarrier for poorly water-soluble drugs; however, micelles also show favorable biocompatibility, longevity, high stability (*in vitro* and *in vivo*), and the ability to accumulate in the target zone based on the enhanced permeability and retention effect (Mukherjee et al., 2016; Movassaghian et al., 2015). Accordingly, micelles have been extensively studied and used to deliver hydrophobic drugs, which are kept in the core of the micelle until dissociation (Overstreet et al., 2011). Moreover, polymeric micelles are widely considered as suitable nanocarriers for a wide range of applications, such

as diagnostic imaging and drug and gene delivery (Movassaghian et al., 2015).

Liposomes are artificial spherical vesicles composed of a curved phospholipid membrane, i.e., an amphiphilic, single or multiple, lipid bilayer(s), suspended in a dispersion medium, which encloses hydrophilic or lipophilic molecules in their core and shell (Aryasomayajula and Torchilin, 2016; Singh et al., 2016; Revathy et al., 2016). Depending on the liposome preparation process and the lipids used, they can be grouped as unilamellar, multilamellar and multivesicular (Dey et al., 2016). The pharmacokinetic parameters of the liposomes mainly rely on its physicochemical properties, such as size, membrane lipid packing, surface charge, steric stabilization, dose and route of administration (Zamboni, 2007). Since the lipid bilayer system is biodegradable, similar to cell membranes, it fuses well with infectious microbes and is quickly removed from the bloodstream by immune cells; hence, these versatile NPs are widely used, for instance, in drug delivery (Revathy et al., 2016; Sawant and Shegokar, 2016). In fact, liposomes have been widely used as NP delivery systems for countless routes of drug administration.

Another well-known organic NP type is the nanogel, which is made up of a nanosized network of cross-linked amphiphilic or hydrophilic polyionic polymers. (Szekely and Didaskalou, 2016; Dang et al., 2017). They can be of natural or synthetic origin, since they can be produced by various methods (including precipitation, emulsion and solution phase polymerization) using natural biopolymers (such as dextran, dextrin, pullulan, mannan, chitosan (CS), poly-l-lysine, heparin, hyaluronic acid and alginate), and synthetic biodegradable and biocompatible polymers (such as poly(methyl methacrylate), PLA, poly(glycolic acid), poly(d,l-lactic-co-glycolic acid) and poly(ε-caprolactone). Nanogels are widely used, especially as nanocarriers for drug delivery and imaging (Szekely and Didaskalou, 2016). Their unique properties include their flexible nanosize, large surface area for multivalent conjugation, biocompatibility, stability and loading capacity (Dang et al., 2017).

Dendrimers are another interesting group of NPs that are monodisperse, unimolecular, micellar nanostructures having a symmetrical and three-dimensional structure of around 20 nm (Singh and Sharma, 2016), as can be seen from Figure 9.3. They are highly branched, monodisperse, three-dimensional macromolecules having an internal space, repeating units and a number of functional end groups at their periphery (Sadjadi, 2016; Singh and Sharma, 2016). Dendrimers have a solvent-filled interior core as well as a homogeneous exterior surface.

The unique physicochemical characteristics of dendrimers are responsible for their increasing uses in various fields, such as catalysis, nanomedicines, nanomaterials and nanodevices (Sadjadi, 2016).

9.2.2 Inorganic

Examples of inorganic NPs are gold nanoparticles (AuNPs or GNPs), superparamagnetic iron oxide nanomaterials (SPIONs), QDots and paramagnetic lanthanide ions (Rajabi and Mousa, 2016).

In the past years, interest in the application of gold nanostructures and NPs in various fields of chemistry has been growing. GNPs possess unique physicochemical properties that make them very useful as scaffold components, some of these properties include their tunable size (1–100 nm), redox properties, functionality and conductivity (Figure 9.4). They show many advantages for cellular imaging, for instance (i) GNPs can scatter light intensely,

being much brighter than chemical fluorophores; (ii) they do not photobleach and (iii) they can be detected at low concentrations (10−16 M) (Chandran and Thomas, 2015). These NPs can be an efficient system for both DNA and RNA delivery, for gene silencing and for therapeutic purposes (Ravichandran et al., 2016). In fact, GNPs are able to carry and deliver not only large biomolecules but also small molecular drugs.

SPIONs show several possibilities for application in medicine. In fact, magnetic NPs have fascinated researchers and medical doctors due to their drug delivery application, which can decrease the amount of drugs necessary for treatment (Pansailom et al., 2017). They are also extensively applied as (i) a robust negative contrast agent for conventional magnetic resonance imaging, (ii) drug transporters in magnetic drug targeting, (iii) heat inducers in alternating electromagnetic fields for magnetic hyperthermia, and also for (iv) magnetization of cells for magnetic tissue engineering (Alipour et al., 2017; Janko et al., 2017).

QDots are a class of semiconductor NPs that are portions of matter (luminescent semiconductor crystals) whose holes and electrons are confined in all three spatial dimensions, behaving as artificial atoms and having the core potential replaced by confinement potential. These NPs have exceptional physicochemical characteristics due to their small size (results in quantum confinement and size-dependent fluorescence properties) and have a highly compact structure (Yadav and Raizaday, 2016; Tom et al., 2016). They can be in the form of semiconductors, metals and metal oxides, and these have been at the vanguard of the most recent studies because of the resulting electronic, optical, magnetic and catalytic properties (Njuguna et al., 2014). It is easy to find literature about different types of QDots, such as those of cadmium selenide, carbon, graphene, phosphorene, copper oxide, nickel (II) oxide, among others. There are several methods to produce QDots, with the wet chemical colloidal processes being the most common (Njuguna et al., 2014).

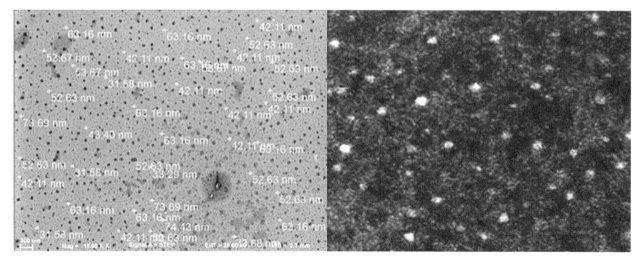

FIGURE 9.3 Images of STEM (left) and AFM (right), of PAMAM 1.5 generation dendrimers functionalized to treat rheumatoid arthritis.

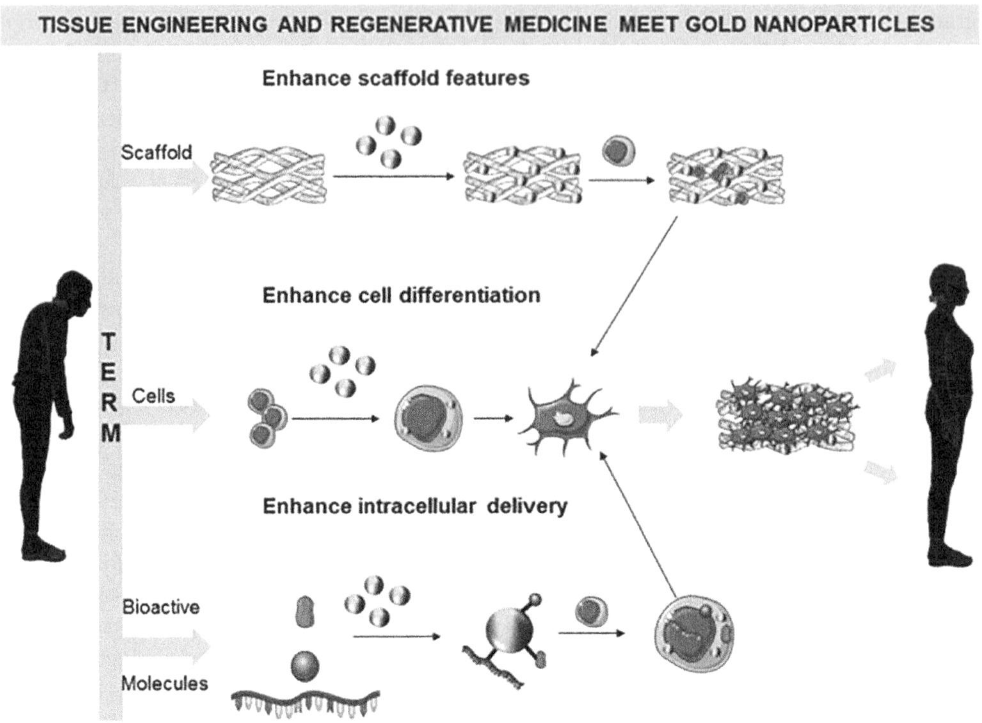

FIGURE 9.4 Scheme representing the importance of introducing AuNPs in TERM realm. TERM combines three elements: scaffold, cells and bioactive molecules for engineered or repaired tissue. The addition of AuNPs in each element aims to enhance scaffold features. (Reprinted by permission from Springer Elsevier: Springer, *Nano Research*, Vial et al., 2017.)

9.3　Application of NPs in BTE

There are several studies showing the potential of NPs in BTE. In this section, some examples of studies made with the NPs in the last 2 years will be given (Table 9.1).

Terranova et al. (2017) studied electrospun polystyrene scaffolds (PS) combined with β-TCP or GNPs to repair calvarial bone defects. Biocompatibility was assessed by implanting the material into a critical-size calvarial defect in mice. Weeks after implantation, the mouse calvaria model showed that β-TCP-PS scaffolds had more newly formed bone tissue as compared to Au-PS. A dense fibrous connective tissue containing blood vessels was identified histologically in both types of scaffolds. Another study analyzed electrospun silk fibroin/nano-HAp/bone morphogenetic protein-2 (BMP-2) scaffolds to promote bone

TABLE 9.1　Different NPs and Some of Their Applications in BTE Strategies

NP		Application in BTE Strategies	References
Organic	Micelles	(a) target fracture and improve healing, (b) targetable bone imaging, (c) treatment of rheumatoid arthritis, (d) osteosarcoma chemotherapy, (e) as templates for the synthesis of porous HAp NPs;	Jia et al. (2015), Park et al. (2002), Liu et al. (2013), Wang et al. (2016)
	Liposomes	(a) as a bone-targeting carrier, (b) controlled bone formation, (c) enhanced bone healing by carring TGF-β1 and BMP-4 /epidermal growth factor (EGF), (d) bone targeting gene delivery regulating mesenchymal stem cell migration;	Yamashita et al. (2018), Lee et al. (2016), Crasto et al. (2016), Marquez et al. (2013), Ferreira et al. (2013), Zheng et al. (2016)
	Nanogels	(a) bone loss prevention by delivering delivery of a TNF-α and RANKL, (b) GF delivery system for bone formation, (c) bone regeneration by BMP-2 direct- and indirect-binding designed nanogels;	Alles et al. (2009), Fujioka-Kobayashi et al. (2012), Nelson et al. (2013)
	Dendrimers	(a) bone imaging, (b) bone-targeting carriers for the treatment of bone diseases, (c) ectopic bone formation;	Pes et al. (2017), Yamashita et al. (2017), Pan et al. (2012), Oliveira et al. (2010)
Inorganic	GNPs	(a) bone regeneration, (b) induce differential cell response on mesenchymal stem cell osteogenesis, (c) stimulate differentiation and mineralization of primary osteoblasts through the ERK/MAPK signaling pathway, (d) repair of calvarial bone defects;	Ribeiro et al. (2017), Kawazoe and Chen (2015), Zhang et al. (2014), Terranova et al. (2017)
	SPIONs	(a) magnetic resonance imaging of bone turnover in early osteoarthritis, (b) to treat metastatic bone tumors, and reduced lesion and visible bone formation, (c) MRI contrast agent	Panahifar et al. (2014), Adamiano et al. (2018), Revia and Zhang (2016)
	QDots	Dentistry and bone-related disease diagnostics	Khan et al. (2014)

extracellular matrix (ECM) production and remodeling, thus being an approach with potential for bone repair and regeneration. The results showed that the scaffolds had good biocompatibility and promoted osteogenic differentiation (Niu et al., 2017).

Yet another group studied scaffolds of HAp NPs disseminated in 1, 6-diisocyanatohexane-extended poly(1, 4-butylene succinate)/poly(methyl methacrylate) for regeneration of bone in the cranio-maxillofacial region. The scaffolds showed high porosity and excellent water retention ability, controlled degradation and sustained drug release ability, thus presenting many advantages for being used as vehicles in drug delivery. Results indicated that the scaffolds had good bone regeneration aptitude and cell growth, having an antimicrobial activity on the surrounding tissues (Kaur et al., 2017).

Bastami et al. (2017) produced β-tricalcium phosphate (β-TCP) scaffolds with BMP2-loaded CS NPs for controlled release of bone morphogenetic protein-2. The obtained outputs demonstrated that the scaffolds can be mechanically and biologically compatible, acting as an osteoinductive graft by enabling controlled delivering of rhBMP2, allowing the differentiation of hBFPSCs toward the osteoblast lineage. Besides, the scaffold model can be used for the delivery of cells and other growth factors, such as vascular endothelial growth factor (VEGF). A study investigated the osteoinductive effects of different compositions of bioactive glass NPs for BTE. The results showed that osteogenic differentiation was significantly highest in alkaline phosphatase activity and biomineralization. Besides the small particle size, the osteoinductivity and negative zeta potential turn this material into a good candidate for BTE applications (Tavakolizadeh et al., 2017). Another research group studied the influence of the antibiotic vancomycin (VANCO) loaded in silk fibroin NPs (SFNPs) and then entrapped it in silk scaffolds, to produce sustained drug delivery systems. These VANCO-loaded SFNPs, entrapped in scaffolds, show reduced bone infections at the defected site with better results than the other treatments. Therefore, this delivery system has good biocompatibility and sustained release properties and can be appropriate for further study in the context of osteomyelitis (Hassani Besheli et al., 2017). Yassin and co-authors (Yassin et al., 2017) investigated the effect f-poly(l-lactide-co-ε-caprolactone) (poly(LLA-co-CL)) scaffolds with nDPs and seed with bone marrow stromal cells (BMSCs) on bone regeneration in a rat calvarial critical size defect. In the calvarial defects implanted with BMSC-seeded poly(LLA-co-CL)/nDPs scaffolds, there was a significant increase in osteogenic metabolic activity, which revealed a substantial increase in bone volume. Furthermore, results showed that compared with conventional poly(LLA-co-CL) scaffolds those functionalized with nDPs promote osteogenic metabolic activity and mineralization capacity. A study by Xie and co-authors (2017) evaluated graphene oxide nanolayers as an anchor for the immobilization of BMP-2-encapsulated bovine serum albumin NPs on HAp and TCP for bone regeneration. Graphene oxide nanolayers impeded

rapid degradation of TCP scaffolds and promoted NP adsorption on these scaffolds, and realized BMP-2 sustained release. The NPs enriched the scaffold surfaces with a nanostructure similar to that of the ECM, improving BMSC attachment. The nanostructure, charge balance and BMP-2 sustained release capability synergistically improved BMSC differentiation and bone regeneration. Another group investigated multilayer nanoscale poly(lactic-co-glycolic acid) and nanoHAp, by means of using poly(allylamine hydrochloride) and poly(sodium 4-styrenesulfonate) as polyelectrolytes encapsulating biofunctional peptides, to enhance bone tissue regeneration *in vivo*. The results showed that the peptide incorporation increased cellular processes, with good viability and significant enhancement of alkaline phosphatase activity, osteopontin and osteocalcin. Furthermore, the functionalized membrane induces a favorable *in vivo* response after implantation in a non-healing rat calvarial defect model (Gentile et al., 2017). Brett et al. (2017) described a prefabricated scaffold integrated with magnetic nanoparticles (MNPs) used to upregulate Bcl-2 expression in implanted adipose-derived stromal cells for bone regeneration. The results showed that there was an increase osteogenic potential, biological resilience and that the magnetofection had an efficiency rate of 30%, which in turn resulted in significantly improved healing as compared with the control group. A study made by Roh et al. (2017) investigated the effects of adding magnesium oxide (MgO) NPs to polycaprolactone (PCL)/HAp composites and treating PCL/HAp/MgO scaffolds with oxygen and nitrogen plasma for improving bone regeneration. The combination of MgO/HAp NPs and plasma treatment increased the adhesion, proliferation and differentiation of pre-osteoblast (MC3T3-E1) cells in the PCL scaffolds. Behera et al., 2017 investigated the use of HAp with RGD-containing silk fibroin composite scaffolds in bone regeneration. The results showed that HAp-deposited fibroin scaffolds provide greater mechanical strength and cytocompatibility, but minimal immune responses to both types of composite scaffolds are observed using osteoblast–macrophage co-culture model (Behera et al., 2017). Hadavi et al. (2017) investigated a novel calcified gum Arabic porous nanocomposite scaffold for bone tissue regeneration. The study demonstrated that the hydroxyapatite/gum Arabic nanocomposite had favorable biocompatibility and a similar structure to natural bone matrix, but showed significant changes in terms of biomechanical properties and osteoconductivity of the nanocomposite scaffold by modulating its mineral content. Nanocomposite scaffolds containing gum and n-HAp of 40%–50% exhibited the highest mechanical properties, as well as supported increased biomineralization. A study made by Huang et al. (2017) investigated the effect of magnetic NP composite scaffold under a pulsed electromagnetic field on bone marrow mesenchymal stromal cells (BMSCs). The results showed that BMSC-seeded magnetic scaffolds when exposed to an electromagnetic field had a positive effect on the osteogenic differentiation of BMSCs. Yet another study evaluated osseointegration of silver NP-coated additive manufactured titanium implants in rat tibial defects. The newly formed bone demonstrated a trabecular morphology, with the bone being located at the implant surface and bone–implant interface showing that silver was present primarily in the osseous tissue and co-localized with sulfur. Other analyses revealed silver sulfide NPs in the newly regenerated bone, presenting solid evidence that the earlier *in vitro* observed biotransformation of silver to silver sulfide occurs *in vivo* (Geng et al., 2017). A study was done by Leena et al. where they fabricated a drug delivery system using CS NPs loaded with Silibinin at different concentrations and then incorporated this into scaffolds containing alginate and gelatin (Alg/Gel) for the sustained and prolonged release of Silibinin for bone formation *in vitro*. These scaffolds were biocompatible with mouse mesenchymal stem cells and promoted osteoblast differentiation, and they found that the Silibinin released from scaffolds regulated miRNAs that control the bone morphogenetic protein pathway. These results suggest the potential for sustained and prolonged release of Silibinin to promote bone formation (Leena et al., 2017). Lee et al. studied the bioceramic nanocomposite thiol-acrylate polyHIPE scaffolds (mixture of trimethylolpropane tris(3-mercaptopropionate) and dipentaerythritol penta-/hexa-acrylate in the presence of HAp or strontium-modified hydroxyapatite (SrHAp) NPs) for increasing osteoblastic cell culture in 3D. The results showed that cells were able to migrate throughout all scaffolds, producing a high density by the end of the culture period (21 days). The presence of HAp, and in particular SrHAp, led to significantly increased cell proliferation; furthermore, *in vitro* studies demonstrated significant mineralization from inclusion of bioceramics. The level of alkaline phosphatase was significantly reduced on HAp- and SrHAp-modified scaffolds by day 7, which agrees with the observed early onset of mineralization in the presence of the bioceramics. The study supported that HAp and SrHAp increase osteoblastic cell proliferation on poly-HIPE scaffolds and promote early mineralization (Lee et al., 2017). Another group reported the use of monodispersed bioactive glass nanoclusters (BGNCs) with ultra-large pores and intrinsic exceptionally high miRNA loading for efficiently increasing bone regeneration. Bioactive glass NPs showed excellent apatite-forming ability and high biocompatibility, and furthermore they demonstrated an almost 19 times higher miRNA loading than those of conventional bioactive glass NPs. Additionally, BGNCs-miRNA nanocomplexes presented significantly high antienzymolysis and increase cellular uptake and miRNA transfection efficiency. BGNCs-mediated miRNA delivery significantly promoted the osteogenic differentiation of bone marrow stromal stem cells *in vitro* and efficiently increased bone formation *in vivo* (Xue et al., 2017). A study by Fang et al. (2017) investigated the effect of MNP-induced hyperthermia on destroying biofilm and promoting bactericidal effects of antibiotics in the treatment of osteomyelitis. In this study, the material was implanted into the bone marrow cavity of distal femur of sixty 12-weeks-old male Wistar rats. The results showed that the implants can be heated up to 75°C by magnetic heating without any significant thermal damage on the surrounding tissue. Systemic administration of VANCO into the femoral canal and the presence of MNP-induced hyperthermia promoted the eradication of bacteria in a biofilm-based colony. Bhowmick et al. (2018) studied the design of biomimetic organically modified montmorillonite clay (OMMT)-supported CS/HAp-zinc oxide (CTS/HAp-ZnO) nanocomposites (ZnCMH I-III) with improved mechanical and biological properties as compared to previously reported CTS/OMMT/HAp composite. The results suggested that addition of 5 wt% of OMMT into CTS/HAP-ZnO (ZnCMH I) gives the best mechanical strength and water absorption capacity, and addition of 0.1 wt% of ZnO NPs into CTS-OMMT-HAp significantly increased the tensile strengths of ZnCMH I–III compared with the previously reported CTS-OMMT-HAp composite. Furthermore ZnCH, without OMMT also showed reduced tensile strength, antibacterial effect and cytocompatibility with osteoblastic cell compared to ZnCMH I. Han et al., 2017 investigated the use CS-coated bovine serum albumin nanoparticles (CBSA NPs) and oxidized alginate (OSA) in a layer-by-layer deposition on Ti scaffolds, for bone regeneration and anti-infection properties. The structure possesses abundant functional groups from CBSA NPs and OSA to improve the surface biocompatibility and provide active sites for stable and efficient immobilization of BMP2 through chemical and physical interactions without compromising its bioactivity. The immobilized BMP-2 on the scaffold improved cell adhesion and proliferation and induced osteogenic differentiation of bone marrow stromal cells *in vitro*. Furthermore, this modification also increased ectopic bone formation, and the scaffolds presented good antibacterial activity which is important during the bone healing process (Han et al., 2017). A study made by Saber-Samandari et al. (2018) studied the role of titanium dioxide (TiO_2) on the morphology, microstructure and bioactivity of grafted cellulose/hydroxyapatite nanocomposites for potential application in bone repair. Results showed that enhancing the amount of TiO_2 decreases the swelling behavior of the nanocomposite scaffolds, and cell culture experiments showed that the scaffold extracts do not have cytotoxicity in any concentration and that 3D nanocomposite scaffolds have a great potential as a bone tissue substitute. Melancon et al. (2017) evaluated the antitumoral effect of irreversible electroporation alone or in combination with doxorubicin (DOX)-loaded superparamagnetic iron oxide (SPIO) NPs (SPIO-DOX), in a VX2 rabbit tibial tumor model and in five porcine vertebral bodies in one pig (IRE electrode placement without ablation for bone metastases). SPIO-DOX led to reduction in signal intensity within the tumor by up to 5 days after treatment and was related to the presence of iron. The percentage of residual viable tumor in bone was significantly less in the combination therapy model compared with control, SPIO-DOX and IRE treatment. The percentage of residual viable tumor in soft tissue was significantly less with IRE and SPIO-DOX than with SPIO-DOX alone. They verified that tibial VX2 tumors treated with a combination of SPIO-DOX and IRE showed increased anti-tumor effect as compared with individual treatments alone.

Mahdavi et al., 2017 investigated the osteogenic differentiation content of equine adipose-derived stem cells (e-ASCs) on nano-bioactive glass (nBGs)-covered poly(l-lactic acid) (PLLA) nanofiber scaffold (nBG-PLLA). Their results indicated that nBGs was covered by PLLA and di not have any negative effect on cell growth rate. Furthermore, e-ASCs that differentiated on nBGs-PLLA scaffold demonstrated higher ALP activity and higher expression of bone-associated genes than that on uncoated PLLA scaffold and more calcium capacity. This work showed that a combination of bioceramics and biopolymeric nanofibers could be a valuable and promising tool to use for BTE application (Mahdavi et al., 2017). A study by Türkkan et al. (2017) evaluated the effect of nanosized CaP-silk fibroin-PCL-PEG-PCL/PCL-based bilayer membranes (incorporating nanocalcium phosphate (SPCA)1 and one layer of PCL membrane) for guided bone regeneration. The mechanical test results showed notable improvement in the tensile strength of membranes with the incorporation of NPs. Higher water affinity of nanoCaP including in the membranes was proved by lower contact angle values and higher percentage of water uptake capacity. A biomineralization assay demonstrated that nucleation and growth of apatites around fibers of SPCA10 and SPCA20 were evident, while on SPCA0 apatite minerals were hardly observed. Human dental pulp stem cells (DPSC) were seeded, and it was observed that there was increasing nanoCaP amount resulting in higher cell adhesion, proliferation, ALP activity and calcium deposition on membranes. Babitha et al. (2018) fabricated biomimetic ZeinPDA nanofibrous scaffold permeated with BMP-2 peptide-linked TiO_2NP for BTE. The results showed that there was increased cell adhesion, maintaining delivery of bioactive signals, mineralization and differentiation that can be associated to its greatly linked nanofibrous matrix with unique material composition. Furthermore, the expression of osteogenic markers showed that the fabricated nanofibrous scaffold possesses better cell–biomaterial interactions. So, they demonstrated that these results are promising when using the composite nanofibrous scaffold as an adequate biomaterial substrate for bone regeneration. A study was carried out by Xia et al. (2018) wherein they developed a novel calcium phosphate cement containing GNPs (GNP-CPC) to increase osteogenic differentiation of hDPSCs. The results demonstrated that the incorporation of GNPs improved hDPSC behavior on CPC; there was better cell adhesion, proliferation and enhanced osteogenic differentiation. GNPs donate CPC with micro-nano-structure, thus improving surface properties for cell adhesion and subsequent behaviours. Furthermore, GNPs released from GNP-CPC were internalized by hDPSCs, thus enhancing cell functions. The results were coherent with and supported the osteogenic induction results of GNP-CPC. The culture media containing GNPs enhanced

the cellular activities of hDPSCs, which is consistent with and supported the osteogenic induction results of GNP-CPC. The study showed that GNP-CPC significantly increased the osteogenic functions of hDPSCs; thus, GNPs are promising candidates to increase bone regeneration. Another group (Johari et al., 2018) studied fluoridated silk fibroin (SF)/TiO_2 nanocomposite scaffolds for BTE, to improve the mechanical properties of scaffolds. Nevertheless, adverse effects were observed by further increase in the TiO_2-F content. Cell cytotoxicity results showed that the SF/TiO_2-F nanocomposite scaffolds are nontoxic to osteoblasts, and cell fixation results after 3 days of incubation reported that the cell attachment and spreading on SF/TiO_2-F nanocomposite scaffolds increase with respect to the SF/TiO_2 nanocomposite scaffolds control sample (Johari et al., 2018). Khorshidi and Karkhaneh (2018) studied hydrogel/fiber scaffold for BTE, and verified that evaluation on pristine hydrogel and hydrogel/fiber showed that inclusion of conducting fibers into hydrogel enhanced elastic modulus, roughness and electrical conductivity, while reducing hydrophilicity. Furthermore, the results demonstrated that the hydrogel/fiber composite better sustained human osteoblast-like cell adhesion, proliferation and morphology compared with the hydrogel alone. In summary, the presence of gel/fiber architecture along with electrical conductivity can allow the scaffold to become a very promising tool for bone regeneration. Chen et al. (2017) analysed dexamethasone-loaded biphasic calcium phosphate NPs/collagen porous composite scaffolds for BTE. *In vitro* and *in vivo* studies showed that the composite scaffolds demonstrated symbiotic effects on the osteogenic differentiation of human mesenchymal stem cells (hMSCs) and bone regeneration. The scaffolds demonstrated helpful effects on the production of capillary blood vessels in the reconstructed bone. This was the first study to elaborate on DEX-loaded BCP NPs/collagen porous composite scaffolds, and the significantly positive results of the composite scaffolds denote that composite scaffolds should be highly useful for BTE. Another study (Kim et al., 2018) analysed a novel injectable methylcellulose (MC) hydrogel with calcium phosphate NPs (CaP NPs) for bone regeneration. The results of the *in vitro* study using mesenchymal stem cells demonstrated that MC-CaP NPs composite hydrogel was biocompatible. The *in vivo* study suggested that the regeneration rate of new mature bone was also superior in the MC-CaP NPs composite hydrogel than in the MC hydrogel alone. The results of this study suggested that injectable MC-CaP NPs composite hydrogel has huge potential for use in bone tissue regeneration (Kim et al., 2018).

Despite all these new research outputs, few have evolved to the stage of clinical trials; nevertheless, it is expected that some will soon reach that stage. Liposomal prednisolone (PEGylated) was found in clinical studies to be useful in rheumatoid arthritis, in two different clinical trials in two different phases: NCT02495662 (Ph II) and NCT02534896 (Ph III)(Anselmo and Mitragotri, 2016, van den Hoven et al., 2011).

9.4 Concluding Remarks and Future Trends

Significant advances have been made concerning the knowledge and understanding of nanotechnology, since the term NP was first introduced. Currently, nanotechnology is globally one of the main research areas of knowledge in therapeutic and industrial applications, which has increasing development and impact.

It is possible to find several uses of NPs from the fields of medicine to technology, since NPs can be designed to fit multiple purposes, due to their extraordinary characteristics such as the ability to create diverse structural forms with the desired size and hierarchical organization. The main applications thus far are related with (targeted) therapeutics delivery and imaging systems. For instance, these systems are able to target fracture and improve the bone healing, deliver growth factor for bone formation or induce differential cell response on mesenchymal stem cell osteogenesis.

However, the main clinical trials are mostly carried out on liposomes and on gene therapy and cancer (e.g., solid tumors, leukemia, lymphoma and also lungs, gastric, pancreatic, liver and breast cancer). Clinical trials are also found for instance for the use of nanotechnology in hepatitis and pneumonia. Liposomal prednisolone (PEGylated) is an important compound that has found use in rheumatoid arthritis and hemodialysis fistula maturation in two different clinical trials: NCT02495662 (Ph II) and NCT02534896 (Ph III). In fact, there are not many NPs that can be applied for bone disorders that have reached the clinical trial stage. The NP complexity requires careful engineering, analysis and reproducible scale-up to reach a reliable product with the envisioned properties. The safety and efficacy of nanomedicines is very volatile and requires very detailed preclinical and clinical study.

There is still plenty of room for new approaches that apply the NPs' potential in tailoring the treatment of bone defects. Several promising studies, as was herein described, are currently being performed and published. Thus, it is expected that a significant increase is needed in clinical trials with requisite widespread efforts made in preclinical, commercial and clinical studies. Moreover, new approaches for BTE will also emerge, with the expected advancement of parallel technologies, interdisciplinary collaboration and increasing interest and knowledge on NPs.

References

Abdullaeva, Z. 2017. *Synthesis of Nanoparticles and Nanomaterials: Biological Approaches.* Springer, Cham.

Adamiano, A., Iafisco, M., Sandri, M., Basini, M., Arosio, P., Canu, T., Sitia, G., Esposito, A., Iannotti, V. & Ausanio, G. 2018. On the use of superparamagnetic hydroxyapatite nanoparticles as an agent for magnetic and nuclear *in vivo* imaging. *Acta Biomaterialia*, 73, 458–469.

Ahmed, S., Sheikh, J. & Ali, A. 2018. A review on chitosan centred scaffolds and their applications in tissue engineering. *International Journal of Biological Macromolecules*, 116, 849–862.

Alipour, A., Soran-Erdem, Z., Utkur, M., Shamra, V. K., Algin, O., Saritas, E. U. & Demir, H. V. 2017. A new class of cubic SPIONs as a dual-mode T1 and T2 contrast agent for MRI. *Magnetic Resonance Imaging*, 49, 16–24.

Allen, G., Aggarwal, S. L. & Russo, S. 1992. *Comprehensive Polymer Science: Supplement.* Pergamon Press, Oxford.

Alles, N., Soysa, N. S., Hussain, M. A., Tomomatsu, N., Saito, H., Baron, R., Morimoto, N., Aoki, K., Akiyoshi, K. & Ohya, K. 2009. Polysaccharide nanogel delivery of a TNF-α and RANKL antagonist peptide allows systemic prevention of bone loss. *European Journal of Pharmaceutical Sciences*, 37, 83–88.

Amini, A. R., Laurencin, C. T. & Nukavarapu, S. P. 2012. Bone tissue engineering: Recent advances and challenges. *Critical Reviews*™ *in Biomedical Engineering*, 40, 363–408.

Anselmo, A. C. & Mitragotri, S. 2016. Nanoparticles in the clinic. *Bioengineering & Translational Medicine*, 1, 10–29.

Aryasomayajula, B. & Torchilin, V. P. 2016. Nanoformulations: A lucrative tool for protein delivery in cancer therapy. In A. M. Grumezescu (ed.), *Nanobiomaterials in Cancer Therapy.* Elsevier, Amsterdam, 307–330.

Azmi, M. A. & Shad, K. F. 2017. Role of nanostructure molecules in enhancing the bioavailability of oral drugs. In D. Ficai & A. M. Gruemzescu (eds.), *Nanostructures for Novel Therapy.* Elsevier, Amsterdam, 375–407.

Babitha, S., Annamalai, M., Dykas, M. M., Saha, S., Poddar, K., Venugopal, J. R., Ramakrishna, S., Venkatesan, T. & Korrapati, P. S. 2018. Fabrication of a biomimetic ZeinPDA nanofibrous scaffold impregnated with BMP2 peptide conjugated TiO$_2$ nanoparticle for bone tissue engineering. *Journal of Tissue Engineering and Regenerative Medicine*, 12, 991–1001.

Bastami, F., Paknejad, Z., Jafari, M., Salehi, M., Rad, M. R. & Khojasteh, A. 2017. Fabrication of a three-dimensional β-tricalcium-phosphate/gelatin containing chitosan-based nanoparticles for sustained release of bone morphogenetic protein-2: Implication for bone tissue engineering. *Materials Science and Engineering: C*, 72, 481–491.

Behera, S., Naskar, D., Sapru, S., Bhattacharjee, P., Dey, T., Ghosh, A. K., Mandal, M. & Kundu, S. C. 2017. Hydroxyapatite reinforced inherent RGD containing silk fibroin composite scaffolds: Promising platform for bone tissue engineering. *Nanomedicine: Nanotechnology, Biology and Medicine*, 13, 1745–1759.

Bhowmick, A., Banerjee, S. L., Pramanik, N., Jana, P., Mitra, T., Gnanamani, A., Das, M. & Kundu, P. P. 2018. Organically modified clay supported chitosan/hydroxyapatite-zinc oxide nanocomposites with enhanced mechanical and biological properties for the application in bone tissue engineering. *International Journal of Biological Macromolecules*, 106, 11–19.

Bonzani, I. C., George, J. H. & Stevens, M. M. 2006. Novel materials for bone and cartilage regeneration. *Current Opinion in Chemical Biology*, 10, 568–575.

Boskey, A. L. 2013. Bone composition: Relationship to bone fragility and antiosteoporotic drug effects. *Bonekey Reports*, 2, 447.

Brett, E., Zielins, E. R., Luan, A., Ooi, C. C., Shailendra, S., Atashroo, D., Menon, S., Blackshear, C., Flacco, J. & Quarto, N. 2017. Magnetic nanoparticle-based upregulation of B-cell lymphoma 2 enhances bone regeneration. *Stem Cells Translational Medicine*, 6, 151–160.

Chandran, P. R. & Thomas, R. T. 2015. Gold nanoparticles in cancer drug delivery. In S. Thomas, Y. Grohens & N. Ninan (eds.), *Nanotechnology Applications for Tissue Engineering*. Elsevier, Amsterdam, 221–237.

Chen, Y., Kawazoe, N. & Chen, G. 2017. Preparation of dexamethasone-loaded biphasic calcium phosphate nanoparticles/collagen porous composite scaffolds for bone tissue engineering. *Acta Biomaterialia*, 67, 341–353.

Choi, H. S. & Frangioni, J. V. 2010. Nanoparticles for biomedical imaging: Fundamentals of clinical translation. *Molecular Imaging*, 9, 291–310..

Clarke, B. 2008. Normal bone anatomy and physiology. *Clinical Journal of the American Society of Nephrology*, 3, S131–S139.

Coleman, R. 2014. Breaking the barriers of resolution. *Acta Histochemica*, 116, 1209.

Crasto, G. J., Kartner, N., Reznik, N., Spatafora, M. V., Chen, H., Williams, R., Burns, P. N., Clokie, C., Manolson, M. F. & Peel, S. A. 2016. Controlled bone formation using ultrasound-triggered release of BMP-2 from liposomes. *Journal of Controlled Release*, 243, 99–108.

Dahman, Y. 2017. *Nanotechnology and Functional Materials for Engineers*. Elsevier, Amsterdam.

Dang, N., Liu, T. & Prow, T. 2017. Nano- and microtechnology in skin delivery of vaccines. In M. Skwarczynski & I. Toth (eds.), *Micro and Nanotechnology in Vaccine Development*. Elsevier, Amsterdam, 327–341.

Dey, S., Datta, S., Dasgupta, S., Mazumder, B. & Pathak, Y. V. 2016. Lipid nanoparticles for topical application of drugs for skin diseases. In A. M. Grumezescu (ed.), *Nanobiomaterials in Galenic Formulations and Cosmetics*. Elsevier, Amsterdam, 327–361.

Fang, C.-H., Tsai, P.-I., Huang, S.-W., Sun, J.-S., Chang, J. Z.-C., Shen, H.-H., Chen, S.-Y., Lin, F. H., Hsu, L.-T. & Chen, Y.-C. 2017. Magnetic hyperthermia enhance the treatment efficacy of peri-implant osteomyelitis. *BMC Infectious Diseases*, 17, 516.

Feng, X. 2009. Chemical and biochemical basis of cell-bone matrix interaction in health and disease. *Current Chemical Biology*, 3, 189–196.

Ferreira, C. L., de Abreu, F. A. M., Silva, G. A. B., Silveira, F. F., Barreto, L. B. A., Paulino, Tde P., Miziara, M. N. & Alves, J. B. 2013. TGF-β1 and BMP-4 carried by liposomes enhance the healing process in alveolar bone. *Archives of Oral Biology*, 58, 646–656.

Fujioka-Kobayashi, M., Ota, M. S., Shimoda, A., Nakahama, K.-I., Akiyoshi, K., Miyamoto, Y. & Iseki, S. 2012. Cholesteryl group- and acryloyl group-bearing pullulan nanogel to deliver BMP2 and FGF18 for bone tissue engineering. *Biomaterials*, 33, 7613–7620.

Ganji, D. D. & Kachapi, S. H. H. 2015. *Application of Nonlinear Systems in Nanomechanics and Nanofluids: Analytical Methods and Applications*. William Andrew, Amsterdam, 1–4.

Geng, H., Poologasundarampillai, G., Todd, N., Devlin-Mullin, A., Moore, K. L., Golrokhi, Z., Gilchrist, J. B., Jones, E., Potter, R. J. & Sutcliffe, C. 2017. Biotransformation of silver released from nanoparticle coated titanium implants revealed in regenerating bone. *ACS Applied Materials & Interfaces*, 9, 21169–21180.

Gentile, P., Ferreira, A. M., Callaghan, J. T., Miller, C. A., Atkinson, J., Freeman, C. & Hatton, P. V. 2017. Multi-layer nanoscale encapsulation of biofunctional peptides to enhance bone tissue regeneration *in vivo*. *Advanced Healthcare Materials*, 6, 1–11.

Hadavi, M., Hasannia, S., Faghihi, S., Mashayekhi, F., Zadeh, H. & Mostofi, S. 2017. Novel calcified gum Arabic porous nano-composite scaffold for bone tissue regeneration. *Biochemical and Biophysical Research Communications*, 488, 671–678.

Han, L., Wang, M., Sun, H., Li, P., Wang, K., Ren, F. & Lu, X. 2017. Porous titanium scaffolds with self-assembled micro/nano hierarchical structure for dual functions of bone regeneration and anti-infection. *Journal of Biomedical Materials Research Part A*, 105, 3482–3492.

Hassani Besheli, N., Mottaghitalab, F., Eslami, M., Gholami, M., Kundu, S. C., Kaplan, D. L. & Farokhi, M. 2017. Sustainable release of vancomycin from silk fibroin nanoparticles for treating severe bone infection in rat tibia osteomyelitis model. *ACS Applied Materials & Interfaces*, 9, 5128–5138.

Henkel, J., Woodruff, M. A., Epari, D. R., Steck, R., Glatt, V., Dickinson, I. C., Choong, P. F., Schuetz, M. A. & Hutmacher, D. W. 2013. Bone regeneration based on tissue engineering conceptions—a 21st century perspective. *Bone Research*, 1, 216–248.

Hill, R. G. 2012. *Drug Discovery and Development-E-Book: Technology in Transition*. Elsevier Health Sciences, Edinburgh.

Hochella Jr, M., Aruguete, D., Kim, B. & Madden, A. 2012. Naturally occurring inorganic nanoparticles: General assessment and a global budget for one of Earth's last unexplored geochemical components. In A. S. Barnard & H. Guo (eds.), *Nature's Nanostructures*. Pan Stanford Publishing, Singapore, 1–42.

Hochella Jr, M. F. 2002. There's plenty of room at the bottom: Nanoscience in geochemistry. *Geochimica et Cosmochimica Acta*, 66, 735–743.

Huang, J., Wang, D., Chen, J., Liu, W., Duan, L., You, W., Zhu, W., Xiong, J. & Wang, D. 2017. Osteogenic differentiation of bone marrow mesenchymal stem cells by magnetic nanoparticle composite scaffolds under a pulsed electromagnetic field. *Saudi Pharmaceutical Journal*, 25, 575–579.

Janko, C., Zaloga, J., Pöttler, M., Dürr, S., Eberbeck, D., Tietze, R., Lyer, S. & Alexiou, C. 2017. Strategies to optimize the biocompatibility of iron oxide nanoparticles–"SPIONs safe by design". *Journal of Magnetism and Magnetic Materials*, 431, 281–284.

Jia, Z., Zhang, Y., Chen, Y. H., Dusad, A., Yuan, H., Ren, K., Li, F., Fehringer, E. V., Purdue, P. E. & Goldring, S. R. 2015. Simvastatin prodrug micelles target fracture and improve healing. *Journal of Controlled Release*, 200, 23–34.

Johari, N., Hosseini, H. R. M. & Samadikuchaksaraei, A. 2018. Novel fluoridated silk fibroin/TiO$_2$ nanocomposite scaffolds for bone tissue engineering. *Materials Science and Engineering: C*, 82, 265–276.

Jurvetson, S. & Jurvetson, D. F. 2007. Transcending Moore's law with molecular electronics and nanotechnology. In M. C. Roco & W. S. Bainbridge (eds.), *Nanotechnology: Societal Implications—Individual Perspectives*. Springer, Dordrecht, 43–55.

Katari, R., Peloso, A. & Orlando, G. 2015. Tissue engineering and regenerative medicine: Semantic considerations for an evolving paradigm. *Frontiers in bioengineering and biotechnology*, 2, 57.

Kaur, K., Singh, K., Anand, V., Bhatia, G., Kaur, R., Kaur, M., Nim, L. & Arora, D. S. 2017. Scaffolds of hydroxyl apatite nanoparticles disseminated in 1, 6-diisocyanatohexane-extended poly (1, 4-butylene succinate)/poly (methyl methacrylate) for bone tissue engineering. *Materials Science and Engineering: C*, 71, 780–790.

Kawazoe, N. & Chen, G. 2015. Gold nanoparticles with different charge and moiety induce differential cell response on mesenchymal stem cell osteogenesis. *Biomaterials*, 54, 226–236.

Khan, I., Saeed, K. & Khan, I. 2019. Nanoparticles: Properties, applications and toxicities. *Arabian Journal of Chemistry*, 12, 908–931.

Khan, M. Z. M., Ng, T. K. & Ooi, B. S. 2014. Self-assembled InAs/InP quantum dots and quantum dashes: Material structures and devices. *Progress in Quantum Electronics*, 38, 237–313.

Khorshidi, S. & Karkhaneh, A. 2018. Hydrogel/fiber conductive scaffold for bone tissue engineering. *Journal of Biomedical Materials Research Part A*, 106, 718–724.

Kim, M. H., Kim, B. S., Park, H., Lee, J. & Park, W. H. 2018. Injectable methylcellulose hydrogel containing calcium phosphate nanoparticles for bone regeneration. *International Journal of Biological Macromolecules*, 109, 57–64.

Krukemeyer, M., Krenn, V., Huebner, F., Wagner, W. & Resch, R. 2015. History and possible uses of nanomedicine based on nanoparticles and nanotechnological progress. *Journal of Nanomedicine & Nanotechnology*, 6, 1.

Lee, A., Langford, C. R., Rodriguez-Lorenzo, L. M., Thissen, H. & Cameron, N. R. 2017. Bioceramic nanocomposite thiol-acrylate polyHIPE scaffolds for enhanced osteoblastic cell culture in 3D. *Biomaterials Science*, 5, 2035–2047.

Lee, S.-G., Gangangari, K., Kalidindi, T. M., Punzalan, B., Larson, S. M. & Pillarsetty, N. V. K. 2016. Copper-64 labeled liposomes for imaging bone marrow. *Nuclear Medicine and Biology*, 43, 781–787.

Leena, R., Vairamani, M. & Selvamurugan, N. 2017. Alginate/Gelatin scaffolds incorporated with Silibinin-loaded Chitosan nanoparticles for bone formation *in vitro*. *Colloids and Surfaces B: Biointerfaces*, 158, 308–318.

Liu, X., Li, X., Zhou, L., Li, S., Sun, J., Wang, Z., Gao, Y., Jiang, Y., Lu, H. & Wang, Q. 2013. Effects of simvastatin-loaded polymeric micelles on human osteoblast-like MG-63 cells. *Colloids and Surfaces B: Biointerfaces*, 102, 420–427.

Mahdavi, F. S., Salehi, A., Seyedjafari, E., Mohammadi-Sangcheshmeh, A. & Ardeshirylajimi, A. 2017. Bioactive glass ceramic nanoparticles-coated poly (L-lactic acid) scaffold improved osteogenic differentiation of adipose stem cells in equine. *Tissue and Cell*, 49, 565–572.

Marquez, L., de Abreu, F. A. M., Ferreira, C. L., Alves, G. D., Miziara, M. N. & Alves, J. B. 2013. Enhanced bone healing of rat tooth sockets after administration of epidermal growth factor (EGF) carried by liposome. *Injury*, 44, 558–564.

Melancon, M. P., Appleton Figueira, T., Fuentes, D. T., Tian, L., Qiao, Y., Gu, J., Gagea, M., Ensor, J. E., Muñoz, N. M. & Maldonado, K. L. 2017. Development of an electroporation and nanoparticle-based therapeutic platform for bone metastases. *Radiology*, 286, 149–157.

Mendes, A. C., Baran, E. T., Reis, R. L. & Azevedo, H. S. 2013. Self-assembly in nature: Using the principles of nature to create complex nanobiomaterials. *Wiley Interdisciplinary Reviews: Nanomedicine and Nanobiotechnology*, 5, 582–612.

Movassaghian, S., Merkel, O. M. & Torchilin, V. P. 2015. Applications of polymer micelles for imaging and drug delivery. *Wiley Interdisciplinary Reviews: Nanomedicine and Nanobiotechnology*, 7, 691–707.

Mukherjee, B., Chakraborty, S., Mondal, L., Satapathy, B. S., Sengupta, S., Dutta, L., Choudhury, A. & Mandal, D. 2016. Multifunctional drug nanocarriers facilitate more specific entry of therapeutic payload into tumors and control multiple drug resistance in cancer. In A. M. Grumezescu (ed.), *Nanobiomaterials in Cancer Therapy*. Elsevier, Amstredam, 203–251.

Nagarajan, R. 2008. Nanoparticles: Building blocks for nanotechnology. *ACS Publications*, 1, 2–14.

Nelson, J. T., Hashmi, S., Lee, S. S., Ghodasra, J. H., Nickoli, M. S., Ashtekar, A., Park, C., Hsu, E. L., Hsu, W. K. & Sonn, K. A. 2013. BMP-2 direct-and indirect-binding nanogels designed for bone regeneration: A comparison of spinal fusion capacity. *The Spine Journal*, 13, S71.

Niu, B., Li, B., Gu, Y., Shen, X., Liu, Y. & Chen, L. 2017. In vitro evaluation of electrospun silk fibroin/nano-hydroxyapatite/BMP-2 scaffolds for bone regeneration. *Journal of Biomaterials Science, Polymer Edition*, 28, 257–270.

Njuguna, J., Pielichowski, K. & Zhu, H. 2014. *Health and Environmental Safety of Nanomaterials: Polymer Nancomposites and Other Materials Containing Nanoparticles*. Elsevier, Amsterdam.

Oliveira, I. M., Gonçalves, C., Reis, R. L. & Oliveira, J. M. 2018. Engineering nanoparticles for targeting rheumatoid arthritis: Past, present, and future trends. *Nano Research*, 11, 4489–4506.

Oliveira, J. M., Kotobuki, N., Tadokoro, M., Hirose, M., Mano, J., Reis, R. & Ohgushi, H. 2010. Ex vivo culturing of stromal cells with dexamethasone-loaded carboxymethylchitosan/poly (amidoamine) dendrimer nanoparticles promotes ectopic bone formation. *Bone*, 46, 1424–1435.

Overstreet, D., Von Recum, H. & Vernon, B. 2011. Drug delivery applications of injectable biomaterials. In B. Vernon (ed.), *Injectable Biomaterials*. Elsevier, Oxford, 96–141.

Pan, J., Wen, M., Yin, D., Jiang, B., He, D. & Guo, L. 2012. Design and synthesis of novel amphiphilic Janus dendrimers for bone-targeted drug delivery. *Tetrahedron*, 68, 2943–2949.

Panahifar, A., Jaremko, J., Mahmoudi, M., Lambert, R., Maksymowych, W. & Doschak, M. 2014. Bone-seeking superparamagnetic iron oxide nanoparticles (SPIONs) for magnetic resonance imaging of bone turnover in early osteoarthritis. *Osteoarthritis and Cartilage*, 22, S256–S257.

Pansailom, N., Rattanamai, S., Leepheng, P., Rattanawarinchai, P., Chattrairat, K. & Phromyothin, D. 2017. Surface modification: PEG/Dextran encapsulation of SPIONs. *Materials Today: Proceedings*, 4, 6306–6310.

Park, Y. J., Lee, J. Y., Chang, Y. S., Jeong, J. M., Chung, J. K., Lee, M. C., Park, K. B. & Lee, S. J. 2002. Radioisotope carrying polyethylene oxide–polycaprolactone copolymer micelles for targetable bone imaging. *Biomaterials*, 23, 873–879.

Pes, L., Kim, Y. & Tung, C.-H. 2017. Bidentate iminodiacetate modified dendrimer for bone imaging. *Bioorganic & Medicinal Chemistry Letters*, 27, 1252–1255.

Pitkethly, M. 2008. Nanotechnology: past, present, and future. *Nano Today*, 3, 6.

Raab, C., Simkó, M., Gazsó, A., Fiedeler, U. & Nentwich, M. What are synthetic nanoparticles? (NanoTrust Dossier No. 002en–February 2011). Institut für Technikfolgen-Abschätzung (ITA), Wien, 1–4.

Rajabi, M. & A. Mousa, S. 2016. Lipid nanoparticles and their application in nanomedicine. *Current Pharmaceutical Biotechnology*, 17, 662–672.

Rapp, B. E. 2016. *Microfluidics: Modeling, Mechanics and Mathematics*. William Andrew, Oxford.

Ravichandran, M., Jagadale, P. & Velumani, S. 2016. Inorganic nanoflotillas as engineered particles for drug and gene delivery. In A. M. Grumezesctu (ed.), *Engineering of Nanobiomaterials*. Elsevier, Amsterdam, 429–483.

Revathy, T., Jayasri, M. A. & Suthindhiran, K. 2016. Antimicrobial magnetosomes for topical antimicrobial therapy. In A. M. Grumezescu (ed.), *Nanobiomaterials in Antimicrobial Therapy*. Elsevier, Amsterdam, 67–101.

Revia, R. A. & Zhang, M. 2016. Magnetite nanoparticles for cancer diagnosis, treatment, and treatment monitoring: Recent advances. *Materials Today*, 19, 157–168.

Ribeiro, M., Ferraz, M. P., Monteiro, F. J., Fernandes, M. H., Beppu, M. M., Mantione, D. & Sardon, H. 2017. Antibacterial silk fibroin/nanohydroxyapatite hydrogels with silver and gold nanoparticles for bone regeneration. *Nanomedicine: Nanotechnology, Biology and Medicine*, 13, 231–239.

Rogers, M. A. 2016. Naturally occurring nanoparticles in food. *Current Opinion in Food Science*, 7, 14–19.

Roh, H.-S., Lee, C.-M., Hwang, Y.-H., Kook, M.-S., Yang, S.-W., Lee, D. & Kim, B.-H. 2017. Addition of MgO nanoparticles and plasma surface treatment of three-dimensional printed polycaprolactone/hydroxyapatite scaffolds for improving bone regeneration. *Materials Science and Engineering: C*, 74, 525–535.

Rotello, V. M. 2004. *Nanoparticles: Building Blocks for Nanotechnology*. Springer Science & Business Media, New York.

Saber-Samandari, S., Yekta, H., Ahmadi, S. & Alamara, K. 2018. The role of titanium dioxide on the morphology, microstructure, and bioactivity of grafted cellulose/hydroxyapatite nanocomposites for a potential application in bone repair. *International Journal of Biological Macromolecules*, 106, 481–488.

Sadik, O. A. 2013. Anthropogenic nanoparticles in the environment. *Environmental Science: Processes & Impacts*, 15, 19–20.

Sadjadi, S. 2016. Dendrimers as nanoreactors. *Organic Nanoreactors*. Elsevier, Amsterdam, 159–201.

Salata, O. V. 2004. Applications of nanoparticles in biology and medicine. *Journal of Nanobiotechnology*, 2, **3**.

Sawant, S. & Shegokar, R. 2016. Bone scaffolds: What's new in nanoparticle drug delivery research? In A. M. Grumezescu (ed.), *Nanobiomaterials in Hard Tissue Engineering*. Elsevier, Amsterdam, 155–187.

Singh, N., Joshi, A. & Verma, G. 2016. Engineered nanomaterials for biomedicine: Advancements and hazards. In A. M. Grumezescu (ed.), *Engineering of Nanobiomaterials*. Elsevier, Amsterdam, 307–328.

Singh, T. G. & Sharma, N. 2016. Nanobiomaterials in cosmetics: Current status and future prospects. In A. M. Grumezescu (ed.), *Nanobiomaterials in Galenic Formulations and Cosmetics*. Elsevier, Amsterdam, 149–174.

Strambeanu, N., Demetrovici, L. & Dragos, D. 2015. Natural sources of nanoparticles. In M. Lungu, A. Neculae, M. Bunoiu & C. Biris (eds.), *Nanoparticles' Promises and Risks*. Springer, Cham, 9–19.

Szekely, G. & Didaskalou, C. 2016. Biomimics of metalloenzymes via imprinting. In S. Li, S. Cao, S. A. Piletsky & A. P. F. Turner (eds.), *Molecularly Imprinted Catalysts*. Elsevier, Amsterdam, 121–158.

Tavakolizadeh, A., Ahmadian, M., Fathi, M. H., Doostmohammadi, A., Seyedjafari, E. & Ardeshirylajimi, A. 2017. Investigation of osteoinductive effects of different compositions of bioactive glass nanoparticles for bone tissue engineering. *Asaio Journal*, 63, 512–517.

Terranova, L., Dragusin, D. M., Mallet, R., Vasile, E., Stancu, I.-C., Behets, C. & Chappard, D. 2017. Repair of calvarial bone defects in mice using electrospun polystyrene scaffolds combined with β-TCP or gold nanoparticles. *Micron*, 93, 29–37.

Tom, S., Jin, H.-E. & Lee, S.-W. 2016. Aptamers as functional bionanomaterials for sensor applications. In A. M. Grumezescu (ed.), *Engineering of Nanobiomaterials*. Elsevier, Amsterdam, 181–226.

Torgbo, S. & Sukyai, P. 2018. Bacterial cellulose-based scaffold materials for bone tissue engineering. *Applied Materials Today*, 11, 34–49.

Türkkan, S., Pazarçeviren, A. E., Keskin, D., Machin, N. E., Duygulu, Ö. & Tezcaner, A. 2017. Nanosized CaP-silk fibroin-PCL-PEG-PCL/PCL based bilayer membranes for guided bone regeneration. *Materials Science and Engineering: C*, 80, 484–493.

Van Den Hoven, J. M., Van Tomme, S. R., Metselaar, J. M., Nuijen, B., Beijnen, J. H. & Storm, G. 2011. Liposomal drug formulations in the treatment of rheumatoid arthritis. *Molecular Pharmaceutics*, 8, 1002–1015.

Vial, S., Reis, R. L. & Oliveira, J. M. 2017. Recent advances using gold nanoparticles as a promising multimodal tool for tissue engineering and regenerative medicine. *Current Opinion in Solid State and Materials Science*, 21, 92–112.

Vieira, S., Vial, S., Reis, R. L. & Oliveira, J. M. 2017. Nanoparticles for bone tissue engineering. *Biotechnology Progress*, 33, 590–611.

Virlan, M. J. R., Miricescu, D., Radulescu, R., Sabliov, C. M., Totan, A., Calenic, B. & Greabu, M. 2016. Organic nanomaterials and their applications in the treatment of oral diseases. *Molecules*, 21, 207.

Voss, D. F. 1994. No ordinary genius: The illustrated Richard Feynman. *Science*, 264, 1617–1618.

Wang, Q., Jiang, J., Chen, W., Jiang, H., Zhang, Z. & Sun, X. 2016. Targeted delivery of low-dose dexamethasone using PCL–PEG micelles for effective treatment of rheumatoid arthritis. *Journal of Controlled Release*, 230, 64–72.

Wolf, E. L. 2015. *Nanophysics and Nanotechnology: An Introduction to Modern Concepts in Nanoscience*. John Wiley & Sons, Weinheim.

Xia, Y., Chen, H., Zhang, F., Bao, C., Weir, M. D., Reynolds, M. A., Ma, J., Gu, N. & Xu, H. H. 2018. Gold nanoparticles in injectable calcium phosphate cement enhance osteogenic differentiation of human dental pulp stem cells. *Nanomedicine: Nanotechnology, Biology and Medicine*, 14, 35–45.

Xie, C., Sun, H., Wang, K., Zheng, W., Lu, X. & Ren, F. 2017. Graphene oxide nanolayers as nanoparticle anchors on biomaterial surfaces with nanostructures and charge balance for bone regeneration. *Journal of Biomedical Materials Research Part A*, 105, 1311–1323.

Xue, Y., Guo, Y., Yu, M., Wang, M., Ma, P. X. & Lei, B. 2017. Monodispersed bioactive glass nanoclusters with ultralarge pores and intrinsic exceptionally high miRNA loading for efficiently enhancing bone regeneration. *Advanced Healthcare Materials*, 6 (20), 1–11.

Yadav, H. K. S. & Raizaday, A. 2016. Inorganic nanobiomaterials for medical imaging. *Nanobiomaterials in Medical Imaging*. Elsevier, 8, 365–401.

Yamashita, S., Katsumi, H., Hibino, N., Isobe, Y., Yagi, Y., Kusamori, K., Sakane, T. & Yamamoto, A. 2017. Development of PEGylated carboxylic acid-modified polyamidoamine dendrimers as bone-targeting carriers for the treatment of bone diseases. *Journal of Controlled Release*, 262, 10–17.

Yamashita, S., Katsumi, H., Hibino, N., Isobe, Y., Yagi, Y., Tanaka, Y., Yamada, S., Naito, C. & Yamamoto, A. 2018. Development of PEGylated aspartic acid-modified liposome as a bone-targeting carrier for the delivery of paclitaxel and treatment of bone metastasis. *Biomaterials*, 154, 74–85.

Yassin, M. A., Mustafa, K., Xing, Z., Sun, Y., Fasmer, K. E., Waag, T., Krueger, A., Steinmüller-Nethl, D., Finne-Wistrand, A. & Leknes, K. N. 2017. A copolymer scaffold functionalized with nanodiamond particles enhances osteogenic metabolic activity and bone regeneration. *Macromolecular Bioscience*, 17, 1–11.

Zamboni, W. C. 2007. Carrier-mediated and artificial-cell targeted cancer drug delivery. In S. Prakash (ed.), *Artificial Cells, Cell Engineering and Therapy*. CRC Press, Boca Raton, 469.

Zhang, D., Liu, D., Zhang, J., Fong, C. & Yang, M. 2014. Gold nanoparticles stimulate differentiation and mineralization of primary osteoblasts through the ERK/MAPK signaling pathway. *Materials Science and Engineering: C*, 42, 70–77.

Zhang, Y., Cao, Y., Luo, S., Mukerabigwi, J. F. & Liu, M. 2016. Nanoparticles as drug delivery systems of combination therapy for cancer. In A. M. Grumezescu (ed.), *Nanobiomaterials in Cancer Therapy*. Elsevier, Amsterdam, 253–280.

Zheng, C., Pan, H. & Zhao, X. 2016. Alendronate-modified liposome for bone targeting gene delivery regulating mesenchymal stem cell migration. *Journal of Orthopaedic Translation*, 100 (7), 128.

DNA-Functionalized Gold Nanoparticles

Anand Lopez and Juewen Liu
University of Waterloo

10.1 Introduction

DNA-functionalized gold nanoparticles (AuNPs) represent a classic example of bio-/nanosystems useful for a diverse range of applications ranging from biosensor development (Rosi and Mirkin 2005) and materials science (Storhoff and Mirkin 1999) to nanobiotechnology (Jones et al. 2015) and nanomedicine (Giljohann et al. 2010). DNA oligonucleotides are highly programmable in terms of creating nanoscale structures. In addition, DNA has molecular recognition and catalytic functions. AuNPs have excellent optical properties, good biocompatibility and high stability. Their hybrids have thus enabled versatile applications. At the same time, this system also possesses interesting fundamental physical chemistry properties such as sharp melting transitions and tighter binding to complementary DNA.

Since the publication of the first papers in 1996 by the labs of Mirkin (Mirkin et al. 1996) and Alivisatos (Alivisatos et al. 1996), the field has grown tremendously with nearly 1,000 papers on this topic each year. Therefore, it is impossible to provide a comprehensive review of the field in this chapter. Within the scope of this book, this chapter starts with the description of the basic properties of DNA and AuNPs related to DNA-directed assembly and sensing. Then, some unique physical properties are highlighted. Finally, the application of these materials in sensing, materials assembly and drug delivery are reviewed with representative examples.

10.1.1 Structural and Chemical Functions of DNA Oligonucleotides

Before describing the conjugate, we will first go through some basic properties of DNA relevant for this discussion. Single-stranded DNA oligonucleotides are most frequently used for functionalizing AuNPs. Such DNA can be synthesized with a low cost (if unmodified), and it is possible to introduce a diverse range of modifications, albeit at a higher cost. The structure of a four-nucleotide DNA 5′-ATCG-3′ is shown in Figure 10.1a. DNA has a phosphate-based backbone and four types of nucleobases. The bases are charge-neutral between pH 5 and 8. Protonated adenine and cytosine have pK_a values of 3.5 and 4.2, respectively. Therefore, the pH needs to drop to ∼3 to protonate these two bases. Thymine and guanine cannot be protonated unless pH is lower than 2, which is outside the typical range of experimentation. Each phosphate carries one negative charge ($pK_a < 2$). As a result, DNA is a highly negatively-charged polymer (i.e. a polyanion).

DNA is known to hybridize with its complementary strand, forming a duplex structure with a diameter of just 2 nm and a persistent length of ∼ 40 nm. Since each turn of a B-form DNA is 10.5 base pairs (∼3.6 nm), 40 nm would require ∼120 base pairs. Therefore, for most of the studies using oligonucleotides, DNA duplexes can be considered as rigid rods. In addition to forming simple duplexes, by introducing branched structures, various sophisticated nanostructures have been prepared, with DNA origami and DNA Lego bricks being representative highlights of the field (Rothemund 2006, Ke et al. 2012).

Finally, certain DNA sequences also have excellent molecular binding properties, and such binding DNAs are known as aptamers. Aptamers are often isolated via a combinatorial biology technique called *systematic evolution of ligands by exponential enrichment* (SELEX) (Tuerk and Gold 1990, Ellington and Szostak 1990). To date, hundreds of aptamers targeting small molecules, proteins and even live cells have been reported. DNA can also perform enzyme-like catalytic functions, and these are called DNAzymes or deoxyribozymes (Joyce 2004). DNAzymes are isolated using a similar selection method, and they require metal ions for catalysis. Such a metal requirement has made DNAzymes useful for metal detection. Therefore, DNA has both excellent structural and functional properties. Combined with its high stability and low cost of synthesis as well as ease of modification, DNA is a highly attractive molecule for interfacing with AuNPs (Rosi and Mirkin 2005, Liu 2012, Tan et al. 2014, Wang et al. 2009, Song et al. 2010, Zhao et al. 2008).

10.1.2 Optical and Colloidal Properties of AuNPs

AuNPs have a unique role in nanoscience. High-quality monodispersed AuNPs can be readily synthesized, and they are stable for years in clean, low-salt buffers. AuNPs have good biocompatibility, low toxicity (Giljohann et al. 2010) and excellent optical properties. With surface plasmon resonance, the extinction coefficients of AuNPs are three to five orders of magnitude higher than the brightest organic dyes, allowing visual observation of low nanomolar and even picomolar concentrations of AuNPs (Daniel and Astruc 2004). In addition, AuNPs have distance-dependent optical properties; dispersed AuNPs are red, while aggregated ones are blue or purple due to coupling of surface plasmons (Storhoff et al. 2000). AuNPs are also strong fluorescence quenchers and can enhance Raman signal (Maxwell et al. 2002). Therefore, AuNPs are very versatile in designing optical biosensors. Finally, it is quite easy to attach ligands to a gold surface in general. Thiol-containing molecules can be readily adsorbed on AuNPs, forming self-assembled monolayers (SAM) (Love et al. 2005, Hakkinen 2012). The most popular and robust method of attaching DNA to AuNPs is by using thiol-modified DNA.

The colloidal properties of AuNPs are strongly affected by its surface ligands. We limit our discussion here to AuNPs prepared by citrate reduction of $HAuCl_4$ (citrate-capped AuNPs) (Park and Shumaker-Parry 2014, Frens 1973, Grabar et al. 1995, Liu and Lu 2006c). Larger AuNPs are produced with a lower concentration of citrate (Handley 1989). The lower size limit using this method is ~13 nm, and such 13 nm AuNPs can be readily prepared with a small size distribution. It is more difficult to prepare AuNPs larger than 40 nm with a narrow size distribution, and distorted

shapes are sometimes obtained if the reaction is not well controlled.

AuNPs have a very weak affinity for citrate, and citrate can be easily displaced by stronger ligands. Such AuNPs have an overall negatively-charged surface, rendering moderate charge stabilization. By adding salt, the Debye length is shortened and AuNPs can approach each other, leading to irreversible aggregation from attractive van der Waals force. Gold has a very large Hamaker constant (i.e. strong van der Waals force) (Bishop et al. 2009), making AuNPs particularly susceptible to aggregation. Figure 10.1d shows a photograph of as-synthesized 13 nm AuNPs, and its intense red (light gray in print) color turns blue (dark gray in print) due to aggregation upon addition of just over 20 mM of NaCl.

10.1.3 Adsorption of DNA by AuNPs

To ensure desired functions of DNA on AuNPs, we need to consider their interactions. Each DNA base can be adsorbed strongly by the gold surface, and the adsorption energy ranks A>C>G>T>>phosphate (Storhoff et al. 2002, Kimura-Suda et al. 2003, Liu et al. 2018). All the bases adsorb on a gold surface with energy greater than 100 kJ/mol in a vacuum, even for the weakest thymine (Demers et al. 2002). Although the adsorption energy is lower in water, it is still very strong. For example, if a DNA is stably adsorbed by AuNPs as shown in Figure 10.1b, adding its complementary DNA (cDNA) cannot desorb it (Zhang et al. 2012b). This indicates that the energy of DNA hybridization is much lower than DNA adsorption energy on AuNPs. For comparison, adsorbed DNA can be easily desorbed by its cDNA from graphene oxide (Lu et al. 2009, Liu et al. 2016). An analytical implication is that unmodified DNA adsorbed onto AuNPs likely loses its molecular recognition function. If DNA adsorption is achieved via a thiol modification to prepare conjugates as shown in Figure 10.1c, the DNA bases do not directly come in contact with AuNP surface and are available for hybridization.

10.2 Preparation of DNA-AuNP Conjugates

While it seems to be quite straightforward to attach a thiolated DNA to AuNP, this has been quite a technical challenge. Studying this conjugation process has also improved understanding of fundamental biointerface sciences.

10.2.1 Attaching DNA to Planar Bulk Gold Surface

Before working on AuNPs, studies have already been carried out to functionalize gold electrodes and other bulk gold surfaces with DNA (Hegner et al. 1993, Hashimoto et al. 1994). The process of thiolated DNA adsorption was carefully studied by Tarlov and co-workers using X-ray photoelectron spectroscopy (XPS) (Peterlinz et al. 1997, Levicky

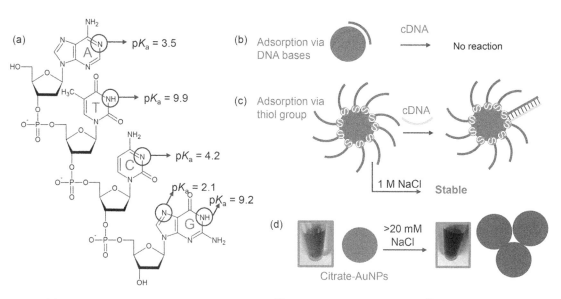

FIGURE 10.1 (a) The structure of a 4-mer DNA and the pK_a values of the bases. The pK_a of the nitrogen atoms without a proton refer to their conjugate acid. The phosphate backbone is negatively-charged. All the bases can strongly adsorb onto AuNPs via coordination interactions. (b) A scheme of an unmodified DNA adsorbing lengthwise on AuNPs using its bases. Since the bases are tightly adsorbed, it cannot react with its cDNA. (c) A thiolated DNA adsorbs via its thiol group, allowing the DNA to hybridize with its cDNA. After DNA attachment, the colloidal stability of AuNPs is significantly enhanced and can often survive 1 M NaCl. (d) Citrate-capped AuNPs are easily aggregated with even a low concentration of salt. (Figure reprinted from Liu and Liu 2017b with permission. Copyright © 2017 Royal Society of Chemistry.)

et al. 1998). In a typical experiment, 1 M buffer was directly used since DNA does not adsorb without salt (their phosphate buffer also served as salt). The hybridization efficiency of as-prepared samples was often quite poor, which was attributed to non-specific adsorption of DNA bases as mentioned above. To solve this problem, a small thiol ligand, such as MCH, was added afterwards to back-fill the gold surface and displace DNA bases. Note that thiol adsorption is stronger than DNA base affinity. This helps the DNA to adopt an upright confirmation for hybridization.

10.2.2 Salt-Aging for Conjugation

While adding 1 M salt directly can effectively screen charge repulsion, it cannot be applied to AuNPs for DNA attachment due to the colloidal stability problem. Adding too much salt all at once induces irreversible aggregation of AuNPs. To address this, a salt-aging method was developed by Mirkin and co-workers (Storhoff et al. 1998, Cutler et al. 2012, Mirkin et al. 1996), which has been reliably reproduced in many labs, and it should work for essentially any DNA sequence. This method still relies on salt to screen charge repulsion, but salt is added gradually in small increments to retain the stability of the AuNPs (Liu and Lu 2006c). Briefly, citrate-capped AuNPs are mixed with thiolated DNA. For the 13 nm AuNPs, DNA is typically mixed with AuNPs at a ratio of ∼ 300:1. For larger AuNPs, this ratio needs to be even higher. After an initial incubation, a concentrated NaCl stock solution is added dropwise with simultaneous shaking to reach a final NaCl of ∼ 50 mM.

After further incubation, another dose of 50 mM NaCl is added until a final of 300 mM or more of NaCl is reached. Then, overnight incubation is typically performed; the whole process takes more than a full day. The final product is highly stable even in 1 M NaCl, and the attached DNA can hybridize to its cDNA.

The reactions during the salt-aging process are presented in Figure 10.2a. After the initial mixing, only a few DNA strands are adsorbed via either the thiol group or via the bases. This is likely to happen since the overall DNA density is very low. These initially adsorbed DNA molecules increased the negative charge density of the AuNPs, and they repel incoming DNA strands more strongly than the bare AuNPs. A new equilibrium is thus reached, and no more DNA is adsorbed. With the adsorbed DNA, the stability of such conjugates is higher than that of bare AuNPs and they can tolerate a slightly higher salt concentration. With more NaCl added, a few more DNA strands are adsorbed until a new electrostatic equilibrium is reached and the product can tolerate even more NaCl. As the density of DNA increases during this salt-aging process, the thiol group gradually displaces the DNA bases adsorbed on AuNPs, resulting in the DNAs in an upright conformation. The density on AuNPs can be higher than that on a planar gold surface. The high curvature of small AuNPs is helpful for such a high density.(Hill et al. 2009) This method has been improved by also adding surfactants, which coat the AuNPs to allow functionalization of larger AuNPs up to about 100 nm (Hurst et al. 2006).

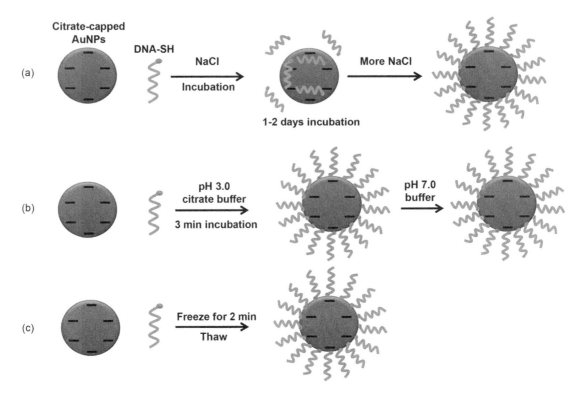

FIGURE 10.2 Schematics of attaching negatively-charged thiolated DNA to negatively-charged AuNPs using (a) the salt-aging method, (b) the low pH-assisted method, and (c) the freezing method.

10.2.3 Low pH-Assisted DNA Attachment

In 2012, Liu and coworkers studied the adsorption of DNA by AuNPs as a function of buffer composition (Zhang et al. 2012a). It was discovered that DNA adsorption was much faster at acidic pH, especially at low salt concentrations. Based on this, they developed an improved method that needs only a few minutes (Zhang et al. 2012a). In a typical experiment, 13 nm AuNPs are mixed with a thiolated DNA, and there is also no need to add excess amount of DNA. By using a 100:1 ratio (about the adsorption capacity), the system works just as well. After a brief incubation, a final aliquot of 10 mM sodium citrate buffer is quickly added. Three minutes later, a stable conjugate is formed and the sample can be brought back to pH 7 (Figure 10.2b). The role of the acid is believed to overcome the kinetic barrier of DNA attachment since adenine and cytosine bases can be protonated at pH 3. Later, it was found that the formation of parallel poly-A duplex at low pH could also be responsible for the ultrafast attachment (Huang et al. 2016). Note, such DNA typically has a poly-A spacer between the thiol group and the DNA intended for hybridization. The quality of the AuNPs was characterized by UV–vis spectroscopy and DLS, and the conjugate is quite comparable with that from the salt-aging method.

10.2.4 Freezing-Assisted DNA Attachment

Recently, a freezing-assisted method was also developed by Liu and coworkers (Figure 10.2c) (Liu and Liu 2017a). After mixing thiolated DNA with AuNPs without adding any

other reagent, the sample is stored in a −20°C freezer. After freezing and thaw, stable conjugates were formed and the density of DNA was even higher by ∼20% than the ones prepared by the typical salt-aging. This is attributed to the exclusion of non-water species, such as salt, AuNPs and DNA during freezing to the micro-pockets formed between ice crystals. In the micro-pockets, a eutectic phase of water exists and DNA could quickly attach in this process. Note that, in this case, an excess amount of DNA is still required to ensure the stability of AuNPs. This method works for all tested DNA sequences.

While NaCl has been the most frequently used salt for attaching DNA to AuNPs, the effect of cations and anions have been systematically studied. By changing cations, (e.g. Li^+, Na^+, K^+, Rb^+, and Cs^+), the rate of DNA adsorption and the final density of DNA can be adjusted (Liu et al. 2014), while changing anions (e.g. F^-, Cl^-, Br^- and I^-) can affect the quality of the finally obtained conjugates. For example, Br^- can selectively displace certain bases and finely control the conformation of DNA on AuNPs (Liu et al. 2018). From the above description, we can see the application of basic colloidal science principles in this particular system to achieve bioconjugation.

10.2.5 Unique Physical Properties of DNA-Functionalized AuNPs

By default, we consider DNA-functionalized AuNPs to be densely functionalized DNA. A direct consequence is highly stable conjugates that can survive 1 M NaCl or even higher salt concentration. With a dense layer of highly negatively-charged DNA on AuNPs, this conjugate has many other interesting and unique physical properties. We just describe two such properties here useful for biosensor development.

Sharp Melting Transitions

When a DNA duplex is heated, the two strands gradually dissociate into single-stranded DNA, a process called DNA melting. The temperature at which 50% of the DNA is in the single-stranded form is called the melting temperature (T_m). A scheme of melting of free DNA (bottom) and DNA-linked AuNPs (top) is shown in Figure 10.3a. A typical DNA melts over a temperature range spanning over 20°C, while in DNA-linked AuNPs, the same transition takes places within just a few degrees. This has been illustrated in the original work by Mirkin and coworkers (Figure 10.3b) (Elghanian et al. 1997). The reason for this was attributed to the co-operative melting of all the DNA strands at the same time, and prematurely melted DNA can quickly rehybridize to the same or neighboring DNA, since the AuNP is still held in place by other DNA linkages to other AuNPs. Only when most of the linkages break, a burst of melting takes place, which is accompanied by a color change. In a particular case, melting of DNA-AuNP aggregates took place within 1°C (Figure 10.3c). This is analytically useful, allowing colorimetric discrimination of single base mismatches. Such a melting-based method has also been used for detecting mercury (Lee et al. 2007) and for screening of DNA binding agents (Han et al. 2006).

Tighter Binding to cDNA

Another feature of this conjugate highlighted here is its higher binding affinity to the complementary DNA. For example, for a 15-mer AT-rich strand, its binding constant to complementary DNA is 1.8×10^{12} M^{-1} (at 298K), while its binding to a densely functionalized AuNP

($\sim$ 100 DNA / AuNP) is 100-fold higher (Lytton-Jean and Mirkin 2005). Control experiments have shown that the increased binding was due to the very high density of DNA enabled by the high curvature of AuNPs (Hill et al. 2009), instead of due to the absolute number of DNA. With these interesting properties, this type of conjugate is termed spherical nucleic acids (SNAs) by Mirkin and coworkers (Cutler et al. 2012).

10.3 Biosensor Applications

After introducing the physical properties of this conjugate, we describe its applications. DNA can recognize a diverse range of analytes including not only complementary nucleic acids, but also metal ions, small molecules and proteins using aptamers and DNAzymes. AuNPs, with excellent optical properties, are ideal for signal transduction. Combining these two can produce interesting and useful biosensors.

10.3.1 Colorimetric DNA Detection

With DNA-functionalized AuNPs, it is quite straightforward to achieve colorimetric detection based on DNA-induced assembly or disassembly of AuNPs. The first examples were reported by Mirkin and coworkers using the target DNA to crosslink two types of DNA-functionalized AuNPs to produce blue or purple (gray in print) colored aggregates (gray in print) as shown in Figure 10.4a (Elghanian et al. 1997). Single-base-mismatch detection was achieved by measuring the melting temperature of the aggregated AuNPs. Due to the presence of multiple DNA linkages and co-operative DNA melting, these aggregated AuNPs possess much sharper melting transitions compared to that for the free DNA (Jin et al. 2003). As a result, mismatched DNAs were detected with a high precision (Figure 10.4b). Since larger AuNPs have higher extinction coefficients, they can be visualized at much lower concentrations. Therefore,

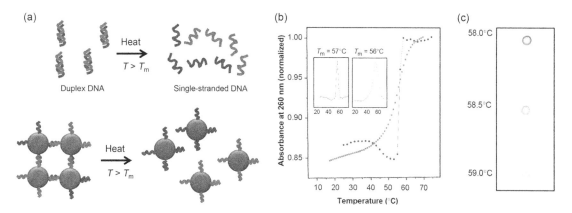

FIGURE 10.3 (a) Scheme of melting of (top row) free DNA strands, and (bottom row) DNA-functionalized AuNPs. Below the melting temperature, most of the DNA strands are in the duplex form. With AuNPs, melting is accompanied by a color change due to surface plasmon coupling. (b) DNA melting curve for free DNA strands (squares) and DNA-AuNP hybrids (circles). (c) The abrupt change in color below and above the melting temperature of the DNA-AuNP hybrid. ((b and c) adapted from Elghanian et al. 1997 with permission. Copyright © 1997 American Association for the Advancement of Science.)

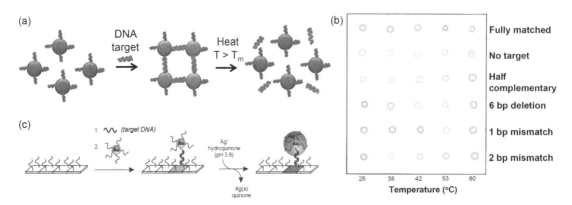

FIGURE 10.4 (a) Scheme of colorimetric DNA detection. (b) DNA detection using DNA-directed assembly of AuNPs with high specificity. To discriminate mismatched targets, a melting experiment is often needed by gradually heating the sample. (Adapted from Elghanian et al. 1997 with permission. Copyright © 1997 American Association for the Advancement of Science.) (c) Signal amplification by growing silver using the immobilized AuNPs as a catalyst. (Adapted from Taton et al. 2000 with permission. Copyright © 2000 American Association for the Advancement of Science.)

sub-nanomolar sensitivity can be achieved with 40–50 nm AuNPs (Reynolds et al. 2000). Further improvement in sensitivity was achieved by immobilizing one of the probe DNAs on a surface, and the target DNA brings AuNPs bearing the other probe DNA to the surface. Signal amplification was achieved using immobilized AuNPs as a catalyst to grow silver (Figure 10.4c), which drastically decreased the detection limit to 50 fM DNA (Taton et al. 2000). In addition, two types of DNA can be immobilized on AuNP surface, with one for hybridization and the other as a barcode. The barcode DNA can later be released as another linker for further signal amplification. With this method, sensitivity close to PCR was achieved to detect just a few copies of target DNA.

10.3.2 Using Aptamers

As mentioned previously, DNA aptamers have been developed to bind specific target molecules in solution. These aptamers are single-stranded DNA and typically fold into well-defined structures in the presence of their target molecules (Hermann and Patel 2000). Interfacing DNA aptamers with AuNPs was first reported Liu and Lu (2006b, c). In their early work, they designed a colorimetric sensor by crosslinking two types of AuNPs using a linker strand containing an aptamer fragment (Figure 10.5a). Upon mixing in solution, the AuNPs are brought close together, and the solution color changes to purple or blue. The target (e.g. adenosine) is then added, causing the aptamer conformation to change and leading to dehybridization of the linker strands from the AuNPs. The AuNPs are separated from each other, and the color of the solution changes from blue to red. This sensing process is fast; reported results were obtained within a few minutes. This was first applied to both adenosine and cocaine, resulting in a detection range between 0.3–2 mM and 50–500 μM, respectively. Using dual aptamer sequences within the linker DNA, multiplexed detection was also achieved (Liu and Lu

2006d). While AuNPs provide the most striking example, this method has also been used for magnetic nanoparticles, quantum dots, and hydrogels, which produced different types of physical signals upon changing the assembly state of these nanomaterials (Liu et al. 2007, Yigit et al. 2007, Yang et al. 2008, Zhu et al. 2010).

While the above methods normally reflect disaggregation of AuNPs (i.e. color shifting from blue to red), the opposite trend (i.e. aggregation for a red-to-blue transition) has also been applied in colorimetric sensing. Ye, Fan and coworkers applied this for the detection of both adenosine and cocaine (Li et al. 2009). They split the aptamer into two pieces, attaching each piece on one population of AuNPs. Then, the two populations were mixed and the target was added. This reunited the two aptamer segments, which also brought the AuNPs closer together, causing a color change from red to blue. They were able to detect targets within 30 minutes and reported a limit of detection of 0.25 mM (adenosine) and 0.1 mM (cocaine). This concept of assembly or disassembly of AuNPs mediated by DNA aptamers has been used for detection of proteins with more than one aptamer binding sites (Pavlov et al. 2004, Huang et al. 2005) and metal ions (Lee et al. 2007, Xue et al. 2008, Liu and Lu 2006d). Furthermore, this type of system has been integrated within lateral flow devices, where simple dipping of the device into the sample solution could complete the detection process (Liu et al. 2006).

10.3.3 Using DNAzymes

DNAzymes are DNA-based catalysts. Since most DNAzymes require specific metal ions to function, they are ideal for metal detection (Zhou et al. 2017). The most commonly used DNAzymes perform the RNA cleavage reaction, which provides a convenient method to disassemble aggregated AuNPs. This method was first reported by Liu and Lu for developing colorimetric biosensors for Pb^{2+} detection (Liu and Lu 2003). A prime example of this

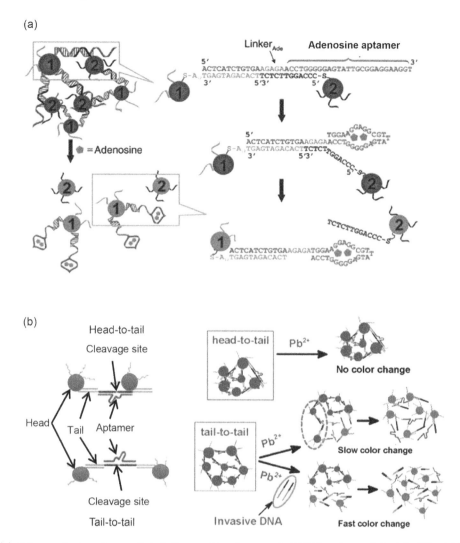

FIGURE 10.5 (a) Scheme of adenosine mediated disassembly of aggregated DNA aptamer-linked AuNPs. (Adapted from Liu and Lu 2006c with permission. Copyright © 2006 Nature Publishing Group.) (b) Scheme of disassembly of tail-to-tail assemblies of AuNPs mediated by cleavage by a Pb^{2+} DNAzyme. (Adapted from Liu and Lu 2005 with permission. Copyright © 2005 American Chemical Society.)

is shown in Figure 10.5b (Liu and Lu 2005). The system consists of DNA-functionalized AuNPs conjugated to the DNAzyme–substrate complex either in a head-to-tail or tail-to-tail conformation. Without the addition of Pb^{2+}, the AuNPs were very close together such that the color of the solution was purple. In the head-to-tail arrangement, the linking DNAzyme was inactive and Pb^{2+} did not cause cleavage or disassembly. This was attributed to the high degree of steric hindrance toward the cleavage site, preventing any reaction from occurring. In the tail-to-tail case, the addition of Pb^{2+} causes cleavage of the substrate strand, resulting in a gradual separation of the AuNPs and a corresponding color change from purple to red. However, the relatively slow cleavage kinetics presented a major limitation for sensing applications. In addition, after cleavage, the cleaved fragments need to be released before the color change can be observed. To overcome this, an invasive DNA strand (which was complementary to the cleaved strand) was added, resulting in a much quicker reaction time (Liu and Lu 2006a).

10.3.4 Fluorescent Biosensors

While only colorimetric sensors have been discussed so far, AuNP-DNA conjugates have also been utilized for detection using fluorescence as a sensor output. Fluorescence methods offer a higher sensitivity compared to colorimetric detectors with a corresponding decrease in the limit of detection (LOD). AuNPs are excellent fluorescence quenchers, and can actually quench fluorophores over a longer distance than conventional molecular quenchers (Yun et al. 2005, Ray et al. 2007). In general, most strategies involve conjugating thiolated DNA with a fluorophore and anchoring it to the AuNP surface, producing an "off" state, where the fluorescence is quenched. With the addition of the target analyte, changes in the DNA conformation (e.g. hybridization or folding) release the fluorophore from the surface of the AuNP, and the sensor changes to the "on" state.

AuNP-DNA fluorescence-based detection was first demonstrated by Libchaber's research group (Dubertret et al. 2001). They designed a DNA hairpin where one end was capped by a fluorophore while the other end was conjugated to a very small AuNP. As a result, the fluorophore was in close proximity to the AuNP and the fluorescence was efficiently quenched (Figure 10.6a). This can also be described as a molecular beacon. Once a DNA complementary to the hairpin structure was added, there was fluorescence recovery due to hybridization and the fluorophore being far enough from the AuNP that quenching could not occur. Soon after, a hairpin-independent design was reported by Nie and coworkers, but the detection scheme was similar (Maxwell et al. 2002). Instead of using the hairpin structure, they relied on simple adsorption of the dye at the other end of DNA strand on the AuNP, effectively quenching the fluorescence. Confirmation of fluorescence quenching was seen upon conjugation of the DNA strand, and subsequently restored with complementary DNA. They were able to detect cDNA concentrations as low as 10 nM.

Two aptamer-based fluorescence detection methods were first proposed by Zhao and co-workers to detect thrombin (Wang et al. 2008). In both cases, a thiolated DNA was conjugated to the surface of the AuNP and hybridized with a fluorophore-conjugated complementary sequence. The first mechanism is shown in Figure 10.6b. The thiolated aptamer sequence was bound to the AuNP and the complementary sequence was conjugated with the fluorophore (signal "off" state). Once the target (in this case, thrombin) was added, the complementary sequence was displaced and the fluorescence increased. In the second mechanism, the thiolated complementary sequence was conjugated to the AuNP hybridized with the fluorophore-labeled aptamer strand (Figure 10.6c). Addition of the target displaced the aptamer and fluorescence increased. The detection limit for the first and second case was 0.14 nM and 3.78 nM, respectively.

The common theme throughout the previous sensor designs is that there is a change in the fluorophore distance from the AuNPs through a binding event and/or conformational change. Another way to change this distance would be to fully break the linkage between the AuNP and the DNA. The proof-of-concept design was reported by Ray, Fortner and Darbha in 2006 using a DNA nuclease (Ray et al. 2006). Essentially, the AuNP is conjugated with a short, thiolated strand bearing a fluorophore and hybridized with a sequence which is complementary to all but the closest base to the AuNP (Figure 10.6d). The distance between the fluorophore and the AuNP is short enough so that quenching

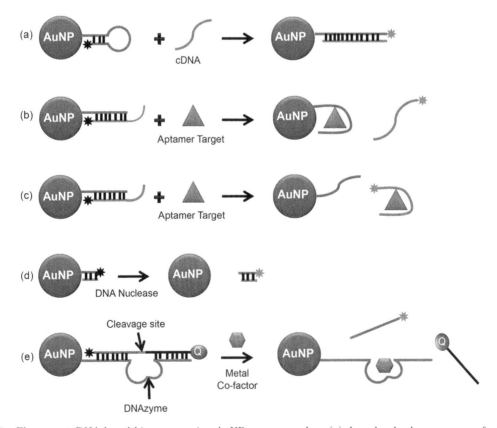

FIGURE 10.6 Fluorescent DNA-based biosensors using AuNPs as a quencher. (a) A molecular beacon sensor for DNA detection. Aptamer-based detection where the fluorophore is labeled on (b) the non-aptamer strand, or (c) the aptamer strand. (d) Detection of nuclease activity by cleaving DNA off the AuNP surface. (e) Detection of metal using DNAzyme-mediated cleavage of a fluorescent substrate strand.

of the fluorescence occurs. In the next step, the nuclease is added and cleaves the DNA at the unmatched base site, releasing the small duplex into solution with a corresponding increase in fluorescence. This nuclease only cleaves single-stranded DNA; the free duplex DNA in solution would not be affected. As such, this system could be used to either detect short DNA strands or assay DNA nuclease activity.

DNAzymes have also been used to detecting other targets using AuNPs as fluorescence quenchers and was first demonstrated by Wu and coworkers (Wu et al. 2013). The basic scheme of their sensor design is shown in Figure 10.6e. In principle, thiolated DNAzymes were conjugated to AuNPs and hybridized with a partially complementary substrate strand. This substrate strand was conjugated with a quencher at one end and a fluorophore at the other end and contained a ribonucleotide site in the middle. While the AuNP provided significant quenching capabilities, the addition of another quencher lowered the background fluorescence to improve the sensitivity of the sensor. This particular DNAzyme catalyzed the cleavage of the ribonucleotide site mediated by a metal co-factor (in their case, the uranyl group). Upon cleavage of the substrate, de-hybridization occurs, and fluorescence increases due to increasing distance from the AuNP quencher. The target for most DNAzyme-based sensors is the metal co-factor and, as such, much research has been directed toward screening toxic metal moieties for DNAzyme activity.

10.4 DNA-Directed Assembly of AuNPs

In addition to preparing biosensors, DNA-functionalized AuNPs have been used for preparing sophisticated and dynamically-controllable nanomaterials.

10.4.1 Colloidal Crystals

For biosensor applications, AuNPs are often randomly assembled by linker DNA and a color change is produced. Most of these products were non-crystalline without long-range order, but this was not a concern for biosensor development. In 2008, it was discovered by the Mirkin and Gang groups that crystalline materials can be obtained by introducing flexibility to the DNA strands (Park et al. 2008, Nykypanchuk et al. 2008). The strategy from Mirkin's group is shown in Figure 10.7a. Linker A represents a strand which can hybridize with the DNA conjugated on the AuNP, and can also hybridize with itself through the short sticky ends. Linkers X and Y can also hybridize with the DNA on the AuNP surface, but the sticky ends can only hybridize with the other type of linker (i.e. linker X can hybridize only with linker Y). By slowly cooling the short duplex from above to below T_m, crystallization occurs through duplex hybridization. If the self-complementary linker A is used, then a face-centered-cubic (FCC) structure is formed, while if the binary

linkers (X and Y) are used, a body-centered-cubic (BCC) structure is formed. For these studies, small angle X-ray scattering (SAXS) was the main tool for characterization of the packing and periodicity of the material. Since this initial discovery, many types of colloidal crystals have been produced by combining different particles of varied size, shape, and composition and DNA length and sequence using the same general concept (Macfarlane et al. 2011, Jones et al. 2010). Gold nanorods were found to assemble into a 2D hexagonal layer with a hexagonal-close-packed lattice structure (Figure 10.7b, left). Triangular nanoprisms stacked in 1 dimension, though no specific long-range ordering was observed (Figure 10.7b, middle). Finally, rhombic dodecahedrons formed FCC structures with a packing efficiency that was even higher than spherical particles (Figure 10.7b, right). Furthermore, through controlling DNA hybridization, it is also possible to dynamically control the inter-particle distance and the lattice structure (Kim et al. 2016, Maye et al. 2010). Short and intermediate inter-particle distance favours the formation of disordered and FCC structures, while longer inter-particle distances favour BCC lattice structures.

10.4.2 AuNPs as "atoms"

So far, we have focused on AuNPs with a densely functionalized layer of DNA, and the advantage of such particles is a very high colloidal stability and unique physical properties. In addition to using densely functionalized DNA, another direction is to control the number of DNA and thus build valency, which was demonstrated first in 1996 by Alivisatos and coworkers (Alivisatos et al. 1996). Typically, such conjugates were prepared by mixing a relatively long thiolated DNA and small AuNPs capped with a stabilizing ligand at a low DNA-to-AuNP ratio. A range of DNA densities is produced following the Poisson distribution. The product is then separated using gel electrophoresis, and AuNPs with 0, 1, 2, 3… copies of DNA are harvested from the gel for application as shown in Figure 10.8a (Zanchet et al. 2001). This is a powerful method, allowing the rational construction of different particles and different geometries. Simple structures containing just two or three AuNPs were demonstrated by Alivisatos' group using head-to-tail arrangements of mono-functionalized AuNPs (Loweth et al. 1999). This was further developed to form tetramers of mono-dispersed (Figure 10.8b, top row) or poly-dispersed (Figure 10.8b, bottom row) AuNPs (Mastroianni et al. 2009). In this case, each population of AuNP was conjugated with a strand complementary to a portion of the strand on two other populations of AuNP, followed by mixing and hybridization to form the nanoparticle clusters. Chiral AuNP-DNA tetramers were assembled using DNA origami as a support by Shen et al (Shen et al. 2013). Using a large viral DNA, they used stapler strands to form a rectangular DNA origami template with sticky ends for the formation of either a left-handed (Figure 10.8c, top row) or right-handed (Figure 10.8c, bottom row) structure. The difference

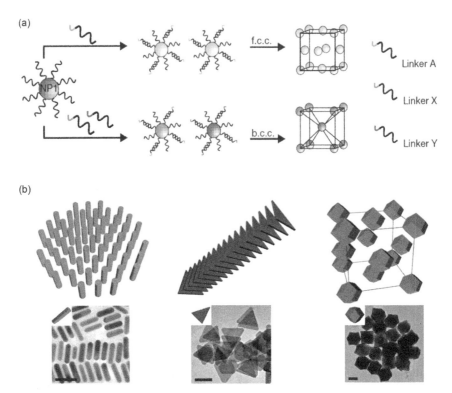

FIGURE 10.7 (a) Scheme of DNA-mediated crystallization of AuNPs into either an FCC or BCC lattice. (Adapted from Park et al. 2008 with permission. Copyright © 2008 Nature Publishing Group.) (b) Complex superstructures formed by DNA-mediated assembly of non-spherical AuNPs. (Adapted from Jones et al. 2010 with permission. Copyright © 2010 Nature Publishing Group.)

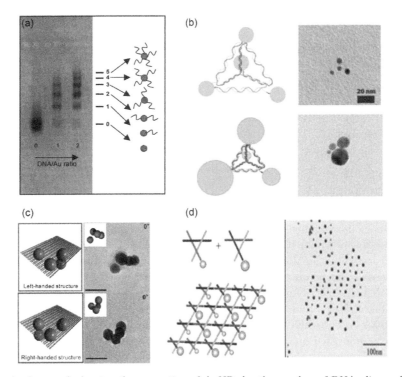

FIGURE 10.8 (a) A gel micrograph showing the separation of AuNPs by the number of DNA oligonucleotide attached. (Adapted from Zanchet et al. 2001 with permission. Copyright © 2001 American Chemical Society.) (b) DNA-linked AuNPs organized into well-defined structures. (Adapted from Mastroianni et al. 2009 with permission. Copyright © 2009 American Chemical Society.) (c) Chiral DNA-AuNP tetramers deposited on DNA origami superstructures. (Adapted from Shen et al. 2013 with permission. Copyright © 2013 American Chemical Society.) (d) Periodic arrays of AuNPs formed by DX-DNA self-assembly. (Adapted from Zheng et al. 2006 with permission. Copyright © 2006 American Chemical Society.)

in chirality of the tetramers was confirmed using circular dichroism experiments, in addition to transmission electron microscopy. In addition to AuNPs, other types of materials including silver and platinum have also been used (Zheng et al. 2012b, Li et al. 2012).

10.4.3 DNA Nanostructures as a Template

A third type of material is prepared by building a DNA nanostructure first. DNA overhangs are engineered at designated positions for hybridization with DNA-functionalized AuNPs. If the DNA nanostructure is a large periodic network, then periodic AuNP arrays can be obtained. For example, Seeman and co-workers constructed arrays of alternating AuNPs of different sizes as shown in Figure 10.8d (Zheng et al. 2006). In this case, the DNA structure used was termed a double-crossover (DX). At one of the blunt ends, AuNPs were conjugated, while the other end was sticky and complementary to the sticky end of another DX structure. Mixing two DX populations, each conjugated with a different size of AuNP, allowed for self-assembled arrays of alternating AuNPs to be obtained. If the structure is limited by complex geometry, smaller AuNP structures can also be prepared. An example of this was seen in 2012 by Kuzyk et al (Kuzyk et al. 2012). They used DNA origami to assemble 24-helix bundles containing multiple sticky ends. The addition of AuNPs conjugated with complementary DNA allowed for precise insertion of these AuNPs into the superhelix. This precision allows for the study of plasmonic and catalytic properties of AuNPs as a function of interparticle distance.

10.5 Cellular Uptake of DNA-Functionalized AuNPs

The growth of the field of nanomedicine in recent decades has led to the development of many unique ways to implement nanoparticles for biomedical applications. AuNPs are no exception to this; their low toxicity and inertness, combined with a relative ease of production, has certainly helped in this regard. It therefore comes as no surprise that research groups have combined the biological functionality of DNA with AuNPs for enhanced function.

DNA is a highly negatively-charged polymer, and it is naturally repelled by the negatively-charged cell surface. Typically, transfection agents are polycationic to allow for electrostatic attraction between the transfection agent and cell membrane. Mirkin and coworkers discovered that DNA-functionalized AuNPs can be effectively internalized by cells (Rosi et al. 2006). Interestingly, a higher loading density on AuNPs is associated with better cell uptake, suggesting that electrostatic forces are unlikely to be the primary cause for uptake.

At the first glance, it is difficult to understand the internalization of negatively-charged DNA/AuNP complex by

cells. To determine the cause of this, the size and zeta potential of the DNA-functionalized AuNP were measured before and after cell uptake (Giljohann et al. 2007). An increase in particle size and a rise in zeta-potential were measured, suggesting the adsorption of extracellular proteins by the DNA strands. Further research has revealed the specific proteins that might mediate cellular uptake of DNA-functionalized AuNPs (Patel et al. 2010). Serum generally contains a wide range of proteins such as bovine serum albumin (BSA), which is traditionally used as a blocking agent for protein adsorption and is found in great abundance in serum. The adsorption of BSA onto the surface of a DNA-functionalized AuNP was demonstrated to be non-effective in aiding cellular internalization. Increased cell uptake is accomplished by the displacement of these serum proteins by scavenger receptors, which then act as mediators for endocytosis of the DNA-functionalized AuNPs (Figure 10.9a).

10.5.1 Nanoflares

Probing products or reagents of intracellular metabolism can prove difficult in real time using antibodies. Typically, if the target is a protein, cells need to be fixed to a surface in order for an assay to take place. Furthermore, the nature of the method requires multiple washing steps and is, in general, quite tedious to get meaningful results. Since delivering protein-based sensors into cells has proven difficult, intracellular expression of fluorescent proteins (such as the Green Fluorescent Protein, GFP) has been the method of choice to detect binding events or other processes within the cell. However, AuNP-DNA hybrids provide a new way to achieve this goal: through the so-called "nanoflare". This concept was developed by Mirkin and co-workers in 2007 for the detection of messenger RNA (mRNA) and illustrated in Figure 10.9b (Seferos et al. 2007). Essentially, AuNPs are conjugated with a recognition sequence complementary to the mRNA being probed. It is then partially hybridized with a smaller, dye-conjugated sequence. Due to the proximity of the fluorescent dye to the AuNP, the fluorescence is quenched and the nanoflare is ready to be utilized. Upon exposure to the target mRNA, the smaller strand is displaced, resulting in an increase in fluorescence (i.e. the "flare" is released). This allows for visual observation of mRNA production using fluorescence microscopy, while quantitative data could be obtained using flow cytometry. This simple idea has been modified to regulate mRNA concentration (Prigodich et al. 2009), as well as detect multiple mRNA with one nanoflare (Prigodich et al. 2012).

10.5.2 Tissue Penetration

Drugs which silence specific signaling pathways within cells (typically protein-based) have led to many developments in treating many diseases. However, these drugs are limited by their toxicity, especially by injection or

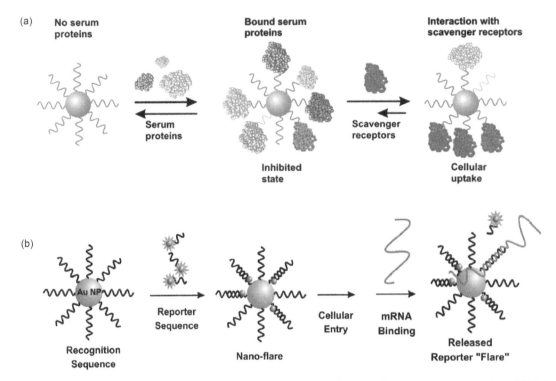

FIGURE 10.9 (a) Adsorption of serum proteins (and their subsequent displacement by scavenger receptors) on DNA-functionalized AuNPs. (Adapted from Patel et al. 2010 with permission. Copyright © 2010 American Chemical Society.) (b) Scheme of nanoflares for mRNA binding in cells. (Adapted from Seferos et al. 2007 with permission. Copyright © 2007 American Chemical Society.)

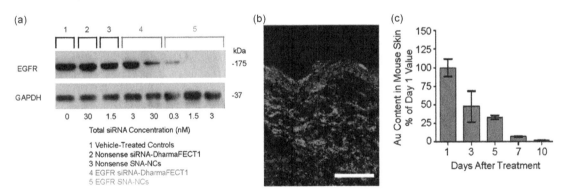

FIGURE 10.10 (a) Western blot micrograph of EGFR and GAPDH expression. Treatment with EGFR-SNA-NCs effectively knocks down any production of this receptor. (b) Confocal microscopy image of rat skin cross-section showing internalization of SNA-NCs through the tissue. Individual cell nuclei stained with Hoescht dye. (c) Clearance of SNA-NCs from rat skin after 10 days of treatment. At the end of the experiment period, only 2% of the initial Au content was measured in the skin. (All the images adapted from Zheng et al. 2012a with permission. Copyright © 2012 United States National Academy of Sciences.)

by the gastrointestinal route. As a result, there has been a movement toward topical delivery of such drugs to bypass these limitations. Topical delivery presents new challenges, namely the requirement of small particle size for effective penetration through the outermost skin layers. Mirkin, Paller and co-workers used small-interfering RNA (siRNA)-functionalized AuNPs (SNA-NCs) to silence the epidermal growth factor receptor (EGFR) in human keratinocytes, as well as investigating its viability as a topical agent (Zheng et al. 2012a). They were able to see effective knockdown of EGFR using small amounts of EGFR SNA-NCs (Figure 10.10a). In comparison,

non-sense RNA-SNAs displayed no knockdown. Furthermore, this knockdown was specific to EGFR; expression of the unrelated glyceraldehyde 3-phosphate dehydrogenase (GAPDH) was unchanged. They mixed the RNA-SNAs with Aquaphor® (a commercial skin moisturizer) and applied it topically to mouse skin (Figure 10.10b). They were able to see effective penetration of the RNA-SNAs (red color; light gray in print) through the skin cells (blue color; dark gray in print). In addition, the RNA-SNAs were easily cleared from the skin area, with only 2% of Au content being detected after 10 days of measurement (Figure 10.10c).

10.6 Summary

In summary, DNA-functionalized AuNPs are highly versatile materials, and they represent a classic example in bionanotechnology. DNA has programmable structures ranging from isolated geometric shapes to sophisticated 2D and 3D periodic networks. At the same time, DNA aptamers and DNAzymes allow molecular recognition beyond simple complementary nucleic acids. AuNPs have excellent stability and optical properties. Thiolated DNA can be readily attached to AuNPs via a few protocols, all screening the charge repulsion between DNA and AuNPs. The conjugate has been used for designed colorimetric biosensors relying on the plasmonic properties of AuNPs. Many fluorescent sensors using the quenching property of AuNPs were also demonstrated. The conjugate was also used to assemble AuNP-based nanomaterials and for delivery of nucleic acids to cells and penetrating tissues for therapeutic applications. Aside from the applications, this conjugate has interesting physical properties such as very sharp melting transitions and higher binding affinity to complementary DNA, which in turn also facilitated biosensor development.

References

Alivisatos, A.P., K.P. Johnsson, X. Peng, T.E. Wilson, C.J. Loweth, M.P. Bruchez, Jr, and P.G. Schultz. "Organization of 'Nanocrystal Molecules' Using DNA." *Nature* 382, no. 6592 (1996): 609–11.

Bishop, K.J.M., C.E. Wilmer, S. Soh, and B.A. Grzybowski. "Nanoscale Forces and Their Uses in Self-Assembly." *Small* 5, no. 14 (2009): 1600–30.

Cutler, J.I., E. Auyeung, and C.A. Mirkin. "Spherical Nucleic Acids." *J. Am. Chem. Soc.* 134, no. 3 (2012): 1376–91.

Daniel, M.-C., and D. Astruc. "Gold Nanoparticles: Assembly, Supramolecular Chemistry, Quantum-Size-Related Properties, and Applications toward Biology, Catalysis, and Nanotechnology." *Chem. Rev.* 104, no. 1 (2004): 293–346.

Demers, L.M., M. Oestblom, H. Zhang, N.-H. Jang, B. Liedberg, and C.A. Mirkin. "Thermal Desorption Behavior and Binding Properties of DNA Bases and Nucleosides on Gold." *J. Am. Chem. Soc.* 124, no. 38 (2002): 11248–49.

Dubertret, B., M. Calame, and A.J. Libchaber. "Single-Mismatch Detection Using Gold-Quenched Fluorescent Oligonucleotides." *Nat. Biotechnol.* 19, no. 4 (2001): 365–70.

Elghanian, R., J.J. Storhoff, R.C. Mucic, R.L. Letsinger, and C.A. Mirkin. "Selective Colorimetric Detection of Polynucleotides Based on the Distance-Dependent Optical Properties of Gold Nanoparticles." *Science* 277, no. 5329 (1997): 1078–80.

Ellington, A.D., and J.W. Szostak. "In Vitro Selection of RNA Molecules That Bind Specific Ligands." *Nature* 346, no. 6287 (1990): 818–22.

Frens, G. "Controlled Nucleation for the Regulation of the Particle Size in Monodisperse Gold Suspensions." *Nature* 241, no. 105 (1973): 20–22.

Giljohann, D.A., D.S. Seferos, W.L. Daniel, M.D. Massich, P.C. Patel, and C.A. Mirkin. "Gold Nanoparticles for Biology and Medicine." *Angew. Chem. Int. Ed.* 49, no. 19 (2010): 3280–94.

Giljohann, D.A., D.S. Seferos, P.C. Patel, J.E. Millstone, N.L. Rosi, and C.A. Mirkin. "Oligonucleotide Loading Determines Cellular Uptake of DNA-Modified Gold Nanoparticles." *Nano Lett.* 7, no. 12 (2007): 3818–21.

Grabar, K.C., R.G. Freeman, M.B. Hommer, and M.J. Natan. "Preparation and Characterization of Au Colloid Monolayers." *Anal. Chem.* 67, no. 4 (1995): 735–43.

Hakkinen, H. "The Gold-Sulfur Interface at the Nanoscale." *Nat Chem* 4, no. 6 (2012): 443–55.

Han, M.S., A.K.R. Lytton-Jean, B.-K. Oh, J. Heo, and C.A. Mirkin. "Colorimetric Screening of DNA-Binding Molecules with Gold Nanoparticle Probes." *Angew. Chem., Int. Ed.* 45, no. 11 (2006): 1807–10.

Handley, D.A. "Methods for Synthesis of Colloidal Gold." In *Coloidal Gold Principles, Methods, and Applications*, edited by M.A. Hayat, 13–32. San Diego: Academic Press, 1989.

Hashimoto, K., K. Ito, and Y. Ishimori. "Sequence-Specific Gene Detection with a Gold Electrode Modified with DNA Probes and an Electrochemically Active Dye." *Anal. Chem.* 66, no. 21 (1994): 3830–33.

Hegner, M., P. Wagner, and G. Semenza. "Immobilizing DNA on Gold Via Thiol Modification for Atomic Force Microscopy Imaging in Buffer Solutions." *FEBS Lett.* 336, no. 3 (1993): 452–56.

Hermann, T., and D.J. Patel. "Adaptive Recognition by Nucleic Acid Aptamers." *Science* 287, no. 5454 (2000): 820–25.

Hill, H.D., J.E. Millstone, M.J. Banholzer, and C.A. Mirkin. "The Role Radius of Curvature Plays in Thiolated Oligonucleotide Loading on Gold Nanoparticles." *ACS Nano* 3, no. 2 (2009): 418–24.

Huang, C.-C., Y.-F. Huang, Z. Cao, W. Tan, and H.-T. Chang. "Aptamer-Modified Gold Nanoparticles for Colorimetric Determination of Platelet-Derived Growth Factors and Their Receptors." *Anal. Chem.* 77, no. 17 (2005): 5735–41.

Huang, Z., B. Liu, and J. Liu. "Parallel Polyadenine Duplex Formation at Low Ph Facilitates DNA Conjugation onto Gold Nanoparticles." *Langmuir* 32, no. 45 (2016): 11986–92.

Hurst, S.J., A.K.R. Lytton-Jean, and C.A. Mirkin. "Maximizing DNA Loading on a Range of Gold Nanoparticle Sizes." *Anal. Chem.* 78, no. 24 (2006): 8313–18.

Jin, R., G. Wu, Z. Li, C.A. Mirkin, and G.C. Schatz. "What Controls the Melting Properties of DNA-Linked Gold Nanoparticle Assemblies?" *J. Am. Chem. Soc.* 125, no. 6 (2003): 1643–54.

Jones, M.R., R.J. Macfarlane, B. Lee, J. Zhang, K.L. Young, A.J. Senesi, and C.A. Mirkin. "DNA-Nanoparticle

Superlattices Formed from Anisotropic Building Blocks." *Nat. Mater.* 9, no. 11 (2010): 913–17.

Jones, M.R., N.C. Seeman, and C.A. Mirkin. "Programmable Materials and the Nature of the DNA Bond." *Science* 347, no. 6224 (2015): 1260901.

Joyce, G.F. "Directed Evolution of Nucleic Acid Enzymes." *Ann. Rev. Biochem.* 73 (2004): 791–836.

Ke, Y., L.L. Ong, W.M. Shih, and P. Yin. "Three-Dimensional Structures Self-Assembled from DNA Bricks." *Science* 338, no. 6111 (2012): 1177–83.

Kim, Y., R.J. Macfarlane, M.R. Jones, and C.A. Mirkin. "Transmutable Nanoparticles with Reconfigurable Surface Ligands." *Science* 351, no. 6273 (2016): 579–82.

Kimura-Suda, H., D.Y. Petrovykh, M.J. Tarlov, and L.J. Whitman. "Base-Dependent Competitive Adsorption of Single-Stranded DNA on Gold." *J. Am. Chem. Soc.* 125, no. 30 (2003): 9014–15.

Kuzyk, A., R. Schreiber, Z. Fan, G. Pardatscher, E.-M. Roller, A. Högele, F.C. Simmel, A.O. Govorov, and T. Liedl. "DNA-Based Self-Assembly of Chiral Plasmonic Nanostructures with Tailored Optical Response." *Nature* 483 (2012): 311.

Lee, J.-S., M.S. Han, and C.A. Mirkin. "Colorimetric Detection of Mercuric Ion (Hg^{2+}) in Aqueous Media by DNA-Functionalized Gold Nanoparticles." *Angew. Chem., Int. Ed.* 46 (2007): 4093–96.

Levicky, R., T.M. Herne, M.J. Tarlov, and S.K. Satija. "Using Self-Assembly to Control the Structure of DNA Monolayers on Gold: A Neutron Reflectivity Study." *J. Am. Chem. Soc.* 120, no. 38 (1998): 9787–92.

Li, F., J. Zhang, X.N. Cao, L.H. Wang, D. Li, S.P. Song, B.C. Ye, and C.H. Fan. "Adenosine Detection by Using Gold Nanoparticles and Designed Aptamer Sequences." *Analyst* 134, no. 7 (2009): 1355–60.

Li, Y., Y. Zheng, M. Gong, and Z. Deng. "Pt Nanoparticles Decorated with a Discrete Number of DNA Molecules for Programmable Assembly of Au-Pt Bimetallic Superstructures." *Chem. Commun.* 48, no. 31 (2012): 3727–29.

Liu, B., E.Y. Kelly, and J. Liu. "Cation-Size-Dependent DNA Adsorption Kinetics and Packing Density on Gold Nanoparticles: An Opposite Trend." *Langmuir* 30, no. 44 (2014): 13228–34.

Liu, B., and J. Liu. "Freezing Directed Construction of Bio/Nano Interfaces: Reagentless Conjugation, Denser Spherical Nucleic Acids, and Better Nanoflares." *J. Am. Chem. Soc.* 139 (2017a): 9471–74.

Liu, B., and J. Liu. "Methods for Preparing DNA-Functionalized Gold Nanoparticles, a Key Reagent of Bioanalytical Chemistry." *Anal. Methods* 9 (2017b): 2633–43.

Liu, B., S. Salgado, V. Maheshwari, and J. Liu. "DNA Adsorbed on Graphene and Graphene Oxide: Fundamental Interactions, Desorption and Applications." *Curr. Opin. Colloid Interface Sci.* 26 (2016): 41–49.

Liu, B., P. Wu, Z. Huang, L. Ma, and J. Liu. "Bromide as a Robust Backfiller on Gold for Precise Control of DNA

Conformation and High Stability of Spherical Nucleic Acids." *J. Am. Chem. Soc.* 140 (2018): 4499–502.

Liu, J. "Adsorption of DNA onto Gold Nanoparticles and Graphene Oxide: Surface Science and Applications." *Phys. Chem. Chem. Phys.* 14 (2012): 10485–96.

Liu, J., J.H. Lee, and Y. Lu. "Quantum Dot Encoding of Aptamer-Linked Nanostructures for One Pot Simultaneous Detection of Multiple Analytes." *Anal. Chem.* 79 (2007): 4120–25.

Liu, J., and Y. Lu. "Stimuli-Responsive Disassembly of Nanoparticle Aggregates for Light-up Colorimetric Sensing." *J. Am. Chem. Soc.* 127, no. 36 (2005): 12677–83.

Liu, J., and Y. Lu. "Design of Asymmetric Dnazymes for Dynamic Control of Nanoparticle Aggregation States in Response to Chemical Stimuli." *Org. Biomol. Chem.* 4, no. 18 (2006a): 3435–41.

Liu, J., and Y. Lu. "Fast Colorimetric Sensing of Adenosine and Cocaine Based on a General Sensor Design Involving Aptamers and Nanoparticles." *Angew. Chem., Int. Ed.* 45, no. 1 (2006b): 90–94.

Liu, J., and Y. Lu. "Preparation of Aptamer-Linked Gold Nanoparticle Purple Aggregates for Colorimetric Sensing of Analytes." *Nat. Protoc.* 1, no. 1 (2006c): 246–52.

Liu, J., and Y. Lu. "Smart Nanomaterials Responsive to Multiple Chemical Stimuli with Controllable Cooperativity." *Adv. Mater.* 18 (2006d): 1667–71.

Liu, J., D. Mazumdar, and Y. Lu. "A Simple and Sensitive 'Dipstick' Test in Serum Based on Lateral Flow Separation of Aptamer-Linked Nanostructures." *Angew. Chem., Int. Ed.* 45, no. 47 (2006): 7955–59.

Liu, J.W., and Y. Lu. "A Colorimetric Lead Biosensor Using DNAzyme-Directed Assembly of Gold Nanoparticles." *J. Am. Chem. Soc.* 125, no. 22 (2003): 6642–43.

Love, J.C., L.A. Estroff, J.K. Kriebel, R.G. Nuzzo, and G.M. Whitesides. "Self-Assembled Monolayers of Thiolates on Metals as a Form of Nanotechnology." *Chem. Rev.* 105, no. 4 (2005): 1103–69.

Loweth, C.J., W.B. Caldwell, X. Peng, A.P. Alivisatos, and P.G. Schultz. "DNA-Based Assembly of Gold Nanocrystals." *Angew. Chem., Int. Ed.* 38, no. 12 (1999): 1808–12.

Lu, C.H., H.H. Yang, C.L. Zhu, X. Chen, and G.N. Chen. "A Graphene Platform for Sensing Biomolecules." *Angew. Chem. Int. Ed.* 48, no. 26 (2009): 4785–87.

Lytton-Jean, A.K.R., and C.A. Mirkin. "A Thermodynamic Investigation into the Binding Properties of DNA Functionalized Gold Nanoparticle Probes and Molecular Fluorophore Probes." *J. Am. Chem. Soc.* 127, no. 37 (2005): 12754–55.

Macfarlane, R.J., B. Lee, M.R. Jones, N. Harris, G.C. Schatz, and C.A. Mirkin. "Nanoparticle Superlattice Engineering with DNA." *Science* 334, no. 6053 (2011): 204–08.

Mastroianni, A.J., S.A. Claridge, and A.P. Alivisatos. "Pyramidal and Chiral Groupings of Gold Nanocrystals Assembled Using DNA Scaffolds." *J. Am. Chem. Soc.* 131, no. 24 (2009): 8455–59.

Maxwell, D.J., J.R. Taylor, and S. Nie. "Self-Assembled Nanoparticle Probes for Recognition and Detection of Biomolecules." *J. Am. Chem. Soc.* 124, no. 32 (2002): 9606–12.

Maye, M.M., M.T. Kumara, D. Nykypanchuk, W.B. Sherman, and O. Gang. "Switching Binary States of Nanoparticle Superlattices and Dimer Clusters by DNA Strands." *Nat. Nanotechnol.* 5, no. 2 (2010): 116–20.

Mirkin, C.A., R.L. Letsinger, R.C. Mucic, and J.J. Storhoff. "A DNA-Based Method for Rationally Assembling Nanoparticles into Macroscopic Materials." *Nature* 382, no. 6592 (1996): 607–09.

Nykypanchuk, D., M.M. Maye, D. van der Lelie, and O. Gang. "DNA-Guided Crystallization of Colloidal Nanoparticles." *Nature* 451, no. 7178 (2008): 549–52.

Park, J.-W., and J.S. Shumaker-Parry. "Structural Study of Citrate Layers on Gold Nanoparticles: Role of Intermolecular Interactions in Stabilizing Nanoparticles." *J. Am. Chem. Soc.* 136, no. 5 (2014): 1907–21.

Park, S.Y., A.K.R. Lytton-Jean, B. Lee, S. Weigand, G.C. Schatz, and C.A. Mirkin. "DNA-Programmable Nanoparticle Crystallization." *Nature* 451, no. 7178 (2008): 553–56.

Patel, P.C., D.A. Giljohann, W.L. Daniel, D. Zheng, A.E. Prigodich, and C.A. Mirkin. "Scavenger Receptors Mediate Cellular Uptake of Polyvalent Oligonucleotide-Functionalized Gold Nanoparticles." *Bioconjug. Chem.* 21, no. 12 (2010): 2250–56.

Pavlov, V., Y. Xiao, B. Shlyahovsky, and I. Willner. "Aptamer-Functionalized Au Nanoparticles for the Amplified Optical Detection of Thrombin." *J. Am. Chem. Soc.* 126, no. 38 (2004): 11768–69.

Peterlinz, K.A., R.M. Georgiadis, T.M. Herne, and M.J. Tarlov. "Observation of Hybridization and Dehybridization of Thiol-Tethered DNA Using Two-Color Surface Plasmon Resonance Spectroscopy." *J. Am. Chem. Soc.* 119, no. 14 (1997): 3401–02.

Prigodich, A.E., P.S. Randeria, W.E. Briley, N.J. Kim, W.L. Daniel, D.A. Giljohann, and C.A. Mirkin. "Multiplexed Nanoflares: Mrna Detection in Live Cells." *Anal. Chem.* 84, no. 4 (2012): 2062–66.

Prigodich, A.E., D.S. Seferos, M.D. Massich, D.A. Giljohann, B.C. Lane, and C.A. Mirkin. "Nano-Flares for Mrna Regulation and Detection." *ACS Nano* 3, no. 8 (2009): 2147–52.

Ray, P., G. Darbha, A. Ray, J. Walker, and W. Hardy. "Gold Nanoparticle Based FRET for DNA Detection." *Plasmonics* 2, no. 4 (2007): 173–83.

Ray, P.C., A. Fortner, and G.K. Darbha. "Gold Nanoparticle Based Fret Asssay for the Detection of DNA Cleavage." *J. Phys. Chem. B* 110, no. 42 (2006): 20745–48.

Reynolds, R.A., III, C.A. Mirkin, and R.L. Letsinger. "Homogeneous, Nanoparticle-Based Quantitative Colorimetric Detection of Oligonucleotides." *J. Am. Chem. Soc.* 122, no. 15 (2000): 3795–96.

Rosi, N.L., D.A. Giljohann, C.S. Thaxton, A.K.R. Lytton-Jean, M.S. Han, and C.A. Mirkin. "Oligonucleotide-Modified Gold Nanoparticles for Intracellular Gene Regulation." *Science* 312, no. 5776 (2006): 1027–30.

Rosi, N.L., and C.A. Mirkin. "Nanostructures in Biodiagnostics." *Chem. Rev.* 105, no. 4 (2005): 1547–62.

Rothemund, P.W.K. "Folding DNA to Create Nanoscale Shapes and Patterns." *Nature* 440, no. 7082 (2006): 297.

Seferos, D.S., D.A. Giljohann, H.D. Hill, A.E. Prigodich, and C.A. Mirkin. "Nano-Flares: Probes for Transfection and mRNA Detection in Living Cells." *J. Am. Chem. Soc.* 129, no. 50 (2007): 15477–79.

Shen, X., A. Asenjo-Garcia, Q. Liu, Q. Jiang, F.J. García de Abajo, N. Liu, and B. Ding. "Three-Dimensional Plasmonic Chiral Tetramers Assembled by DNA Origami." *Nano Lett.* 13, no. 5 (2013): 2128–33.

Song, S.P., Y. Qin, Y. He, Q. Huang, C.H. Fan, and H.Y. Chen. "Functional Nanoprobes for Ultrasensitive Detection of Biomolecules." *Chem. Soc. Rev.* 39, no. 11 (2010): 4234–43.

Storhoff, J.J., R. Elghanian, C.A. Mirkin, and R.L. Letsinger. "Sequence-Dependent Stability of DNA-Modified Gold Nanoparticles." *Langmuir* 18, no. 17 (2002): 6666–70.

Storhoff, J.J., R. Elghanian, R.C. Mucic, C.A. Mirkin, and R.L. Letsinger. "One-Pot Colorimetric Differentiation of Polynucleotides with Single Base Imperfections Using Gold Nanoparticle Probes." *J. Am. Chem. Soc.* 120, no. 9 (1998): 1959–64.

Storhoff, J.J., A.A. Lazarides, R.C. Mucic, C.A. Mirkin, R.L. Letsinger, and G.C. Schatz. "What Controls the Optical Properties of DNA-Linked Gold Nanoparticle Assemblies?" *J. Am. Chem. Soc.* 122, no. 19 (2000): 4640–50.

Storhoff, J.J., and C.A. Mirkin. "Programmed Materials Synthesis with DNA." *Chem. Rev.* 99, no. 7 (1999): 1849–62.

Tan, L.H., H. Xing, and Y. Lu. "DNA as a Powerful Tool for Morphology Control, Spatial Positioning, and Dynamic Assembly of Nanoparticles." *Acc. Chem. Res.* 47, no. 6 (2014): 1881–90.

Taton, T.A., C.A. Mirkin, and R.L. Letsinger. "Scanometric DNA Array Detection with Nanoparticle Probes." *Science* 289, no. 5485 (2000): 1757–60.

Tuerk, C., and L. Gold. "Systematic Evolution of Ligands by Exponential Enrichment: RNA Ligands to Bacteriophage T4 DNA Polymerase." *Science* 249, no. 4968 (1990): 505–10.

Wang, H., R.H. Yang, L. Yang, and W.H. Tan. "Nucleic Acid Conjugated Nanomaterials for Enhanced Molecular Recognition." *ACS Nano* 3, no. 9 (2009): 2451–60.

Wang, W.J., C.L. Chen, M.X. Qian, and X.S. Zhao. "Aptamer Biosensor for Protein Detection Using Gold Nanoparticles." *Anal. Biochem.* 373, no. 2 (2008): 213–19.

Wu, P., K. Hwang, T. Lan, and Y. Lu. "A DNAzyme-Gold Nanoparticle Probe for Uranyl Ion in Living Cells." *J. Am. Chem. Soc.* 135, no. 14 (2013): 5254–57.

Xue, X.J., F. Wang, and X.G. Liu. "One-Step, Room Temperature, Colorimetric Detection of Mercury (Hg^{2+}) Using DNA/Nanoparticle Conjugates." *J. Am. Chem. Soc.* 130, no. 11 (2008): 3244–45.

Yang, H.H., H.P. Liu, H.Z. Kang, and W.H. Tan. "Engineering Target-Responsive Hydrogels Based on Aptamer – Target Interactions." *J. Am. Chem. Soc.* 130, no. 20 (2008): 6320–21.

Yigit, M.V., D. Mazumdar, H.-K. Kim, J.H. Lee, B. Odintsov, and Y. Lu. "Smart 'Turn-on' Magnetic Resonance Contrast Agents Based on Aptamer-Functionalized Superparamagnetic Iron Oxide Nanoparticles." *ChemBioChem* 8, no. 14 (2007): 1675–78.

Yun, C.S., A. Javier, T. Jennings, M. Fisher, S. Hira, S. Peterson, B. Hopkins, N.O. Reich, and G.F. Strouse. "Nanometal Surface Energy Transfer in Optical Rulers, Breaking the Fret Barrier." *J. Am. Chem. Soc.* 127, no. 9 (2005): 3115–19.

Zanchet, D., C.M. Micheel, W.J. Parak, D. Gerion, and A.P. Alivisatos. "Electrophoretic Isolation of Discrete Au Nanocrystal/DNA Conjugates." *Nano Lett.* 1, no. 1 (2001): 32–35.

Zhang, X., M.R. Servos, and J. Liu. "Instantaneous and Quantitative Functionalization of Gold Nanoparticles with Thiolated DNA Using a pH-Assisted and Surfactant-Free Route." *J. Am. Chem. Soc.* 134, no. 17 (2012a): 7266–69.

Zhang, X., M.R. Servos, and J. Liu. "Surface Science of DNA Adsorption onto Citrate-Capped Gold Nanoparticles." *Langmuir* 28 (2012b): 3896–902.

Zhao, W., M.A. Brook, and Y. Li. "Design of Gold Nanoparticle-Based Colorimetric Biosensing Assays." *ChemBioChem* 9, no. 15 (2008): 2363–71.

Zheng, D., D.A. Giljohann, D.L. Chen, M.D. Massich, X.-Q. Wang, H. Iordanov, C.A. Mirkin, and A.S. Paller. "Topical Delivery of siRNA-Based Spherical Nucleic Acid Nanoparticle Conjugates for Gene Regulation." *Proc. Natl. Acad. Sci. U.S.A.* 109, no. 30 (2012a): 11975–80.

Zheng, J., P.E. Constantinou, C. Micheel, A.P. Alivisatos, R.A. Kiehl, and N.C. Seeman. "Two-Dimensional Nanoparticle Arrays Show the Organizational Power of Robust DNA Motifs." *Nano Lett.* 6, no. 7 (2006): 1502–04.

Zheng, Y., Y. Li, and Z. Deng. "Silver Nanoparticle-DNA Bionanoconjugates Bearing a Discrete Number of DNA Ligands." *Chem. Commun.* 48, no. 49 (2012b): 6160–62.

Zhou, W., R. Saran, and J. Liu. "Metal Sensing by DNA." *Chem. Rev.* 117 (2017): 8272–325.

Zhu, Z., C.C. Wu, H.P. Liu, Y. Zou, X.L. Zhang, H.Z. Kang, C.J. Yang, and W.H. Tan. "An Aptamer Cross-Linked Hydrogel as a Colorimetric Platform for Visual Detection." *Angew. Chem. Int. Ed.* 49, no. 6 (2010): 1052–56.

11

Undoped Tetrahedral Amorphous Carbon (ta-C) Thin Films for Biosensing

Anja Aarva, Miguel Caro, and
Tomi Laurila
Aalto University

11.1 Introduction

Diamond-like carbon (DLC) is a metastable form of amorphous carbon characterized by its high fraction of sp^3-bonded carbon atoms. This "diamond-likeness" arising from the sp^3 bonded carbon atoms gives DLC many of its unique properties. Thus, applications of DLC include a wide range of fields, such as machine parts, biomedical coatings, microelectromechanical (MEMS) devices, sunglasses and so forth [1–5]. DLC is also a promising bioelectrode material owing to its several attractive electrochemical properties, such as (i) chemical inertness and the resulting (ii) wide potential window as well as (iii) low background current. Thus, DLC has been used recently in several analytical applications ranging from biomolecule detection [1–4] to trace analysis of heavy metals [5]. The basic electrochemical properties and response of DLC to several redox systems have been investigated [6–9] and also recently reviewed [10]. Many of the unique electrochemical properties of DLC originate from the specific structural features of the material in thin film form (thickness range from a few nanometers to a few tens of nanometers) as recently discussed in Ref. [11].

A special form of highly sp^3-bonded DLC that has electrochemical properties reasonably close to those of diamond is known as tetrahedral amorphous carbon (ta-C). In this chapter, we will concentrate on the physical, chemical and electrochemical properties of this type of high sp^3-containing DLC films. We will not consider other types of DLC films, e.g., those with high hydrogen content or the ones that have been doped with nitrogen or alloyed with other elements. We will start by introducing briefly the (i) fabrication of ta-C thin films. We will shortly give the three main techniques to obtain high sp^3 carbon–containing ta-C films. We will then continue with an outlook of the proposed growth mechanisms for ta-C films. The new insight put forward by recent simulation activities and their implications to the established growth models are subsequently discussed. After obtaining some understanding about the film formation and especially about the resulting sp^3 to sp^2 bonded carbon ratio, we will start to examine the (ii) physical properties of ta-C thin films, and how the abovementioned sp^3/sp^2 ratio affects the various material features. Again, an approach based on a tight combination of experimental and computational information is utilized. Next, we will address the issue of (iii) surface chemical properties of these films and look at recent experimental and computational evidence of an occurrence of various (mainly) oxygen containing functional groups on these surfaces. After we have formed a more or less coherent picture of the physical and surface chemical properties of ta-C thin films, we will proceed to examine the (iv) electrochemical properties of these films. We will cover the fundamental electrochemical properties, such as potential window, double-layer capacitance and heterogeneous electron transfer (HET) kinetics as well as study the adsorption of various inner sphere couples on ta-C thin films. As before, a tight combination of experimental and computational work is maintained. We will finish the chapter by giving a (v) summary and outlook for the future.

11.2 Fabrication and Growth of ta-C and the Role of Simulation

High sp^3-containing a-C films (namely tetrahedral amorphous carbon ta-C) are typically fabricated with three main approaches: (i) pulsed DC sputtering, (ii) different types of arc processes and (iii) various laser ablation methods. We will not go into details of any one of these methods as these have been summarized in detail, for example in Ref. [12,13]. We just point out here that one of the main features or challenges in arc processes is the unavoidable presence of macroparticles in the films even if different types of filters are utilized. These will of course have an effect on the film properties and material performance in many applications. Pulsed DC sputtering on the other hand is relatively free of any kind of macroparticles. However, at the same time arc techniques are the methods of choice if one really targets a high sp³-containing a-C film, which cannot be obtained with most of the other methods. Whatever is the method used to fabricate the ta-C thin films, the following parameters should be determined and controlled as they heavily influence the properties of the resulting films: (i) distribution of energetic species involved, (ii) their energy and incidence angle, (iii) background pressure during deposition, (iv) trace impurities in the system, (v) substrate temperature and (vi) deposition rate. As the study of the atomistic picture of film growth experimentally is next to impossible, many efforts have been made to provide an explanation for the film growth phenomena based on computational simulations. Unfortunately, due to the limitations of classical potentials, which are computationally efficient, but lack accuracy in describing bond formation, and the high computational cost associated with ab initio methods, which could describe the flexible bonding environment, these efforts have failed to provide any definitive answers.

We have recently utilized a new computational approach using machine learning-based Gaussian Approximation potential (GAP) trained from local density approximation density functional theory (DFT) data [14] to simulate the growth of ta-C thin films. With this new method, we have been, for the first time, able to reproduce the experimental results with very high accuracy (without any experimental input), as shown below, and provided new insight into the atomic-level picture of the growth of ta-C thin films. A detailed description of the method and the results obtained can be found in Ref. [15]; here, we just provide a concise summary of the main items. The atomistic details of the growth of an a-C film were investigated by the deposition of C atoms onto a carbon substrate, one atom at a time, using molecular dynamics (MD). A large [111]-oriented diamond substrate, terminated by the stable 2 × 1 reconstructed surface, was used, containing a total of 3,240 atoms in periodic boundary conditions, which means an initial structure of approximate dimensions 38 Å × 38 Å in plane and 16 Å thickness. A total of 2,500 single monoenergetic C atoms were shot with kinetic energy of 60 eV from the top of the simulation box onto the diamond substrate, to create an initial a-C template. Subsequently, an additional 5,500 atoms, each with a kinetic energy corresponding to the different deposition energies studied (20, 60 or 100 eV), were deposited. This sums up to a total of 8,000 impact events per investigated energy. All MD simulations were carried out with Large-scale Atomic/Molecular Massively Parallel Simulator (LAMMPS) [16,17] using the aforementioned GAP potential [14].

Figure 11.1 shows the main features of the deposited a-C films. The figure shows the in-plane averaged mass density profile of the films grown at different deposition energies. Mass density is one of the most straightforward characterization criteria for the high sp³ content of ta-C as well as experimentally quite easily accessible with X-ray reflection measurements. Thus, it provides a feasible connection between simulations and experiments. As can be seen, very high densities and accordingly high sp³ fractions are obtained in the interior of the film. The simulated deposition at 60 eV, which is the ion energy at which sp³ content is expected to peak based on experimental observations [15], shows sp³ fractions of up to 90%. Previous simulations [18–21], either based on deposition or alternative methods such as liquid quenching, have systematically failed to reproduce these high sp³ fractions, regardless of the level of theory employed. The highest sp³ fractions (slightly less than 85%) reported previously for computational studies were based on DFT geometry optimization followed by pressure correction [18,21]. Explicit deposition simulations (based on the so-called Carbon- Environment-Dependent Interatomic Potential (C-EDIP)) had not been able to produce a-C structures with sp³ fractions exceeding ∼60% [19]. In fact, as can be seen from Figure 11.1, the films deposited with energies of 20 eV, 60 eV and 100 eV reach mass densities of around 3.5 g/cm³, which is very close to the experimental value for diamond. Another important feature seen in Figure 11.1 is the identical sp²-rich surface (the thickness of which depends on deposition energy) on all samples despite the different deposition energies. This is fully consistent with

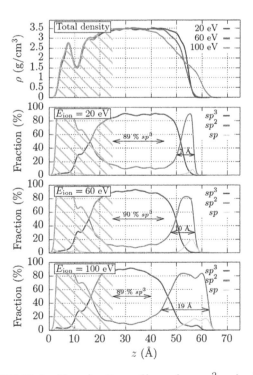

FIGURE 11.1 Mass density profiles and sp, sp² and sp³ fractions in the bulk of the film, for the different deposition regimes studied [15].

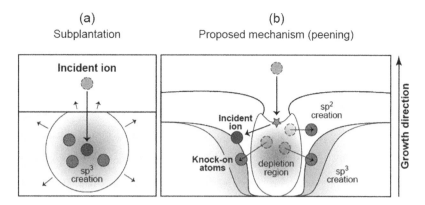

FIGURE 11.2 (a) Previously accepted growth mechanism in ta-C and (b) growth mechanism proposed in Ref. [15].

the experimental results as will be seen later on and provides yet another strong evidence that this model can reproduce all the important experimental features of ta-C films with a high precision.

In addition to providing high-quality results consistent with the experiments, the model also provided a new atomistic picture of the ta-C growth process. The consensus in the literature has long been that the "subplantation" mechanism is behind the growth phenomenon of ta-C films [22]. This mechanism is illustrated in Figure 11.2a and relates the increase in bonding coordination to the packing of atoms in too small a volume, as newly arrived atoms are being deposited/forced in. It is the relaxation of the surrounding matrix that then explains film growth and the formation of sp^3-coordinated carbon atoms. However, based on the atomistic view provided by our simulations, this view is not the correct one. Instead, the computational data show that, in fact, (i) each single impact induces coordination changes for roughly 80 atoms and that (ii) local destruction of sp^3 carbon at and around the impact site occurs. However, (iii) the dynamical balance between sp^3 creation and annihilation builds up laterally and *away from the impact region* to yield *net generation of sp^3* carbon as a result (Figure 11.2b).

11.3 Structural and Electrical Properties of ta-C Thin Film

The ratio of sp^3-bonded carbon to the sp^2-bonded one is perhaps the most important feature that determines many properties of hydrogen-free amorphous carbon films as pointed out above. In order to study the ta-C bonding structure, it is typical to use Raman spectroscopy. Visible Raman (vis-Raman) spectroscopy only detects sp^2-related vibrations, and so its information about the sp^3 component is indirect. For direct sp^3 detection UV-Raman should be used. However, owing to the much more extensive amount of vis-Raman results in literature we will use results obtained with that technique in the following discussion. Figure 11.3a shows the change in the vis-Raman I(D)/I(G) ratio of amorphous carbon thin films as a function of film thickness.

It is common practice to use the ratio of 0.25 as a threshold below which the film is considered highly sp^3-bonded carbon and, therefore, can be taken to be of ta-C type [12]. On the other hand, as the ratio increases, the sp^2 fraction grows and the film becomes more amorphous carbon (a-C) in its properties. As shown in Section 11.2, the deposition energy used for the film fabrication influences this ratio heavily. However, a much less known fact is that in the case of nanometer-scale thin films the thickness of the film also affects the apparent ratio. As Figure 11.3a illustrates, the I(D)/I(G) ratio is a strong function of the thickness of the film, with the thinner films being overall much more sp^2 rich than the thicker films. This can be expected to have an effect especially on the electrical properties of the films, which are to a large degree determined by the sp^2 fraction. Indeed, as shown in Figure 11.3b, the average current flowing through the electrode structure as a result of a given bias voltage heavily depends on the film thickness. Figure 11.3b illustrates that as the thickness of the carbon thin film increases, the current drops down significantly. The maximum current flowing through the structure is seen to occur around a film thickness of 7 nm. The decrease in the current values as the thickness of the carbon film becomes 4 nm or less is most likely related to the pin holes in these very thin carbon films and the resulting formation of Ti-oxide at the interface [23]. Further evidence of this will be seen in Section 11.5 when electrochemical measurements are discussed. This Ti-oxide originates from the Ti adhesion layer used underneath the ta-C layer on top of Si wafer.

To rationalize the fundamental physical mechanisms that determine the properties of a-C and ta-C (depending on sp^3 fraction) observed experimentally, a series of computational studies based on DFT calculations were performed. A detailed description of the approach employed to generate computational a-C samples of different densities, as well as an assessment of its accuracy, is given in Ref. [24]. Subsequent computational band gap [22] and X-ray absorption spectroscopy (XAS) spectra [25] results confirmed that the properties calculated using this DFT-based model of a-C could accurately reproduce the experimental findings. The evolution of the Tauc gap, which has been proposed as an estimate of optical gap in amorphous

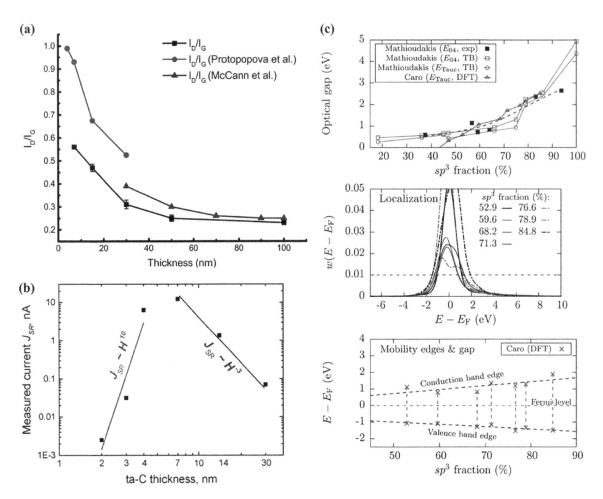

FIGURE 11.3 (a) I(D)/I(G) ratio determined with vis-Raman spectroscopy showing the change in the ratio as a function of thickness [11], (b) change in the average current through the structure as a function of film thickness (H) [24] and (c) computational results showing the Tauc gap, state localization and change in the mobility gap as a function of sp^3 fraction. Tauc gap and mobility gap data are new but the procedure is the same as in Ref. [11]. TB in (c) indicated tight binding. (Composite figure from Ref. [54].)

semiconductors [26–28], with sp^3 fraction is shown in Figure 11.3c. The procedure that was followed to estimate Tauc gaps is the same as reported in Refs. [24,29]. While the absolute values illustrated in Figure 11.3c with the calculations using the Heyd–Scuseria–Ernzerhof (HSE) hybrid functional are much more meaningful than our previous results based on the Perdew–Burke–Ernzerhof (PBE) generalized-gradient approximation (GGA) reported in Ref. [24], we obtained a very similar trend with composition, +0.058 eV/sp^3 % (HSE06) compared to +0.051 eV/sp^3 % (PBE). Thus, the gap size is a strong function of sp^3 fraction and increases as the ratio of sp^3 to sp^2 increases. The mobility gap is probably a more useful quantity than the Tauc gap, at least with respect to characterization of the electrical properties of a-C. In order to define the mobility gap, we relied on an electronic state localization criterion, as used also for instance for a-Si in Ref. [30]. In our case, we employed the localization function w, which was based on a projected (or local) DOS analysis [31]. The mobility edges were estimated as the energy values in the valence and conduction bands at which the transition between the extended and localized states based on a localization threshold criterion [31] occurred. The results for different sp^3 fractions are shown in Figure 11.3c. A clear increase of the mobility gap with sp^3 content was observed, in good agreement with the experimental observations from scanning tunneling spectroscopy (STS) [23].

As shown above, the increase in the sp^2 fraction heavily influences the electrical properties of amorphous carbon films. Thus, the spatial distribution of sp^2-bonded carbon within the films should be determined to have a more comprehensive understanding of the electrical properties of these materials. In Ref. [32] it was reported, by utilizing a combination of transmission electron microscopy (TEM) and electron energy loss spectroscopy that an sp^2-rich carbon layer was formed on the film surfaces regardless of the sp^3 fraction of the underlying film. It was also pointed out that the thickness of this surface layer was dependent on the deposition conditions used to realize the films and especially on the energy of the incoming carbon ions. As seen from Figure 11.1, these features also arise naturally from the GAP/DFT simulations.

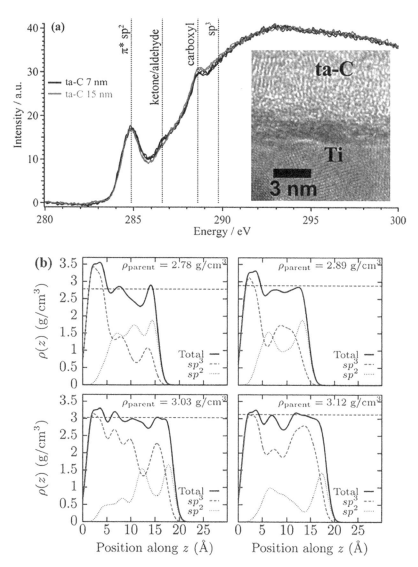

FIGURE 11.4 (a) AEY spectra from the XAS measurements for the 7 and 15 nm ta-C films and high-resolution TEM image of the structure of the 7 nm thick ta-C film on Ti [11]. (b) Computational results showing the presence of identical sp²-rich surface layer in all samples regardless of the bulk sp³-fraction [21]. (Composite figure from Ref. [54].)

To further experimentally elaborate this issue, we carried out a detailed spectroscopic study utilizing XAS in total electron yield (TEY) and Auger electron yield (AEY) modes on different types of ta-C films [25]. In this study, we showed that (i) there were clear differences in the surface chemistry of the two different ta-C films with different sp³ fractions that were investigated (especially in the amount and nature of oxygen functionalities), despite the fact that (ii) the sp² fraction of the surface region was identical in both cases (Figure 11.4a). The latter type of feature was also verified with computational studies based on density functional theory (Figures 11.1 and 11.4b), which showed that, independently of the average mass-density of the samples, the surfaces always exhibited a similar sp²-rich region. The simulations further pointed out that the increase in the sp² fraction at the surface resulted in the introduction of surface-localized electronic states of π character [21,33]. This was accompanied by a pronounced increase in the DOS

within the pseudogap region [21,34]. Hybrid-functional DFT calculations also showed that a-C surfaces should be able to retain a very narrow electronic band gap despite the observed increase in mid-gap states [29]. The simulated work function from DFT has been estimated at around 5 eV irrespective of bulk density [35] (compared with the experimental value of 4–5 eV [36])— a further indication that the properties of a-C surfaces are not strongly influenced by properties of the underlying bulk material.

Thus, we can conclude that: (a) electrical properties of the ta-C thin films are heavily dependent on the thickness of the films and consequently on the overall sp² fraction. (ii) The surface of the ta-C thin films is always sp² rich despite the level of sp³ fraction in the underlying material. Hence, structure of the surfaces of the amorphous carbon thin films can be expected to be almost identical as well as possess electrical properties, which are widely different from those of the "bulk" of the film.

11.4 Surface Chemical Properties of ta-C Films

Tetrahedral amorphous carbon (ta-C) thin films exhibit a rich surface chemistry even after the deposition. Further, there exist a plethora of different recipes to modify the surface functionalization with different physical, chemical and electrochemical treatments. It has been shown that various (mainly oxygen based) functional groups at the ta-C surface play a major role in many applications [18]. However, atomic-level information about these groups is not readily available, and accordingly most of the treatments used to change the surface chemistry of the ta-C (or a-C) films are completely based on trial-and-error approaches. As was shown above, high-quality atomistic simulations can be of great help in acquiring a very detailed view of the atomistic processes taking place on a-C surfaces. Thus, to probe reactivity of a-C surfaces, we carried out a series of adsorption energy calculations for the functional groups, which are expected to be present most abundantly on a-C surfaces [7].

The geometry of the slabs created with GAP molecular dynamics was relaxed by performing geometry optimization within the framework of DFT. After this step, the surfaces were functionalized with hydrogen (–H), oxygen (–O), hydroxyl (–OH) and carboxylic acid groups (–COOH), and the geometry of the system was allowed to relax. The adsorption sites for the groups were chosen according to the Smooth Overlap of Atomic Positions (SOAP)-based clustering scheme presented in detail in Ref. [37]. This way, a connection between the geometrical trends of the different adsorption sites and their reactivity could be established. In order to compare the chemistry and binding properties of these sites, adsorption energies of all the functional groups considered were computed for each set of surface atomic motifs. The simulations were carried out following the methodology outlined in Ref. [37]. Examples of the geometries of the each identified atomic motifs discussed in this work are depicted in Figure 11.5.

Motifs 2 and 3 are sp sites typically contained along a carbon chain that forms rings on the surface. Motifs 4 and 5 are sp^2 sites, whereas motif 6 corresponds to diamond-like sp^3 sites. About 20 adsorption sites per motif were selected for further studies. For more details, see Ref. [37]. The distributions of calculated adsorption energies are presented in Figure 11.6. It can be seen that for motifs 2 and 3 (sp) and 4 and 5 (sp^2), sites corresponding to different motifs displayed markedly different adsorption energies. sp sites were clearly more reactive than sp^2 sites, with a difference in adsorption energies between motifs 4 and 3 ranging between −2 (H adsorption) and −3.5 eV (O adsorption). While the reactivity of the different adsorption sites toward –H and –OH groups was similar, –O adsorption showed a significant increase in energies, with some sites showing adsorption energies as large as −6.5 eV. The interaction of the different motifs with the –COOH group was the weakest among the tested functionalizations. In all cases, the ordering of adsorption energies was the same.

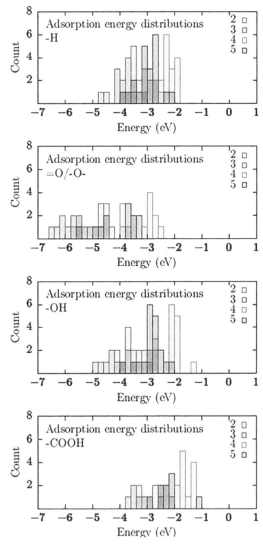

FIGURE 11.6 Adsorption energy distributions of the functional groups for different clusters [37].

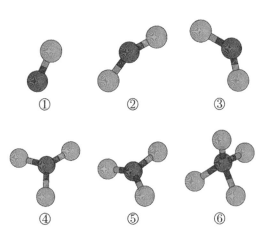

FIGURE 11.5 Ball and stick representation of typical motifs encountered in a-C surfaces.

To gain further insight into the connection between geometrical features, electronic structure and reactivity, in Figure 11.7 adsorption energies are plotted against the local density of states (LDOS) integrated around the Fermi level. It is highly probable that these states will be involved in chemisorption of functional groups, and therefore the number of states (as given by the integrated LDOS) can be expected to act as a good descriptor for site reactivity. The interval chosen for integration was from −3 to 3 eV. The average LDOS for each cluster is shown in Figure 11.8. From Figure 11.7, we see that the integrated LDOS value correlate strongly with the adsorption energies as expected. Figure 11.7 clearly shows that, when LDOS around the Fermi level is high, adsorption energies are more negative, and vice versa. In a similar way, transition metal d-band occupation and overlap with the adsorbates energy bands has previously been shown to determine the characteristics of hydrogen chemisorption as well as to play a significant role in electrocatalysis [38,39]. Furthermore, sites in a

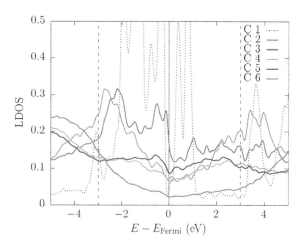

FIGURE 11.8   Average local density of states (LDOS) of each cluster. Vertical dashed lines describe the integration interval.

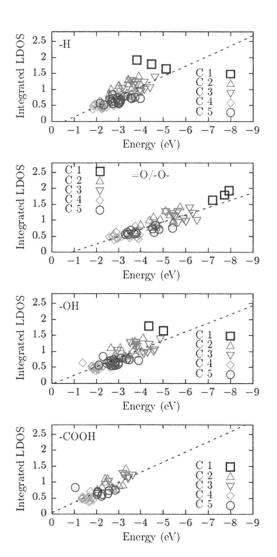

FIGURE 11.7   Adsorption energies of the functional groups vs. integrated local density of states (LDOS) for each site in each cluster.

certain cluster are aggregated around similar energy values (Figures 11.6 and 11.7). However, as Figures 11.6 and 11.7 show, the interaction between −O and the a-C surface is more complicated than the interaction between a-C and −H, −OH and −COOH. In the case of oxygen, the adsorption energies are much more scattered. Oxygen behaves differently than other groups, most likely because it can become bonded to the C site in various ways. Based on the present results, oxygen tends to form mostly either ketone or epoxide type of bonds.

In addition to exploring the chemical reactivity of the surfaces, the above-described models also enable quantitative deconvolution of spectroscopic data and thus acquisition of atomic-level chemical information from the actual ta-C thin films. As discussed above, XAS, or near-edge X-ray absorption fine structure (NEXAFS), is a very powerful technique for analyzing the surface characteristics and for identifying functional groups present on the surfaces. In fact, as an instrument, XAS can provide extremely detailed information about the electronic structure of the surface and possible adsorbates. However, considering the complexity of the materials discussed here, interpretation of the experimental data is very difficult, and thus sophisticated simulation tools are again required to aid in rationalization of the results. Thus, DFT-based simulations of XAS spectra of carbonaceous materials were utilized in order to rationalize the experimental results. Simulations were carried out with the GPAW code [40] where the method to calculate the spectra has been implemented by Ljungberg et al. [41]. To examine the roles of differently bonded carbon, three types of carbon materials were included: graphene, diamond and amorphous carbon. Furthermore, these surfaces were functionalized with hydrogen, oxygen, hydroxyl group and carboxylic acid in order to obtain the fingerprint spectrum of each group on a certain site type. Finally, the simulation outcome was compared with the experimental XAS results to provide more in-depth knowledge of the effects of these groups on the XAS spectrum.

Details of the simulation methods and models used can be found in Ref. [42]. In brief, all simulations were carried out using the supercell approach, where "infinite" surfaces are reconstructed from finite-size slabs by employing periodic boundary conditions [43]. In case of grapheme, two types of surfaces were used: pristine and defective. The size of the graphene sheet was $9 \times 5 \times 1$ primitive unit cells; thus, there were 180 carbon atoms in a pristine graphene sheet. Defective graphene was simulated by introducing a single vacancy defect [44] into a graphene sheet of the same size as the pristine sample. In both cases, k-space integration was performed using Monkhorst Pack (MP) grid [45] with $2 \times 2 \times 1$ k-point sampling. Binding energies of the functional groups on graphene, diamond and a-C are presented in Table 11.1.

On pristine graphene, only oxygen had a binding energy that was strong enough to indicate covalent bonding. Introduction of the single vacancy defect created a favorable site for attachment of the functional group, and the binding energies on that site were thus much stronger. A diamond surface was simulated by employing 3×5 primitive unit cells of the 2×1 reconstruction of the diamond (111) surface [46] with a thickness of 10 Å. Two types of termination were used at the bottom of the slab—hydrogen and carbon termination—corresponding to 330 and 360 atom systems, respectively. The k-point sampling in case of diamond was $2 \times 2 \times 1$. The a-C surface was created by Caro et al., and the method is explained in detail in Ref. [29]. Briefly, an a-C surface sample was constructed via random initialization followed by geometry optimization and "anchoring" of the slab by binding it to a diamond substrate. However, this full surface sample was considerably too large for carrying out numerous DFT calculations needed in this work, and thus a smaller slab was created by cleaving off the bottom of the full surface structure. The bottom of the smaller surface sample was passivated with hydrogen. The slab used in this work was 7 Å thick and consisted of 326 atoms. In a-C calculations, k-point sampling of $2 \times 1 \times 1$ was used. Unlike the case of graphene and diamond, there are no equivalent sites, or carbon atoms, in the a-C sample. Accordingly, in order to functionalize the a-C surface, the most favorable sites need to be identified. This was done by comparing the local density of states (LDOS) of the various surface sites as it provides a feasible descriptor for site reactivity as mentioned earlier. The functional groups were placed on these sites one site and one group at a time, to avoid creating excessively complicated systems that could lead to results open to various interpretations. In all cases, an adequate amount of vacuum was set above the system to ensure that the energy is converged

to the correct value, and in binding energy calculations, dipole corrections [47] were applied. The PBE GGA [48] was used as the exchange-correlation density functional. Van der Waals corrections were included via the method developed by Tkatchenko and Schefler [49]. System sizes and k-point samplings were tested to ensure sufficient convergence with respect to binding energies and XAS core levels. Because of the existence of local (atomic) magnetic moments in a-C and because a defect can introduce magnetization to the system [50], all calculations were carried out with spin polarization. The self-consistent Kohn–Sham (KS) density functional theory [51,52] calculations were performed with the GPAW suite [40].

Figure 11.9 shows an experimental C K-edge XAS spectrum from an a-C thin film sample. As can be seen, there are a few distinct features (peaks) in the spectrum, but otherwise it is relatively featureless. Thus, the interpretation of this kind of spectrum is very difficult, and fitting to single peaks typically becomes at least somewhat arbitrary. In what follows, we will show that again the use of atomistic simulations carried out as described above and the concept of "fingerprint" spectra for a given functional group can greatly help in resolving atomic-level information from the experimental spectra.

In Figure 11.10a–d we present examples of the calculated fingerprint spectra for different functional groups. As can be seen, each functional group has its own spectra instead of a single peak as is typically taken to be the case when XAS (or XPS) spectra are fit. Figure 11.11 shows the sum spectrum constructed based on the fingerprint spectra for various functional groups on a-C surfaces. As can be seen, the main features of the experimental spectrum are reproduced remarkably well. The main advantages of the present method are as follows: (i) the "exact" positions of the main peaks of the functional groups are given, and thus one does not need to rely on intuition and vague literature information during fitting, and (ii) for each functional group a fingerprint spectrum instead of a single peak is given. This greatly helps in interpretation

TABLE 11.1 Calculated Adsorption Energies for Different Functional Groups on Various Carbonaceous Surfaces [42]

	–H	–O–I = O	–OH	–COOH
Pristine graphene	−0.850	−2.151	−0.772	−0.207
Defective graphene	−4.331	−6.314	−3.853	−3.520
a-C (averaged)	−4.100	−5.249	−4.374	−3.973
Reconstructed diamond	−2.533	−4.057	−2.435	−1.900

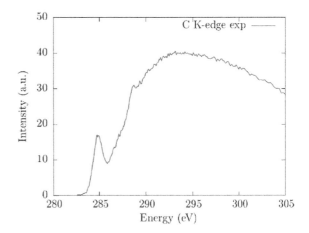

FIGURE 11.9 XAS K-edge C spectrum from a typical ta-C thin film sample.

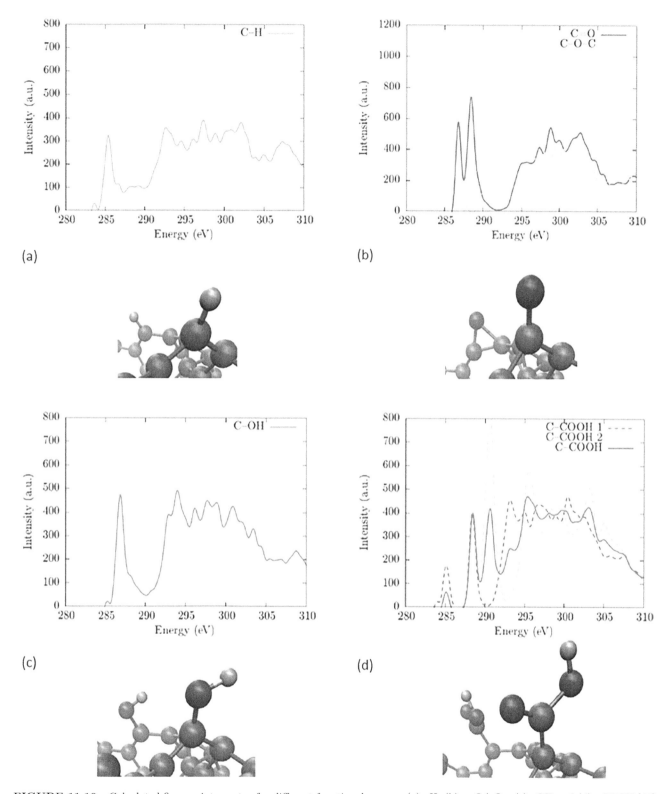

FIGURE 11.10 Calculated fingerprint spectra for different functional groups: (a) –H, (b) = O/–O–, (c) –OH and (d) –COOH [42].

of the spectrum, especially in the more or less "feature-less" region, as now all the fine structures of the fingerprint spectra of the functional groups are available. This method enables one to quantitatively deconvolute the experimental spectra and to acquire atomic-level chemical information about the real a-C surfaces. This can then be used to:

(i) guide and give feedback to other computational studies (such as adsorption calculations), (ii) assess the outcome of various functionalization treatments and (iii) help in establishing connections between chemistry of the a-C surfaces and their electrochemical behavior, as seen in the next section.

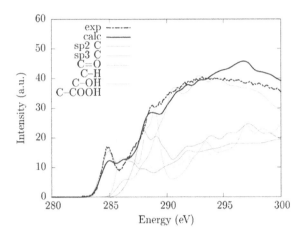

FIGURE 11.11 The calculated sum spectrum and its comparison to the measured one.

11.5 Electrochemical Properties of ta-C Thin Films

Some of the electrochemical properties of interest from a sensor point of view include the water window of the electrode material, double-layer capacitance as well as HET kinetics. The significance of a wide water window stems from the fact that when water starts to decompose, either producing oxygen at the anodic end or hydrogen at the cathodic end, the currents arising from these reactions are much higher than those coming from any analyte signals, thus overrunning all other oxidation/reduction reactions in water-based media. Thus, by pushing the practical (kinetically determined) potentials of these reactions as far apart as possible (thus increasing the water window), a broader measurement range and thus possible access to a wider range of analytes is made possible. The importance of the double-layer capacitance is easily understood by considering the background current, which becomes larger the higher the double-layer capacitance is, thus compromising our ability to sense small current peaks related to small analyte concentrations.

HET kinetics gives an indication about the facility of electron transfer under specific circumstances and is commonly investigated by utilizing so-called outer sphere redox couples. These are (more or less) insensitive to the surface chemistry of the electrodes. Thus, they do not adsorb on the surface and there is always at least one solvent layer between the redox probe and the electrode surface. In addition, there are no changes in the chemical bonds. Therefore, they give information about the electronic interaction between the probe and the electrode surface—ideally without chemical complications.

In Ref. [24], basic electrochemical properties of the ta-C thin film electrodes were studied by cyclic voltammetry by measuring the water window and apparent double-layer capacitance in sulfuric acid. A current threshold of 200 mA was used as the limit to determine the width of the water window. The sample structure used throughout this section,

unless otherwise stated, is as follows: highly doped Si (100) substrate on which a Ti adhesion promotion layer about 20 nm thick has been deposited, followed by the actual ta-C thin film electrode of various thicknesses. Figure 11.12 shows the cyclic voltammograms for the bare Ti layer (20 nm) as well as for the electrodes with various thicknesses on top of ta-C layer (of which thickness varies from 2 to 30 nm) measured at 400 mV/s cycling speed. One can see that the potential windows measured on electrodes with the 7–30 nm thick ta-C layers were much wider when compared to electrodes with the 2–4 nm thick ta-C layer. Further, the background current was also much lower with the thicker (7–30 nm) electrodes. It is to be noted that the shape of cyclic voltammograms for electrodes with the thinnest ta-C (2–4 nm) are very similar to those of the bare Ti layer. This indicates that with the thinnest ta-C layers the underlying Ti layer has been partly exposed to the electrolyte, and oxidization and reduction reactions of titanium took place during cycling. This is as expected, since in very thin ta-C films there are mostly likely at least some pinholes and other defects that make the layer non-uniform. Implications of this have already explained in Section 11.3 where electrical measurements were discussed. In the case of electrodes with the top ta-C layer having a thickness of 7–30 nm, oxidization or reduction of the Ti underlayer was not observed (Figure 11.12). These samples were stable and showed a wide water window as well as reasonably low apparent double-layer capacitance.

As discussed in Section 11.3, increase in the thickness of ta-C films leads to a decrease in their overall volume sp^2 fraction, increase in the mobility gap and decrease in the average current flowing through the film [21,23]. Further, it was verified in Sections 11.2 and 11.3 that the surface of the ta-C thin films is always sp^2 rich and possesses electrical properties distinctly different from the underlying bulk. These features should be reflected, in addition to the water window discussed earlier, even in the HET rates, when outer-sphere redox probes are utilized. Thus, in Ref. [11] the reaction

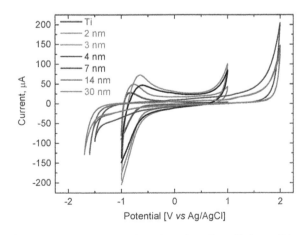

FIGURE 11.12 Potential windows for a bare Ti layer (20 nm) and electrodes with different thicknesses of top ta-C layers (2–30 nm) in sulfuric acid measured at 400 mV/s cycling speed [24].

kinetics on ta-C electrodes as a function of film thickness was assessed. The results are shown in Figures 11.13a and b.

Based on the electrochemical impedance spectroscopy (EIS) measurements, there was a clear increase in the charge transfer resistance for $Ru(NH_3)_6^{3+/2+}$ redox couple, as the film thickness increased from 7 to 100 nm. This observation is consistent with the changes in the electrical properties discussed in Section 11.3. This behavior is most probably related to the electron *transport* through the film, as opposed to the electron *transfer* from a solution species to the surface of ta-C. This argument is supported by the experimental (x-ray absorption spectroscopy, XAS, Figure 11.4a) and computational data (Figures 11.1 and 11.4b) showing that the surfaces of ta-C thin films are sp^2-rich and practically identical in all cases, despite the sp^3 content of the underlying bulk film [18,21,23]. Further, with the inclusion of a Ti interlayer between the Si and ta-C thin film, there was a clear decrease in the apparent charge transfer resistance (compare Figures 11.13a and b) indicating that the reaction kinetics became more facile. As it is expected that Ti interlayer provides a better electrical contact to the Si substrate, this provides further support for the view that the overall reaction kinetics of the outer-sphere redox systems is controlled by the electron transport through the ta-C/Ti/Si electrode stack.

The EIS data was subsequently fit with a modified Randles circuit model (Figure 11.13c) to gain numerical data. The significant new feature in the modified model was that the charge transfer resistance was divided into two parts connected in series (Figure 11.13c). The first one corresponded to the electron transfer from the redox species in the solution to the surface of the ta-C and the second one to the subsequent charge transport through the ta-C film. The fitting naturally only gives a value for the total resistance R_{ct}. However, based on the arguments presented above, it can be expected that the resistance to electron transfer is about the same for all the films, whereas the resistance for the electron transport through the film increases significantly as a function of film thickness. The rise of the latter part of the charge transfer resistance then gives us the observed behavior. An estimate of the electron *transfer* part can be obtained from the value of "total" charge transfer resistance for the 7 nm thick ta-C films, as they are highly conductive (see Section 11.3), and thus the electron *transport* part of the charge transfer resistance can be considered to be negligible.

The electrical double-layer capacitance C_{dl} and the a-values determined from the EIS measurements also showed significant thickness dependencies. The observed decrease in C_{dl} with increasing film thickness can be explained by the increasing volume fraction of the sp^3-bonded carbon in the films. The double-layer capacitance can be represented by two capacitances, the solution capacitance and the space charge capacitance of ta-C, connected in series. When the ta-C film is thin, the contribution from both capacitances is roughly equal. However, when the thickness (and sp^3 content) increases, the contribution from the space charge capacitance to the total capacitance diminishes (because the

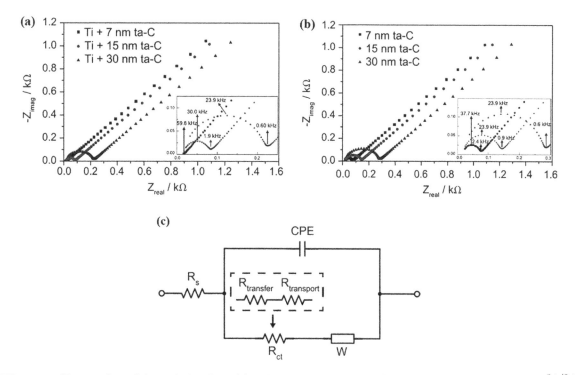

FIGURE 11.13 Nyquist plots of the ta-C thin films (a) with the Ti interlayer and (b) without it in 5 mM $Ru(NH_3)_6^{3+/2+}$ in 1 M KCl. The insets show a magnification of the high-frequency range. (c) Randles equivalent circuit used to fit the Nyquist plots of the ta-C films. The charge transfer resistance is represented by an electron transfer resistance (from solution to the surface) connected in series with an electron transport resistance (through the ta-C film) [11]. (Composite figure from Ref. [54].)

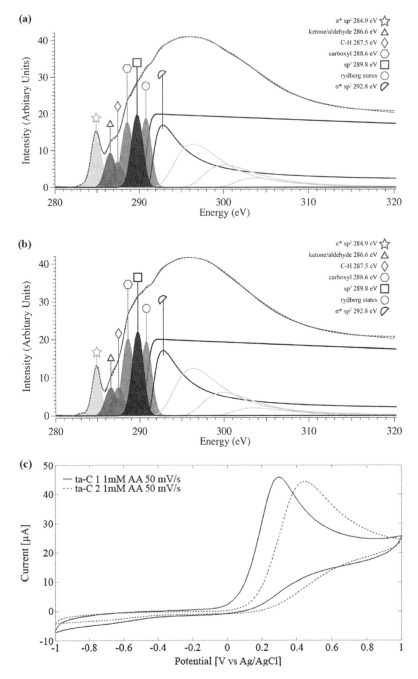

FIGURE 11.14   (a) ta-C 1 C 1s TEY spectra. Peaks fitted to the actual XAS-curve, (b) ta-C 2 C 1s TEY spectra. Peaks fitted to the actual XAS-curve and (c) ta-C 1 and ta-C 2 electrodes measured with cycling voltammetry in ascorbic acid with 50 mV/s cycling speed [25]. (Composite figure from Ref. [54].)

space charge capacitance increases). This is shown by the increase of the value of the variable a, which describes the behavior of the constant phase elements used in the model. When the value of a approaches one, the films start to behave more like an ideal capacitor.

Typical target molecules in sensor applications do not react via the outer sphere route on electrode surfaces. Instead, the reactions take the inner sphere route, and the overall kinetics therefore becomes heavily dependent on the chemistry of the electrode surface. While investigating the surface chemistries of different types of ta-C thin films,

we have observed clear variations in the amount and nature of surface oxygen functional groups (Section 11.3). Thus, one can use these differences as a starting point to look for reasons behind the observed differences in the electrochemical behavior of different inner-sphere redox probes on these materials [18]. In Ref. [25], we carried out a detailed spectroscopic study on two different types of ta-C thin film electrodes (ta-C1 and ta-C2), which had different surface functionalities. The main differences between the materials were that (i) slightly higher oxygen content was expected for the ta-C1 film and (ii) somewhat larger measured sp^3

fraction was seen for the ta-C2 film. Based on the C1s spectra (Figure 11.14a and b), the largest difference was found for the aldehyde/ketone groups, which were more abundant on the ta-C1 sample surface. Further differences were found for the amount of carboxyl groups between the samples. Moreover, the oxygen spectrum from the ta-C1 sample verified the expectation that it contained in total more oxygen than the ta-C2 sample.

This higher amount of oxygen surface functional groups on the ta-C1 surface should be manifested as a higher affinity toward redox probes with high polar nature and known preference to oxygen. One such species is ascorbic acid [53]. When the CVs were run to detect ascorbic acid (AA) with these two different types of ta-C electrodes, the results showed that (i) the peak oxidation potential for AA was less anodic and (ii) the current increased more steeply in the case of ta-C1 than that of ta-C2. Both of these features indicated that the ta-C1 electrode had a higher activity toward AA (Figure 11.14c) than ta-C2. As the amount and nature of the oxygen functional groups on ta-C surfaces are influenced by processing, chemical and electrochemical differences, and pH, this provides another factor that must not only be taken into account when rationalizing results but can also be utilized to modify the electrochemical behavior of ta-C thin films.

11.6 Summary and Outlook for the Future

We have shown in this chapter that ta-C thin films possess various interesting properties that make them potential materials for many application fields. In addition, we demonstrated that it is possible to understand (at least to some degree) the properties of these complex materials by combining computational techniques with experimental work. We showed that this method not only helps in rationalization of the observed behavior but also gives us an atomic-level picture of the various physical processes occurring on ta-C surfaces and can help us in giving guidelines for tailoring the surfaces of these materials to reach certain desired properties as was pointed out in the end of Section 11.5. This brings us to the concept of application-specific nanomaterials that, in our opinion, will play a crucial enabling role in many future technologies, especially those relying on detailed control of surface chemistry. The idea is relatively simple: by combining computational work (in addition to physical simulations also including machine learning and related methods) with the experimental one, we can construct a comprehensive understanding of our materials structure and chemistry, relate them to materials performance under various conditions and finally, based on this, provide a recipe for a tailored material for a specific application. This would construct a basis of so-called rational data-driven design of nanomaterials. This would change the way people approach materials design as earlier it was quite extensively been based on trial-and-error

type of methods, which are not only unreliable but also time consuming and therefore expensive. We expect that as our understanding of these complex carbon materials grows during the upcoming years, we will start to see their applications in much wider scope of fields than is the case currently.

References

1. T. Laurila, V. Protopopova, S. Rhode, S. Sainio, T. Palomäki, M. Moram et al., New electrochemically improved tetrahedral amorphous carbon films for biological applications, *Diam. Relat. Mater.* 49 (2014) 62–71.
2. R.A. Medeiros, R. Matos, A. Benchikh, B. Saidani, C. Debiemme-Chouvy, C. Deslouis et al., Amorphous carbon nitride as an alternative electrode material in electroanalysis: Simultaneous determination of dopamine and ascorbic acid, *Anal. Chim. Acta* 797 (2013) 30–39.
3. D. Sopchak, B. Miller, R. Kalish, Y. Avyigal, X. Shi, Dopamine and ascorbate analysis at hydrodynamic electrodes of boron doped diamond and nitrogen incorporated tetrahedral amorphous carbon, *Electroanalysis* 14 (2002) 473–478.
4. G. Yang, E. Liu, N.W. Khun, S.P. Jiang, Direct electrochemical response of glucose at nickel-doped diamond like carbon thin film electrodes, *J. Electroanal. Chem.* 627 (2009) 51–57.
5. L.X. Liu, E. Liu, Nitrogenated diamond-like carbon films for metal tracing, *Surf. Coat. Technol.* 198 (2005) 189–193.
6. Y. Tanaka, M. Furuta, K. Kuriyama, R. Kuwabara, Y. Katsuki, T. Kondo et al., Electrochemical properties of N-doped hydrogenated amorphous carbon films fabricated by plasma-enhanced chemical vapor deposition methods, *Electrochim. Acta* 56 (2011) 1172–1181.
7. T. Palomäki, S. Chumillas, S. Sainio, V. Protopopova, M. Kauppila, J. Koskinen et al., Electrochemical reactions of catechol, methylcatechol and dopamine at tetrahedral amorphous carbon (ta-C) thin film electrodes, *Diam. Relat. Mater.* 59 (2015) 30–39.
8. X. Yang, L. Haubold, G. DeVivo, G.M. Swain, Electroanalytical performance of nitrogen-containing tetrahedral amorphous carbon thin-film electrodes, *Anal. Chem.* 84 (2012) 6240–6248.
9. K. Yoo, B. Miller, R. Kalish, X. Shi, Electrodes of nitrogen-incorporated tetrahedral amorphous carbon A novel thin-film electrocatalytic material with diamond-like stability, *Electrochem. Solid-State Lett.* 2 (1999) 233–235.
10. A. Zeng, V.F. Neto, J.J. Gracio, Q.H. Fan, Diamond-like carbon (DLC) films as electrochemical electrodes, *Diam. Relat. Mater.* 43 (2014) 12–22.
11. T. Palomäki, N. Wester, M.A. Caro, S. Sainio, V. Protopopova, J. Koskinen, T. Laurila, Electron

transport determines the electrochemical properties of tetrahedral amorphous carbon (ta-C) thin films, *Electrochim. Acta* 225 (2017) 1–10.

12. J. Robertson, Diamond-like amorphous carbon, *Mater. Sci. Eng. R Rep.* 37 (2002) 129–281.

13. Y. Lifshitz, Diamond-like carbon—present status, *Diam. Relat. Mater.* 8 (1999) 1659–1676.

14. V.L. Deringer, G. Csanyi, Machine learning based interatomic potential for amorphous carbon, *Phys. Rev. B* 95 (2017) 094203.

15. M. Caro, V. Deringer, J. Koskinen, T. Laurila, G. Csanyi, Growth mechanism and origin of high sp^3 content in tetrahedral amorphous carbon, *Phys. Rev. Lett.* 120 (2018) 166101.

16. S. Plimpton, Fast parallel algorithms for short-range molecular dynamics, *J. Comput. Phys.* 117 (1995) 1.

17. http://lammps.sandia.gov.

18. T. Laurila, S. Sainio, M.A. Caro, Hybrid carbon based nanomaterials for electrochemical detection of biomolecules, *Prog. Mater. Sci.* 88 (2017) 499.

19. N.A. Marks, Thin film deposition of tetrahedral amorphous carbon: a molecular dynamics study, *Diam. Relat. Mater.* 14 (2005) 1223.

20. D.G. McCulloch, D.R. McKenzie, C.M. Goringe, Ab initio simulations of the structure of amorphous carbon, *Phys. Rev. B* 61 (2000) 2349.

21. M.A. Caro, R. Zoubko, O. Lopez-Acevedo, T. Laurila, Atomic and electronic structure of tetrahedral amorphous carbon surfaces from density functional theory: Properties and simulation strategies, *Carbon* 77 (2014) 1168.

22. J. Robertson, Plasma deposition of diamond-like carbon, *Jpn. J. Appl. Phys.* 50 (2011) 01AF01.

23. V. Protopopova, A. Iyer, N. Wester, A. Kondrateva, S. Sainio, T. Palomäki et al., Ultrathin undoped tetrahedral amorphous carbon films: The role of the underlying titanium layer on the electronic structure, *Diam. Relat. Mater.* 57 (2015) 43–52.

24. V.S. Protopopova, N. Wester, M.A. Caro, P.G. Gabdullin, T. Palomaki, T. Laurila et al., Ultra-thin undoped tetrahedral amorphous carbon films: thickness dependence of the electronic structure and implications for their electrochemical behavior, *Phys. Chem. Chem. Phys.* 17 (2015), 9020–9031.

25. S. Sainio, D. Nordlund, M.A. Caro, R. Gandhiraman, J. Koehne, N. Wester et al., Correlation between sp3-to-sp2 ratio and surface oxygen functionalities in tetrahedral amorphous carbon (ta-C) thin film electrodes and implications of their electrochemical properties, *J. Phys. Chem. C* 120 (2016) 8298–8304.

26. J. Tauc, R. Grigorovici, A. Vancu, Optical properties and electronic structure of amorphous germanium, *Phys. Status Solidi (b)* 15 (1966) 627–637.

27. S. Knief, W. von Niessen, Disorder, defects, and optical absorption in a-Si and a-Si:H, *Phys. Rev. B* 59 (1999) 12940–12946.

28. S.K. O'Leary, S.R. Johnson, P.K. Lim, The relationship between the distribution of electronic states and the optical absorption spectrum of an amorphous semiconductor: An empirical analysis, *J. Appl. Phys.* 82 (1997) 3334–3340.

29. M.A. Caro, J. Määttä, O. Lopez-Acevedo, T. Laurila, Energy band alignment and electronic states of amorphous carbon surfaces in vacuo and in aqueous environment, *J. Appl. Phys.* 117 (2015) 034502.

30. M. Legesse, M. Nolan, G. Fagas, Revisiting the dependence of the optical and mobility gaps of hydrogenated amorphous silicon on hydrogen concentration, *J. Phys. Chem. C* 117 (2013) 23956–23963.

31. G.A. Tritsaris, C. Mathioudakis, P.C. Kelires, E. Kaxiras, Optical and elastic properties of diamond-like carbon with metallic inclusions: A theoretical study, *J. Appl. Phys.* 112 (2012) 103503.

32. C. Davis, G. Amaratunga, K. Knowles, Growth mechanism and cross-sectional structure of tetrahedral amorphous carbon thin films, *Phys. Rev. Lett.* 80 (1998) 3280.

33. J. Dong, D.A. Drabold, Ring formation and the structural and electronic properties of tetrahedral amorphous carbon surfaces, *Phys. Rev. B* 57 (1998) 15591.

34. K.J. Koivusaari, T.T. Rantala, J. Levoska, S. Leppävuori, Surface electronic density of states of tetrahedral amorphous carbon investigated by scanning tunneling spectroscopy and ab initio calculations, *Appl. Phys. Lett.* 76 (2000) 2794–2796.

35. M.A. Caro, R. Zoubkoff, O. Lopez-Acevedo, T. Laurila, Corrigendum to "Atomic and electronic structure of tetrahedral amorphous carbon surfaces from density functional theory: Properties and simulation strategies"[Carbon 77 (2014) 1168–1182]. *Carbon* (2015) 612–613.

36. A. Ilie, A. Hart, A. Flewitt, J. Robertson, W. Milne, Effect of work function and surface microstructure on field emission of tetrahedral amorphous carbon, *J. Appl. Phys.* 88 (2000) 6002–6010.

37. M. Caro, A. Aarva, V. Deringer, G. Csányi, T. Laurila, Reactivity of amorphous carbon surfaces: rationalizing the role of structural motifs on functionalization using machine learning, *Chem. Mater.* 30 (2018) 7446–7455.

38. D.M. Newns, Self-consistent model of hydrogen chemisorption, *Phys. Rev.* 178 (1969) 1123.

39. E. Santos, W. Schmickler, Electrocatalysis of hydrogen oxidation theoretical foundations, *Angew. Chem. Int. Ed.* 46 (2007) 8262.

40. https://wiki.fysik.dtu.dk/gpaw/.

41. M.P. Ljungberg, J.J. Mortensen, L.G.M. Pettersson, An implementation of core level spectroscopies in a real space projector augmented wave density functional theory code, *J. Electron. Spectrosc.* 184 (2011) 427–439.

42. A. Aarva, V.L. Deringer, S. Sainio, T. Laurila, M. Caro, Understanding X-ray spectroscopy of carbonaceous

materials by combining experiments, density functional theory and machine learning. Part I: fingerprint spectra, *Chem. Mater.* (accepted), (2019).

43. G. Makov, M.C. Payne, Periodic boundary conditions in ab initio calculations, *Phys. Rev. B* 51 (1995) 4014.

44. F. Banhart, J. Kotakoski, A.V. Krasheninnikov, Structural defects in graphene, *ACS Nano* 5 (2010) 26–41.

45. H.J. Monkhorst, J.D. Pack, Special points for brillouin-zone integrations, *Phys. Rev. B* 13 (1976) 5188.

46. S. Iarlori, G. Galli, F. Gygi, M. Parrinello, E. Tosatti, Reconstruction of the diamond (111) surface, *Phys. Rev. Lett.* 69 (1992) 2947.

47. J. Neugebauer, M. Schefler, Adsorbate-substrate and adsorbate-adsorbate interactions of Na and K adlayers on Al (111), *Phys. Rev. B* 46 (1992) 16067.

48. J.P. Perdew, K. Burke, M. Ernzerhof, Generalized gradient approximation made simple, *Phys. Rev. Lett.* 77 (1996) 3865.

49. A. Tkatchenko, M. Schefler, Accurate molecular van der Waals interactions from ground-state electron density and free-atom reference data, *Phys. Rev. Lett.* 102 (2009) 073005.

50. H. Xia, W. Li, Y. Song, X.Yang, X. Liu, M. Zhao, Y. Xia, C. Song, T.-W. Wang, D. Zhu, J. Gong, Z. Zhu, Tunable magnetism in carbon-ion-implanted highly oriented pyrolytic graphite, *Adv. Mater.* 20 (2008) 4679–4683.

51. W. Kohn, L.J. Sham, Self-consistent equations including exchange and correlation effects, *Phys. Rev.* 140 (1965) A1133.

52. R.M. Martin, *Electronic Structure: Basic Theory and Practical Methods* (Cambridge University Press, Cambridge, UK, 2004).

53. R.W. Berg, Investigation of L(+)-ascorbic acid with raman spectroscopy in visible and UV light, *Appl. Spectrosc. Rev.* 50 (2015) 193–239.

54. T. Laurila, M. Caro, Special features of the electrochemistry of undoped tetrahedral amorphous carbon (ta-C) thin films, in *Encyclopedia of Interfacial Chemistry: Surface Science and Electrochemistry*, Editors-in-Chief: Klaus Wandelt, (Elsevier, Netherlands, 2017), pp.856–862.

Charge and Spin Dynamics in DNA Nanomolecules: Modeling and Applications

Samira Fathizadeh and
Sohrab Behnia
Urmia University of Technology, Urmia, Iran

12.1 Charge and Spin Transfer in DNA

Several groups have attempted theoretical modeling of the charge transfer in DNA based on transport through coherent tunneling [1]. They include classical diffusion under the condition of temperature-driven fluctuations [2], incoherent phonon-assisted hopping [3], variable range hopping between localized states [4], solitons [5], and polarons [6]. Here, we have tried to consider some practical models for studying the charge and spin dynamics in DNA.

12.2 PBH Model

DNA is easily deformable, and its structural deformations are important for an accurate description of its charge transport properties [7]. Charge coupling with such DNA distortions can create a polaron and enhance its mobility. The interaction of the charge carriers with a deformable DNA chain forms polaronic excitations. Available results have indeed been interpreted through mechanisms involving standard polaronic motion [8–10]. The tight-binding Peyrard–Bishop–Holstein (PBH) [11,12] and Su–Scherifer–Heeger (SSH) [13,14] are two effective models which are based on the polaron. In both models, overlapping π orbitals of the DNA base-pairs are responsible for migration of charge in it. In the PBH model, the coupling is between the charge and the hydrogen bond stretching of complementary bases. Here, lattice points are identified as bases rather than base pairs

and the graph resembles a ladder rather than a chain. There is a classical interaction between sites which are linearly coupled to a tight-binding Hamiltonian in the PBH model. In the current study, we have used an extended PBH model to describe the coupled structural and electronics aspects of DNA. In this model, a tight-binding linear chain is also used for the description of charge transfer in a one-dimensional channel. The model characterizes the lattice dynamics as classic PBD model and charge transfer phenomena with a nearest-neighbor tight-binding approach. The interplay of charge and lattice is studied through the Holstein-like approach with a charge–lattice interaction term. We start with the consideration of N base-pairs charge–lattice system with the extended PBH Hamiltonian [15]

$$H = H_{\text{lat}} + H_{\text{car}} + H_{\text{int}}. \quad (12.1)$$

H_{lat} is the Hamiltonian due to the lattice distortion described by nonharmonic PBD model [11]

$$H_{\text{lat}} = \sum_n \left[\frac{1}{2} m \dot{y}_n^2 + V(y_n) + W(y_{n+1}, y_n) \right], \quad (12.2)$$

where $V(y_n) = D_n (e^{-a_n y_n} - 1)^2$ is the Morse potential that provides effective interaction between complementary bases and $W(y_{n+1}, y_n) = \frac{k}{2}(1 + \rho e^{-b(y_{n+1}+y_n)})(y_{n+1} - y_n)^2$ is the interaction of neighboring stacked base pairs. The electronic part of the Hamiltonian is given by

$$H_{\text{car}} = \sum_n \left[\epsilon_n c_n^\dagger c_n - V_{n,n+1} \left(c_n^\dagger c_{n+1} + c_{n+1}^\dagger c_n \right) \right], \quad (12.3)$$

where $c_n^\dagger$ and c_n are the charge's creation and annihilation operators, respectively. ϵ_n is the onsite energy for base pair in nth site. In this model, charge hopping is restricted to nearest-neighbor base-pairs and given by $V_{n,n+1}$. The transfer matrix is supposed to depend on the relative distance between two consecutive molecules on the chain in the following exponential fashion [16], $V_{n,n+1} = V_0 e^{-\beta_n(y_{n+1}-y_n)}$. V_0 determines the constant of hopping integral and the quantity β regulates how strongly $V_{n,n+1}$ is influenced by the distance $r = y_{n+1} - y_n$. In the limit of $\beta = 0$, the hopping integral reduces to $V_0 = constant$ and the Hamiltonian yields the standard PBH model. A limiting case is recovered by keeping only the linear term in an expansion of the exponential function, i.e., $V_{n,n+1} = V_0[1 - \beta_n(y_{n+1} - y_n)]$, which is justified only for small arguments $\beta_n(y_{n+1} - y_n)$.

The other way to take into account the effects of charge lattice interaction is via the linear coupling of the onsite energy with the lattice displacements y_n, as proposed in Holstein model. This interaction is represented by a Hamiltonian term

$$H_{\text{int}} = \chi \sum_n y_n c_n^\dagger c_n, \tag{12.4}$$

where χ is the electron-lattice coupling constant.

In the current study, we propose the effect of the electrical field on charge transfer in DNA. In this regard, the corresponding general Hamiltonian has the following form

$$H_{\text{field}} = -eE \sum_n n d c_n^\dagger c_n, \tag{12.5}$$

where d is the distance between the base pairs in the lattice.

We have attempted to examine the effect of both DC and AC fields on DNA conductivity. In the DC condition, it is supposed that a constant field E is applied to DNA. In the other case, a time-periodic field is applied. So, it provides an extra degree of freedom (frequency of field) in addition to the field intensity to study the response of DNA to the external field. In this regard, the external field could be $E = E_0\cos(\omega t)$, where E_0 and ω are the amplitude and the frequency of the electrical field, respectively.

To study the temperature effect, we have examined the dynamics of the system in contact with a thermal bath through the Nosé method. We should point out that if a real canonical ensemble with correct fluctuations is needed, it is better to use the Nosé–Hoover thermostat. In Hoover's reformulation of Nose's method, the equations of motion are formulated by [17]

$$\ddot{y}_n = \frac{2a_n D_n}{m} e^{-a_n y_n}\left(e^{-a_n y_n} - 1\right)$$
$$+ \frac{kb\rho}{2m}\left[e^{-b(y_n+y_{n-1})}(y_n - y_{n-1})^2\right.$$
$$\left. + e^{-b(y_{n+1}+y_n)}(y_{n+1} - y_n)^2\right]$$
$$- \frac{k}{m}\left[\left(1 + \rho e^{-b(y_n+y_{n-1})}\right)(y_n - y_{n-1})\right.$$
$$\left. - \left(1 + \rho e^{-b(y_{n+1}+y_n)}\right)(y_{n+1} - y_n)\right]$$

$$- \frac{V_0\beta_n}{m}\left(c_{n-1}^\dagger c_n + c_n^\dagger c_{n-1} - c_n^\dagger c_{n+1} - c_{n+1}^\dagger c_n\right)$$
$$- \frac{\chi}{m}c_n^\dagger c_n - \xi\dot{y}_n, \tag{12.6}$$

$$\dot{\xi} = \frac{1}{M}\left[\sum_n m\dot{y}_n^2 - Nk_B T\right]. \tag{12.7}$$

Here, ξ is the thermodynamics friction coefficient which interacts with the particles, T is the temperature maintained by heat bath and M is the constant of Nosé–Hoover thermostat that has been set to $M = 1,000$.

To obtain the equations governing the electronic part, we have used the Heisenberg approach $\dot{c}_n = -\frac{i}{\hbar}[c_n, H]$. Then, we have

$$\dot{c}_n = -\frac{i}{\hbar}\left\{[\chi y_n - \text{eEnd}]c_n - V_0[1 - \beta_n(y_n - y_{n-1})]c_{n-1}\right.$$
$$\left. - V_0[1 - \beta_n(y_{n+1} - y_n)]c_{n+1}\right\}. \tag{12.8}$$

12.2.1 Selection of the Parameters of Model

It is clear that the calculations, validity, and approaching logical values require specification of the parameters of the system more accurately. The study procedure of charge transfer mechanism in DNA is mostly affected by the selection of the parameters of the model [18]. These parameters have been previously obtained experimentally and analytically [11]. All calculations are carried out for the sequence $L60B36$ [19,20] according to parameters published in Tables 12.1, 12.2, and 12.3. But, the sequence dependence conductivity is compared for $L60B36$ and $CH22$ sequences with $N = 60\ bp$ [21] and $N = 40\ bp$, respectively [22]. In this work, we have tried to choose the best range of parameters by using the Mean Lyapunov exponent (MLE) theory. It is worth mentioning that the MLE takes the minimum values when the system is orderly. MLE as a measure which expresses the complexity of the spatiotemporal patterns obtained from the time series of dynamical systems [23]. In this regard, it is convenient to transform the second-order differential equations into an autonomous system of first-order differential equations. Then, we can obtain the $(3N + 1) \times (3N + 1)$ Jacobian matrix written as:

$$B_{k,N} = \begin{pmatrix} Y_Y & Y_U & Y_C & Y_\xi \\ U_Y & U_U & U_C & U_\xi \\ C_Y & C_U & C_C & C_\xi \\ \xi_Y & \xi_U & \xi_C & \xi_\xi \end{pmatrix}. \tag{12.9}$$

The full matrix $B_{k,N}$ may be considered as a 4×4 matrix, each element of which is itself a block $Y_Y, Y_U, \dots$ that they are the derivatives with respect to the elements. Accordingly, the Lyapunov exponent is defined as:

$$\lambda_k = \frac{1}{3N+1}\ln|B_{k,N}| \tag{12.10}$$

where $|B_{k,N}|$ means the determinant of matrix $B_{k,N}$.

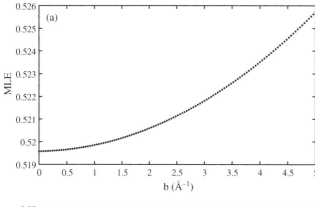

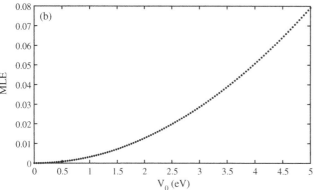

FIGURE 12.1 Mean Lyapunov exponent with respect to the (a) damping coefficient, and (b) hopping parameter [15].

TABLE 12.1 Constant parameters used in the extended PBH model [11,24]

Symbol	Description	Units	Value
m	Base-pairs mass	amu	300
a_{AT}	Width of the Morse potential for AT base pairs	Å^{-1}	6.9
a_{GC}	Width of the Morse potential for GC base pairs	Å^{-1}	4.2
D_{AT}	Depth of the Morse potential for AT base pairs	eV	0.05
D_{GC}	Depth of the Morse potential for GC base pairs	eV	0.075
k	Coupling constant	eV/Å^2	0.04
ρ	Stiffness parameter		0.5
b	Damping coefficient	Å^{-1}	0.35
β_n	Coupling parameter of hopping integral	Å^{-1}	$\simeq [1-1.7]a_n$
χ	Electron-lattice coupling	eV/Å	$[0.1-0.6]$

TABLE 12.2 The electron hopping constants for different base-pairs in meV [25]

base-pair sequence	electron hopping constant
AA, TT	-29
AT	0.5
AG, CT	3
AC, GT	32
TA	2
TG, CA	17
TC, GA	-1
GG, CC	20
GC	-10
CG	-8

TABLE 12.3 The onsite energies for the two possible base-pairs AT and GC in eV [25]

B-DNA base-pair	onsite energy
A-T	-4.9
G-C	-4.5

The best range of parameters is matched to the lowest values of MLE (Figure 12.1a and b). Here, we have examined the MLE to determine the range of b and V_0. In both of them, it seems that increasing the parameter corresponds to the growth of the MLE, and hence the system irregularity (Figure 12.1a and b). So, we have chosen the smallest value previously offered to be closer to the actual results. The obtained values are in agreement with the previous experimental and theoretical studies. The same procedure was carried out to determine the other parameters. The parameter values established in Refs. [11,24] are represented in Table 12.1. The onsite energies and electron hopping constants used in this work are according to Ref. [25], represented in Table 12.2 and Table 12.3.

12.2.2 Electrical Current

The charge transport properties of DNA can be investigated through the flowing electrical current. We could use the definition of the particle density operator in Heisenberg picture: $n_i(t) = e^{iHt} n_i e^{-iHt}$, where $n_i = c_i^+ c_i$ is the charge density. Then, we calculate the time-dependent current operator as follows:

$$I(t) = \frac{d(en_i(t))}{dt} = \frac{ie}{\hbar} \sum_n V_0 [1 - \beta(y_{n+1} - y_n)]$$
$$\times (c_n^\dagger c_{n+1} - c_{n+1}^\dagger c_n). \quad (12.11)$$

According to the obtained relation, the electrical current is dependent on the relative position of the base pairs and probability amplitude to find the charge carriers every time. So, the electrical current shows the oscillatory behavior over time and oscillates with irregular periods sometimes (Figure 12.2a and b). Several factors contribute to the flowing current through DNA. The temperature is a key agent that has a considerable effect on the evolution of the system. On the other hand, it is thought that denaturation temperature is a critical temperature for DNA. It seems reasonable to expect that the passing current behavior is affected by the denaturation temperature. To study this effect, we have investigated the passing electrical current of DNA over time at (a) room temperature and (b) denaturation temperature. It could be shown that at both temperatures, the electrical current behaves in an irregular oscillatory manner. But, at the denaturation temperature, the electrical current through the DNA is more irregular. We have obtained similar results for time-dependent (AC) fields (Figure 12.2c and d).

12.2.3 I–V Characteristic Diagram in DC Fields

Measurements of electrical current as a function of the potential applied across single DNA could indicate metallic-like behaviors and efficient conduction and conversely

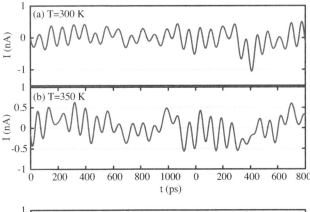

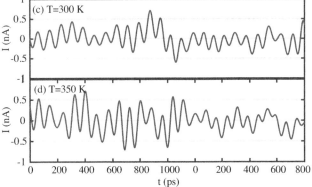

FIGURE 12.2 Electrical current time series in presence of a DC field at (a) $T = 300$ K, and (b) $T = 350$ K, and in the presence of a AC field at (c) $T = 300$ K, and (d) $T = 350$ K [15].

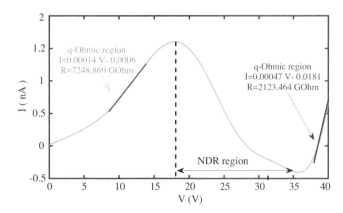

FIGURE 12.3 I-V characteristic diagram in the constant field [15].

12.2.4 I–V Characteristic Diagram in AC Fields

In order to clarify the effect of variation of frequency on electrical current flowing through DNA, a large domain range is intended [33,34]. By considering ranges of about $[0.01 - 0.8]$ THz, we have observed different output characteristics (see Figure 12.4a–c). It could be shown that when the frequency of the field is increased, the quasi-ohmic regions (with a linear gradient) and NDR regions are distinguished. Accordingly, by the enhancement of the frequency, one could see the appearance of the neoteric quasi–ohmic regions. On the other hand, the quasi-ohmic areas change to the NDR regions, and thus novel quasi-ohmic and the negative slope areas are created. It is worth mentioning that the slope of the I–V diagram for quasi-linear ranges is proportional to the quasi–ohmic resistance of the conductors as in the DC case. The quasi–ohmic resistance value of DNA is specified on the characteristic diagrams (Figure 12.4a–c). It is shown that at a constant frequency, quasi–ohmic resistance varies by increasing the applied potential. On the other hand, the width of the NDR regions is reduced by increasing the voltage.

12.2.5 The Effect of DNA Sequence

It should be noted that the sequence length and its content affect the conductivity properties of DNA (see Figure 12.5). Previous studies have demonstrated the effect of chain variation on electronic transport and localization length [35,36].

In the current study, we examined the sequence-dependent conductivity for sequences represented in Table 12.4. Here, only the sequence along one strand is presented, and that for the other strand can be derived according to Watson–Crick base-pairing rules. The first sequence ($L60B36$) is the most well-known DNA molecule which is widely adopted in DNA experiments such as melting, etc [19,20,37]. $CH22$ sequences are human chromosome 22–based sequences with 60 bp and 40 bp, respectively [22]. It is clear that sequence variation does not change the general shape of the I–V characteristic but

insulator-like behaviors [26]. For a more detailed study of DNA conductivity, we could investigate the current–voltage characteristic of DNA in the present model, see (Figure 12.3). It is clear that DNA shows different behaviors on increasing the applied field and thus through increasing the voltage. The I–V diagram characterizes the quasi-linear regions with quasi-ohmic behaviors. The slope of I–V diagram in these regions corresponds to the ohmic resistance. The equation of the line and the ohmic resistance in quasi-linear areas investigated in Figure 12.3. On the other hand, there are regions with negative slope among the linear areas as the differential variation of the current with respect to the applied potential is negative ($\frac{dI}{dV} < 0$). When the voltage continuously increases, the current through DNA increases at the beginning and then decreases, resulting in a negative differential resistance (NDR) peak. NDR has been observed experimentally in DNA previously [27,28]. In electronics, NDR devices are used to make bistable switching circuits and electronic oscillators [29]. Also, it opens the possibility to develop molecular electronic switches and memory devices [30]. We could also see the appearance of negative value current in a positive bias. Such an event is perhaps due to the Bloch oscillations in DNA superlattice. According to this phenomenon, the motion of electrons in a perfect crystal or superlattice under the action of even a constant electric field would be oscillatory instead of uniform [31,32].

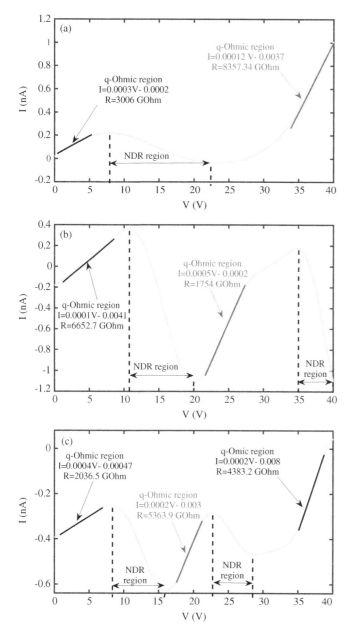

FIGURE 12.4 I-V characteristic diagram in time variable field and (a) $\omega = 0.01\ THz$, (b) $\omega = 0.6\ THz$, and (c) $\omega = 0.8\ THz$ [15].

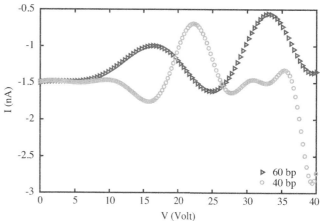

FIGURE 12.5 The sequences length dependent I-V characteristic diagram [15].

TABLE 12.4 The sequences of the DNA molecules

Name	Number of base pairs	DNA sequence
$L60B36$	60 bp	CCGCCAGCGGCGTTATTACA TTTAATTCTTAAGTATTATA AGTAATATGGCCGCTGCGCC
$CH22$	60 bp	AGGGCATCGCTAACGAGGTC GCCGTCCACAGCATCGCTAT CGAGGACACCACACCGTCCA
$CH22$	40 bp	CAATGCAGTCTATCCACCTG ACGGACCCCGACCCGCCTTT

helix. Here, we have modeled the charge transfer phenomena based on the SSH approach [13,38]. In this model, the non-diagonal matrix element depends on intersite displacements is considered [39]. It is based on the assumption of a classical harmonic interaction between sites, which is linearly coupled to a tight-binding Hamiltonian. To understand how electron transport is affected by the environmental effect, we used the coupling between the chain and phonon bath. However, quantum fluctuations from the external interferences of e-bath interaction are important and inevitable. Hence, in this study, the quantum effects of these interactions in the environment and also in transport are investigated.

In this work, we assume that the nucleobases with their high–energy frontier molecular orbitals are responsible for the charge transport properties and neglect the deoxyribose and the phosphate units which play a scaffolding role for the base pairs. The system Hamiltonian is comprised of four parts, which are as follows [40]:

$$H = H_{\mathrm{SSH}} + H_{\mathrm{bath}} + H_{\mathrm{e-bath}} + H_{\mathrm{field}}. \quad (12.12)$$

The first term is used to simulate the electronic and lattice dynamics of the system called the SSH Hamiltonian [41]:

$$\begin{aligned}
H_{\mathrm{SSH}} = & \sum_n \frac{1}{2} m\dot{x}_n^2 + \sum_n \frac{k_s}{2}(x_{n+1} - x_n)^2 \\
& - \sum_n \left[t_0 - \alpha(x_{n+1} - x_n) \right] \left[c_{n+1}^\dagger c_n + c_n^\dagger c_{n+1} \right] \\
& + \sum_n \epsilon_0 c_n^\dagger c_n, \quad (12.13)
\end{aligned}$$

different regions are slightly displaced. The major effect of variation of DNA length and its type could be a further deviation from ohmic behavior and appear nonlinear effects. Another effect of variation of the chain could be the displacement of different areas (quasi-ohmic or NDR region) in I–V characteristic (see Figure 12.5).

12.3   SSH Model

Let us consider a DNA charge transfer model through a single, simple, flexible and chemically specific model Hamiltonian that allows for full variational control. The coupling between the charge and lattice can be regarded along the where m is the base pair mass and $x_{n+1} - x_n$ is the deviation of the distance between the neighboring base pairs from their equilibrium state. The energy ε_0 represents the onsite energy and $c_n^\dagger$ and c_n are creation and annihilation operators of an electron at the nth site. t_0 denotes the hopping integral, α is the electron-lattice coupling constant and k_s is the harmonic potential constant.

The next two terms in Hamiltonian represent the vibrational mode of a phonon bath at frequency ω_0 coming from external sources and the local external e–bath interaction term, respectively.

$$H_{\text{bath}} + H_{\text{e-bath}} = \omega_0 \sum_n b_n^\dagger b_n + \gamma_0 \sum_n c_n^\dagger c_n (b_n^\dagger + b_n),$$
(12.14)

where $b_n^\dagger$ and b_n are creation and annihilation operators of a phonon at site n and γ_0 is the external e–bath coupling constant.

It is worth noting that temperature plays an important role in conductivity properties in DNA [42]. To study the temperature effect, the dynamics of the molecule in contact with a thermal bath is investigated via molecular dynamics simulation using Nosé–Hoover method [17]. Consequently, the coupled nonlinear equations of the system are calculated.

12.3.1   Electrical Current

Electrical field can be applied as either constant field E_0 (DC) or time-varying (AC) field ($E = E_0 \sin(\omega t)$). Therefore, the time-dependent current operator is as follows:

$$I(t) = \frac{ie}{\hbar} \sum_n [t_0 - \alpha(x_{n+1} - x_n)] \left(c_n^\dagger c_{n+1} - c_{n+1}^\dagger c_n \right).$$
(12.15)

We have investigated the flowing electrical current of DNA over time from when it is at room temperature to when it reaches the threshold of denaturation temperature. It could be shown that at both temperatures, the electrical current behaves in an oscillatory manner. Also, secondary peaks appear in the current versus time diagram, but the number and amplitude of peaks are more when DNA is denatured while the generic shape of the diagrams is not changed (see Figure 12.6). It should be noted that the electrical current demonstrates a significant transition by increasing the temperature to the denaturation temperature in a short time interval (about 600 ps) (Figure 12.6). In these conditions, the current intensity increases suddenly and the period of its oscillation becomes shorter in time. The sharp changes in the current diagram can be considered as a signature of alternation in the DNA. Perhaps it can be said that DNA has begun to unfold. Ref. [43] reports the time (τ), referred to as the characteristic time for unwinding or as the time required for strand separation. This characteristic time is dependent on the chain length N (the number of base pairs in double-stranded DNA) in both studies. Denaturation temperature has been investigated in the previous studies, and it has been in agreement with the

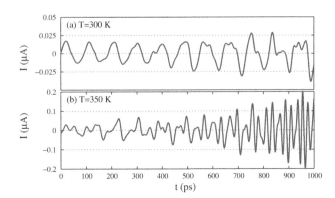

FIGURE 12.6   Electrical current time series at (a) $T = 300$ K, and (b) $T = 350$ K [40].

empirical studies [44,45]. It is found that flowing a weak electrical current through DNA and its detection is presumably a method to estimate the start time of the denaturation phenomenon in denaturation temperature.

Here, the I–V characteristic of DNA is investigated and compared with previous works (Figure 12.7) [46–48]. Measurements of electrical current as a function of the potential applied across single DNA could be indicated by the metallic-like behaviors and efficient conduction and conversely insulator-like behaviors [26]. Figure 12.7 compares the SSH results with empirical findings at low voltages and the results are in agreement with each other. However, by increasing the applied potential, the other form of the conductivity of DNA can be identified. In this circumstance, Ohmic (quasi-Ohmic) regions (with a linear gradient) and the regions with a negative gradient are observable (see Figure 12.8). On the other hand, in AC fields and lower frequencies the current only shows a quasi-linear behavior with respect to the electrical potential difference (see Figure 12.9a). Hence, at the small frequencies, the increasing the electrical potential difference corresponds to the greater passing current. Then, it was mentioned that the observed behavior together with the fact that DNA molecules of specific composition and length, ranging from a few nucleotides to several tens of micrometers can be

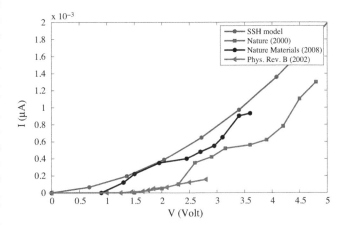

FIGURE 12.7   I-V characteristic diagram to compare SSH model with experimental data [40].

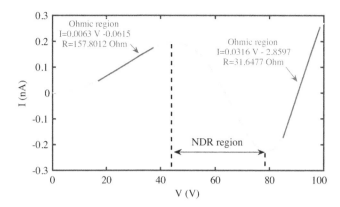

FIGURE 12.8 I-V characteristic diagram at DC fields [40].

prepared, making DNA ideally suited for the construction of mesoscopic electronic devices. This situation would be different when the frequency of the field is increased. In this case, for different values of frequency, the I–V characteristic diagram shows the different phenomena (Figure 12.9b and c). Here, one can consider the quasi–ohmic and NDR regions in the I–V diagram. In DNA, by the enhancement of frequency, one could see the appearance of neoteric quasi–ohmic regions. On the other hand, the quasi-Ohmic areas change to the NDR regions, and thus the novel quasi–ohmic and negative slope areas are created. It is clear that the slope of I–V diagram for quasi–ohmic ranges is proportional to the ohmic resistance of the conductors. Then, the resistance of DNA in different conditions in the quasi–ohmic regions are specified on the characteristic diagram. Previous studies have demonstrated the effect of chain length variation on electronic transport and localization lengths [35]. It is clear that the major effect of variation of DNA length is a further deviation from ohmic behavior and leads to the appearance of nonlinear effects (see Figure 12.10). Another effect of length variation of the chain is the displacement of different areas (quasi–ohmic or NDR regions) in the I–V characteristic curve. It means that in shorter sequences NDR occurs at lower voltages while with increasing DNA chain length, this effect is delayed. Here, we have considered the environmental effect as a phonon bath and included its effect on the calculations. The most important finding is the transition between the (quasi)- ohmic and NDR behaviors in DNA. These behaviors are revealed by the I–V characteristics at room temperature. Also, the variation in DNA length leads to further deviation from ohmic behavior and displacement of different areas (quasi–ohmic or NDR regions) in the I–V characteristic curve.

12.4 Spin Transfer in DNA Chains: Modeling and Results

Increasing the rate and security of information transport is one of the major challenges in information theory. Modern information transport is principally based on the manipulation of the spin of the electron in addition to the

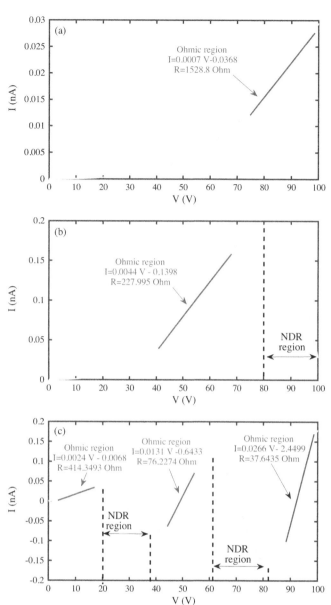

FIGURE 12.9 I-V characteristic diagram at AC fields, (a) $\omega = 0.01\ MHz$, (b) $\omega = 0.2\ MHz$, and (c) $\omega = 5\ MHz$ [40].

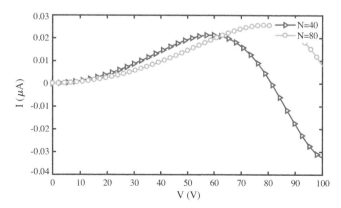

FIGURE 12.10 Sequence-dependent I-V characteristic diagram [40].

electron charge. It could be used to create devices with novel functionalities and potentially superior performance, such as new types of memory [49] or low power consumption devices. Therefore, using of electron spin state as a quantum of information for storage, sensing, and computing has generated considerable interest in the context of data storage and communication devices [50,51] opening avenues for developing multifunctional spintronics [52]. Spintronics aims to utilize the spin degree freedom of electrons for new forms of information storage and logic devices [53].

The direct injection and transport of spin carriers in metallic and semiconducting media have been reported previously [54,55]. On the other hand, organic electronics has grown into a vast research activity that has already seen success in commercial applications with organic light-emitting diodes (OLEDs). In particular, a flurry of research has recently surrounded the study of magneto-optoelectronic properties, an emerging research area referred to as 'organic spintronics' that has produced a range of device concepts for magnetic-field sensors [56], spin valves [57] and spin-OLEDs [58,59]. The term organic spintronics has been used to describe either the purported exchange of spin for a charge as the information carrier in devices or, more generally, for a spin and thus the magnetic field-dependent charge-transport processes [60].

Also, chiral organic molecules open up new possibilities for using the molecule as an information carrier in next-generation electronics and optoelectronics [61,62]. The spin polarization effect is also shown in DNA molecules as a chiral organic molecule [63]. It could be said that DNA molecules find use in spintronics tools too. It would be interesting to explore its potential for a wide variety of technological applications such as read heads, magnetic information storage [64], low– and high–field magnetic sensors, IR detectors [65], and numerous other spintronic applications in which carrier spins play a key role in transmitting, processing and storing information [66]. It could be a promising medium to transport spin-encoded information due to low spin-orbit interaction and thus be an ideal spin-polarized current source.

12.4.1 Spin Selectivity in the DNA/Gold Interface

We look for a secure channel for data transmission with minimum dissipation. Therefore, we have to try to generate a pure spin current. The pure spin current could be used as information carrier in new memory devices. In spintronics, pure spin currents play a key role in transmitting, processing and storing information. A pure spin current is a flow of electron spin angular momentum without a simultaneous flow of charge current [67]. It can be carried by conduction of electrons or magnons and has been studied in many inorganic metals, semiconductors, and insulators, but not yet in organic semiconductors [68]. A nearly pure spin current has been reported in the electrical current flowing through DNA molecules previously [69]. Charge carriers in conjugated organic materials are the localized spin-1/2 polarons

which move by hopping, but the mechanisms of their spin transport and relaxation are not well understood.

The aim of this work is to investigate the conditions which allow forming a pure spin current for information transmission in biomatterials. In this regard, we have considered DNA molecules as a data transmission channel, so studied the spin transfer mechanism in DNA molecules absorbed on metal. Inasmuch as one of the characteristic properties of organic materials is the strong electron-lattice interaction leading to the formation of polarons, we take into account the spin coupling with radicals and the electron-lattice interaction. Therefore, we use PBH model [11,15] for studying the spin polaron transfer in DNA by considering the spin degree of freedom and spin-orbit coupling. We have taken inspiration from the work of Göhler et al, in which the spin polarization of photoelectrons ejected from the gold substrate and passed through DNA is studied [63]. We would try to determine the initial conditions and the regions in which the pure spin current is created and study the effect of external agents on these regions. Consider N base-pairs system with the Hamiltonian [70]

$$H_{\text{chiral}} = H_{\text{PBH}} + H_{\text{so}}, \qquad (12.16)$$

where H_{PBH} is the extended PBH model Hamiltonian (Eq. 12.1) [11] that describes the DNA lattice, the charge carrier, and the interaction between the DNA lattice and charge. The electronic part of the Hamiltonian is written by adding the spin effect as follows:

$$H_{\text{car}} = \sum_n \sum_\sigma \left[\epsilon_n c_n^{\sigma\dagger} c_n^\sigma - V_{n,n+1}(c_n^{\sigma\dagger} c_{n+1}^\sigma + c_{n+1}^{\sigma\dagger} c_n^\sigma) \right], \qquad (12.17)$$

where $c_n^{\sigma\dagger}$ and c_n^σ are the creation and annihilation operators of a charge with spin σ, respectively, and ϵ_n is the onsite energy for each base-pair.

$$H_{\text{lat-car}} = \chi \sum_n y_n c_n^{\sigma\dagger} c_n^\sigma, \qquad (12.18)$$

where χ is the coupling between the charge and DNA lattice. We have considered a Rushba-like spin-orbit interaction for moving an electron with spin σ in helical DNA molecules [71]. The spin-orbit interaction in the second quantization representation using the creation and annihilation operators is represented as follows:

$$\begin{aligned}
H_{\text{so}} = \sum_n \Big[& 2it_{\text{so}} \cos\theta \Big(c_n^{\dagger\uparrow} c_{n+1}^\uparrow - c_n^{\dagger\uparrow} c_{n-1}^\uparrow \\
& - c_n^{\dagger\downarrow} c_{n+1}^\downarrow + c_n^{\dagger\downarrow} c_{n-1}^\downarrow \Big) \\
& + D_{n,n+1} c_n^{\dagger\uparrow} c_{n+1}^\downarrow - D_{n,n+1}^* c_n^{\dagger\downarrow} c_{n+1}^\uparrow \\
& + D_{n-1,n}^* c_n^{\dagger\downarrow} c_{n-1}^\uparrow - D_{n-1,n} c_n^{\dagger\uparrow} c_{n-1}^\downarrow \Big], \qquad (12.19)
\end{aligned}$$

where $D_{n,n+1} = it_{\text{so}} \sin\theta \{\sin[n\Delta\phi] + \sin[(n+1)\Delta\phi] + i\cos[n\Delta\phi] + i\cos[(n+1)\Delta\phi]\}$. Also, t_{so} represents the spin-orbit coupling constant, θ is the helix angle and $\phi = n\Delta\phi$ is the cylindrical coordinate with ϕ being the twist angle.

On the other hand, preservation of time–reversal symmetry leads to $D_{n,n-1} = D_{n-1,n}^*$.

Now, by considering the laser radiation and gold substrate as the electron pumping machine, one could obtain the following Hamiltonian:

$$H_{\text{pump}} = H_{\text{Gold}} + H_{\text{Gold--DNA}} + H_{\text{Laser}}, \qquad (12.20)$$

where H_{Gold} and $H_{\text{Gold--DNA}}$ are the Hamiltonian of gold metal and Hamiltonian which represents the connection of DNA to metal, respectively. These terms are written as:

$$H_{\text{Gold}} + H_{\text{Gold--DNA}} = \sum_{k,\sigma} \epsilon_k a_k^{\sigma\dagger} a_k^{\sigma}$$
$$+ t \sum_{k,\sigma} \left(a_k^{\sigma\dagger} c_1^{\sigma} + c_1^{\sigma\dagger} a_k^{\sigma} \right) \qquad (12.21)$$

The laser photon Hamiltonian consists of two perpendicular electrical and magnetic fields irradiated on the gold and has the following form [69]:

$$H_{\text{Laser}} = \sum_k \left[\left(-eEkd - \mu_B B \right) a_k^{\uparrow\dagger} a_k^{\uparrow} \right.$$
$$\left. + \left(-eEkd + \mu_B B \right) a_k^{\downarrow\dagger} a_k^{\downarrow} \right] \qquad (12.22)$$

It is assumed that $\vec{E} = E\hat{x}$ is the electrical field along the x axis and d is the distance between the lattice points of gold, $\vec{B} = B\hat{z}$ is the magnetic field along the z axis and $\mu_B = \frac{e\hbar}{2mc} = 5.78838 \times 10^{-5}$ eV.T^{-1} is the Bohr magneton.

We use the nonlinear dynamical systems theory for analyzing the system. Therefore, evolution equations of the system can be derived.

For studying information transport using the pure spin current, we have to obtain the electrical currents corresponding to spin up and down electrons. The currents could be obtained directly from motion equations. We could define the current operator as $I^{\sigma} = e\frac{dn_i^{\sigma}}{dt}$, where $n_i^{\sigma} = c_i^{\sigma\dagger} c_i^{\sigma}$. Therefore, the current operators are written as [72]:

$$I^{\uparrow}(t) = \frac{-ie}{\hbar} \left\{ \sum_n \left[W_{n,n+1} c_n^{\uparrow\dagger} c_{n+1}^{\uparrow} + W_{n-1,n}^* c_n^{\uparrow\dagger} c_{n-1}^{\uparrow} \right. \right.$$
$$\left. + D_{n,n+1} c_n^{\uparrow\dagger} c_{n+1}^{\downarrow} - D_{n-1,n} c_n^{\uparrow\dagger} c_{n-1}^{\downarrow} \right]$$
$$\left. + \sum_k [c_1^{\uparrow\dagger} a_k^{\uparrow} - a_k^{\uparrow\dagger} c_1^{\uparrow}] \right\} \qquad (12.23)$$

$$I^{\downarrow}(t) = \frac{-ie}{\hbar} \left\{ \sum_n \left[W_{n,n+1}^* c_n^{\downarrow\dagger} c_{n+1}^{\downarrow} + W_{n-1,n} c_n^{\downarrow\dagger} c_{n-1}^{\downarrow} \right. \right.$$
$$\left. - D_{n,n+1}^* c_n^{\downarrow\dagger} c_{n+1}^{\uparrow} + D_{n-1,n}^* c_n^{\downarrow\dagger} c_{n-1}^{\uparrow} \right]$$
$$\left. + \sum_k [c_1^{\downarrow\dagger} a_k^{\downarrow} - a_k^{\downarrow\dagger} c_1^{\downarrow}] \right\} \qquad (12.24)$$

Now, we define the net charge, $I_c = I^{\uparrow} + I^{\downarrow}$, and net spin, $I_s = I^{\uparrow} - I^{\downarrow}$, currents. The other quantity for comparing the net charge and spin currents is the ratio of net spin current

(I_s) to net charge current (I_c) represented as $P = \frac{I_s}{I_c}$. Using the nonlinear dynamics methods and obtained equations for the current operators, we could easily assess the effect of various factors on spin-dependent currents in DNA. In order to achieve to pure spin currents to allow its application in different spintronics fields, we should consider the effect of limiting internal and external factors on spin filtering of DNA wires. The effective range of laser power, external magnetic field and the temperature effect caused by internal fluctuation and laser radiation could be some of the determining agents in spin transfer.

The effect of laser fluence

Here, an organic chiral layer is absorbed on a nonmagnetic metal surface which is not self-magnetized, and the photoelectrons ejected from such a layer would not be polarized. The ionization energy of gold is less than DNA, and photoelectrons originate from the gold substrate. Therefore, the electrical current is due to photoelectrons emitted from Gold via an ultraviolet (UV) laser radiation. Since the gold work function and the bandgaps in DNA are greater than 4 eV, in order to release an electron in these materials, the photon energy must be greater than 4 eV, indicating an ultraviolet light source is needed. Then, laser power is one of the parameters most affected in the polarization of the spin current. We have examined the variation of laser power from the photon energy of 5.1 eV corresponding to work function of gold up to the threshold of energy required for gold melting. But, we try to study only the spin filtering mechanism of DNA for electrons ejected from the gold substrate and not electrons originating from DNA. Then, the laser energy set to less than ≈ 8.4 eV, that is lower than the ionization energy of base-pairs [73,74]. We have applied a laser with a pulse duration of about 200 ps at a 20-kHz repetition rate and varied the laser fluence from 131 pJ/cm^2 up to 9 mJ/cm^2. Then, we have defined a critical fluence (f_c) for laser power where at fluences higher than it, gold melts and the structure is destroyed. On the other hand, we have determined the effect of the incident laser power with energies lower than base-pairs ionization.

In all calculations, we have used the parameter values established in Refs. [11,24,71,75] represented in Tables 12.1, 12.2, and 12.5. On the other hand, the onsite energies and electron hopping constants are different for distinct base-pairs in DNA sequences. By considering a segment from human chromosome 22 (CH22)–based sequences with $N = 60$ *bp* [76] (see the Table 12.6), one could obtain spin-up and spin-down currents and corresponding I_s and I_c. Figure 12.11 shows the spin currents with respect to the energy densities of laser. It determines the values of laser power in which the maximum net spin current, and therefore the minimum net charge current flow through DNA molecules. As can be seen from Figure 12.11, one could investigate the laser powers with the highest efficiency to create the pure spin current. On the other hand, we have determined the best regions for information coding in laser power.

TABLE 12.5   Constant parameters used in the combined extended PBH model [11,24,71]

Symbol	Description	Units	Value
θ	helix angle	rad	0.66
$\Delta\varphi$	twist angle		$\pi/5$
t_{so}	spin-orbit coupling constant	eV	0.01
d	distance between lattice point of Au	A°	4.065
ϵ_k	onsite energy of Au	eV	7.75
t	hopping constant between DNA and gold	eV	0.42

TABLE 12.6   The DNA sequence used in the calculation [22,76]

Name	Number of base pairs	DNA sequence
$CH22$	60 bp	AGGGCATCGCTAACGAGGTC GCCGTCCACAGCATCGCTAT CGAGGACACCACACCGTCCA
$hc1$	40 bp	TAAATAAATAAATAAATAAA TAAAATAAATAAAAGCCTTT
$hc3$	40 bp	AGCTGGGGAGCAGGGCTCCA CTCTGGGAGGGGGGGCAGCCT

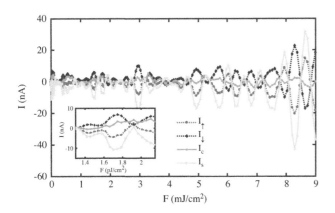

FIGURE 12.11   Spin currents with respect the laser fluence [70].

12.4.2   The Effect of Laser Field Frequency

By considering the laser power in which the maximal net spin current flows through DNA ($\approx$1.7 pJ/cm^2), we have studied the variation of incident radiation frequency on spin-polarized currents through DNA. It is worth mentioning that the repetition rate of laser pulse could have an effective impression on spin selectivity. To this extent, one would be able to determine the best range of laser frequency for approaching the pure spin currents (see Figure 12.12a). It is worth noting that the pure spin current is a good candidate for information transfer with minimal dissipation [77,78]. Also, the results based on the ratio of net spin current (I_s) to net charge current (I_c), (P), shows significant peaks at certain frequencies (see Figure 12.12b). It is clear that these frequency values could be important in magnetic information transfer based on spin current. Figure 12.12b shows polarization peaks for different laser fluences. It is clear that there were no noticeable change in intensity of peaks.

12.4.3   External Magnetic Field Effect

It is clear that the magnetic field has a considerable effect on spin arrangement, and as a result spin polarization of materials [79,80]. Then, we have tried to discuss the

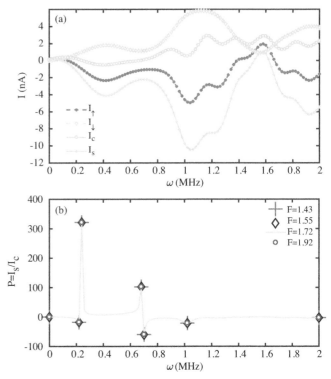

FIGURE 12.12   (a) Spin currents, and (b) the ratio of net spin current to net charge current, with respect to the frequency of the fields [70].

magnetic field effect on DNA spin filtering. According to the relation $B = E/c$, if E is the electrical field of the radiated photon, its magnetic field would be weakened by c order of magnitude, where c is the velocity of light. Then, the spin filtering will be virtually unaffected by the laser magnetic field. We have tried to examine the effect of the periodic magnetic field which varies with frequency equal to the laser frequency on DNA spin currents. It is assumed that the external magnetic field is applied to DNA chain in the same direction as the laser magnetic field. To this end, we have added Hamiltonian H_{field} to the total Hamiltonian

$$H_{\mathrm{field}} = \sum_n \left[-\mu_B B c_n^{\uparrow\dagger} c_n^{\uparrow} + \mu_B B c_n^{\downarrow\dagger} c_n^{\downarrow} \right] \quad (12.25)$$

and corrected the evolution equations. For examining the effect of AC field on the DNA spin transport and considering the field frequency degree of freedom in addition of field intensity, we have considered the general form of field as $B = B_0 \cos(\omega t)$, where B_0 is the amplitudes of magnetic field and ω is the frequency of the field. As can be seen from Figure 12.13, the effect of external magnetic is appeared as creating the quasi–periodic I_s current. In this way, one could determine the intervals in the external magnetic field for creating nearly pure spin currents. There are magnetic field values in which I_c is almost zero, and one could report nearly pure spin current. In contrast, for some field intensities, I_s has zero value. The region [7.4–9.5 μT] corresponds to magnetic field intensities of incident radiation

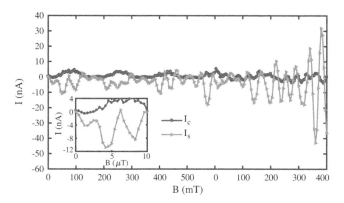

FIGURE 12.13 Spin currents with respect to the external magnetic field intensity [70].

which are very weak intensities. Here, the DNA current is the spontaneous spin filter and not induced via external driving.

12.4.4 The Simultaneous Effects of Laser Fluence and External Magnetic Field

It is reported that the laser energy and the applied external magnetic field are important factors in spin filtering efficiency of DNA. By studying the laser effect and external field simultaneously, some island are emerged in some values which they have reported the nearly pure spin current regions (Figure 12.14).

12.4.5 DNA in Thermal Bath

It is significant that temperature has a considerable effect on spin current [81]. In this work, the temperature effect occurs via internal fluctuations and laser radiation. DNA lattice fluctuation and the photon flux induced by laser could transfer the thermal energy to the system. For studying temperature effect and its simulation, we have modeled DNA in contact with a thermal bath [17]. A deterministic and time-reversible bulk thermostating based on introducing

a momentum-dependent friction coefficient in the equations of motion is the mechanism of the Nosé–Hoover thermostat. In Hoover's reformulation of Nosé's method, the evolution equation of the thermostat is formulated using Eq. 12.7. Figure 15(a) shows the variation of spin currents with respect to temperature. By considering the constant values of laser power, one could choose the best ranges of system temperature for the highest performance of DNA wires in information transfer.

It is clear from Figure 12.15b that the parameter P (the ratio of net spin to net charge current) shows significant characteristic peaks in [300–350 K] known as the pre-melting region of DNA thermodynamics.

12.4.6 Sequence–Dependent Spin Filtering in DNA Segments

In most works on studying the charge transfer in DNA, its sequence–dependent electrical conductivity is determined [76]. However, the spin filtering efficiency in DNA segments depends on the sequence and its length [22]. We have examined the different DNA blocks from human chromosome 22's (CH22) sequence to study the effect of DNA sequence type and length on the net spin current to determine a better sequence for information transport. Table 12.7 reports the spin currents and spin filtering for different DNA segments. It is clear that the AT-rich oligomers are a better candidate than CG-rich for creating the net spin current

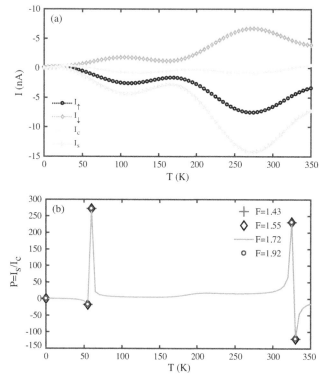

FIGURE 12.15 (a) The spin currents, and (b) the ratio of net spin current to net charge current, with respect to the temperature [70].

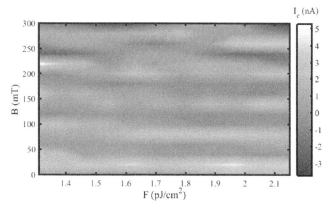

FIGURE 12.14 The simultaneous effects of laser fluence and external magnetic field on the spin current [70].

TABLE 12.7 The effect of different DNA oligomers on spin filtering, (T=300 K)

Number of blocks	Sequence	Number of base-pairs	$I_s(nA)$	$I_c(nA)$	$P = I_s/I_c$
1	CH22	60 (bp)	−12.58	−0.57	21.95
1	hc 3	40 (bp)	−15.85	−0.30	53.42
1	hc 1	40 (bp)	−52.09	−0.48	108.61
2	CH22	120 (bp)	−12.23	−0.54	22.61
3	hc 3	120 (bp)	−14.30	−0.28	50.20
3	hc1	120 (bp)	−50.43	−0.42	120.70
3	hc 1-hc 3 -hc 1	120 (bp)	−50.41	−0.40	124.75
3	hc 3-hc 1 -hc 3	120 (bp)	−14.44	−0.25	58.13

(Figure 12.16). Also, increasing of the AT-rich blocks has a positive impact on the ratio of the net spin current to net charge current.

12.4.7 Spin Hall Effect

Internal interactions and their effects could have substantial effect on the spin–polarized currents. One of the phenomena observed in this case is the spin-Hall effect (SHE). SHE, driven by the spin–orbit interaction, converts a charge current into a pure spin current [82,83]. Spin Hall effect often find use in various applications such as magnetization switching, domain wall motion, spin current detection, etc. On the basis of studying the roles of these effects, it is essential that we pay attention to spin-Hall effect in the DNA chain attached on gold.

As an electron moves along a chiral molecule, it experiences an electric field which is also chiral (E_{chiral}). E_{chiral} acting on the moving electron arises from the electrons and nuclei that comprise the chiral molecule [63]. Therefore, a moving charge in a helical DNA molecule experience an electrostatic potential where $E_{\text{chiral}} = -\nabla V$. For the dsDNA, the the potential difference is usually bigger along the radial direction $\hat{r}$ than that along the helix axis (here, the helix axis is $\hat{z}$) [84]. On the other hand, dV/dr is very large at the boundary $r = R$ with R denoting the helix radius. Then, it is reasonable to consider the r component of E only. It could be said that there is a internal transverse electrical field (E_r) in DNA. An internal field is necessary to break

inversion symmetry and generate significant spin-orbit interaction. Here, a spin Hall effect is mainly determined by the magnitude of spin-orbit coupling parameter.

One could consider the spin-orbit interaction as follows:

$$H_{\text{so}} = -\frac{\alpha}{\hbar}\hat{\sigma}.(\hat{r} \times \hat{P}), \qquad (12.26)$$

with $\alpha = -(\frac{\hbar}{2mc})^2 E_r$, where m is the electron mass, c the speed of light, σ the Pauli matrices and P the momentum operator.

On the other hand, using the second quantization, we could obtain the previous SOC Hamiltonian (Eq. 12.19). Now, one could define t_{so} as spin-orbit constant as [71]

$$t_{\text{so}} = -\frac{\alpha}{4l_a}$$

In DNA molecules, l_a is the arc length that satisfies $l_a \cos(\theta) = R\Delta\phi$ and $l_a \sin(\theta) = \Delta h$, with Δh being the stacking distance between neighboring base-pairs. By considering parameter values $\Delta h \simeq 0.34$ nm, $R = 0.7$ nm, one could obtain $l_a \simeq 0.56$ nm.

Now, we could obtain the transversal electrostatic field and calculate the transverse charge current density as follows:

$$E_r = \left(\frac{2mc}{\hbar}\right)^2 l_a t_{\text{so}}$$

and

$$J_c = \sigma E_r,$$

here σ is the charge conductivity of the DNA [33].

The efficiency of SHE (conversion of charge current to spin current) is characterized by a single material-specific parameter, the spin-Hall angle, γ_{SH}, given by the ratio of spin to charge current density [85]. Then, one could define γ_{SH} for DNA molecules with the above condition as follows:

$$\gamma_{\text{SH}} = \frac{J_s}{J_c} = \frac{I^\uparrow - I^\downarrow}{\pi R^2 \sigma E_r}. \qquad (12.27)$$

However, the spin Hall conductivity (σ_{SH}) is related to γ_{SH} according to

$$\sigma_{\text{SH}} = \left(\frac{\hbar}{2e}\right)\gamma_{\text{SH}}\sigma.$$

So, we have obtained the hall conductivity of DNA oligomers as:

$$\sigma_{\text{SH}} = \left(\frac{\hbar}{2e}\right)\frac{I^\uparrow - I^\downarrow}{\pi R^2 E_r}. \qquad (12.28)$$

Using the above relations, one can find the others quantifiers for studying the pure spin current in DNA.

The increasing of the spin Hall angle, and therefore the spin Hall conductivity is corresponding to high net spin current.

Figure 12.17a and b show the variation of spin hall conductivity with respect to the temperature and external magnetic field intensity, respectively. It is clear that its variation is similar to the variation of the net spin current. Then, the increasing of the spin hall conductivity is directly related to amplification of the net spin current through DNA.

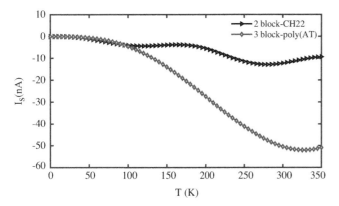

FIGURE 12.16 The net spin current with respect to the temperature for different sequences.

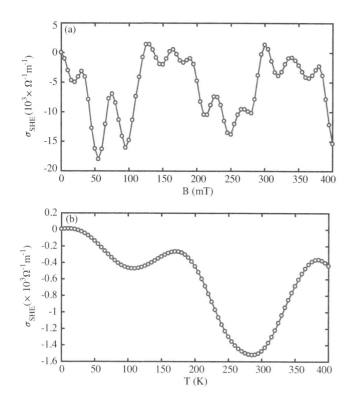

FIGURE 12.17 The spin Hall conductivity with respect to the (a) external magnetic field intensity, and (b) temperature.

In conclusion, we have established the current model for photoelectrons ejected from a gold substrate due to laser radiation and passed through DNA. By varying the laser energy from the work function of gold until to DNA ionization energy is reached as well as radiation frequency, one could characterize the best functional range of laser. These areas are where the most net spin current and, by contrast, the least net charge current flows through DNA. In order to increase the polarization of spin current without affecting on the system structure, we have applied an external magnetic field in addition to the field of incident radiation. Applying the external magnetic field has amplified the nearly pure spin current. Temperature as another important factor defines peaks in the ratio of the net pin current to net charge current. By studying the effect of laser fluence and external magnetic field, simultaneously, one can report islands with maximal net spin current. By investigating the spin Hall effect in DNA and obtaining its related quantity, we could determine the direct relevance between the net spin current and the spin Hall angle as well as spin Hall conductivity. Then, one can quantify the net spin current with spin Hall conductivity.

12.5 Conclusion

We have theoretically observed the transition between the quasi-ohmic and NDR behaviors in DNA through two theoretical approaches based on PBH and SSH models. Reproducible and stable NDR behaviors are revealed by the

I–V characteristics at room temperature. NDR behavior originates the active redox reactions of transition metal ions in DNA. It is found that there is a relation between the width of the quasi-ohmic and NDR areas and the frequency of the applied field. Also, DNA sequence variation leads to further deviation from ohmic behavior and displacement of different areas (quasi-ohmic or NDR region) in the I–V characteristic curve.

On the other hand, spin transport has technological and functional application in spintronics and particularly secure information transfer. We have studied the spin transfer mechanism in DNA based on a spin-polaronic PBH model. By varying the laser energy from the work function of gold until to DNA ionization energy as well as radiation frequency, one can characterize the best functional range of the laser. Applying the external magnetic field led to amplification of the nearly pure spin current. Temperature as another important factor defines peaks in the ratio of the net pin current to net charge current. By investigating the spin Hall effect in DNA, we could determine the direct relevance of the net spin current and the spin Hall conductivity.

References

1. D. D. Eley, and D. I. Spivey, *Trans. Faraday Soc.* **58**, 405 (1962).
2. R. Bruinsma, G. Grüner, M. R. Dorsogna, and J. Rudnick, *Phys. Rev. Lett.* **85**, 4393 (2000).
3. D. Ly, Y. Kan, B. Armitage, and G. B. Schuster, *J. Am. Chem. Soc.* **118**, 8747 (1996).
4. Z. G. Yu, and X. Song, *Phys. Rev. Lett.* **86**, 6018 (2001).
5. Z. Hermon, S. Caspi, and E. Ben-Jacob, *Europhys. Lett.* **43**, 482 (1998).
6. K.-H. Yoo, D. H. Ha, J.-O. Lee, J. W. Park, J. Kim, J. J. Kim, H.-Y. Lee, T. Kawai, and H. Y. Choi, *Phys. Rev. Lett.* **87**, 198102 (2001).
7. E. M. Conwell, and S. V. Rakhmanova, *Proc. Natl. Acad. Sci. USA* **97**, 4556 (2000).
8. E. Conwell, *Long-Range Charge Transfer in DNA II* (Springer Berlin Heidelberg), 2004.
9. D. Hennig, and J. F. R. Archilla, *Modern Methods for Theoretical Physical Chemistry of Biopolymers* (Elsevier), 2006, 429.
10. G. Kalosakas, K. Ø. Rasmussen, and A. R. Bishop, *Synth. Met.* **141** 93 (2004).
11. S. Komineas, G. Kalosakas, and A. R. Bishop, *Phys. Rev. E* **65**, 061905 (2002).
12. J. Zhu, K. Ø. Rasmussen, A. V. Balatsky, and A. R. Bishop, *J. Phys. Condens. Matter* **19**, 136203 (2007).
13. W. P. Su, J. R. Schrieffer, and A. J. Heeger, *Phys. Rev. Lett.* **42**, 1698 (1979).
14. W. P. Su, J. R. Schrieffer, and A. J. Heeger, *Phys. Rev. B* **22**, 2099 (1980).
15. S. Behnia, and S. Fathizadeh, *Phys. Rev. E* **91**, 022719 (2015).

16. D. Hennig, C. Neissner, M. G. Velarde, and W. Ebeling, *Phys. Rev. B* **73**, 024306 (2006).

17. T. Dauxois, M. Peyrard, and A. R. Bishop, *Phys. Rev. E*, **47**, 684 (1993).

18. E. Díaz, R. P. A. Lima, and F. Domínguez-Adame, *Phys. Rev. B* **78**, 134303 (2008).

19. B. Alexandrov, N. K. Voulgarakis, K. Ø. Rasmussen, A. Usheva, and A. R. Bishop, *J. Phys. Condens. Matter* **21**, 034107 (2009).

20. Y. Zeng, A. Montrichok, and G. Zocchi, *J. Mol. Biol.* **339**, 67 (2004).

21. S. Roche, *Phys. Rev. Lett.* **91**, 108101 (2003).

22. A. M. Guo, and Q. F. Sun, *Phys. Rev. B* **86**, 115441 (2012).

23. H. Shibata, *Physica A* **292**, 182 (2001).

24. D. Hennig, *Phys. Rev. E* **62**, 2846 (2000).

25. C. Simserides, *Chem. Phys.* **440**, 31 (2014).

26. L. M. Bezerril, D. A. Moreiraa, E. L. Albuquerque, U. L. Fulcob, E. L. de Oliveira, and J. S. de Sousa, *Phys. Lett. A* **373**, 3381 (2009).

27. P. C. Jangjian, T. F. Liu, M. Y. Li, M. S. Tsai, and C. C. Chang, *Appl. Phys. Lett.* **94**, 043105 (2009).

28. N. Kang, A. Erbe, and E. Scheer, *Appl. Phys. Lett.* **96**, 023701 (2010).

29. T. C. L. G. Sollner, W. D. Goodhue, P. E. Tannenwald, C. D. Parker, and D. D. Peck, *Appl. Phys. Lett.* **43**, 588 (1983).

30. J. Chen, M. A. Reed, A. M. Rawlett, and J. M. Tour, *Science*, **286**, 1550 (1999).

31. T. Dekorsy, R. Ott, H. Kurz, and K. Kohler, *Phys. Rev. B*, **51**, 17275 (1995).

32. J. M. Zhang, and W. M. Liu, *Phys. Rev. A* **82**, 025602 (2010).

33. P. Maniadis, G. Kalosakas, K. O. Rasmussen, and A. R. Bishop, *Phys. Rev. E* **72**, 021912 (2005).

34. P. Maniadis, G. Kalosakas, K. O. Rasmussen, and A. R. Bishop, *Phys. Rev. B* **68**, 174304 (2003).

35. R. Gutiérrez, S. Mandal, and G. Cuniberti, *Nano Lett.* **5**, 1093 (2005).

36. R. Gutiérrez, S. Mandal, and G. Cuniberti, *Phys. Rev. B* **71**, 235116 (2005).

37. M. Peyrard, S. Cuesta-Lopez, and D. Angelov, *J. Phys. Condens. Matter* **21**, 034103 (2009).

38. J. Lei, and Y. Shimoi, *J. Phys. Soc. Jpn.* **80**, 034702 (2011).

39. D. A. Tikhonov, N. S. Fialko, E. V. Sobolev, and V. D. Lakhno, *Phys. Rev. E* **89**, 032124 (2014).

40. S. Behnia, S. Fathizadeh, and A. Akhshani, *J. Phys. Soc. Jpn.* **84**, 084002 (2015).

41. D. M. Basko, and E. M. Conwell, *Phys. Rev. E* **65**, 061902 (2002).

42. P. Tran, B. Alavi, and G. Gruner, *Phys. Rev. Lett.* **85**, 1564 (2000).

43. G. F. Calvo, and R. F. Alvarez-Estrada, *J. Phys: Condens. Matter* **20**, 035101 (2008).

44. S. Behnia, A. Akhshani, M. Panahi, A. Mobaraki, and M. Ghaderian, Phys. Rev. E **84**, 031918 (2011).

45. S. Behnia, A. Akhshani, M. Panahi, A. Mobaraki, and M. Ghaderian, *Phys. Lett. A* **376**, 2538 (2012).

46. D. Porath, A. Bezryadin, S. de Vries, and C. Dekker, *Nature* **403**, 635 (2000).

47. G. Cuniberti, L. Craco, D. Porath, and C. Dekker, *Phys. Rev. B* **65**, 241314 (2002).

48. E. Shapir, H. Cohen, A. Calzolari, C. Cavazzoni, D. A. Ryndyk, G. Cuniberti, A. Kotlyar, R. D. Felice, and D. Porath, *Nat. Mater.* **7**, 68 (2008).

49. K. V. Raman, A. M. Kamerbeek, A. Mukherjee, N. Atodiresei, T. K. Sen, P. Lazic, V. Caciuc, R. Michel, D. Stalke, S. K. Mandal, S. Blügel, M. Münzenberg, and J. S. Moodera, *Nature* **493**, 509 (2013).

50. M. Verdaguer, *Science* **272**, 698 (1996).

51. T. Miyamachi, M. Gruber, V. Davesne, M. Bowen, S. Boukari, L. Joly, F. Scheurer, G. Rogez, T. K. Yamada, P. Ohresser, E. Beaurepaire, and W. Wulfhekel, *Nat. Commun.* **3**, 938 (2012).

52. Y. Tserkovnyak, *Nat. Nanotechnol.*, **8**, 706 (2013).

53. I. Zutic, J. Fabian, and S. Das Sarma, *Rev. Mod. Phys.* **76**, 323 (2004).

54. I. Appelbaum, B. Q. Huang, and D. J. Monsma, *Nature* **447**, 295 (2007).

55. B. T. Jonker, G. Kioseoglou, A. T. Hanbicki, C. H. Li, and P. E. Thompson, *Nat. Phys.* **3**, 542 (2007).

56. S. Pleasants, *Nat. Photonics* **8**, 168 (2014).

57. Z. H. Xiong, D. Wu, Z. V. Vardeny, and J. Shi, *Nature* **427**, 821 (2004).

58. T. D. Nguyen, E. Ehrenfreund, and Z. V. Vardeny, *Science* **337**, 204 (2012).

59. W. J. Baker, K. Ambal, D. P. Waters, R. Baarda, H. Morishita, K. van Schooten, D. R. McCamey, J. M. Lupton, and C. Boehme, *Nat. Commun.* **3**, 898 (2012).

60. C. Boehme, and J. M. Lupton, *Nat. Nanotechnol.* **8**, 612 (2013).

61. R. Naaman, and D. H. Waldeck, *J. Phys. Chem. Lett.* **3**, 2178 (2012).

62. Sh. Ohkoshi, Sh. Takano, K. Imoto, M. Yoshikiyo, A. Namai, and H. Tokoro, *Nat. Commun.* **5**, 3757 (2014).

63. B. Göhler, V. Hamelbeck, T. Z. Markus, M. Kettner, G. F. Hanne, Z. Vager, R. Naaman, and H. Zacharias, *Science* **331**, 894 (2011).

64. G. M. Church, Y. Gao, and S. Kosuri, *Science* **337**, 1628 (2012).

65. E. Tchekanda, D. Sivanesan, and S. W. Michnick, *Nat. Methods* **11**, 641 (2014).

66. M. Cinchetti, *Nat. Nanotechnol.* **9**, 965 (2014).

67. A. Manchon, *Nat. Phys.* **10**, 340 (2014).

68. S. Watanabe, K. Ando, K. Kang, S. Mooser, Y. Vaynzof, H. Kurebayashi, E. Saitoh, and H. Sirringhaus, *Nat. Phys.* **10**, 308 (2014).

69. D. Rai, and M. Galperin, *J. Phys. Chem. C* **117**, 13730 (2013).

70. S. Behnia, S. Fathizadeh, A. Akhshani, *Chem. Phys.* **477**, 61 (2016).

71. A. M. Guo, and Q. F. Sun, *Phys. Rev. Lett.* **108**, 218102 (2012).

72. S. Behnia, S. Fathizadeh, and A. Akhshani, *J. Phys. Chem. C* **120**, 2973 (2016).

73. S. G. Ray, S. S. Daube, and R. Naaman, *Proc. Natl. Acad. Sci. USA* **102**, 15 (2005).

74. C. E. Crespo-Hernández, R. Arce, and Y. Ishikawa, *J. Phys. Chem. A* **108**, 6373 (2004).

75. S. Malakooti, E. R. Hedin, Y. D. Kim, and Y. S. Joe, *J. Appl. Phys.* **112**, 094703 (2012).

76. S. Roche, D. Bicout, E. Maciá, and E. Kats, *Phys. Rev. Lett.* **91**, 228101 (2003).

77. G. A. Prinz, *Science* **282**, 1660 (1998).

78. L. Bogani and W. Wernsdorfer, *Nat. Mater.* **7**, 179 (2008).

79. T. Kimura, T. Goto, H. Shintani, K. Ishizaka, T. Arima, and Y. Tokura, *Nature* **426**, 6962 (2003).

80. S. M. Frolov, A. Venkatesan, W. Yu, and J. A. Folk, *Phys. Rev. Lett.* **102**, 116802 (2009).

81. Z. M. Liao, Y. D. Li, J. Xu, J. M. Zhang, K. Xia, and D. P. Yu, *Nano Lett.* **6**, 1087 (2006).

82. J. E. Hirsch, *Phys. Rev. Lett.* **83**, 1834 (1999).

83. Y. K. Kato, R. C. Myers, A. C. Gossard, and D. D. Awschalom, *Science* **306**, 1910 (2004).

84. D. Hochberg, G. Edwards, and T. W. Kephart, *Phys. Rev. E* **55**, 3765 (1997).

85. L. Liu, C.-F. Pai, Y. Li, H. W. Tseng, D. C. Ralph, and R. A. Buhrman, *Science* **336**, 555 (2012).

13

A Nanomaterials Genome

Chenxi Qian
California Institute of Technology

Geoffrey A. Ozin
University of Toronto

13.1 Introduction

What do we mean by genome? The term genome was originally coined by the botanist Hans Winkler in 1920.[1] He combined the words gene and chromosome to describe the origin of a species in terms of its constituent set of chromosomes and protoplasm. These days, it is an informatics descriptor used in molecular biology and genetics to describe how the heredity of an organism is encoded in its DNA. While it is true that the elements of Mendeleev's periodic table of elements are not the same as the nucleotide base code carriers of DNA, they do carry all the information required to enable the design and discovery of molecules and materials with specific functionality and purposeful utility through chemical control of composition, structure, and property relations. While the number of permutations and combinations of more than a 100 elemental building blocks provides infinite molecules and materials synthetic possibilities, it is only through the rational and systematic understanding and manipulation of the chemical and physical properties of the elemental building blocks that the molecules and materials discovery process becomes tractable.

On this basis and in the context of chemistry, one could argue that Mendeleev's periodic table is the most fundamental of all genomes, the "element genome" although in the framework of physics, all matter ultimately evolves from a "sub-atomic particle genome". In a chemical sense, the "element genome" is the basic building code for all matter, biological or abiological, inorganic or organic, soft or hard, including matter intermediate between molecules and materials, namely nanomaterials. In the language of chemical informatics, the element genome stores all the information required to make molecules and materials of any kind by design, and the human genome and materials genome build

on this platform to categorize and store patterns of structure and properties in biological and abiological matter that expose function and manifest utility.

By analogy, the nanomaterials genome (NMG) can be contemplated to emerge naturally into the genomic paradigm through the information content programmed into a periodic table of nanomaterials[2] coupled with the coded interactions that direct nanomaterials self-assembly into functional and useful structures.[3] On this basis, maybe it is time to expand and transform the term genome to include the information content embedded in all forms of matter, the "omninome", Human/Nature's all-purpose genome. By expanding the informatics descriptors for the term genome, we can encompass a broader spectrum of examples of animate and inanimate matter, from the atom-up over all scales, broadly and deeply applied from idea to innovation, research to development, and industry to business. This raises a fundamental question: how are two seemingly different physical and biological systems similar to one another? That question stands at the crux of the scientific challenge to connect various forms of matter, and to understand the creative potential of these connections.

13.2 What Is a NMG?

13.2.1 Background

One breakthrough insight into the relationship between genes and the Human Genome project is that the amount of protein produced for a given amount of mRNA not only depends on the gene it is transcribed from but also on the biological environment in which it exists. In simpler words, human genome is much more than just genes.[4−6] A recent finding also indicates that even among genes themselves, cooperative mechanisms are playing important roles in cell regulation.[7] Researchers now understand that the protein is not correlated with the gene as closely and exclusively as once assumed. From this deep insight emerged the field of proteomics, which manages a large-scale study of the structures and functions of proteins. The proteome refers to an entire set of proteins produced or modified by a definite system, such as an organism.

Both genomics and proteomics are based on massive studies of the biological interactions between genes and proteins. These studies are becoming increasingly more extensive and systematized.[8–10] Recently, the Human Proteome Organization has launched an international collaborative Human Proteome Project, which aims to experimentally observe all of the proteins produced by the sequences translated from the Human Genome.[11,12]

As global activity in nanomaterials research and development continues to grow in intensity and its practical implementation becomes more pronounced, the use of new nanomaterials and new methods for making and examining them and exploring their potential uses are becoming more systematized and layered with complexity. Despite the huge diversity and complexity we have reached, simple rules that regulate the development and evolution of nanomaterials need to be re-addressed and understood better. This will help to further codify, characterize, categorize, and utilize nanomaterials.

We endeavor to systematically build on this natural pattern that has been revealed, aiming to empirically confirm and reproduce this in a predictable, and reliable, manner with the aid of the proposed NMG and its building block representations in the form of an envisioned periodic table of nanomaterials.[13,14] In this context, the National Nanomanufacturing Network in the U.S. in 2010 started a community-owned program called "Nanoinformatics" and announced the "Nanoinformatics 2020 Roadmap". The goal of this initiative is to identify, collect, validate, store, share, analyze, model, and apply nanoinformation that is deemed pertinent to the science, engineering, and medical community, to enable and enhance connections between researchers in academia, industry, and government agencies. It shows a community-wide resolution in building up a shared network and a tool that fosters efficient scientific discoveries.[15]

It is worth mentioning here that the groups of Mirkin and Tomalia[16,17] have introduced the concept of a Periodic Table of Nanomaterials previously but point out the difficulty of calling it a "Periodic Table" as long as the intrinsic imperfections of the nanomaterial building blocks with respect to their variations in their size, shape, and surface exist. However, with continuing improvements in nanochemistry synthetic methods and nanomaterial separation techniques, the ubiquitous problem of nanomaterial polydispersity could be resolved and the use of the term periodic would become more apt. Such a Periodic Table of Nanomaterials with improvements in building block perfection could ultimately prove to be a convenient instrument to help organize ones thoughts and provide a guide to what is possible.

In this chapter, we aim to provide an overview and general description of an atlas of nanomaterials, created and categorized by means of the NMG. By naming it as the NMG, we believe it is more than just a tool in the mode "nanoinformatics" will function. It urges one to understand the inner nature of the nanomaterials, sort out the connections of their chemical and physical parameters, and assess how this relates to function and how recognition of that function enables the development of value in advanced materials and biomedical applications.

13.2.2   Concept

The Federal Government of the United States used the term materials genome in its report "Materials Genome Initiative for Global Competitiveness".[18] To complete this work and realize its full potential, we propose a comprehensive, complementary program "Nanomaterials Genome Initiative (NMGI)," which operates with an equally important, urgent, and tandem mission: namely, studying the inherited behavior of complex nanometer-scale matter from its constituent nanomaterial building blocks that consist of nanocrystals, nanowires, nanotubes, and nanosheets. Essentially, these basic building blocks of nanomaterials can be used to synthesize a world of materials targeted for advanced materials and biomedical technologies.[19]

There are no new materials in the known universe that are not explicitly represented in the Periodic Table of Elements. These include an infinite number of nanomaterials that can be created and, in principle, represented in a Periodic Table of Nanomaterials.[16] Our primary goal aims to both sort out and build upon the interconnections between materials and nanomaterials. These interconnections underscore the creative act of making nanomaterials themselves, as well as overseeing their self-assembly into higher order advanced structural materials. By comparing and connecting gene and genome to nanomaterials, we mean to highlights these two important pieces of nano-information: (i) the order and hierarchy within the system from the building blocks to higher order structures, and (ii) the representations and connections between the elements (or descriptors) that define the identification of each type of nanomaterial.

In the context of biology, a gene denotes a molecular unit of heredity of a living organism, or, described in modern terms, "a locatable region of genomic sequence, corresponding to a unit of inheritance, which is associated with regulatory regions, transcribed regions, and or other functional sequence regions".[20,21] The term NMG is used here in a non-biological context.

In terms of the unit itself, a gene is defined by four basic construction subunits that encode information into DNA (adenine, guanine, cytosine, thymine [AGCT]),[22] while a nanomaterials gene is identified by the descriptors' elemental composition, structure, size, shape, surface, degree of imperfection, self-assembly, and how these are connected function and utility.

Note that in the NMG we are mixing together structure and function. Also, note that we list self-assembly as a descriptor. We think of it as a descriptor insofar as it describes the differences in the way a particular "nanomaterials gene" self-assembles. And, nanomaterials don't self-assemble in one-and-the-same way.[23-29] They all have different rates and characteristics in the process of self-assembly, and there are five major classes of self-assembly, with overlap between them: static self-assembly, dynamic self-assembly, co-assembly, hierarchical self-assembly, and directed self-assembly.[30]

In addition, by comparing materials with living organisms, we imply the formation of complex nanoscale matter that evolves from the pre-programmed assembly of nanoscale building blocks. Their functional architecture and actions behave, more or less, in a naturalistic or organic way that resembles the process in which a living organism grows and evolves. The change of a given phenotypic trait observed for an organism owes its origin to the change of the nucleic acid sequence. Similarly, the inherited characteristics and behaviors of complex nanomatter vary as the information encoded in a genomic format varies. This information flow starts from the combination and sequence of all key construction units – namely the element composition, size, shape, surface, degree of imperfection, self-assembly, function, and utility.

In other words, similar to the central principles of molecular biology, a given combination and sequence of those key construction units regulates the formation of a corresponding form of complex nanomatter with pre-programmed relations between its structure and properties, desired function, and ultimate use.

Like all analogies, regardless of how they seem to possess a "ring of truth",[31] we must rely on our skepticism and critical thinking as we move through the door of this conceptual connection. It is a truism that molecules are, for a large majority, in thermodynamic equilibrium. They are stable. In a good solvent, they do not dissociate or reassociate. If they do, we know exactly how much, and how they do it. With nanomaterials, their state is dependent on their environment. The surface ligand concentration will depend on the concentration of ligands in the solvent around them.[32-34] Are you going to standardize that as well? In high-purity conditions, on top of the problem of polydispersity, many nanoparticle systems will ripen and change size, shape, concentration, and so forth.[35-38] Now that the information you need to fully characterize one nanomaterial is beyond the means of most laboratories, how is it that anyone will be able to associate their nanomaterial to an entry in the database? Do you account for incomplete and incorrect entries? How do you do so when any of the missing characteristic can be a determinant of the property you are looking to optimize? These are all good points, especially the reality that nanomaterials do not have the perfection of molecules, and it is this that should be posed as a question and explored in-depth. These are seminal questions beckoning insightful responses by the nanoscience community.

We plan to use the NMG as a generative tool for controlling the combinatory creations of nanomaterials in a reliable and predictable manner. This tool enables us to see the order and hierarchy contained within complex nanomatter; meaning, it helps see through the apparent complexities of nanomatter, spotting the "simplicity within the complexity" of the NMG. This approach is similar to molecular biologists and geneticists using their tools to see the simple basic building blocks in the complex biological matter manifested in the Human Genome.

Expressed another way, we can glean some simple organizing principles nested within the complexity of nanomatter. It fosters an explicit and intuitive way of designing, engineering, and deploying nanomaterials, without losing its scientific precision and elegance, to observe a particular group of nanomaterials and its inner nature as a whole.

13.2.3 Realization

The Beginning: Database of Nanomaterials

In order to bring the idea to reality, academia and industry need to work collaboratively to build a central database of nanomaterials. This database might bear a resemblance to an encyclopedia or atlas, where each particular type of nanomaterial is assigned one entry, which is written according to a common standard. Each entry should be utility-directed, since the ultimate function of any material is to serve humankind. For every particular type of nanomaterial with single or multiple utility, we link with it a unique Reference Number as an identifier after we register that nanomaterial in this universal database. "Universal" means more than "central", which implies that, in the foreseeable future when this database is brought to reality, we can use the Reference Number of the nanomaterials to carry out any cross-database search. This Number can be used by international organizations, national institutions, and governmental departments such as WHO, UNESCO, NIST, NIH, FDA, DOE, DOF, DARPA, etc.

For example, an entry of titania nanoparticles would look as follows, but might be more detailed than the following lines of descriptors:

 I. {Reference Number}: BRP100XXXXXXXXX
 II. {Elemental Composition}: TiO_2
 III. {Structural Information}: Anatase
 IV. {Size (min, max /nm)}: 20, 35
 V. {Shape}: Elongated Rhombic (specific geometrical parameters, face index, etc.)
 VI. {Surface}: OA capped
 VII. {Degree of Imperfection}: (Polydispersity Index, crystallinity, defects, etc.)
 VIII. {Utility and Function}

Other descriptors can be added, as well, for purposes of clarity and innovation. Note that this information can be presented as a matrix in which all the key descriptors are named and described for easy identification, similar to how a pharmacy codifies prescribed medications. To standardize every descriptor or term and to avoid scientific mistakes and unnecessary contradictions, introducing a systematic nomenclature might be helpful as well.[39] How to implement this is still under discussion and development, and already the International Standards Organization has been working on the creation of a useful nomenclature system.[40,41]

The initiation of this database would require considerable effort since every research group in academia, government, and industry would have to help edit the entries of nanomaterials reported from their previous work (from publications, patents, etc.). It could be an open-source, operating system like Linux.[42] Once this first step is done, the rest will be a standard procedure whereby getting a Reference Number of the newly reported nanomaterials is required for publication purpose, just like crystallographers will have to get a CCDC (Cambridge Crystallographic Data Centre[43]) number before their crystal structure gets published.

It's important to mention here that every effort will be made to ensure that the NMG Database will not be used as a governance system that determines the publication of one's research. More to the point: many research scientists share a growing concern that Big Data may inadvertently create masses of useless bureaucratic workload that can hinder human creativity rather than foster it. Already in academia there is an assessment and evaluation process in place that requires researchers to simulate every trace of one's experimental idea. If the simulations don't support one's idea, then the researchers are not funded.

This Database, when reduced to practice, would be not only the central connector of all research databases and patents libraries but would act as the center of the NMGI. The peripherals of the system would take the advantage of the central database, serving all the researchers, companies, and institutions, facilitating their research and production (see the following sections). However, the building of such a database will not be easy and short-term work – it took hundreds of researcher's years to analyze a few percent of the Human Genome – we believe the start of a NMG would take longer, but in view of the burgeoning activity in the field it should be initiated in haste. Though the discussions in this handbook are based on a proposed central database, a network of de-centralized databases may also be of great help for materials scientists to share their data, as long as every node supports open-access and shares the same standards for exchanging data. Those with knowledge and expertise in computer science would be needed to build such a network.

Evolution: Beyond Database

The working principle of the NMG that endows it with scientific assets beyond a nanomaterials registry can be expressed in a type of Venn scheme. At its center point is featured a periodic table of nanomaterials, which comprises

a multi-dimensional searchable library that compiles and organizes all known forms of nano-matter in terms of its basic building block, size, shape, and surface descriptors. Each one of these building block descriptors interlaces at the next level of complexity, with all known information concerning their properties (e.g., electronic,[44−48] optical,[49−54] magnetic,[55−61] surface reactivity[62−68]), their imperfections (e.g., dopants,[69−73] non-stoichiometry,[74−79] defects,[80−90] impurities,[91−95] polydispersity,[96−100] surface irregularity[101,102]), and their interactions (e.g., self-assembly, co-assembly, directed assembly, hierarchical assembly).[23−30]

This panoply of integrated nanomaterials information provides a platform to the next level of the decision-making process where connections to nanomaterials function are made and deciphered at the highest level in terms of nanomaterials utility in advanced materials and biomedical technologies.

In this way, the NMG if programmed judiciously can be seen to be much more than just a data base. It is the way the makers, characterizers, testers, and users of nanomaterials working across the borders of the disciplines of chemistry, materials science, engineering, biology and medicine, actually 'think' and 'act' in their quest to solve a nanomaterials-based problem.

It is the creative thought process based on connecting an umbrella of nano-information that defines the various pathways to achieve an objective, and it is this entire way of thinking that can in principle be encapsulated in the concept of the NMG as illustrated in the multidimensional Venn diagram (Figure 13.1).

Cautions

As mentioned earlier, nanomaterials are not molecules. The inherent imperfection of nanomaterial building blocks with respect to variations in their size, shape, and surface will complicate the exactitude of the data base and limit

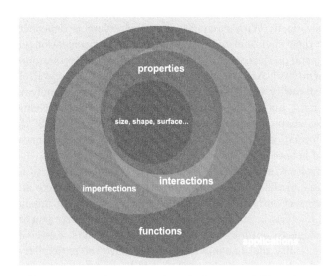

FIGURE 13.1 A conceptual, multidimensional Venn diagram showing the construction of the NMG.

its usefulness until synthetic methods have improved to the point that the perfection of nanomaterials begins to approach that of molecules. Is this a realistic expectation?

Currently, the heterogeneity of a nanomaterial makes it virtually impossible to define a precise code or even an accurate building block entry; so, it is not yet productive to make gross generalizations with so many variables contributing to nanomaterials' properties which will ultimately influence functionality and utility. And policing the suitability of an entry will be challenging in order to avoid the generation of misinformation and prevent incorrect predictions.

However, this situation is expected to improve with advances in the continuing quest to "perfect the imperfection" of nanomaterial building blocks. It is worth noting that there is an analogous problem of imperfection with the biological genome which influences and impacts the behavior of genes; so in that regard, nanomaterials are really no exception.

13.2.4 Visualization

It has been routine practice in genomic studies to use data manipulation tools and associated visualization functions. Herein, we describe an applicable method among many others.

Circos is a data visualizing tool, widely used in various fields, especially the field of bioinformatics by cell biologists. In scientific journals, it is the standard of genome plotting.[103] That said, there are plenty of other available and proven tools for plotting multi-variant data, which may also be very useful. To be clear, the NMG is totally different from the Human Genome both by definition and by nature.

How do we represent the NMG with the Circos diagram? How do we present the qualitative or quantitative descriptions of each determinant (composition, structure, size, shape, surface, degree of imperfection, function, utility) and their interconnections? Different from plotting the Human Genome, NMG plotting is multi-dimensional rather than just two-dimensional. So the challenge appears to be, how to display this multi-dimensional space in a plain, comprehensible, and user-friendly way to researchers, governments, and the public? In recent years, other related models with similar geometries have been proposed, exploring new tools for mapping the "genometries" of nanomaterials. The multi-dimensional Periodic Table of Nanomaterials and NMG are aimed to visualize, describe, and demonstrate the interconnectivity of these determinants.[14,104] These alternative models and approaches provide valuable insights into the nature of nanomaterials, and their creation involving morphogenesis and morphosynthesis.[104]

One possible layout of the Circos diagram presenting the NMG is designed as shown. Use the link (as seen in the Human Genome diagram) to connect the composition–structure pair and size–shape pair. For example, the composition–structure pairs of a group of nanoparticles can be presented as in Figure 13.2.

As shown in Figure 13.2, points on the circle stand for different elemental compositions or different structure (amorphous or various crystal structures). The grayscale lines (links) connecting composition–structure pairs stand for all nanomaterials featuring the corresponding composition–structure combinations. Similarly, we can plot a Circos diagram for size–shape combinations, as shown by Figure 13.3.

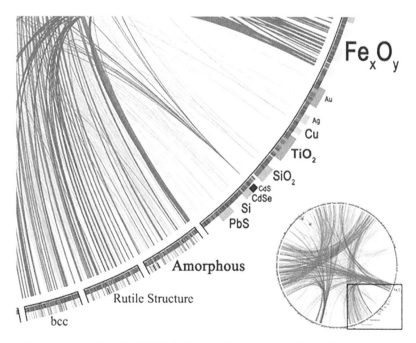

FIGURE 13.2 A Circos diagram presenting the NMG. Points on the circle stand for different elemental compositions or different structure (amorphous or various crystal structures). The grayscale lines (links) connecting composition–structure pairs stand for all nanomaterials featuring the corresponding composition–structure combinations.

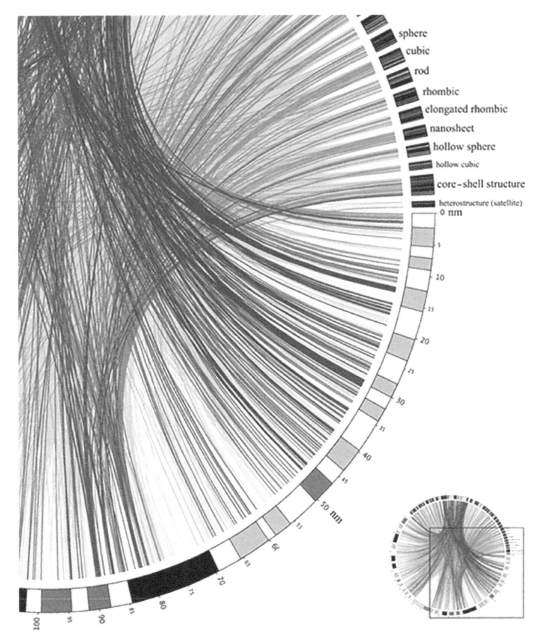

FIGURE 13.3   Another example of NMG presentation in a Circos diagram. In this case, a link is used to connect the size–shape pair.

By assembling those two circles, we form a cylinder like that shown in Figure 13.4. Imagine a 2-dimensional sheet determined by a given composition–structure–size–shape combination. This sheet contains an infinite amount of nano-materials, which are further described with different surface conditions and different functions – those two variables make these two additional dimensions.

Another variable shown in Figure 13.4 is the color (repre-sented as a grayscale gradient). It is a scale of degree of imperfection. As we know in the world of nanomaterials, the degree of imperfection is a significant property of almost all kinds of nanomaterials. In a narrow sense, it measures the fine quality of a nanomaterial, providing information about its purity, its compositional precision, and structural exactitude. In a broader sense, it stands for a group of other

descriptors, e.g., defects (including surface defects), dopants, non-stoichiometry, impurities, polydispersity index (PDI), etc. An inclusion of all these determinants depicts a gener-alized visualization of NMG for a specific range of nanoma-terials (shown in Figure 13.5).

With an infinite amount of combinations of these deter-minants, we can pinpoint any specific nanomaterial, e.g., an aggregation of rhombic-shaped titanium dioxide nanocrys-tals with an anatase structure [Tetragonal $(4/m\ 2/m\ 2/m)$], sized from 20 to 30 nm with a PDI of 1.15, passivated with oleylamine on the surface. It is presented as a point, located on the corresponding composition–structure–size–shape sheet within the Circos cylinder, with a particular color.

What we have proposed above is an idea, an example of the visualization of representative data in the

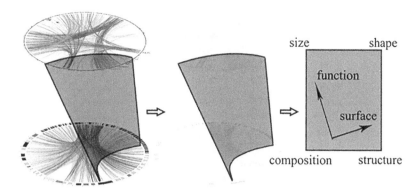

FIGURE 13.4   A 3-dimensional NMG presentation, where we can extract a 2-dimensional sheet showing composition–structure–size–shape information of a given material.

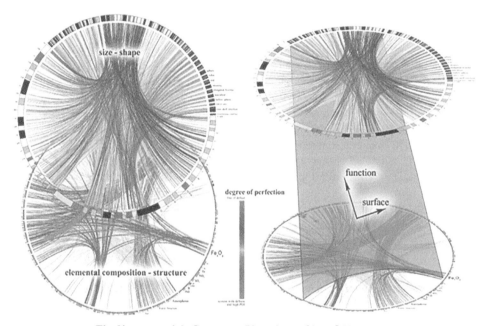

The Nanomaterials Genome—Plotted as a Circos® Kaleidoscope

FIGURE 13.5   A generalized visualization of NMG for a specific range of nanomaterials. A complete group of descriptors are included, including defects, dopants, non-stoichiometry, impurities, poly-dispersity index (PDI), etc.

NMG database. Others might think of better ways of visualization.[104] With the development of information technology and biotechnology and the promotion of open-access, the NMG system bodes well for a bright future. Tools similar to Cytoscape could be made possible.[105–108] (Cytoscape is an open source bioinformatics software platform for visualizing molecular interaction networks and integrating with gene expression profiles and other state data.)

13.3   Applications

13.3.1   More than a Database

The NMG has the potential to serve the nanoscience and materials science communities with an opportunity to speed up the development continuum of nanomaterials through the steps of discovery, structure determination and property

optimization, functionality elucidation, system design and integration, certification, etc.

When we take its application into consideration, what first comes into our mind is providing platforms for data management, analysis, and synthesis – that is usually the second step after we have a database. The world's leading sources of intelligent information can provide access portals of the database,[109] data management software, and data analysis software to customers, most of who are from academia or businesses.

How does this serve researchers and technology developers? A good example can be given in the field of materials research for sustainable energy. Suppose we have both the NMG database and the omnipotent platform provided by an intelligent source at hand (by "omnipotent", we mean the platform has integrated functions of database access, data management, and data analysis). A user in this field is capable of accessing the database, doing a fast and comprehensive search by filtering selected keywords, and drawing intuitive graphs for analysis and decision-making. These keywords might be, for example, given composition, given size range, given electronic band-gap range (if they are semiconductors for photocatalysis), required minimum quantum efficiency, required maximum cost of materials, and even device parameters like power conversion efficiency, device cost, etc. These essential details would be useful and necessary for scoping out the feasibility of manufacturing the nanomaterials for this energy-related application. They are also important for securing venture capital funds and financial support from governmental and non-governmental investment groups that require this information for their stakeholders.

Another benefit is that by filtering data as a keyword we can even see research trends on every aspect of these energy materials, including the development of intellectual property that's grown along with these innovative materials. It is true that most of these details are embedded in scientific papers and patents; ferreting them out takes some serious effort. However, if such a database of nanomaterials and a multifunctional platform are provided to researchers, it will facilitate their research and profoundly enhance their understanding in this field. Users will learn to appreciate its power just as they are grateful for things like reference management software and citation analysis reports. This will inspire researchers to choose the best candidate energy nanomaterials for evaluation and optimization. Equally important, both researchers and developers will be able to leverage the collective wisdom of this collaborative enterprise in interpreting the rich data in such a way that leads to "making uncommon connections", a process of inventing and innovating practiced by leading organizations such as The Lemelson-MIT Program.[110]

Unfortunately, the most compelling argument for this NMGI cannot just be the basic pursuit of scientific knowledge. In today's globally market-driven environment – in the life of business – there's an unspoken belief that knowledge is worthless unless it is somehow productized and monetized. This leads us to consider the commercial use of the NMG in everyday life. There are many applications for this new knowledge if we take advantage of the NMG. One example is the employment of the QR code scanner and reader for identifying a commercial product. This QR code is a unique identifier linked with the Reference Number of the corresponding nanomaterial in the NMG Database. It works as a technological brand, as important as (if not more than) the commercially registered brand. Customers with portable devices such as smartphones can access the technological background of the nanomaterial used in the product in seconds. The day when shoemakers print patent numbers on their shoes has passed with the QR code reader linking to the central database of the nanomaterials world. A buyer of water-proof shoes can easily be informed with all the knowledge he needs, e.g., technological principles, working conditions, preserving conditions, etc. Advertisements from the media can thus provide more information to interested readers by simply placing a small QR code, without taking too much costly space.

Noting the practical commercialization of this new knowledge in no way is meant to imply that we've run out of solid scientific ideas on how to make our materials better. If anything, the above earmarked example serves merely to invite the communities of potential users to weigh in on the most important products to highlight. Clearly, there are a handful of significant and urgent products to consider, such as nano-related energy products for artificial photosynthesis devices applied to climate control or health-related examples, such as nanomaterials designed for treating neurodegenerative diseases,[111] or improving oncological treatments for cancers.[112,113]

No doubt, the number of nanomaterials that will be permitted to enter the marketplace at the level we envision will be relatively small, because each material and preparation will be subjected to FDA/EPA approval; this carefully controlled process can still be made more effective and efficient. Needless to say, this approval system is essential, since it guarantees there will be strict criteria for maintaining the high quality of nanomaterials.

Simply put: It's neither necessary nor advantageous to produce a million different types of non-steroidal anti-inflammatory drugs such as Motrin and Advil. The same may hold true for a million different types of nanomaterials. No one really knows, any more than the specialists managing the Materials Genome Initiative (MGI) know what they're going to do with the massive data they're aiming to collect and analytically mine. However, this current reality shouldn't exclude the fact that there may well be many compelling examples proposed, as our international community of scientists begin to earnestly percolate on determining and prioritizing the best examples. Here and now, we openly invite our readers to suggest their outstanding examples. We welcome individuals or teams of researchers to show and tell us why their examples are important. Perhaps, there are specific problems or challenges that they see a way of solving by accessing the NMG Database.

One particularly useful application for industry is the employment of the NMG Database in market intelligence. In the area of nanomaterials and nanotechnology, data and citing from this database can enrich the background of a market intelligence report, making it more convincing for the purpose of guiding accurate and confident decision-making in determining market opportunity, market penetration strategy, and market development metrics. It saves consulting groups huge amounts of money and saves producers huge amount of time and opportunity costs as well.

We are confident that smart businessmen will come up with elegant and responsible ways of utilizing the NMG Database. Once the human ecosystem is built and the data-mining tools are working synchronously, researchers, industries, and service providers bound to it will be all in a virtuous circle.

13.3.2 Invent, Discover, Connect, and Apply

One of the major challenges for practitioners of the art of making nanomaterials with a designated purpose is the stream of consciousness that is required to actually go into the laboratory and literally know where and how to begin. It takes extraordinary background knowledge, experimental and theoretical expertise, within and across the science, engineering and biomedical disciplines, to be able to intelligently sift through increasingly large mounds of chemical, physical, and biological information on masses of nanomaterials and connect and envision how to discover the best nanomaterial targeted for a specific task. It is not just a matter of classifying nanomaterials with a particular composition, size, shape, and surface to know what it is good for, it also entails understanding how this particular combination of nanomaterial parameters work synergistically to create the properties required to provide the desired function and use of the namomaterial.

A rudimentary attempt to express this way of thinking in the form of an NMG flow diagram (Figures 13.6 and 13.7) for two classes of nanomaterials, one for a targeted cancer therapy and another for a water splitting photocatalyst, is illustrated in the following schemes. The information included in these schemes would represent the first elementary steps on a staircase of increasing complexity that would include synthesis and characterization details with associated information on chemical, physical, and biological properties. The key to the success of the NMG will be to elevate its capability beyond that of just a searchable database of codified nanomaterials to one that instead has the capacity to stimulate integrative thinking that connects, relates, explores, analyzes, and transforms diverse information, knowledge, ideas, and experiences and makes and discovers connections between them for new meaning and uses in advanced materials and biomedical science and technology. It is foreseen that the NMG will eventually be able to generate blueprints for making and assembling nanometer-scale advanced materials and biomedical systems.

13.4 Benefits and Impact

The significance of materials chemistry and materials science, materials engineering, and materials technology in both our daily life and throughout the human history is self-evident – every industrial revolution in history is preceded with or accompanied by advances in materials science. That's why the US government proposed the MGI. Our proposed NMGI starts where the other stops so to speak. It builds on the work and results of the MGI, rather than being built into it as a subdivision, which is also a possibility. The point is we aim to go beyond that point.

We have passed the era when humankind depends only on natural materials and metallurgy. Apart from further advancement and improvement of the study of traditional materials and metallurgy, materials scientists today also deal with advanced materials, paying close attention to the morphology, surface chemistry, doping, defects, and self-assembly on a micro-scale.

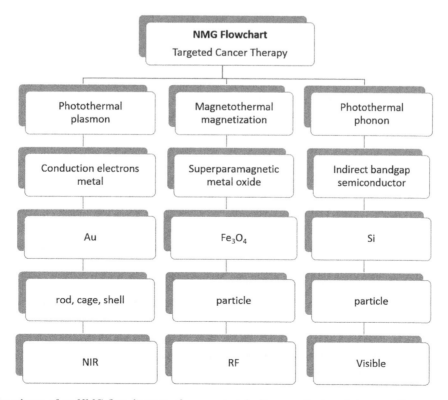

FIGURE 13.6 The scheme of an NMG flow diagram of nanomaterials discovery for targeted cancer therapy.

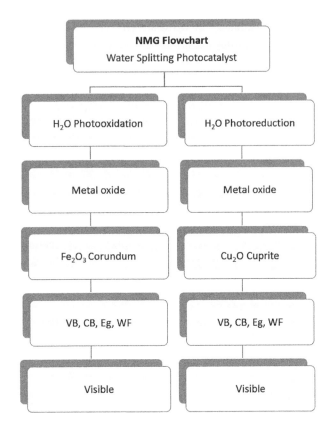

FIGURE 13.7 The scheme of an NMG flow diagram of nanomaterials discovery for water splitting.

We have entered the era of solid-state chemistry and physics where there is increasing demand for structured nanophase materials with stringent requirements of size, shape, and dimensionality, as well as the type and concentration of dopants, defects, and impurities.[114] In a broader sense, a modern view of materials should encompass nanoscale or mesoscale building blocks which self-assemble at different length scales to form hierarchical structures, and finally, to perform their functions as an integrated system. It is a general preconception but should be true from both a materials chemist's and a condensed-matter physicist's point of view. It also makes sense no matter what the approach is used: either bottom–up chemical synthesis or top–down lithography, or a creative integration of both.

Today, thinking small is the new way of thinking big. That's also why the US government in recent years has readdressed the significance of the National Nanotechnology Initiative (NNI) started during the Clinton Administration.[115] Recently, there has been some feedback on how the MGI and NNI have been changing the way materials scientists work, which infers that researchers are beginning to adopt data sharing, with collaboration between them possibly improving in the future.[116] Now that researchers in the US are going through the early stage of NNI and GMI supported by the US government, we feel it is important to establish an international collaborative cross-disciplinary program, based on our idea of mapping the NMG as delineated in this book chapter.

In April 2013, the Obama administration announced the BRAIN (Brain Research through Advancing Innovative Neurotechnologies) Initiative,[117] which challenges scientists to map the brain to better understand how we think, learn, and remember. It is regarded as the government's latest ambitious Big Science initiative. There have been three of these before the launch of the BRAIN initiative – most notably the Manhattan project, the Apollo program, and the Human Genome program, which changed the course of history. We believe that implementing the NMGI, as a comparable endeavor, could make a similarly important contribution in the history of innovation. It would benefit humankind and the globally emerging nanotechnology enterprise by promoting public health, easing the energy crisis, and contributing to economic growth and the quality of life.

Granted, the Manhattan Project changed the world because it yielded the unique capabilities of Nature by unlocking the potential of energy stored in the atom. Keep in mind that it wasn't clear or certain that nuclear energy would result from that remarkable achievement. Moreover, there were serious concerns about controlled nuclear fission (or fusion energy, for that matter). The commercialized, civilian use of nuclear energy was in many respects an afterthought rather than forethought. In fact, there were plenty of die-hard skeptics who sided with the view of the 1908 Nobel laureate in Chemistry Ernest Rutherford: "The energy produced by the breaking down of the atom is a very poor kind of thing. Anyone who expects a source of power from the transformation of these atoms is talking nonsense." And there were scientists who accepted the view of the 1923 Nobel laureate in Physics, Robert Milliken: "There is no likelihood that man can ever tap the power of the atom."

The Manhattan project suggested the possibility of realizing the Holy Grail of Energy: tapping a limitless energy supply from a completely new source that few experts knew anything about. To this day, we're still groping to figure out that stellar technology, especially controlled nuclear fusion energy systems. These continue to confound plasma physicists and fusion specialists who are building the next generation of magnetic confinement machines,[118,119] inertial confinement devices,[120] and hybrid fusion–fission reactors[121] with the prospect of producing high-temperature, commercial-grade plasmas in a self-sustaining system. And one day soon, we're counting on them succeeding.

At this nascent or seedling state of development, no one can say with any authority that the NMGI will be able to show and tell us exactly what the absolute best nanomaterial for a specific application will be, especially if that material doesn't even exist yet. There simply is no crystal ball or magic crucible that can show conclusively how to best do this and get the desired results in a reliable and efficient manner. To make good on that promise, we'd all have to be living aboard Gene Roddenbury's Star Trek: Enterprise, controlling a universe of virtual matter composed in the boundless space of the Holodeck.

The NMGI would also have a great impact on materials research, targeting two of the most urgent challenges confronting our world: the search for a sustainable source of carbon-neutral renewable energy and the improvement and protection of public health. No matter what NMGI aims to achieve or is tasked to do at any given time (e.g., designing more efficient solar cells and hydrogen fuel cells, or researching cures for common cancers, or devising new medical devices for treating neurodegenerative diseases), university, industry, and government scientists around the world will benefit from the existence of a NMGI and the scientific and technological platform that supports it.

This platform will enable researchers and developers to engage in more effective and creative collaborations that search, connect, analyze, synthesize, interpret, and share information. It will help facilitate the whole process of ideating and innovating, which drives creative and critical thinking from research to development to application. The NMG promises a highly integrated form of collaboration, communication, and cooperation throughout the global nanoscience community.

13.5 Challenges

The fundamental challenges that NMG is facing are described in Section 13.2.3.3 (Cautions), which includes advances to be made in fundamental understanding and development of applied tools. Some challenges in terms of real-world applications will be discussed here.

The first challenge that the NMG system would face is establishment. The first steps are always the most difficult to take. One of the very first steps is the initiation and sustainable maintenance of its universal database. If we recall the start of the Protein Data Bank (PDB),[122–124] we should clearly see what elements a usable database should have. 1. There are growing sets of well-defined, high-quality, complete, insightful data; 2. There are applicable and user-friendly data manipulation tools. To achieve the first objective requires researchers in the field of nanoscience share with the database a detailed description and quantitative data set for every synthetic or functional nanomaterial.

Second, international collaboration and sustained funding support from specified organizations are needed after the initial establishment of this database. Because of the fast growing need for storage and usage of this database, individuals or small organizations will soon not be able to afford such an operation. Cooperation between different research institutions and funding agencies is necessary. That requires a worldwide recognition of the NMG concept and system, its value be seen, and its principles be understood. This would be not easy for the first decade of building the NMG, with possible lack of common recognition or collaboration.

Another challenge that NMG will face after successful establishment and application of the database is further development, application, and inventions based on the database. These would require the herculean effort of NMG

developers, database managers, researchers, and administrators. Commercialization issues, regulatory affairs, and global harmonization are all involved in this long and winding road ahead. Future development could also include possible merger with the MGI, which should rely on the support from the government as well.

13.6 What Is Next?

Mapping the NMG and responsibly exploiting its attributes are core processes of building a "human ecosystem", one that respects "The Human Element," as one popular Dow Chemical advertisement wisely heralds. In this human ecosystem, researchers, product and service providers, universities, and governments are all playing an important role, forming a complex web of information with a flexible knowledge-creation and sharing system of communication. The entire nanoscience community is obliged to add to its diversity, growing this communication system while maintaining its stability and viability. By "diversity", we mean the construction of more and more applicable peripherals of the NMG, along with the products and services those peripherals provided. Once started, this human ecosystem will be able to produce a massive output in terms of scientific, technological, and socioeconomic benefits.

The NMG provides a new window through which researchers, governments, and the public can better observe, understand, and utilize the nano-world. The immediate task and responsibilities of its builders entail first figuring out how to make this window on the nano-world transparent, accessible, and meaningful for the public so they are informed about this work and learn to embrace it, rather than fear it. That is one critical reason why optimization of this system is so important and must include some educational outreach programs that engage the public in the whole discovery process.

We hope the NMG depicted in this book chapter, even in its embryonic form with its existing imperfections and limitations, will be enhanced and enriched by future generations of innovators whose mission will be to make the NMG as useful and versatile a tool as the Human Genome, which is proving to be an indispensable tool for enhancing the health and well-being of the human race.

In our view, this Big Data work is not revolutionary science. But it certainly helps accelerate and facilitate an abundance of scientific discoveries, leading to a wealth of technological innovations. For example, recent development in machine learning and its application in nanomaterials science and application-oriented engineering has started to show great promises.[125–128] With more and more data available offered by the envisioned NMG system, the technological advances and advances in fundamental understanding will be significantly accelerated. It is worth noting that early visionaries in computer sciences assumed that building computers would free our minds from the burden of doing mindlessly repetitive work, and that they would enable everyone to concentrate on what the brain does best and enjoys most: imaginative work that works our imagination. Ironically, the opposite seems to be happening, as some pioneer scientists have pointed out not as a caveat but as a cautionary measure. So we need to be cognizant of this fact and factor it into the creation of a new innovative system that cultivates human creativity and fosters wonderment – one that safeguards our sense of humanity by avoiding accidentally turning this database into little more than another ambitious information-collecting platform for serving a world of consumer automatons.

References

1. Winkler, H. Verbreitung Und Ursache Der Parthenogenesis Im Pflanzen-Und Tierreiche. **1920**.

2. Shevchenko, E. V.; Talapin, D. V.; Kotov, N. A.; O'Brien, S.; Murray, C. B. Structural diversity in binary nanoparticle superlattices. *Nature* **2006**, *439* (7072), 55–59.

3. Chenxi, Q.; Todd, S.; Ozin, G. A. Exploring the possibilities and limitations of a nanomaterials genome. *Small* **2015**, *11* (1), 64–69.

4. ENCODE. Identification and analysis of functional elements in 1% of the human genome by the ENCODE pilot project. *Nature* **2007**, *447* (7146), 799–816.

5. Ecker, J. R.; Bickmore, W. A.; Barroso, I.; Pritchard, J. K.; Gilad, Y.; Segal, E. Genomics: ENCODE explained. *Nature* **2012**, *489*, 52–55.

6. ENCODE. An integrated encyclopedia of DNA elements in the human genome. *Nature* **2012**, *489*, 57–74.

7. Kuzmin, E.; VanderSluis, B.; Wang, W.; Tan, G.; Deshpande, R.; Chen, Y.; Usaj, M.; Balint, A.; Mattiazzi Usaj, M.; van Leeuwen, J. et al. Systematic analysis of complex genetic interactions. *Science (80-.)* **2018**, *360* (6386).

8. Tyers, M.; Mann, M. From genomics to proteomics. *Nature* **2003**, *422*, 193–197.

9. de Hoog, C. L.; Mann, M. Proteomics. *Annu. Rev. Genomics Hum. Genet.* **2004**, *5*, 267–293.

10. Cox, J.; Mann, M. Is proteomics the new genomics? *Cell* **2007**, *130*, 395–398.

11. Project, H. P. The call of the human proteome. *Nat. Methods* **2010**, *7* (9), 661.

12. Legrain, P.; Aebersold, R.; Archakov, A.; Bairoch, A.; Bala, K.; Beretta, L.; Bergeron, J.; Borchers, C. H.; Corthals, G. L.; Costello, C. E. et al. The human proteome project: Current state and future direction. *Mol. Cell. Proteomics* **2011**, *10*, 1–5.

13. Amato, I. A periodic table of nanoparticles. *Chem. Eng. News* **2006**, *84*, 45–46.

14. Macfarlane, R. J.; O'Brien, M. N.; Petrosko, S. H.; Mirkin, C. A. Nucleic acid-modified nanostructures as programmable atom equivalents: Forging a new "table of elements." *Angew. Chem. Int. Ed.* **2013**, *52*, 5688–5698.

15. nanoinformatics.org http://nanoinformatics.org/nanoinformatics/index.php/Main_Page.

16. Mirkin, C. A.; Rathmann, G. B. The polyvalent gold nanoparticle conjugate-materials synthesis, biodiagnostics, and intracellular gene regulation. *MRS Bull.* **2010**, *35*, 532–539.

17. Tomalia, D. A. In quest of a systematic framework for unifying and defining nanoscience. *J. Nanoparticle Res.* **2009**, *11*, 1251–1310.

18. Council, U. S. N. S. and T. Materials Genome Initiative for Global Competitiveness. 2011. www.mgi.gov/ accessed September, 2019.

19. Ozin, G. A; Arsenault, A.; Cademartiri, L. Nanochemistry: A chemical approach to nanomaterials – 2nd Ed. *Mater. Today* **2009**, *12* (4), 46.

20. Pearson, H. What is a gene? *Nature* **2006**, *441*, 398.

21. Pennisi, E. DNA study forces rethink of what it means to be a gene. *Science (80-.)* **2007**, *316* (5831), 1556–1557.

22. Watson, J. D.; Crick, F. H. C. Molecular structure of nucleic acids. *Nature* **1953**, *171*, 737–738.

23. Li, M.; Schnablegger, H.; Mann, S. Coupled synthesis and self-assembly of nanoparticles to give structures with controlled organization. *Nature* **1999**, *402* (6760), 393–395.

24. Boal, A. K.; Ilhan, F.; Derouchey, J. E.; Thurn-Albrecht, T.; Russell, T. P.; Rotello, V. M. Self-assembly of nanoparticles into structured spherical and network aggregates. *Nature* **2000**, *404* (6779), 746–748.

25. Rabani, E.; Reichman, D. R.; Geissler, P. L.; Brus, L. E. Drying-mediated self-assembly of nanoparticles. *Nature* **2003**, *426* (6964), 271–274.

26. Grzelczak, M.; Vermant, J.; Furst, E. M.; Liz-Marzán, L. M. Directed self-assembly of nanoparticles. *ACS Nano* **2010**, *4*, 3591–3605.

27. Shenton, W.; Davis, S. A.; Mann, S. Directed self-assembly of nanoparticles into macroscopic materials using antibody-antigen recognition. *Adv. Mater.* **1999**, *11* (6), 449–452.

28. Duan, H.; Wang, D.; Kurth, D. G.; Möhwald, H. Directing self-assembly of nanoparticles at water/oil interfaces. *Angew. Chem. Int. Ed.* **2004**, *43* (42), 5639–5642.

29. Roldughin, V. I. Self-assembly of nanoparticles at interfaces. *Russ. Chem. Rev.* **2004**, *73* (2), 115–145.

30. Cademartiri, L.; Ozin, G. A.; Lehn, J. M. *Concepts of Nanochemistry*; Wiley (Weinheim, Germany), 2009.

31. Morrison, P.; Morrison, P. *The Ring of Truth*; Random House Value Publishing (New York), 1993.

32. Sperling, R. A.; Parak, W. J. Surface modification, functionalization and bioconjugation of colloidal inorganic nanoparticles. *Philos. Trans. R. Soc. A Math. Phys. Eng. Sci.* **2010**, *368* (1915), 1333–1383.

33. Lin, Y.; Skaff, H.; Emrick, T.; Dinsmore, A. D.; Russell, T. P. Nanoparticle assembly and transport at liquid-liquid interfaces. *Science (80-.)* **2003**, *299* (5604), 226–229.

34. Pfeiffer, C.; Rehbock, C.; Huhn, D.; Carrillo-Carrion, C.; de Aberasturi, D. J.; Merk, V.; Barcikowski, S.; Parak, W. J. Interaction of colloidal nanoparticles with their local environment: The (ionic) nanoenvironment around nanoparticles is different from bulk and determines the physico-chemical properties of the nanoparticles. *J. R. Soc. Interface* **2014**, *11* (96), 20130931.

35. Voorhees, P. W. The theory of ostwald ripening. *J. Stat. Phys.* **1985**, *38* (1–2), 231–252.

36. Yang, H. G.; Zeng, H. C. Preparation of hollow anatase TiO_2 nanospheres via ostwald ripening. *J. Phys. Chem. B* **2004**, *108* (10), 3492–3495.

37. Yin, Y.; Rioux, R. M.; Erdonmez, C. K.; Hughes, S.; Somorjai, G. A.; Alivisatos, A. P. Formation of hollow nanocrystals through the nanoscale kirkendall effect. *Science (80-.)* **2004**, *304* (5671), 711–714.

38. Wang, W.; Dahl, M.; Yin, Y. Hollow nanocrystals through the nanoscale kirkendall effect. *Chem. Mater.* **2013**, *25*, 1179–1189.

39. Gentleman, D. J.; Chan, W. C. W. A systematic nomenclature for codifying engineered nanostructures. *Small* **2009**, *5* (4), 426–431.

40. Chan, W. C. W. Complexities abound. *Nat. Nanotechnol.* **2013**, *8* (2), 72.

41. International Standards Organization ISO-TC229/TR11360:2010 www.iso.org/iso/home/store/catalogue_tc/catalogue_detail.htm?csnumber=55967&commid=381983.

42. Tuomi, I. *Networks of Innovation: Change and Meaning in the Age of the Internet*; Oxford University Press (New York), 2006.

43. Allen, F. H.; Bellard, S.; Brice, M. D.; Cartwright, B. A.; Doubleday, A.; Higgs, H.; Hummelink, T.; Hummelink-Peters, B. G.; Kennard, O.; Motherwell, W. D. S. The Cambridge Crystallographic Data Centre: Computer-based search, retrieval, analysis and display of information. *Acta Crystallogr. Sect. B Struct. Crystallogr. Cryst. Chem.* **1979**, *35* (10), 2331–2339.

44. McConnell, W. P.; Novak, J. P.; Brousseau, L. C.; Fuierer, R. R.; Tenent, R. C.; Feldheim, D. L. Electronic and optical properties of chemically modified metal nanoparticles and molecularly bridged nanoparticle arrays. *J. Phys. Chem. B* **2000**, *104*, 38.

45. Zhang, P.; Sham, T. K. X-ray studies of the structure and electronic behavior of alkanethiolate-capped gold nanoparticles: The interplay of size and surface effects. *Phys. Rev. Lett.* **2003**, *90* (24), 245502.

46. Liqiang, J.; Honggang, F.; Baiqi, W.; Dejun, W.; Baifu, X.; Shudan, L.; Jiazhong, S. Effects of Sn dopant on the photoinduced charge property and photocatalytic activity of TiO_2 nanoparticles. *Appl. Catal. B Environ.* **2006**, *62* (3–4), 282–291.

47. Peng, Z.; Yang, H. Designer platinum nanoparticles: Control of shape, composition in alloy, nanostructure and electrocatalytic property. *Nano Today* **2009**, *4* (2), 143–164.

48. Chen, X.; Wu, G.; Chen, J.; Chen, X.; Xie, Z.; Wang, X. Synthesis of "clean" and well-dispersive Pd nanoparticles with excellent electrocatalytic property on graphene oxide. *J. Am. Chem. Soc.* **2011**, *133* (11), 3693–3695.

49. Elghanian, R.; Storhoff, J. J.; Mucic, R. C.; Letsinger, R. L.; Mirkin, C. A. Selective colorimetric detection of polynucleotides based on the distance-dependent optical properties of gold nanoparticles. *Science (80-.)* **1997**, *277* (5329), 1078–1081.

50. Kelly, K. L.; Coronado, E.; Zhao, L. L.; Schatz, G. C. The optical properties of metal nanoparticles: The influence of size, shape, and dielectric environment. *J. Phys. Chem. B* **2003**, *107*, 3668–3677.

51. Rechberger, W.; Hohenau, A.; Leitner, A.; Krenn, J. R.; Lamprecht, B.; Aussenegg, F. R. Optical properties of two interacting gold nanoparticles. *Opt. Commun.* **2003**, *220* (1–3), 137–141.

52. Schaadt, D. M.; Feng, B.; Yu, E. T. Enhanced semiconductor optical absorption via surface plasmon excitation in metal nanoparticles. *Appl. Phys. Lett.* **2005**, *86* (6), 63106.

53. Nehl, C. L.; Liao, H.; Hafner, J. H. Optical properties of star-shaped gold nanoparticles. *Nano Lett.* **2006**, *6* (4), 683–688.

54. Huang, X.; Jain, P. K.; El-Sayed, I. H.; El-Sayed, M. A. Gold nanoparticles: Interesting optical properties and recent applications in cancer diagnostics and therapy. *Nanomedicine (Lond)* **2007**, *2* (5), 681–693.

55. Varadan, V. K.; Chen, L.; Xie, J. *Nanomedicine: Design and Applications of Magnetic Nanomaterials, Nanosensors and Nanosystems*; John Wiley & Sons (Chichester, UK), 2008.

56. Zhang, S.; Niu, H.; Cai, Y.; Zhao, X.; Shi, Y. Arsenite and arsenate adsorption on coprecipitated bimetal oxide magnetic nanomaterials: $MnFe_2O_4$ and $CoFe_2O_4$. *Chem. Eng. J.* **2010**, *158* (3), 599–607.

57. Kumar, C. S. S. R.; Mohammad, F. Magnetic nanomaterials for hyperthermia-based therapy and controlled drug delivery. *Adv. Drug Deliv. Rev.* **2011**, *63* (9), 789–808.

58. Suber, L.; Peddis, D. *Approaches to Synthesis and Characterization of Spherical and Anisometric Metal Oxide Magnetic Nanomaterials*; Wiley (Weinheim, Germany), 2011.

59. Badruddoza, A. Z. M.; Shawon, Z. B. Z.; Rahman, M. T.; Hao, K. W.; Hidajat, K.; Uddin, M. S. Ionically modified magnetic nanomaterials for arsenic and

chromium removal from water. *Chem. Eng. J.* **2013**, *225*, 607–615.

60. Rossi, L. M.; Costa, N. J. S.; Silva, F. P.; Wojcieszak, R. Magnetic nanomaterials in catalysis: Advanced catalysts for magnetic separation and beyond. *Green Chem.* **2014**, *16* (6), 2906–2933.

61. Evans, R. F. L.; Fan, W. J.; Chureemart, P.; Ostler, T. A.; Ellis, M. O. A.; Chantrell, R. W. Atomistic spin model simulations of magnetic nanomaterials. *J. Phys. Condens. Matter* **2014**, *26* (10), 103202.

62. Aballe, L.; Barinov, A.; Locatelli, A.; Heun, S.; Kiskinova, M. Tuning surface reactivity via electron quantum confinement. *Phys. Rev. Lett.* **2004**, *93* (19), 196103.

63. Gong, X.-Q.; Selloni, A. Reactivity of anatase TiO_2 nanoparticles: The role of the minority (001) surface. *J. Phys. Chem. B* **2005**, *109* (42), 19560–19562.

64. Koh, S.; Strasser, P. Electrocatalysis on bimetallic surfaces: Modifying catalytic reactivity for oxygen reduction by voltammetric surface dealloying. *J. Am. Chem. Soc.* **2007**, *129* (42), 12624–12625.

65. Tian, N.; Zhou, Z.-Y.; Sun, S.-G. Platinum metal catalysts of high-index surfaces: From single-crystal planes to electrochemically shape-controlled nanoparticles. *J. Phys. Chem. C* **2008**, *112* (50), 19801–19817.

66. Yang, H. G.; Sun, C. H.; Qiao, S. Z.; Zou, J.; Liu, G.; Smith, S. C.; Cheng, H. M.; Lu, G. Q. Anatase TiO_2 single crystals with a large percentage of reactive facets. *Nature* **2008**, *453* (7195), 638.

67. Huang, K.; McNab, I. R.; Polanyi, J. C.; Yang, J. S. Y. Adsorbate alignment in surface halogenation: Standing up is better than lying down. *Angew. Chem. Int. Ed.* **2012**, *124* (36), 9195–9199.

68. Agarwal, S.; Lefferts, L.; Mojet, B. L.; Ligthart, D. A. J.; Hensen, E. J. M.; Mitchell, D. R. G.; Erasmus, W. J.; Anderson, B. G.; Olivier, E. J.; Neethling, J. H. Exposed surfaces on shape-controlled ceria nanoparticles revealed through AC-TEM and water–gas shift reactivity. *ChemSusChem* **2013**, *6* (10), 1898–1906.

69. Stouwdam, J. W.; van Veggel, F. C. J. M. Near-infrared emission of redispersible Er3+, Nd3+, and Ho3+ doped LaF3 nanoparticles. *Nano Lett.* **2002**, *2* (7), 733–737.

70. Xu, A.-W.; Gao, Y.; Liu, H.-Q. The preparation, characterization, and their photocatalytic activities of rare-earth-doped TiO_2 nanoparticles. *J. Catal.* **2002**, *207* (2), 151–157.

71. Burda, C.; Lou, Y.; Chen, X.; Samia, A. C. S.; Stout, J.; Gole, J. L. Enhanced nitrogen doping in TiO_2 nanoparticles. *Nano Lett.* **2003**, *3* (8), 1049–1051.

72. Chen, D.; Yang, D.; Wang, Q.; Jiang, Z. Effects of boron doping on photocatalytic activity and microstructure of titanium dioxide nanoparticles. *Ind. Eng. Chem. Res.* **2006**, *45* (12), 4110–4116.

73. Wang, F.; Liu, X. Upconversion multicolor fine-tuning: Visible to near-infrared emission from lanthanide-doped NaYF4 nanoparticles. *J. Am. Chem. Soc.* **2008**, *130* (17), 5642–5643.

74. Neagu, D.; Tsekouras, G.; Miller, D. N.; Ménard, H.; Irvine, J. T. S. In situ growth of nanoparticles through control of non-stoichiometry. *Nat. Chem.* **2013**, *5* (11), 916.

75. Hernández-Alonso, M. D.; Hungría, A. B.; Martínez-Arias, A.; Coronado, J. M.; Conesa, J. C.; Soria, J.; Fernández-García, M. Confinement effects in quasi-stoichiometric CeO_2 nanoparticles. *Phys. Chem. Chem. Phys.* **2004**, *6* (13), 3524–3529.

76. Baranchikov, A. E.; Polezhaeva, O. S.; Ivanov, V. K.; Tretyakov, Y. D. Lattice expansion and oxygen non-stoichiometry of nanocrystalline ceria. *CrystEngComm* **2010**, *12* (11), 3531–3533.

77. Dai, M.; Ogawa, S.; Kameyama, T.; Okazaki, K.; Kudo, A.; Kuwabata, S.; Tsuboi, Y.; Torimoto, T. Tunable photoluminescence from the visible to near-infrared wavelength region of non-stoichiometric $AgInS_2$ nanoparticles. *J. Mater. Chem.* **2012**, *22* (25), 12851–12858.

78. Hu, W.; Tong, W.; Li, L.; Zheng, J.; Li, G. Cation non-stoichiometry in multi-component oxide nanoparticles by solution chemistry: A case study on $CaWO_4$ for tailored structural properties. *Phys. Chem. Chem. Phys.* **2011**, *13* (24), 11634–11643.

79. Dzhagan, V.; Lokteva, I.; Himcinschi, C.; Jin, X.; Kolny-Olesiak, J.; Zahn, D. R. T. Phonon raman spectra of colloidal CdTe nanocrystals: Effect of size, non-stoichiometry and ligand exchange. *Nanoscale Res. Lett.* **2011**, *6* (1), 79.

80. Gubicza, J. *Defect Structure in Nanomaterials*; Woodhead Publishing (Sawston, UK), 2012.

81. Smyth, D. M. *The Defect Chemistry of Metal Oxides*. Foreword by DM Smyth. Oxford University Press (New York). ISBN-10 0195110145. ISBN-13 9780195110142, 2000, p. 304.

82. González, G. B.; Mason, T. O.; Quintana, J. P.; Warschkow, O.; Ellis, D. E.; Hwang, J.-H.; Hodges, J. P.; Jorgensen, J. D. Defect structure studies of bulk and nano-indium-tin oxide. *J. Appl. Phys.* **2004**, *96* (7), 3912–3920.

83. Robinson, J. A.; Snow, E. S.; Bǎdescu, Ş. C.; Reinecke, T. L.; Perkins, F. K. Role of defects in single-walled carbon nanotube chemical sensors. *Nano Lett.* **2006**, *6* (8), 1747–1751.

84. Nowotny, M. K.; Sheppard, L. R.; Bak, T.; Nowotny, J. Defect chemistry of titanium dioxide. Application of defect engineering in processing of TiO2-based photocatalysts. *J. Phys. Chem. C* **2008**, *112* (14), 5275–5300.

85. Gai, P. L.; Harmer, M. A. Surface atomic defect structures and growth of gold nanorods. *Nano Lett.* **2002**, *2* (7), 771–774.

86. Hartmann, P.; Brezesinski, T.; Sann, J.; Lotnyk, A.; Eufinger, J.-P.; Kienle, L.; Janek, J. Defect chemistry of oxide nanomaterials with high surface area: Ordered mesoporous thin films of the oxygen storage catalyst CeO2–ZrO2. *ACS Nano* **2013**, *7* (4), 2999–3013.

87. Bom, D.; Andrews, R.; Jacques, D.; Anthony, J.; Chen, B.; Meier, M. S.; Selegue, J. P. Thermogravimetric analysis of the oxidation of multiwalled carbon nanotubes: Evidence for the role of defect sites in carbon nanotube chemistry. *Nano Lett.* **2002**, *2* (6), 615–619.

88. Yu, Z.; Pan, Y.; Shen, Y.; Wang, Z.; Ong, Z.-Y.; Xu, T.; Xin, R.; Pan, L.; Wang, B.; Sun, L. Towards intrinsic charge transport in monolayer molybdenum disulfide by defect and interface engineering. *Nat. Commun.* **2014**, *5*, 5290.

89. Switzer, J. A.; Hung, C.-J.; Breyfogle, B. E.; Shumsky, M. G.; Van Leeuwen, R.; Golden, T. D. Electrodeposited defect chemistry superlattices. *Science (80-.)* **1994**, *264* (5165), 1573–1576.

90. McDonnell, S.; Addou, R.; Buie, C.; Wallace, R. M.; Hinkle, C. L. Defect-dominated doping and contact resistance in MoS2. *ACS Nano* **2014**, *8* (3), 2880–2888.

91. Pumera, M.; Ambrosi, A.; Chng, E. L. K. Impurities in graphenes and carbon nanotubes and their influence on the redox properties. *Chem. Sci.* **2012**, *3* (12), 3347–3355.

92. Addou, R.; McDonnell, S.; Barrera, D.; Guo, Z.; Azcatl, A.; Wang, J.; Zhu, H.; Hinkle, C. L.; Quevedo-Lopez, M.; Alshareef, H. N. Impurities and electronic property variations of natural MoS_2 crystal surfaces. *ACS Nano* **2015**, *9* (9), 9124–9133.

93. Chen, D.; Wang, Y. Impurity doping: A novel strategy for controllable synthesis of functional lanthanide nanomaterials. *Nanoscale* **2013**, *5* (11), 4621–4637.

94. Hull, M. S.; Kennedy, A. J.; Steevens, J. A.; Bednar, A. J.; Weiss Charles, A. J.; Vikesland, P. J. Release of metal impurities from carbon nanomaterials influences aquatic toxicity. *Environ. Sci. Technol.* **2009**, *43* (11), 4169–4174.

95. Lu, F.; Zhou, M.; Li, W.; Weng, Q.; Li, C.; Xue, Y.; Jiang, X.; Zeng, X.; Bando, Y.; Golberg, D. Engineering sulfur vacancies and impurities in $NiCo_2S_4$ nanostructures toward optimal supercapacitive performance. *Nano Energy* **2016**, *26*, 313–323.

96. Ohara, P. C.; Leff, D. V.; Heath, J. R.; Gelbart, W. M. Crystallization of opals from polydisperse nanoparticles. *Phys. Rev. Lett.* **1995**, *75* (19), 3466.

97. Gaumet, M.; Vargas, A.; Gurny, R.; Delie, F. Nanoparticles for drug delivery: The need for precision in reporting particle size parameters. *Eur. J. Pharm. Biopharm.* **2008**, *69* (1), 1–9.

98. Lim, J.; Eggeman, A.; Lanni, F.; Tilton, R. D.; Majetich, S. A. Synthesis and single-particle optical detection of low-polydispersity plasmonic-superparamagnetic nanoparticles. *Adv. Mater.* **2008**, *20* (9), 1721–1726.

99. Xia, Y.; Nguyen, T. D.; Yang, M.; Lee, B.; Santos, A.; Podsiadlo, P.; Tang, Z.; Glotzer, S. C.; Kotov, N. A. Self-assembly of self-limiting monodisperse supraparticles from polydisperse nanoparticles. *Nat. Nanotechnol.* **2011**, *6* (9), 580.

100. Tomaszewska, E.; Soliwoda, K.; Kadziola, K.; Tkacz-Szczesna, B.; Celichowski, G.; Cichomski, M.; Szmaja, W.; Grobelny, J. Detection limits of DLS and UV-Vis spectroscopy in characterization of polydisperse nanoparticles colloids. *J. Nanomater.* **2013**, *2013*, 60.

101. Nie, S.; Emory, S. R. Probing single molecules and single nanoparticles by surface-enhanced raman scattering. *Science (80-.)* **1997**, *275* (5303), 1102–1106.

102. Sun, Y.; Xia, Y. Shape-controlled synthesis of gold and silver nanoparticles. *Science (80-.)* **2002**, *298* (5601), 2176–2179.

103. Circos. http://circos.ca/ accessed September, 2019.

104. Ozin, G. A. Nanomaterials Kaleidoscope: Building a Nanochemistry Periodic Table. www.advancedsciencenews.com/nanomaterials-kaleidoscope-building-a-nanochemistry-periodic-table/ accessed September, 2019.

105. Cytoscape Consortium. What is Cytoscape? https://cytoscape.org/ accessed September, 2019.

106. Saito, R.; Smoot, M. E.; Ono, K.; Ruscheinski, J.; Wang, P. L.; Lotia, S.; Pico, A. R.; Bader, G. D.; Ideker, T. A travel guide to cytoscape plugins. *Nat. Methods* **2012**, *9*, 1069–1076.

107. Shannon, P.; Markiel, A.; Ozier, O.; Baliga, N. S.; Wang, J. T.; Ramage, D.; Amin, N.; Schwikowski, B.; Ideker, T. Cytoscape: A software environment for integrated models of biomolecular interaction networks. *Genome Res.* **2003**, *13* (11), 2498–2504.

108. Cline, M. S.; Smoot, M.; Cerami, E.; Kuchinsky, A.; Landys, N.; Workman, C.; Christmas, R.; Avila-Campilo, I.; Creech, M.; Gross, B. et al. Integration of biological networks and gene expression data using cytoscape. *Nat. Protoc.* **2007**, *2* (10), 2366–2382.

109. Zhu, D.; Porter, A. L. Automated extraction and visualization of information for technological intelligence and forecasting. *Technol. Forecast. Soc. Change* **2002**, *69* (5), 495–506.

110. Thursby, J. G. INTELLECTUAL PROPERTY: Enhanced: University Licensing and the Bayh-Dole Act. *Science* **2003**, *301*, 1052.

111. Gilmore, J. L.; Yi, X.; Quan, L.; Kabanov, A. V. Novel nanomaterials for clinical neuroscience. *J. NeuroImmune Pharmacol.* **2008**, *3* (2), 83–94.

112. Yezhelyev, M. V.; Gao, X.; Xing, Y.; Al-Hajj, A.; Nie, S.; O'Regan, R. M. Emerging use of nanoparticles in diagnosis and treatment of breast cancer. *Lancet Oncol.* **2006**, *7* (8), 657–667.

113. Davis, M. E.; Shin, D. M. Nanoparticle therapeutics: An emerging treatment modality for cancer. *Nat. Rev. Drug Discov.* **2008**, *7* (9), 771.

114. Ozin, G. A. Nanochemistry: Synthesis in diminishing dimensions. *Adv. Mater.* **1992**, *4* (10), 612–649.

115. Roco, M. C.; Mirkin, C. A.; Hersam, M. C. Nanotechnology research directions for societal needs in 2020: Summary of International Study. *J. Nanoparticle Res.* **2011**, *13*, 897–919.

116. Tinkle, S.; McDowell, D. L.; Barnard, A.; Gygi, F.; Littlewood, P. B. Sharing data in materials science. *Nature* **2013**, *503*, 463–464.

117. Insel, T. R.; Landis, S. C.; Collins, F. S. The NIH brain initiative. *Science (80-.)* **2013**, *340* (6133), 687–688.

118. Furth, H. P. Magnetic confinement fusion. *Science (80-.)* **1990**, *249* (4976), 1522–1527.

119. Ongena, J.; Koch, R.; Wolf, R.; Zohm, H. Magnetic-confinement fusion. *Nat. Phys.* **2016**, *12* (5), 398.

120. Duderstadt, J. J.; Moses, G. A. *Inertial Confinement Fusion*; John Wiley & Sons (New York), 1982.

121. Leonard Jr, B. R. A review of fusion-fission (hybrid) concepts. *Nucl. Technol.* **1973**, *20* (3), 161–178.

122. Berman, H. M.; Westbrook, J.; Feng, Z.; Gilliland, G.; Bhat, T. N.; Weissig, H.; Shindyalov, I. N.; Bourne, P. E. The Protein Data Bank. *Nucleic Acids Res.* **2000**, *28* (1), 235–242.

123. Bernstein, F. C.; Koetzle, T. F.; Williams, G. J. B.; Meyer, E. F.; Brice, M. D.; Rodgers, J. R.; Kennard, O.; Shimanouchi, T.; Tasumi, M. The Protein Data Bank: A computer-based archival file for macromolecular structures. *Arch. Biochem. Biophys.* **1978**, *185* (2), 584–591.

124. Berman, H.; Henrick, K.; Nakamura, H. Announcing the worldwide Protein Data Bank. *Nat. Struct. Biol.* **2003**, *10*, 980.

125. Copp, S. M.; Bogdanov, P.; Debord, M.; Singh, A.; Gwinn, E. Base motif recognition and design of DNA templates for fluorescent silver clusters by machine learning. *Adv. Mater.* **2014**, *26* (33), 5839–5845.

126. Ramakrishnan, R.; Dral, P. O.; Rupp, M.; Von Lilienfeld, O. A. Big data meets quantum chemistry approximations: The Δ-machine learning approach. *J. Chem. Theory Comput.* **2015**, *11* (5), 2087–2096.

127. Mitchell, J. B. O. Machine learning methods in chemoinformatics. *Wiley Interdiscip. Rev. Comput. Mol. Sci.* **2014**, *4* (5), 468–481.

128. Mueller, T.; Kusne, A. G.; Ramprasad, R. Machine learning in materials science. *Rev. Comput. Chem.* **2016**, *29* (1), 186–273.

14

Nanoscience of Large Immune Proteins

Alexey Ferapontov,
Kristian Juul-Madsen, and
Thomas Vorup-Jensen
Aarhus University

14.1 Introduction

The immune system protects the body against many infectious threats through the function of a group of large, soluble proteins in the blood. A critical part of immunity is the select recognition of potential pathogens while avoiding recognition of the body's own structures. To understand how this balance is kept, research in recent years has indicated that many methodologies and concepts of nanoscience are helpful, and indeed, might currently be our only way to handle structural information on the relevant protein structures. Many immunoactive plasma proteins reach cross-sectional diameters of approximately 50 nm, and even beyond (Vorup-Jensen and Boesen 2011). Often their structure is made from several identical structural units, each containing a ligand-binding site. By oligomerization of these units into a single assembly particle, the size becomes comparable to a nanoparticle (NP). In this range, methodologies such as light scattering and certain types of scanning probe microscopy can be used to characterize these large structures in a functionally meaningful way. Indeed, our understanding of the function has been greatly enabled by combining the earlier provided structural information at the atomic level with modeling of the entire molecular structure from, in particular, small-angle X-ray scattering (SAXS) data. Atomic force microscopy (AFM) is one of only a few techniques permitting direct insight into the structure of surface-bound proteins in their hydrated, physiologic state. Under these conditions, insights on conformational regulation associated with surface binding can be obtained from comparison with solution structures obtained by, e.g., SAXS. A perhaps more surprising aspect is the observation that nanometer-scale structure of these proteins,

notably their symmetries, can be related in a non-trivial way to their target recognition (Vorup-Jensen 2012, 2016). Chemical binding of oligomeric proteins involves complex thermodynamic aspects (Kitov and Bundle 2003), which enable almost irreversible binding to ligand-dense surfaces (Gjelstrup et al. 2012). Biosensor technologies such as surface plasmon resonance (SPR) easily record such binding, but only recent advances have permitted a clearer interpretation of the complex ligand binding kinetics (Vorup-Jensen 2016).

Here, we take the reader through various techniques from the nanoscience toolbox to show how they have enabled detailed insights on the immunobiology of large proteins. In each case, we make an emphasis of both technological advances as well as scientific achievements pertinent to immunology.

14.2 Overview of Select Immunoactive Proteins and Challenges in Understanding Their Function

It is, of course, beyond the scope of the present chapter to present an exhaustive analysis of the proteins and protein structures of the immune system, and the reader is referred to leading textbooks in the field for a more comprehensive coverage (Murphy et al. 2017). Mainly to help the understanding of the topics brought up in Section 14.3, a brief introduction to certain soluble plasma proteins of the immune system is made below, namely the large proteins of the complement system, immunoglobulins, and lectins.

As a very broad generalization, one of the main challenges of the immune system is distinguishing between chemical structures of biological macromolecules of a "self" origin and those of a "non-self" origin, and hence belonging to microbial synthesis by, e.g., viruses or bacteria, or even mutated cancer cells.

The sources of immune recognition of the non-self structures have been divided into innate and adaptive immunity.

Innate immunity relies on proteins encoded by germline genes with a structure that permits these proteins to bind motifs distinctive of microbial organisms. A classic example is the human receptor expressed for bacterial flagellin, known as Toll-like receptor (TLR)-5, on the surface of leukocytes. This principle is known as "pattern recognition" (Janeway 1989). Certain yeast species express a high level of carbohydrate, e.g., mannose, in their cell wall. As a result, it is possible to identify receptors and soluble molecule of the innate immune system, e.g., macrophage mannose receptor and mannan-binding lectin (MBL), respectively, which will recognise the microbial carbohydrate structure, in particular those typical of fungi (Hoffmann et al. 1999). However, as an important caveat, mannose is also expressed on human cells, although typically in much lower amounts than seen in fungal cells. While the TLRs were evolutionary selected for the ability to bind microbial non-self molecules, with no equivalents expressed by human cells, for instance MBL seems to rely on a structural aspect of non-self, where the structural presentation, but not the synthesis, differs between friend and foe. In a way and as discussed further below, this makes the notion of pattern recognition even more justified, now with the meaning of a certain geometric presentation of ligands. This type of immune recognition is tightly linked with chemical binding by large immune proteins, notably those with a homooligomeric structure (Vorup-Jensen 2012). Research into the significance of "topological pattern recognition" is appropriately addressed with instrumentation and, perhaps more importantly, the principles of nanoscience.

As suggested by the name, the adaptive immunity response is formed specifically to target molecules that have microbes intruding or, for the matter, any chemical structure not being an endogenous product of the body. The cellular basis of these mechanisms is the lymphocytic compartment of the immune system, namely B and T lymphocytes. Both cell types express receptors for recognising foreign substances, here referred to as antigens. The gene *loci* responsible for encoding the antigen-recognising part of these receptors undergo so-called somatic recombination in a *quasi*-random way, which enables the encoding of a very large number of antigen receptors. Each lymphocyte carries only one type of antigen receptor. Those cells successfully recognizing a non-self antigen are then subsequently expanded to produce a number of cells with identical antigen-recognizing properties sufficient for an effective immune response toward the carriers, e.g., microbes,

of that antigen. By prior selection, in particular the T lymphocytes are prevented from reacting with self-antigens, which otherwise would cause autoimmune disease, as is nevertheless sometimes encountered in patients. The selection process of B lymphocytes is less clearly understood. However, a major role of B lymphocytes is the differentiation to plasma cells, which, in turn, are cellular "factories" of antibodies. For many antigens, albeit not all, such differentiation requires the help of T lymphocytes, which, in principle, would limit antibody development to only non-self antigens. Unlike the innate immune response, adaptive immune response contains a memory of past exposures in the form of certain lymphocyte subsets. Upon reexposure to the same antigen, these cellular compartments will rapidly enable once again a cellular response and formation of antibodies to the incoming threat.

14.2.1 The Complement System

The complement system comprises of a group of soluble (plasma) proteins, which largely act by forming a deposit on target surfaces such as those presented by microbes, including both bacteria and viruses (Vorup-Jensen and Boesen 2011). As also discussed in Sections 14.2.2 and 14.2.3, the complement system involves components of the adaptive immune system through the involvement of antibodies, but may also be activated through the recognition of carbohydrate patterns expressed by certain microbes. In several steps, the complement deposition is initiated by proteolytic cleavage of the complement. Upon such deposition, the microbe is labelled for phagocytic uptake (sometimes referred to as opsonisation) through complement receptors (CR) expressed on, for instance, macrophages and neutrophil granulocytes. Another outcome is lysis through pore formation in the microbial membrane by complement proteins, which may also serve to clear the bloodstream of infectious pathogens. Finally, complement activation produces small protein fragments named anaphylatoxins. Again through binding of CRs, the anaphylatoxins mediate several proinflammatory responses such as activation of the endothelium and mast cells, leading to extravasation of leukocytes and vasodilation. The detailed enzymatic mechanisms and medical implications have been the topic of several excellent reviews (Ricklin et al. 2010; Ricklin and Lambris 2007; Ricklin et al. 2018). Briefly, the activation of the complement system proceeds through three pathways. The classic pathway follows deposition of either immunoglobulins of the IgG or IgM isotype on target surfaces expressing antigens recognized by these antibodies. The enzymatic machinery responsible for the activation involves the C1 complex, with the C1q subcomponent directly binding the antibodies and C1q-associated C1serine proteases (C1r and C1s) responsible for further downstream activation of C4 and C2, which form an enzymatic convertase that will enable the covalent binding of C3 to the surface. In several proteolytic forms, fragments of C3 serve as ligands for CRs that enables phagocytosis.

The alternative pathway proceeds by the spontaneous deposition of C3 onto surfaces, followed by convertase formation through the association with a fragment of complement factor B. Most host cells will be able rapidly to degrade this convertase, preventing more complement activation. By contrast, some microbial surfaces are not able to clear the convertase, permitting a positive-loop amplification of C3 deposition. Finally, a third route of complement activation is the so-called lectin pathway. Originally, this pathway was thought to involve MBL with its associated proteases in a way resembling the C1 complex, i.e., with activation of C2 and C4 to form C3 convertases. More recent research has demonstrated, however, that this pathway involves several distinct initiator complexes; some, like MBL, with carbohydrates as ligands, others, such as the ficolins, with acetylated compounds as their major ligands (Holmskov et al. 2003). Pattern recognition remains a unifying principle for complement activation through these molecules with the assumption that the mentioned ligands are presented on microbial surfaces in ways different from their location on host cells. Below, a more detailed description is given of the initiation of the complement cascade through these molecules.

14.2.2   Immunglobulins: The Case of IgM

The structural units of any immunoglobulin in the human body are two heavy chains and two light chains, so named owing to their molecular masses, and covalently associated with disulfide bridges between the heavy and light chains, and between the heavy chains. Both chains contain a variable domain, which contributes to forming the antigen-binding sites. The constant domains are responsible for several immunological effector mechanisms of the immunoglobulins, including complement activations. The body produces different heavy-chain isoforms of immunoglobulins, organized into the five families of IgA, IgD, IgE, IgG and IgM immunoglobulins. Although the structural frame of heavy and light chains is conserved throughout these families, considerable differences exist in both tissue distribution and immunological effector mechanisms between the isotype families. Even within the family of IgG immunoglobulins (IgG1-4), significant differences exist with regard to what effector mechanism is supported. The versatile IgG1 activates the complement system and is the ligand for several receptors on leukocytes. By contrast, IgG4 has been reported to be incapable of activating the complement system and fails to bind receptors on leukocytes (Murphy et al. 2017).

IgG1, among the smallest soluble immunoglobulins with two heavy chains and two lights chains, has an M_r of 150,000. In comparison with this and the other immunoglobulins, IgM contains at least 5 times this number of chains, with the further addition of the J chain (not found in IgG) and possesses an M_r of ~800,000, thereby being the largest of the immunoglobulins. It exists also in a hexameric form with an M_r of ~950,000. Already in the

1960s it was clear from EM studies that the cross-sectional diameter is about 35–40 nm (Feinstein and Munn 1966). From further EM studies, it became clear that the molecule in its antigen-bound state make take a "staple" conformation (Feinstein and Munn 1969). A subsequent study by small-angle neutron scattering established that the soluble form, by contrast, was flat. Moreover, with the binding sites for C1q known, it was possible to rationalize that the C1 complex would bind the staple conformation, but not the flat conformation in solution (Perkins et al. 1991). This elegantly explains why soluble IgM will not activate the complement system, while antigen-bound IgM will. In Section 14.3.1.1, we discuss several applications of nanotechnology to address investigations into the conformational regulation of IgM complement activation.

14.2.3   The Concept of Protein Ultrastructure of Large Immunoactive Proteins

Both in the case of IgM and several soluble plasma lectins such as MBL, the proteins are homooligomers. This implies that they contain multiple identical binding sites (or, in the case of IgM, properly the antigen-combining sites). Understanding the chemical binding reaction of such molecules has been a research topic in immunology for almost a century (Vorup-Jensen 2012). Briefly, two aspects are logically important, namely the ultrastructure of large immunoactive proteins and the topology of the ligand-presenting surface. As will be explained in Section 14.3, both aspects are tightly linked with nanoscience as part of protein chemistry. The concepts of ultrastructure and topology are, however, by no means standard topics in immunology and need, at least for this reason, some definition.

Protein structure has typically been divided into primary structure, or sequence of amino acid residues in the protein, secondary structure, essentially reflecting the propensity of that sequence to take either an alpha helical, beta stranded, or unfolded structure of the polypeptide chain, tertiary structure, describing spatial organization of the polypeptide chain into, e.g., a globular domain structure, and finally, quaternary structure, reflecting the organization of more polypeptide chains into a single structure (Linderstrøm-Lang 1952). While the borders of the quaternary structure, in principle, can cover even very large protein complexes, the structure of both IgM and MBL highlights some limitations of the classic definition. As noted above, IgM is built from units of the fundamental immunoglobulin structure of two heavy and two light chains. The unit itself contains a protein architecture with all levels of protein structure, from the primary to the quaternary structure. MBL also contains structural units made from the folded MBL chain, which at the quaternary level a forms trimeric structure, referred to as MBL_3 (Gjelstrup et al. 2012). The fully assembled, polydisperse molecules apparently contain at least 3–8 of these units, and maybe even large assemblies (Gjelstrup et al. 2012; Jensenius et al. 2009). These assemblies are referred to as $3 \times MBL_3$–$8 \times MBL_3$, in this way clearly highlighting the nature of the oligomeric structure (Vorup-Jensen 2012). It would seem awkward to mix the quaternary level of the MBL_3 structure with the oligomerization of these units. Hence, we prefer to refer to the oligomerization of the structural unit as forming the ultrastructure of the MBL and IgM.

The role of surface topology in the binding of large, polyvalent proteins is rarely mentioned in immunology textbooks. This is disappointing since the field of nano-microbiology is uncovering a surprising level of detail in microbial surfaces, with structural features at the nanometer scale (Xiao and Dufrene 2016). In line with the definitions above, these structures could well be termed ultrastructures as they often originate from the bottom–up synthesis of the cell wall of the fungi and bacteria, involving polymerization of biomacromolecules. As one well-characterized example, the fungus, and human pathogen, *Aspergillus fumigatus* creates a highly intricate pattern of so-called rodlets in its cell wall, resembling the texture of woven fabric. This pattern is dynamic and disappears in germinating conidia (Dague et al. 2008). With a spacing of ∼10 nm between the repeating units, it seems likely that binding by immunoactive proteins could be affected by this pattern. The bacterium *Staphylococcus aureus* also creates a cell wall with a distinct topology. In this case, the cell wall synthesis is seen as concentric circles, which can be visualized by AFM (Touhami et al. 2004). For both of these examples, it seems a reasonable expectation that the presentation of ligands, or epitopes, to molecules such as MBL or IgM on the microbial surfaces follows the ultrastructural features of the surfaces. Polyvalent interactions are sensitive to the geometry of the ligand presentation in a non-trivial way, which could have ramifications with regard to the patterns in the microbial surfaces now seen with AFM.

A simpler, yet not less important aspect of surface topology in biology, is curvature. In planktonic cultures of *S. aureus*, it possible to demonstrate the presence of *S. aureus*-derived particles with a mean radius, $<r>$, of ∼30 nm (Pedersen et al. 2010). It seems likely that such material is also released to the surroundings when the bacteria grows *in vivo*, but it has remained an enigma what benefit the bacterium gains from producing this material. One topological parameter of perfect sphere surfaces is the Gaussian curvature of circles defined as $1/<r>$. With the almost spherical appearance of intact *S. aureus* cells and a radius of 500 nm, the curvature of the released material is 17-fold steeper. In the case of both the released particles and intact *S. aureus*, the assumption of a perfect spherical topology is only a useful approximation. Nevertheless, it seems clear that the consequence of binding large proteins to these two types of surfaces impacts the ultrastructure differently in consequence of the difference in curvature. As just one example of a principle probably widely applicable, the consequences of IgM binding to nanoparticulate material is discussed further below.

14.3 Case Studies in Unravelling the Function of Large Immune Active Proteins by Nanotechnology

Below, an outline is provided of studies making use of SAXS, AFM, and SPR for researching connections between nanometer-scale protein ultrastructures and biological functions of large immune proteins. The examples from inter-disciplinary studies highlight that a novel understanding of such large proteins in the immune system is well explored and conceptualized by taking a nanoscience approach. None of these methodologies is entirely new, but nor are their potentials exhausted. For this reason, in each case a section is included briefly outlining some potential future developments and uses.

14.3.1 Using SAXS for Probing Large and Small Oligomeric Complexes

Biological SAXS experiments make use of the interaction between X-ray photons and the outermost electrons of large molecules such as proteins (Oliveira et al. 2009). The momentum of photons is transferred to electrons, which leads to Compton scattering of the X-ray beam with a mathematical relation between the intensity of the scattering and momentum transfer. Since the electrons most often are asymmetrically distributed, the scattering intensity plotted against the momentum transfer becomes a source of information on the topography of the protein envelope. From constructive and destructive interference between the scattered X-ray waves, the SAXS spectrum reflects the internal distance between the electrons in the envelope.

The use of SAXS to retrieve the ultrastructure of large proteins relevant to the functions of the complement system is far from new. In particular, Perkins et al. have published several reports, including some on the structure of IgM and the C1 component (Perkins et al. 1991; Vorup-Jensen and Boesen 2011). Certainly, these reports have helped the understanding of the conformational regulation of especially IgM in terms of activating the complement system and are some examples on how technological developments have been key to understanding immunology at the molecular level (Sim et al. 2016). Most of this the initial work with SAXS and large immune proteins was only helped by low-resolution EM images and lack of computational power limited the precision of *ab initio* structure construction. Even so, much of the data has stood the test of time (Vorup-Jensen and Boesen 2011). Meanwhile, of course, SAXS analysis has been greatly helped by the availability of protein structure determined by high-resolution X-ray crystallography. Here, especially the software packages by Svergun et al. have helped make a direct connection between the crystallographic structures and the molecular envelopes determined by SAXS (Svergun 2010).

Examples of SAXS-Based Solution Structure of Large Immunoactive Protein Oligomers

Structural investigations on MBL oligomers serve as a good example of the powers of SAXS to deal with large protein structures.

In a paper by Dong et al., recombinant human MBL was subjected to size fractionation by gel permeation chromatography to isolate smaller ($\sim 3 \times MBL_3$) oligomers (Dong et al. 2007). An initial *ab initio* modeling revealed a highly ramified structure with a striking rotation symmetrical organization of the structural units. The analysis was taken further with a few assumptions on the collagenous region, modeled from available crystallographic data, and similarly for the trimers of carbohydrate recognition domains, it was possible to generate a reliable model of the structural unit. With subsequent relaxation of some of the structure as well as an imposed rotational symmetry of the structural subunits, the analysis produced a convincing structure of the oligomers and probably the first experimentally based three-dimensional structure of an intact MBL oligomer. Later work on rat MBL, which is considerably less polydisperse, largely seemed to confirm this structural arrangement (Miller et al. 2012). Unfortunately, here the fitting of the SAXS spectrum was in a narrow scattering range. It was contended that the imposed symmetrical organization was chosen *ad hoc* for the human MBL (Dong et al. 2007), but both of these reports largely agreed on such an arrangement.

Use of the low-resolution SAXS, the structure of $3 \times MBL_3$ was also mapped by applying the same material to ligand-coated mica surfaces, followed by AFM analysis of the resulting sizes of the bound MBL oligomers (Dong et al. 2007). Compared with the SAXS-determined structure in solution, the ligand surface-bound oligomers showed a larger cross-sectional diameter at 60 nm, still in liquid, while the diameter in solution was only 40 nm. This is consistent with a conformational change stretching the oligomers upon binding surfaces (Figure 14.1). While this in principle may offer mechanistic insight into how such binding triggers complement activation, these studies did not provide any direct link to function. As discussed in Section 14.3.2 on conformational changes in IgM, other techniques are necessary to make such a link.

It is of considerable interest to the usefulness of SAXS that methodologies permitting analyses of structural variation have been introduced. In the first place, SAXS is a solution-based measurement, which obviously asks the question if analysis of fluctuations in the soluble state of large molecules is permitted. Gjelstrup et al. (2012) extended the analysis by Dong et al., mainly by using the algorithm developed by Bernado et al. (2007). By creating an ensemble of $3 \times MBL_3$ and $6 \times MBL_3$, it was possible to estimate their relative abundance. It was also possible to gain information on the flexibility of the oligomers, by permitting a best fit of the oligomeric organization of the ramified structure. These data showed some variation in the relative positions of the structural units in each oligomer.

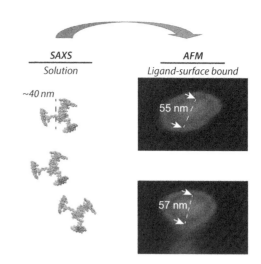

FIGURE 14.1 Conformational change in MBL following binding to a ligand-coated surface. The cross-sectional diameter of $3 \times MBL_3$ oligomers was determined in solution by SAXS and on a mannose-amine coated mica surface by AFM. Comparison of this simple structural information showed a widening of the surface-bound oligomer. (Data from Dong et al. 2007.)

Future Analyses with SAXS of Polydisperse Oligomeric Proteins

As described above, within the past decade it has been shown several time that SAXS is an appropriate tool for getting information on the structure of polydisperse oligomeric proteins. In particular, the algorithm by Bernado et al. (2007) seems to extend the technical possibilities significantly and deserves to be used more extensively for such characterization. The work on MBL is illustrative with regard to the actual potentials of this methodology. On one hand, the distribution of smaller and larger oligomers was quantified. On the other hand, it is clear, with the highly heterogeneous composition of the oligomers, that the analysis was not able to provide an exhaustive analysis. There are few competing technologies for making similar analyses. In principle, the tried and tested gel permeation chromatography could provide knowledge on size distribution; however, to our knowledge no column would adequately fractionate the larger MBL oligomers.

The technical possibilities and limitations become all the more interesting when considering the large number of proteins where SAXS could be applied to oligomeric proteins. The number of known members of the collectin family, i.e., proteins such as MBL with a collagen-like region and a lectin domain, was increasing until very recently (Hansen et al. 2016). Also, their polydisperse character has been under investigations, albeit predominantly with more standard techniques in the protein biochemistry field. We think there are ample opportunities for making headway in understanding the nanometer-scaled ultrastructures of these large molecules of the immune system and hence their functions, as further discussed in Section 14.3.3.

14.3.2 Conformational Changes in IgM and Complement Activation: NPs as Tools for Probing the Influence of Curvature

In the field of nanomedicine, NPs are often grown for their capacity to serve as engineered vehicles to enable better drug delivery or act as contrast reagents for *in vivo* scanning and, in this sense, they are tools. One exception is the concern over so-called nanotoxicity, where the submicron size of these particles for some has raised concerns over the unknown toxicological parameters, including those originating from provocations of the immune system (Moghimi et al. 2012). Both as tools and health threats, however, the rational for explaining why especially nanoparticulate materials possess noteworthy properties has been somewhat lacking, although the field of nanotoxicity is now gradually expanding (Vorup-Jensen and Peer 2012). It clear that materials made from certain heavy metals are toxic, almost irrespective of their size or form when they enter the human body. Likewise, positively charged particles are often incompatible with human plasma proteins and cells in the sense that they will cause widespread coagulation and disruption of the cells, again with few indications that their size in itself is responsible for these properties.

IgM and Its Interaction with NPs

The complement system provides an interesting opportunity to link a biological outcome, namely complement activation, directly to the size of the particulate material inducing it. Pedersen et al. (2010) used a classic and simple procedure for measuring complement activation using dextran-coated iron oxide particles: human serum was incubated with the particles under conditions permitting complement activation, followed by a test of their ability to lyse rabbit erythrocytes through the inability of these cells to defend themselves against human complement components. If the particles activated complement, a loss was observed in the subsequent ability to lyse erythrocytes. Particles with sizes of 50, 250, and 600 nm were used. With serum from some donors, it was clear that the particles did not activate complement equally, not even when corrected for the available surface area. Of significance to the interpretation, activation by the 250-nm particles was stronger than that by either the 50 or 600-nm particles. Considering that the surface chemistry of all of the particles was identical, an explanation for the difference in complement activation seems to rely on a structural property of the particles. At the outset, trivial explanations such as an influence of aggregation were quickly ruled out by performing light scattering experiments. Upon closer scrutiny, it appeared that the complement activation correlated with IgM titers to dextran, but not with IgG titers. From these observations, the authors rationalized that the known conformational flexibility of IgM could be

involved in explaining the outcome. Indeed, from structural modeling, it became clear that the curvature of the 250-nm particles would be sufficient to bring the IgM molecule into its "staple" conformation, which enables complement activation. By contrast, the 600-nm particles had a too low curvature to do so, essentially being a flat surface compared to the size of the IgM molecules. The 50-nm one obviously had a steeper curvature than the 250-nm particles; however, it was too steep to accommodate binding of a sufficient number of antigens to enable the conformational change required to induce complement activation. In this way, the work by Pedersen et al. provides an interesting link between polyvalent interaction and sensitivity in binding to surface topography. With the link to complement activation, this work supports the view that size differences between particulate materials can induce significant differences in biological reaction to the material. In effect, at the nanometer-scale of particulates and interacting large immune proteins, the link between size and biological effects are non-trivial because they critically involve structural – i.e., size – properties of both entities.

Future Perspectives: From Protein–NP Interactions to a Structural Understanding of Nanomicrobiology

Both specific and non-specific adsorptive events of protein interactions with NPs are of significance to their fate and action in the human body. With the promising perspective of NPs as drug delivery vehicles or contrast reagent in several types of *in vivo* scanner technologies, the work by Pedersen et al. points to important principles in understanding the outcome of such interactions as acknowledge by others (Chen et al. 2017).

The administration of NPs in medicine would rarely seek complement activation as an outcome, which indeed could pose a significant health risk by inducing anaphylactic shock. IgM-mediated complement activation is, of course, not the only route of such activation. Nevertheless, it is an important part of hypersensitivity reaction to materials, for instance to those carrying a carbohydrate-covered surface. Pedersen et al.'s work suggests that anaphylactic responses may be avoidable by appropriate tuning of the particulate size distribution, either in the smaller or larger direction. A similar approach could also be useful for regulating the spontaneous, adsorptive-type interaction between NPs and proteins. It is now well known that many NPs in protein-rich body fluids such as plasma form both a firmly attached and a softer corona of select protein (Miclaus et al. 2016). Evidence from studies on fibrinogen suggest NP size, and hence curvature, could be a factor in structural stability of the attached protein, probably due to curvature-related deflection of the ultrastructure of the adsorbed fibrinogen (Deng et al. 2009). Again, it would seem a fair suggestion that such forces gain their impact from fibrinogen being a protein sufficiently large to be sensitive to the curvature of the applied NPs, quite similar to the more specific binding of IgM to carbohydrate-covered NP surfaces (Pedersen et al. 2010).

From research on the nanostructure of microbial surfaces, it is gradually becoming clearer that these surfaces present structural features on the nanometer scale. A likely explanation for this phenomenon is the bottom–up synthesis of materials such as a Gram-positive cell wall, which, as demonstrated in AFM studies by Touhami et al. (2004), has a pattern of concentric circles. As noted by Pedersen et al. (2010), such a structure fits remarkably well with the observation that many large immune proteins have a structure with some element of rotational symmetry. With a curvature sharper than for peptidoglycan found on intact *S. aureus*, the shed peptidoglycan was the prime activator of complement compared with the intact cells, as discussed in Section 14.2.3. Of course, this suggest a kind of protection mechanism, where the *S. aureus* cell shed nanoparticulate fragments with structural features, i.e., curvature, that enables complement activation at a distance from the intact bacteria. We have recently extended this finding by the discovery that exposure of *S. aureus* to antimicrobial peptides induces a prominent shedding of nanoparticulate material, again probably protecting *S. aureus* against the peptide by providing decoys for adsorbing the peptides (Christiansen et al. 2017). Altogether, this work suggests that the nanometer-scaled properties of microbial materials could be a significant factor contributing to the virulence of these organisms, which clearly calls for further investigations.

14.3.3 SPR as a Tool in the Nanoscience of Large Immune Proteins

SPR technology is widely used to study the interactions between large biomolecules and quantify their binding kinetics. SPR utilizes the phenomenon of total internal reflection, which occurs when propagating light meets an interface located in-between media of high and low refractive index (Schuck 1997). The light is fully reflected back into the media with high refractive index at the interface, generating an evanescent wave in the media with low refractive index in the process. The interface layer is made up of a conducting, non-magnetic metal such as gold. In this setting, the surface coating on the gold part of the chip makes up the low-refractive index part together with the surrounding solution. The evanescent wave will penetrate the gold layer and interact with the surface plasmon waves that are present within the conductor, thus exciting them. The interaction will lead to a reduction in intensity of the reflected light at a specific angle of reflection, known as the SPR angle. This angle is in direct correlation with the refractive index of the coating of the gold chip, therefore representing changes due to interaction of biomolecules on the surface.

Recent Advances in Interpreting the Binding Kinetics of Large Immune Proteins

By measuring the changes in the SPR angle, which is linearly correlated to changes in refractive index, R or sometimes denoted S, of the coating due to bimolecular interactions, it is possible to estimate chemical binding parameters of the binding. Most systems for measuring SPR work by injection of analyte over a ligand-coated surface, thereby enabling a steady-state equilibrium with one of the reactants (the analyte) in vast stoichiometric excess over the other. For a simple 1:1 binding reaction between a protein (P) and its ligand (L), L + P $\leftrightarrow$ LP, the dissociation constant can be estimated from $K_D = [L]_{Eq}[P]_{Eq}/[LP]_{Eq}$. The Langmuir binding isotherm, $K_D = (1 - \theta)[P]_{Eq}/\theta$ will apply, where θ is the fractional coverage of binding ligands immobilized on the gold layer. In reality, far from all binding reactions will reach a steady-state equilibrium within a reasonable time frame (or consumption of analyte). Experiments are conducted with a contact phase, with injection of analyte, and a dissociation phase, where the injection of analyte is stopped and the decline in signal follows. The differential equations originally derived by Langmuir for a 1:1 binding can sometimes be used to derive the association constant, k_a, and dissociation constant, k_d, rates. For the topic discussed in the present chapter, it should be noted that Langmuir 1:1 binding kinetics unfortunately often in the literature are forced onto experimental data where neither theoretical considerations nor goodness of fit justifies such an approach. Polyvalent interactions are one such type where it has even been demonstrated that the simple 1:1 binding scheme grossly neglects significant differences in binding kinetics as a function of the polyvalency (Gjelstrup et al. 2012). A solution to this particular problem now exists, partly based on the significant work by Kitov and Bundle, Schuck et al., and Vorup-Jensen et al. Briefly, Kitov and Bundle demonstrated that the binding of IgM is a sum interaction, each contributing to the sum characterized by the number of bonds formed (Kitov and Bundle 2003). A polyvalent molecule with ten binding sites can form anywhere between 1 and 10 bonds, provided sufficient ligands are present. Since the number of bonds formed determines the Gibbs free energy of the interaction, it possible to consider polyvalent interactions as an ensemble of interactions, or Gibbs free energies, with each their respective bindings kinetics (Vorup-Jensen 2016). To determine the contribution of each ensemble to the overall binding, a methodology developed by Schuck et al. (Gorshkova et al. 2008; Svitel et al. 2003) was employed to determine the minimal ensemble required to account for the experimental data of MBL oligomers binding to the ligand. As shown by Vorup-Jensen (Vorup-Jensen 2012, 2016), the principles derived by Kitov and Bundle justify this approach.

The value of characterizing the binding of large, polyvalent immune proteins becomes clear from also bringing in the ability of SPR to account for the amount of immobilized ligand. It is simple to transform the SPR signal (in arbitrary response units, RU) to a coating density of protein in pico-moles of ligand per square millimeter. This, in turn, makes an estimate of the interligand distance possible, and hence enables recording correlating changes in binding as related to changes in this distance. The outcome was a 100,000-fold change in affinity between surfaces with interligand distances ranging only between 6 and 14 nm (Figure 14.2) (Gjelstrup et al. 2012). It is astonishing that this relatively marginal change in average interligand distance produced such a difference in affinity for the surfaces. As noted elsewhere, these observations greatly support the concept of topological patterns as an important part of innate immunity (Vorup-Jensen 2012).

Future Perspectives: SPR Biosensors and Polyvalent Interactions

MBL belongs to the family of collectins, the numbers of which have recently been expanded by addition of newly identified members. In addition, the ficolins share significant structural similarities with the collectins, especially the oligomeric structure (Hansen et al. 2016; Holmskov et al. 2003). Insights on the ligand-binding kinetics of these molecules are remarkably sparse, probably because of the analytical challenges mentioned above. However, with the methods implemented by Gjelstrup et al. (2012), new possibilities are available. It is interesting that other oligomeric

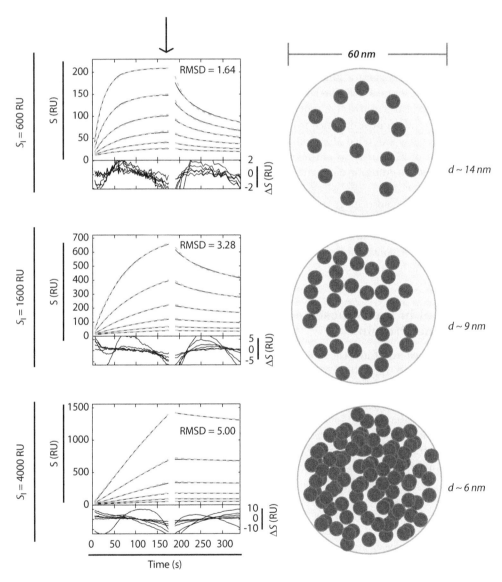

FIGURE 14.2 Influence of ligand density on MBL oligomer binding (Gjelstrup et al. 2012). The binding of fractionated MBL oligomers (with approximately 94% (w/w) 6 × MBL$_3$ as well as some larger species) to three surfaces with increasing amount of ligands. In the sensorgrams, experimental binding curves (sensorgrams) are indicated in black and the fitted, minimum-ensemble model in gray. The root-mean square deviation (RMSD) indicates the goodness of fit. Smaller panels below the sensorgrams also indicate the difference (ΔS) between the experimental data and the model. The oligomers were applied at concentrations of 0.5, 1, 2, 4, 8, and 16 nM of the structural unit (c_{MBL_3}). The arrow indicates the time when the injection of the analyte was stopped. The panels to the right show the approximate density of ligands, drawn to scale relative to the cross-sectional diameter of MBL at 60 nm.

protein interactions, albeit not with other collectins, have been studied with this approach and have worked well in this situation (Kapinos et al. 2014). This probably implies that the findings by Gjelstrup et al. can be extended to a broad range of polyvalent interactions. As noted several times earlier, oligomeric proteins account for a considerable proportion of the large proteins, which are relevant for studies in the realm of nanoscience. We envisage further enquiries into the binding properties of these proteins to extend the significance of nanometer-scaled topological features in immunology.

A rarely used aspect of SPR technology involves the probing depth. This is dependent on the evanescent waves, hereunder the incident light used. The amplitude of the generated evanescent wave is decaying exponentially as a function of the distance from the surface of the chip. This means that the depth of the wave that is typically useful for measurements is around 300 nm, corresponding to 37% of the maximum intensity of the wave. However, since the evanescent waves and incident light used are correlated, different light wavelength can be used to adjust the measurements. This option has mostly been used for cellular studies with SPR (Yang et al. 2015). On one hand, current commercial instruments operate at a fixed wavelength, not permitting such changes. On the other hand, by considering the information to be gained from more flexibility in the choice of wavelength, maybe there is a case for such options in the study of large, oligomeric proteins. Longer wavelengths allow probing further into the solution. In principle, this may lead to capturing weaker, non-specific binding interactions with more background signal. However, we know little about these weaker interactions. In case of the large protein oligomers, they may contain information on the initial steps of polyvalent binding. Indeed, Kitov and Bundle proposed in their model that the first contact between polyvalent molecules and ligand-presenting surfaces is a single, intermolecular bond formation. In particular, with long carbohydrates as ligand, but maybe also in other cases, it is conceivable that such bond formation occurs at a distance from the SPR surfaces greater than the conventional probing depth. By comparing with shorter-wavelength SPR, which allows capturing the binding interactions in the close proximity to the gold surface, insights on the binding dynamics can possibly be derived this way, enabling a better understanding of the events in polyvalent interactions.

14.4 Conclusions

Here, we have presented a select number of cases that use instrumentation and analyses typical of nanoscientific enquiries for protein characterization. At the outset, these investigations were prompted by the large size of some immune proteins, which for many more standard techniques make retrieving structural information a challenge. By nature, SAXS and AFM easily return information for

large complexes, which can then be aided by techniques such as X-ray crystallography and EM to further model the atomic structure of the complexes. It is an important feature of SAXS and AFM that these techniques permit investigations in liquid environment, both ensuring physiological conditions or a choice to change these conditions to study molecular forces responsible for protein contacts, as we have recently demonstrated with SAXS analysis of polypeptide complex formation. We believe these uses can be greatly extended in the field of the large proteins of the immune system.

Protein structure has been described as organized at four levels, from the primary structure embodying the sequence of amino acid residues to the quaternary level describing the non-covalent or covalent association of folded polypeptide chains (Linderstrøm-Lang 1952). By the use of the nanoscientific analysis of proteins, it has become clear that a fifth level can be thought of as constituted by the assembly of structural units in oligomeric proteins. We name this level the "ultrastructure" of proteins owing to the relatively large size such proteins typically will attain. The MBL oligomers are but one example where a unit of three MBL chains (MBL_3) assemble into oligomers of anywhere between 3 and at least 8 such units ($3-8 \times MBL_3$). The methodologies for structural investigations here described seem quite efficient at dealing with unraveling this level of structure, unlike many other techniques more focused at dealing with protein structure at the atomic level. Changes in the structure at this level, for instance the deflection of the IgM molecule or widening of the MBL oligomers upon surface binding, are functionally significant as particularly shown by the complement activation properties of IgM on the surface of particles. Perhaps more surprisingly, the ultrastructure appears to be linked with the symmetrical organization of certain microbial surfaces, emphasizing the need to better understand this part of microbial biology for explaining the ultrastructural properties of large immune proteins. Taken together, this points to the significance of analyzing the large immune proteins from the perspective of nanoscience, and with the tools from nanoscience, to contribute a fuller picture of these versatile biological macromolecules.

References

Bernado, P., Mylonas, E., Petoukhov, M.V., Blackledge, M., and Svergun, D.I. 2007. Structural characterization of flexible proteins using small-angle X-ray scattering. *J Am Chem Soc* 129: 5656–5664.

Chen, F., Wang, G., Griffin, J.I. et al. 2017. Complement proteins bind to nanoparticle protein corona and undergo dynamic exchange in vivo. *Nat Nanotechnol* 12: 387–393.

Christiansen, S.H., Murphy, R.A., Juul-Madsen, K. et al. 2017. The immunomodulatory drug glatiramer acetate is also an effective antimicrobial agent that kills gram-negative bacteria. *Sci Rep* 7: 15653.

Dague, E., Alsteens, D., Latge, J.P., and Dufrene, Y.F. 2008. High-resolution cell surface dynamics of germinating Aspergillus fumigatus conidia. *Biophys J* 94: 656–660.

Deng, Z.J., Mortimer, G., Schiller, T., Musumeci, A., Martin, D., and Minchin, R.F. 2009. Differential plasma protein binding to metal oxide nanoparticles. *Nanotechnology* 20: 455101.

Dong, M., Xu, S., Oliveira, C.L. et al. 2007. Conformational changes in mannan-binding lectin bound to ligand surfaces. *J Immunol* 178: 3016–3022.

Feinstein, A., and Munn, E.A. 1966. An electron microscopic study of the interaction of macroglobulin (IgM) antibodies with bacterial flagella and of the binding of complement. *J Physiol* 186: 64P–66P.

Feinstein, A., and Munn, E.A. 1969. Conformation of the free and antigen-bound IgM antibody molecules. *Nature* 224: 1307–1309.

Gjelstrup, L.C., Kaspersen, J.D., Behrens, M.A. et al. 2012. The role of nanometer-scaled ligand patterns in polyvalent binding by large mannan-binding lectin oligomers. *J Immunol* 188: 1292–1306.

Gorshkova, II, Svitel, J., Razjouyan, F., and Schuck, P. 2008. Bayesian analysis of heterogeneity in the distribution of binding properties of immobilized surface sites. *Langmuir* 24: 11577–11586.

Hansen, S.W., Ohtani, K., Roy, N., and Wakamiya, N. 2016. The collectins CL-L1, CL-K1 and CL-P1, and their roles in complement and innate immunity. *Immunobiology* 221: 1058–1067.

Hoffmann, J.A., Kafatos, F.C., Janeway, C.A., and Ezekowitz, R.A. 1999. Phylogenetic perspectives in innate immunity. *Science* 284: 1313–1318.

Holmskov, U., Thiel, S., and Jensenius, J.C. 2003. Collections and ficolins: humoral lectins of the innate immune defense. *Annu Rev Immunol* 21: 547–578.

Janeway, C.A., Jr. 1989. Approaching the asymptote? Evolution and revolution in immunology. *Cold Spring Harb Symp Quant Biol* 54 Pt 1: 1–13.

Jensenius, H., Klein, D.C., van Hecke, M., Oosterkamp, T.H., Schmidt, T., and Jensenius, J.C. 2009. Mannan-binding lectin: structure, oligomerization, and flexibility studied by atomic force microscopy. *J Mol Biol* 391: 246–259.

Kapinos, L.E., Schoch, R.L., Wagner, R.S., Schleicher, K.D., and Lim, R.Y. 2014. Karyopherin-centric control of nuclear pores based on molecular occupancy and kinetic analysis of multivalent binding with FG nucleoporins. *Biophys J* 106: 1751–1762.

Kitov, P.I., and Bundle, D.R. 2003. On the nature of the multivalency effect: a thermodynamic model. *J Am Chem Soc* 125: 16271–16284.

Linderstrøm-Lang, K.U. (1952). *Lane Medical Lectures: Proteins and Enzymes* (Redwood City, CA: Stanford University Press).

Miclaus, T., Beer, C., Chevallier, J. et al. 2016. Dynamic protein coronas revealed as a modulator of silver nanoparticle sulphidation in vitro. *Nat Commun* 7: 11770.

Miller, A., Phillips, A., Gor, J., Wallis, R., and Perkins, S.J. 2012. Near-planar solution structures of mannose-binding lectin oligomers provide insight on activation of lectin pathway of complement. *J Biol Chem* 287: 3930–3945.

Moghimi, S.M., Hunter, A.C., and Andresen, T.L. 2012. Factors controlling nanoparticle pharmacokinetics: an integrated analysis and perspective. *Annu Rev Pharmacol Toxicol* 52: 481–503.

Murphy, K., Travers, P., and Walport, M.J. (2017). *Janeway's Immunobiology* (New York: Garland Science).

Oliveira, C.L.P., Vorup-Jensen, T., Andersen, C.B.F., Andersen, G.R., and Pedersen, J.S. (2009). Discovering new features of protein complexes structures by small-angle X-ray scattering. In *Applications of Synchrotron Light to Scattering and Diffraction in Materials and Life Sciences*, T.A. Ezquerra, M.C. Garcia-Gomez, and M. Gomez, eds. (Heidelberg, Germany: Springer Verlag), pp. 231–244.

Pedersen, M.B., Zhou, X., Larsen, E.K. et al. 2010. Curvature of synthetic and natural surfaces is an important target feature in classical pathway complement activation. *J Immunol* 184: 1931–1945.

Perkins, S.J., Nealis, A.S., Sutton, B.J., and Feinstein, A. 1991. Solution structure of human and mouse immunoglobulin M by synchrotron X-ray scattering and molecular graphics modelling. A possible mechanism for complement activation. *J Mol Biol* 221: 1345–1366.

Ricklin, D., Hajishengallis, G., Yang, K., and Lambris, J.D. 2010. Complement: a key system for immune surveillance and homeostasis. *Nat Immunol* 11: 785–797.

Ricklin, D., and Lambris, J.D. 2007. Complement-targeted therapeutics. *Nat Biotechnol* 25: 1265–1275.

Ricklin, D., Mastellos, D.C., Reis, E.S., and Lambris, J.D. 2018. The renaissance of complement therapeutics. *Nat Rev Nephrol* 14: 26–47.

Schuck, P. 1997. Use of surface plasmon resonance to probe the equilibrium and dynamic aspects of interactions between biological macromolecules. *Annu Rev Biophys Biomol Struct* 26: 541–566.

Sim, R.B., Schwaeble, W., and Fujita, T. 2016. Complement research in the 18th-21st centuries: Progress comes with new technology. *Immunobiology* 221: 1037–1045.

Svergun, D.I. 2010. Small-angle X-ray and neutron scattering as a tool for structural systems biology. *Biol Chem* 391: 737–743.

Svitel, J., Balbo, A., Mariuzza, R.A., Gonzales, N.R., and Schuck, P. 2003. Combined affinity and rate constant distributions of ligand populations from experimental surface binding kinetics and equilibria. *Biophys J* 84: 4062–4077.

Touhami, A., Jericho, M.H., and Beveridge, T.J. 2004. Atomic force microscopy of cell growth and division in Staphylococcus aureus. *J Bacteriol* 186: 3286–3295.

Vorup-Jensen, T. 2012. On the roles of polyvalent binding in immune recognition: perspectives in the nanoscience of immunology and the immune response to nanomedicines. *Adv Drug Deliv Rev* 64: 1759–1781.

Vorup-Jensen, T. (2016). The nanoscience of polyvalent binding by proteins in the immune response. In *Nanomedicine*, K.A. Howard, T. Vorup-Jensen, and D. Peer, eds. (New York: Springer Nature), pp. 53–76.

Vorup-Jensen, T., and Boesen, T. 2011. Protein ultrastructure and the nanoscience of complement activation. *Adv Drug Deliv Rev* 63: 1008–1019.

Vorup-Jensen, T., and Peer, D. 2012. Nanotoxicity and the importance of being earnest. *Adv Drug Deliv Rev* 64: 1661–1662.

Xiao, J., and Dufrene, Y.F. 2016. Optical and force nanoscopy in microbiology. *Nat Microbiol* 1: 16186.

Yang, C.T., Mejard, R., Griesser, H.J., Bagnaninchi, P.O., and Thierry, B. 2015. Cellular micromotion monitored by long-range surface plasmon resonance with optical fluctuation analysis. *Anal Chem* 87: 1456–1461.

15

Nanozymes and Their Applications in Biomedicine

Qian Liang, Ruofei Zhang, and Xiyun Yan

CAS Engineering Laboratory for Nanozyme, Institute of Biophysics, Chinese Academy of Sciences

University of Chinese Academy of Sciences

Kelong Fan

CAS Engineering Laboratory for Nanozyme, Institute of Biophysics, Chinese Academy of Sciences

15.1 Introduction

Natural enzymes are proteins or RNA produced by living cells that are highly specific and catalytic to their substrates, which are so important that almost all metabolic processes in the cell need enzyme catalysis (J. M. Berg 2002). The high catalytic activity of natural enzymes has attracted many researchers to try to use them in catalytic applications. However, the applications of enzymes are limited by their lack of stability in non-physiological conditions, such as in organic solvents or at high temperatures. For instance, horseradish peroxidase (HRP) can only catalyze the synthesis of water-soluble conducting polyaniline in the very narrow pH range of 4.0–4.65. Otherwise, only non-conductive polyaniline can be prepared (pH > 4.65) or HRP loses its catalytic activity (pH < 4.0) (Wei Liu 1999). In addition, the content of natural enzymes in organisms is too low to obtain them in large quantities, which increases the production cost. As a consequence, artificial enzyme-mimetic synthesis has become an active area of research aimed at creating new enzymes with novel properties, either through total-synthetic or semi-synthetic methods.

Before the discovery of nanozymes, artificial enzyme-mimetics generally referred to chemical complexes made by attaching external catalytic elements to chemically synthesized materials. To mimic the active site of natural enzymes, classical artificial enzymes typically employ organic polymers (e.g. cyclodextrin, crown ether, or calixarene) as substrate-binding receptors and use characteristic molecular moieties (e.g. amino acids and peptides) as catalytic functional groups (Morten Meldal 2005, Mikael Bols 2010). Compared with natural enzymes, these enzyme-mimetics are more stable and cheaper, thus enabling industrial production and application. Nevertheless, despite the extensive development of these additive-based materials, their process complexity and limited stability have impeded their wide application.

While scientists were still struggling to find new extrinsic catalytic ligands for coupling to nanomaterials, nanozymes with intrinsic enzyme-like activities were discovered in 2007. Without any ligands, inorganic Fe_3O_4 nanoparticles (NPs) were found to possess intrinsic catalytic activity similar to HRP with comparable catalytic efficiency and the same mechanism (Lizeng Gao 2007). Since then, nanozymes have gradually become a hot topic of research in the study of enzyme-mimetics because of their unique properties, such as high catalytic activity, good stability and easy large-scale preparation. Nowadays, research on nanozymes has made remarkable progress in many fields, especially in biomedicine. This chapter gives a comprehensive review of the discovery and development of nanozymes and outlines the progress of nanozyme applications in biomedicine, such as biosensing, bioimaging, antibiofouling, disease diagnosis and therapy.

15.2　The Discovery and Development of Nanozymes

As the size of inorganic materials is reduced from the macroscale to the nanoscale, materials may exhibit some unexpected physical and chemical properties, such as surface effects, small size effects, quantum size effects and macroscopic quantum tunneling effects. Still, to the best of our knowledge, there were few reports that inorganic nanomaterials possessed intrinsic biological effects before the discovery of nanozymes. Therefore, to conventionally endow nanomaterials with biological functions, such as enzyme-like catalytic activity, surface modification has been a commonly used method, in which enzymes or other catalytic groups are coupled to the surface of nanomaterials via covalent or non-covalent bonds. For example, in order to mimic the function of natural RNase, Scrimin et al. modified gold NPs with an azacrown on their surface, which was then chelated with zinc ion to obtain the catalytic function of shearing phosphodiester bonds (Jack L.-Y. Chen 2016). Although these modified gold NPs are also called nanozymes in this study, their catalytic activity derives from the surface-modified moieties rather than the properties of the nanomaterials themselves.

Before nanozymes were systematically studied and applied as biological enzyme-mimetics, there was some evidence suggesting that nanomaterials may possess enzyme-like catalytic functions. In 1996, Dugan et al. reported that fullerene derivatives are effective free radical scavengers and could reduce excitotoxic and apoptotic death in cultured neurons (Laura L. Dugan 1996, Laura L. Dugan 1997). In 2006, CeO_2 NPs were found to possess the ability of scavenging reactive oxygen species (ROS) and used for preventing ROS-induced retinal degeneration (Junping Chen 2006, Gabriel A. Silva 2006). These studies imply that inorganic nanomaterials may possess intrinsic enzyme-like activities, but still do not provide strong experimental evidence and theoretical support for the catalytic mechanisms.

It was not until 2007 that the intrinsic catalytic activity of Fe_3O_4 was first systematically studied and illustrated from the perspective of enzymology. In the pioneering work of Gao et al., the peroxidase-like activity and catalytic mechanism of Fe_3O_4 magnetite NPs were systematically compared with those of natural HRP enzymes (Lizeng Gao 2007). This study found that magnetite NPs can catalyze the oxidation of various substrates (tetramethylbenzidine (TMB), diaminobenzidine (DAB), and o-phenylenediamine (OPD), three representative HRP substrates) of HRP in the presence of hydrogen peroxide (H_2O_2) and produce the same colored products, exactly the same as those catalyzed by HRP (Figure 15.1). In addition, the catalytic efficiency, kinetic parameters and catalytic mechanism of Fe_3O_4 nanozymes are also similar to those of natural HRP enzymes. Like HRP, the catalytic efficiency of the Fe_3O_4 nanozyme is dependent on pH, temperature and substrate concentration. Under the conditions of pH 3.5 and 40°C, Fe_3O_4 nanozymes show the highest catalytic activity.

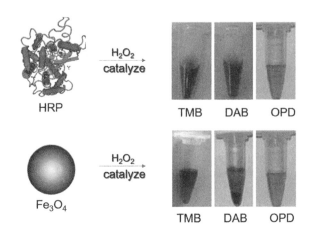

FIGURE 15.1　Fe_3O_4 nanozyme demonstrates peroxidase-like activity by catalyzing H_2O_2 and various substrates with color reactions. (Reprinted with permission from Lizeng Gao 2007. Copyright (2007) Springer Nature.)

Moreover, Fe_3O_4 nanozymes exhibit a characteristic ping-pong catalytic mechanism over a range of TMB and H_2O_2. Based on the above evidence, Gao et al. suggested that Fe_3O_4 magnetite NPs are a kind of peroxidase mimetic, which represented the discovery of a nanozyme for the first time.

The emergence of nanozymes has changed the traditional concept that inorganic nanomaterials are bio-inert substances, revealed the inherent biological effects and new characteristics of nanomaterials, and expanded the study of artificial enzyme-mimetics from organic polymers to inorganic nanomaterials (Figure 15.2) (Hui Wei 2013). Because nanozymes possess both high enzyme-like catalytic activity and more stable structures than those of traditional artificial enzymes, their research and application have gradually expanded into a wide range of fields, including biomedicine, chemical industry, food industry, agriculture and environmental governance. To date, there have been more than 245 laboratories from 26 countries focusing on the research of nanozymes. The number of papers published about nanozymes has also increased explosively in recent years (Figure 15.3). From these publications, we draw conclusion that the research on nanozymes mainly focuses on the following three aspects: (i) discovering new nanomaterials with enzyme-like catalytic activity; (ii) revealing the catalytic mechanism of nanozymes, optimizing their catalytic efficiency and substrate specificity; (iii) continuing to explore and expand the application of nanozymes in biology, medicine, the environment, chemical industry, and other fields. More detailed advances in these areas will be discussed in the following sections.

15.3　Discovering Novel Nanozymes

After the discovery of the Fe_3O_4 peroxidase-mimic nanozyme, various nanozymes composed of different materials or exhibiting novel enzyme-mimic activities have

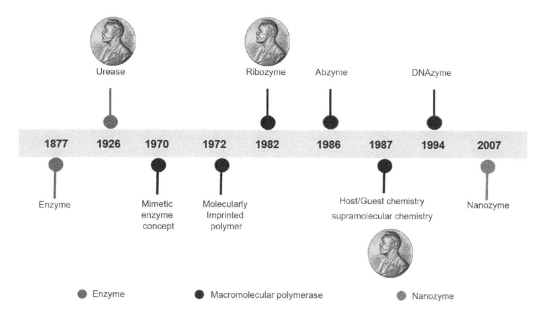

FIGURE 15.2 Nanozymes are a new type of enzyme-mimetics that are different from traditional artificial enzymes. (Reprinted with permission from Hui Wei 2013. Copyright (2013) Royal Society of Chemistry.)

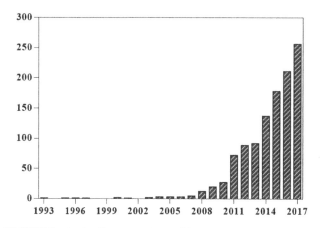

FIGURE 15.3 Progress in publications in the field of nanozymes. (Data Source: Web of Science, deadline: December 31, 2017. Key Words: nanozyme or nano and ase-like or ase-mim or enzyme mim or enzyme like.)

$$2AH + H_2O_2 \xrightarrow{\text{peroxidase}} 2A + 2H_2O$$

$$2AH + ROOH \xrightarrow{\text{peroxidase}} 2A + ROH + H_2O$$

$$AH + O_2 \xrightarrow{\text{oxidase}} A + H_2O$$

$$AH + O_2 + H_2O \xrightarrow{\text{oxidase}} A + H_2O_2$$

$$2O_2^{\bullet-} + 2H^+ \xrightarrow{\text{SOD}} H_2O_2 + O_2$$

$$H_2O_2 \xrightarrow{\text{catalase}} O_2 + 2H_2O$$

FIGURE 15.4 Enzyme-like reactions catalyzed by nanozymes. (Reprinted with permission from Qingqing Wang 2018. Copyright (2018) Elsevier.)

been reported one after another. So far, more than 100 types of nanomaterials involving 18 elements have been confirmed to possess enzymatic activities, mainly including metal oxides (e.g. V_2O_5, CeO_2, Co_3O_4 and Mn_3O_4), metal sulfides (e.g. FeS, MoS_2), metal-based (e.g. Au, Ag, Pt and their alloy) and carbon-based (e.g. carbon tube, GO) nanomaterials, and various kinds of composite materials. The enzyme-like activities shown by these new nanozymes include, but are not limited to, peroxidase, oxidase, catalase, and superoxide dismutase (SOD) mimic activities (Figure 15.4) (Qingqing Wang 2018).

15.3.1 Metal Oxide Nanozymes

Metal oxide nanomaterials are a large class of nanozymes and were the first ones discovered. As mentioned previously,

Fe_3O_4 magnetite NPs were the first reported nanozymes that exhibited peroxidase-like properties. Before that, iron oxide nanomaterials had already been extensively used in biomedical fields due to their unique magnetic properties. For instance, Fe_3O_4 NPs are commonly used as contrast agents in magnetic resonance imaging (MRI) for disease diagnosis. The magnetic properties of Fe_3O_4 NPs are also used for the separation and purification of biological samples (Ajay Kumar Gupta 2005). Moreover, the Fe_3O_4 NP property of generating heat in the presence of an external magnetic field has been employed to develop tumor therapy strategies. In recent years, enzymatic catalysis by iron oxide nanomaterials has attracted wide attention. Initial studies mainly focused on the effects of their size, morphology, and surface modification on catalytic efficiency. Beyond their peroxidase-mimic activity in acidic conditions,

Fe_3O_4 nanozymes also exhibit catalase-mimic properties under physiological conditions, and can decompose H_2O_2 into water and oxygen (Zhongwen Chen 2012, Yi Zhang 2016). Subsequent studies showed that some mixed oxide nanomaterials formed by iron and other metals could also exhibit peroxidase-like behavior, such as $BiFeO_3$ (Wei Luo 2010) and $MnFeO_4$ (V. Figueroa-Espí 2011). Interestingly, Zhang et al. reported that $CoFe_2O_4$ NPs could catalyze the oxidation of luminol by dissolved oxygen to produce an intensified chemiluminescence signal, indicating that these NPs have intrinsic oxidase-like activity (Xiaodan Zhang 2013).

In addition to iron-based nanozymes, CeO_2 NPs (nanoceria) are another representative metal oxide nanozyme that has been comprehensively studied. The surface cerium ions of nanoceria can easily transform between Ce^{4+} and Ce^{3+} rapidly; thus, nanoceria can easily accept or lose electrons based on the environment. The most important application of nanoceria in industrial catalysis is to treat automobile pollution (hydrocarbons, CO, NO, etc.) as an active component of the three-way catalyst (Alessandro Trovarelli 1999). Self et al. first discovered and studied the ability of nanoceria to catalyze the dismutation of superoxide in vitro (Cassandra Korsvik 2007). By using two kinds of nanoceria with different Ce^{3+}/Ce^{4+} ratios, they found that the surface Ce^{3+}/Ce^{4+} ratio determines the SOD-like activity of nanoceria. Subsequently, catalase-like activity was also reported by Self et al. (Talib Pirmohamed 2010). However, in contrast to the SOD-like mechanism, nanoceria with lower Ce^{3+}/Ce^{4+} ratio exhibits higher catalase-like activity. In order to explain the phenomenon of nanoceria's multi-enzyme mimetic catalysis, Celardo et al. have put forward a detailed reaction mechanistic model for the SOD-mimic (Figure 15.5) and catalase-mimic (Figure 15.6) activities of nanoceria. They demonstrated that the transformation of Ce^{3+}/Ce^{4+} and oxygen vacancies on the surface of nanoceria determine the multi-enzymatic activities of nanoceria.

Similar to Fe_3O_4, the peroxidase-mimic activity of transition-metal oxides is generally like that of the Fenton reaction (Ivana Celardo 2011). Self et al. found that Ce^{3+} could also participate in a Fenton-like reaction in the presence of H_2O_2, which provides the possibility for the discovery of the peroxidase-like activity of nanoceria. In a later study, Lv et al. showed that nanoceria synthesized by a hydrothermal method actually possesses peroxidase-like activity. Based on this characteristic of nanoceria, they developed a simple colorimetric method for glucose detection (Xue Jiao 2012). Moreover, a kind of water-soluble nanoceria synthesized by an aqueous phase method and modified with L-glucoside and polyacrylic acid was found to exhibit oxidase-like activity (Atul Asati 2009). This activity is related to pH, thickness of the polymer layer, and particle size. The nanoceria with a thin polymer layer and small size exhibit the best oxidase-like activity under acidic conditions.

Most interestingly, nanoceria has also been reported to have phosphatase-mimic activity. Kuchma et al. reported that nanoceria could hydrolyze many biologically related molecules with phosphate bonds, such as p-nitrophenylphosphate (pNPP), Adenosine triphosphate (ATP), and o-phospho-l-tyrosine (Melissa Hirsch Kuchma 2010). In addition, Qian et al. suggested that nanoceria can efficiently catalyze the dephosphorylation of phosphopeptides without temperature limitations. Because phosphorylation and dephosphorylation of organisms play important roles in signal transduction in many biological processes,

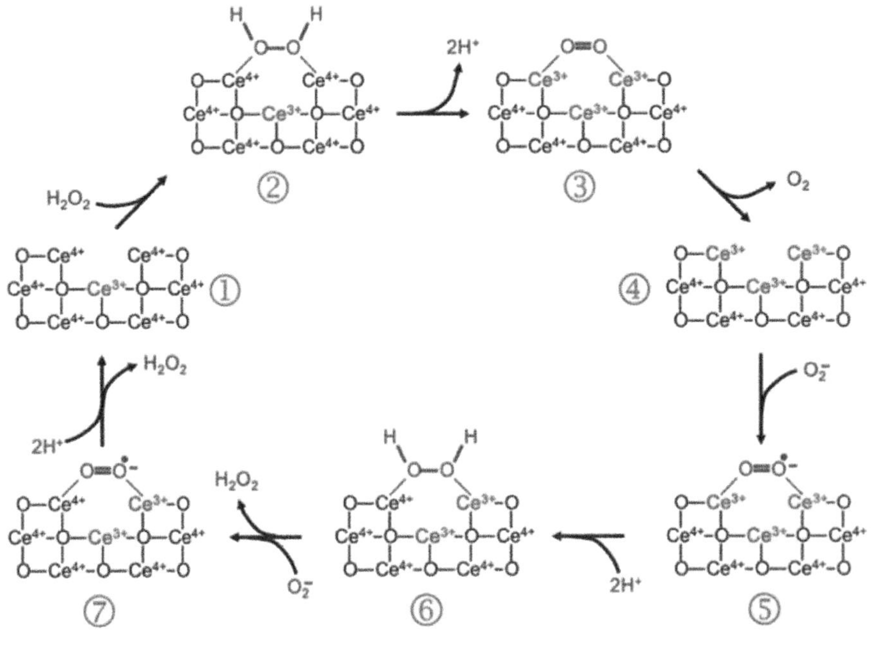

FIGURE 15.5 A reaction mechanism model of the SOD-mimic activity of nanoceria. (Reprinted with permission from Ivana Celardo 2011. Copyright (2011) Royal Society of Chemistry.)

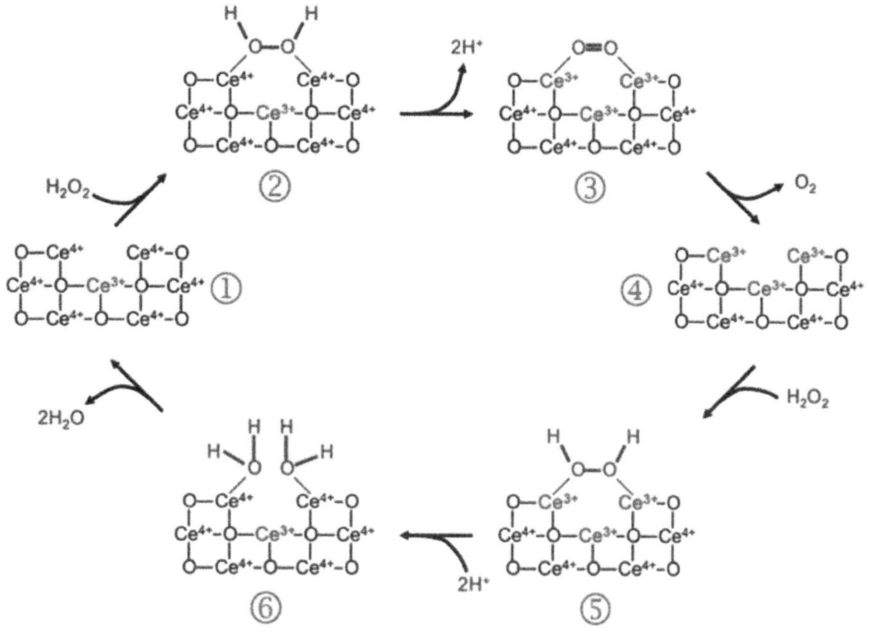

FIGURE 15.6 A reaction mechanism model of the catalase-mimic activity of nanoceria. (Reprinted with permission from Ivana Celardo 2011. Copyright (2011) Royal Society of Chemistry.)

the phosphatase-mimic activity of nanoceria may offer the potential to expand its biomedical applications.

Tremel et al. found that V_2O_5 NPs catalyze the oxidation reaction of 2,2′-Azinobis-(3-ethylbenzthiazoline-6-sulfonate) (ABTS) and TMB in the presence of H_2O_2, indicating that nano-V_2O_5 is a kind of peroxidase-mimetic nanozyme (Rute André 2011). Subsequently, Mugesh et al. reported that V_2O_5 NPs could maintain the balance of redox in cells and protect cells by reducing oxidative stress owing to their glutathione peroxidase (GPx)-like activity (Amit A. Vernekar 2014).

In addition, the metal oxides Co_3O_4 and Mn_3O_4 are also novel nanozymes that possess multiple catalytic activities. Co_3O_4 nanozyme has been reported to exhibit the enzyme-like activities of peroxidase, catalase and SOD (Jianshuai Mu 2012, Jinlai Dong 2014), and Mn_3O_4 exhibits the three enzyme-like activities of catalase, SOD and GPx simultaneously (Jia Yao 2013, Namrata Singh 2017). Similar to these metal oxide nanozymes, some metal sulfide nanomaterials such as FeS and MoS_2 were also found to exhibit peroxidase-like activity (Zhihui Dai 2009, Amit Kumar Dutta 2012, Tianran Lin 2014).

15.3.2 Metal-Based Nanozymes

In addition to the metal oxide nanozymes mentioned above, noble metal nanomaterials (e.g. Au, Pt, Ag, etc.) and composite nanomaterials (e.g. Au-Pt, Ag-Pt, Au-Bi, etc.) formed by a variety of metals are also found to exhibit enzyme-like activities.

Au NPs possess high electron density, dielectric properties and catalytic activity. And they are able to bind to a variety of biological macromolecules without affecting their biological functions. Thus, before the discovery of their enzyme-like properties, Au NPs had already been widely used in the biomedical field, for example as biological labeling tracers. Despite there being many studies about the catalytic activity of nanogold, studies focused on their enzyme-like activity were only initiated in the past decade (Massimiliano Comotti 2004). Cao et al. reported that positively charged Au NPs possess intrinsic peroxidase-like activity that can catalyze oxidation of the peroxidase substrate TMB by H_2O_2 (Yun Jv 2010). Moreover, nanogold modified with bovine serum albumin (BSA) to improve its biocompatibility was found to still possess highly peroxidase-like activity (Xian-Xiang Wang 2011). Huang et al. studied the impact of different metal ions on the enzyme mimetic activity of nanogold and found that Hg^0 can stimulate the peroxidase-like activity. Based on this study, a label-free colorimetric assay for Hg^{2+} was developed (Yi Juan Long 2011). Moreover, nanogold could also catalyze the rapid decomposition of H_2O_2 and superoxide based on its catalase-mimic and SOD-mimic activities, respectively (Weiwei He 2013).

Platinum NPs can effectively scavenge superoxide anion radicals and hydroxyl radicals, as suggested in an early study (Takeki Hamasaki 2008). Later on, Nie et al. synthesized a kind of platinum NPs using apoferritin (Ft) as a nucleation substrate and found that these protein-coated Pt-Ft NPs possess both catalase-like and peroxidase-like activities under different conditions (Jia Fan 2011). The enzyme-like properties of the Pt-Ft nanozyme showed a significant increase in catalase-mimic activity with increasing pH and temperature, while its peroxidase-mimic activity was optimal at physiological temperature and slightly acidic conditions.

Moreover, monodispersed cubic platinum nanocrystals have also been reported to act as a peroxidase-like nanozyme (Ming Ma 2011). However, this kind of Pt nanozyme needs stabilizers to prevent aggregation, which will distinctly affect its catalytic activity if it occurs. In addition, it is also worth mentioning that silver NPs are another noble-metal nanozyme that has been found to possess peroxidase-like properties (Huan Jiang 2012).

Based on the enzymatic activities of noble-metal nanozymes, the question arises: would hybrids of different noble metals affect the enzyme-like activities? Amazingly, Au@Pt nanorods are reported to possess four kinds of enzyme-like activities: oxidase, peroxidase, catalase and SOD (Weiwei He 2011, Jo-Won Lee 2014). As shown in Figure 15.7, the oxidase and peroxidase activities of Au@Pt nanorods could be applied to the same immunoassay reaction at the same time to improve the accuracy of detection (Weiwei He 2011). Au@Ag NPs can enhance peroxidase-like activity a hundredfold compared to Au NPs (Chen-I Wang 2012). Ternary precious metal complexes of Au@Ag@Pt also have the same four enzyme activities and are affected by the Ag components (Xiaona Hu 2013).

15.3.3 Carbon-Based Nanozymes

Many non-metallic materials, especially carbon-based nanomaterials (e.g. carbon nanotubes, graphene oxide, carbon nanodots, etc.) have also been reported to have enzyme-like activities.

Qu et al. first reported that single-walled carbon nanotubes (SWCNTs) have intrinsic peroxidase-like activity and applied them in the label-free colorimetric detection of single nucleotides (Yujun Song 2010). Similar to natural enzymes, the catalytic activity of SWCNTs is affected by pH, temperature and substrate concentration. After removing residual cobalt from SWCNTs by ultrasonic treatment in concentrated sulfuric acid and nitric acid, SWCNTs still exhibit enzyme-like catalytic activity.

These results confirmed that the catalytic activity comes from the SWCNTs themselves. Moreover, in addition to SWCNTs, multi-walled carbon nanotubes and helical carbon nanotubes also possess peroxidase activity. However, the catalytic activity of helical carbon nanotubes is related to their iron content, and the catalytic activity increased with the increase of iron content (Rongjing Cui 2011).

Graphene oxide is an oxide of graphene which still has a structure consisting of a single molecular layer of graphite, but many oxygen functional groups are introduced in it, such as carboxyl, hydroxyl, epoxy and so on. The peroxidase-like activity of graphene oxide was first discovered by Yang et al. (Fengli Qu 2011). The catalytic mechanism of graphene oxide is based on its ability to accelerate the electron transfer between H_2O_2 and donor molecules. In addition, graphene oxide possesses a high specific surface area. Its special molecular structure makes it exhibit a high affinity for TMB. In the TMB color reaction catalyzed by graphene oxide, the substrate affinity toward TMB is higher than that of HRP, but the substrate affinity toward H_2O_2 is lower than that of HRP, which is very similar to the other peroxidase nanozymes.

The water-soluble derivatives of fullerenes exhibit SOD-like activity (Sameh S. Ali 2004). Although their catalytic activity is relatively low, their catalytic mechanism is consistent with the ping-pong reaction mechanism of natural SOD enzyme.

15.3.4 Other Nanozymes

The nanozymes mentioned above can be assembled together to form complexes of integrated nanozymes with multiple enzyme-like catalytic activities. Glucose oxidase (GOx)@zif-8 (Ni/Pd) exhibits both the peroxidase-like activity of Ni/Pd NPs and the catalytic activity of the coupled natural glucose oxidase (Qingqing Wang 2017). Rationally designed multiple nanozyme complexes could also catalyze enzyme cascade reactions (Hanjun Cheng 2016,

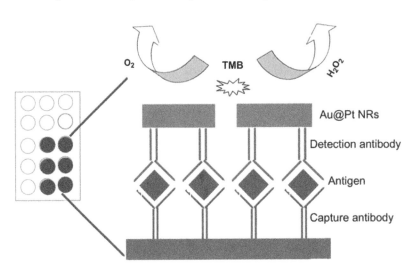

FIGURE 15.7 Au@Pt NRs-based enzyme-linked immunosorbent assay. (Reprinted with permission from Weiwei He 2011. Copyright (2011) Elsevier.)

Minfeng Huo 2017, Qingqing Wang 2017, Secheon Jung 2017). For example, Au nanozyme combined with urate oxidase could initiate a one-pot enzyme–nanozyme cascade reaction, which could be used in the treatment of hyperuricemia (Secheon Jung 2017). In this cascade reaction, uric acid oxidase degrades uric acid caused by hyperuricemia, and then the resulting H_2O_2 could be decomposed by the catalase activity of Au nanozymes. During the whole reaction process, the uric acid is rapidly removed without releasing the damaging H_2O_2, thus eliminating oxidative damage.

At present, the reported enzymatic activities of nanozymes are basically limited to redox enzyme-like activities. Recently, several nanozymes with novel types of enzyme-like activities were reported; for example, copper metal-organic frameworks (MOFs) were reported to possess intrinsic protease activities and could hydrolyze BSA and casein (Bin Li 2014). Moreover, confining passivated Au NPs with multiple cerium (IV) complexes on the surface of colloidal magnetic Fe_3O_4/SiO_2 core/shell particles produced a nanozyme exhibiting DNase-like activity (Zhaowei Chen 2016). Nanozymes with more natural enzyme-like activities, such as lyase, ligase, isomerase and synthetic enzyme-mimic activities, need to be explored.

15.4 Regulation of the Activities of Nanozymes

Compared with natural enzymes, one significant advantage of nanozymes is that their activities are easy to modulate and optimize. Because the catalytic reaction of nanozymes mainly occurs on the surface of NPs, factors affecting the surface properties (e.g. size, structure, modification, etc.) can be used to regulate the enzymatic activities of nanozymes.

15.4.1 Size-Dependent Effects

The surface area of the NPs increases significantly with decreasing size, and the electrons on their surfaces become very active simultaneously. This size-dependent effect can indirectly affect the catalytic activity of a nanozyme by influencing the active sites on the surface of the NPs. Taking Fe_3O_4 as an example, the catalytic efficiency of the nanozyme increases with decreasing particle size under the same conditions (Figure 15.8) (Lizeng Gao 2007).

15.4.2 Morphology-Dependent Effects

Besides the size effect, the morphology of a nanozyme also affects its enzymatic activity. The effects of the structure of Fe_3O_4 nanozymes on peroxidase-like activity have been studied by comparing the activity of three distinct nanoclusters with different structures (cluster spheres, octahedra, and triangular plates) (Shanhu Liu 2011). Among these three nanocrystals, the cluster spheres showed the highest activity, while the octahedral particles showed the lowest

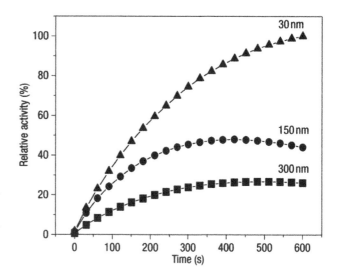

FIGURE 15.8 Under the same conditions, smaller Fe_3O_4 magnetite NPs show higher peroxidase-like activity. (Reprinted with permission from Lizeng Gao 2007. Copyright (2007) Springer Nature.)

activity. The mechanism of this difference may be based upon the distinct arrangement of iron atoms on the surface of the different nanostructures.

In addition, Zhang et al. compared the peroxidase-like and catalase-like activities of five types of Co_3O_4 nanozymes with different shapes (nanoplates, nanopolyhedron-CP, nanopolyhedron-ST, nanorods, nanocubes) and found that Co_3O_4 nanoplates possessed the highest catalytic activity (Wei Zhang 2017). They concluded that the (112) plane, which contains more Co^{3+} on its surface, plays a chief role in the catalytic behavior of Co_3O_4 nanozymes. Similarly, the GPx-like activity of orthorhombic V_2O_5 nanozymes is also influenced by the morphologies of NPs. Because of the different crystal facets exposed, V_2O_5 nanospheres exhibit remarkably higher catalytic activity than V_2O_5 nanowires (Sourav Ghosh 2018). Moreover, Mn_3O_4 nanozymes, which possesses three types of enzyme-mimic activities (SOD, catalase and GPx), also exhibit morphology-dependent enzyme mimetic behaviors (Namrata Singh 2018). Mugesh et al. showed that nanoflowers (NFs) possess the highest catalytic activity in all three enzyme-mimic systems compared with other shapes of Mn_3O_4 NPs. The multi-enzymatic activities of these particles followed the order: flowers > flakes > hexagonal plates ≈ polyhedrons ≈ cubes.

As the morphology is so important to the behaviors of nanozyme, it is worth studying the selection of a nanozyme with appropriate morphology to maintain high catalytic activity and stability.

15.4.3 Surface Microenvironment Effects

In addition to the size and morphology, the enzymatic activities of nanozymes are also affected by the surface microenvironment, such as surface charge, hydrophilicity, hydrophobicity and other weak interactions

(Zhongwen Chen 2012, Kelong Fan 2018). For example, iron oxide NPs exhibit catalase-like catalytic activity under neutral conditions but peroxidase-like catalytic activity under acidic conditions (Zhongwen Chen 2012). Moreover, some modifying functional groups, ions or molecules can be added to the surface of nanozymes to change their microenvironment, thereby regulating their affinity toward substrates and improving their catalytic capacities (Figure 15.9) (Biwu Liu 2017).

Modifying nanozymes with charged groups was found to significantly affect their enzymatic properties. Fe_3O_4 modified with positively charged molecules, such as glycine, polylysine, poly(ethyleneimine) (PEI), etc., showed enhanced peroxidase-like activity for catalyzing ABTS by improvement of the affinity of the Fe_3O_4 nanozyme to this kind of substrate. By contrast, negatively charged modification of Fe_3O_4 with citrate, carboxymethyl dextran and heparin could increase TMB binding and facilitate the oxidation reaction of TMB in the presence of H_2O_2 (Faquan Yu 2009). Moreover, a study suggested that Fe_3O_4 nanozymes modified with negatively charged DNA showed similar charge-related enzymatic activities, in which the TMB oxidation was greatly improved, while the oxidation of ABTS was inhibited (Biwu Liu 2013). Similarly, an amino-modified gold nanozyme possessing peroxidase-like activity showed higher activity toward ABTS, while citrate-capped gold nanozymes showed higher activity for catalyzing TMB (Sheng Wang 2012).

Modification by ions could also affect the enzymatic activity of nanozymes. Huang et al. have studied the peroxidase-, oxidase-, and catalase-like activity of gold nanozyme in the presence of various metal ions (Ag^+, Bi^{3+}, Pb^{2+}, Pt^{4+} and Hg^{2+}). They found that the gold NPs showed enhanced peroxidase-like activity after deposition of Ag^+, Bi^{3+} or Pb^{2+} ions on their surface. In addition, the oxidase- and catalase-like activities were also improved

by reacting with Ag^+/Hg^{2+} and Hg^{2+}/Bi^{3+} ions, respectively. In a later study, these researchers suggested that the high peroxidase-like activity of gold NPs modified with Bi^{3+} may be due to the various valence (oxidation) states of Bi^{3+} and Au (Au^+/Au^0) atoms on the surface. However, when Bi^{3+} and Hg^{2+} coexisted, the peroxidase-like activity of the gold NPs decreased greatly because of the strong Hg-Au amalgamation (Chia-Wen Lien 2013, Chia-Wen Lien 2014). Besides the abovementioned metal ions, some biologically relevant ions, such as carbonate or sulfate anions, have also been used to regulate the peroxidase-like activity of gold nanozymes. However, the carbonate or sulfate anions showed little influence on the catalytic activity of gold NPs (Juhi Shah 2015). In addition, sulfide ions (S^{2-}) were found to suppress the peroxidase-like activity of gold NPs (Hao-Hua Deng 2014). The peroxidase-like catalytic activity of Fe_3O_4 nanozymes is thought to come from the iron atoms on their surface, and is greatly weakened if the surface Fe^{3+} is blocked. Phosphate ion adsorbed on Fe^{3+} on the surface of the Fe_3O_4 nanozyme was found to strongly decrease the peroxidase-like activity of these particles (Chuanxia Chen 2014). In the same study, the authors also modified Fe_3O_4 nanozymes with some metallic ions, including Na^+, K^+, Ca^{2+}, Mg^{2+}, Al^{3+}, Ba^{2+}, Cu^{2+}, Co^{2+}, Ni^{2+}, Zn^{2+} and Pb^{2+}, but found that none of these ions significantly affected the peroxidase-like activity of Fe_3O_4 nanozymes. A study carried out by Willner et al. suggested that Cu^{2+}-modified GO composite nanomaterials exhibit heterogeneous enzyme-mimicking functions of HRP and of NADH peroxidase, catalyzing the oxidation of dopamine or the oxidation of NADH in the presence of H_2O_2, respectively, while unmodified GO NPs did not show any such activity (Shan Wang 2017). Moreover, these researchers also found that other metal ion modifications such as Ni^{2+}, Co^{2+}, Pd^{2+} and Cd^{2+} did not affect the catalytic activity of GO NPs for the oxidation of NADH. While the catalytic activities of nanozymes have been found to be related to the ions deposited on their surface, the specific underlying mechanisms still need to be elucidated.

Besides the abovementioned modifications, some other molecules may also be used to affect the catalytic activity of nanozymes, including macropolymers and some micromolecules. It has been found that the effect of molecular modification, such as by polyethylene glycol (PEG), amino, dextran, and silica, could inhibit the peroxidase-like activity of Fe_3O_4 (Lizeng Gao 2007). When catecholamine binds to Fe_3O_4, the resulting complex weakens the catalytic activity of Fe_3O_4 NPs, which is caused by the catechol moiety of catecholamines binding to the surface iron ions. Therefore, any substance with a catechol moiety, such as dopamine, could affect the activity of Fe_3O_4 nanozymes (Cheng-Hao Liu 2012). The complex formed by lipase and Fe_3O_4 significantly decreases the catalytic activity of Fe_3O_4 nanozyme because the dense protein layer formed on the surface of Fe_3O_4 blocks the substrate binding sites. In addition, dextran-coated nanoceria showed oxidase-like activity and quickly oxidized a series of substrates without H_2O_2

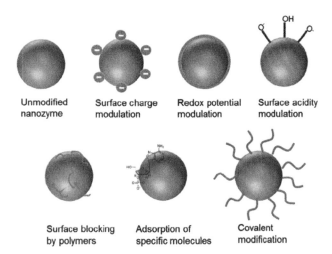

FIGURE 15.9 The strategies of modulating the enzyme-like actives of nanozymes via surface modification. (Reprinted with permission from Biwu Liu 2017. Copyright (2017) Tsinghua University Press and Springer-Verlag Berlin Heidelberg.)

under acidic conditions, which is not shown by non-coated nanoceria (Atul Asati 2009).

ATP is known to form charge transfer complexes with aromatic molecules and to participate in free radical redox reactions. Qu et al. found that ATP significantly enhanced the peroxidase-like activity of Au nanozymes toward ABTS, and greatly expanded the temperature range of its catalytic activity (Youhui Lin 2014). The authors concluded that ATP can stabilize the oxidation product ABTS and serve as an H_2O_2 activator and electron transfer energizer. Sanjay et al. also found that compared with natural enzymes, ATP modification could only increase the peroxidase-like activity of Au nanozyme without affecting other catalytic activities (Juhi Shah 2015). Sanjay et al. further studied the mechanism by which ATP enhances Au nanozyme peroxidase-like activity (Juhi Shah 2018). They found that the catalytic activity of the Au nanozyme modified by ATP is not enhanced by producing more hydroxyl radicals. The likely mechanism of activity improvement by ATP is the stabilization of oxidized TMB. With a similar mechanism, Au nanozyme coupled with melamine also showed increased catalytic capacity toward TMB in the presence of H_2O_2 (Pengjuan Ni 2014).

In order to improve the enzyme activity, nanozymes can be modified with active moieties corresponding to the active centers or structures of natural enzymes. For instance, Fe_3O_4 was modified with histidine to simulate the active center of natural HRP, and showed significantly enhanced peroxidase-like activity (Kelong Fan 2016). Interestingly, some studies also showed that modification by histidine-containing peptides could accelerate the transesterification of the p-nitrophenyl ester of N-carboxybenzylphenylalanine catalyzed by Au NPs (Davide Zaramella 2012).

As we all know, substrate selectivity is one of the important properties of enzymes. Although there have been many studies on the regulation of nanozymes' catalytic activity, there is still a lack of research on their selectivity. Methods like polymer imprints have been used to improve substrate selectivity, but more effective methods still need to be developed (Zijie Zhang 2017).

15.5 The Applications of Nanozymes in Biomedicine

Natural enzymes have precisely regulated catalytic activity and high substrate selectivity. These characteristics make natural enzymes valuable for applications in industry, agriculture, medicine, environmental management and life sciences. However, the applications of natural enzymes have been limited by their disadvantages such as having poor stability and being easy to inactivate and difficult to prepare and preserve. Although many attempts have been made to create artificial enzyme mimetics using nanotechnology, most of them are still limited by the complicated and tedious preparation processes required. Compared with traditional artificial enzymes, nanozymes with intrinsic enzyme-like activities are easier to synthesize and functionalize while exhibiting higher stability and catalytic activity. The emergence of nanozymes facilitates the wider use of nanotechnology in biomedical fields. Their high catalytic activity makes it possible to replace natural enzymes in biological detection, disease diagnosis and treatment. This section mainly introduces the current research on nanozymes in the field of life sciences and their applications in biomedicine.

15.5.1 Biosensors Based on Nanozymes

Based on the effective catalytic activity of nanozymes, several novel strategies have been developed for the detection of various biological entities, including bioactive small molecules (e.g., H_2O_2, glucose, lactic acid) and biological macromolecules (e.g., nucleic acids, proteins, polysaccharides). Therefore, nanozymes can play an important role in environmental conservation and disease detection.

Environmental Monitoring

One of the important tasks of environmental monitoring is to monitor the concentration of peroxide in raindrops. Nanozymes can replace natural enzymes for the detection of peroxide in raindrops (Figure 15.10).

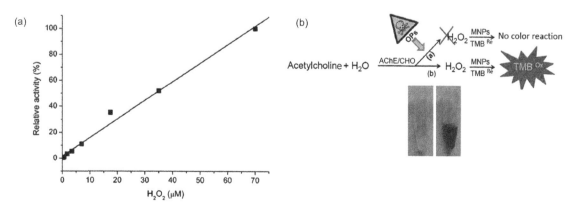

FIGURE 15.10 (a) The Fe_3O_4 magnetite NPs are efficient catalysts for the determination of H_2O_2 in rainwater. (Reprinted with permission from Jie Zhuang 2008. Copyright (2008) Elsevier.) (b) Fe_3O_4 magnetite NPs colorimetric assay for the rapid detection of organophosphate neurotoxins. (Reprinted with permission from Minmin Liang 2013. Copyright (2013) American Chemical Society.)

Researchers have developed a new acid rain–detection method using the peroxidase-mimic property of Fe_3O_4 nanozymes (Figure 15.10a). In acid rain detection, acidic conditions may affect the activity of biological enzymes, thus affecting the final detection results. The Fe_3O_4 nanozyme-based method can quickly detect the concentration of H_2O_2 in rainwater, in which the activity of Fe_3O_4 nanozyme is positively correlated with the concentration of H_2O_2. Moreover, Fe_3O_4 nanozymes can be reused after recycling by separation under a magnetic field and always maintain a high catalytic activity (Jie Zhuang 2008).

In addition, a study combining Fe_3O_4 nanozymes with acetylcholinesterase (AChE), and choline oxidase (CHO) demonstrated a colorimetric method for the rapid detection of organophosphorus pesticides and nerve agents (Figure 15.10b). In this method, H_2O_2 can be formed under the catalysis of the enzymes AChE and CHO in the presence of acetylcholine, which then stimulates Fe_3O_4 nanozymes to catalyze the oxidation of colorimetric substrates to produce a color reaction. Moreover, the organophosphorus neurotoxins inhibit the enzymatic activity of AChE, which reduces the content of H_2O_2 and then results in decreased peroxidase-like catalytic activity of the Fe_3O_4 nanozymes, accompanied by a decrease in color intensity after the oxidation of colorimetric substrates. Using this method, as low as 1 nM Sarin, 10 nM methyl-paraoxon and 5 μM acephate could be easily detected (Minmin Liang 2013).

Detection of Biomarkers for Disease

Glucose detection plays an important role in biomedical analysis and diabetes monitoring. At present, glucose oxidase colorimetry is mainly used in the clinical detection of glucose, and its principle is based on a dual-enzyme system, which is HRP and GOx combined to produce a color reaction. Since the Fe_3O_4 nanozyme exhibits the catalytic function of peroxidase, it can not only replace HRP in colorimetry but also conjugate GO_X directly onto the surface of Fe_3O_4 NPs. While GOx catalyzes glucose to produce H_2O_2, the nanozyme can directly exert its peroxidase catalytic activity, and then produce a color reaction with the substrate ABTS. This provides a concept whereby one type of nanozyme can also be used in combination with other kinds of natural enzymes for specific biomarker detection (Figure 15.11a) (Hui Wei 2008). Taking advantage of this feature, a nanozyme-based biomarker diagnosis method can also be used *in vivo* to monitor the dynamic changes of biomolecules in real time to monitor disease progression after drug administration. Wei et al. designed and developed an integrated nanozyme (INAzymes) containing the molecular catalyst hemin and natural enzyme glucose oxidase inside the zeolitic imidazolate framework (ZIF-8) nanocomposite structure, which showed high peroxidase-like activity, to achieve *in vivo* glucose measurements. This nanocomposite can carry out an enzymatic cascade reaction, to achieve products of the first reaction that can be used immediately as substrates for the second reaction,

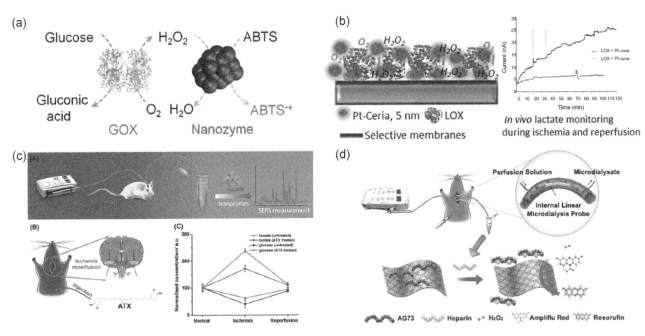

FIGURE 15.11 (a) Combining the catalytic reaction of glucose with GOx and the Fe_3O_4 magnetite NPs catalytic reaction, a colorimetric method for glucose detection was developed. (Reprinted with permission from Hui Wei 2008. Copyright (2008) American Chemical Society.) (b) A platinum-doped ceria-based biosensor enables real-time *in vivo* monitoring of lactic acid during hypoxia. (Reprinted with permission from Naimish Pandurang Sardesai 2015. Copyright (2015) American Chemical Society.) (c) Real-time monitoring of glucose and lactic acid in living rats' brains with an integrative nanozyme-based detection platform. (Reprinted with permission from Yihui Hu 2017. Copyright (2017) American Chemical Society.) (d) Monitoring of the heparin elimination process in live rats using 2D MOF nanozymes. (Reprinted with permission from Hanjun Cheng 2017. Copyright (2017) American Chemical Society.)

forming a progressive amplification system to improve detection sensitivity. The combination of the multiple functions of INAzymes and the advantages of synergistically enhanced catalytic activity may also make this one of the future development directions of nanozymes (Hanjun Cheng 2016).

A double-enzyme continuous reaction system can also be used to detect lactic acid and heparin. Lactic acid is an indicator for tissue oxygen levels, and elevated levels of lactic acid are associated with various disorders such as shock and myocardial ischemia. A platinum-doped ceria-based biosensor designed by Andreescu et al. enables real-time *in vivo* monitoring of lactic acid during hypoxia. The catalase-like catalytic activity of these NPs decompose H_2O_2 to generate oxygen, and these released oxygen molecules can then be used by lactic acid oxidase (LOx) in the conversion of lactic acid to regenerate H_2O_2 at the bubble surface. This allows the benign cycle of substrate–product formation to be continuously monitored for 2 h (Figure 15.11b) (Naimish Pandurang Sardesai 2015). Wei et al. also designed a peroxidase-mimicking nanozyme by growing AuNPs *in situ* in a highly porous and thermally stable MOF, coupled with GOx and LOx, which realized the respective monitoring of glucose and lactic acid in living brains (Figure 15.11c) (Yihui Hu 2017). In addition to monitoring glucose and lactic acid in real time, these researchers also extended the real-time monitoring method of nanozymes to the detection of heparin. They designed a 2D MOF nanosheet modified with tetrakis (4-carboxyphenyl) porphyrin (TCPP) and metallized with Zn^{2+} metal ions, which showed enhanced peroxidase-mimicking activities compared with their 3D bulk analogues. As a proof-of-concept bioassay, heparin-specific peptides AG73 were physically adsorbed onto these MOF nanosheets, reducing their enzymatic activity by blocking the active sites. The catalytic activity of the MOF nanosheets is released after heparin reacts with the peptides AG73, which is positively related to the concentration of heparin, and was used to monitor the dynamic changes of heparin in the arteries of live rats after drug administration. These 2D MOF nanosheets provide a platform for biological detection and may be improved to meet multiple detection requirements (Figure 15.11d) (Hanjun Cheng 2017).

In addition to sensing small molecules in disease, nanozymes are widely used in detecting macromolecular biotargets (nucleotides, proteins, cells, bacteria and viruses, etc.). A rapid and simple DNA detection assay, based on the mechanism where DNA binding inhibits the peroxidase-like activity of the magnetite NP nanozyme, was reported to be accomplished using naked-eye detection within 30 mins, which overcomes the time-consuming disadvantages of the polymerase chain reaction (PCR) and reverse transcription-polymerase chain reaction (RT-PCR). This is the first time that DNA has been detected using nanozymes (Ki Soo Park 2011).

Many antigens can be detected rapidly by using an immunological detection method employing nanozymes. *In situ* growth of porous platinum NPs on graphene oxide (Pt NPs/GO) can act as a colorimetric assay for the direct detection of cancer cells. Through antibody targeting, this system can specifically detect tumor cells expressing corresponding antigens. After specific binding with tumor cells, it can catalyze the substrate TMB to produce a color reaction, thus achieving the goal of cancer detection. Doping with Pt greatly improves the peroxidase-like activity of GO, which showed a higher affinity for TMB than HRP (Figure 15.12a) (Ling-Na Zhang 2014).

Subsequently, researchers developed a label-free chemiluminescent (CL) immunosensor by using cupric oxide nanorods (CuO NRs) as peroxidase mimics to directly detect tumor antigens more accurately. Antigens could be introduced into the sensing system, forming large immunocomplexes with biotinylated anti-antigen, streptavidin and the CuO NRs-chitosan support that prevents the CL substrate from accessing the surface, thereby reducing the CL signal in a concentration-dependent fashion. The proposed label-free immunosensor was able to rapidly determine carcinoembryonic antigen (CEA), which was used as a model analyte, with a wide linear range of 0.1–60 ng/mL and a low detection limit of 0.05 ng/mL (Figure 15.12b) (Juan Li 2018).

Nanozymes also play a role in application for the detection of infectious diseases. Yan et al. fabricated a nanozyme probe by conjugating the anti-EBOV antibody to the Fe_3O_4 nanozyme, which possessed intrinsic nanozyme activity and generated a color reaction with substrates to enhance the signal in a strip. The sensitivity of the nanozyme-based strip for EBOV detection and diagnostic accuracy was comparable with ELISA, but much faster (within 30 min) and simpler (without the need for specialized facilities) (Figure 15.12c) (Demin Duan 2015).

15.5.2 Tumor Imaging

Tumor imaging in early progression stages remains a key requirement for efficient cancer diagnosis and treatment. Researchers demonstrated the application of a nanozyme in tumor imaging by using the high ROS characteristics in tumor regions to induce the catalytic activity of the nanozyme. Gu et al. explored the application of the catalase-like activity of Prussian Blue NPs (PBNPs) (KFe^{3+} $[Fe^{2+}(CN)_6]$) in ultrasound imaging. PBNPs can catalyze H_2O_2 to generate O_2, and these gas bubble–forming molecules can be used as an ultrasound contrast agent to enhance ultrasound signals. In tissues that overproduce radical oxidants, such as tumors, PBNPs can make them more easily detectable by ultrasound imaging (Fang Yang 2012).

MnO NPs functionalized with catechol-PEG can significantly enhance magnetic resonance imaging (MRI) contrast when exposed to superoxide radicals, such as in the tumor microenvironment. The MRI relaxation times of MnO increase due to the SOD reaction and lead to a significant enhancement (Figure 15.13) (R. Ragg 2013).

Moreover, taking advantage of the enzymatic properties of nanozymes, several colored products of natural

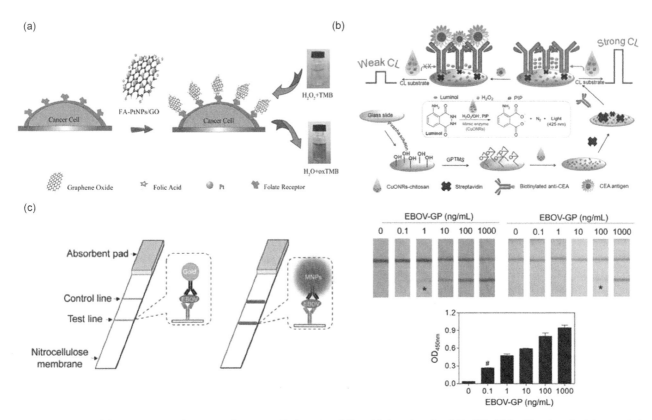

FIGURE 15.12 (a) Colorimetric direction of cancer cells by using folic acid-functionalized Pt NPs/GO. (Reprinted with permission from Ling-Na Zhang 2014. Copyright (2014) American Chemical Society.) (b) Dual-functional CuO NRs as peroxidase mimics developed as a label-free chemiluminescent immunosensor for biosensing applications. (Reprinted with permission from Juan Li 2018. Copyright (2018) Elsevier.) (c) Standard colloidal gold strip and nanozyme-strip employing magnetite NPs in place of colloidal gold to form a novel nanozyme probe with higher sensitivity. (Reprinted with permission from Demin Duan 2015. Copyright (2015) Elsevier.)

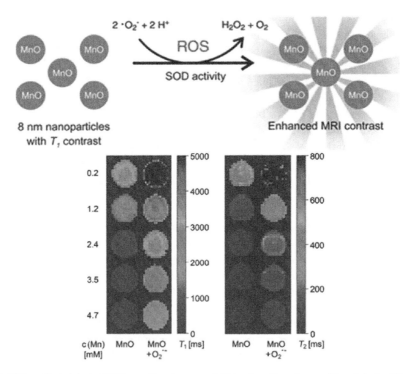

FIGURE 15.13 MnO NPs with intrinsic SOD reaction increase MRI relaxation times. (Reprinted with permission from R. Ragg 2013. Copyright (2013) Royal Society of Chemistry.)

enzyme HRP-based immunohistochemistry (IHC) could also be generated for tumor imaging. With the development of nanozymes, researchers have tried to replace HRP in conventional IHC with peroxidase-like nanozymes. Magnetoferritin NPs (M-HFn), a well-defined peroxidase-mimicking Fe_3O_4 nanozyme synthesized within recombinant human heavy-chain ferritin shells, exhibit peroxidase-like activity toward typical peroxidase substrates. Relying on the specific targeting activity toward tumor high-expression antigen transferrin receptor protein 1 (TfR1) and the peroxidase-like catalytic behavior of M-HFn for the substrate DAB, the imaging of tumors could be achieved through a one-step incubation, showing an intense brown color for visualizing the tumor tissues. More than 1,400 clinical specimens with ten types of cancer were tested, and the results showed that the nanozyme imaging platform can discriminate cancerous cells from normal cells with 98% sensitivity and 95% specificity (Figure 15.14a) (Kelong Fan 2012).

Gu et al. have carried out a series of works on the use of nanozymes for IHC detection of tumors (Jinlai Dong 2014, Lin Fan 2016, Wei Zhang 2017). In a 2014 paper published by these authors, IHC was performed using a Co_3O_4 conjugate antibody. Co_3O_4 NPs were modified by dimercaptosuccinic acid (DMSA), and then immobilized with an antibody on the surface through activation and an amide reaction. The antibody can be captured by antigens on the tumor slice. With the addition of H_2O_2 and DAB, Co_3O_4 then acts as a peroxidase, oxidizing DAB into brown products on the surface of the clinical tumor tissues. Their staining ability was comparable to that of natural HRP (Figure 15.14b) (Jinlai Dong 2014). Later, they employed the heavy chain of the antibody, which is prevented from aggregating, to couple with Co_3O_4 nanozymes. The nanoprobes were successfully applied in the detection of epidermal growth factor receptor (EGFR) expression in non-small-cell lung cancer tissues (Wei Zhang 2017).

In order to achieve more sensitive imaging for IHC, HRP-labeled secondary antibodies conjugated with Au NPs were constructed. Au NPs can be loaded with more HRP enzymes, therefore resulting in enhanced DAB chromogenicity. Simultaneously, Au NPs also act as a synergistically enhanced agent due to their peroxidase-like enzyme catalysis (Figure 15.14c) (Lin Fan 2016).

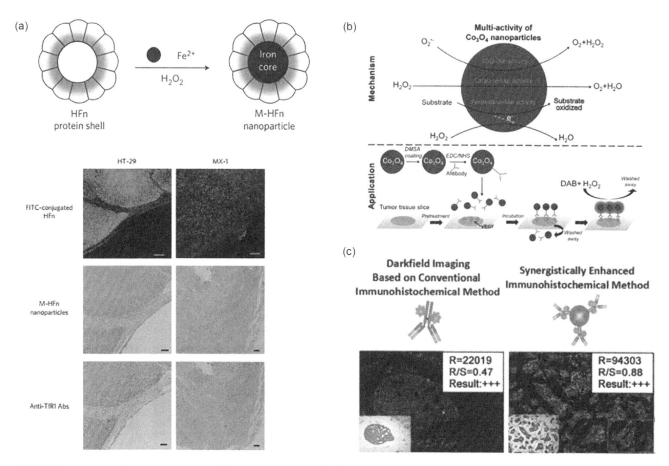

FIGURE 15.14 (a) The synthesis of M-HFn and the usage of M-HFn as peroxidase mimic for targeting and visualizing tumor tissues. (Reprinted with permission from Kelong Fan 2012. Copyright (2012) Springer Nature.) (b) Multi-enzyme mimetics of Co_3O_4 NPs and IHC detection process based on the peroxidase-like activity. (Reprinted with permission from Jinlai Dong 2014. Copyright (2014) American Chemical Society.) (c) Synergistically enhanced immunohistochemical method based on HRP-labeled secondary antibodies conjugated with Au NPs. (Reprinted with permission from Lin Fan 2016. Copyright (2016) Royal Society of Chemistry.)

15.5.3 Anti-bacterial activity

Nanozymes can destroy the biofilms on the surface of bacteria, so they have great application in wound cleaning, dental caries treatment, ship dredging and so on. Moreover, the anti-bacterial activity of nanozymes mainly relies on the production of ROS, for which bacteria have difficulty developing drug resistance (Yuyun Zhao 2013, Xiaoning Li 2014, Li-Sheng Wang 2016). Gao et al. found that Fe_3O_4 NPs with peroxidase-like activity enhanced the cleavage of biofilm components (model nucleic acids, proteins and oligosaccharides) in the presence of H_2O_2, because the Fe_3O_4 nanozymes could catalyze H_2O_2 to make free radicals that attack these biomolecules. In addition, the Fe_3O_4–H_2O_2 system efficiently broke down the existing biofilm and prevented new biofilms from forming (Figure 15.15a) (Lizeng Gao 2014).

Based on this, Qu et al. fabricated multiple composite nanozymes, confining passivated gold NPs with multiple cerium (IV) complexes on the surface of colloidal magnetic Fe_3O_4/SiO_2 core/shell particles, and found that this hybrid nanozyme possesses DNase-mimetic activities and exhibits high cleavage ability toward extracellular DNA to kill biofilm-encased bacteria and biofilms (Zhaowei Chen 2016). Later, Gao et al. used a Fe_3O_4 nanozyme to inhibit the occurrence and development of dental caries. Fe_3O_4

nanozymes, known to exhibit peroxidase-like activity under acidic conditions, can catalyze the substrate H_2O_2 to produce free radicals, kill bacteria on teeth and degrade extracellular polysaccharides until the attached biofilm matrix is destroyed (Figure 15.15b) (Lizeng Gao 2016, Lizeng Gao 2017).

The antibacterial properties of nanozymes can also prevent wound infection and promote wound healing. Zhao et al. synthesized PEG-functionalized molybdenum disulfide nanoflowers (PEG-MoS$_2$ NFs). They found that PEG-MoS$_2$ NFs have peroxidase-like activity and can efficiently catalyze decomposition of low-concentration H_2O_2 to generate hydroxyl radicals (·OH). With this excellent antibacterial activity, PEG-MoS$_2$ NFs could be conveniently used for wound disinfection. Moreover, the PEG-MoS$_2$ NFs-based anti-bacterial strategy also avoids the toxicity of high concentrations of H_2O_2 (Figure 15.15c) (Wenyan Yin 2016). Employing the enzymatic activity of Graphene quantum dots (GQDs), GQDs-Band-Aids were prepared and showed excellent antibacterial properties with the assistance of H_2O_2 at low dose *in vivo* (Hanjun Sun 2014).

Moreover, V_2O_5 nanowires act like naturally occurring vanadium haloperoxidases to thwart biofilm formation. Therefore, this can be a green and efficient method to remove biological deposition on ship hulls. In the presence

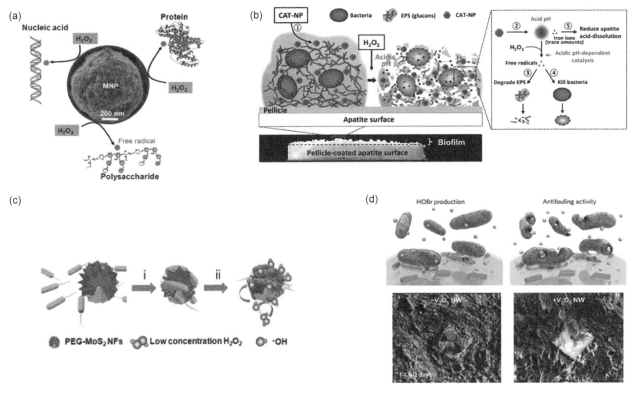

FIGURE 15.15 (a) Fe_3O_4 NP-enhanced cleavage of a nucleic acid, a protein and a polysaccharide, for biofilm elimination. (Reprinted with permission from Lizeng Gao 2014. Copyright (2014) Royal Society of Chemistry.) (b) Biofilm disruption under acidic conditions by Fe_3O_4-NP/H_2O_2. (Reprinted with permission from Lizeng Gao 2016. Copyright (2016) Elsevier.) (c) PEG-MoS$_2$ as a peroxidase catalyst eliminating bacteria in wound disinfection. (Reprinted with permission from Wenyan Yin 2016. Copyright (2016) American Chemical Society.) (d) V_2O_5 nanowires act like vanadium haloperoxidases to prevent biofilm formation by bacteria on ships' hulls. (Reprinted with permission from Filipe Natalio 2012. Copyright (2012) Springer Nature.)

of substrates such as Br^- and H_2O_2, small amounts of hypobromous acid (HOBr) are produced continuously. The released HOBr prevents adhesion of the bacteria and biofilm formation without being toxic to marine biota (Figure 15.15d) (Filipe Natalio 2012).

15.5.4 Disease Therapy

By effectively regulating the levels of ROS *in vivo*, the different enzymatic activities of nanozymes can be well applied in the treatment of various diseases, such as tumors, inflammatory diseases and other types of diseases caused by ROS imbalance (Toren Finkel 2000, Xiaoqiang Chen 2011).

Cancer Therapy

ROS-induced apoptosis is a popular strategy for cancer therapy (Dunyaporn Trachootham 2009, Lakshmi Raj 2011, Zijian Zhou 2016). The tumor therapy strategies utilizing nanozymes mainly act through prompting the production of ROS. Fe_3O_4 nanozymes can simulate peroxidase, and thereby could catalyze the decomposition of H_2O_2 to generate ROS efficiently to inhibit tumors *in vivo*. Tumor cell viability was significantly decreased after treatment by Fe_3O_4 nanozymes and H_2O_2. Moreover, the results of an experiment in tumor-bearing mice also showed that the tumor growth rate was greatly delayed and even cured after treatment with Fe_3O_4 nanozymes and H_2O_2 (Di Zhang 2013). Linoleic acid hydroperoxide tethered on Fe_3O_4

nanozymes was also able to induce efficient apoptotic cancer cell death both *in vitro* and *in vivo* by producing ROS in a specific tumor microenvironment (Zijian Zhou 2017).

Recently, Shi et al. fabricated an enzymatic cascade reaction system, which was composed of large-pore-size, biodegradable dendritic silica NPs (DMSNs), decorated natural GOx and an ultra-small Fe_3O_4 nanozyme (designated as GOx-Fe_3O_4@DMSNs nanocatalyst), to efficiently treat tumors *in vivo*. In this enzymatic system, the GOx could effectively deplete the glucose in tumor cells, resulting in tumor nutrient deprivation, meanwhile producing considerable amounts of H_2O_2. Fe_3O_4 nanozymes then exhibit peroxidase catalysis under the acidic microenvironment of the tumor, to generate abundant highly toxic hydroxyl radicals. Through these sequential catalytic processes, consequent tumor-cell apoptosis and death were finally achieved (Figure 15.16a) (Minfeng Huo 2017).

Interestingly, nanozymes with peroxidase-like activities also exhibit synergistic efficacy in combination with ROS-generating anticancer agents for cancer therapy. In NAD(P)H:quinone oxidoreductase 1 (NQO1)-overexpressing cancer cells, β-lapachone (β-lap) can specifically increase the oxidative stress in tumor cells to efficiently kill tumors. Huang et al. reported that superparamagnetic iron oxide NPs with peroxidase-like activity synergistically enhanced the ROS stress in β-lap-exposed cancer cells, thereby enhancing the therapeutic index of β-lap by 10-fold (Gang Huang 2013).

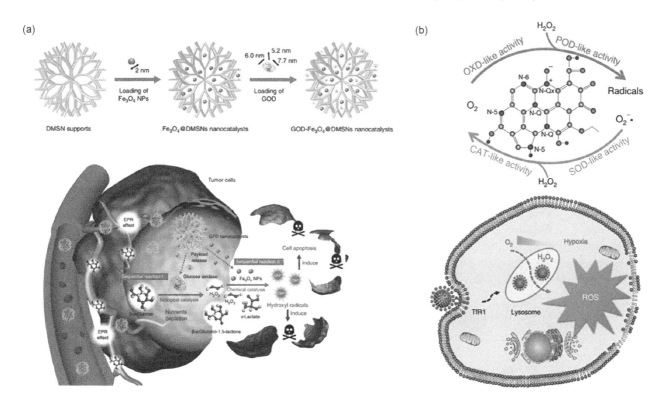

FIGURE 15.16 (a) Synthetic procedure for GOx-Fe_3O_4@DMSNs nanocatalysts and the sequential catalytic therapeutic mechanism of the generation of hydroxyl radicals for cancer therapy. (Reprinted with permission from Minfeng Huo 2017. Copyright (2017) Nature Publishing Group.) (b) N-PCNSs induced tumor cell destruction by generating ROS via ferritin-mediated specific tumor delivery and lysosome localization. (Reprinted with permission from Kelong Fan 2018. Copyright (2018) Nature Publishing Group.)

In addition to Fe_3O_4 nanozymes, carbon nanozymes have also been used to develop anti-tumor strategies. In 2018, Yan's group fabricated a novel type of nanozyme by using nitrogen-doped porous carbon nanospheres (N-PCNSs). This nanozyme possesses four enzymelike activities (oxidase, peroxidase, catalase and SOD) depending on pH, all of which are involved in ROS regulation in biological systems. Guided by the conjugated ferritin, N-PCNSs nanozymes specifically target the tumor, and localize in the lysosome of tumor cells. The acidic microenvironment of lysosomes allows N-PCNSs to perform oxidase-like and peroxidase-like activities, resulting in abundantly increasing ROS levels to destroy tumor cells (Figure 15.16b) (Kelong Fan 2018).

Since the multi-enzymatic activities of nanozymes are pH dependent, the sub-locations of nanozymes after entering the cell are important. However, most nanozymes are typically localized in the lysosome after entering cells. Thus, the predominant enzymatic activities in different conditions determine the *in vivo* applications of nanozymes. The catalase-like catalytic activity of CeO_2 nanozymes was inhibited in acidic environments, while the SOD catalytic activity was not affected. Therefore, CeO_2 nanozymes could be used as a sensitizer for radiotherapy. The SOD-like activity of CeO_2 nanozymes can continuously transform superoxide radical in acidic pancreatic cancer cells into H_2O_2. The accumulated H_2O_2 can enhance the apoptosis effect in radiation-induced tumor cells (Melissa S. Wason 2013).

Anti-oxidative property

In inflammatory diseases, nanozymes are primarily employed to scavenge ROS to relieve symptoms of diseases. Nanozymes with SOD, catalase or GPx-like activities, such as platinum NPs (Lianbing Zhang 2010, Mauro Moglianetti 2016), cerium oxide NPs (Francesca Pagliari 2012), V_2O_5 nanowires (Amit A. Vernekar 2014), Mn_3O_4 (Namrata Singh 2017), PBNPs (Wei Zhang 2016, Jiulong Zhao 2018)

and nitrogen-doped carbon nanodots (Zi-Qiang Xu 2015) can remove ROS to resist both endogenous and exogenous oxidative pressure, and reshape the balance of oxidative reduction, thus protecting cells.

The overproduction of ROS is the key to the initiation and progression of inflammatory bowel disease. In mice with the colitis model, polyvinylpyrrolidone (PVP)-modified PBNPs nanozymes reduced intestinal inflammation, bleeding and infiltration of immune cells, and alleviated epithelial damage to mucosal areas via eliminating ROS. This treatment exhibits improved clinical outcomes (Figure 15.17a) (Jiulong Zhao 2018).

Nanozymes are also used in other types of diseases induced by the pathologic status of ROS, including degenerative diseases, senility, cardiovascular disease and so on. The catalase-like activity of Fe_3O_4 nanozymes can be used to protect cells from H_2O_2-induced oxidative stress in Parkinson disease (PD) and Alzheimer's disease (AD). Dietary Fe_3O_4 nanozymes have already been proved to have anti-aging effects and were recently shown to extend the lifespan of *Drosophila* (Yi Zhang 2016).

Mn_3O_4 nanozymes possesses SOD, catalase and GPx-like activities, which are important enzymes to reduce ROS *in vivo*. In an MPP^+(1-methyl-4-phenylpyridinium) induced PD-like cellular model, Mn_3O_4 nanozymes significantly regulated the redox balance in cells and successfully helped cells resist oxidative stress (Namrata Singh 2017).

The overabundance of ROS is one of the most critical mechanisms responsible for causing ischemic injury (C. L. Allen 2009), which can lead to ischemic stroke or cardiovascular diseases. Taeghwan Hyeon et al. demonstrated that nanoceria could eradicate ROS to protect against ischemic stroke *in vivo*. Animal experiments showed that nanoceria with optimal doses at 0.5 and 0.7 mg/kg significantly reduced the brain infarct volume (Chi Kyung Kim 2012). 2, 3-dimercaptosuccinic acid (DMSA)-modified Fe_2O_3 NPs (Fe_2O_3@DMSA NPs) in a range of small sizes exhibited

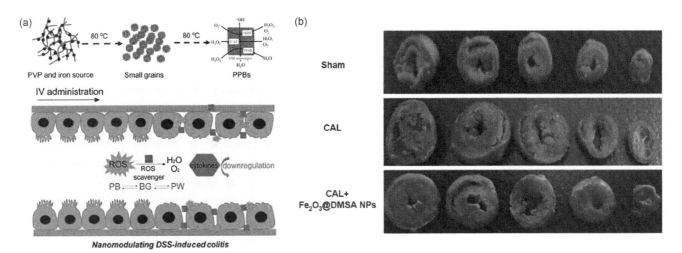

FIGURE 15.17 (a) PBNPs nanozymes could modulate colitis by eliminating ROS. (Reprinted with permission from Jiulong Zhao 2018. Copyright (2018) American Chemical Society.) (b) Fe_2O_3@DMSA NPs can protect against cardiovascular disease caused by ischemia. (Reprinted with permission from Fei Xiong 2015. Copyright (2015) Nature Publishing Group.)

cardioprotective activity *in vitro* and *in vivo*. Fe_2O_3@DMSA NPs treatment could significantly repair coronary artery ligature-induced injury in rats. The authors concluded that Fe_2O_3@DMSA NPs may enhance the cellular SOD activity, thus decreasing ROS levels. Even compared with clinical drugs, the therapeutic effect is appreciable, showing its potential application in clinical treatment of cardiovascular diseases (Figure 15.17b) (Fei Xiong 2015).

Nanozymes entering the body can act as enzymes to treat diseases, but the safety of nanozymes in the body has been a concern. The safety of nanozymes *in vivo* has been controversial and needs further exploration (Feng Zhao 2011, Hamed Arami 2015, Wei Zhang 2017).

15.6 Challenges and Perspectives of Nanozymes

The emergence of nanozymes not only changes the traditional concept that nanomaterials are biologically inert materials but also provides a new perspective for the study of the biological effects of nanomaterials. The decade of development of nanozymes has depended on the efforts of many scientists. In 2007, the enzymatic characteristics of inorganic nanomaterials, which can be used as simulated enzymes for disease detection, were first reported (Lizeng Gao 2007). In 2013, the first review on nanozymes was published, and the term "nanozyme" has been widely accepted since then (Hui Wei 2013). In 2016, a special website on "nanozymes" was established (Hui Wei 2016). In 2018, the standardization of nanozymes was formulated (Bing Jiang 2018). There are still many challenges ahead for nanozymes, but this also represents an opportunity.

To sum up, we have outlined the development of nanozymes over the past decade, including the discovery and design of nanozymes, activity control and optimization, and biomedical applications of nanozymes. Although nanozymes have unparalleled advantages in stability and large-scale preparation, they are not perfect.

The catalytic activity of nanozymes is still not comparable to that of natural enzymes in most cases, which makes it more difficult to completely replace natural enzymes in application. In addition, the substrate specificity of nanozymes is still poor. So far, single-substrate-specific nanozymes have rarely been reported. High substrate specificity is especially important for the catalytic activity of natural enzymes, and it may be the same for nanozymes. Therefore, finding ways to improve the substrate specificity of nanozymes may play an equally important role in improving their catalytic activity.

To compensate the imperfections of nanozymes in catalytic activity and substrate specificity, deep understanding their catalytic mechanism is required. At present, most of the studies on the catalytic mechanism of nanozymes still remain in the hypothesis stage, lacking reliable evidence to explain the specific catalytic process of nanozymes. The main reason for this situation is the lack of technical means

needed to accurately detect the rapid molecular changes in the process of nanozyme catalysis. Nevertheless, because there are many similarities between the catalytic behavior of nanozymes and that of natural enzymes, the catalytic mechanism of natural enzymes may draw some lessons for the study of the mechanism of nanozymes. By learning from natural enzymes, scientists may design nanozymes more elaborately to achieve the best catalytic effect. For instance, some natural enzymes use cofactors including metal ions and organic compounds to improve their catalytic activity. Nanozymes may also use similar cofactors to promote their activity (as mentioned in Section 15.4.3).

Moreover, compared with natural enzymes that have a wide variety of catalytic prowess (e.g., oxidoreductases, transferases, hydrolases, cleavages, isomerases and ligases), most nanozymes can only catalyze redox reactions. The limited reaction types of nanozymes may be related to their relatively rough structure, which also implies that more sophisticated design of nanozymes is needed in future studies.

In addition to enzyme-like activities, nanozymes have unique physical and chemical properties of nanomaterials, such as magnetic, photothermal and photoacoustic properties. Combing these characteristics of nanozymes with their highly efficient catalytic activity is a hotspot in the application of nanozymes, which may lead to them exerting some functions that natural enzymes do not have (as mentioned in Section 15.5). However, because nanozymes are exogenous substances to the body, the toxicity of nanozymes *in vivo* has to be carefully considered. Although many studies have confirmed that nanozymes have significant anti-oxidative or pro-oxidative effects, the overall and long-term effects of these nanomaterials in vivo remain unknown.

So far, *in vivo* studies of nanozymes mainly focused on the catalytic activity of nanozymes in disease microenvironment. However, these exogenous substances may also have important effects on the metabolism and immune system of organisms which cannot be ignored. Also, as enzyme mimetics, nanozymes can replace natural enzymes in applications *in vitro*, but whether they can be used as substitutes for natural enzymes *in vivo* is unidentified. As we can see, the *in vivo* research of nanozymes has just started, and there are still many problems to be solved in the process from bench to bedside.

Acknowledgments

This work was supported in part by the National Natural Science Foundation of China (Grant No. 31871005, 31530026, 31900981), the Strategic Priority Research Program, CAS (Grant No. XDPB29040101), Chinese Academy of Sciences (Grant No. YJKYYQ20180048), the Key Research Program of Frontier Sciences, CAS (Grant No. QYZDB-SSW-SMC013), National Key R&D Program of China (2017YFA0205501), and Youth Innovation Promotion Association CAS (Grant No. 2019093).

References

Ajay Kumar Gupta, Mona Gupta (2005). "Synthesis and surface engineering of iron oxide nanoparticles for biomedical applications." *Biomaterials* **26**(18): 3995–4021.

Alessandro Trovarelli, Carla de Leitenburg, Marta Boaro, Giuliano Dolcetti (1999). "The utilization of ceria in industrial catalysis." *Catalysis Today* **50**(2): 353–367.

Amit A. Vernekar, Devanjan Sinha, Shubhi Srivastava, Prasath U. Paramasivam, Patrick D'Silva, Govindasamy Mugesh (2014). "An antioxidant nanozyme that uncovers the cytoprotective potential of vanadia nanowires." *Nature Communications* **5**: 5301.

Amit Kumar Dutta, Swarup Kumar Maji, Divesh N. Srivastava, Anup Mondal, Papu Biswas, Parimal Paul, Bibhutosh Adhikary (2012). "Synthesis of FeS and FeSe nanoparticles from a single source precursor: a study of their photocatalytic activity, peroxidase-like behavior, and electrochemical sensing of H_2O_2." *ACS Applied Materials & Interfaces* **4**(4): 1919–1927.

Atul Asati, Santimukul Santra, Charalambos Kaittanis, Sudip Nath, J. Manuel Perez (2009). "Oxidase-like activity of polymer-coated cerium oxide nanoparticles." *Angewandte Chemie International Edition* **48**(13): 2308–2312.

Bin Li, Daomei Chen, Jiaqiang Wang, Zhiying Yan, Jinping Zhang, Fagui Yuan (2014). "MOFzyme: Intrinsic protease-like activity of Cu-MOF." *Scientific Reports* **4**: 6759.

Bing Jiang, Demin Duan, Lizeng Gao, Guohui Nie, Minmin Liang, Xiyun Yan (2018). "Standardized assays for determining the catalytic activity and kinetics of peroxidase-like nanozymes." *Nature Protocols* **13**(7): 1506–1520.

Biwu Liu, Juewen Liu (2013). "Accelerating peroxidase mimicking nanozymes using DNA." *Nanoscale* **7**(33): 13831–13835.

Biwu Liu, Juewen Liu (2017). "Surface modification of nanozymes." *Nano Research* **10**(4): 1125–1148.

C. L. Allen, U. Bayraktutan (2009). "Oxidative stress and its role in the pathogenesis of ischaemic stroke." *International Journal of Stroke* **4**(6): 461–470.

Cassandra Korsvik, Swanand Patil, Sudipta Seal, William T. Self (2007). "Superoxide dismutase mimetic properties exhibited by vacancy engineered ceria nanoparticles." *Chemical Communications* **10**(10): 1056–1058.

Chen-I Wang, Wen-Tsen Chen, Huan-Tsung Chang (2012). "Enzyme mimics of Au/Ag nanoparticles for fluorescent detection of acetylcholine." *Analytical Chemistry* **84**(22): 9706–9712.

Cheng-Hao Liu, Cheng-Ju Yu, Wei-Lung Tseng (2012). "Fluorescence assay of catecholamines based on the inhibition of peroxidase-like activity of magnetite nanoparticles." *Analytica Chimica Acta* **745**: 143–148.

Chi Kyung Kim, Taeho Kim, In-Young Choi, Byung-Woo Yoon, Seung-Hoon Lee, Taeghwan Hyeon (2012). "Ceria nanoparticles that can protect against ischemic stroke." *Angewandte Chemie International Edition* **51**(44): 1–6.

Chia-Wen Lien, Ying-Chieh Chen, Huan-Tsung Chang, Chih-Ching Huang (2013). "Logical regulation of the enzyme-like activity of gold nanoparticles by using heavy metal ions." *Nanoscale* **5**(17): 8227–8234.

Chia-Wen Lien, Yu-Ting Tseng, Chih-Ching Huang, Huan-Tsung Chang (2014). "Logic control of enzyme-like gold nanoparticles for selective detection of lead and mercury ions." *Analytical Chemistry* **86**(4): 2065–2072.

Chuanxia Chen, Lixia Lu, Yu Zheng, Dan Zhao, Fan Yang, Xiurong Yang (2014). "A new colorimetric protocol for selective detection of phosphate based on the inhibition of peroxidaselike activity of magnetite nanoparticles." *Analytical Methods* **7**(1): 161–167.

Davide Zaramella, Paolo Scrimin, Leonard J. Prins (2012). "Self-assembly of a catalytic multivalent peptide−nanoparticle complex." *Journal of the American Chemical Society* **134**(20): 8396−8399.

Demin Duan, Kelong Fan, Dexi Zhang, Gary P. Kobinger, George Fu Gao, Xiyun Yan (2015). "Nanozyme-strip for rapid local diagnosis of Ebola." *Biosensors and Bioelectronics* **74**: 134–141).

Di Zhang, Ying-Xi Zhao, Yun-Shan Fan, Xiao-Jun Li, Zhong-Yu Duan, Hao Wang (2013). "Anti-bacteria and in vivo tumor treatment by reactive oxygen species generated by magnetic nanoparticles." *Journal of Materials Chemistry B* **1**: 5100–5107.

Dunyaporn Trachootham, Jerome Alexandre, Peng Huang (2009). "Targeting cancer cells by ROS-mediated mechanisms: a radical therapeutic approach." *Nature Reviews Drug Discovery* **8**(7): 579–591.

Fang Yang, Sunling Hu, Yu Zhang, Xiaowei Cai, Gaojun Teng, Ning Gu (2012). "A hydrogen peroxide-responsive O_2 nanogenerator for ultrasound and magnetic-resonance dual modality imaging." *Advanced Materials* **24**(38): 5205–5211.

Faquan Yu, Yongzhuo Huang, Adam J. Cole, Victor C. Yang (2009). "The artificial peroxidase activity of magnetic iron oxide nanoparticles and its application to glucose detection." *Biomaterials* **30**(27): 4716–4722.

Fei Xiong, Hao Wang, Yidong Feng, Lina Song, Yu Zhang, Ning Gu (2015). "Cardioprotective activity of iron oxide nanoparticles." *Scientific Reports* **5**: 8579.

Feng Zhao, Ying Zhao, Ying Liu, Xueling Chang, Chunying Chen, Yuliang Zhao (2011). "Cellular uptake, intracellular trafficking, and cytotoxicity of nanomaterials." *Small* **7**(10): 1322–1337.

Fengli Qu, Ting Li, Minghui Yang (2011). "Colorimetric platform for visual detection of cancer biomarker based on intrinsic peroxidase activity of graphene oxide." *Biosensors and Bioelectronics* **26**(9): 3927–3931.

Filipe Natalio, Rute André, Aloysius F. Hartog, Klaus Peter Jochum, Ron Wever, Wolfgang Tremel (2012). "Vanadium pentoxide nanoparticles mimic vanadium haloperoxidases and thwart biofilm formation." *Nature Nanotechnology* **7**(8): 530–535.

Francesca Pagliari, Corrado Mandoli, Giancarlo Forte, Eugenio Magnani, Paolo Di Nardo, Enrico Traversa (2012). "Cerium oxide nanoparticles protect cardiac progenitor cells from oxidative stress." *ACS Nano* **6**(5): 3767–3775.

Gang Huang, Huabing Chen, Ying Dong, Erik A. Bey, David A. Boothman, Jinming Gao (2013). "Superparamagnetic iron oxide nanoparticles: amplifying ROS stress to improve anticancer drug efficacy." *Theranostics* **3**(2): 116–126.

Hamed Arami, Amit Khandhar, Denny Liggitt, Kannan M. Krishnan (2015). "In vivo delivery, pharmacokinetics, biodistribution and toxicity of iron oxide nanoparticles." *Chemical Society Reviews* **44**(23): 8576–8607.

Hanjun Cheng, Lei Zhang, Jian He, Wenjing Guo, Shuming Nie, Hui Wei (2016). "Integrated nanozymes with nanoscale proximity for in vivo neurochemical monitoring in living brains." *Analytical Chemistry* **88**(10): 5489–5497.

Hanjun Cheng, Yufeng Liu, Yihui Hu, Yubin Ding, Hang Xing, Hui Wei (2017). "Monitoring of heparin activity in live rats using metal organic framework nanosheets as peroxidase mimics." *Analytical Chemistry* **89**(21): 11552–11559.

Hanjun Sun, Nan Gao, Kai Dong, Jinsong Ren, Xiaogang Qu (2014). "Graphene quantum dots-band-aids used for wound disinfection." *ACS Nano* **8**(6): 6202–6210.

Hao-Hua Deng, Shao-Huang Weng, Shuang-Lu Huang, Ai-Lin Liu, Xin-Hua Lin, Wei Chen (2014). "Colorimetric detection of sulfide based on target-induced shielding against the peroxidase-like activity of gold nanoparticles." *Analytica Chimica Acta* **852**: 218–222.

Huan Jiang, Zhaohui Chen, Haiyan Cao, Yuming Huang (2012). "The peroxidase-like activity of chitosan stabilized silver nanoparticles for visual and colorimetric detection of glucose." *Analyst* **137**(23): 5560–5564.

Hui Wei, Erkang Wang (2008). "Fe_3O_4 magnetic nanoparticles as peroxidase mimetics and their applications in H_2O_2 and glucose detection." *Analytical Chemistry* **80**(6): 2250–2254.

Hui Wei, Erkang Wang (2013). "Nanomaterials with enzyme-like characteristics (nanozymes): next-generation artificial enzymes." *Chemical Society Reviews* **42**(24): 6060–6093.

Ivana Celardo, Jens Z. Pedersen, Enrico Traversa, Lina Ghibelli (2011). "Pharmacological potential of cerium oxide nanoparticles." *Nanoscale* **3**(4): 1411–1420.

Jack L.-Y. Chen, Cristian Pezzato, Paolo Scrimin, Leonard J. Prins (2016). "Chiral nanozymes–gold nanoparticle-based transphosphorylation catalysts capable of enantiomeric discrimination." *Chemistry-A European Journal* **22**(21): 7028–7032.

Jia Fan, Jun-Jie Yin, Bo Ning, Jingyan Wei, Yuliang Zhao, Guangjun Nie (2011). "Direct evidence for catalase and peroxidase activities of ferritineplatinum nanoparticles." *Biomaterials* **32**(6): 1611–1618.

Jia Yao, Yuan Cheng, Min Zhou, Jiangjiexing Wu, Sirong Li, Hui Wei (2013). "ROS scavenging Mn_3O_4 nanozymes for in vivo anti-inflammation." *Chemical Science* **9**(11): 2927–2933.

Jianshuai Mu, Yan Wang, Min Zhao, Li Zhang (2012). "Intrinsic peroxidase-like activity and catalase-like activity of Co_3O_4 nanoparticlesw." *Chemical Communicayions* **48**(19): 2540–2542.

Jie Zhuang, Jinbin Zhang, Lizeng Gao, Yu Zhang, Ning Gu, Xiyun Yan (2008). "A novel application of iron oxide nanoparticles for detection of hydrogen peroxide in acid rain." *Materials Letters* **62**(24): 3972–3974.

Jinlai Dong, Jun-Jie Yin, Weiwei He, Yihang Wu, Ning Gu, Yu Zhang (2014). "Co_3O_4 nanoparticles with multi-enzyme activities and their application in immunohistochemical assay." *ACS Applied Materials & Interfaces* **6**(3): 1959–1970.

Jiulong Zhao, Xiaojun Cai, Wei Gao, Yuanyi Zheng, Zhaoshen Li, Hangrong Chen (2018). "Prussian blue nanozyme with multi-enzyme activity reduces colitis in mice." *ACS Applied Materials & Interfaces* **10**(31): 26108–26117.

J. M. Berg, J. L. Tymoczko, L. Stryer (2002). *Biochemistry*. 5th edition. New York: W H Freeman. ISBN-10: 0-7167-3051-0.

Jo-Won Lee, Jihye Son, Kyung-Mi Yoo, Y. Martin Loc, BoKyung Moon (2014). "Characterization of the antioxidant activity of gold@platinum nanoparticles." *RSC Advances* **4**(38): 19824.

Juan Li, Yue Cao, Samuel S. Hinman, Kristy S. McKeating, Quan Cheng, Zhanjun Yang (2018). "Efficient label-free chemiluminescent immunosensor based on dual functional cupric oxide nanorods as peroxidase mimics." *Biosensors and Bioelectronics* **100**: 304–311.

Juhi Shah, Rahul Purohit, Ragini Singh, Ajay Singh Karakoti, Sanjay Singh (2015). "ATP-enhanced peroxidase-like activity of gold nanoparticles." *Journal of Colloid and Interface Science* **456**: 100–107.

Juhi Shah, Sanjay Singh (2018). "Unveiling the role of ATP in amplification of intrinsic peroxidase-like activity of gold nanoparticles." *3 Biotech* **8**(1): 67.

Junping Chen, Swanand Patil, Sudipta Seal, James F. McGinnis (2006). "Rare earth nanoparticles prevent retinal degeneration induced by intracellular peroxides." *Nature Nanotechnology* **1**(2): 142–150.

Kelong Fan, Changqian Cao, Yongxin Pan, Di Lu, Minmin Liang, Xiyun Yan (2012). "Magnetoferritin nanoparticles for targeting and visualizing tumour tissues." *Nature Nanotechnology* **7**(7): 459–464.

Kelong Fan, Hui Wang, Juqun Xi, Demin Duan, Lizeng Gao, Xiyun Yan (2016). "Optimization of Fe_3O_4 nanozyme activity via single amino acid modification mimicking an enzyme active site." *Chemical Communications* **53**(2): 424–427.

Kelong Fan, Juqun Xi, Lei Fan, Peixia Wang, Xiyun Yan, Lizeng Gao (2018). "In vivo guiding nitrogen-doped carbon nanozyme for tumor catalytic therapy." *Nature Communications* **9**(1): 1440.

Ki Soo Park, Moon Il Kim, Dae-Yeon Cho, Hyun Gyu Park (2011). "Label-free colorimetric detection of nucleic acids based on target-induced shielding against the peroxidase mimicking activity of magnetic nanoparticles." *Small* **7**(11): 1521–1525.

Lakshmi Raj, Takao Ide, Aditi U. Gurkar, Anna Mandinova, Stuart L. Schreiber, Sam W. Lee (2011). "Selective killing of cancer cells by a small molecule targeting the stress response to ROS." *Nature* **475**(7355): 231–234.

Laura L. Dugan, Dorothy M. Turetsky, Cheng Du, Tien-Yau Luh, Dennis W. Choi, Tien-Sung Lin (1997). "Carboxyfullerenes as neuroprotective agents." *Proceedings of the National Academy of Sciences of the United States of America* **94**(17): 9434–9439.

Laura L. Dugan, Joseph K. Gabrielsen, Shan P. Yu, Tien-Sung Lin, Dennis W. Choi (1996). "Buckminsterfullerenol free radical scavengers reduce excitotoxic and apoptotic death of cultured cortical neurons." *Neurobiology of Disease* **3**(2): 129–135.

Li-Sheng Wang, Akash Gupta, Vincent M. Rotello (2016). "Nanomaterials for the treatment of bacterial biofilms." *ACS Infectious Diseases* **2**(1): 3–4.

Lianbing Zhang, Linda Laug, Wolfram Münchgesang, Ulrich Gösele, Matthias Brandsch, Mato Knez (2010). "Reducing stress on cells with apoferritin-encapsulated platinum nanoparticles." *Nano Letters* **10**(1): 219–223.

Lin Fan, Yanyan Tian, Rong Yin, Doudou Lou, Ning Gu, Yu Zhang (2016). "Enzyme catalysis enhanced darkfield imaging as a novel immunohistochemical method." *Nanoscale* **8**(16): 8553–8558.

Ling-Na Zhang, Hao-Hua Deng, Feng-Lin Lin, Xiong-Wei Xu, Xing-Hua Xia, Wei Chen (2014). "In situ growth of porous platinum nanoparticles on graphene oxide for colorimetric detection of cancer cells." *Analytical Chemistry* **86**(5): 2711−2718.

Lizeng Gao, Jie Zhuang, Leng Nie, Jinbin Zhang, Sarah Perrett, Xiyun Yan (2007). "Intrinsic peroxidaselike activity of ferromagnetic nanoparticles." *Nature Nanotechnology* **2**(9): 577–583.

Lizeng Gao, Koo H (2017). "Do catalytic nanoparticles offer an improved therapeutic strategy to combat dental biofilms?" *Nanomedicine* **12**(4): 275–279.

Lizeng Gao, Krista M. Giglio, Jacquelyn L. Nelson, Holger Sondermann, Alexander J. Travisa (2014). "Ferromagnetic nanoparticles with peroxidase-like activity enhance the cleavage of biological macromolecules for biofilm elimination." *Nanoscale* **6**(5): 2588–2593.

Lizeng Gao, Yuan Liu, Dongyeop Kim, Yong Li, David P. Cormode, Hyun Koo (2016). "Nanocatalysts promote Streptococcus mutans biofilm matrix degradation and enhance bacterial killing to suppress dental caries in vivo." *Biomaterials* **101**: 272–284.

Mikael Bols (2010). "From enzyme models to model enzymes." *ChemBioChem* **11**(4): 581–583.

Massimiliano Comotti, Cristina Della Pina, Roberto Matarrese, Michele Rossi (2004). "The catalytic activity of 'naked' gold particles." *Angewandte Chemie International Edition* **43**(43): 5812–5815.

Mauro Moglianetti, Elisa De Luca, Deborah Pedone, Heinz Amenitsch, Saverio Francesco Retta, Pier Paolo Pompa (2016). "Platinum nanozymes recover cellular ROS homeostasis in oxidative stress-mediated disease model." *Nanoscale* **8**(6): 3739–3752.

Morten Meldal (2005). "Artificial Enzymes." *Angewandte Chemie International Edition* **44**(48): 7829–7830.

Melissa S. Wason, Jimmie Colon, Sudipta Seal, James Turkson, Jihe Zhao, Cheryl H. Baker (2013). "Sensitization of pancreatic cancer cells to radiation by cerium oxide nanoparticle-induced ROS production." *Nanomedicine* **9**(4): 558–569.

Melissa Hirsch Kuchma, Christopher B. Komanski, Jimmie Colon, Sudipta Seal, Justin Summy, Cheryl H. Baker (2010). "Phosphate ester hydrolysis of biologically relevant molecules by cerium oxide nanoparticles." *Nanomedicine: Nanotechnology, Biology, and Medicine* **6**(6): 738–744.

Minfeng Huo, Liying Wang, Yu Chen, Jianlin Shi (2017). "Tumor-selective catalytic nanomedicine by nanocatalyst delivery." *Nature Communications* **8**(1): 357.

Ming Ma, Yu Zhang, Ning Gu (2011). "Peroxidase-like catalytic activity of cubic Pt nanocrystals." *Colloids & Surfaces A Physicochemical & Engineering Aspects* **373**(1–3): 6–10.

Minmin Liang, Kelong Fan, Yong Pan, Hui Jiang, Liu Yang, Xiyun Yan (2013). "Fe_3O_4 magnetic nanoparticle peroxidase mimetic-based colorimetric assay for the rapid detection of organophosphorus pesticide and nerve agent." *Analytical Chemistry* **85**(1): 308−312.

Naimish Pandurang Sardesai, Mallikarjunarao Ganesana, Anahita Karimi, James C Leiter, Silvana Andreescu (2015). "Platinum-doped ceria based biosensor for in vitro and in vivo monitoring of lactate during hypoxia." *Analytical Chemistry* **87**(5): 2996–3003.

Namrata Singh, Geethika Motika, Sandeep M. Eswarappa, Govindasamy Mugesh (2018). "Manganese-based nanozymes: multienzyme redox activity and effect on the nitric oxide produced by endothelial nitric oxide synthase." *Chemistry-A European Journal* **24**(33): 8393–8403.

Namrata Singh, Mohammed Azharuddin Savanur, Shubhi Srivastava, Patrick D'Silva, Govindasamy Mugesh (2017). "Redox modulatory Mn_3O_4 nanozyme with multi-enzyme activity provides efficient cytoprotection to human cells in Parkinson's Disease model." *Angewandte Chemie International Edition* **56**(45): 14267–14271.

Pengjuan Ni , Haichao Dai, Yilin Wang , Yan Shi, Jingting Hu , Zhuang Li (2014). "Visual detection of melamine based on the peroxidase-like activity enhancement of bare gold nanoparticles." *Biosensors and Bioelectronics* **60**: 286–291.

Qingqing Wang, Xueping Zhang, Liang Huang, Zhiquan Zhang, Shaojun Dong (2017). "GOx@ZIF-8(NiPd) nanoflower: an artificial enzyme system for tandem catalysis." *Angewandte Chemie International Edition* **56**(50): 16082–16085.

Qingqing Wang, Hui Wei, Zhiquan Zhang, Erkang Wang, Shaojun Dong (2018). "Nanozyme: An emerging alternative to natural enzyme for biosensing and immunoassay." *TrAC-Trends in Analytical Chemistry* **105**: 218–224.

R. Ragg, A. M. Schilmann, K. Korschelt, Blümler, M. N. Tahir, Natalio, W. Tremel (2013). "Intrinsic superoxide dismutase activity of MnO nanoparticles enhances magnetic resonance imaging contrast." *Journal of Materials Chemistry B* **4**(46): 7423–7428.

Rongjing Cui, Zhida Han, Jun-Jie Zhu (2011). "Helical carbon nanotubes: intrinsic peroxidase catalytic activity and its application for biocatalysis and biosensing." *Chemistry-A European Journal* **17**(34): 9377–9384.

Rute André, Filipe Natálio, Madalena Humanes, H.-C. Schröder, Werner E. G. Müller, Wolfgang Tremel (2011). "V_2O_5 nanowires with an intrinsic peroxidase-like activity." *Advanced Functional Materials* **21**(3): 501–509.

Sameh S. Ali, Joshua I. Hardt, Kevin L. Quick, Ting-Ting Huang, Charles J. Epstein, Laura L. Dugan (2004). "A biologically effective fullerene (C60) derivative with superoxide dismutase mimetic properties." *Free Radical Biology & Medicine* **37**(8): 1191–1202.

Secheon Jung, Inchan Kwon (2017). "Synergistic degradation of a hyperuricemia-causing metabolite using one-pot enzyme-nanozyme cascade reactions." *Scientific Reports* **7**: 44330.

Shan Wang, Rémi Cazelles, Wei-Ching Liao, Margarita Vázquez-González, Raed Abu-Reziq, Itamar Willner (2017). "Mimicking horseradish peroxidase and NADH peroxidase by heterogeneous Cu^{2+}-modified graphene oxide nanoparticles." *Nano Letters* **17**(3): 2043–2048.

Shanhu Liu, Feng Lu, Ruimin Xing, Jun-Jie Zhu (2011). "Structural effects of Fe_3O_4 nanocrystals on peroxidase-like activity." *Chemistry-A European Journal* **17**(2): 620–625.

Sheng Wang, Wei Chen, Ai-Lin Liu, Lei Hong, Hao-Hua Deng, Xin-Hua Lin (2012). "Comparison of the peroxidase-like activity of unmodified, amino-modified, and citrate-capped gold nanoparticles." *ChemPhysChem* **13**(5): 1199–1204.

Gabriel A. Silva (2006). "Nanomedicine: seeing the benefits of ceria." *Nature Nanotechnology* **1**(2): 92–94.

Sourav Ghosh, Punarbasu Roy, Naiwrit Karmodak, Eluvathingal D. Jemmis, Govindasamy Mugesh (2018). "Nanoisozymes: crystal-facet-dependent enzyme-mimetic activity of V_2O_5 nanomaterials." *Angewandte Chemie International Edition* **57**(17): 4510–4515.

Takeki Hamasaki, Taichi Kashiwagi, Toshifumi Imada, Noboru Nakamichi, Yoshio Hisaeda, Sanetaka Shirahata (2008). "Kinetic analysis of superoxide anion radical-scavenging and hydroxyl radical-scavenging activities of platinum nanoparticles." *Langmuir* **24**(14): 7354–7364.

Talib Pirmohamed, Janet M. Dowding, Sanjay Singh, Brian Wasserman, Sudipta Seal, William T. Self (2010). "Nanoceria exhibit redox state-dependent catalase mimetic activityw." *Chemical Communications* **46**(16): 2736–2738.

Tianran Lin, Liangshuang Zhong, Liangqia Guo, Fengfu Fu, Guonan Chen (2014). "Seeing diabetes: visual detection of glucose based on the intrinsic peroxidase-like activity of MoS_2 nanosheets." *Nanoscale* **6**(20): 11856–11862.

Toren Finkel, Nikki J. Holbrook (2000). "Oxidants, oxidative stress and the biology of ageing." *Nature* **408**(6809): 239–247.

V. Figueroa-Espí, A. Alvarez-Paneque, M. Torrens, A.J. Otero-González, E. Reguera (2011). "Conjugation of manganese ferrite nanoparticles to an anti sticholysin monoclonal antibody and conjugate applications." *Colloids & Surfaces A Physicochemical & Engineering Aspects* **387**(1–3): 118–124.

Wei Liu, Jayant Kumar, Sukant Tripathy, Kris J. Senecal, Lynne Samuelson (1999). "Enzymatically synthesized conducting polyaniline." *Journal of the American Chemical Society* **121**(1): 71–78.

Wei Luo, Lihua Zhu, Nan Wang, Heqing Tang, Meijuan Cao, Yuanbin She (2010). "Efficient removal of organic pollutants with magnetic nanoscaled $BiFeo_3$ as a reusable heterogeneous fenton-like catalyst." *Environmental Science & Technology* **44**(5): 1786–1791.

Wei Zhang, Jinlai Dong, Yang Wu, Ming Ma, Ning Gu, Yu Zhang (2017). "Shape-dependent enzyme-like activity of Co_3O_4 nanoparticles and their conjugation with his-tagged EGFR single-domain antibody." *Colloids and Surfaces B: Biointerfaces* **154**: 55–62.

Wei Zhang, Sunling Hu, Jun-Jie Yin, Weiwei He, Ning Gu, Yu Zhang (2016). "Prussian blue nanoparticles as multi-enzyme mimetics and ros scavengers." *Journal of the American Chemical Society* **138**(18): 5860–5865.

Wei Zhang, Ying Sun, Zhichao Lou, Lina Song, Ning Gu, Yu Zhang (2017). "In vitro cytotoxicity evaluation of graphene oxide from the peroxidase-like activity perspective." *Colloids and Surfaces B: Biointerfaces* **151**: 215–223.

Weiwei He, Jun-Jie Yin, Xiaochun Wu, Chunying Chen, Yinglu Ji, Yuting Guo (2011). "Au@Pt nanostructures as oxidase and peroxidase mimetics for use in immunoassays." *Biomaterials* **32**(4): 1139–1147.

Weiwei He, Yu-Ting Zhou, Wayne G. Wamer, Xiaona Hu, Xiaochun Wu, Jun-Jie Yin (2013). "Intrinsic catalytic activity of Au nanoparticles with respect to hydrogen peroxide decomposition and superoxide scavenging." *Biomaterials* **34**(3): 765–773.

Wenyan Yin, Jie Yu, Fengting Lv, Liang Yan, Zhanjun Gu, Yuliang Zhao (2016). "Functionalized nano-MoS2 with peroxidase catalytic and near-infrared photothermal activities for safe and synergetic wound antibacterial applications." *ACS Nano* **10**(12): 11000–11011.

Xian-Xiang Wang, Qi Wu, Zhi Shan, Qian-Ming Huang (2011). "BSA-stabilized au clusters as peroxidase mimetics for use in xanthine detection." *Biosensors & Bioelectronics* **26**(8): 3614–3619.

Xiaodan Zhang, Shaohui He, Zhaohui Chen, Yuming Huang (2013). "CoFe$_2$o$_4$ nanoparticles as oxidase mimic-mediated chemiluminescence of aqueous luminol for sulfite in white wines." *Journal of Agricultural & Food Chemistry* **61**(4): 840–847.

Xiaona Hu, Aditya Saran, Shuai Hou, Weiwei He, Jun-Jie Yin, Xiaochun Wu (2013). "Au@PtAg core/shell nanorods: tailoring enzyme-like activities via alloying3." *RSC Advances* **3**(17): 6095–6105.

Xiaoning Li, Sandra M. Robinson, Ali Sahar, Margaret A. Riley, Vincent M. Rotello (2014). "Functional gold nanoparticles as potent antimicrobial agents against multi-drug-resistant bacteria." *ACS Nano* **8**(10): 10682–10686.

Xiaoqiang Chen, Xizhe Tian, Injae Shin, Juyoung Yoon (2011). "Fluorescent and luminescent probes for detection of reactive oxygen and nitrogen species." *Chemical Society Reviews* **40**(9): 4783–4804.

Xiaoyu Wang, Wenjing Guo, Yihui Hu, Jiangjiexing Wu, Hui Wei *"Nanozymes: Next Wave of Artificial Enzymes."* Verlag GmbH Berlin Heidelberg, Springer. 2016. pp. 1–127.

Xue Jiao, Hongjie Song, Huihui Zhao, Wei Bai, Lichun Zhang, Yi Lv (2012). "Well-redispersed ceria nanoparticles: promising peroxidase mimetics for H$_2$O$_2$ and glucose detection." *Analytical Methods* **4**(10): 3261–3267.

Yi Juan Long, Yuan Fang Li, Yue Liu, Jia Jia Zheng, Jie Tanga, Cheng Zhi Huang (2011). "Visual observation of the mercury-stimulated peroxidase mimetic activity of gold nanoparticles." *Chemical Communications* **47**(43): 11939–11941.

Yi Zhang, Zhuyao Wang, Xiaojiao Li, Nan Chen, Chunhai Fan, Haiyun Song (2016). "Dietary iron oxide nanoparticles delay aging and ameliorate neurodegeneration in drosophila." *Advanced Materials* **28**(7): 1387–1393.

Yihui Hu, Hanjun Cheng, Xiaozhi Zhao, Zhengyang Zhou, Shuming Nie, Hui Wei (2017). "Surface-enhanced raman scattering-active gold nanoparticles with enzyme mimicking activities for measuring glucose and lactate in living tissues." *ACS Nano* **11**(6): 5558–5566.

Youhui Lin, Yanyan Huang, Jinsong Ren, Xiaogang Qu (2014). "Incorporating ATP into biomimetic catalysts for realizing exceptional enzymatic performance over a broad temperature range." *NPG Asia Materials* **6**: e114.

Yujun Song, Xiaohui Wang, Chao Zhao, Konggang Qu, Jinsong Ren, Xiaogang Qu (2010). "Label-free colorimetric detection of single nucleotide polymorphism by using single-walled carbon nanotube intrinsic peroxidase-like activity." *Chemistry-A European Journal* **16**(12): 3617–3621.

Yun Jv, Baoxin Li, Rui Cao (2010). "Positively-charged gold nanoparticles as peroxidase mimic and their application in hydrogen peroxide and glucose detection." *Chemical Communications* **46**(42): 8017–8019.

Yuyun Zhao, Xingyu Jiang (2013). "Multiple strategies to activate gold nanoparticles as antibiotics." *Nanoscale* **5**(18): 8340–8350.

Zhaowei Chen, Haiwei Ji, Chaoqun Liu, Wei Bing, Zhenzhen Wang, Xiaogang Qu (2016). "A multinuclear metal complex based Dnase-mimetic artificial enzyme: matrix cleavage for combating bacterial biofilms." *AngewandteÜhemie International Edition* **55**(36): 10732–10736.

Zhihui Dai, Shaohua Liu, Jianchun Bao, Huangxian Ju (2009). "Nanostructured FeS as a mimic peroxidase for biocatalysis and biosensing." *Chemistry-A European Journal* **15**(17): 4321–4326.

Zhongwen Chen, Jun-JieYin, Yu-Ting Zhou,Yu Zhang, Sunling Hu, Ning Gu (2012). "Dual enzyme-like activities of iron oxide nanoparticles and their implication for diminishingcytotoxicity." *ACS Nano* **6**(5): 4001–4012.

Zi-Qiang Xu, Jia-Yi Lan, Jian-Cheng Jin, Ping Dong, Feng-Lei Jiang, Yi Liu (2015). "Highly photoluminescent nitrogen-doped carbon nanodots and their protective effects against oxidative stress on cells." *ACS Applied Materials & Interfaces* **7**(51): 28346–28352.

Zijian Zhou, Jibin Song, Liming Nie, Xiaoyuan Chen (2016). "Reactive oxygen species generating systems meeting challenges of photodynamic cancer therapy." *Chemical Society Reviews* **45**(23): 6597–6626.

Zijian Zhou, Jibin Song, Rui Tian, Zhen Yang, Liming Nie, Xiaoyuan Chen (2017). "Activatable singlet oxygen generation from lipid hydroperoxide nanoparticles for cancer therapy." *AngewandteÜhemie International Edition* **56**(23): 6492–6496.

Zijie Zhang, Xiaohan Zhang, Biwu Liu, Juewen Liu (2017). "Molecular imprinting on inorganic nanozymes for hundred-fold enzyme specificity." *Journal of the American Chemical Society* **139**(15): 5412–5419.

Self-Assembling Protein Nanomaterials – Design, Production and Characterization

Bhuvana K. Shanbhag,
Victoria S. Haritos, and
Lizhong He
Monash University

16.1 Introduction

The last decade has seen immense development in the field of nanomaterials with wide application to food, cosmetics, energy, catalysis and medical industries. Protein-based nanomaterials have gained importance in the medical field due to their biocompatibility in drug delivery applications (Herrera Estrada and Champion, 2015; Lee, Lee and Kim, 2016), gene therapy (Ferrer-Miralles et al., 2015) and tissue engineering (Zhang, 2003b). Here we consider three general classes of protein nanomaterials including naturally occurring, computationally designed and, thus, well-defined protein nanostructures and peptide-driven self-assembled protein nanomaterials.

In all three approaches, the basic monomeric protein unit is vital because it is the building block that determines the final protein nanostructure formed. In the case of natural protein nanoparticles, the scope of template is limited owing to constraints of naturally occurring sequence and properties. However, both computationally designed and peptide-based approaches offer potential to tailor the monomeric unit to suit the intended application of these protein nanomaterials. The protein nanostructures formed using one of the above methods may be used to create nanomaterials that primarily function as a scaffold or can be integrated with other types of functional proteins to incorporate multiple functions within the same nanomaterial.

As a scaffold, the monomeric unit is designed either to form enclosed structures such as a compartment for use as a drug carrier or as a framework for tissue regeneration. Important design considerations for scaffold-forming monomeric units are their structure-forming abilities, stability and size of the final nanomaterial as these parameters influence their performance in intended applications. For example, the size of protein nanocarriers used in drug delivery must allow penetration into specific tissues for cellular uptake of the drug, (Shang, Nienhaus and Nienhaus, 2014) which, in turn, depends on the type of target tissue or cells. For protein nanomaterials that integrate functional and self-assembly properties, monomeric units must have the ability to self-assemble into the desired nanostructure without hampering the functionality of the integrated protein; ideally, self-assembly and protein functions should be independent of each other.

The following sections of the chapter describe the three approaches available for creating protein nanomaterials with emphasis on the design principles and properties of the basic monomeric units. An overview of the various design strategies used for different types of protein nanomaterials is illustrated in Figure 16.1. The influence of the monomeric unit design on protein expression and purification is also discussed to enable the end user to obtain high biological production yields and ready characterization of functional protein nanomaterials.

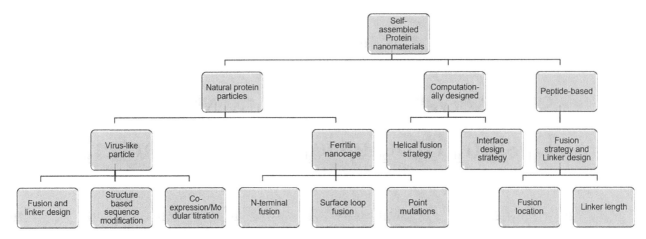

FIGURE 16.1 Overview of the three approaches to self-assembled protein nanomaterials and design strategies employed for each approach.

16.2 Naturally Occurring Protein Nanoparticles

In nature, protein monomers interact with each other in specific ways to form larger assemblies that deliver advanced functions. Some examples of natural protein assemblies are viral capsids, ferritin cages, tubular flagella, and bacterial microcompartments. Self-assembly of these proteins is driven by the symmetric structures of the monomeric units containing low-energy interfaces that favor nanostructure formation. Since the self-assembly feature is an intrinsic property of these protein subunits, design modifications usually focus on improving their *in vitro* assembly for large-scale production or for functionalization purposes depending on the specific application. Using molecular biology techniques and existing crystal structure data combined with computational tools, new protein subunit designs can be developed. For instance, the monomeric PduA microcompartment shell protein that naturally forms layers was re-engineered to form dodecahedral nanocages for encapsulating cargo (Jorda et al., 2016). Here we will focus on two natural protein assemblies namely virus-like particles (VLPs) and ferritin cages and discuss the various design modifications of their monomeric subunits that improve their properties and functions.

16.2.1 Virus-Like Particles (VLP)

Viruses protect their genetic material by encapsulating them within protein capsids that are composed of protein subunits called capsomeres, which self-assemble *in vivo* during viral replication. VLPs are formed from capsomeres, themselves produced by recombinant expression, and retain their self-assembling ability under *in vitro* conditions but lack virus genetic material. As they lack viral genetic material, VLPs are non-pathogenic and the hollow capsids are ideal for encapsulating substances within or displaying desired molecules on the surface. Due to these attributes, VLPs find wide application in the field of vaccine development

and drug delivery (Ding et al., 2018). Therefore, the typical design considerations for VLP subunits include interface design for controlled *in vitro* assembly, improving stability of VLPs, surface modifications for antigen presentation, and mutations to interior and exterior surfaces for functionalisation (Frietze, Peabody and Chackerian, 2016; Hill et al., 2017).

In vitro assembly of VLPs occurs in two stages: first is the assembly of capsomeres or subunits into oligomers, followed by assembly of oligomers into VLP. Optimal conditions must be met in both stages to obtain consistent VLP structure otherwise the result may be simple protein aggregation rather than ordered VLP. To minimize unwanted aggregation, design strategies are used to promote favorable interactions between subunits for controlled *in vitro* assembly. Introduction of long flanking linker segments with complementary charged residues on either sides of the hydrophobic rotavirus peptide prevented protein aggregation and resulted in stable murine polyomavirus VP1 capsomeres (Tekewe et al., 2016). Controlled assembly of hybrid human papillomavirus VLP was achieved by swapping the helix4 region components of HPV16 and HPV18 (Jin et al., 2016). Furthermore, *in vitro* self-assembly may be tuned to achieve desired VLP morphology. Cowpea chlorotic mottle virus VLP was tuned to form different shapes (spherical or tubular) based on the type of platinum (PtII) amphiphilic monomer used as a template for VLP assembly (Sinn et al., 2018). An alternative approach to improve *in vitro* assembly is by visualizing the process *in silico* by molecular dynamic simulations and modifying the sequence designs accordingly (L. Zhang et al., 2015).

For antigen presentation on VLP surface, the antigenic component is fused to one of the viral coat protein subunits to be presented on the capsid surface. A key factor here for the protein design is ensuring sufficient distance of the antigen component from the capsid surface and avoiding steric hindrance during self-assembly. The standard approach involves the fusion of the antigenic module to either the N- or C-terminus of the coat protein subunits.

A linking peptide is inserted in the design between the antigenic module and viral coat protein to provide flexibility and distance between each protein domain to ensure stable particle formation and efficient antigen presentation. The avian influenza antigen (HA1) was fused to the C-terminus of murine polyomavirus (MuPyV) VP1 capsomere *via* the GSA linkers (GSAGSAAGSGEF) (Waneesorn et al., 2018). Furthermore, in order to accommodate the large size of the HA1 antigen, truncation of the VP1 capsomere was necessary. Based on structural information, a C-terminal truncation was preferred over that in the N-terminus to maintain VLP stability. In a study displaying rotavirus antigen on MuPyV VLP, Q25 (25 amino acid residues) and P6 (6 residues) linkers resulted in stable capsomere formation due to their flexibility and length in comparison to the shorter G4S (4 residues) linker, which provided insufficient space between antigen and capsomere. (Lua et al., 2015).

Modular titration is another strategy to achieve highest number of antigen molecules presented on VLP surface without steric hindrance from inter-capsomere interactions. This involves the co-expression of unmodified VP subunits along with antigen-fused VP subunits in the same host cell to obtain an optimal ratio of the unmodified and modified forms of VP subunits (Lua et al., 2015). This method minimizes unwanted capsomere interactions and prevents steric hindrance leading to stable VLP formation (Tekewe et al., 2016). A combination of computational tools and experimental methods has also been employed to design chimeric VLP subunits. The interactions between subunits have been examined at the atomic level using homology modeling to understand the influence of the fused antigen in comparison with the unmodified subunit (Arcangeli et al., 2014). Other *in silico* tools such as computationally optimized broadly reactive antigens (COBRAs) are used to design optimal antigen structure for improved antigen presentation and protective response.

Apart from vaccine development, VLPs have also been designed with other functions such as a molecular cargo container although the molecular design for such VLPs differs from those presented for vaccine development. Enzymes such as cytochrome P450 have been encapsulated into the interior of the bacteriophage P22 capsid by fusing the enzyme to the N-terminus of the truncated version of the scaffold proteins, prepared as a single gene construct, without hampering the capsid assembly process (Sánchez-Sánchez et al., 2015). Alternatively, *in vivo* packaging of RNA into VLP has been achieved by co-expression of a RNA scaffold gene containing the Qβ hairpin region that has high affinity to the coat protein of bacteriophage Qβ, thereby resulting in VLP-RNAi particles assembled within *Escherichia coli* (Fang et al., 2016).

16.2.2 Ferritin Protein Nanocages

Ferritins are proteins that are produced by both prokaryotes and eukaryotes with the primary function of iron storage and transport. They are composed of 24 subunits that self-assemble into spherical nanocages with an inner and outer diameter of 8 and 12 nm, respectively (Truffi et al., 2016). Particle formation is reversible and responsive to pH changes. Their defined structure is ideal as a scaffold, and the hollow interior has a fixed volume that provides a contained environment for encapsulating desired cargo. Modification of the ferritin subunits has been carried out to suit specific applications. For instance, computational design of the interior of the ferritin cage has yielded hydrophobic cavities that extend the nature of the encapsulating products to hydrophobic molecules (Swift et al., 2006). However, the most common modification is the fusion strategy, where the desired protein or antigen is fused to the N-terminus of the ferritin subunit (H-chain or L-chain) leading to the presentation of the protein/antigen on the exterior surface of the ferritin cage. For example, a single genetic construct designed with the human H-chain ferritin (HFn) fused to a cell-penetrating peptide (RGD-4C) at the N-terminus resulted, following recombinant expression, in the production of ferritin cages with RGD-4C peptide on the surface (Uchida et al., 2006). Similarly, large proteins such as the native-like HIV-1 envelope glycoprotein trimers were fused to the N-terminus of *Helicobacter pyroli* ferritin *via* a GSG linker to display multiple copies of the antigen on the ferritin cage surface (Sliepen et al., 2015).

Computational 3D models also help to identify further fusion locations that may be used to present antigen on the ferritin cage surface. Wang et al. (2017) expressed antigenic peptides of *Neisseria gonorrhoeae* by fusing them between surface-exposed loops of helices αA and αB of *H. pyroli* ferritin (Figure 16.2a). This strategy resulted in the successful peptide presentation over the cage surface (Figure 16.2b) similar to the N-terminal fusion method with no perturbation on either cage formation or immunogenic response.

An alternative strategy involves deletion of a section of amino acids in the individual ferritin subunits to create space for fusing multiple antigenic components in a geometric pattern. Deletion of the first 18 and 29 N-terminal residues of insect-derived ferritin heavy and light chains, respectively, was performed, and the truncated forms were fused with two different trimeric influenza antigens via a GS-linker at the same positions (Georgiev et al., 2018). The choice of amino acid deletion at specific locations has the benefits of multiple antigen presentation at predetermined locations over the ferritin particle surface in a repetitive fashion. Other design strategies have involved the introduction of point mutations in the genetic code for insertion of unnatural amino acids that allow site-specific conjugation purposes. An example of this is the mutation of human ferritin light chain to insert an unnatural amino acid, 4-azidophenylalanine (4-AzF), which allowed site-specific conjugation of small molecules or affinity ligands on the surface of the ferritin cage (Khoshnejad et al., 2018).

In summary, the widely used strategy for functional ferritin cage design is modifying the gene sequence to incorporate N-terminal fusion of epitopes that will be displayed

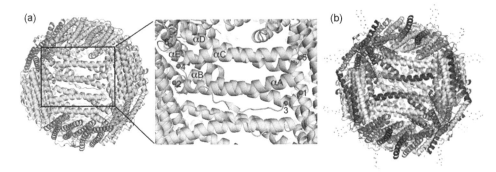

FIGURE 16.2 Design strategy for antigen display over *H. pyroli* ferritin particle surface. (a) Structure of ferritin particle with zoomed image showing one subunit comprising of 5 helices (αA-αE), with possible insertion sites in the loop regions (number 1–6) between helical structures (αA–αE). (b) Cartoon representation of the twofold symmetry of the spherical ferritin cages displaying multiple copies of *N. gonorrhoeae* antigenic peptides (cyan dotted lines) on the exterior surface. (Reproduced from Wang et.al. (2017) 'Structure-based design of ferritin nanoparticle immunogens displaying antigenic loops of Neisseria gonorrhoeae', *FEBS Open Bio*, 7(8), pp. 1196–1207.)

on the surface of the final nanocage. While the genetic sequence and cage formation attributes of the ferritin particle are conserved, flexibility to extend the design features can be obtained via adoption of different sources of ferritins with reduced complexity. Other aspect to consider when designing ferritin particles is whether the iron sequestration function of the ferritin particle is necessary for the intended application, since mutating or suppressing this feature may provide further flexibility to improve particle design tailored to specific applications.

16.3 Computationally Designed and Defined Protein Nanostructures

The defined design approach has recently gained pace using small protein subunits with interface attributes as the starting building block for the formation of larger protein assemblies (King and Lai, 2013). Using computational approaches, the interfaces or orientations of the individual blocks are modified to promote ordered spatial organization of the subunits resulting in symmetric protein assemblies of defined shape and size. There are two typical strategies used to design such protein nanomaterials; they are: the helical linker fusion and the interface design strategies. In the helical linker fusion strategy (Padilla, Colovos and Yeates, 2001), two protein monomers P_A and P_B that are naturally capable of forming their respective oligomers are chosen from sources such as Protein Data Bank (www.rcsb.org/). Only those monomers that form oligomeric structures having an α-helix either at their N- or C-terminus may be used in this strategy. Using a computer program, the two monomers P_A and P_B are fused via a short helical linker to create the hypothetical fusion protein P_A–P_B. An important fusion criteria is that the orientation of the monomers P_A and P_B must have their α-helix regions directed towards the N- or C-terminal. In other words, the fusion protein P_A–P_B is designed to have α-helix

structure at its N-terminus, linker, and C-terminus regions. Such a design ensures that the monomers are linked in a rigid fashion to produce the starting protein building block/subunit having a predictable orientation. Further, the geometry of the symmetry elements such as angle and axes is defined based on certain construction rules (Padilla, Colovos and Yeates, 2001) that direct the protein building blocks to self-assemble into a particular type of nanostructure. Padilla et al. (2001) produced protein cages that were 15 nm wide using 12 subunits formed by fusion of trimeric bromoperoxidase and dimeric M1 matrix proteins of the influenza virus linked by a helical region, nine amino acids in length. In this case, the construction rules applied were a tetrahedral symmetry with intersecting axes at an angle of 54.7°. Alternatively, cube-shaped cages were formed from 24 subunits of the fusion protein made from trimeric KDPGal aldolase with dimeric domain of FkpA protein via four residue helical linker, which was defined with a octahedral symmetry and angle of 36.5° (Lai et al., 2014).

Despite their defined shape, it was observed that the resultant protein assemblies were heterogeneous in nature causing problems during crystallization and characterization studies. Single amino acid changes in individual components of the fusion protein can address potential flaws in the subunit design and reduce heterogeneity of the protein nanostructures. The triple mutant version of bromoperoxidase-M1 matrix fusion protein addressed steric hindrance arising from a lysine residue and minimized exposure of hydrophobic residues to solvent resulting in a subunit design that was successfully crystallized (Lai et al., 2013). These modifications demonstrated that interactions within the subunit influence its orientation, and optimization of residues to promote favorable interactions is important for obtaining a homogenous population of protein assemblies. Furthermore, generating a range of protein nanostructures is possible by varying the combination of protein monomers, their oligomeric states, and the length of the helical linker. An interesting application of this strategy was the creation of a protein scaffolding system, which allowed the display of the very small protein DARPin (∼17 kDa) for its structural analysis via single cryo-electron microscopy; an imaging technique that is currently limited to larger proteins (>50kDa) (Liu et al., 2018).

The second strategy also uses protein monomers that have a defined oligomeric state similar to the previous strategy, but instead of using linkers to define subunit orientation, there is a direct manipulation of the surfaces of the oligomeric subunits for self-assembly. The design has two stages where, in the first stage, symmetric docking of the subunits is performed to achieve a defined nanostructure. In the next stage, protein–protein interfaces between the subunits are identified and manipulated to have low energies that promote self-assembly (King et al., 2012). The computer programs required for docking and interface design in this strategy have been developed using the Rosetta3 software (DiMaio et al., 2011). Based on this strategy, a single-component trimeric protein building block was designed to form protein complexes with octahedral (13 nm diameter) and tetrahedral (11 nm diameter) structures (King et al., 2012). Later, this was extended to two-component protein assemblies using subunits having distinct trimeric arrangements resulting in a 24 subunit tetrahedral protein cage-like structure (King et al., 2014). Functional protein nanocages displaying green and yellow fluorescent proteins have been developed by fusing the functional proteins to the subunit I3-01 (Hsia et al., 2016). The protein nanocage also incorporated a designed pentamer on its pentagonal face that provided control over the size of the channel that can regulate the passage of molecules in and out of the nanocage. Synthetic nucleocapsids that encapsulate RNA within the protein cages have been created using modified subunits whereby the interior of the cage is rich in positively charged amino acid residues or by fusing a Tat RNA-binding peptide to the subunit (Butterfield et al., 2017). Such features are vital for the application of protein nano-assemblies in the field of imaging, drug delivery, and synthetic biology.

A generalized stepwise procedure for designing protein nanostructures using computational approaches is presented in Figure 16.3. A variety of protein architectures may be achieved using different subunit designs using predefined geometric rules (Yeates, Liu and Laniado, 2016). A more detailed discussion on the underlying design principles for both helical linker fusion and the interface design strategies has been covered in a recent review (Yeates, 2017), which provides the foundation for designing protein nanomaterials with novel architectures. Currently, the defined design approach has been successful in creating protein nanostructures with a defined shape and size in a predictable manner, and a high level of correlation has been achieved between the computational model and experimental crystal structures. Areas of practical improvement would be the ability to control assembly and disassembly processes through manipulation of solution conditions. These aspects provide the scope for further improvement of computationally designed protein nanomaterials with useful and practical applications.

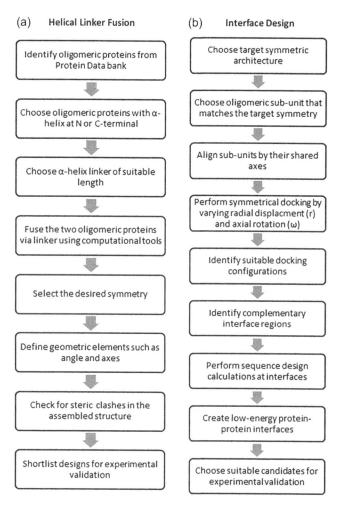

FIGURE 16.3 Stepwise procedure of two different computational approaches to design protein nanostructures. (a) Helical linker fusion strategy and (b) interface design strategy.

16.4 Peptide-Driven Self-Assembly of Protein Nanomaterials

This approach uses peptide sequences to promote self-assembly into structures that can incorporate larger moieties like proteins. Unlike the former approaches, the peptide-based method has the advantage of ready self-assembly due to their smaller size, quick response to changes in solution conditions, and the potential to switch between structured and unstructured forms. The basis of many naturally occurring self-assembly systems depends on this potential for switching states between ordered and monomer. Early inspiration for peptide design came from naturally occurring segments from larger proteins or domains with no significant modifications in the sequence. These peptides, such as the K24 peptide derived from the transmembrane domain of IsK protein (A. Aggeli, et al, 1997), were used to study self-assembly characteristics. Later significant improvements were carried out in peptide design and sequence to incorporate features such as stimuli responsiveness to achieve controlled and predictable self-assembly.

Today these designed and responsive peptides form a large and special family of peptides called the 'self-assembling peptides' (SAP). Another benefit of this approach is the ability to fuse small peptides to larger functional proteins to create functional protein assemblies as opposed to fusing large protein subunits (defined design approach) which may impair the functionality of the protein of choice or the self-assembly ability of the subunit itself. Furthermore, the peptide-based fusion approach can be an easier and computationally less demanding. However, the protein nanostructures formed by this approach lack comparable symmetry and shape precision that is achieved by the computationally design approach.

16.4.1 Overview on Self-Assembling Peptides (SAP)

An interesting peptide that assembles spontaneously into larger structures was reported by Zhang and co-workers while working with the yeast protein Zuotin. They observed a mixture of short 16-residue peptides named EAK16 form a stable, transparent membrane upon addition of salt (Zhang et al., 1993). Subsequently, 'self-assembly' was used to explain the spontaneous organization of such peptides into ordered structures without human intervention (Zhang et al., 2002). As a general definition, 'Self-Assembling Peptides (SAP)' are short peptides with sequence length ranging from 2 to 20 amino acids that spontaneously assemble to form structures of various orders following the principles of molecular recognition.

The family of SAPs contains different types of peptides each with its unique self-assembling conditions and structures formed. There are several classifications of SAPs that have been suggested based on differing criteria. The simplest classification is based on the secondary structure of the peptide, which divides SAPs into α-helix, β-sheet and peptide amphiphiles (Tsutsumi and Mihara, 2013). Another criterion for classification is based on the type of self-assembled structure formed. For example, Gazit differentiated SAPs based on nanostructures namely nanotubes, nanospheres, vesicles and sheets (Gazit, 2007). Zhang also suggests a similar classification of SAPs as 'molecular lego', 'molecular carpet', 'molecular cargo' and 'molecular switch', by correlating the nanostructure formed with the intended application of the final self-assembled assembled material (Zhang, 2003a). The third type of classification is based on the responsiveness of the peptide to stimuli to form self-assembled structures (Löwik et al., 2010). The SAPs, which have the property of self-assembling or disassembling by changes in external stimuli, such as altered solution conditions, are also known as 'stimuli-responsive peptides' or 'stimuli sensitive peptides'.

Regardless of the different types of SAPs, a certain minimal peptide concentration is required for self-assembly to occur. Additionally, the presence of other external stimuli (e.g., temperature, light) and solution conditions such as pH and metal ions can promote protein–protein interactions among peptide monomers favoring the self-assembly process. In some cases, such as the $P_{11}4$ peptide, the stimuli trigger a structural change of the peptide from unstructured to β-sheet that favors monomer interaction leading to nanofiber formation. Under specific conditions, the protein interactions between nanostructures may favor the formation of larger macroscopic forms such as hydrogels, fibers and aggregates (Dasgupta, Mondal and Das, 2013). Furthermore, the formation of assembled nanostructures may be reversible, and this process controlled by changing solution conditions.

16.4.2 Enhancing and Controlling Peptide Self-Assembly

Two factors are important for the initiation or increase in peptide self-assembly: (i) the concentration of the self-assembling peptide and (ii) an external stimulus. Typically, in a self-assembly process, it is essential that the assembling component is present at a critical concentration that increases the probability of chemical interactions between the monomeric units. This may also be known as the critical micelle concentration since the earliest studies on self-assembly began using micellar materials and were based on the principles of colloid chemistry. However, the limitations of this approach are that there is no precise control over the self-assembly process apart from varying the component concentration and the process is relatively slow due to the lack of driving force within the system, and hence, the self-assembly requires time to achieve the final nanostructure rather than being spontaneous.

On the other hand, the addition of an external stimulus can trigger a quicker and spontaneous self-assembly by providing the required driving force. From a fundamental point of view, the free energy of a spontaneous process is negative, and for this, the enthalpy must be negative and exceed the entropy term (O'Mahony et al., 2011). Therefore, in a self-assembly system, external stimuli such as pH and salt can increase the enthalpy of the system through electrostatic screening or salting in/out effects to favor intermolecular interactions between monomeric units (Mason et al., 2017). As a result, the external stimuli reduce the energy barrier leading to spontaneous or faster self-assembly process. A detailed review on the self-assembly mechanism has been reported by Löwik et al. (2010); examples of stimuli and their mechanisms are summarized in Table 16.1. Depending on the type of interactions involved, the assembly process may be reversible or irreversible. For example, the P_{11} peptide family is based on non-covalent interactions and can be switched between monomeric and assembled state by a change of pH. On the contrary, application of enzymatic stimulus in the form of peptide bond cleavage is a permanent change and thus irreversible.

The type of external stimuli not only influence self-assembly process but also size of the nanostructure formed. Also, the intensity/concentration of the stimuli can affect the final size of the protein nanoparticles formed. To achieve

TABLE 16.1 Summary of Stimuli and Their Response Mechanism in SAPs

Stimuli	Mechanism	Effect	Example of SAP
pH	Protonation/deprotonation of charged residues	Switch/change secondary structure; changes net charge of peptide	P_{11}-2,4,8,9
Temperature	Folding/unfolding of peptides	Initiate hydrophobic interactions	Elastin-like polypeptide
Metal ions	Co-ordination interaction between metal ion and peptide amino acid residue	Induce conformational change; charged interactions	Alkyl-C4G3S(P)RGD-COOH
Enzyme-catalyzed reactions	Phosphorylation/dephosphorylation; cleavage of peptide bonds	Induce conformational change; imbalance in ratio of hydrophobicity/hydrophilicity	HAAHHELH
Light	Moieties that absorb light	Induce conformational change	MAX7CNB

a controlled and desired particle size in a predictable fashion, it is essential to find the optimal conditions of the influencing factors. A straightforward approach to optimize assembly conditions is by varying one factor and keeping all other factors constant at any given instance, known as 'One-Factor At a Time – OFAT' experiments. When number of factors are few, this method is simple and quick to determine the optimal settings. However, interaction between different factors may be significant, making this method less predictable under realistic conditions.

A more sophisticated approach is the 'Design of experiments – DoE' method using statistical tools to study various factors and their interactions. The greatest advantage of this method is maximum information that may be obtained with minimum number of experiments. It reduces time, improves reliability, and enhances process robustness. In addition, the method can generate an empirical model for prediction and experimental validation. Due to these benefits, DoE has been widely used to optimize various industrial process and chemical reactions (Weissman and Anderson, 2015). As an example for protein nanoparticle formation, a two-level full-factorial design has been used to optimize the key factors, i.e., pH, $MgCl_2$ concentration, and protein concentration that influence the BCA-P_{11}4 nanoparticle sizes formed (Shanbhag et al., 2018). The factorial design revealed an interaction between factors of pH and $MgCl_2$ as illustrated by a 3D-contour plot (Figure 16.4a), suggesting that a combination of pH and $MgCl_2$ concentration may be used to better control BCA-P_{11}4 nanoparticle with a desired size in a predictable fashion. Using the equation, BCA-P_{11}4 nanoparticle sizes were predicted with more than 90% accuracy (Figure 16.4b). Therefore, the DoE approach provides highly valuable information for better understanding of self-assembly processes. Knowledge of interacting parameters can provide new self-assembly protocols to obtain the desired nanoparticle size in a controlled and predictable manner. DoE has also been used to optimize particle sizes of inactivated influenza virus particles (Kanojia et al., 2016) for vaccine development.

16.4.3 Design Strategy Using SAPs for Nanomaterials

Peptide-based assembly is a relatively straightforward approach to combine functional proteins with SAPs to create functional protein nanomaterials with self-assembly attributes. The first step towards protein nanoparticles

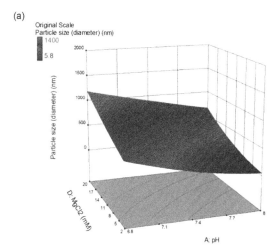

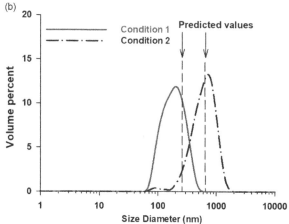

FIGURE 16.4 Two-level full-factorial DoE approach for the controlled self-assembly of BCA-P_{11}4 nanoparticles; (a) 3D-contour plot showing the interaction effect between the factors pH and $MgCl_2$ concentration. (b) Comparison of observed particle size distribution data determined by dynamic light scattering (DLS) with the predicted particle size using empirical equation at two different conditions: (1) 0.5 mg/mL BCA-P_{11}4, pH 7.5, 6 mM $MgCl_2$ and (2) 1.0 mg/mL BCA-P_{11}4, pH 7.0, 8 mM $MgCl_2$ (Reprinted with permission from Shanbhag et al. (2018) 'Understanding the interplay between self-assembling peptides and solution ions for tunable protein nanoparticle formation', *ACS Nano*, 12, pp. 6956–6967. Copyright (2018) American Chemical Society.)

is the careful design of the building block or the monomeric unit, to ensure that protein activity and peptide self-assembly function independently. The monomeric unit typically requires three components: a) functional protein, b) self-assembling peptide and c) linker molecule. The fusion of the linker and peptide component may be performed at either the N- or C-terminal end of the protein or at both ends (Figure 16.5). An example where the peptide is fused to both N- and C-termini of a functional protein is alcohol dehydrogenase with an α-helix peptide at both termini which together is capable of self-assembling into a hydrogel (Figure 16.5d) (Wheeldon et al, 2009). The choice of location of the peptide may be based on the distance of the terminus from the protein's active site which can be estimated from the enzyme structure, where available. Positioning the peptide and linker away from the active site would help to avoid any steric hindrance of protein activity. For example, the self-assembling peptide $P_{11}4$ was fused to the C-terminus of the bovine carbonic anhydrase enzyme to ensure that the $P_{11}4$ peptide was away from the entrance of the active site, which is closer to the N-terminus than the C-terminus (Figure 16.6).

The number of peptide units fused to the protein is another variable to be considered in the design. The peptide number per protein monomer may be increased by fusing the peptides at both termini (Figure 16.5c) or having multiple peptide repeats at one terminus (Figure 16.5e). As peptide concentration is an important factor for self-assembly, increasing the number of repeated units can increase the overall peptide concentration relative to the functional protein (Kyle et al., 2009). However, there may be repercussions for recombinant expression as high concentration of SAPs within the host cell may

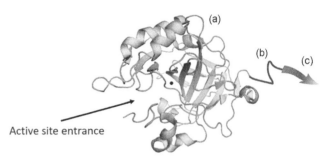

FIGURE 16.6 Components of the enzyme–peptide fusion system. (a) Enzyme–bovine carbonic anhydrase, (b) linker and (c) self-assembling peptide $P_{11}4$. (Reprinted with permission from Shanbhag et al. (2016). Self-assembled enzyme nanoparticles for carbon dioxide capture', *Nano Letters*, 16, pp. 3379–3384. Copyright (2016) American Chemical Society.)

cause spontaneous aggregation leading to the formation of inclusion bodies or aggregates (Wu et al., 2011). The crystal structure of the protein of interest if available can provide useful information to decide the preferred fusion location of the peptide with respect to the protein's termini. Crystallographic data can provide insights into the secondary structure confirmation of the termini and low-energy surfaces; to prevent undesired intramolecular interaction between protein and peptide within a single fusion system.

The next design consideration is the linker peptide, which connects the protein and the SAP peptide components of the fusion protein. The main purpose of the linker peptide is to

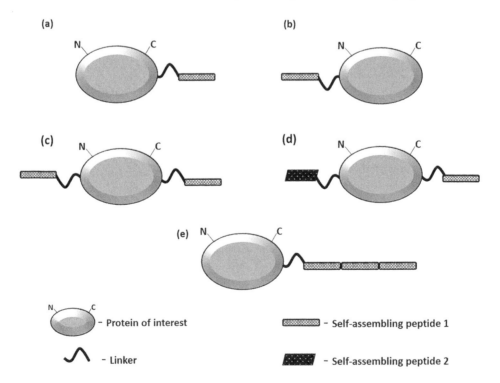

FIGURE 16.5 Different fusion protein designs based on the fusion location of the self-assembling peptide relative to the termini of the protein of interest. (a) C-terminus, (b) N-terminus, (c) same peptide at both N- and C-termini, (d) different peptides at each N and C-termini and (e) multiple repeats of the same peptide.

provide distance and flexibility between the two molecules as, in most cases, the peptide is much smaller than the protein counterpart. A lack of distance between the two entities may introduce steric hindrance and interfere with peptide self-assembly. While there are no specifications for the length and constitution of a linker sequence for preparing protein nanoparticles, cues can be obtained from other types of fusion systems for their design.

The amino acid glycine is commonly used in linker peptides because the primary carbon atom has a single hydrogen as its side chain, offering free rotational movement. Consequently, glycine-based linkers provide flexibility and have been used to attach short helical coils for improved movement and distance (Bromley et al., 2009). Glycine residues may be alternated/interspersed with serine residues to form a GS-linker, a random coil with no defined secondary structure. Flexible, unstructured linkers are useful in application to protein nanomaterials as SAPs undergo structural changes during the transition between monomeric and self-assembled form, and flexibility in the linker peptide ensures self-assembly is unimpeded (Reddy Chichili, Kumar and Sivaraman, 2013). The GS-linker has been successfully used in enzyme–peptide fusion system BCA-$P_{11}4$ to form enzyme protein particles (Shanbhag et al., 2016). A further example of a flexible linker is the serine-glutamate-glycine (SEG) linker, which has been used to link *de novo* designed peptides capable of forming self-assembling peptide-based cages (SAGEs) (Ross et al. 2017).

Another important factor is the length of the linker, which may be varied to suit the functional protein of interest and/or the self-assembled form desired for a specific application. Based on other successful fusion systems, it is suggested that the minimum linker length to be equal or greater than the length of one protein domain based on X-ray crystallography data (Wong et al. 2009). In another case, the linker length was designed based on the size of the substrate molecule to allow lipase enzyme displayed on yeast cells to efficiently interact with the substrate (Washida et al., 2001).

There are a variety of SAPs available to choose from for the design of protein nanomaterials (as elaborated in Section 16.4.1). Alternatively, it is also possible to custom design peptides or modify existing peptides to suit the functional protein or intended application, using the principles of amino acid chemistry and their fundamental interactions. Examples of SAPs used in conjunction with proteins to generate functional protein nanomaterials are summarized in Table 16.2.

16.5 Bioproduction and Purification

Research and applications into protein nanomaterials require a method of production of the protein building blocks or subunits. The most straightforward and practical approach for scalable production of the building blocks is to use living microorganisms as host/factories to produce the proteins followed by downstream processing to obtain purified building blocks. A typical production process involves generation and insertion of the DNA sequence coding for the target protein building block encompassing the protein, associated linker and assembly peptide elements into a DNA plasmid. The DNA plasmid is transformed into a suitable host cell followed by growth in either shake flasks or a bioreactor. The cells are harvested and subjected to one or more purification methods to yield purified proteins.

There are several options for host cells for expression of protein building blocks ranging from bacteria and yeast to insect and mammalian cell lines. Expression in a bacterial system such as the *E. coli* is favored due to its relative simplicity, fast doubling time and high protein production, and it is one of the most commonly used hosts for protein expression for protein nanomaterials discussed in this chapter. Therefore in this section, bioproduction and purification procedures using microbial systems, mainly *E.coli*, with a focus on obtaining high level of soluble proteins will be emphasized. An overview of the various process steps and some important considerations in bioproduction and purification of protein nanomaterial building blocks are presented in Figure 16.7. In case of VLPs, although the basic process steps are similar to production of other protein nanomaterials building blocks, there are certain process variations unique to VLPs and these are discussed separately.

TABLE 16.2 Examples of SAPs Used to Form Functional Protein Nanomaterials

SAP	Functional Partner	Self-Assembly Initiator	Self-Assembled Structure	References
β-Tail tag and Q11 peptide	Green fluorescent protein Cutinase	Peptide concentration	Nanofibers	Hudalla et al. (2014)
EAK-16-II	Fluorogen-activating protein (FAP)	Peptide concentration	Membrane	Saunders et al. (2013)
RTX – β roll peptide	Green fluorescent protein β-lactamase Alcohol dehydrogenase	Addition of calcium	Aggregates Hydrogel	Shur et al. (2013), Dooley, Bulutoglu and Banta (2014)
$P_{11}4$ peptide	Bovine carbonic anhydrase	pH	Nanoparticles	Shanbhag et al. (2016)
Tryptophan-phenylalanine dipeptide	Aptamer	Zinc	Nanoparticles	Fan et al. (2016)
Glycylglycine bolaamphiphile	Lipase	pH and concentration	Peptide nanotubes	Yu et al. (2005)
Polyarginine peptide (R9)	Green fluorescent protein	Salt concentration	Nanoparticles	Unzueta et al. (2012)

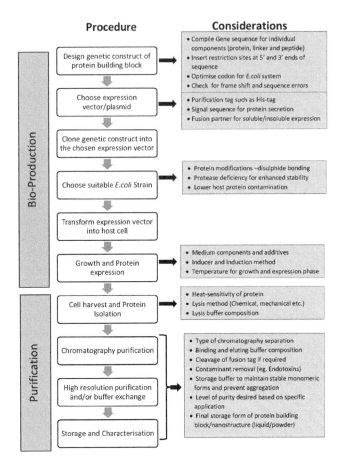

Procedure

- Design genetic construct of protein building block
- Choose expression vector/plasmid
- Clone genetic construct into the chosen expression vector
- Choose suitable *E.coli* Strain
- Transform expression vector into host cell
- Growth and Protein expression
- Cell harvest and Protein Isolation
- Chromatography purification
- High resolution purification and/or buffer exchange
- Storage and Characterisation

Considerations

- Compile Gene sequence for individual components (protein, linker and peptide)
- Insert restriction sites at 5' and 3' ends of sequence
- Optimise codon for *E.coli* system
- Check for frame shift and sequence errors

- Purification tag such as His-tag
- Signal sequence for protein secretion
- Fusion partner for soluble/insoluble expression

- Protein modifications –disulphide bonding
- Protease deficiency for enhanced stability
- Lower host protein contamination

- Medium components and additives
- Inducer and Induction method
- Temperature for growth and expression phase

- Heat-sensitivity of protein
- Lysis method (Chemical, mechanical etc.)
- Lysis buffer composition

- Type of chromatography separation
- Binding and eluting buffer composition
- Cleavage of fusion tag if required
- Contaminant removal (eg. Endotoxins)
- Storage buffer to maintain stable monomeric forms and prevent aggregation
- Level of purity desired based on specific application
- Final storage form of protein building block/nanostructure (liquid/powder)

FIGURE 16.7 Overview of various process steps and the corresponding factors to be considered during the bioproduction and purification of protein nanomaterial building blocks using *E.coli* as a host system.

16.5.1   Strategies to Increase Expression of Protein in Soluble Form

During the production process, achieving expression of the protein building blocks in the soluble fraction of the cell lysate is desired and beneficial. Firstly, having soluble protein reduces uncontrolled aggregation, thereby reducing complexity in the subsequent purification and characterization steps. Secondly, a higher protein yield can be obtained because aggregated protein formed within the host cell often elicits protease activity and reduces overall protein yield. However, as the protein nanomaterial building blocks are designed to self-assemble, it is challenging to achieve high level of expression in soluble form. There are many factors that affect the solubility of a protein namely amino acid sequence, protein folding, and expression conditions to name a few. Each of these factors may be addressed at different stages of the bioproduction process if there has been difficulty in achieving suitable expression yields.

During the design of the protein subunit, the amino acid sequence may be altered to promote soluble expression without significantly affecting the self-assembly behavior. For instance, the soluble yield of the T33-09 subunit A forming tetrahedral nanostructures was increased after it was re-engineered with five mutations on its surface based on computational characterization (Bale et al., 2015). In the case of fusion proteins containing a linker, the linker region may also influence solubility. For example, protein–peptide systems fused via the rigid PT-linker formed inclusion bodies under recombinant expression (Wu et al., 2011; Zhou et al., 2012); this may be attributed to the rigid nature of the linker which contains high percentage of proline residues (Schuler et al., 2005), thereby hampering protein folding leading to misfolded proteins that are insoluble. Misfolded protein building blocks could be overcome by selecting a flexible GS-linker instead of the PT-linker which is a random coil with no defined secondary structure. This has been demonstrated in VLP production where the addition of longer GS-linker and a sequence change with introduction of four additional glutamic acids on either sides of the hydrophobic element improved soluble yield of VP1 (Tekewe et al., 2016).

Recombinant protein accumulated as inclusion bodies in bacteria may be a result of incorrect folding. Where there is a limitation to changing the amino acid sequence of the protein subunit, alternative methods to improving protein folding and enhancing solubility may be required. One readily available method is to use solubility tags such as glutathione S-transferase (GST-tag), thioredoxin (Trx-tag) and maltose-binding domain (MBP-tag) (Esposito and Chatterjee, 2006) fused to the protein of interest. The Trx-tag was used to obtain soluble expression of protein–peptide fusion system composed of a photo-responsive protein and self-assembling peptide (X. Zhang et al., 2015). Also GST-tag is a popular choice to express the VP1 and VP8 capsomere subunits in soluble form (Tekewe et al., 2017).

More complex protein-folding processes such as those involving disulfide bonds within proteins may be hampered by expression in *E .coli* as the organism has a reduced environment within the cell which does not support the retention of disulfide bonds. Co-expression of chaperons or protein-folding components may be incorporated into the same or as a separate DNA construct and introduced into the host simultaneously. This method has improved the soluble expression of the human L-chain ferritin, which is normally prone to inclusion body formation (Zou et al., 2016). Alternatively, newer *E. coli* strains (SHuffle ® T7 express) have been developed to support disulfide bond formation in the cytosol (Lobstein et al., 2012).

Expression of soluble protein may be achieved by optimizing growth and culture conditions. High protein production in soluble form can be achieved by growing cells to high density cultures at induction temperatures as low as 18°C (Studier, 2005; Sivashanmugam et al., 2009). The low temperature is expected to slow down the growth of cells and the rate of protein synthesis allowing more time for complex protein folding to occur (Schein and Noteborn, 1988). Other factors to be considered include culture media components, nature of the expression inducing agent used and concentration, and these may be optimized using statistical methods to obtain maximum soluble expression (Papaneophytou and Kontopidis, 2014). In case of failure to express soluble protein, there is further advice available via the troubleshooting strategies recommended by Rosano and Ceccarelli (2014).

16.5.2 Specific Considerations for Downstream Processing

The primary goal of downstream processing is to achieve high purity proteins with minimum processing steps. Purification tags may be fused with the gene sequence of the protein building blocks as part of the DNA construct design as this can simplify purification. Purification tags aid in selective separation of the desired protein from other host contaminants due to their affinity towards a specific ligand or matrix and hence also known as 'affinity tags'. The most popular and commonly used purification tag is the His6-tag, which contains six repeats of histidine amino acid, which has high affinity to certain metal ions and can be selectively separated by immobilized metal affinity chromatography (IMAC). The His6-tag has been used to purify computationally designed protein building blocks as well as protein–peptide fusion systems. Other purification tags that may be used in the *E. coli* production system include Poly-Arg, FLAG® peptide (DYKDDDDK **tag**), cellulose-binding domain and chitin-binding domain (Terpe, 2003).

An important aspect to use of purification tags is any possible interference on self-assembly or protein function

and whether cleavage of the tag prior to formation of protein nanoparticles is required which introduces an additional process step. Tags can be cleaved by introducing a sequence coding for protease cleavage site between the tag and protein nanomaterial building block in the translated sequence. Most vector sequences incorporate the thrombin cleavage site after the His6-tag that can be cleaved using thrombin protease. This method has been used to remove the GST-tag from the VP1 capsomere subunit (Wibowo et al., 2013) and His6-tag from the computationally designed I53-47 and I53-50 protein subunits (Butterfield et al., 2017). However, where the protein nanomaterial is to be used for medical applications, it is not recommended to use thrombin for cleavage in the process due to its impact on human physiology. As such, the VLP production vector was updated to introduce the tobacco etch virus protease (TEVp) sequence rather than thrombin, since this protease has no known effect in humans (Connors et al., 2014).

Alternatively, column-free purification may be achievable by selecting a specific SAP that can also act as a precipitation partner to induce precipitation of the protein nanomaterial building block. This method is more suitable for the protein–peptide approach as it provides flexibility to select a SAP that can not only self-assemble but also aid in the purification steps. The elastin-like polypeptide (ELP) and β-roll tag have been employed for purification purpose based on their precipitation behavior in response to temperature and calcium salt, respectively. Furthermore, a self-cleaving sequence has been incorporated along with these SAPs to obtain a protease-free cleavable process (Fan et al., 2018). Alternatively, proteins may be targeted to be secreted out of the host cell to simplify purification steps. While protein secretion is not undertaken in *E. coli* expression, proteins may be targeted to the periplasmic space, close to the outer membrane, by the addition of well-known signal sequences for secretion. Protein secretion is a characteristic trait of yeasts and mammalian cells; their systems can efficiently secrete proteins into the medium eliminating the need to lyse the cells and reducing the purification required. *Pichia pastoris* has been used as host for the soluble expression of the P_{11}-14 self-assembling peptide (Moers et al., 2010) and *Saccharomyces cerevisiae* as the host for human papillomavirus capsid protein.

16.5.3 Process Variations Specific to Virus-Like Particle Production

Typical application of VLPs is in the development of vaccines, and therefore, the VLPs are used as a scaffold to present specific antigens. Hence, the first step in the production process requires the construction of an expression vector containing DNA of the VLP subunit fused to the antigenic protein (Figure 16.8a) resulting in chimeric protein subunits. Although the overall production process is similar to recombinant protein production, there are distinct process steps such as aggregate removal and assembly of VLPs that are critical in its production (Figure 16.8b).

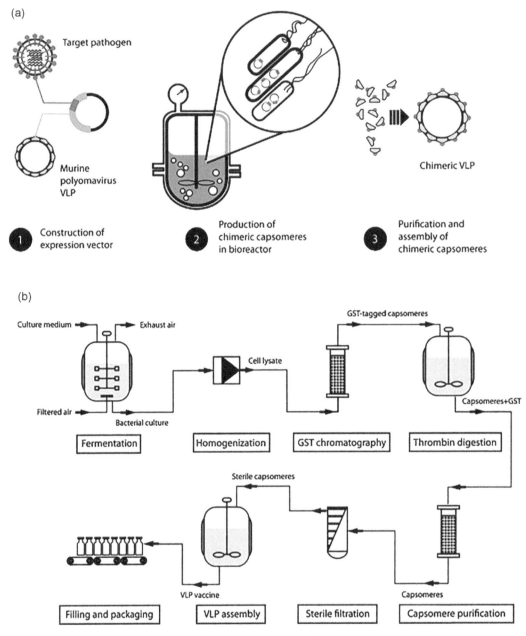

FIGURE 16.8 Microbial platform for bioproduction and purification of VLP. (a) Overview of production and assembly of chimeric VP1 capsomeres. (b) Standard process flow diagram for vaccine production using VLP. (Reprinted from Middelberg et al. (2011). 'A microbial platform for rapid and low-cost virus-like particle and capsomere vaccines', *Vaccine*, 29 (41), pp. 7154–7162. Copyright (2011) with permission from Elsevier).

Assembly of VLPs requires a controlled environment, otherwise they are prone to form aggregates that would have adverse effects on their immunogenic properties. Therefore, it is essential to understand and control the kinetics of the VLP formation to reduce unwanted aggregation (Ding et al., 2010). VLP assembly process is performed during the formulation phase of the production and has two key steps which are: the removal of stabilizing agents added to prevent self-association of the capsomeres and promoting controlled VLP assembly by reducing pH and addition of salts such as calcium and ammonium sulfate. Drug molecules that need to be loaded into the VLP are also added at this stage which increases the complexity of the assembly process. Various methods such as dialysis against suitable buffers and varying folds of dilution have been explored for VLP assembly of which an optimized twofold dilution process resulted in 22% increase in VLP yield (Liew, Chuan and Middelberg, 2012a). By using a reactive constant volume diafiltration process, high-quality VLP with sizes similar to that of a virus was obtained with a 42% increase in the yield of VLP (Liew, Chuan and Middelberg, 2012b). Other improvements include high-throughput miniaturized cell culture processes that allow quick screening and optimization of fermentation (Ladd Effio et al., 2016) and formulation steps (Mohr et al., 2013) and computationally designed peptides as a ligand for one-step affinity purification of capsomeres of VLP from murine polyomavirus (MuPyV) (Li et al., 2014).

16.6 Characterization of Protein Nanomaterials

Protein nanomaterial characterization occurs at two levels of production process: the protein subunit level to ensure that the expressed and purified protein is as expected. This is vital because any variations at the subunit level will influence not only protein self-assembly but also their function. Routine techniques such as polyacrylamide gel electrophoresis can confirm the molecular weight of the expressed protein and samples can be removed for mass spectrometry analysis for more accurate molecular weight determination and amino acid sequence verification. The second stage of characterization is at the protein particle level which provides comprehensive information about the protein assembly in terms of particle size and shape. The structural data is vital to understanding the efficiency of the assembly process and to differentiate between assembled structures, aggregates, and unassembled subunits. Many techniques are available to assess protein nanomaterials and to obtain both quantitative and qualitative information; these are summarized in Table 16.3 with their benefits and limitations. Furthermore, the approaches are categorized into groups based on their working principle and discussed in later sections.

16.6.1 Light Scattering Techniques

Both static light scattering (SLS) and DLS can be used for determination of particle sizes, and they are relatively easy and non-destructive. For example, SLS has been used to determine the molecular weights of modular VLP-RV10 capsomeres (Tekewe et al., 2016), and DLS has been used to determine particle size distributions of BCA-$P_{11}4$ nanoparticles as shown in Figure 16.9. More recently, the company Nano Sight Ltd. in UK developed nanoparticle tracking analysis (NTA) method, whose working principle is the same as DLS but uses a video to track particle motion as opposed to the digital correlator used in DLS (Filipe, Hawe and Jiskoot, 2010).

16.6.2 Size-Exclusion Chromatography

Size-exclusion chromatography (SEC) also known as 'gel permeation chromatography' uses a solid porous matrix to separate molecules based on their size. Typically, when a sample containing a mixture of particle sizes is applied to the matrix, the larger molecules pass through the void volume of the matrix and elute first, whereas smaller molecules are impeded by the matrix and pass through the pores of the

matrix, which takes longer to elute. Appropriate matrix selection having the desired fractionation range must be selected to achieve the desired level of separation. If a suitable standard molecule of known molecular weight is used, then the elution volume of each species may be correlated to determine their molecular weights and sizes. SEC was used to differentiate the assembled form of the I3-01 60-subunit protein cage from its trimeric building block (Hsia et al., 2016). Often SEC is used in combination with other techniques, for instance, SEC having an in-line multi-angle scattering analysis (MALLS) (Figure 16.10a) can provide real-time information on size, shape and molecular weight distribution even if the species are not resolved completely (Patterson et al., 2011).

16.6.3 Sedimentation Methods

Ultracentrifugation operates at high rotational speeds and can separate particles of different sizes and densities due to centrifugal forces that act on a mixture. As an analytical tool, sedimentation experiments can be performed to determine shape, molar mass and particle size distribution of the sample. From the resultant sedimentation traces (Figure 16.10b), assembled protein cages were differentiated from their monomeric building blocks based on the calculated sedimentation coefficients which were 10–25 and 5.2–5.8 S, respectively (Patterson et al., 2011).

16.6.4 Microscopic Methods

Transmission electron microscopy (TEM) and atomic force microscopy (AFM) are the most widely used microscopy methods for the characterization of protein nanomaterials. TEM produces high-resolution images due to the smaller wavelength of the electrons. The final assembled structures of protein nanomaterials can be visualized using TEM, by subjecting the protein sample to a negative staining procedure using uranyl acetate/uranyl formate. An example of a TEM image of assembled VLP and protein nanoparticles formed by BCA-$P_{11}4$ is shown in Figure 16.11a and b, respectively. For higher resolution images, 2D and 3D reconstruction of protein nanostructures can be performed by subjecting the electron micrographs to computer-aided image processing and refining processes in an iterative fashion. This procedure is often used to validate the experimental structure of the protein assembly with the computational model (King et al., 2012, 2014). Cryo-EM, a variation of TEM, is also used to view protein assemblies (Liu et al., 2018) where the sample is prepared using liquid nitrogen allowing visualization of the biological structures close to their native state.

AFM measurements can provide a topographic image of the sample surface (Figure 16.11c). AFM was used to study the probable differences in the nanomorphology of dipeptide nanoparticles due to changed amino acid positions (Trp-Phe or Phe-Trp) in the sequence (Fan et al., 2016). Single particle measurements of self-assembled peptide cages

TABLE 16.3 Summary of Various Analytical Methods and Their Utility in Protein Nanomaterials Characterization

Method	Output Data	Benefits	Limitation
Dynamic light scattering (DLS)	Particle size distribution	• Low barrier analysis • Quick results • Non-disruptive	• Errors due to multiple scattering from sample
Static light scattering (SLS)	Weight average molecular weight	• Simple, User friendly • Quick results • Non-disruptive • *In situ* method	• Errors due to multiple scattering from sample
Nanoparticle tracking analysis (NTA)	Particle size distribution	• Improved accuracy than DLS • Sample visualization possible • Aggregate architecture information • Non-disruptive	• Requires trained personnel • Time consuming • Less reproducible
Size-exclusion chromatography	Size-based separation Molecular weight	• User-friendly • Gentle method • Can be combined with other detection methods	• Cannot separate aggregates and particles of the same size • Time consuming depending on column size
Analytical ultracentrifuge	Shape and size distribution	• User-friendly • Gentle method • Can be combined with optical detection system for real-time monitoring	• Time consuming • Expensive • High maintenance
Transmission electron microscopy (TEM)	Qualitative information on shape and size	• Near atomic resolution • 2D and 3D constructed images • Accurate shape and size details • Low sample volumes	• Trained personnel required • Expensive • High maintenance • Only a small section of the sample population is visualized
Atomic force microscopy	Qualitative information on surface topology	• High resolution • Low sample volumes • No vacuum or electron source required and can be operated under ambient conditions	• Trained personnel required • Expensive • High maintenance • Low scanning speed and area
X-ray crystallography	Qualitative information on protein nanostructure	• Atomic-level accuracy • Validation tool	• Restricted to proteins that can be crystallized • Low throughput • Time consuming • Trained personnel with expertise
HT-SAXS	Information on protein assembly in solution state	• High throughput • Can evaluate the effect of solution conditions on protein assembly	• Trained personnel
AFFF (AF4)	Quantitative information on particle size	• Analysis of a broad size range • Gentle method • Small sample volume and easy preparation • No solid matrix	• Trained personnel with expertise • Expensive
ES-DMA	Quantitative information on particle size and molecular weight	• Analysis of a broad size range • Resolution higher than DLS • Sensitive technique • Study kinetics of protein aggregation	• Aerosol formation may affect protein assemblies • Trained personnel with expertise • Expensive

(SAGEs) using AFM were performed to obtain information on particle dimensions such as diameter, height and height-to-diameter ratio (Ross et al., 2017). In addition, AFM provides information on the mechanical properties of protein assemblies. For instance, the stretching behavior of coiled-coil-mediated self-assembly of Ig27 molecules (Dietz et al., 2007) and the mechanical stability of VLPs based on the interacting and breaking patterns of VLP subunits (Llauró et al., 2016) were determined using AFM.

16.6.5   X-ray Based Methods

X-ray Crystallography

X-ray crystallography is an essential tool for computationally designed protein assemblies because initial data on symmetry calculations and space groups are obtained from existing crystal structures of protein subunits (DiMaio et al., 2011; Yeates, 2017). It is also a validation method to confirm the atomic-level accuracy of the shape and structure of protein assemblies with the original design models (Harcus et al., 2016). In conjunction with TEM images, crystal structure data provides valuable feedback about the design of the protein assemblies for future modifications. However, this technique is limited to those proteins that can form crystal structures. The uncertain and time-consuming attributes of protein crystallization limit the use of this method to characterize protein assembles formed through various routes.

Small-Angle X-ray Scattering (SAXS)

To overcome the main limitation of X-ray crystallography, researchers have applied small-angle X-ray scattering (SAXS) technique, which analyze proteins in solution

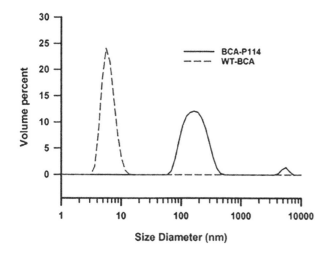

FIGURE 16.9 DLS data comparing the particle size distribution of BCA-P$_{11}$4 nanoparticle with monomeric wild-type protein subunit. (Reprinted with permission from Shanbhag et al. (2016). 'Self-assembled enzyme nanoparticles for carbon dioxide capture', *Nano Letters*, 16, pp. 3379–3384. Copyright (2016) American Chemical Society.)

rather than crystalline and study the structure of protein nanoparticles. Recently, a combination of TEM and SAXS data was used to validate the icosahedral structures formed using two protein components with the design models (Harcus et al., 2016). SAXS analysis was also used to determine the efficiency of protein cage formation based on the results of weight percentages of the various species observed in solution (Lai et al., 2014). In addition, it was possible to evaluate the effectiveness of thermal annealing on the assembly of protein cages using SAXS. Furthermore, oligomerization states (6-mer, 12-mer or 18-mer) of possible protein assemblies can be determined (Kobayashi et al., 2015). The dependence of NaCl concentration and pH on

the superlattice formation of cowpea chlorotic mottle virus (CCMV) and recombinant ferritin protein cage (apoFT) was determined by SAXS, showing the presence of amorphous gel-like form, crystalline and free particle forms (Kostiainen et al., 2013). New improvements to this technique include using high-throughput SAXS (HT-SAXS) which improved the understanding of the assembly process of PC*trip* and PC*quad* subunits as a function of pH and salt concentration, enabling control over the conformation of the assembled protein cage (Lai et al., 2016).

16.6.6 Flow-Based Separations

The asymmetric field flow fractionation (AFFF or AF4) method is a cross-flow separation channel with a semipermeable membrane along the lower wall of the channel causing the particles to separate based on differences in the diffusion coefficients and particle size. AF4 has been a valuable tool in characterizing protein aggregates and separating them from their monomers (Cao, Pollastrini and Jiang, 2009). The technique can be coupled with multiangle light scattering to obtain consistent VLP production of desired size such as where AF4-MALLS was used to compare the size distributions and changes in quaternary structure of VLPs formed by *in vivo* (insect cells) and *in vitro* assembly (Chuan et al., 2008). A different flow-based technique is electrospray-differential mobility analysis (ES-DMA) where the sample is aerosolized and its electrical mobility measured and correlated with particle size (Guha et al., 2012). This method has also been used to characterize VLPs. Both AF4 and ES-DMA were used to differentiate between empty and packaged VLP and also those VLPs with surface modifications (Pease et al., 2009), illustrating the sensitivity of these techniques to subtle differences in the particle morphology.

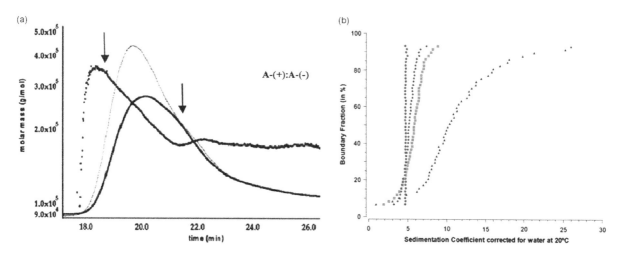

FIGURE 16.10 Analysis of protein cage formation by 1:1 mixture of A-(+):A-(−) KDPG aldolase trimers via coiled-coil linker domains. (a) SEC-MALLS traces showing the presence of more than one species indicated by arrows. (b) Sedimentation coefficient distributions for protein cages (triangles) and the individual unassembled monomers components (circles, diamonds and squares). (Reproduced from Patterson et al. (2011). 'Evaluation of a symmetry-based strategy for assembling protein complexes', *RSC Advances*, 1(6), pp. 1004–1012 with permission of The Royal Society of Chemistry.)

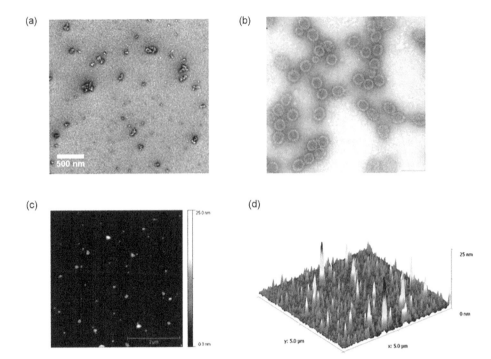

FIGURE 16.11 TEM and AFM data for different types of protein assemblies. (a) TEM image of BCA-P$_{11}$4 nanoparticle. (b) TEM image of purified VLP-V8. (c) AFM image of protein cage formed by 1:1 mixture of A-(+):A-(−) KDPG aldolase trimers via coiled-coil linker domains. (d) Force measurements shown as X–Y plot for AFM image in (c). ((a) Reprinted with permission from Shanbhag et al. (2016). 'Self-assembled enzyme nanoparticles for carbon dioxide capture', *Nano Letters*, 16, pp. 3379–3384. Copyright (2016) American Chemical Society. (b) Reprinted from Lua et al. (2015). 'Synthetic biology design to display an 18kDa rotavirus large antigen on a modular virus-like particle', *Vaccine*, 33 (44), pp. 5937–5944. Copyright (2015) with permission from Elsevier. (c and d) Reproduced from Patterson et al. (2011). 'Evaluation of a symmetry-based strategy for assembling protein complexes', *RSC Advances*, 1(6), pp. 1004–1012 with permission of The Royal Society of Chemistry.)

16.7 Perspectives and Conclusion

Recent research in protein engineering and nanotechnology have made significant progress in the field of protein nanomaterials, providing different routes to prepare these with desired attributes. While the field initially started by mimicking naturally occurring protein nanoparticles (e.g., VLPs), it has developed to include the engineering of brand new nanoparticles by *de novo* design, providing exciting opportunities to bring new functionalities. On the other hand, self-assembly peptide-based protein nanomaterials have advantages of the simplicity and variety although the field is at an early stage for the delivery of nanoparticles with defined structures. It is anticipated that the design, production and purification approaches reviewed in this chapter will be further developed into generalized methods but that the field will continue to develop to deliver advanced materials based on a broad range of functional protein nanoparticles in near future.

References

Aggeli, A. et al. (1997) 'Responsive gels formed by the spontaneous self-assembly of peptides into polymeric beta-sheet tapes', *Nature*, 386, pp. 259–262.

Arcangeli, C. et al. (2014) 'Structure-based design and experimental engineering of a plant virus nanoparticle for the presentation of immunogenic epitopes and as a drug carrier', *Journal of Biomolecular Structure and Dynamics*, 32(4), pp. 630–647. doi:10.1080/07391102.2013.785920.

Bale, J. B. et al. (2015) 'Structure of a designed tetrahedral protein assembly variant engineered to have improved soluble expression', *Protein Science*, 24(10), pp. 1695–1701. doi:10.1002/pro.2748.

Bromley, E. H. C. et al. (2009) 'Designed r -helical tectons for constructing multicomponent synthetic biological systems', *Journal of American Chemical Society*, 131, pp. 928–930.

Butterfield, G. L. et al. (2017) 'Evolution of a designed protein assembly encapsulating its own RNA genome', *Nature*, 552(7685), pp. 415–420. doi:10.1038/nature25157.

Cao, S., Pollastrini, J. and Jiang, Y. (2009) 'Separation and characterization of protein aggregates and particles by field flow fractionation', *Curr Pharm Biotechnol*, 10(4), pp. 382–390. Available at: www.ncbi.nlm.nih.gov/pubmed/19519413.

Chuan, Y. P. et al. (2008) 'Quantitative analysis of virus-like particle size and distribution by field-flow fractionation',

Biotechnology and Bioengineering, 99(6), pp. 1425–1433. doi:10.1002/bit.21710.

Connors, N. K. et al. (2014) 'Improved fusion tag cleavage strategies in the downstream processing of self-assembling virus-like particle vaccines', *Food and Bioproducts Processing*, 92(2), pp. 143–151. doi:10.1016/j.fbp.2013.08.012.

Dasgupta, A., Mondal, J. H. and Das, D. (2013) 'Peptide hydrogels', *RSC Advances*, 3(24), p. 9117. doi:10.1039/c3ra40234g.

Dietz, H. et al. (2007) 'Programming protein self assembly with coiled coils', *New Journal of Physics*, 9. doi:10.1088/1367-2630/9/11/424.

DiMaio, F. et al. (2011) 'Modeling symmetric macromolecular structures in Rosetta3', *PLoS ONE*, 6(6). doi:10.1371/journal.pone.0020450.

Ding, X. et al. (2018) 'Virus-like particle engineering: From rational design to versatile applications', *Biotechnology Journal*, 13(5), pp. 1–7. doi:10.1002/biot.201700324.

Ding, Y. et al. (2010) 'Modeling the competition between aggregation and self-assembly during virus-like particle processing', *Biotechnology and Bioengineering*, 107(3), pp. 550–560. doi:10.1002/bit.22821.

Dooley, K., Bulutoglu, B. and Banta, S. (2014) 'Doubling the cross-linking interface of a rationally designed beta roll peptide for calcium-dependent proteinaceous hydrogel formation', *Biomacromolecules*, 15(10), pp. 3617–3624. doi:10.1021/bm500870a.

Esposito, D. and Chatterjee, D. K. (2006) 'Enhancement of soluble protein expression through the use of fusion tags', *Current Opinion in Biotechnology*, 17(4), pp. 353–358. doi:10.1016/j.copbio.2006.06.003.

Fan, Y. et al. (2018) 'Column-free purification methods for recombinant proteins using self-cleaving aggregating tags', *Polymers*, 10(5), p. 468. doi:10.3390/polym10050468.

Fan, Z. et al. (2016) 'Bioinspired fluorescent dipeptide nanoparticles for targeted cancer cell imaging and real-time monitoring of drug release', *Nature Nanotechnology*, 11(January), pp. 388–394. doi:10.1038/nnano.2015.312.

Fang, P. Y. et al. (2016) 'Functional RNAs: Combined assembly and packaging in VLPs', *Nucleic Acids Research*, 45(6), pp. 3519–3527. doi:10.1093/nar/gkw1154.

Ferrer-Miralles, N. et al. (2015) 'Engineering protein self-assembling in protein-based nanomedicines for drug delivery and gene therapy', *Critical Reviews in Biotechnology*, 35(2), pp. 209–221. doi:10.3109/07388551.2013.833163.

Filipe, V., Hawe, A. and Jiskoot, W. (2010) 'Critical evaluation of nanoparticle tracking analysis (NTA) by NanoSight for the measurement of nanoparticles and protein aggregates', *Pharmaceutical Research*, 27(5), pp. 796–810. doi:10.1007/s11095-010-0073-2.

Frietze, K. M., Peabody, D. S. and Chackerian, B. (2016) 'Engineering virus-like particles as vaccine platforms', *Current Opinion in Virology*, 18, pp. 44–49. doi:10.1016/j.coviro.2016.03.001.

Gazit, E. (2007) 'Self-assembled peptide nanostructures: The design of molecular building blocks and their technological utilization', *Chemical Society Reviews*, 36(8), pp. 1263–1269. doi:10.1039/b605536m.

Georgiev, I. S. et al. (2018) 'Two-component ferritin nanoparticles for multimerization of diverse trimeric antigens', *ACS Infectious Diseases*, 4(5), pp. 788–796. doi:10.1021/acsinfecdis.7b00192.

Guha, S. et al. (2012) 'Electrospray-differential mobility analysis of bionanoparticles', *Trends in Biotechnology*, 30(5), pp. 291–300. doi:10.1016/j.tibtech.2012.02.003.

Harcus, T. E. et al. (2016) 'Accurate design of megadalton-scale two-component icosahedral protein complexes', *Science*, 353(6297), pp. 389–395. doi:10.5061/dryad.8c65s.

Herrera Estrada, L. P. and Champion, J. A. (2015) 'Protein nanoparticles for therapeutic protein delivery', *Biomaterials Science*, 3(6), pp. 787–799. doi: 10.1039/C5BM00052A.

Hill, B. D. et al. (2017) 'Engineering virus-like particles for antigen and drug delivery', *Current Protein & Peptide Science*, 19(1), pp. 112–127. doi:10.2174/1389203718666161122113041.

Hsia, Y. et al. (2016) 'Design of a hyperstable 60-subunit protein icosahedron', *Nature*, 535(7610), pp. 136–139. doi:10.1038/nature18010.

Hudalla, G. A. et al. (2014) 'Gradated assembly of multiple proteins into supramolecular nanomaterials', *Nature Materials*, 13(8), pp. 829–836. doi: 10.1038/nmat3998.

Jin, S. et al. (2016) 'Controlled hybrid-assembly of HPV16/18 L1 Bi VLPs *in vitro*', *ACS Applied Materials and Interfaces*, 8(50), pp. 34244–34251. doi:10.1021/acsami.6b12456.

Jorda, J. et al. (2016) 'Structure of a novel 13 nm dodecahedral nanocage assembled from a redesigned bacterial microcompartment shell protein', *Chemical Communications*, 52(28), pp. 5041–5044. doi:10.1039/c6cc00851h.

Kanojia, G. et al. (2016) 'A design of experiment approach to predict product and process parameters for a spray dried influenza vaccine', *International Journal of Pharmaceutics*, 511(2), pp. 1098–1111. doi:10.1016/j.ijpharm.2016.08.022.

Khoshnejad, M. et al. (2018) 'Ferritin nanocages with biologically orthogonal conjugation for vascular targeting and imaging', *Bioconjugate Chemistry*, 29(4), pp. 1209–1218. doi:10.1021/acs.bioconjchem.8b00004.

King, N. P. et al. (2012) 'Computational design of self-assembling protein nanomaterials with atomic level accuracy', *Science*, 336(6085), pp. 1171–1174. doi:10.1126/science.1219364.

King, N. P. et al. (2014) 'Accurate design of co-assembling multi-component protein nanomaterials', *Nature*, 510(7503), pp. 103–108. doi:10.1038/nature13404.

King, N. P. and Lai, Y.-T. (2013) 'Practical approaches to designing novel protein assemblies', *Current Opinion in Structural Biology*, 23(4), pp. 632–638. doi:10.1016/j.sbi.2013.06.002.

Kobayashi, N. et al. (2015) 'Self-assembling nano-architectures created from a protein nano-building block using an intermolecularly folded dimeric de novo protein', *Journal of the American Chemical Society*, 137(35), pp. 11285–11293. doi:10.1021/jacs.5b03593.

Kostiainen, M. A. et al. (2013) 'Electrostatic assembly of binary nanoparticle superlattices using protein cages', *Nature Nanotechnology*, 8(1), pp. 52–56. doi:10.1038/nnano.2012.220.

Kyle, S. et al. (2009) 'Production of self-assembling biomaterials for tissue engineering', *Trends in Biotechnology*, 27(7), pp. 423–433. doi:10.1016/j.tibtech.2009.04.002.

Ladd Effio, C. et al. (2016) 'High-throughput process development of an alternative platform for the production of virus-like particles in Escherichia coli', *Journal of Biotechnology*, 219, pp. 7–19. doi:10.1016/j.jbiotec.2015.12.018.

Lai, Y. T. et al. (2013) 'Structure and flexibility of nanoscale protein cages designed by symmetric self-assembly', *Journal of the American Chemical Society*, 135(20), pp. 7738–7743. doi:10.1021/ja402277f.

Lai, Y. T. et al. (2014) 'Structure of a designed protein cage that self-assembles into a highly porous cube', *Nature Chemistry*, 6(12), pp. 1065–1071. doi:10.1038/nchem.2107.

Lai, Y. T. et al. (2016) 'Designing and defining dynamic protein cage nanoassemblies in solution', *Science Advances*, 2(12). doi:10.1126/sciadv.1501855.

Lee, E. J., Lee, N. K. and Kim, I. S. (2016) 'Bioengineered protein-based nanocage for drug delivery', *Advanced Drug Delivery Reviews*, 106, pp. 157–171. doi:10.1016/j.addr.2016.03.002.

Li, Y. et al. (2014) 'Biomimetic design of affinity peptide ligand for capsomere of virus-like particle', *Langmuir*, 30(28), pp. 8500–8508. doi:10.1021/la5017438.

Liew, M. W. O., Chuan, Y. P. and Middelberg, A. P. J. (2012a) 'High-yield and scalable cell-free assembly of virus-like particles by dilution', *Biochemical Engineering Journal*, 67, pp. 88–96. doi:10.1016/j.bej.2012.05.007.

Liew, M. W. O., Chuan, Y. P. and Middelberg, A. P. J. (2012b) 'Reactive diafiltration for assembly and formulation of virus-like particles', *Biochemical Engineering Journal*, 68, pp. 120–128. doi:10.1016/j.bej.2012.07.009.

Liu, Y. et al. (2018) 'Near-atomic cryo-EM imaging of a small protein displayed on a designed scaffolding system', *Proceedings of the National Academy of Sciences*, 115(13), p. 201718825. doi:10.1073/pnas.1718825115.

Llauró, A. et al. (2016) 'Tuning viral capsid nanoparticle stability with symmetrical morphogenesis', *ACS Nano*, 10(9), pp. 8465–8473. doi:10.1021/acsnano.6b03441.

Lobstein, J. et al. (2012) 'SHuffle, a novel Escherichia coli protein expression strain capable of correctly folding disulfide bonded proteins in its cytoplasm', *Microbial Cell Factories*, 11(1), p. 56. doi:10.1186/1475-2859-11-56.

Löwik, D. W. P. M. et al. (2010) 'Stimulus responsive peptide based materials', *Chemical Society Reviews*, 39(9), pp. 3394–3412. doi:10.1039/b914342b.

Lua, L. H. L. et al. (2015) 'Synthetic biology design to display an 18kDa rotavirus large antigen on a modular virus-like particle', *Vaccine*, 33(44), pp. 5937–5944. doi:10.1016/j.vaccine.2015.09.017.

Mason, T. O. et al. (2017) 'Thermodynamics of polypeptide supramolecular assembly in the short-chain limit', *Journal of the American Chemical Society*, 139(45), pp. 16134–16142. doi:10.1021/jacs.7b00229.

Moers, A. P. H. A. et al. (2010) 'Secreted production of self-assembling peptides in Pichia pastoris by fusion to an artificial highly hydrophilic protein', *Journal of Biotechnology*, 146(1–2), pp. 66–73. doi:10.1016/j.jbiotec.2010.01.010.

Mohr, J. et al. (2013) 'Virus-like particle formulation optimization by miniaturized high-throughput screening', *Methods*, 60(3), pp. 248–256. doi:10.1016/j.ymeth.2013.04.019.

O'Mahony, T. C. et al. (2011) 'The thermodynamics of defect formation in self-assembled systems', in J. C. Moreno-Piraján (ed.), *Thermodynamics – Systems in Equilibrium and Non-Equilibrium*. InTech. doi:10.5772/20145.

Padilla, J. E., Colovos, C. and Yeates, T. O. (2001) 'Nanohedra: Using symmetry to design self assembling protein cages, layers, crystals, and filaments', *Proceedings of the National Academy of Sciences*, 98(5), pp. 2217–2221. doi:10.1073/pnas.041614998.

Papaneophytou, C. P. and Kontopidis, G. (2014) 'Statistical approaches to maximize recombinant protein expression in Escherichia coli: A general review', *Protein Expression and Purification*, 94, pp. 22–32. doi:10.1016/j.pep.2013.10.016.

Patterson, D. P. et al. (2011) 'Evaluation of a symmetry-based strategy for assembling protein complexes', *RSC Advances*, 1(6), pp. 1004–1012. doi:10.1039/C1RA00282A.

Pease, L. F. et al. (2009) 'Quantitative characterization of virus-like particles by asymmetrical flow field flow fractionation, electrospray differential mobility analysis, and transmission electron microscopy', *Biotechnology and Bioengineering*, 102(3), pp. 845–855. doi:10.1002/bit.22085.

Reddy Chichili, V. P., Kumar, V. and Sivaraman, J. (2013) 'Linkers in the structural biology of protein-protein interactions', *Protein Science*, 22(2), pp. 153–167. doi:10.1002/pro.2206.

Rosano, G. L. and Ceccarelli, E. A. (2014) 'Recombinant protein expression in Escherichia coli: Advances and challenges', *Frontiers in Microbiology*, 5, pp. 1–17. doi:10.3389/fmicb.2014.00172.

Ross, J. F. et al. (2017) 'Decorating self-assembled peptide cages with proteins', *ACS Nano*, 11, pp. 7901–7914. doi:10.1021/acsnano.7b02368.

Sánchez-Sánchez, L. et al. (2015) 'Design of a VLP-nanovehicle for CYP450 enzymatic activity delivery', *Journal of Nanobiotechnology*, 13(1), pp. 1–10. doi:10.1186/s12951-015-0127-z.

Saunders, M. J. et al. (2013) 'Engineering fluorogen activating proteins into self-assembling materials', *Bioconjugate Chemistry*, 24(5), pp. 803–810. doi:10.1021/bc300613h.

Schein. C. H. and Noteborn, M. H. M. (1988) 'Formation of soluble recombinant proteins in Escherichia coli is favored by lower growth temperature', *Nature Biotechnology*, 6, pp. 291–294.

Schuler, B. et al. (2005) 'Polyproline and the "spectroscopic ruler" revisited with single-molecule fluorescence', *Proceedings of the National Academy of Sciences of the United States of America*, 102(8), pp. 2754–2759. doi:10.1073/pnas.0408164102.

Shanbhag, B. K. et al. (2016) 'Self-assembled enzyme nanoparticles for carbon dioxide capture', *Nano Letters*, 16(5), pp. 3379–3384. doi:10.1021/acs.nanolett.6b01121.

Shanbhag, B. K. et al. (2018) 'Understanding the interplay between self-assembling peptides and solution ions for tunable protein nanoparticle formation', *ACS Nano*. doi:10.1021/acsnano.8b02381.

Shang, L., Nienhaus, K. and Nienhaus, G. U. (2014) 'Engineered nanoparticles interacting with cells: size matters', *Journal of Nanobiotechnology*, 12(1), p. 5. doi:10.1186/1477-3155-12-5.

Shur, O. et al. (2013) 'A designed, phase changing RTX-based peptide for efficient bioseparations', *BioTechniques*, 54(4), pp. 197–198, 200, 202, 204, 206. doi:10.2144/000114010.

Sinn, S. et al. (2018) 'Templated formation of luminescent virus-like particles by tailor-made pt(II) amphiphiles', *Journal of the American Chemical Society*, 140(6), pp. 2355–2362. doi:10.1021/jacs.7b12447.

Sivashanmugam, A. et al. (2009) 'Practical protocols for production of very high yields of recombinant proteins using Escherichia coli', *Protein Science: A Publication of the Protein Society*, 18(5), pp. 936–948. doi:10.1002/pro.102.

Sliepen, K. et al. (2015) 'Presenting native-like HIV-1 envelope trimers on ferritin nanoparticles improves their immunogenicity', *Retrovirology*, 12(1), pp. 1–5. doi:10.1186/s12977-015-0210-4.

Studier, F. W. (2005) 'Protein production by auto-induction in high-density shaking cultures', *Protein Expression and Purification*, 41(1), pp. 207–234. doi:10.1016/j.pep.2005.01.016.

Swift, J. et al. (2006) 'Design of functional ferritin-like proteins with hydrophobic cavities', *Journal of the American Chemical Society*, 128(20), pp. 6611–6619. doi:10.1021/ja057069x.

Tekewe, A. et al. (2016) 'Design strategies to address the effect of hydrophobic epitope on stability and *in vitro* assembly of modular virus-like particle', *Protein Science*, 25, pp. 1507–1516. doi:10.1002/pro.2953.

Tekewe, A. et al. (2017) 'Integrated molecular and bioprocess engineering for bacterially produced immunogenic modular virus-like particle vaccine displaying 18 kDa

rotavirus antigen', *Biotechnology and Bioengineering*, 114(2), pp. 397–406. doi:10.1002/bit.26068.

Terpe, K. (2003) 'Overview of tag protein fusions: from molecular and biochemical fundamentals to commercial systems', *Applied Microbiology and Biotechnology*, 60(5), pp. 523–533. doi:10.1007/s00253-002-1158-6.

Truffi, M. et al. (2016) 'Ferritin nanocages: A biological platform for drug delivery, imaging and theranostics in cancer', *Pharmacological Research*, 107, pp. 57–65. doi:10.1016/j.phrs.2016.03.002.

Tsutsumi, H. and Mihara, H. (2013) 'Soft materials based on designed self-assembling peptides: From design to application', *Molecular BioSystems*, 9(4), pp. 609–617. doi:10.1039/c3mb25442a.

Uchida, M. et al. (2006) 'Targeting of cancer cells with ferrimagnetic ferritin cage nanoparticles', *Journal of the American Chemical Society*, 128(51), pp. 16626–16633. doi:10.1021/ja0655690.

Unzueta, U. et al. (2012) 'Non-amyloidogenic peptide tags for the regulatable self-assembling of protein-only nanoparticles', *Biomaterials*, 33(33), pp. 8714–8722. doi:10.1016/j.biomaterials.2012.08.033.

Waneesorn, J. et al. (2018) 'Structural-based designed modular capsomere comprising HA1 for low-cost poultry influenza vaccination', *Vaccine*, 36(22), pp. 3064–3071. doi:10.1016/j.vaccine.2016.11.058.

Wang, L. et al. (2017) 'Structure-based design of ferritin nanoparticle immunogens displaying antigenic loops of Neisseria gonorrhoeae', *FEBS Open Bio*, 7(8), pp. 1196–1207. doi:10.1002/2211-5463.12267.

Washida, M. et al. (2001) 'Spacer-mediated display of active lipase on the yeast cell surface', *Applied Microbiology and Biotechnology*, 56(5–6), pp. 681–686. doi:10.1007/s002530100718.

Weissman, S. A. and Anderson, N. G. (2015) 'Design of Experiments (DoE) and process optimization. A review of recent publications', *Organic Process Research and Development*, 19(11), pp. 1605–1633. doi:10.1021/op500169m.

Wheeldon, I. R., Campbell, E. and Banta, S. (2009) 'A chimeric fusion protein engineered with disparate functionalities-enzymatic activity and self-assembly', *Journal of Molecular Biology*, 392(1), pp. 129–142. doi:10.1016/j.jmb.2009.06.075.

Wibowo, N. et al. (2013) 'Modular engineering of a microbially-produced viral capsomere vaccine for influenza', *Chemical Engineering Science*, 103(September 2009), pp. 12–20. doi:10.1016/j.ces.2012.04.001.

Wong, C. T. S. et al. (2009) 'Two-component protein-engineered physical hydrogels for cell encapsulation', *Proceedings of the National Academy of Sciences of the United States of America*, 106(52), pp. 22067–22072.

Wu, W. et al. (2011) 'Active protein aggregates induced by terminally attached self-assembling peptide ELK16 in Escherichia coli', *Microbial Cell Factories*, 10(1), p. 9. doi:10.1186/1475-2859-10-9.

Yeates, T. O. (2017) 'Geometric principles for designing highly symmetric self-assembling protein nanomaterials', *Annual Review of Biophysics*, 46(1), pp. 23–42. doi:10.1146/annurev-biophys-070816-033928.

Yeates, T. O., Liu, Y. and Laniado, J. (2016) 'The design of symmetric protein nanomaterials comes of age in theory and practice', *Current Opinion in Structural Biology*, 39, pp. 134–143. doi:10.1016/j.sbi.2016.07.003.

Yu, L. et al. (2005) 'Fabrication and application of enzyme-incorporated peptide nanotubes', *Bioconjugate Chemistry*, 16(6), pp. 1484–1487. doi:10.1021/bc050199a.

Zhang, L. et al. (2015) 'Biomolecular engineering of virus-like particles aided by computational chemistry methods', *Chemical Society Reviews*, 44(23), pp. 8608–8618. doi:10.1039/C5CS00526D.

Zhang, S. et al. (1993) 'Spontaneous assembly of a self-complementary oligopeptide to form a stable macroscopic membrane', *Proceedings of the National Academy of Sciences of the United States of America*, 90, pp. 3334–3338.

Zhang, S. et al. (2002) 'Design of nanostructured biological materials through self-assembly of peptides and proteins', *Current Opinion in Chemical Biology*, 6(6), pp. 865–871. doi:10.1016/S1367-5931(02)00391-5.

Zhang, S. (2003a) 'Building from the bottom up', *Materials Today*, 6, pp. 20–27.

Zhang, S. (2003b) 'Fabrication of novel biomaterials through molecular self-assembly', *Nature Biotechnology*, 21(10), pp. 1171–1178. doi:10.1038/nbt874.

Zhang, X. et al. (2015) 'Rational design of a photoresponsive UVR8-derived protein and a self-assembling peptide–protein conjugate for responsive hydrogel formation', *Nanoscale*, pp. 16666–16670. doi:10.1039/C5NR05213K.

Zhou, B. et al. (2012) 'Small surfactant-like peptides can drive soluble proteins into active aggregates', *Microbial Cell Factories*, 11(1), p. 10. doi:10.1186/1475-2859-11-10.

Zou, W. et al. (2016) 'Expression, purification, and characterization of recombinant human L-chain ferritin', *Protein Expression and Purification*, 119, pp. 63–68. doi:10.1016/j.pep.2015.11.018.

17

Nanoengineering Neural Cells for Regenerative Medicine

Christopher F. Adams and
Stuart I. Jenkins
*Neural Tissue Engineering group: Keele
(NTEK)*
Keele University

17.1 Tissue Engineering and Regenerative Medicine for the Central Nervous System

This chapter covers a broad range of techniques for improving neural repair by utilizing various cutting-edge nanotechnology strategies. Initially, we will provide a brief description of CNS (brain and spinal cord) disease/injury and why repair is difficult to achieve. Subsequent sections will outline different nanotechnologies and discuss how these materials can be used to address the varied and complex challenges associated with CNS injury and repair. We will specifically focus on nanoengineering neural cells using nanoparticles to improve transplantation outcomes and finish with a discussion examining future areas of research combining multiple advances in nanotechnology for combinatorial therapies.

17.1.1 CNS Disease Is a Major and Growing Problem, Lacking Effective Treatments

Disease and injury of the CNS severely affect the quality of life of patients and their families/carers, as well as representing an enormous and growing healthcare burden. This is partly due to broader successes in reducing mortality, contributing to the increasing numbers of older people, the population most prone to developing neurodegenerative conditions. Due to increasing life expectancies, the prevalence of neurodegenerative diseases is predicted to become one of the leading causes of disability and death by 2030 (World Health Organisation, 2013). Efforts to find therapies which can restore function to the damaged CNS are vital both for human well-being and to alleviate the cost-burden for global healthcare services. However, this constitutes a major challenge as CNS tissue has a complex cytoarchitecture and a poor regenerative capacity, posing considerable obstacles to the development of neuroregenerative strategies.

17.1.2 The Native and Injured CNS Is Highly Complex, with Various Important Cell Types

In terms of its cytoarchitecture, the CNS consists of two major classes of cells: the neurons and their supporting glia. Neurons transmit electrical signals and reside in groups forming multiple connections with other neurons to make up neural circuits, which perform a common function, for example, vision or movement (Bear, Connors and Paradiso, 2015). Neurons extend axons which are highly specialized structures unique to neuronal cells and adapted to relay information within the body. Axons are ensheathed by layers of an electrically-insulating fatty deposit called myelin. CNS myelin is made and maintained by the oligodendrocytes, each of which can myelinate multiple axons. Astrocytes are the major supporting cell type within the CNS, with new roles regularly being discovered. Their currently accepted functions include: maintaining CNS homeostasis; clearance and recycling of neurotransmitters; providing metabolic support to neurons; roles in synapse formation and maintenance (Bear, Connors and Paradiso, 2015; Liddelow and Barres, 2015; Verkhratsky and Butt, 2013).

Other important CNS cell types include the microglia which are the resident immune cells (involved in phagocytosis and cell recruitment to sites of injury) and stem/precursor cells – both neural stem cells (NSCs) and oligodendrocyte progenitor cells (OPCs) (Burda and Sofroniew, 2014). NSCs are multipotent precursor cells to three major cell types of the CNS (neurons, oligodendrocytes and astrocytes) with the capacity to self-renew (Kang

et al., 2010; Burda and Sofroniew, 2014). Finally, a mixture of astrocytes, endothelial cells and pericytes form the blood brain barrier (BBB) which strictly controls the entry of molecules to the CNS from the bloodstream. In this manner, it functions to maintain the specific molecular environment (e.g. optimal ion concentrations for signal transduction) crucial for CNS function (Bear, Connors and Paradiso, 2015; Verkhratsky and Butt, 2013).

Insult to the CNS generally results in a complex, multifactorial pathology which can be summarized by three stages of response: (i) Cell death and breakdown of the BBB – typically involving a wave of neuronal cell death followed by the death of oligodendrocytes (which normally receive pro-survival signaling from the neurons). Cell death coupled with breakdown of the BBB produces inflammatory and other cell signaling molecules which can be toxic to surrounding CNS cells. (ii) Reactive gliosis – where glia are mobilized in response to this environment. This includes recruitment of microglia (microgliosis) which attempt to clear debris, proliferation and migration of OPCs, to replace lost oligodendrocytes, and proliferation and hypertrophy of local astrocytes (astrogliosis) (Pekny and Pekna, 2016; Gao et al., 2013). (iii) Tissue remodeling – which involves formation of the so-called *astrocyte scar* – a combination of hypertrophied astrocytes and secreted extracellular matrix (ECM) molecules including repair-inhibiting chondroitin sulfate proteoglycans (CSPGs) (Brosius Lutz and Barres, 2014; Burda and Sofroniew, 2014; Kim, Sajid and Trakhtenberg, 2018; Kyritsis, Kizil and Brand, 2014). Figure 17.1 and Box 17.1 illustrate and describe the main pathological features of CNS damage. Disrupting the astrocyte scar, removing inhibitory signaling molecules, promoting intrinsic neuronal growth and replacing lost neurons are all exciting therapeutic avenues which if targeted simultaneously could promote repair in the CNS. It should be noted that the

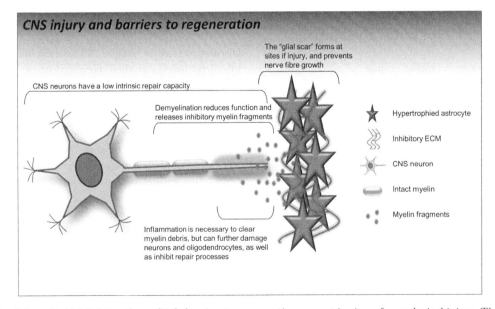

FIGURE 17.1 Schematic highlighting the multiple barriers to regeneration present in sites of neurological injury. The diagram shown depicts a traumatic injury which has severed the axon projecting from the nerve cell.

precise inhibitory nature of the adult, aged and injured CNS is a complex and evolving area of research, with concepts such as inhibition of axonal growth by the glial scar being challenged as simplistic (Liddelow and Barres, 2016).

Box 17.1: Glossary of Terms for CNS Damage and Repair

Astrogliosis: Astrocytes are the homeostatic cells of the CNS. Following injury, they often become hypertrophic, altering their morphology, and physically isolate lesion sites (e.g. cavities following spinal cord injury), preventing regeneration.

Axonal regeneration: When severed, nerve fibers (axons) can sometimes regrow. However, the CNS milieu is usually inhibitory to this process, especially in lesion sites. If axons can be encouraged to regenerate, they could re-innervate cells, restoring synaptic connections and possibly function. Some nanomaterials can promote axonal extension and even provide directional cues.

Blood–brain barrier: Specific cellular arrangements in capillaries that limit movement of molecules from bloodstream to CNS.

CNS: Brain and spinal cord.

Demyelination: Loss of the insulating, protective coating around nerve fibers, which is produced by oligodendrocytes.

Glial scar: Astrogliosis at sites of injury, inhibitory to regeneration.

Microgliosis: Pro-inflammatory microglial response; neuroinflammation.

Remyelination: One of the few repair processes in the CNS; generation of new oligodendrocytes and myelin, following demyelination.

Synapse: Connection between neurons, where neurotransmitters mediate communication.

Wallerian degeneration: Where nerve fibers (axons) are severely damaged, the section no longer connected to the cell body degenerates.

17.1.3 Repairing the Damaged CNS Requires Complex Therapies

Current clinical treatments for neurological injury generally aim to reduce symptoms and slow disease progression, for example, by administering drugs to modulate inflammation in spinal cord injury (SCI; Plemel, Wee Yong and Stirling, 2014) or reduce blood clotting in stroke (Frank et al., 2013). Physiotherapy is also vital to improve neurological function and has shown benefits for patients experiencing stroke, traumatic brain injury (TBI) and SCI (Harvey, 2016; Reinkensmeyer, Lum and Winters, 2002). However, currently there are no treatments which are truly regenerative.

In this context, numerous challenges are stalling advances in the field of neuroregenerative medicine, including the BBB, which severely restricts or prevents ingress of most drugs into the CNS, the complex environment of CNS lesions, which inhibits regenerative events and necessitates combinatorial therapies, and the lack of understanding of etiology for many CNS diseases.

For example, drug delivery into the CNS is hampered by limited BBB permeability and diffusion of drugs within the blood and cerebrospinal fluid (CSF). This can mean that high concentrations of the drug have to be administered, potentially resulting in unwanted side-affects, or multiple deliveries are required increasing expense and exacerbating damage at the target site.

Stem cell transplantation has shown significant potential in early pre-clinical experiments. However, addressing a single factor by utilizing cell replacement (or gene delivery) is unlikely to be sufficient to restore the function of the CNS. To promote effective neural repair, 'combinatorial' therapies, such as cell replacement coupled with therapeutic gene delivery, are thought to be the most realistic approach to achieve functional repair. Supporting this view, studies have demonstrated enhanced benefits when transplanting genetically manipulated neural cells compared to transplantation of cells alone (Abati et al., 2019; Gransee et al., 2015; Moradian et al., 2017), a topic which will be examined in more detail later.

17.1.4 Nanomaterials and Nanoengineering of Neural Tissue Offer Many Solutions

Given the significant challenges within the field of neuroregeneration, new methods need to be developed which can support and improve existing treatments and provide new platforms for neural therapy. In particular, nanosized tools are supporting novel approaches for studying neuropathology and developing therapeutic interventions. The key aim of this chapter is to explore the potential of combining the rapidly emerging field of nanotechnology with neural cell engineering to address these barriers. An overview of various nanotechnologies will be given in the subsequent sections and details on how they can be used to **nanoengineer** both endogenous and exogenous (for cell transplantation) neural cells to potentially achieve successful repair of the damaged CNS.

17.2 Tools for Nanoengineering

Here, any material with a dimension in the nanoscale (1–1,000 nm) will be considered 'nano', although some authorities suggest restricting the term to at least one dimension in the range 1–100 nm (Saner and Stoklosa, 2013; Boverhof et al., 2015). This not only includes nanoparticles (all three dimensions <1,000 nm) but also nanotubes and nanowires (which may have length much greater than 1,000 nm), and sheets of materials such as graphene (where a single dimension is <1,000 nm). With respect to materials, many substances exhibit different behaviors/properties at the nanoscale, compared to the same material in bulk (Lynch et al., 2007). This can affect the traversal/penetration of cell or other biological membranes, affect material responses to magnetic fields and vastly increase the available surface area. Exploiting these physicochemical properties allows the design of multifunctional nanoplatforms, with great potential for neural engineering. Some nanoengineering approaches are unlikely to be translated to the clinic, yet still have practical applications in basic research. For example, nanoparticles with fluorescent tags, or low-level toxicity, or applications requiring activation signals/imaging techniques not yet approved for clinical use (e.g. high-grade MRI fields, plasmon resonance imaging).

17.2.1 Nanoparticles for Drug/Gene Delivery, Tumor Labeling

Nanoparticles are small enough to be internalized by individual cells, facilitating drug delivery, non-viral genetic engineering, and labeling of cells for manipulation or tracking (Figure 17.2). Non-viral gene delivery has clinically translational benefits over viral delivery. Viral delivery can be associated with insertional mutagenesis (where genetic material is inserted into unfavorable places within the target cell genome), activation of the host's immune response and significant manufacturing hurdles. Plasmid-based engineering can overcome some of these challenges as plasmids are not incorporated into the genome, can be engineered without immunogenic DNA and can be manufactured relatively easily (Mayrhofer, Schleef and Jechlinger, 2009; Williams and Kingston, 2011).

Nanoparticles can be fabricated from a wide range of materials, including different materials arranged in layers, and both surface and core may be functionalized, e.g. with drugs or labels for tracking. This versatility means that particles can be designed for a huge variety of applications, including drug delivery, gene delivery and non-invasive tracking.

Magnetic NanoParticles (MNPs)

MNPs, generally synthesized through addition of iron oxide to the nanoparticle formulation, are attracting great interest due their capacities for remote manipulation by magnetic fields and non-invasive imaging/tracking inside patients. These particles are typically superparamagnetic, meaning that they only exhibit magnetic properties in the presence of a magnetic field – important to reduce particle aggregation (due to magnetic attraction between particles), which may impair their utility, and even produce toxic effects. Along with acting as a contrast enhancer, the iron core can be detected histologically using Perls' stain for identification of labeled cells (Dunning et al., 2004).

The additional functionality provided by the magnetic element makes MNPs particularly attractive for use as multifunctional tools potentially capable of achieving a range of biomedical goals within one formulation (Figure 17.2). MNPs for nanoengineering neural cell transplants will be discussed in more detail later.

17.2.2 Nanomaterials as Guidance and Differentiation Cues

Nanomaterials can be generated as sheets or fibers, possibly with topographic cues, that promote adhesion of neural cells, and/or directional guidance of cell migration or nerve fiber growth (Baiguera et al., 2014; Entekhabi et al., 2016; Xue et al., 2018). These cell-material interactions can be exploited to guide axonal outgrowth, for example, across lesion sites following SCI. Nanotopography can also influence cellular differentiation, with graphene reportedly promoting neuronal differentiation from NSCs (Park et al., 2011; Gardin, Piattelli and Zavan, 2016; Kamudzandu et al., 2015).

Electrospinning is a technique that can produce nanofibers from a range of liquid polymers, including gelatin, poly lactic-co-glycolic acid (PLGA) and polycaprolactone (PCL), which may be aligned largely parallel with each other (Kamudzandu, Roach et al., 2015; Kamudzandu, Yang et al., 2015; Meco and Lampe, 2018). Astrocytes, OPCs and neurites (fiber outgrowths from neurons) have been shown to align with poly-L,D-lactic acid nanofibers, with laminin coating of the fibers inducing neurite extension across an in vitro spinal cord lesion (Weightman, Jenkins et al., 2014; Weightman, Pickard et al., 2014).

An alternative material for neural cell adherence/guidance is carbon. Carbon nanotubes (CNTs) may

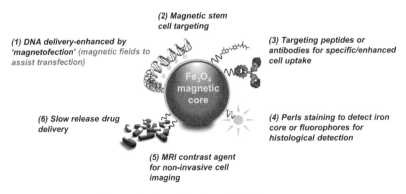

FIGURE 17.2 Schematic detailing multiple biomedical applications of MNPs.

be single-walled (SWNTs; prepared by rolling graphene sheets into tubes), or multi-walled (MWNTs), consisting of tubes within tubes (concentric; may be 25+ layers), or a single-sheet 'Swiss roll' format. CNTs are extremely strong, can be produced with large aspect ratios (length:diameter) and can be filled with cargo such as drugs (Mallapragada et al., 2015). These materials are also electrically conductive, offering further functionality in terms of electrical stimulation and/or recording, which will be discussed next.

17.2.3   Electrical Nanomaterials

Electrical signaling is known to be vital in neurodevelopment and in wound healing with electrical stimulation of nervous system tissue for therapeutic purposes being a centuries-old practice (Thompson et al., 2014). However, only recently, more sophisticated materials are being developed which mean electrical stimuli can be delivered (and recorded) using nanoscale devices (Sanghavi et al., 2015; Young, Cornwell and Daniele, 2018; Zhang et al., 2018). Conducting polymers have been used to fabricate electroactive nanowires and 3D hydrogel-based systems (Balint, Cassidy and Cartmell, 2014; Thompson et al., 2014). Other electrically conductive nanomaterials (e.g. CNTs, graphene) can be added to non-electroactive materials as layers, or embedded within 3D constructs, to render them electroactive (Fairfield, 2018; Kolarcik et al., 2015). Engineering electroactive materials using both approaches has been shown to be compatible with neural cell growth in vitro. Further, electrical stimulation of neural cells grown on these materials can improve regenerative processes such as increasing stem cell differentiation into neurons and enhancing nerve fiber outgrowth (Lee et al., 2009, 2012; Bechara, Wadman and Popat, 2011; Chang et al., 2011; Kobelt et al., 2014; Xu et al., 2018). Electrical stimulation affects not only neurons but also glia. For example, astrocytes will align perpendicular to an electrical current (Alexander et al., 2006). Aligned astrocytes are thought to mimic developmental processes within the brain which direct neural progenitor migration and also provide a guide for developing neuronal tracts (Weightman, Jenkins et al., 2014; Winter et al., 2016). By replicating this process, it may be possible to create a cellular bridge to repair gaps in the damaged CNS. In addition, Schwann cells (myelinating cells of the peripheral nervous system) will increase production of neurotrophic factors, such as nerve growth factor (Hardy et al., 2015), in response to electrical stimulation. Oligodendrocytes also respond to electrical stimulation, with enhanced survival demonstrated in mixed neural cell cultures (Gary et al., 2012). Perhaps more excitingly, oligodendrocytes have been shown to increase their myelination capacity for neurons which are actively 'firing' (Mitew et al., 2018) suggesting a potential to simulate this scenario using electrical stimulation. These studies raise the possibility of combining topographical/nanoengineered control over neural cell populations with the additional benefit of

electrical stimulation for enhancing therapeutic outcomes. While promising, it must be noted that most studies have used 'neural like' cell lines, which may not be physiologically representative of native cells. In addition, experiments are often performed on one cell type in isolation rather than mimicking the complex cellular and cytoarchitecture of the nervous system. To address this, a few recent studies have demonstrated that organotypic slice cultures of neural tissue can be cultured on electroactive substrates (Fabbro et al., 2012; Usmani et al., 2016). Culturing the slices on these substrates appeared to enhance nerve fiber growth and neural network formation, providing promise that these techniques may be more broadly applied within the in vivo CNS.

17.3   In Vitro and In Vivo Clinical Applications: Challenges of CNS Engineering

In vitro engineering applications are more straightforward than in vivo, as cells can be readily isolated and manipulated. For example, engineering cells before transplantation into a patient, which will be the focus of the subsequent section. First, we discuss challenges of delivering nanomaterials (specifically nanoparticles) to the in vivo CNS.

17.3.1   The Protein Corona

On contact with biological fluids (e.g. blood, CSF, or culture medium), nanomaterials become coated in a 'corona' of biomolecules, particularly proteins (Cedervall et al., 2007; Ke et al., 2017; Walczyk et al., 2010). Propensity for aggregation, and cellular interactions then depend on the properties of this nanomaterial-corona composite, which may differ substantially from the bare nanomaterial (Carillo-Carrion, Carril and Parak, 2017; Lynch et al., 2007). Altering the size and surface chemistry through corona formation can affect the fate of these materials, especially in terms of immune responses, including clearance/destruction of the nanomaterial (Xu et al., 2018).

17.3.2   The Mononuclear Phagocyte System and 'Stealth' Coatings

Typically, materials introduced into the bloodstream will be phagocytosed by the mononuclear phagocyte system (MPS; formerly termed the reticuloendothelial system, RES), especially by Kupffer cells in the liver and macrophages in the spleen (Dobrovolskaia, Shurin and Shvedova, 2016). Clearance is increased by corona formation, particularly if opsonins are present. However, nonspecific absorption of biomolecules can be reduced by 'stealth' surface chemistry, including polyethylene glycol (PEG) coatings (Pelaz et al., 2015). This increases blood circulation time for nanoparticles and is widely used for drug/gene delivery

(Zhang et al., 2008; Wahajuddin and Arora, 2012; Lehner et al., 2013). However, a further hurdle for systemically delivered nanomaterials is access to the CNS.

17.3.3 Crossing the Blood–Brain Barrier

As described in Section 17.1.2, the BBB strictly regulates access from the bloodstream to the brain parenchyma. Some nanoparticles are small enough to cross the BBB, but functionalized particles are typically too large to cross in large numbers and/or are substantially cleared by the MPS. In addition to reducing MPS clearance, PEGylation of nanoparticles enhances their ability to cross the BBB (Calvo et al., 2001; Kim et al., 2007; Nance et al., 2013). Of interest, regarding corona formation around nanoparticles, it has been reported that corona composition can be altered as part of the process of crossing the BBB (Cox et al., 2018). The majority of nanoparticle-based strategies for the CNS rely on particles efficiently crossing the BBB and persisting extracellularly within the parenchyma (Wohlfart, Gelperina and Kreuter, 2012; Jenkins et al., 2016). For example, dalargin, an enkephalin analog that acts on membrane receptors (Schroeder, Sommerfeld and Sabel, 1998; Misra et al., 2003). However, many applications may require that particles are internalized by cells, perhaps specific cell types for gene therapy. We have documented the competitive nanoparticle uptake dynamics that exist between neural cell types, illustrating the clinical challenges posed by the dominance of microglial uptake (Jenkins et al., 2013; Jenkins et al., 2016).

17.3.4 Microglia as a Barrier to Nanoparticle-Based Engineering

Microglia rapidly sequester nanoparticles in vitro (Jenkins et al., 2013) and in vivo (Fleige et al., 2001), typically degrading them, preventing therapeutic action. PEGylation limits, but does not abolish, this clearance (Jenkins et al., 2016). Microglial degradation can be anticipated for other nanomaterials, and identifying techniques to limit or evade microglial degradation of nanomaterials would be of great utility for CNS therapeutic applications.

17.4 Genetic Engineering In Vitro, Including for Transplantation

Given the challenges associated with nanoparticle delivery to the in vivo CNS, many researchers have argued that an alternative use could be in engineering cells before transplantation. This section acts as a precursor to the following sections and will focus on improving cell transplantation through genetic engineering of cells destined for transplantation. Subsequent sections will then argue that MNPs could be utilized to perform this role, while also offering further clinical benefits in cell imaging and localization at injury sites.

17.4.1 Why Genetically Engineer Transplant Cell Populations?

Transplanted NSCs (a key clinical neural transplant population) engineered to release NT-3 and GDNF have shown enhanced neurite extension of host axons into lesion sites and greater evidence of neurogenesis from the transplanted NSCs when compared to non-engineered controls (Lu et al., 2003; Park et al., 2006; Kameda et al., 2007). In a similar manner, NSC-mediated delivery of molecules which target underlying disease pathologies have also been developed. For example, NSCs can increase synaptic connectivity and neuronal survival and consequently cognitive function in a rat model with Alzheimer's. Subsequent engineering of the NSCs to release neprilysin was shown to further address the underlying pathology of Alzheimer's disease by breaking down amyloid plaques in diseased areas (Blurton-Jones et al., 2014). These studies demonstrate the great potential in combining neuroprotective efficacy of NSCs with biomolecule delivery to address several therapeutic goals in one step (also see Section 17.1.3).

17.4.2 Challenges to safe and Efficient Genetic Engineering

Currently, the most popular method of genetic manipulation is the use of viral vectors (used almost exclusively when engineering neural cells for combinatorial therapy), which offer high transduction efficiencies. However, viruses have several drawbacks including: safety issues, associated with toxicity to the target cells and oncogenicity of the transduced population due to insertional mutagenesis (especially relevant when using stem cells which have a capacity to self-renew, with risk of tumor formation); a limit to plasmid size in the most versatile vectors; and a complex, time-consuming method of application (Lentz, Gray and Samulski, 2011; Rogers and Rush, 2012). Additionally, genes introduced by retroviral transduction have been shown to undergo downregulation over time (Palmer et al., 1991). All of these disadvantages pose a considerable barrier to clinical and commercial adoption of stem cells transduced with viruses. Although considered safer than viruses, non-viral transfection is generally associated with low transfection efficiencies, especially in the case of lipofection and electroporation (Cesnulevicius et al., 2006; Tinsley, Faijerson and Eriksson, 2006) or high cell death, in the case of nucleofection (Cesnulevicius et al., 2006).

17.4.3 Magnetofection

Magnetofection is the process of delivering genetic material (complexed to MNPs) into cells under the influence of magnetic force. This is commonly achieved by application of a static magnetic field beneath the cells, which concentrates the particles at the cell surface (Figure 17.3). The Dobson group were the first to report that an oscillating magnetic field (horizontal oscillation of the magnet beneath the

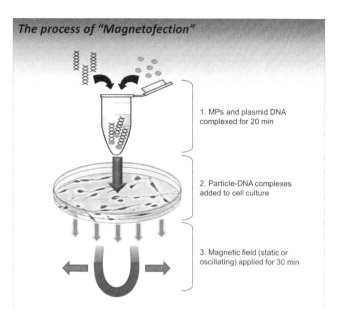

FIGURE 17.3 'Magnetofection'. Schematic detailing the procedural steps to achieve MP-mediated transfection of target cells using applied magnetic fields.

culture plate [Figure 17.3]) can further enhance MNP-mediated gene delivery over and above that of static field transfection (McBain et al., 2008). The mechanism for this enhanced transfection efficiency remains to be elucidated, but several theories have been proposed: (i) the magnetic field induces lateral motion of particles, making contact with a cell more likely (possibly overcoming particle distribution biases inherent to static magnetic fields); (ii) the oscillation of cell-associated particles stimulates the cell membrane (possibly increasing the likelihood of endocytosis in general or promoting a specific endocytotic mechanism [there are several forms of micropinocytosis] more likely to result in nucleic acid transport to the nucleus); (iii) uptake may be unaffected, but intracellular processing of MNPs could be altered, for example oscillation may enable MNP escape from intracellular vesicles (McBain et al., 2008; Pickard and Chari, 2010; Jenkins et al., 2011; Tickle et al., 2014).

Since then, numerous publications have shown that several neural cell populations can be genetically modified using magnetofection technology, often with a frequency-dependent increase in transfection efficiency observed, specific to each cell-type (Pickard and Chari, 2010; Jenkins et al., 2011; Sapet et al., 2011; Adams, Pickard and Chari, 2013; Tickle et al., 2014; Pickard et al., 2015; Adams et al, 2016a; Delaney et al., 2017). Both 2D monolayers of cells and 3D cell culture systems (as floating balls of cells, termed neurospheres, and in 3D matrix culture systems) have been successfully engineered in this manner demonstrating the utility of this approach. The technique has been shown to be safe with respect to several key regenerative properties of the cells, including proliferation, 'stemness' and generation of daughter cells for potential cell replacement strategies (Figures 17.4 and 17.5).

Crucially for clinical application, magnetic nanoparticle (MNP)-mediated *therapeutic* gene delivery to NSCs has been shown with plasmids encoding NeuroD1 (reported to promote neuronal differentiation from NSCs (Sapet et al., 2011)), FGF-2 (enhances stem cell proliferation/survival and angiogenesis [Pickard et al., 2015]) and BDNF (enhances neurite outgrowth [Fernandes and Chari, 2016b]). However, a negative correlation of transfection efficiency with plasmid size (which inevitably increases as therapeutic sequences are included in the construct) was observed (Pickard et al., 2015). A similar observation has been made when using nanoparticle and lipid-based gene delivery techniques in other cell types (Kreiss et al., 1999; Yin, Xiang and Li, 2005; Xu, Capito and Spector, 2008). Although the mechanism for the phenomenon is currently unknown, this could be a major barrier to therapeutic gene delivery as achieving the transfection efficiency required for a therapeutic effect may be compromised (successful pre-clinical trials with engineered cells report a transfection efficiency of between 20% and 80%).

17.4.4 Advances in Plasmid Engineering for Application with MNPs

To address this issue, the Chari laboratory has tested minicircle DNA as an alternative to plasmids for neural cell transfection (Fernandes and Chari, 2016a, 2016b). These are expanded initially as parental plasmids in specially modified bacteria. Upon addition of a chemical to the bacterial growth media, the parental plasmid undergoes a recombination event to produce two daughter plasmids. One contains the transgene and the other contains all the bacterial expansion elements and is subsequently degraded. This dramatically reduces plasmid size, enabling incorporation

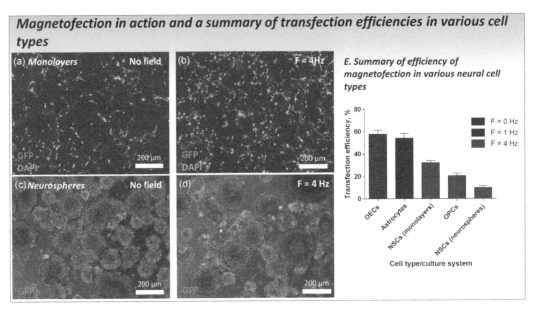

FIGURE 17.4 Neural cell magnetofection. (a) NSCs grown as 2D monolayers on the culture surface that have been magnetofected with MNPs carrying a gene encoding GFP. This causes the cells to turn green upon successful gene delivery and expression by the cells. (a) The MNPs have been added in the absence of a field. (b) The dramatic enhancement in the proportions of cells successfully transfected in the presence of an oscillating magnetic field. (c and d) A similar effect in NSCs grown as 'neurospheres' or balls of cells, showing that magnetofection can also be applied to 3D suspension cell cultures. (e) A summary of the transfection efficiencies, and the magnetic field condition used to obtain them, in a number of neural transplant population demonstrating the versatility of the technique. ((a and b) Reproduced under the Creative Commons Attribution License from Pickard et al., 2015. (c and d) Reprinted from Adams, Pickard and Chari, 2013, with permission from Elsevier. (e) Adapted from Delaney et al., 2017 with permission from The Royal Society of Chemistry.)

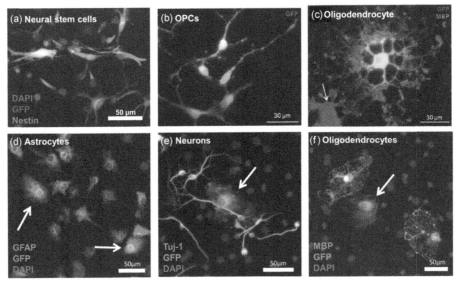

FIGURE 17.5 Magnetofection is effective with multiple neural cell types, without affecting key regenerative properties. (a) NSCs expressing GFP (indicating successful transfection) and positive for NSC marker nestin. Note the cells also display normal morphology and adherence to substrates indicating a lack of effect on NSC 'stemness'. Similar observations were also made after magnetofection of OPCs, shown in (b). Subsequently, these progenitor populations were allowed to differentiate. (c) A genetically engineered oligodendrocyte generated from a magnetofected OPC and (d–f) astrocytes, neurons and oligodendrocytes generated as part of a mixed culture from genetically engineered NSCs. In the studies, it was demonstrated that magnetofection had no effect on differentiation of OPCs or NSCs. It is worth noting though that genetically engineered neurons and oligodendrocytes were hardly ever observed after magnetofection of NSCs and subsequent differentiation – the arrows point to cells with the morphological appearance of astrocytes which appear to be the predominant cell type to 'inherit' the transfected gene. ((a and d–f) Reprinted from Adams, Pickard and Chari, 2013, with permission from Elsevier. (b and c) Reprinted with permission from Jenkins et al., 2011. Copyright 2011 American Chemical Society.)

of large or multiple therapeutic genes into one construct without compromising transfection efficiency due to size limitations. Further, removal of bacterial DNA from plasmids reduces immune responses and mammalian silencing of foreign DNA, important in clinical applications. When combining magnetofection technology with the advanced minicircle vector, 54% of NSCs were successfully engineered, which represents the highest safe, non-viral transfection efficiency reported to date in NSCs (Fernandes and Chari, 2016a). Further, expression and secretion of therapeutic protein were demonstrated after magnetofection-mediated delivery of minicircles encoding BDNF, adding considerable value to this technology as a therapeutic genetic engineering strategy (Fernandes and Chari, 2016b). Subsequently, this technique has been shown to be safe and effective for genetically engineering canine olfactory ensheathing cells (OECs) (Delaney et al., 2017), a clinically-relevant transplant population previously used as part of a veterinary clinical trial in dogs with naturally occurring spinal injuries (Granger et al., 2012). In the trial, around half of the dogs responded to treatment with cells alone. It is hoped that engineering these cells using MNP technology to express neurotrophic factors will further improve clinical outcomes, with significant promise for human application given the genetic and injury pathology similarity between dogs and humans.

17.5 Engineering Cells for Transplantation Using Nanotechnologies

17.5.1 Cell Transplantation

In the context of replacing lost tissue and restoring neurological function, cell transplantation (including genetically engineered populations) is a promising therapeutic strategy. The scientific rationale for cell transplantation involves two main objectives: (i) to replace lost or damaged cells and (ii) to release molecules to promote repair, such as neurotrophic factors. Many cell types are being investigated for the treatment of neurological injury, including stem and progenitor cells that owing to their self-renewal capacity, can be expanded to clinically-required numbers through ex vivo cell culture (Trounson and McDonald, 2015). In addition, differentiated populations of cells such as OECs (Granger et al., 2012; Tabakow et al., 2014) and astrocytes (Davies et al., 2011; Chen et al., 2015; Li et al., 2015) have been transplanted into areas of SCI and shown to promote functional recovery. These can be generated from autologous biopsies and propagated in culture or potentially derived from induced pluripotent stem cells (Li et al., 2015).

17.5.2 Safe and Efficient Delivery of Cells to Injury Sites

Accumulation of cells at the desired site is a key requirement for successful cell therapy. Direct transplantation of cells generally results in a larger accumulation of cells than systemic (non-local) injection. However, this approach is associated with secondary pathology, breakdown of the BBB and embolism (Li et al., 2010). Systemic injection can alleviate these risks; however, cells can be cleared by the MPS, often accumulating in the lung, spleen or liver so that few cells reach the desired site (Kraitchman et al., 2005; Hauger et al., 2006; Zheng et al., 2016). Ramifications of this include the need to inject more cells (which will be costly to produce) to compensate for cell loss together with the possible need for multiple injections, thereby, increasing risk to the patient.

17.5.3 Cell Tracking Post-Transplantation

Non-invasive imaging: Cell transplant populations need to be monitored to correlate cell behavior with functional outcome and assess integration upon transplantation. Real-time, non-invasive imaging of transplant populations in vivo is essential to monitor correct engraftment, both position and viability of the graft, cell migration and ideally cell differentiation. This is commonly achieved by using several techniques such as: magnetic resonance imaging (MRI) where the cells are labeled with a contrast agent; bioluminescence imaging (BLI) where cells express photon-producing enzymes; single photon emission computed tomography (SPECT) and positron emission tomography (PET), which both detect radioactive labels; or near-infrared fluorescence (NIRF) microscopy which detects fluorophore-labeled cells (Hong et al., 2010; Zheng et al., 2017). Each technique has its own advantages and disadvantages, for example, MRI has high spatial resolution and tissue penetration, whereas BLI and PET have a high sensitivity (the ability to detect lower numbers of cells) and BLI has the possibility of providing information on differentiation by detecting luciferase expression that is under the control of cell-specific promoters (Hong et al., 2010; Zheng et al., 2017).

Histological examination: Ultimately, the tissue has to be analyzed post-mortem for thorough histopathological examination of integration, migration and differentiation of the transplanted cells and, also, the effect of the transplant on the host tissue. This means that the transplanted population needs to be distinguished from the host cells. This can be accomplished by using cells from another species which can be detected with antibodies against that species, using cells from mutant animals which constitutively express green fluorescent protein (GFP), labeling cells with lipophilic fluorescent dyes (Lu et al., 2011) or genetically modifying cells to express a detectable marker (Park et al., 2006). The first two techniques, although useful in a laboratory setting, are not readily translatable. Labeling cells with dyes can result in nonspecificity due to 'leaky' dyes being taken up by non-target cells post-transplantation (Lepore et al., 2006) and virally transducing a cell population to express a detectable gene is undesirable from a translational perspective as discussed above.

To address the challenges outlined above, there is a large body of basic research concerning the use of iron nanoparticles to label transplant cells that can be imaged non-invasively by MRI in real time (Fang and Zhang, 2009; Cromer Berman et al., 2011; Jasmin et al., 2017). Feridex, a dextran-coated iron-oxide nanoparticle, was approved by the FDA to act as a contrast agent in the clinic, highlighting these particles' safety (Cromer Berman et al., 2011). Feridex was originally designed as a liver contrast agent as it accumulates there but has since been used to label various stem cell populations (Bulte, 2009; Bulte and Daldrup-Link, 2018). Using a detailed analysis, Cohen et al. have shown that neurospheres isolated from GFP$^+$ mice and labeled with Feridex can be tracked in vivo using MRI (Cohen et al., 2010). In this study, labeled NSCs were transplanted into the cerebral ventricles of mice with an induced form of multiple sclerosis (experimental autoimmune encephalomyelitis). Using in vivo MRI, a hypointense region, where the cells had been injected, was observed on day 1 and was followed by progressive migration of the cells into the corpus callosum over 7 days, thought to result from NSC pathotropism towards lesion sites. After sacrificing the animal, ex vivo MRI confirmed the widespread hypointense signal, which correlated well with both GFP fluorescence and Perls' staining, indicating that the signal was derived from the labeled transplant population. It is worth noting that although Feridex had approval it has since been discontinued (along with a number of other approved MNPs) and there remains a lack of clinically approved MNPs for potential cell tracking purposes (Jasmin et al., 2017).

17.5.4 Transplant Cell Targeting/Localization Strategies

Interestingly, once labeled, localizing magnetically labeled cells using magnetic force may overcome the cell delivery issues mentioned above. This approach could be especially beneficial for indirect methods of cell transplantation such as injection into the bloodstream or the CSF, reducing the risks of secondary pathology due to direct injection into CNS tissue. Applying external magnets after transplantation of MNP-labeled cells by both intravenous and lumbar puncture delivery has been shown to enhance cell retention post-transplantation in the brain and spinal cord. In one study, GFP$^+$ bone marrow stromal cells (BMSCs) were labeled with Feridex complexed to poly-L-lysine (PLL) and transplanted into a rat contusion model of SCI via lumbar puncture into the CSF. When transplanted in the presence of a magnet over the lesion, the area occupied by the MNP-labeled BMSCs was measured as being 20 times that of non-labeled cells. Localization of MNP-labeled cells appeared to reduce cavity formation at the site of injury and improved hind-limb function (Sasaki et al., 2011). Cells transplanted under magnetic field application have also been monitored via MRI, demonstrating the potential multifunctionality of MNP labeling for clinical applications.

17.5.5 Challenges Facing Nanoparticle-Labeling of Cells

Although promising, magnetolabeling of cells raises several considerations. Rapid, efficient and safe labeling of the majority of (ideally all) cells is required, and progress here will be aided by better understanding of what particle properties are most suited to specific cell types. However, even if initial labeling is effective (e.g. sufficient iron per cell to mediate MRI or manipulation), the regenerative capacity of most transplant populations depends on their proliferative capacity. This raises the importance of label dilution, as dividing cells must split their particle load (Kim et al., 2012). Particle loss can also occur by exocytosis and intracellular degradation (Jin et al., 2009). Therefore, consideration must be given to ensuring that a sufficient percentage of a transplant population will retain a sufficient quantity of particles to mediate the intended effects. Research must address the physicochemical and biological parameters that determine this for neural cells which differ greatly in terms of proliferative and endocytotic capacity, have cell-specific processes for uptake and intracellular processing and vary in susceptibility to toxic effects (Jenkins et al., 2013; Tickle et al., 2016).

The few studies addressing this report particle inheritance as largely asymmetric (uneven between daughter cells), although the reasons for this are unknown (Errington et al., 2010; Kim et al., 2012; Summers et al., 2013; Rees et al., 2014; Tickle et al., 2016). Although asymmetric inheritance modeling predicts that some cells will drop below the efficacy-threshold (e.g. for MRI) earlier than for symmetric inheritance, it also predicts that a subset of the population would retain efficacious levels of particles for longer than symmetric modeling would predict. This representative subpopulation may be useful for tracking purposes.

17.6 Multifunctional Nanomaterials for Advanced Therapy

Genetic engineering and labeling of cells for non-invasive tracking of transplanted cells may be vital for successful cell therapies. To achieve these goals currently, multiple steps and techniques have to be employed, greatly increasing cell loss and infection risks associated with cell manipulation. Using MNPs could facilitate these multiple biomedical outcomes using simple one-step procedures, reducing both the risk and cost associated with current approaches. The clinical value of MNPs is based in their ability to mediate multiple biomedical functions, for example, drug/gene delivery combined with cellular detection post-transplantation (Lin et al., 2018; Lo Giudice et al., 2016; Posadas, Monteagudo and Ceña, 2016; Tomitaka et al., 2019; Vissers, Ming and Song, 2019).

Bi-functional particles with the ability to mediate gene delivery and subsequent cell tracking by MRI have been tested in mesenchymal stem cells (Yang et al., 2011; Park et al., 2014) and mouse embryonic fibroblasts (MEFs; Kami et al., 2014). However, despite their significant clinical and commercial promise, there is a critical lack of neurocompatible nanoparticles tailored for such a purpose. We have recently described one such dual mode MNP for application in NSCs based on a maghemite core and coated with polyethyleneimine (PEI) and a novel oxidized PEI/alginic acid coating (Adams et al., 2016b). About 75% of NSCs could be labeled with the MNPs, and contrast from labeled NSCs was generated in MRI. However, transfection efficiency was relatively low (ca 6%), a finding which is also observed for multifunctional MNP application in MEFs (reported as ca 2%; Kami et al., 2014). To generate sufficient contrast in MRI, multifunctional MNPs typically have higher iron content than commercially available MNP-based transfection agents (e.g. 20% iron is present in our described formulations versus 0.5% iron in a widely used commercial MNP (Jenkins et al., 2014)). This could point towards an intracellular barrier to transfection when using multifunctional MNPs with elevated levels of iron oxide and transfection polymer causing toxicity before adequate transfection efficiency can be obtained. Therefore, there remains a key hurdle in developing safe multifunctional particles for efficient transfection and contrast generating capability.

Nanoparticle engineering of neural cells and subsequent incorporation into biomaterial scaffolds is also an ongoing area of research designed to take benefit from both approaches. For example, it has been shown that incorporation of transplant cells into protective matrices before implantation can improve cellular survival and distribution through injury sites (Macaya and Spector, 2012; Pakulska, Ballios and Shoichet, 2012; Führmann, Anandakumaran and Shoichet, 2017). It might be predicted that MNPs can be used as an additional nanoengineering step before neural cell incorporation to substantially enhance the translational relevance of the construct. This could be through non-invasive bioimaging of intraconstruct cells, especially important as the construct may be intended to degrade, releasing cells into the injury site, or to serve as a biofactory, in which genetically engineered intraconstruct cells produce and release therapeutic molecules. While hurdles remain with developing optimal protocols for the incorporation of nanoengineered cells into biomaterial matrices, preliminary steps have been taken demonstrating magnetofection in intraconstruct NSCs (Adams et al., 2016a) and other cell types (Zhang et al., 2010). In addition, magnetolabeled astrocytes (Tickle et al., 2018) and OECs (Adams et al., 2018) have been detected by MRI while within collagen scaffolds, demonstrating the feasibility of tracking cells implanted within supportive matrices using a nanoparticle-based approach.

Exciting opportunities still remain, particularly by supplementing therapeutic strategies with appropriate nanotechnologies. For example, nanoengineering of hydrogels, to alter mechanical or electrical properties, without substantially impairing porosity/perfusion (Mehrali et al., 2017). Such modifications may aid applications involving construct implantation into neural tissue. For example, nanofibers have been incorporated into hydrogels and shown to promote and maintain neural cell alignment (Weightman, Jenkins et al., 2014). Adding an electroactive component to the nanofibers and combining with subsequent electrical stimulation may encourage nerve fiber elongation through these biological bridges. Accordingly, multifactorial therapies based on neural nanoengineering offer sophisticated tools for addressing the complex goals of CNS repair and regeneration.

References

Abati, E., Bresolin, N., Comi, G. P. and Corti, S. (2019) 'Preconditioning and cellular engineering to increase the survival of transplanted neural stem cells for motor neuron disease therapy', *Molecular Neurobiology*, 56(5), pp. 3356–3367. doi:10.1007/s12035-018-1305-4.

Adams, C. F., Delaney, A. M., Carwardine, D. R., Tickle, J., Granger, N. and Chari, D. M. (2018) 'Nanoparticle-based imaging of clinical transplant populations encapsulated in protective polymer matrices', *Macromolecular Bioscience*, 19(2), pp. e1800389. doi:10.1002/mabi.201800389.

Adams, C. F., Dickson, A. W., Kuiper, J. and Chari, D. M. (2016a) 'Nanoengineering neural stem cells on biomimetic substrates using magnetofection technology', *Nanoscale*. doi:10.1039/c6nr05244d.

Adams, C. F, Israel, L. L., Ostrovsky, S., Taylor, A., Poptani, H., Lellouche, J. P. and Chari, D. (2016b) 'Development of multifunctional magnetic nanoparticles for genetic engineering and tracking of neural stem cells', *Advanced Healthcare Materials*, 5(7), pp. 841–849. doi:10.1002/adhm.201500885.

Adams, C. F., Pickard, M. R. and Chari, D. M. (2013) 'Magnetic nanoparticle mediated transfection of neural stem cell suspension cultures is enhanced by applied oscillating magnetic fields', *Nanomedicine: Nanotechnology, Biology, and Medicine*, 9(6), pp. 737–741. doi:10.1016/j.nano.2013.05.014.

Alexander, J. K., Fuss, B. and Colello, R. J. (2006) 'Electric field-induced astrocyte alignment directs neurite outgrowth', *Neuron Glia Biology*, 2(2), pp. 93–103. doi:10.1017/S1740925X0600010X.

Baiguera, S., Del Gaudio, C., Lucatelli, E., Kuevda, E., Boieri, M., Mazzanti, B., Bianco, A. and Macchiarini, P. (2014) 'Electrospun gelatin scaffolds incorporating rat decellularized brain extracellular matrix for neural tissue engineering', *Biomaterials*, 35(4), pp. 1205–1214. doi:10.1016/j.biomaterials.2013.10.060.

Balint, R., Cassidy, N. J. and Cartmell, S. H. (2014) 'Conductive polymers: Towards a smart biomaterial for tissue engineering', *Acta Biomaterialia*, 10(6), pp. 2341–2353. doi:10.1016/j.actbio.2014.02.015.

Bear, M. F., Connors, B. W. and Paradiso, M. A. (2015) *Neuroscience: Exploring the Brain.* Fourth edition. Edited by E. Lupash and P. C. Williams. Baltimore: Lippincott Williams & Wilkins.

Bechara, S., Wadman, L. and Popat, K. C. (2011) 'Electroconductive polymeric nanowire templates facilitates in vitro C17.2 neural stem cell line adhesion, proliferation and differentiation', *Acta Biomaterialia*, 7(7), pp. 2892–2901. doi:10.1016/j.actbio.2011.04.009.

Cromer Berman, S. M., Walczak, P. and Bulte, J. W. M. (2011) 'Tracking stem cells using magnetic nanoparticles', *Wiley Interdisciplinary Reviews. Nanomedicine and Nanobiotechnology*, 3(4), pp. 343–355. doi:10.1002/wnan.140.

Blurton-Jones, M., Spencer, B., Michael, S., Castello, N. A., Agazaryan, A. A, Davis, J. L., Mller, F.-J., Loring, J. F., Masliah, E. and LaFerla, F. M. (2014) 'Neural stem cells genetically-modified to express neprilysin reduce pathology in Alzheimer transgenic models', *Stem cell research & therapy*, 5(2), p. 46. doi:10.1186/scrt440.

Boverhof, D. R., Bramante, C. M., Butala, J. H., Clancy, S. F., Lafranconi, W. M., West, J. and Gordon, S. C. (2015) 'Comparative assessment of nanomaterial definitions and safety evaluation considerations', *Regulatory Toxicology and Pharmacology*, 73(1), pp. 137–150. doi:10.1016/j.yrtph.2015.06.001.

Brosius Lutz, A. and Barres, B. A. (2014) 'Contrasting the glial response to axon injury in the central and peripheral nervous systems', *Developmental Cell*, 28(1), pp. 7–17. doi:10.1016/j.devcel.2013.12.002.

Bulte, J. W. (2009) 'In vivo MRI cell tracking: Clinical studies', *AJR. American Journal of Roentgenology*, 193(2), pp. 314–325. doi:10.2214/AJR.09.3107.In.

Bulte, J. W. M. and Daldrup-Link, H. E. (2018) 'Clinical tracking of cell transfer and cell transplantation: Trials and tribulations', *Radiology*, 289(3), pp. 604–615. doi:10.1148/radiol.2018180449.

Burda, J. E. E. and Sofroniew, M. V. V (2014) 'Reactive gliosis and the multicellular response to CNS damage and disease', *Neuron*, 81(2), pp. 229–248. doi:10.1016/j.neuron.2013.12.034.

Calvo, P., Gouritin, B., Chacun, H., Desmaile, D., D'Angelo, J., Noel, J. P., Georgin, D., Fattal, E., Andreux, J. P. and Couvreur, P. (2001) 'Long-circulating pegylated polycyanoacrylate nanoparticles as new drug carrier for brain delivery', *Pharmaceutical Research*, 18(8), pp. 1157–1166. doi:10.1023/A:1010931127745.

Carillo-Carrion, C., Carril, M. and Parak, W. J. (2017) 'Techniques for the experimental investigation of the protein corona', *Current Opinion in Biotechnology*, 48, pp. 106–113. doi:10.1016/j.copbio.2017.02.009.

Cedervall, T., Lynch, I., Lindman, S., Berggård, T., Thulin, E., Nilsson, H., Dawson, K. A. and Linse, S. (2007) 'Understanding the nanoparticle-protein corona using methods to quantify exchange rates and affinities of proteins for nanoparticles', *Proceedings of the National Academy of Sciences of the United States of America*, 104(7), pp. 2050–2055. doi:10.1073/pnas.0608582104.

Cesnulevicius, K., Timmer, M., Wesemann, M., Thomas, T., Barkhausen, T. and Grothe, C. (2006) 'Nucleofection is the most efficient nonviral transfection method for neuronal stem cells derived from ventral mesencephali with no changes in cell composition or dopaminergic fate', *Stem Cells*, 24(12), pp. 2776–2791. doi:10.1634/stemcells.2006- 0176.

Chang, K. A., Kim, J. W., Kim, J. A., Lee, S., Kim, S., Suh, W. H., Kim, H. S., Kwon, S., Kim, S. J. and Suh, Y. H. (2011) 'Biphasic electrical currents stimulation promotes both proliferation and differentiation of fetal neural stem cells', *PLoS ONE*, 6(4). doi:10.1371/journal.pone.0018738.

Chen, C., Chan, A., Wen, H., Chung, S.-H., Deng, W. and Jiang, P. (2015) 'Stem and progenitor cell-derived astroglia therapies for neurological diseases', *Trends in Molecular Medicine*, 21(11), pp. 715–729. doi:10.1016/j.molmed.2015.09.003.

Cohen, M. E., Muja, N., Fainstein, N., Bulte, J. W. M. and Ben-Hur, T. (2010) 'Conserved fate and function of ferumoxides-labeled neural precursor cells in vitro and in vivo', *Journal of Neuroscience Research*, 88(5), pp. 936–944.

Cox, A., Andreozzi, P., Dal Magro, R., Fiordaliso, F., Corbelli, A., Talamini, L., Chinello, C., Raimondo, F., Magni, F., Tringali, M., Krol, S., Jacob Silva, P., Stellacci, F., Masserini, M. and Re, F. (2018) 'Evolution of nanoparticle protein corona formation across the blood-brain barrier', *ACS Nano*, 12(7), pp. 7292–7300. doi:10.1021/acsnano.8b03500.

Davies, S. J. A., Shih, C.-H., Noble, M., Mayer-Proschel, M., Davies, J. E. and Proschel, C. (2011) 'Transplantation of specific human astrocytes promotes functional recovery after spinal cord injury', *PLoS ONE*, 6(3), p. e17328. doi:10.1371/journal.pone.0017328. Edited by C. Combs.

Delaney, A. M., Adams, C. F., Fernandes, A. R., Al-Shakli, A. F., Sen, J., Carwardine, D. R., Granger, N. and Chari, D. M. (2017) 'A fusion of minicircle DNA and nanoparticle delivery technologies facilitates therapeutic genetic engineering of autologous canine olfactory mucosal cells', *Nanoscale*, 9(25), pp. 8560–8566. doi:10.1039/C7NR00811B.

Dobrovolskaia, M. A., Shurin, M. and Shvedova, A. A. (2016) 'Current understanding of interactions between nanoparticles and the immune system', *Toxicology and Applied Pharmacology*, 299, pp. 78–89. doi:10.1016/j.taap.2015.12.022.

Dunning, M. D., Lakatos, A., Loizou, L., Kettunen, M., Ffrench-Constant, C., Brindle, K. M. and Franklin, R. J. M. (2004) 'Superparamagnetic iron oxide-labeled Schwann cells and olfactory ensheathing cells can be traced in vivo by magnetic resonance imaging and retain functional properties after transplantation into the CNS', *The Journal of Neuroscience*, 24(44), pp. 9799–9810. doi:10.1523/jneurosci.3126-04.2004.

Entekhabi, E., Haghbin Nazarpak, M., Moztarzadeh, F. and Sadeghi, A. (2016) 'Design and manufacture of neural tissue engineering scaffolds using hyaluronic acid and polycaprolactone nanofibers with controlled porosity', *Materials Science and Engineering C*, 69, pp. 380–387. doi:10.1016/j.msec.2016.06.078.

Errington, R. J., Brown, M. R., Silvestre, O. F., Njoh, K. L., Chappell, S. C., Khan, I. A., Rees, P., Wilks, S. P., Smith, P. J. and Summers, H. D. (2010) 'Single cell nanoparticle tracking to model cell cycle dynamics and compartmental inheritance', *Cell Cycle, (Georgetown, Tex.)*, 9(1), pp. 121–130.

Fabbro, A., Villari, A., Laishram, J., Scaini, D., Toma, F. M., Turco, A., Prato, M. and Ballerini, L. (2012) 'Spinal cord explants use carbon nanotube interfaces to enhance neurite outgrowth and to fortify synaptic inputs', *ACS Nano*, 6(3), pp. 2041–2055. doi:10.1021/nn203519r.

Fairfield, J. A. (2018) 'Nanostructured materials for neural electrical interfaces', *Advanced Functional Materials*, 28(12), pp. 1701145. doi:10.1002/adfm.201701145.

Fang, C. and Zhang, M. (2009) 'Multifunctional magnetic nanoparticles for medical imaging applications', *Journal of Materials Chemistry*, 19, pp. 6258–6266.

Fernandes, A. R. and Chari, D. M. (2016a) 'Part I: Minicircle vector technology limits DNA size restrictions on ex vivo gene delivery using nanoparticle vectors: Overcoming a translational barrier in neural stem cell therapy', *Journal of Controlled Release*, 238, pp. 289–299. doi:10.1016/j.jconrel.2016.06.024.

Fernandes, A. R. and Chari, D. M. (2016b) 'Part II: Functional delivery of a neurotherapeutic gene to neural stem cells using minicircle DNA and nanoparticles: Translational advantages for regenerative neurology', *Journal of Controlled Release*, 238, pp. 300–310. doi:10.1016/j.jconrel.2016.06.039.

Fleige, G., Nolte, C., Synowitz, M., Seeberger, F., Kettenmann, H. and Zimmer, C. (2001) 'Magnetic labeling of activated microglia in experimental gliomas', *Neoplasia*, 3(6), pp. 489–499. doi:10.1038/sj/neo/7900176.

Frank, B., Grotta, J. C., Alexandrov, A. V, Bluhmki, E., Lyden, P., Meretoja, A., Mishra, N. K., Shuaib, A., Wahlgren, N. G., Weimar, C. and Lees, K. R. (2013) 'Thrombolysis in stroke despite contraindications or warnings?', *Stroke; A Journal of Cerebral Circulation*, 44(3), pp. 727–733. doi:10.1161/STROKEAHA. 112.674622.

Führmann, T., Anandakumaran, P. N. and Shoichet, M. S. (2017) 'Combinatorial therapies after spinal cord injury: How can biomaterials help?', *Advanced Healthcare Materials*, p. 1601130. doi:10.1002/adhm.201601130.

Gao, Z., Zhu, Q., Zhang, Y., Zhao, Y., Cai, L., Shields, B. and Cai, J. (2013) 'Reciprocal modulation between microglia and astrocyte in reactive gliosis following the CNS injury', *Molecular Neurobiology*, 48(3), pp. 690–701. doi:10.1007/s12035-013-8460-4.

Gardin, C., Piattelli, A. and Zavan, B. (2016) 'Graphene in regenerative medicine: Focus on stem cells and neuronal differentiation', *Trends in Biotechnology*, 34(6), pp. 435–437. doi:10.1016/j.tibtech.2016.01.006.

Gary, D. S., Malone, M., Capestany, P., Houdayer, T. and Mcdonald, J. W. (2012) 'Electrical stimulation promotes the survival of oligodendrocytes in mixed cortical cultures', *Journal of Neuroscience Research*, 90(1), pp. 72–83. doi:10.1002/jnr.22717.

Granger, N., Blamires, H., Franklin, R. J. M. and Jeffery, N. D. (2012) 'Autologous olfactory mucosal cell transplants in clinical spinal cord injury: A randomized double-blinded trial in a canine translational model', *Brain: A Journal of Neurology*, 135(Pt 11), pp. 3227–3237. doi:10.1093/brain/aws268.

Gransee, H. M., Zhan, W-Z., Sieck, G. C. and Mantilla, C. B. (2015) 'Localized delivery of brain-derived neurotrophic factor-expressing mesenchymal stem cells enhances functional recovery following cervical spinal cord injury', *Journal of Neurotrauma*, 32(3), pp. 185–193. doi:10.1089/neu.2014.3464.

Hardy, J., Cornelison, R., Sukhavasi, R., Saballos, R., Vu, P., Kaplan, D. and Schmidt, C. (2015) 'Electroactive tissue scaffolds with aligned pores as instructive platforms for biomimetic tissue engineering', *Bioengineering*, 2(1), pp. 15–34. doi:10.3390/bioengineering2010015.

Harvey, L. (2016) 'Physiotherapy rehabilitation for people with spinal cord injuries', *Journal of Physiotherapy*, 62(1), pp. 4–11. doi:10.1016/j.jphys.2015.11.004.

Hauger, O., Frost, E. E., van Heeswijk, R., Deminire, C., Xue, R., Delmas, Y., Combe, C., Moonen, C. T. W., Grenier, N. and Bulte, J. W. M. (2006) 'MR evaluation of the glomerular homing of magnetically labeled mesenchymal stem cells in a rat model of nephropathy', *Radiology*, 238(1), pp. 200–210. doi:10.1148/radiol.2381041668.

Hong, H., Yang, Y., Zhang, Y. and Cai, W. (2010) 'Non-invasive cell tracking in cancer and cancer therapy', *Current Topics in Medicinal Chemistry*, 10(12), pp. 1237–1248.

Jasmin, Souza, G. T., Louzada, R. A., Rosado-de-Castro, P. H., Mendez-Otero, R., Campos de Carvalho, A. C. (2017) 'Tracking stem cells with superparamagnetic iron oxide nanoparticles: perspectives and considerations' *International Journal of Nanomedicine*, 12, pp. 779–793. doi:10.2147/IJN.S126530.

Jenkins, S. I., Pickard, M. R., Furness, D. N., Yiu, H. H. and Chari, D. M. (2013) 'Differences in magnetic particle uptake by CNS neuroglial subclasses: Implications for neural tissue engineering', *Nanomedicine (Lond.)*, 8(6), pp. 951–968. doi:10.2217/nnm.12.145.

Jenkins, S. I., Pickard, M. R., Granger, N. and Chari, D. M. (2011) 'Magnetic nanoparticle-mediated gene transfer to oligodendrocyte precursor cell transplant populations is enhanced by magnetofection strategies', *ACS Nano*, 5(8), pp. 6527–6538. doi:10.1021/nn2018717.

Jenkins, S. I., Weinberg, D., Al-Shakli, A. F., Fernandes, A. R., Yiu, H. H. P., Telling, N. D., Roach, P. and Chari, D. M. (2016) '"Stealth" nanoparticles evade neural immune cells but also evade major brain cell populations: Implications for PEG-based neurotherapeutics', *Journal of Controlled Release*, 224, pp. 136–145. doi:10.1016/j.jconrel.2016.01.013.

Jenkins, S. I., Yiu, H. H. P., Rosseinsky, M. J. and Chari, D. M. (2014) 'Magnetic nanoparticles for oligodendrocyte precursor cell transplantation therapies: Progress and challenges', *Molecular and Cellular Therapies*, 2(1), p. 23. doi:10.1186/2052-8426-2-23.

Jin, H., Heller, D. A., Sharma, R. and Strano, M. S. (2009) 'Size-dependent cellular uptake and expulsion of single-walled carbon nanotubes: Single particle tracking and a generic uptake model for nanoparticles', *ACS Nano*, 3(1), pp. 149–158. doi:10.1021/nn800532m.

Kameda, M., Shingo, T., Takahashi, K., Muraoka, K., Kurozumi, K., Yasuhara, T., Maruo, T., Tsuboi, T., Uozumi, T., Matsui, T., Miyoshi, Y., Hamada, H. and Date, I. (2007) 'Adult neural stem and progenitor cells modified to secrete GDNF can protect, migrate and integrate after intracerebral transplantation in rats with transient forebrain ischemia', *European Journal of Neuroscience*, 26(6), pp. 1462–1478.

Kami, D., Kitani, T., Kishida, T., Mazda, O., Toyoda, M., Tomitaka, A., Ota, S., Ishii, R., Takemura, Y., Watanabe, M., Umezawa, A. and Gojo, S. (2014) 'Pleiotropic functions of magnetic nanoparticles for ex vivo gene transfer', *Nanomedicine: Nanotechnology, Biology, and Medicine*, 10(6), pp. 1165–1174. doi:10.1016/j.nano.2014.03.018.

Kamudzandu, M., Roach, P., Fricker, R. A. and Yang, Y. (2015) 'Nanofibrous scaffolds supporting optimal central nervous system regeneration: An evidence-based review', *Journal of Neurorestoratology*, 3, pp. 123–131. doi:10.2147/JN.S70337.

Kamudzandu, M., Yang, Y., Roach, P. and Fricker, R. A. (2015) 'Efficient alignment of primary CNS neurites using structurally engineered surfaces and biochemical cues', *RSC Advances*, 5(28), pp. 22053–22059. doi:10.1039/C4RA15739G.

Kang, S. H., Fukaya, M., Yang, J. K., Rothstein, J. D. and Bergles, D. E. (2010) 'NG2+ CNS glial progenitors remain committed to the oligodendrocyte lineage in postnatal life and following neurodegeneration', *Neuron*, 68(4), pp. 668–681. doi:10.1016/j.neuron.2010.09.009.

Ke, P. C., Lin, S., Parak, W. J., Davis, T. P. and Caruso, F. (2017) 'A decade of the protein corona', *ACS Nano*, 11, pp. 11773–11776. doi:10.1021/acsnano.7b08008.

Kim, H. R., Andrieux, K., Gil, S., Taverna, M., Chacun, H., Desmaële, D., Taran, F., Georgin, D. and Couvreur, P. (2007) 'Translocation of poly(ethylene glycol-co-hexadecyl)cyanoacrylate nanoparticles into rat brain endothelial cells: Role of apolipoproteins in receptor-mediated endocytosis', *Biomacromolecules*, 8(3), pp. 793–799. doi:10.1021/bm060711a.

Kim, J. A., Åberg, C., Salvati, A. and Dawson, K. A. (2012) 'Role of cell cycle on the cellular uptake and dilution of nanoparticles in a cell population', *Nature Nanotechnology*, 7(1), pp. 62–68. doi:10.1038/nnano.2011.191.

Kim, J., Sajid, M. S. and Trakhtenberg, E. F. (2018) 'The extent of extra-axonal tissue damage determines the levels of CSPG upregulation and the success of experimental axon regeneration in the CNS', *Scientific Reports*, 8(1), p. 9839. doi:10.1038/s41598-018-28209-z.

Kobelt, L. J., Wilkinson, A. E., McCormick, A. M., Willits, R. K. and Leipzig, N. D. (2014) 'Short duration electrical stimulation to enhance neurite outgrowth and maturation of adult neural stem progenitor cells', *Annals of Biomedical Engineering*, 42(10), pp. 2164–2176. doi:10.1007/s10439-014-1058-9.

Kolarcik, C. L., Catt, K., Rost, E., Albrecht, I. N., Bourbeau, D., Du, Z., Kozai, T. D. Y., Luo, X., Weber, D. J. and Cui, X. T. (2015) 'Evaluation of poly(3,4-ethylenedioxythiophene)/carbon nanotube neural electrode coatings for stimulation in the dorsal root ganglion', *Journal of Neural Engineering*, 12(1). doi:10.1088/1741-2560/12/1/016008.

Kraitchman, D. L., Tatsumi, M., Gilson, W. D., Ishimori, T., Kedziorek, D., Walczak, P., Segars, W. P., Chen, H. H., Fritzges, D., Izbudak, I., Young, R. G., Marcelino, M., Pittenger, M. F., Solaiyappan, M., Boston, R. C., Tsui, B. M. W., Wahl, R. L. and Bulte, J. W. M. (2005) 'Dynamic imaging of allogeneic mesenchymal stem cells trafficking to myocardial infarction', *Circulation*, 112(10), pp. 1451–1461. doi:10.1161/CIRCULATIONAHA.105.537480.

Kreiss, P., Cameron, B., Rangara, R., Mailhe, P., Aguerre-Charriol, O., Airiau, M., Scherman, D., Crouzet, J. and Pitard, B. (1999) 'Plasmid DNA size does not affect the physicochemical properties of lipoplexes but modulates gene transfer efficiency', *Nucleic Acids Research*, 27(19), pp. 3792–3798.

Kyritsis, N., Kizil, C. and Brand, M. (2014) 'Neuroinflammation and central nervous system regeneration in vertebrates', *Trends in Cell Biology*, 24(2), pp. 128–135. doi:10.1016/j.tcb.2013.08.004.

Lee, J. Y., Bashur, C. A., Goldstein, A. S. and Schmidt, C. E. (2009) 'Polypyrrole-coated electrospun PLGA nanofibers for neural tissue applications', *Biomaterials*, 30(26), pp. 4325–4335. doi:10.1016/j.biomaterials.2009.04.042.

Lee, J. Y., Bashur, C. A., Milroy, C. A., Forciniti, L., Goldstein, A. S. and Schmidt, C. E. (2012) 'Nerve growth factor-immobilized electrically conducting fibrous scaffolds for potential use in neural engineering applications', *IEEE Transactions on NanoBioscience*, 11(1), pp. 15–21. doi:10.1109/TNB.2011.2159621.

Lehner, R., Wang, X., Marsch, S. and Hunziker, P. (2013) 'Intelligent nanomaterials for medicine: Carrier platforms and targeting strategies in the context of clinical application', *Nanomedicine: Nanotechnology, Biology, and Medicine*, 9(6), pp. 742–757. doi:10.1016/j.nano.2013.01.012.

Lentz, T. B., Gray, S. J. and Samulski, R. J. (2011) 'Viral vectors for gene delivery to the central nervous system', *Neurobiology of Disease*, 48(2), pp. 179–188. doi:10.1016/j.nbd.2011.09.014.

Lepore, A. C., Walczak, P., Rao, M. S., Fischer, I. and Bulte, J. W. M. (2006) 'MR imaging of lineage-restricted neural precursors following transplantation into the adult spinal cord', *Experimental Neurology*, 201(1), pp. 49–59. doi:10.1016/j.expneurol.2006.03.032.

Li, K., Javed, E., Scura, D., Hala, T. J., Seetharam, S., Falnikar, A., Richard, J.-P., Chorath, A., Maragakis, N. J., Wright, M. C. and Lepore, A. C. (2015) 'Human iPS cell-derived astrocyte transplants preserve respiratory function after spinal cord injury', *Experimental Neurology*, 271(1), pp. 479–492. doi:10.1016/j.expneurol.2015.07.020.

Li, L., Jiang, Q., Ding, G., Zhang, L., Zhang, Z. G., Li, Q., Panda, S., Lu, M., Ewing, J. R. and Chopp, M. (2010) 'Effects of administration route on migration and distribution of neural progenitor cells transplanted into rats with focal cerebral ischemia, an MRI study', *Journal of Cerebral Blood Flow and Metabolism*, 30(3), pp. 653–662. doi:10.1038/jcbfm.2009.238.

Liddelow, S. and Barres, B. (2015) 'Snapshot: Astrocytes in health and disease', *Cell* 162(5), pp. 1170. doi:10.1016/j.cell.2015.08.029.

Liddelow, S. and Barres, B. (2016) 'Not everything is scary about a glial scar', *Nature* 532(7598), pp. 182–183. doi:10.1038/nature17318.

Lin, G., Li, L., Panwar, N., Wang, J., Tjin, S. C., Wang, X. and Yong, K.-T. (2018) 'Non-viral gene therapy using multifunctional nanoparticles: Status, challenges, and opportunities', *Coordination Chemistry Reviews*, 374, pp. 133–152.

Lo Giudice, M. C., Meder, F., Polo, E., Thomas, S. S., Alnahdi, K., Lara, S. and Dawson, K. A. (2016) 'Constructing bifunctional nanoparticles for dual targeting: Improved grafting and surface recognition assessment of multiple ligand nanoparticles', *Nanoscale*, 8(38), pp. 16969–16975. doi:10.1039/c6nr05478a.

Lu, H.-X., Hao, Z., Jiao, Q., Xie, W., Zhang, J., Lu, Y.-F., Cai, M., Parker, T., Liu, Y., Wang, Y. and Yang, Z. (2011) 'Neurotrophin-3 gene transduction of mouse neural stem cells promotes proliferation and neuronal differentiation in organotypic hippocampal slice cultures', *Medical Science Monitor*, 17(11), pp. 305–311.

Lu, P., Jones, L., Snyder, E. and Tuszynski, M. (2003) 'Neural stem cells constitutively secrete neurotrophic factors and promote extensive host axonal growth after spinal cord injury', *Experimental Neurology*, 181(2), pp. 115–129. doi:10.1016/S0014-4886(03)00037-2.

Lynch, I., Cedervall, T., Lundqvist, M., Cabaleiro-Lago, C., Linse, S. and Dawson, K. A. (2007) 'The nanoparticle-protein complex as a biological entity; a complex fluids and surface science challenge for the 21st century', *Advances in Colloid and Interface Science*, 134–135, pp. 167–174. doi:10.1016/j.cis.2007.04.021.

Macaya, D. and Spector, M. (2012) 'Injectable hydrogel materials for spinal cord regeneration: A review', *Biomedical Materials*, 7(1), p. 012001. doi:10.1088/1748-6041/7/1/012001.

Mallapragada, S. K., Brenza, T. M., McMillan, J. M., Narasimhan, B., Sakaguchi, D. S., Sharma, A. D., Zbarska, S. and Gendelman, H. E. (2015) 'Enabling nanomaterial, nanofabrication and cellular technologies for nanoneuromedicines', *Nanomedicine: Nanotechnology, Biology, and Medicine*, 11(3), pp. 715–729. doi:10.1016/j.nano.2014.12.013.

Mayrhofer, P., Schleef, M. and Jechlinger, W. (2009) 'Use of minicircle plasmids for gene therapy', *Methods in Molecular Biology (Clifton, N.J.)*, 542, pp. 87–104. doi:10.1007/978-1-59745-561-9˙4.

McBain, S. C., Griesenbach, U., Xenariou, S., Keramane, A., Batich, C. D., Alton, E. W. F. W. and Dobson, J. (2008) 'Magnetic nanoparticles as gene delivery agents: Enhanced transfection in the presence of oscillating magnet arrays', *Nanotechnology*, 19(40), p. 405102. doi:10.1088/0957-4484/19/40/405102.

Meco, E. and Lampe, K. J. (2018) 'Microscale architecture in biomaterial scaffolds for spatial control of neural cell behavior', *Frontiers in Materials*, 5, p. 2. doi:10.3389/fmats.2018.00002.

Mehrali, M., Thakur, A., Pennisi, C. P., Talebian, S., Arpanaei, A., Nikkhah, M. and Dolatshahi-Pirouz, A. (2017) 'Nanoreinforced hydrogels for tissue engineering: Biomaterials that are compatible with load-bearing and electroactive tissues', *Advanced Materials*, 29(8). doi:10.1002/adma.201603612.

Misra, A., Ganesh, S., Shahiwala, A. and Shah, S. P. (2003) 'Drug delivery to the central nervous system: A review', *Journal of Pharmacy & Pharmaceutical Sciences*, 6(2), pp. 252–273. Available at: www.ncbi.nlm.nih.gov/pubmed/12935438.

Mitew, S., Gobius, I., Fenlon, L. R., McDougall, S. J., Hawkes, D., Xing, Y. L., Bujalka, H., Gundlach, A. L., Richards, L. J., Kilpatrick, T. J., Merson, T. D. and Emery, B. (2018) 'Pharmacogenetic stimulation of neuronal activity increases myelination in an axon-specific manner', *Nature Communications*, 9(1), pp. 1–16. doi:10.1038/s41467-017-02719-2.

Moradian, H., Keshvari, H., Fasehee, H., Dinarvand, R. and Faghihi, S. (2017) 'Combining NT3-overexpressing MSCs and PLGA microcarriers for brain tissue engineering: A potential tool for treatment of Parkinson's disease', *Materials Science and Engineering: C*, 76, pp. 934–943. doi:10.1016/j.msec.2017.02.178.

Nance, E. A., Woodworth, G. F., Sailor, K. A., Shih, T.-Y., Xu, Q., Swaminathan, G., Xiang, D., Eberhart, C. and Hanes, J. (2013) 'A dense poly(ethylene glycol) coating improves penetration of large polymeric nanoparticles within brain tissue', *Science Translational Medicine*, 4(149), p. 149ra119. doi:10.1126/scitranslmed.3003594.A.

Pakulska, M. M., Ballios, B. G. and Shoichet, M. S. (2012) 'Injectable hydrogels for central nervous system therapy', *Biomedical Materials*, 7(2), p. 024101. doi:10.1088/1748-6041/7/2/024101.

Palmer, T. D., Rosman, G. J., Osborne, W. R. and Miller, A. D. (1991) 'Genetically modified skin fibroblasts persist long after transplantation but gradually inactivate introduced genes', *Proceedings of the National Academy of Sciences of the United States of America*, 88(4), pp. 1330–1334.

Park, K. I., Himes, B. T., Stieg, P. E., Tessler, A., Fischer, I. and Snyder, E. Y. (2006) 'Neural stem cells may be uniquely suited for combined gene therapy and cell replacement: Evidence from engraftment of Neurotrophin-3- expressing stem cells in hypoxic-ischemic brain injury', *Experimental Neurology*, 199(1), pp. 179–190. doi:10.1016/j.expneurol.2006.03.016.

Park, S. Y., Park, J., Sim, S. H., Sung, M. G., Kim, K. S., Hong, B. H. and Hong, S. (2011) 'Enhanced differentiation of human neural stem cells into neurons on graphene', *Advanced Materials*, 23(36), pp. 263–267. doi:10.1002/adma.201101503.

Park, W., Yang, H. N., Ling, D., Yim, H., Kim, K. S., Hyeon, T., Na, K. and Park, K.-H. (2014) 'Multi-modal transfection agent based on monodisperse magnetic nanoparticles for stem cell gene delivery and tracking', *Biomaterials*, 35(25), pp. 7239–7247. doi:10.1016/j.biomaterials.2014.05.010.

Pekny, M. and Pekna, M. (2016) 'Reactive gliosis in the pathogenesis of CNS disease', *Biochimica et Biophysica Acta*, 1862(3), pp. 483–491. doi:10.1016/j.bbadis.2015.11.014.

Pickard, M. R., Adams, C. F., Barraud, P. and Chari, D. M. (2015) 'Using magnetic nanoparticles for gene transfer to neural stem cells: Stem cell propagation method influences outcomes', *Journal of Functional Biomaterials*, 6(2), pp. 259–276. doi:10.3390/jfb6020259.

Pickard, M. R. and Chari, D. M. (2010) 'Enhancement of magnetic nanoparticle-mediated gene transfer to astrocytes by "magnetofection": Effects of static and oscillating fields', *Nanomedicine (Lond.)*, 5(2), pp. 217–232. doi:10.2217/nnm.09.109.

Plemel, J. R., Wee Yong, V. and Stirling, D. P. (2014) 'Immune modulatory therapies for spinal cord injury – past, present and future', *Experimental Neurology*, 258, pp. 91–104. doi:10.1016/j.expneurol.2014.01.025.

Posadas, I., Monteagudo, S. and Ceña, V. (2016) 'Nanoparticles for brain-specific drug and genetic material delivery, imaging and diagnosis', *Nanomedicine (Lond.)*, 11(7), pp. 833–849. doi:10.2217/nnm.16.15.

Rees, P., Wills, J. W., Brown, M. R., Tonkin, J., Holton, M. D., Hondow, N., Brown, A. P., Brydson, R., Millar, V., Carpenter, A. E. and Summers, H. D. (2014) 'Nanoparticle vesicle encoding for imaging and tracking cell populations', *Nature Methods*, pp. 1–7. doi:10.1038/nmeth.3105.

Reinkensmeyer, D., Lum, P. and Winters, J. (2002) 'Emerging technologies for improving access to movement therapy following neurologic injury', In J.M. Winters, C. Robinson, R. Simpson and G. Vanderheiden (eds.), *Emerging and Accessible Telecommunications, Information and Healthcare Technologies: Engineering Challenges in Enabling Universal Access*, pp. 136–150. Arlington, VA: RESNA Press.

Rogers, M.-L. and Rush, R. A. (2012) 'Non-viral gene therapy for neurological diseases, with an emphasis on targeted gene delivery', *Journal of Controlled Release*, 157(2), pp. 183–189. doi:10.1016/j.jconrel.2011.08.026.

Saner, M. and Stoklosa, A. (2013) 'Commercial, societal and administrative benefits from the analysis and clarification of definitions: The case of nanomaterials', *Creativity and Innovation Management*, 22(1), pp. 26–36. doi:10.1111/caim.12014.

Sanghavi, B. J., Wolfbeis, O. S., Hirsch, T. and Swami, N. S. (2015) 'Nanomaterial-based electrochemical sensing of neurological drugs and neurotransmitters', *Microchimica Acta*, 182, pp. 1–41. doi:10.1007/s00604-014-1308-4.

Sapet, C., Laurent, N., de Chevigny, A., Le Gourrierec, L., Bertosio, E., Zelphati, O. and Béclin, C. (2011) 'High transfection efficiency of neural stem cells with magnetofection', *BioTechniques*, 50(3), pp. 187–189. doi:10.2144/000113628.

Sasaki, H., Tanaka, N., Nakanishi, K., Nishida, K., Hamasaki, T., Yamada, K. and Ochi, M. (2011) 'Therapeutic effects with magnetic targeting of bone marrow stromal cells in a rat spinal cord injury model', *Spine (Phila Pa 1976)*, 36(12), pp. 933–938. doi:10.1097/BRS.0b013e3181eb9fb0.

Schroeder, U., Sommerfeld, P. and Sabel, B. A. (1998) 'Efficacy of oral dalargin-loaded nanoparticle delivery across the blood-brain barrier', *Peptides*, 19(4), pp. 777–780. doi:10.1016/S0196-9781(97)00474-9.

Summers, H. D., Brown, M. R., Holton, M. D., Tonkin, J. A., Hondow, N., Brown, A. P., Brydson, R. and Rees, P. (2013) 'Quantification of nanoparticle dose and vesicular inheritance in proliferating cells', *ACS Nano*, 7(7), pp. 6129–6137. doi:10.1021/nn4019619.

Tabakow, P., Raisman, G., Fortuna, W., Czyz, M., Huber, J., Li, D., Szewczyk, P., Okurowski, S., Miedzybrodzki, R., Czapiga, B., Salomon, B., Halon, A., Li, Y., Lipiec, J., Kulczyk, A. and Jarmundowicz, W. (2014) 'Functional regeneration of supraspinal connections in a patient with transected spinal cord following transplantation of bulbar olfactory ensheathing cells with peripheral nerve bridging', *Cell Transplantation*, pp. 1–70. doi:10.3727/096368914X685131.

Thompson, D. M., Koppes, A. N., Hardy, J. G. and Schmidt, C. E. (2014) 'Electrical stimuli in the central nervous system microenvironment', *Annual Review of Biomedical Engineering*, 16, pp. 397–430. doi:10.1146/annurev-bioeng-121813-120655.

Tickle, J. A., Jenkins, S. I., Pickard, M. R. and Chari, D. M. (2014) 'Influence of amplitude of oscillating magnetic fields on magnetic nanoparticle-mediated gene transfer to astrocytes', *Nano LIFE*, 5(1), p. 1450006. doi:10.1142/S1793984414500068.

Tickle, J. A., Jenkins, S. I., Polyak, B., Pickard, M. R. and Chari, D. M. (2016) 'Endocytotic potential governs magnetic particle loading in dividing neural cells: Studying modes of particle inheritance', *Nanomedicine (Lond.)*, 11(4), pp. 345–358. doi:10.2217/nnm.15.202.

Tickle, J. A., Poptani, H., Taylor, A. and Chari, D. M. (2018) 'Noninvasive imaging of nanoparticle-labeled transplant populations within polymer matrices for neural cell therapy', *Nanomedicine (Lond.)*, 13(11), pp. 1333–1348. doi:10.2217/nnm-2017-0347.

Tinsley, R. B., Faijerson, J. and Eriksson, P. S. (2006) 'Efficient non-viral transfection of adult neural stem/progenitor cells, without affecting viability, proliferation or differentiation', *The Journal of Gene Medicine*, 8(1), pp. 72–81. doi:10.1002/jgm.823.

Tomitaka, A., Kaushik, A., Kevadiya, B. D., Mukadam, I., Gendelman, H. E., Khalili, K., Liu, G. and Nair, M. (2019) 'Surface-engineered multimodal magnetic nanoparticles to manage CNS diseases', *Drug Discovery Today*, 24(3), pp. 873–882. doi:10.1016/j.drudis.2019.01.006.

Trounson, A. and McDonald, C. (2015) 'Stem cell therapies in clinical trials: Progress and challenges', *Cell Stem Cell*, 17(1), pp. 11–22. doi:10.1016/j.stem.2015.06.007.

Usmani, S., Aurand, E. R., Medelin, M., Fabbro, A., Scaini, D., Laishram, J., Rosselli, F. B., Ansuini, A., Zoccolan, D., Scarselli, M., De Crescenzi, M., Bosi, S., Prato, M. and Ballerini, L. (2016) '3D meshes of carbon nanotubes guide functional reconnection of segregated spinal explants', *Science Advances*, 2(7), pp. 1–10. doi:10.1126/sciadv.1600087.

Verkhratsky, A. and Butt, A. (2013) *Glial physiology and pathophysiology*. First edition. Chichester: John Wiley & Sons, Ltd.

Vissers, C., Ming, G.-L. and Song, H. (2019) 'Nanoparticle technology and stem cell therapy team up against neurodegenerative disorders', *Advanced Drug Delivery Reviews*, Epub ahead of print, p. S0169-409X(19)30023-7. doi:10.1016/j.addr.2019.02.007.

Wahajuddin and Arora, S. (2012) 'Superparamagnetic iron oxide nanoparticles: Magnetic nanoplatforms as drug carriers', *International Journal of Nanomedicine*, pp. 3445–3471. doi:10.2147/IJN.S30320.

Walczyk, D., Bombelli, F. B., Monopoli, M. P., Lynch, I. and Dawson, K. A. (2010) 'What the cell "sees" in bionanoscience', *Journal of the American Chemical Society*, 132(16), pp. 5761–5768. doi:10.1021/ja910675v.

Weightman, A. P., Jenkins, S. I., Pickard, M., Chari, D. M. and Yang, Y. (2014) 'Alignment of multiple glial cell populations in 3D nanofiber scaffolds: Toward the development of multicellular implantable scaffolds for repair of neural injury', *Nanomedicine: Nanotechnology, Biology, and Medicine*, 10(2), pp. 291–5. doi:10.1016/j.nano.2013.09.001.

Weightman, A. P., Pickard, M. R., Yang, Y. and Chari, D. M. (2014) 'An in vitro spinal cord injury model to screen neuroregenerative materials', *Biomaterials*, 35(12), pp. 3756–3765. doi:10.1016/j.biomaterials.2014.01.022.

Williams, P. D. and Kingston, P. A. (2011) 'Plasmid-mediated gene therapy for cardiovascular disease', *Cardiovascular Research*, 91(4), pp. 565–576. doi:10.1093/cvr/cvr197.

Winter, C. C., Katiyar, K. S., Hernandez, N. S., Song, Y. J., Struzyna, L. A., Harris, J. P. and Cullen, D. K. (2016) 'Transplantable living scaffolds comprised of micro-tissue engineered aligned astrocyte networks to facilitate central nervous system regeneration', *Acta Biomaterialia*, pp. 1–15. doi:10.1016/j.actbio.2016.04.021.

Wohlfart, S., Gelperina, S. and Kreuter, J. (2012) 'Transport of drugs across the blood-brain barrier by nanoparticles', *Journal of Controlled Release*, 161(2), pp. 264–273. doi:10.1016/j.jconrel.2011.08.017.

World Health Organization (2013) *Global Health Estimates Summary Tables*, Geneva: World Health Organization.

Xu, X., Capito, R. M. and Spector, M. (2008) 'Plasmid size influences chitosan nanoparticle mediated gene transfer to chondrocytes', *Journal of Biomedical Materials Research. Part A*, 84(4), pp. 1038–1048. doi:10.1002/jbm.a.31479.

Xu, Q., Jin, L., Li, C., Kuddannayai, S. and Zhang, Y. (2018) 'The effect of electrical stimulation on cortical cells in 3D nanofibrous scaffolds', *RSC Advances*, 8(20), pp. 11027–11035. doi:10.1039/C8RA01323C.

Xue, C., Zhu, H., Tan, D., Ren, H., Gu, X., Zhao, Y., Zhang, P., Sun, Z., Yang, Y., Gu, J., Gu, Y. and Gu, X. (2018) 'Electrospun silk fibroin-based neural scaffold for bridging a long sciatic nerve gap in dogs', *Journal of Tissue Engineering and Regenerative Medicine*, 12(2), pp. e1143–e1153. doi:10.1002/term.2449.

Yang, J., Lee, E.-S., Noh, M.-Y., Koh, S.-H., Lim, E.-K., Yoo, A.-R., Lee, K., Suh, J.-S., Kim, S. H., Haam, S. and Huh, Y.-M. (2011) 'Ambidextrous magnetic nanovectors for synchronous gene transfection and labeling of human MSCs', *Biomaterials*, 32(26), pp. 6174–6182. doi:10.1016/j.biomaterials.2011.04.007.

Yin, W., Xiang, P. and Li, Q. (2005) 'Investigations of the effect of DNA size in transient transfection assay using dual luciferase system', *Analytical Biochemistry*, 346(2), pp. 289–294. doi:10.1016/j.ab.2005.08.029.

Young, A. T., Cornwell, N. and Daniele, M. A. (2018) 'Neuro-nano interfaces: Utilizing nano-coatings and nanoparticles to enable next-generation electrophysiological recording, neural stimulation, and biochemical modulation', *Advanced Functional Materials*, 28(12), p. 1700239. doi:10.1002/adfm.201700239.

Zhang, H., Lee, M., Hogg, M. G., Dordick, J. S. and Sharfstein, S. T. (2010) 'Gene delivery in three-dimensional cell cultures by superparamagnetic nanoparticles', *ACS Nano*, 4(8), pp. 4733–4743.

Zhang, L., Gu, F. X., Chan, J. M., Wang, A. Z., Langer, R. S. and Farokhzad, O. C. (2008) 'Nanoparticles in medicine: Therapeutic applications and developments', *Clinical Pharmacology & Therapeutics*, 83(5), pp. 761–769. doi:10.1038/sj.clp.

Zhang, S., Song, Y., Wang, M., Xiao, G., Gao, F., Li, Z., Tao, G., Zhuang, P., Yue, W., Chan, P. and Cai, X. (2018) 'Real-time simultaneous recording of electrophysiological activities and dopamine overflow in the deep brain nuclei of a non-human primate with Parkinson's disease using nano-based microelectrode arrays', *Microsystems & Nanoengineering*, 4(1), p. 17070. doi:10.1038/micronano.2017.70.

Zheng, B., von See, M. P., Yu, E., Gunel, B., Lu, K., Vazin, T., Schaffer, D. V., Goodwill, P. W. and Conolly, S. M. (2016) 'Quantitative magnetic particle imaging monitors the transplantation, biodistribution, and clearance of stem cells in vivo', *Theranostics*, 6(3), pp. 291–301. doi:10.7150/thno.13728.

Zheng, Y., Huang, J., Zhu, T., Li, R., Wang, Z., Ma, F. and Zhu, J. (2017) 'Stem cell tracking technologies for neurological regenerative medicine purposes', *Stem Cells International*, 2017, pp. 2934149. doi:10.1155/2017/2934149.

Plasmonic Nanoparticles for Cancer Bioimaging, Diagnostics and Therapy

Bridget Crawford and
Tuan Vo-Dinh
Duke University

18.1 Introduction

Cancer has been a significant threat to human health with more than eight million cancer-related deaths each year worldwide [1]. Therefore, there is a clear clinical need for novel technologies to effectively detect and treat cancer, ultimately reducing cancer recurrences, treatment costs and mortality. A promising diagnostic and therapeutic platform is offered by nanoparticle-mediated assays, imaging and therapies. The development of plasmonic-active nanosystems has received increasing interest for use in a wide variety of applications ranging from molecular manufacturing, environmental monitoring and chemical sensing to medical diagnostics and therapy. Plasmonics refers to the study and application of enhanced electromagnetic properties of metallic nanostructures. The term has its origin in the word "plasmon," the quanta associated with longitudinal waves propagating in matter through the collective motion of large numbers of electrons. Intense absorption of near-infrared (NIR) radiation by the metallic nanostructures also causes photothermal heating of their surroundings and the generation of radicals that can be used for therapeutic applications. Noble metal plasmonic nanostructures, such as gold and silver nanoparticles, are most commonly used due to their relative inertness and visible and NIR absorption bands, with gold nanoparticles popular for its biocompatibility, while silver nanoparticles have larger molar extinction coefficients, providing more sensitive assays. Taking advantage of gold's high biocompatibility, gold nanoparticles (AuNP) can be injected intravenously and accumulate preferentially in cancer cells due to the enhanced permeability and retention (EPR) effect. Various gold nanoplatforms, including nanospheres, nanoshells, nanorods, nanocages and nanostars, have been used for effective photothermal treatment of various cancers. Nanoparticle-mediated photothermal therapy (PTT) has demonstrated the potential for precise tumor cell ablation, radiosensitization of hypoxic regions, enhancement of drug delivery, activation of thermosensitive agents, and enhancement of the immune system.

The earlier cancer can be detected, the better the chance of a cure. Enhancement of cancer diagnostics based on imaging and biomarker detection has been a focus of current scientific and medical research. Biosensors can be designed to detect emerging cancer biomarkers (i.e. antibodies or proteins, DNA, RNA) and have the potential to provide fast and accurate detection, reliable imaging of cancer cells, and monitoring of angiogenesis and cancer metastasis, and the ability to determine the effectiveness of anticancer therapy agents. Plasmonic nanoparticles offer a unique platform for cancer diagnostics as surface plasmons generated following irradiation have been associated with important practical applications in surface plasmon resonance (SPR), surface-enhanced Raman scattering (SERS) and surface-enhanced luminescence (also referred to as metal-enhanced luminescence). Discovery of the SERS effect in the 1970s [2,3] indicated that Raman scattering efficiency could be enhanced by factors of up to 10^6 when the sample is located on or near nano-textured surfaces of plasmonic-active metals such as silver, gold and transition metals. However, this initial enthusiasm for SERS declined from the mid-1970s to the early 1980s due to the irreproducibility of preparing metal nanoparticles in colloidal solution and the applicability of SERS has been questionable. In 1984, we first reported the general applicability of SERS as an analytical technique and demonstrated that the SERS phenomenon is in fact a general effect and that it can be applied to a wide variety of chemicals including homocyclic and heterocyclic polyaromatic compounds [4]. Research since has yielded a wide variety of nanostructures for efficient and reproducible plasmonic-active media, which has led to a wide variety of SERS platforms for environmental monitoring and biomedical diagnostics. This chapter provides a review of the various plasmonic-active nanoparticles developed for a wide variety of applications ranging from biomedical diagnostics to disease treatment.

18.2 Plasmonic Nanoparticle-Enabled Bioimaging

Bioimaging remains one of the most relied-upon methods in disease diagnosis. Gold nanoparticles (AuNP) have been used extensively in bioimaging applications due to their tunable optical properties, i.e. their scattering properties and the frequency of their localized surface plasmon resonance (LSPR) bands. The intense scattering properties exhibited by AuNPs are readily detected and easily quantified using standard optical microscopes. Additionally, modifying the AuNP size and shape allows for tuning of the LSPR wavelength within the "biological window" (650–1,100 nm) where there is minimal attenuation by tissue. This allows for both optical imaging using elastically scattered light, spectroscopic imaging using inelastically scattered light and photoacoustic (PA) imaging using absorbed light. The tunable scattering and absorption properties of AuNPs, combined with their excellent photostability, have led to their use as probes for cellular imaging, both *in vitro* and *in vivo*.

18.2.1 Optical Bioimaging

Optical bioimaging utilizes an optical contrast provided by a difference in the optical properties, such as light scattering, absorption, luminescence, and nonlinear frequency mixing, between the region to be imaged and the background. The first method of distinguishing between healthy and cancerous cells using AuNPs was demonstrated in 2003 by Sokolov et al. through the use of laser scanning confocal reflectance microscopy and bioconjugated AuNPs [5]. AuNPs were functionalized with anti-epidermal growth factor receptor (anti-EGFR) antibody to bind to EGFR, a transmembrane glycoprotein that is overexpressed in epithelial cancers, allowing selective identification of cancer cells. Using similarly functionalized spherical AuNPs, El-Sayed and coworkers utilized dark-field microscopy to distinguish between cancerous and noncancerous cells [6]. Dark-field microscopy utilizes a white light source and a condenser that prevent centered light rays from reaching the sample of interest, therefore only allowing scattered light to enter the objective. Image collection is accomplished by coupling the microscopy to a color charge-coupled device (CCD) camera. Within this study, SPR scattering results were confirmed by SPR absorption spectroscopy, which showed an increase at 545 nm for cancerous cells due to the binding of AuNPs to the cell surface. Antibody-functionalized gold nanorods (GNRs) and gold nanoshells have also been used in dark-field microscopy assays with the added potential for selective PTT using near-IR light irradiation [7,8].

The combination of dark-field microscopy and colorimetric analysis of AuNP scattered light has been used for simultaneous visualization and quantification of cell surface biomarkers. Recently, live cell dark-field microscopy has been used to monitor biomolecular interactions at the nanoscale in real time. By incorporating a temperature and atmosphere-controlled incubator and using an angled incident white light source in place of a dark-field condenser, El-Sayed and coworkers designed a system to image the cellular interactions of AuNPs in real time. AuNPs were functionalized with cell-penetrating peptide, arginylglycylaspartic acid (RGD) and a nuclear localizing sequence (NLS) peptide and cellular uptake, and localization was monitored by observing changes in scattering intensity and wavelength. The larger the red shift in scattering wavelength and the greater the scattering intensity, the denser the nanoparticle cluster [9]. This method was then applied to evaluating the efficacy of three anticancer drugs that induce apoptotic cell death: cisplatin, 5-fluorouracil (5FU) and camptothecin. Following the scattering increase over time yielded a profile of drug efficacy, similar to that determined using cell viability kits [10].

18.2.2 Photoacoustic (PA) Imaging

PA imaging offers a proven *in vivo* imaging technique with micron spatial resolution using light that is absorbed by AuNPs, rather than scattered. The absorbed energy is then converted to heat through electron–electron, electron–phonon, and phonon–phonon oscillations within the AuNP [11]. This relaxation process occurs very rapidly and results in intense heating at the nanoparticle surface. When the local nanoparticle environment is rapidly heated and the heat cannot dissipate quickly enough, a thermal expansion occurs. This expansion quickly collapses on itself and generates a PA wave, which is detected using an ultrasonic transducer and converted into an image [12]. Because of the strong absorption of hemoglobin and inherent weakness of PA signals, PA imaging has been relegated to imaging vascular structures; however, this technique does not offer sufficient contrast to distinguish between health and malignant vasculature. To resolve this issue, AuNPs have been used to increase the absorption for tumor imaging applications, the first of which was reported by Copeland et al., who studied the contrast enhancement of 40 nm AuNPs conjugated with Herceptin to selectively target breast cancer cells [13]. With this method, nanoparticle concentrations as low as 1 pM at depths up to 6 cm were detected. This technique, termed photoacoustic tomography (PAT), was used to successfully image malignant populations resulting in a 63% contrast enhancement using gold nanocages (AuNCs) and 81% contrast enhancement using gold nanoshells. Other AuNP geometries have also been used for PAT contrast enhancement, including nanorods [14–16], nanoshells [17], nanocages [18], nanostars [19–22], hollow nanospheres [23,24] and hybrid magnetic core shell particles (gold/iron oxide) [25,26].

18.2.3 Magnetic Resonance Imaging (MRI)

Magnetic resonance imaging (MRI) typically relies on the spin–lattice relaxation (T_1 contrast) or the spin–spin relaxation (T_2 contrast) times of protons contained in different microstructures of the organs to create imaging contrast. To increase contrast, various nanoparticles and complexes, i.e. contrast agents, are administered prior to imaging. Shortening T_1 and T_2 increases the corresponding relaxation rates ($1/T_1$ and $1/T_2$), producing hyperintense and hypointense signals, respectively. The most popular T_1 contrast agent used in MRI is Gd^{III}-containing nanoprobes, because of their large magnetic moment. Chelates of Gd with diethyltriamine pentaacetic acid (DTPA) or tetraazacyclododecane tetraacetic acid (DOTA) are the most clinically used contrast agents, albeit with related major concern of toxicity of the released Gd ions [27,28].

Gold nanoparticles coated by numerous Gd chelates are expected to enhance the positive contrast of magnetic resonance images as compared to single gadolinium chelate species. Since DTPA:Gd is a very stable complex and is currently used for clinical diagnosis, gold nanoparticles have been synthesized in the presence of dithiolated DTPA (DTDTPA) derivatives. The modification of DTPA by two thiol moieties is aimed at ensuring the quasi-irreversible grafting of the ligand onto the gold nanoparticles, consequently providing for enhanced colloidal stability. Debouttière et al. reported the synthesis of gold nanoparticles (2–2.5 nm in size) functionalized by DTDTPA:Gd chelates leading to multilayered shell containing about 150 ligands [29]. These particles exhibit a high relaxivity (r_1 = 585 mM^{-1} s^{-1} as compared to 3.0 mM^{-1} s^{-1} for DTPA:Gd), rendering them very attractive as contrast agents for MRI. Moriggi et al. developed 1–13 nm gold nanoparticles functionalized with thiol-DTTA with paramagnetic gadolinium or diamagnetic yttrium rare-earth ions [30]. Nuclear magnetic relaxation dispersion (NMRD) profiles show very high relaxivities up to 1.5 T. Holbrook et al. developed dithiolane-Gd(III)-conjugated gold nanoparticles with significant contrast enhancement allowing clear identification of the pancreas with contrast-to-noise ratios exceeding 35:1 [31].

18.2.4 X-Ray Computed Tomography (CT)

X-ray computed tomography (CT) is an imaging procedure that uses computer-processed X-ray scans to produce tomographic images of a specific area of the body. Regarded as one of the most reliable imaging techniques, CT has been widely used owing to its micron spatial and density resolution. Imaging contrast agents are generally required to increase the density of the area to be imaged to improve the diagnostic accuracy. Gold nanoparticles are excellent X-ray contrast agents superior to traditional iodinated contrast agents, since gold has a higher atomic number (79) and higher value k-edge energy (80.7 keV) [32]. In addition, the X-ray attenuation of iodine is decreased by water, while that of gold is not.

Reuveni et al. reported the use of intravenously injected antibody-conjugated gold nanoparticles (30 nm diameter) for *in vivo* tumor targeting and molecular CT imaging [33]. Peng et al. reported the synthesis of dendrimer-stabilized AuNPs through use of amine-terminated fifth-generation poly(adidoamine) (PAMAM) dendrimers modified by diatrizoic acid as stabilizers for enhanced CT imaging [34].

Zhou et al. applied PEGylated (polyethylene glycol modified) gold nanoparticles for blood pool and tumor imaging using X-ray CT. PEG is commonly used to minimize aggregation of nanoparticles and allow for increased blood circulation and half-life *in vivo*. A prolonged half-decay circulation time of 11.2 h in rats was observed [35]. A commercial formulation of these gold nanoparticles as X-ray contrast agents for *in vivo* CT imaging is available from Nanoprobes (Yaphank, NY) under the trade name Auro-Vist. Zhang et al. introduced the design and synthesis of polyethyleneimine (PEI)-stabilized gold nanoparticles (Au PSNPs) modified with PEG for blood pool, lymph node and tumor CT imaging [36]. The PEGylated Au PSNPs with

Au core size of 5.1 nm have a relatively long half-decay time (7.8 h) and display better X-ray attenuation properties than conventionally used iodine-based CT contrast agents (e.g., Omnipaque) in imaging blood pool and major organs of rats, lymph node of rabbits and the xenografted tumor model of mice. The PEGylated Au PSNPs could be excreted out of the body with time and showed *in vivo* stability.

Additionally, gold nanoparticles can absorb X-ray, which consequently generate photoelectric and Compton effects. These effects lead to formation of secondary electrons and reactive oxygen species (ROS). Misawa et al. [37] demonstrated that AuNPs contribute to enhanced generation of ROS under irradiations of x-rays in the diagnostic range and that smaller diameter AuNPs with larger surface area lead to a greater yield of ROS.

18.2.5 Radioisotope Imaging

Positron emission tomography (PET) imaging utilizes pairs of Éč rays emitted by the positron-emitting radionuclide in the imaging contrast agents to produce a three-dimensional image in the body. Coupling radiometals to gold nanomaterials via a chelator faces the challenges of possible detachment of the radiometals as well as surface property changes of the nanomaterials. Sun et al. developed a simple and chelator-free ^{64}Cu radiolabeling method by chemically reducing ^{64}Cu onto arginineglycine-aspartic acid (RGD) peptide-modified GNRs [38]. These GNRs showed high tumor-targeting ability in a U87MG glioblastoma xenograft model and were successfully used for PET image-guided PTT owing to the high photon-to-heat conversion of GNRs [38]. In 2016, Frellsen et al. took advantage of the ability of ^{64}Cu to bind nonspecifically to gold surfaces to develop a methodology to embed it within gold nanospheres as a chelator-free PET contrast agent [39]. ^{64}Cu-AuNPs were prepared by incubating AuNP seeds with ^{64}Cu^{2+}, followed by the entrapment of the radionuclide by grafting on a second layer of gold, resulting in radiolabeling efficiencies of 53% $\pm$ 6%. The radiolabel showed excellent stability and showed no loss of ^{64}Cu when incubated with 50% mouse serum for 2 days.

18.2.6 Surface-Enhanced Raman Scattering (SERS)

SERS takes advantage of a unique phenomenon on certain metal nanoparticles, SPR. Surface plasmon refers to oscillating electrons within the conduction band when the metallic nanostructure surface is excited by an external electromagnetic field. The oscillating electrons can generate a secondary electromagnetic field, which is added to the external electromagnetic field to result in SPR. Incident photon energy, when in resonance with the surface plasmon, magnifies the local electromagnetic field that dramatically enhances the intrinsically weak Raman signal. The SERS enhancement factor is typically 10^6–10^8-fold and can be up to 10^{15}-fold in hot spots, where the electromagnetic field is extremely intense [40].

SERS imaging offers high spatial resolution and the capability to perform multiplexing, facilitated by the narrow spectral characteristics of the vibrational footprint of Raman molecules [41–43]. Unlike fluorescence imaging, SERS imaging is not limited by tissue autofluorescence and photobleaching [44].

Human epidermal growth factor receptor 2, HER2, is known to be overexpressed on the surface membrane of breast cancer cells, thus making this receptor protein an important biomarker for breast cancer detection and diagnosis. SERS-based detection of HER2 on breast cancer cells has been reported where antibody-conjugated GNRs [45] and nanospheres [46] coupled with Raman reporter molecules were used as the SERS-active nanoprobes for imaging and detection of HER2 on the surface of MCF-7 cells. More recently, Lee et al. reported a SERS-based multiplexed cellular imaging method for simultaneous imaging, detection and quantification of different breast cancer phenotypic markers [47]. These SERS nanotags were developed using antibody-conjugated silica-encapsulated hollow gold nanospheres (SEHGNs) coupled with different Raman reporters for three different biomarkers, epidermal growth factor (EGF), ErbB2, and insulin-like growth factor-1 (IGF-1) receptors on the surface of MDA-MB-468, KPL4, and SK-BR-3 human breast cancer cell lines, respectively.

Among nanoparticles used for light-induced detection (SERS) or PTT, gold nanostars (GNS) are of particular interest as they offer a wide range of optical tunability by engineering subtle changes in their geometry. The multiple sharp branches on GNS create a "lightning rod" effect that enhances local electromagnetic (EM) field dramatically [48,49]. Transmission electron microscopy (TEM) images and theoretical calculations of the magnitude of the electric field indicated that the largest E-field enhancement occurs at the tips of the branches of the star (Figure 18.1). Intracellular detection of silica-coated SERS-encoded nanostars was also demonstrated in breast cancer cells; the non-aggregated field enhancement makes the GNS ensemble a promising agent for SERS bioapplications [50].

Local pH environment has been considered to be a potential biomarker for tumor diagnosis since solid tumors contain highly acidic environments. A pH-sensing nanoprobe based on SERS using nanostars under NIR excitation has been developed for potential biomedical applications [51]. The SERS peak position of pMBA at ~1,580 cm^{-1} was identified to be a novel pH-sensing index, which has small but noticeable down-shift with pH increase.

GNS SERS nanoprobes have been applied for immune-SERS microscopy of the tumor suppressor p63, imaged in prostate biopsies [52]. More recently, Webb et al. developed particles similar to GNS termed multibranched gold nanoantennas (MGNs) for simultaneously targeting immune checkpoint receptor programmed death ligand 1 (PD-L1) and EGFR overexpressed in triple negative breast cancer (TNBC) cells *in vitro* [53]. A mixture of MGNs functionalized with anti-PDL1 antibodies and Raman tag 5,5-dithio-bis-(2-nitrobenzoic acid) (DTNB),

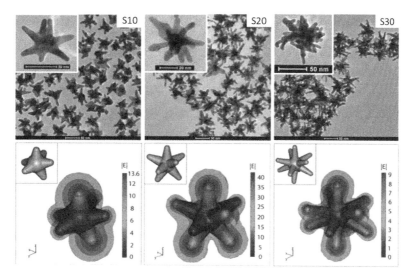

FIGURE 18.1 Structures and "lighting rod" effect of GNSs. (Top) TEM images of nanostars formed under different Ag+ concentrations (S10: 10 μM, S20: 20 μM, S30: 30 μM). (Bottom) Simulation of |E| in the vicinity of the nanostars in response to a z-polarized plane wave incident E-field of unit amplitude, propagating in the y-direction, and with a wavelength of 800 nm. The insets depict the 3D geometry of the stars. Diagrams are not to scale [49].

and MGNs functionalized with anti-EGFR antibodies and para-mercaptobenzoic acid (pMBA) were incubated with cells enabled targeted diagnosis of both biomarkers. In this study, it was also demonstrated that cells incubated with anti-EGFR-pMBA-MGNs and illuminated with an 808 nm laser for 15 min at 4.7 W/cm^2 exhibited photothermal cell death only within the laser spot suggesting the potential theranostic use of such nanoparticles.

In vivo detection of sarcoma in murine model was achieved using SERS-labeled GNS as contrast agents [54]. *In vivo* SERS spectra of 30 nm and 60 nm GNS nanoprobes with pMBA reporter exhibit characteristic SERS peaks at 1,067 and 1,588 cm^{-1} in the tumor but not in the normal muscle of the mice injected with GNS. We have also reported *in vivo* detection of SERS-encoded GNS in small (mouse) and large (pig) animal models [55].

By combining resonance enhancement, surface-enhanced resonance Raman scattering (SERRS) shows even greater Raman signal enhancement than SERS alone. Harmsen et al. developed silica-coated GNS SERRS nanoprobes that have been used for visualizing brain tumor margins and microscopic tumor invasion [56].

18.2.7 Two-Photon Luminescence (TPL) Imaging

Optically nonlinear two-photon excitation is a process involving simultaneous absorption of two photons through a virtual intermediate state, which is quadratically dependent on the intensity of light being absorbed, and is thus very localized near the focus of the light beam. Wang et al. demonstrated that the TPL signal from a single nanorod is 58 times more than that from a single rhodamine molecule allowing for the visualization of single nanorods

flowing through mouse ear blood vessels [57]. Vo-Dinh and coworkers have developed a novel seed-mediated method to synthesize GNS without using toxic surfactant, which is suitable for *in vivo* applications [48,49]. GNS with tip-enhanced plasmonics exhibit extremely high two-photon cross section (4 × 10^6 Goeppert-Mayer (GM) units), several orders of magnitude higher than organic fluorophores (10^2–10^3 GM) and quantum dots (10^4–10^5 GM), allowing real-time tracking of a single GNS nanoparticle using high-resolution two-photon photoluminescence (TPL) imaging [49,58]. The two-photon excitation action cross-sections of GNS has been reported to be 100 times higher than that of GNRs and 8,000 times higher than gold nanocubes [58], making GNS excellent candidates for real-time *in vivo* tracking as well as for sensitive tumor detection [59]. GNS have been successfully applied for *in vivo* imaging such as brain cancer [20]. This imaging platform has been extended for combined cellular tracking, imaging and PTT, demonstrating their potential for biomedical applications [54,59–63].

18.2.8 Multimodal Imaging

In recent years, a growing interest in the development of multimodal imaging, combining multiple medical imaging techniques in one platform to provide complementary information, has fueled interest in developing multifunctional nanoparticles. Arifin et al. developed microcapsules that immunoprotect pancreatic islet cells for treatment of type I diabetes and enabled multimodal cellular imaging by MRI/ultrasound of transplanted islet cells by functionalizing gold nanoparticles with DTDTPA (dithiolated diethylenetriaminepentaacetic acid):gadolinium chelates (GC) [64]. In 2012, a triple-modality gold nanosphere probe for the combination of MRI, PAT contrast, and Raman imaging was developed for both pre-operative detection and intraoperative tumor margin delineation of brain tumors *in vivo*. Using a spherical AuNPs core (for PAT), coated with a Raman active layer and surrounded by a gadolinium shell (for MRI), the probe was developed to combine the benefits of all three imaging modalities into a single, multimodal nanoparticle probe while enhancing the stability of each individual method [65].

More recently, GNS exhibit cross-platform flexibility as multimodal contrast agents for macroscopic X-ray computer tomography (CT), MRI, PET, as well as nanoprobes for PAT, TPL, and SERS. Since star-shaped geometry exhibits greater surface area than the spherical counterpart of equivalent size, GNS can be an improved sensitizer for radiation therapy. For MRI imaging, Gd3+ ions have been linked to the surface of GNSs through (DOTA) chelator. Tumor cells loaded with multifunctional GNS probes display high intensity under CT and MRI examination [60]. PET scan provides an extremely sensitive 3D imaging method. The sensitivity could reach picomolar for PET compared to micromolar for MRI. PET imaging with ^{64}Cu-labeled GNS nanoprobes allowed for dynamic imaging up to 24 h. *In vivo* tracking results showed that the developed multimodal GNS nanoprobes accumulate gradually in tumor with a 3:1 tumor-to-muscle ratio at the end of 24 h [40]. GNS are also suitable to be used under PAT with high extinction coefficient (10^{10} M^{-1}cm^{-1}) [20,22]. Through cranial window chambers implanted with tumor cells in live animals, GNS have been used for angiography with high spatial resolution cerebral capillaries that were clearly visible with minimal tissue autofluorescence background. Unlike conventional contrast agents showing signal decay within 30 min, GNS can be coated with PEG to achieve a much longer intravascular signal stability due to its extended serum half-life (several hours) and lower degree of extravasation [60]. These studies exemplify the potential of GNS for whole-body multimodal imaging.

Furthermore, these nanoparticle-based contrast agents can carry various biotargeting agents (i.e. antibodies) for targeting specific disease markers/sites. These targeted multimodal nanoparticle platforms will advance the sensitivity and selectivity of medical imaging technology, thus allowing for early detection of various diseases.

18.3 Plasmonic Nanoparticle-Based Diagnostics and Sensing

As is the case for many diseases, earlier detection is crucial for increasing cancer patient survival. With their relatively straightforward synthesis and tunable optical and physical properties, AuNPs offer a promising alternative to traditional fluorescence and radiolabeled assays. Due to their strong light scattering, AuNPs offer greater simplicity in assay design while maintaining or even lowering the limit of detection (LOD).

18.3.1 Localized Surface Plasmon Resonance (LSPR) Colorimetric Assays

Colorimetric-based assays are regarded as simple, high-throughput, and cost-effective. Due to aggregation of bioconjugated AuNPs after introduction of a target solution, a solution color change occurs which can often be seen by the naked eye. The interparticle plasmonic coupling that occurs upon AuNP aggregation causes their LSPR band to red shift to longer wavelengths, leading to the adoption of a more purple color in solution as opposed to the red color of colloidal AuNP. Generally, the recognition elements that are applied into colorimetric sensors based on plasmonic nanoparticles are antibodies or proteins for immunoreaction or aptamers with conformational change. Halas, West, and coworkers used gold nanoshells with nanoshell-conjugated antibodies to demonstrate a rapid immunoassay, capable of detecting analytes within a complex blood biological media [66]. Analyte-induced aggregation of these multiple metallic nanoparticles produces a red shift in the LSPR peak, along with a decrease in the extinction coefficient at 720 nm. Hirsch et al. demonstrated the detection of immunoglobulins. An interesting concept used for monitoring DNA conformation changes is that of a "plasmonic molecular ruler," introduced by Alivisatos' group [67]. In a plasmonic ruler, the change in separation between two metallic nanoparticles, the change in separation between two metallic nanoparticles due to DNA hybridization, produced a color change as well as a change in the plasmonic scattering intensity.

18.3.2 Localized Surface Plasmon Resonance (LSPR) Fluorescence Sensors

LSPR fluorescence sensors utilize analyte-induced enhancement of the luminescent intensity of a fluorophore when its distance from a metallic nanoparticle is tuned by analyte interaction. Direct contact of fluorophore with the metal surface will result in dipole-induced quenching of fluorescence, whereas when they are separated, the fluorescence will be fully recovered or enhanced due to elimination of quenching. It is possible that the enhanced local plasmonic field at the metal surface enhances the fluorescence and the enhancement is maximum at an optimum nanoscale distance [68]. Lakowicz et al. reported the development of metal-enhanced fluorescence to probe the hybridization of DNA [69]. Thiolated oligonucleotides were bound to silver particles on a glass substrate. Addition of a complementary fluorescein-labeled oligonucleotide resulted in a dramatic time-dependent 12-fold increase in the fluorescence intensity during the process of hybridization. This hybridization brought the fluorescein molecule close to the silver nanoparticles at an appropriate distance for surface plasmon enhancement of fluorescein's intensity. Mirkin's group has developed an aptamer nano-flare, a gold nanoparticle core functionalized with a dense mono-layer of nucleic acid aptamers, which can directly quantify an intracellular analyte in a living cell [70]. In the initial state, the fluorescence of the flare strand is quenched by the AuNP, but when the ATP molecules bind to the aptamer, it causes a new folded secondary structure causing the release of fluorescent reporter molecules. More recently, Rotello and coworkers used a fluorescence displacement protein sensor they had previously developed for applications of high-throughput multichannel sensor platform that can profile the mechanisms of various chemotherapeutic drugs in minutes [71,72]. This sensor is made of a gold nanoparticle complexed with three different fluorescent proteins that can sense drug-induced physiochemical changes on cell surfaces. The presence of cells will rapidly displace fluorescent proteins from the gold nanoparticle surface and restore the fluorescence, therefore, generating patterns that identify specific mechanisms of cell death induced by various chemotherapeutic drugs.

18.3.3 Surface-Enhanced Raman Scattering (SERS)-based Biosensors and Assays

SERS has gained increasing interest as a powerful analytic and diagnostic tool in recent years. The sensitivity of SERS is demonstrated through its ability to detect single molecules [73–76]. Compared to fluorescence, SERS offers several advantages. Not only does SERS exhibit emission peaks two orders of magnitude narrower than those of fluorescence, making SERS more suitable for multiplex detection, but SERS is also less sensitive to photobleaching than fluorescence. Resonant SERS (SERRS) has a lower LOD than that of fluorescence, i.e. SERRS LOD was three orders of magnitude lower than that of fluorescence with the same commercially available fluorophore [77,78]. Finally, SERS can be excited in the NIR region where absorption and fluorescence background of biological matrices are SERS-based methods and have shown potential for detection of biomarkers [79–81].

Plasmonic nanoparticles have been widely used in SERS-based immunoassays of biomolecules such as proteins, nucleic acids and cells. Antibody detection using SERS was demonstrated by Li et al. with the development of a SERS immune sensor for detection of vascular endothelial growth factor (VEGF) biomarker using a capture antibody immobilized on an Au triangle nanoarray SERS-active chip to which the capture antibodies (anti-VEGF) were immobilized [82]. SERS-active sandwich nanoparticles were constructed by conjugating a Raman reporter (malachite green isothiocyanate, MGITC) to a GNS followed by SiO_2 coating (Au@RamanReporter@SiO_2 sandwich NP). The detection antibodies were then conjugated onto these sandwich NPs and in the presence of the target biomarker protein, the VEGF analyte gets sandwiched between the capture antibody on the nanoarray platform and the detection

antibody on the SERS probes, thereby creating a three-dimensional hierarchical structure with a confined plasmonic field, yielding "hot spots" under light excitation.

A promising demonstration of SERS-based protein detection *in vivo* used actively targeted particles for multiplexed detection of EGFR, CD44 and TGFbRII and led to longer retention times within tumors as compared to passively targeted particles, such as in the 2008 study by Qian et al. [83,84]. Dinish et al. observed that the antibody-coated nanotags exhibited SERS signal for as long as 2 days, while antibody-free nanotags had detectable signals only for 4–6 h.

The current gold standard of nucleic acid-based molecular diagnostics using real-time reverse transcription polymerase chain reaction (qRT-PCR) has displayed high sensitivity and specificity; however, qRT-PCR is quite labor-intensive, is time consuming, and requires relatively bulky and expensive equipment. The development of rapid, easy-to-use, cost-effective, high-accuracy biomarker detection technologies and devices for applications at the point of care and in resource-limited settings remains a goal of current nanotechnology research. Our group demonstrated a target label-free SERS-based approach for nucleic acid detection with the development of an approach referred to as plasmonic coupling interference (PCI) in which the plasmonic coupling effect between neighboring silver nanoparticles (AgNPs) is interrupted upon hybridization of target nucleic acids [85]. In this approach, AgNPs are functionalized with thiolated oligonucleotides and used as the capture probes (Capture-NPs) and complementary Raman-labeled reporter probes (Reporter-NPs). To induce the plasmonic coupling effect, Capture-NPs and Reporter-NPs are mixed in a hybridization buffer solution to assemble nanoaggregates where the Raman label is located between closely spaced nanoparticles, leading to an intense SERS signal upon laser excitation. Target sequences are therefore the competitors of the reporter probes in a competitive binding process as they share the same sequence as the reporter probes. Therefore, target sequences prevent the formation of nanoaggregates, i.e. there is interference with the plasmonic coupling effect, thereby producing a weak SERS signal. This technique was demonstrated in detecting miR-21, which is involved in a variety of different cancer and other diseases.

To generate a strong SERS signal and to identify specific target sequences, many existing methods require either multiple incubation/wash steps or target labeling, thus increasing the assay complexity. Over two decades ago, our group first introduced the SERS technique for gene diagnostics [86]. Over the past several years, we have further developed a wide variety of SERS plasmonic platforms for chemical and biological sensing and have been extending the strength of the SERS effect as part of a nucleic acid sensing nanoprobes using homogenous assay formats. Particularly, we have demonstrated that highly sensitive, specific and multiplexed detection of nucleic acids can be achieved by utilizing SERS-active silver nanoparticles in the development of a unique SERS-based probe for DNA detection, referred to as the "molecular sentinel" (MS) [87–89]. The MS probe consists of a DNA strand having a Raman label molecule at one end and a metal nanoparticle at the other end. The plasmonic nanoprobe uses a hairpin-like stem-loop structure to recognize target DNA sequences. Note that hairpin DNA structures, first developed by Tyagi and Kramer in 1996, have been used in "molecular beacon" systems that are based on optical and electrochemical detection [90]. The sensing principle of MSs, however, is quite different from that of molecular beacons. With MS systems, in the normal configuration (i.e. in the absence of target DNA), the DNA sequence forms a hairpin loop, which maintains the Raman label in close proximity of the metal nanoparticle, thus inducing an intense SERS signal of the Raman label upon laser excitation. Upon hybridization of a complementary target DNA sequence to the nanoprobe hairpin loop, the Raman label molecule is physically separated from the metal nanoparticle, thus leading to a decreased SERS signal.

More recently, we have reported a novel "turn-on" plasmonic-based nanobiosensor, referred to as "inverse Molecular Sentinel (iMS)" nanoprobes, for the detection of DNA sequences of interest with an "OFF-to-ON" signal switch [91–93]. The term "inverse" is used to distinguish this new technology from our previously developed "ON-to-OFF" MS nanoprobes. Figure 18.2 shows a schematic diagram of the operating principle of the iMS detection modality. The "OFF-to-ON" signal switch is based on a non-enzymatic strand-displacement process and the conformational change of stem-loop (hairpin) oligonucleotide probes upon target binding. The demonstration of using

iMS nanoprobes to detect miRNAs in real biological samples was performed with total small RNA extracted from breast cancer cell lines [89]. MicroRNAs (miRNAs) have demonstrated great promise as a novel class of biomarkers for early detection of various cancers. The multiplex capability of the iMS technique was demonstrated using a mixture of the two differently labeled nanoprobes to detect miR-21 and miR-34a miRNA biomarkers for breast cancer. The results of this study illustrate the transformative potential of the iMS technique as a simple and rapid diagnostic tool for multiplexed detection of miRNA biomarkers. Furthermore, our group demonstrated the proof-of-principle of *in vivo* detection of nucleic acid targets using iMS nanosensor implanted in the skin of a large animal model (pig) [94]. The results of this study illustrate the usefulness of SERS iMS nanosensors as a skin-based *in vivo* biosensing platform, laying the groundwork for a pathway to future continuous health status sensing, disease biomarker monitoring and other clinical translation applications.

Over 30 years ago, our group first introduced a unique type of plasmonic-active substrate, which consists of a metal film over nanospheres (MFONs), as an efficient and reproducible SERS-active substrate [4]. Using this substrate, also termed "Nanowave" [95], our group developed a label-free multiplex DNA biosensor based on MS immobilized on a plasmonic Nanowave chip [96,97]. For this initial work, the MS sensing mechanism was utilized which is based upon the decrease of the SERS intensity upon target incubation. Later, the iMS sensing mechanism, providing off-to-on signal change upon target hybridization, was used and afforded an LOD of six attomoles [98]. The MS and iMS

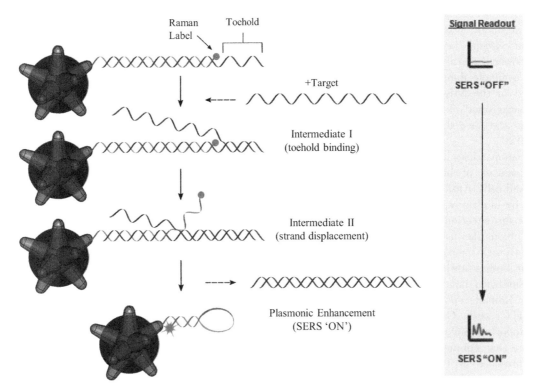

FIGURE 18.2 Detection scheme of the SERS iMS nanoprobes.

sensing mechanisms, whether in solution or immobilized on substrates, do not require secondary hybridization or post-hybridization washing, thus resulting in rapid assay time and low reagent usage.

Our group has developed a sensitive nucleic acid detection method using sandwich hybridization of magnetic beads functionalized with capture probes, target sequences, and ultrabright SERS nanorattles functionalized with reporter probes [99]. Compared to previous sandwich hybridization-based nucleic acid detection methods using SERS gold and silver nanoparticles [100,101], our method utilizes SERS nanorattles with core-gap-shell structure, thus exhibiting ultrabright SERS signals and improving nucleic acid detection sensitivity. Most recently, we utilized the ultrabright SERS nanorattle-based sandwich assay for the development of a low-cost, resource efficient diagnostic as an alternative to standard histopathology to distinguish squamous cell carcinoma from other cell lines [102]. The assay was designed to detect the cytokeratin-14 (CK14) gene, an effective nucleic acid marker for head and neck SCC micrometastases in lymph nodes. In a blinded trial, we demonstrated that the presented method can accurately distinguish the presence of SCC cell lines from alternative cell lines with high sensitivity and specificity, supporting its potential use as a diagnostic tool in head and neck fine-needle aspirations.

Studies have revealed that cancers are associated with mutations or changes in sequences of specific genes that lead to the development and progression of the disease [103]. These mutations can either be a single-nucleotide polymorphism (SNP) or in multiple bases. Alternative splice variants of the breast cancer susceptibility, BRCA1, gene has been reported using a multiplex SERS-based sandwich assay platform [104]. In this study, a gold-coated glass slides onto which capture strands (CS) for four different splice variants of BRCA1 were immobilized. The four alternative splice variants of BRCA1 were used as target strands (TS) that hybridized with their complementary CS forming a complex on the gold substrate. Following hybridization, SERS-active probes (DNA-AuNP-RamanTag) consisting of AuNPs conjugated with specific probe strands and four different Raman reporter molecules were used that specifically hybridized with corresponding CS-TS complex through overhang complementary sequences resulting in the formation of a sandwich complex, yielding enhancement of the Raman signal of the reporter molecules, thereby facilitating multiplexed detection of four different splice variants of BRCA1 simultaneously in a single-assay platform with a detection sensitivity of 1 fM.

Detection of circulating tumor cells (CTCs) in the blood of cancer patients is significant for early cancer diagnosis, treatment efficacy evaluation, and choice of treatment options. In 2011, Wang et al. developed SERS AuNP functionalized with a Raman reporter and EGF peptide as a targeting ligand to directly measure targeted CTCs in the presence of white blood cells [105]. The LOD ranged from 5 to 50 CTCs in 1 mL blood. More recently, Wu et al. developed SERS AuNPs for the direct detection of CTCs in blood

with excellent specificity and sensitivity [106]. These AuNPs were functionalized with a Raman reporter molecule, pMBA and reductive bovine serum albumin (rBSA) to stabilize the AuNPs and decrease nonspecific interaction with blood cells. Finally, the AuNPs were conjugated to folic acid (FA) as a targeting ligand to be recognized by CTCs, which overexpress FA receptor alpha (FRα).

A major problem in simultaneous detection of cancer biomarkers is that in the presence of high-concentration biomarkers, the signals of lower-expression biomarkers are overlapped and may go unseen. Ye et al. demonstrated an asymmetric signal amplification method for simultaneous SERS detection of multiple cancer biomarkers with significantly different levels [107]. Coupling SERS technology with hybridization chain reaction (HCR) amplification enabled the simultaneous detection of microRNA-203 (miR-203) and ATP. The presence of the target ATP results in linear signal amplification, whereas the presence of the target microRNA (miRNA) initiated quadratic signal amplification. Through this, 0.15 fM miRNA and 20 nM ATP were simultaneously detected with detection concentration difference over 11 orders of magnitude.

18.4 Photothermal Therapy (PTT)

Hyperthermia (HT) is a treatment where heat is applied to a tumor or organ [108]. While high-temperature HT (>55°C) can actually induce immediate thermal death (ablation) to targeted tumors, mild fever-range HT (40°C–43°C) can be used to (i) improve drug delivery to tumors, (ii) improve cancer cell sensitivity to anticancer therapy and (iii) trigger potent systemic anticancer immune responses [109,110]. The use of nanoparticles for HT in medicine has attracted increased attention for their improved safety, efficacy and specificity [111]. Nanoparticle (NP)-mediated thermal therapy offers the potential to combine the advantages of precise cancer cell ablation [112] with many benefits of mild HT in the tumor microenvironment, namely radiosensitization of hypoxic regions [113], enhancement of drug delivery [114], activation of thermosensitive agents [115] and boosting the immune system [116]. Key features to consider when selecting a nanoparticle for PTT include the maximum absorption wavelength (λ_{max}) and absorption cross section, as well as the size of the particle. The following sections highlight the use of various nanoparticle platforms for PTT.

18.4.1 Gold-Silica Nanoshells

Developed by Halas and coworkers [117], gold-silica nanoshells are composed of a silica core (∼100 nm) and a thin (∼10 nm) shell of gold. The gold shell is formed by aging small gold clusters attached to the silica core. It is of note that forming a uniform shell by this method is very challenging. The hybridization of the plasmonics of the inner sphere and the outer cavity is believed to cause the red shift in LSPR, which can be controlled by changing the shell thickness [118]. Decreasing the thickness of the gold shell from 20 to 5 nm was shown to provide a red shift from 700 to 1,000 nm, which is attributed to the increased coupling between the inner and outer shell surface plasmons for thinner shell particles [119]. Discrete dipole approximation (DDA) simulations have shown that the SPR frequency depends on the ratio of the shell-to-core thickness in the near-exponential relationship which is independent of particle size, core and shell material and even surrounding medium [120].

For PTT of tumors deeply seated within tissue, NIR light is optimal due to minimal absorption in tissues in this spectral region. Because of the tunable absorption in the NIR region, nanoshells have the capability to allow for *in vivo* PTT of such tumors. Gold-silica nanoshells with 120 nm diameter cores and 15 nm thick shells provide a maximum absorption at 800 nm and are therefore typically used for PTT. In 2003, West, Halas and coworkers demonstrated the use of PEGylated gold-silica nanoshells for both *in vitro* and *in vivo* NIR PTT [121]. *In vitro*, irreversible photothermal damage occurred after exposure to continuous wave (CW) NIR laser (820 nm, 35 W/cm^2, 7 min), while cells treated with laser only did not show cell death. In *in vivo* mouse studies, nanoshells were injected directly into a 1 cm tumor and irradiated at intensity of 4 W/cm^2 for 6 min. Magnetic resonance temperature imaging showed temperature increases of 37.4°C ± 6.6°C, as compared to 9.1°C ± 4.7°C with laser alone, leading to tissue damage, cell shrinkage and loss of nuclear staining. However, fast diffusion of the nanoshells throughout the tumor and high tissue absorption resulted in only 4–6 mm maximum depth of treatment. In later experiments, they performed tail vein injections of PEGylated gold-silica nanoshells to introduce the particles into the blood stream of mice with murine colon tumors in the dorsal flanks [122]. The particles accumulated into the tumor via the EPR effect within 6 h post injection due to leaky vasculature and impaired lymphatic drainage. In this study, all tumors treated with both nanoparticles and laser (808 nm, 4 W/cm^2, 3 min) experienced temperatures of around 50°C, resulting in complete necrosis without regrowth over 90 days. Conversely, all controls demonstrated significant tumor growth and a mean survival time of only 10 days.

Silica-gold nanoshells have demonstrated promise *in vitro* as PTT agents for human breast [8,123,124], prostate [125,126], brain [127] and liver cancers [128]. Along with *in vivo* efficacy demonstrated for the treatment of subcutaneous tumors in mice as described above, silica-gold nanoshells have also been shown to be effective PTT agents against allografted tumors in dogs [129].

An attractive approach that has been recently studied is the use of immune cells as a targeted delivery system by loading macrophages, T cells or DCs with nanoparticles. Macrophages containing gold nanoshells with a core radius of 60 nm and shell thickness of 27 nm have been shown to successfully deliver nanoshells to the hypoxic center of tumor spheroids and allowed for enhanced photo-induced cell death and tumor reduction [130].

18.4.2 Spherical Gold Nanoparticles (AuNP)

PTT using spherical gold nanoparticles (AuNP) can be achieved with pulsed or CW visible lasers due to the SPR absorption in the visible region, making such treatment suitable for shallow cancers. The first comprehensive study using pulsed laser and gold nanospheres was performed in 2003 for targeted photothermolysis of lymphocytes [131]. With this method, cell death is attributed mainly to cavitation damage induced by the micro-scale bubbles that are generated around the nanoparticles. Similar experiments studied the dynamics of laser-induced bubbles during treatment using a pump–probe photothermal technique [132], and this technique was later applied *in vitro* on several cancer types using laser-induced bubbles under nanosecond laser pulses [133–135]. *In vivo* treatment studies using this modality have shown that intracellular bubble formation results in individual tumor cell damage for tumor ablation in rats [136].

To shift the absorbance of gold colloidal nanospheres from visible into the NIR region, particles can be clustered closely together, as demonstrated by Zharov et al. in 2005, or aggregated, as demonstrated by El-Sayed et al. in 2007 [134,137]. It has been shown that when non-aggregated gold colloid internalize within the endosome of cancer cells, the particles form a nanocluster, leading to a shift of the absorbance to the NIR. Upon irradiation with a pulsed NIR laser, photothermal microbubbles cause irreversible damage to cells. This technique has been demonstrated for both leukemia [138] as well as breast cancer cells [134]. Using a short-pulsed NIR laser on aggregates of 30 nm gold colloidal particles, El-Sayed and coworkers were able to selectively ablate oral cancer cells with approximately 20 times lower laser power than what can destroy normal cells.

While PTT using nanosecond-pulsed lasers allows for small tumor eradication due to its highly selective and localized damage, the heating efficiency is relative low because of the heat loss during single pulse excitation. Therefore, the use of CW lasers is favorable for effective heat accumulation to induce cell death in a larger area, mainly via HT and possibly coagulation. As a tradeoff, PTT using CW lasers is more time consuming compared to using pulsed lasers. El-Sayed demonstrated imaging and selective cancer cell photothermal ablation using a CW laser and 40 nm gold nanospheres ($\lambda_{\max} \sim 530$ nm) conjugated to anti-EGFR antibodies to target two types of malignant human oral squamous cell carcinoma [139]. These nanoparticles allowed for cancer cell damage upon irradiation (514 nm, 13–76 W/cm^2, 4 min), while healthy cells showed no loss of cell viability under the same treatment. It was determined that compared to noncancerous cells, cancerous cells targeted with nanoparticles were destroyed with 2–3 times lower laser power.

El-Sayed et al. also investigated how localization of gold nanospheres to different locations inside cancer cells affected photothermal treatment [140]. Gold nanospheres (30 nm, $\lambda_{max} \sim 530$ nm) were functionalized to either localize in the cytoplasm or the nucleus. Human oral squamous cell carcinoma cells were incubated with the nanoparticles for 24 h followed by irradiation using either a CW laser (532 nm, 25 W/cm^2, 5 min) or a nanosecond-pulsed laser (532 nm, 0.3–0.8 mJ, 6–7 ns). It was determined that the energy threshold to induce cell death using pulsed lasers was 0.8 mJ for cytoplasm-targeted and 0.3 mJ for nuclear-targeted nanospheres. In contrast, the energy threshold for cell death using CW laser irradiation was 60 J for cytoplasm-targeted and 210 J for nuclear-targeted nanospheres. A potential explanation for this difference is that nanoparticles were more evenly distributed in the cytoplasm but more aggregated in the nucleus, therefore allowing for enhanced heating efficiency of pulsed lasers at the nucleus due to nonlinear absorption of aggregated gold nanoparticles, whereas slower CW laser heating was more effective in the cytoplasm.

Additionally, human T cells loaded with 45 nm AuNP *ex vivo* have shown to migrate to tumor sites in a human tumor xenograft mouse model [141]. It was found that loaded T cells did not lose function or viability and that the *in vivo* tumor targeting was improved fourfold with the T cell delivery system as compared intravenous injection of free PEGylated AuNPs. Such a system has potential for enhanced delivery of AuNP for tumor PTT.

18.4.3 Gold Nanorods (GNR)

As predicted by Gan theory in 1915, when the shape of AuNPs changes from spheres to rods, the SPR band is split into two bands: a strong band in NIR region and a weak band in the visible region [44]. Unlike the visible band, which is at a wavelength similar to that of gold nanospheres and is insensitive to size changes, the NIR band corresponds to electron oscillations along the long axis and is greatly red shifted with increasing aspect ratios (length/width). Nanorods are formed by asymmetric growth of small gold spheres in the presence of shape-forming surfactants, weak reducing agents and catalysts [142]. The aspect ratio can be precisely controlled by changing the experimental parameters [143,144]. As the aspect ratio increases, the SPR maximum is linearly red shifted. Such behavior is different from that of spheres for which the SPR only slightly red shifts with increasing particle size; however, similar to gold nanospheres, when the aspect ratios increase, the scattering efficiency increases.

In vitro experiments by El-Sayed and coworkers used GNRs (AR 3.9, $\lambda_{max} \sim 790$ nm) functionalized with anti-EGFR antibody to specifically target malignant human oral squamous cell carcinoma (HSC and HOC) [7]. When irradiated with a CW laser (800 nm, 4 min), normal cells required laser powers of 20 W/cm^2 to cause photothermal damage, whereas cancer cells required half the laser energy

(10 W/cm^2) at the same treatment timing. Compared to nanoshells, the use of GNRs enabled effective treatment at more than three times lower laser intensity due to the higher absorption efficiency of GNRs compared to nanoshells with SPR at the same wavelength [145]. When using pulsed laser irradiation, Niidome et al. discovered that GNRs induced cell death; however, successive irradiation caused reshaping of the nanorods into nanospheres, which prevented further cell death [146,147].

Additionally, GNRs have been developed for *in vivo* PTT oral squamous cell carcinoma and colon cancer xenografted into mice [148,149]. PEGylated GNRs were injected into mice both intravenously and subcutaneously and allowed to accumulate within xenografted oral squamous cell tumors [148]. Due to the NIR absorption by the nanorods within the tumor, not only could the tumor be imaged but after exposure to a CW NIR laser (808 nm, 1–2 W/cm^2), tumor growth was significantly inhibited for both treated groups. It was discovered that intravenous injection of nanorods caused lower nanoparticle uptake within the tumor, and therefore, treated tumors showed lower photothermal efficiency. In another study, single intravenous injections of PEGylated GNRs enabled complete eradication of all irradiated tumors in mice with a melanoma cancer xenograft without regrowth over 50 days [150].

As compared to gold nanocages and nanoshells, GNRs have the significant advantages of facile synthesis and superior blood circulation time due to their anisotropic geometry. Studies have shown that PEGylated GNRs have a blood half-life of over 17 h in mice [151]. However, a significant disadvantage is the susceptibility to reshaping to gold nanospheres under intense laser irradiation, which results in the loss of the NIR resonance.

18.4.4 Hollow Gold Nanoshells (HGNS)

Hollow gold nanoshells (HGNS) of 30 nm diameter with a hollow center and thin gold shell have been synthesized by oxidation of cobalt or silver nanoparticles with the addition of chloroauric acid to form a thin shell of gold ($\sim$8 nm) with no solid material at the core [152]. Li et al. demonstrated successful *in vitro* targeting and ablation of EGFR-overexpressing cells by using HGNS conjugated to anti-EGFR antibodies as well as *in vivo* ablation of xenografted subcutaneous murine melanoma tumors by using HGNS conjugated to a melanocyte-stimulating hormone (MSH) analog [153,154]. In *in vitro* experiments, HGNS (30 nm) were conjugated to anti-EGFR antibodies to target A431 cells, and almost all nanoshell-treated cells were damaged when exposed to NIR laser (808 nm, 40 W/cm^2, 5 min), an effect not observed for control groups [153]. *In vivo* biodistribution studies indicated that anti-EGFR tagged HGNS were delivered to EGFR-positive tumors at 6.8% ID/g. In a later study, they demonstrated successful photothermal ablation of a melanoma xenograft using peptide-conjugated 43 nm HGNS and NIR laser irradiation at a low-dose energy of 30 J/cm^2 [154].

In addition to activing as a PTT agent, HGNS may be used as delivery vehicles by loading cargo within the shell [155]. Jin and coworkers recently demonstrated *in vivo* combined PTT and imaging/diagnosis by ultrasound and X-ray CT using 300 nm PEG-stabilized gold nanoshelled perfluorooctoyl bromide nanocapsules [156]. By this modality, they were able to enhance both the imaging contrast in U87 human glioblastoma tumor-bearing nude mice as well as the photothermal heating (808 nm, 1.3 W/cm^2, 10 min) with a 35°C temperature increase and decreased tumor volume by 68% after 16 days.

18.4.5   Gold Nanocages

Xia and coworkers developed gold nanocages, a type of hollow and porous gold nanostructure formed by a galvanic replacement reaction between silver nanocubes and auric acid in aqueous solution [157]. Simultaneous deposition of gold atoms and depletion of silver atoms results in gold nanoshells, which then anneal to generate hollow and porous structures with edge lengths ranging from 30 to 200 nm. By controlling the amount of auric acid solution, the wall thickness can be controlled to tune the absorption peak between 400 and 1,200 nm.

Gold nanocages have been investigated for their utility in *in vitro* PTT experiments using NIR femtosecond pulsed lasers [158,159]. Gold nanocages (45 nm) for HER-2 targeting were shown to generate cell damage to 55% of breast cancer cells when irradiated with pulsed laser (805 nm, 6.4 W/cm^2, 5 min). Chen et al. performed a successful *in vivo* study using gold nanocages to ablate murine models with subcutaneous glioblastoma. Treating mice with gold nanocages and NIR laser therapy resulted in 70% reduced tumor metabolism after 24 h as compared to saline-injected control mice, which showed no difference in metabolic activity by PET [160].

18.4.6   Gold Nanostars (GNS)

Our group has developed a novel seed-mediated method to synthesize GNS without using toxic surfactant, which is suitable for *in vivo* applications [48,49]. The synthesized GNS nanoparticles have multiple sharp branches, resulting in tip-enhanced plasmonics for imaging and PTT. GNS functionalized with cell-penetrating peptide, TAT, demonstrate enhanced intracellular delivery and have been used as agents for combined TPL imaging, cellular tracking and photothermal ablation using NIR laser *in vitro* within various cell types, including 3D inflammatory breast cancer (IBC) tumor emboli [61–63]. We have demonstrated that GNS are endocytosed into multiple cancer cell lines, irrespective of receptor status or drug resistance, and that these nanoparticles penetrate IBC tumor embolic core in 3D culture [61]. Upon NIR laser irradiation, GNS not only photothermally ablate cancer cells in monolayer cultures but also allow effective photothermal ablation of entire IBC tumor emboli, as shown in Figure 18.3.

The size of GNS can be controlled so that they passively accumulate in the tumors due to the EPR effect of tumor vasculature. We have demonstrated GNS selective accumulation in tumor following intravenous injection of GNS in the tail veins in a murine sarcoma model [54]. Following GNS uptake into tumors, the superior photothermal conversion efficiency of GNS was shown through the application of focused NIR laser to the lesion. This PTT using NIR laser power below the maximum permissible exposure (MPE) led to ablation of aggressive tumors containing GNS but had no effect in the absence of GNS.

Brain cancer is one of the most aggressive cancers and how to deliver therapeutic agents into brain tumors is a challenge due to blood–brain barrier (BBB). With GNS, we have found that photothermal treatment with NIR laser can open BBB and GNS penetrate through blood vasculature and get into brain parenchyma [20]. Due to the high photothermal transduction efficiency of GNS, a locally triggered vascular permeabilization can be achieved through a cranial window *in vivo* at low nanoprobe dose (<1 pmole) and laser power (35 mW) (Figure 18.4), thus avoiding unwanted hemorrhagic infarction under high-power PTT. On the contrary, for the region without laser treatment, there is minimal GNS leaking through brain vasculature. Our study, for the first time, achieved a photothermal-triggered and spatially-controlled delivery of GNS nanoprobes into brain tumor under a much lower laser power. Therefore, GNS combined with NIR laser provides a controllable method to selectively open BBB for therapeutic drug delivery into brain tumor for therapy.

18.5   Drug and Gene Delivery

Gold nanoparticles exhibit a combination of unique physicochemical and optical properties that permit their use in various biomedical applications, such as diagnostic imaging and drug and gene delivery [161]. Various shapes, including nanospheres, nanorods, nanostars and nanoparticle clusters, have been explored for drug delivery. Gold nanoparticles are routinely surface functionalized with ligands to achieve increased selectivity in tumors and to specifically deliver their payload. These nanoparticles are generally modified with polymers or PEG-containing linkers to conjugate or complex with drug or siRNA/DNA. Hence, they are increasingly being used for drug or gene delivery purposes by exploiting the advantages of increased drug/gene loading, low toxicity, efficiency in cell uptake and stability in the circulation [162].

18.5.1   Plasmonic Nanoparticles as Drug Delivery Vehicles

The photothermal effect has been widely used to disrupt non-covalent interactions between nanoparticles and drug, resulting in the release of encapsulated drugs [163,164]. For example, drug release can be triggered by the photothermal

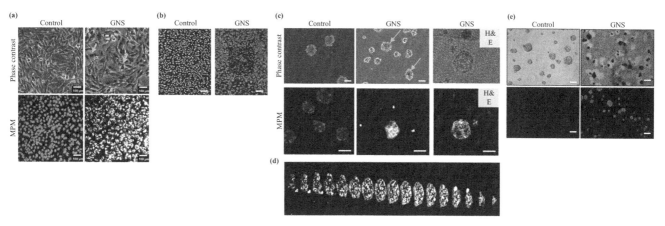

FIGURE 18.3 (a) GNS uptake in IBC cell line, SUM149. The cells were incubated with 0.15 nM GNS for 24 h, washed to remove free particles, stained with Hoechst 33342 and imaged via multiphoton microscopy (MPM). The GNS appear as black particles within the cells under phase-contrast imaging (light gray arrows). Under MPM, the GNS in the cell cytoplasm appear white in color. (b) Photothermolysis of SUM149 cells by irradiation with 800 nm multiphoton pulsed laser for 3 min at 3.7 mW. Following laser exposure, cells were stained with fluorescein diacetate and propidium iodide to determine relative cell viability. The control dish showed no cell death while the GNS dish shows a square region of dead cells in the location of the applied laser spot. The control samples (cells without GNS, i.e., unlabeled) displayed full viability following photothermal treatment. All samples were plated in 35 mm Petri dishes. (c) GNS are endocytosed by SUM149 tumor emboli, after which they penetrate into the tumor embolic core. Tumor emboli shown here are imaged after 96 h of embolic maturation. The control groups represent cells that are free of nanoparticles. For GNS-labeled emboli, nanoparticles were added after 96 h of embolic maturation. Following 24 h of GNS incubation, samples were stained with Hoechst 3342 and imaged with MPM. The black portions, as shown by arrows, represent GNS as observed by phase-contrast microscopy. GNS as observed by MPM is shown by white luminescence. H&E staining displays the cross-sectional area of the emboli, and MPM imaging of the cross section further demonstrates the depth of GNS penetrance into the tumor embolic core (d); 2D projections of the GNS-labeled tumor emboli depicted in (c). Shown here are the individual Z slices of the stacked composite image. This projection demonstrates the depth of GNS penetrance into the tumor emboli. (e) Photothermal treatment of SUM149 tumor emboli. After 96 h of tumor emboli maturation, 0.15 nM of GNS was incubated with tumor emboli for 24 h. Following GNS incubation, samples were exposed to an 808 nm continuous laser at various laser intensities. After treatment, the experimental samples were stained with propidium iodide to illustrate any cellular death that occurred following laser irradiation. All emboli that were incubated with GNS-exhibited cell death following laser exposure. Control samples were treated with the same level of irradiation, which demonstrated virtually no cell death. Representative phase-contrast images of each sample following treatment are shown. Empty spaces seen under phase-contrast for GNS-labeled emboli are due to the repositioning of the emboli in the wells following photothermal treatment. The GNS-labeled emboli appear in black color under phase-contrast imaging due to GNS uptake, compared to the unlabeled controls. (Scale bars =100 μm) [61].

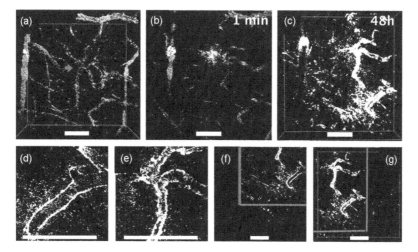

FIGURE 18.4 Photothermal-triggered and spatially-controlled delivery of GNS nanoprobes into brain tumor. MPM images of photothermal-triggered tumor BBB permeation examined through a cranial window. Tumor vessels prior to laser irradiation (a), 1 min after (b), and 48 h after (c–g) laser irradiation. Following irradiation, PEG-GNS (white) were found residing near the blood vessels as well as extravasating deep into the parenchyma (d and e) but not outside the irradiation zone (f and g). Arrow in (a) denotes vascular tortuosity and lines denote the border of irradiation. Scale bar: 100 μm [20].

effect by affecting the hydrophilic-hydrophobic balance of thermally responsive polymers conjugated to nanoparticles. Temperature-responsive polymers and hydrogels exhibit a volume phase transition at their lower critical solution temperature (LCST), below which the polymers are miscible in aqueous solutions, but above which they are insoluble [165]. When thermo-responsive polymers are incorporated into nanoparticle systems, a temperature increase above LCST renders the polymers hydrophobic, causing the polymers to collapse and subsequently release their drug cargo. For example, Yavuz et al. developed gold nanocages covalently functionalized with a thermo-responsive polymer, poly(N-isopropylacrylamideco-acrylicamide) (pNIPAAm-co-pAAm), with an LCST of 37°C [166]. At body temperature (37°C), the polymers were in a conformation that sealed the nanocage pores, preventing dye release. At temperatures above the LCST, the polymers collapsed, allowing the release of the dyes. When nanocages loaded with the doxorubicin (DOX), a chemotherapeutic drug, were incubated with breast cancer cells, irradiation at NIR wavelengths (730–820 nm, 20 mW/cm^2, 2 min) resulted in >30% loss in cell viability. Similarly, Li et al. demonstrated that using doxorubicin-loaded HGNS for combined PTT and chemotherapy was more effective than either treatment alone *in vivo* and also reduced the systemic toxicity of doxorubicin [167]. Because doxorubicin has been shown to cause immunogenic cell death in cancer cells, the inflammatory response, antigen release and immunogenic apoptosis caused by combination PTT and chemotherapy could cause a potent immune response against metastatic cancer sites [168].

Srivastava et al. extensively studied pH-sensitive DOX release using nanosomes as carriers, consisting of gold nanoparticles and exosomes [169]. DOX was conjugated to gold nanoparticles through a hydrazone-based, pH-sensitive drug delivery linker. Isolated exosomes from normal MRC9 lung fibroblast cells were loaded with these nanoparticles. The highest DOX release was observed in conditions that mimicked the tumor microenvironment. Increased cell uptake and cytotoxicity were observed in lung cancer cell lines, while toxicity to normal fibroblast cells and human coronary artery smooth muscle cells were reduced.

18.5.2 Plasmonic Nanoparticles in Gene Therapy

Regulation of the expression of genes that are involved in cancer-related pathways is the goal of various experimental cancer treatment approaches. In the 1990s, a novel method of controlling gene expression was discovered where a short double-stranded RNA was used to specifically and selectively suppress the expression of the target gene [170]. This phenomenon was called RNAi or RNA interference and the double-stranded RNA was termed small interfering RNA, or siRNA. Due to the short *in vivo* half-life of naked siRNA, siRNA-based cancer gene therapy is often degraded by nucleases in biological fluids [171]. Naked siRNA cannot penetrate endothelial cells and tissues due to its size (~15 kDa) and is further prevented from diffusing efficiently into the cellular membrane by the extracellular matrix. In addition to these biological barriers, siRNA's negative charge prevents it from penetrating through negatively charged cell membranes and even if entry into the cell is successful, the siRNA encounters the harsh environment produced by endosomes, which often results in clearing of siRNA through endosomal pathways.

To circumvent these limitations and exploit the benefits of post-transcription gene silencing of cancer-related genes by siRNA, nanoparticle-based delivery vehicles have been developed to transport siRNA into cells. Sardo et al. demonstrated that transfection efficiency and low bioavailability can be considerably improved by using polymer-modified GNS [172]. The GNS were coated with layers of hydrophilic and amphiphilic polymers that contained thiol linker groups to anchor the siRNA layers and NH$_2$ groups, which can be protonated to obtain a positive surface for successive siRNA layering. In doing so, transfection efficiency and colloidal stability were improved for *in vitro* studies on breast cancer cells. The flexibility of gold- and silver-based nanoparticles when developing hybrid nanoparticles offers increasing possibilities to further develop metallic nanoparticles for potential of siRNA delivery in clinics as well as combination therapies when exploiting their NIR photothermal response.

In addition, irradiation of plasmonic nanoparticles can cause PA effects, leading to the formation of vapor microbubbles that disrupt lipid membranes [173,174]. Lipid membrane disruption in liposomes can trigger drug release, while in endosomes, this disruption can enhance endosomal escape [173,175–177]. For pulsed NIR laser-dependent release of siRNA by vapor microbubble formation, Reich and coworkers developed HGNS conjugated to siRNA by Au-s bonds [175]. After irradiation with a femtosecond pulsed laser, endosomal escape and detachment of siRNA were both observed. Endosomal escape was attributed to membrane disruption by cavitation while release of siRNA was attributed to the breaking of Au-S bonds. An *in vitro* study within human epithelial prostate cancer cells (PPC-1) exposed to gold nanoshells carrying siRNA targeting polo-like kinase (PLK1) showed a 60%–70% decrease in PLK1 expression upon irradiation (800 nm, 2.4 W/cm^2, 10 s).

Halas and coworkers demonstrated the use of the photothermal effect to release antisense oligonucleotides from plasmonic nanoparticles via thermal dehybridization [178]. Single-stranded DNA (ssDNA) was bound to complementary DNA sequences immobilized on gold nanoparticles. Upon irradiation, particle heating led to the release of ssDNA. Irradiation (800 nm, 2.5 W/cm^2, 2 min) of nanoparticles loaded with GFP antisense ssDNA caused a 47% downregulation of GFP expression in human lung cancer cells [179]. Irradiation-induced release of other molecules selectively bound to DNA was also demonstrated [180].

18.6 Photodynamic Therapy (PDT)

Photodynamic therapy (PDT) is a promising treatment for a variety of diseases, including cancer, where when a photosensitizer (PS) is activated with the appropriate wavelength of light, it can undergo intersystem crossing to an excited triplet state. In the presence of molecular oxygen ($^3O_2$), the photosensitzer in its excited triplet state can transfer its energy, producing singlet oxygen ($^1O_2$) and other cytotoxic ROS, destroying cells in the immediate vicinity [181]. One of the main obstacles to applying PDT clinically is the inactivation or poor solubility of the PS in biological environments [182]. Loading of the PS onto a nanoparticle can overcome these problems by both protecting the molecule from degradation and acting as a carrier for delivery to the intended target. Due to LSPRs of gold nanoparticles, field enhancement of the incident light around AuNPs could be used to increase the excitation efficiency of the accompanying PS.

Russell et al. were the first to report the synthesis of a three-component (PS/gold/phase transfer reagent) system, where a phase transfer reagent (tetraoctylammonium bromide; TOAB) was utilized to stabilize phthalocyanines (Pc) on AuNPs [183]. The developed 2–4 nm construct generated $^1O_2$ with enhanced quantum yields compared to the free Pc. They later reported *in vitro* uptake and photodynamic efficiency of these AuNPs in breast cancer cells and found that the AuNP conjugates resulted in greater than twice the amount of cell death (57%) compared to free Pc (26%), which could probably be attributed to the 50% enhancement of $^1O_2$ quantum yield observed for the Pc-AuNP [184]. In an *in vivo* study of PDT efficacy at $600-700$ nm (157 J cm^{-2}) for the treatment of amelanotic melanoma in mice, they found that compared to free phthalocyanine, AuNP-Pc conjugates target cancer tissues more selectively and induce more extensive PDT response by promoting an antiangiogenic response by causing extensive damage to the blood capillaries and the endothelial cells [185]. Although promising, the AuNP-Pc conjugates were taken up by the liver and spleen with a prolonged persistence in the liver without any apparent decrease of the PS. Therefore, strategies to limit the accumulation of PS-AuNP in organs such as liver and spleen were investigated by the same group. Conjugation of AuNP-Pc to PEG and anti-HER2 monoclonal antibodies afforded selective targeting to breast cancer cells that overexpress HER2 EGF receptor *in vitro* [185].

Fales et al. demonstrated intracellular Raman imaging and PDT on breast cancer cells by coupling the PS protoporphyrin IX (PpIX) to Raman-labeled plasmonic GNSs, coated with a cell-penetrating peptide, TAT. This nanoconstruct afforded enhanced photodynamic effects and increased enhancement of Raman signals [186]. Shi et al. demonstrated the use of GNRs functionalized with AlPcS4 PSs for the specific PDT of cancer cells [187]. It was found that with NIR irradiation, the AlPcS4 photosensitizer was released; therefore, a high power density coupled with low power density mode was adopted as the initial short duration of high power density laser allows for speeding up of the release of AlPcS4 from the GNR for specific killing of cells containing the GNR-AlPcS4, while the subsequent low power allows for prolonged PDT therapeutic effects through the gradual release of AlPcS4. It was found that due to synergistic plasmonic photothermal properties of the complexes, the high/low PDT mode demonstrated improved efficacy over using single-wavelength continuous laser irradiation and may lead to a safer and more efficient PDT for superficial tumors.

Interestingly, GNRs, which are used as PTT agents with high laser doses, usually 808 nm, 1–48 W/cm^2 irradiations, have also been reported to themselves act as PDT therapeutic agents when the irradiation wavelength is optimized [188]. The researchers demonstrated that GNRs alone can sensitize the formation of singlet oxygen as PDT affects inhibition of tumors in mice under very low LED/laser doses of single-photon NIR (915 nm, <130 mW/cm^2) light excitation. By changing the NIR light excitation wavelengths, Au NRs-mediated phototherapeutic effects can be switched from PDT to PTT or a combination. Kumar et al. also demonstrated PDT without the use of PS by using gold core-petal nanoparticles (CPNs) [189]. It was found that CPNs have both photothermal effects, as shown from its high photothermal conversion efficiency of 32%, and high photodynamic effect, as shown from the large generation of singlet oxygen. It was found that petal length and density is essential for singlet oxygen production. The percentage contribution of ROS in causing cell death was found to be 88%, while the remaining 12% was due to photothermal effects.

18.7 Multimodal and Synergistic Therapies

Development of multifunctional AuNP offering synergistic therapeutic approaches has been a trend in this field. Several studies have examined the combination and synergistic effect of PDT and PTT. Unlike PDT, which is dependent on the availability of tissue oxygen, the PTT effect is oxygen-independent, and hence, the hypoxia environment induced by PDT reaction does not affect PTT. Early reports used two different excitation wavelengths to accomplish PTT and PDT. Jang et al. demonstrated a 95% reduction in tumor growth *in vivo* using a GNR-AlPcS4 complex excited by 810 and 670 nm lasers [190]. Khlebtsov et al. developed composite nanoparticles consisting of a gold-silver nanocage core and a mesoporous silica shell functionalized with the PS (Yb-2,4-dimethoxyhematoporphyrin) for *in vivo* PDT applications [191]. The synthesized nanocomposites generated $^1O_2$ under excitation at 630 nm and produced heat under laser irradiation at the plasmon resonance wavelength ($750-800$ nm). To simultaneously activate the two treatment approaches, Gao et al. developed a nanocomplex using a hypocrellin B-loaded lipid to coat gold nanocages [192]. Using a two-photon laser at 790 nm allowed for simultaneous excitation of both the gold nanocage, for conversion of light to heat energy and the accompany PS to generate ROS for PDT. Wang et al. hypothesized that by adjusting irradiation time, early-phase PDT effects can be combined with late-phase PTT effects to achieve optimal synergistic treatment efficacy [193].

An immunologically modified single-walled carbon nanotube (SWNT) system has been developed by Chen et al. [194] for *in vitro* and *in vivo* cancer treatment by coating the surface of SWNT with an immunoadjuvant, glycated chitosan (GC). It was found that the SWNT-GC system could enter cells and retaine both the optical properties of the carbon nanotubes as well as the immunological function of the chitosan. This system allows for damage via PTT upon irradiation of SWNTs as well as for cellular GC to serve both as a damage-associated molecular patterns (DAMP) and pathogen-associated molecular pattern (PAMP) molecule to enhance tumor immunogenicity and enhanced uptake and presentation of tumor antigens, leading to antitumor response.

Our group has demonstrated a novel cancer therapy, synergistic immune photo nanotherapy (SYMPHONY) on bladder cancer with mice animal model [195]. The SYMPHONY therapy combines GNS-mediated PTT and immune checkpoint inhibitor [196]. In recent years, immunotherapy with specific immune checkpoint inhibitor has provided a promising way to break tumor immunosuppressive environment. PD-L1, a protein overexpressed on cancer cell membrane, contributes to the suppression of the immune system. To modulate T cell function, PD-L1 binds to its receptor, PD-1, found on activated T cells, B cells and myeloid cells. The therapeutic anti-PD-L1 antibody is designed to block the PD-L1/PD-1 interaction and reverse tumor-mediated immunosuppression. Blocking the PD-L1/PD-1 axis has been shown to be highly beneficial in many human tumors [197]. However, only modest clinical response to single-agent activity of PD-L1 and PD-1 antibodies is achieved in patients [198].

The GNS-mediated PTT can be used to activate the immune system and generate a synergistic effect by combining with anti-PD-L1 immunotherapy (Figure 18.5). For the SYMPHONY group, 80 mg/kg GNS was administrated through tail vein followed by laser treatment on the primary tumor but not for the distant tumor 24 h later. Anti-PD-L1 antibody was IP administered after laser irradiation and every 3 days following until mice were tumor-free. On day 49, only SYMPHONY group mice survived. One of these mice was free of both primary and distant tumor and no tumor recurred in the following 3 months. The rechallenge was performed by injection of cancer cells and no tumor development occurred, suggesting a memory immunoresponse, or "cancer vaccine effect." It was also observed that not only the laser treated primary tumor showed therapeutic response from SYMPHONY but the distant tumor was also treated. Immune cell phenotyping was performed to investigate immnuoresponse from SYMPHONY, and it was found that the combination treatment of GNS PTT and anti-PD-L1 significantly increased

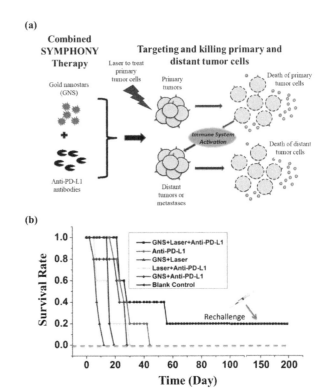

FIGURE 18.5 SYMPHONY: Synergistic immuno photo nanotherapy. (a) By disabling the tumor immune resistance and simultaneously ablating individual cancer cells, we can trigger a powerful thermally enhanced systemic immune response to rapidly eradicate locally aggressive and metastatic cancer. (b) Kaplan–Meier (K-M) survival curve of mice in SYMPHONY study. The tumor-free mouse in SYMPHONY group was monitored for 3 months and no tumor recurred. After that, the rechallenge was performed and no tumor developed [195].

the percentage of total T cells, CD4, CD8 T cells and B cell. On the other hand, both the percentage and cell number of myeloid-derived suppressor cells (MDSCs) were significantly reduced in the combination treatment group. The efficacy of SYMPHONY relies on several synergistic processes. First, localized PTT with GNS and NIR irradiation kills primary tumor cells. Second, dying tumor cells release tumor-specific antigens and DAMPs, including heat shock proteins (HSPs). Third, DAMPs actively stimulate the immune response by modulating maturation, activation and presentation of antigen-presenting cells (APCs). Lastly, these activated APCs present tumor antigens to T cells, activating adaptive immunity. Therefore, blocking PD-L1/PD-1 signaling during T cell activation amplifies immune responses.

In addition to PD-1/PD-L1, CTLA-4 immune checkpoint inhibition has also been combined with PTT to generate immune response to treat cancer. The PTT has also been combined with anti-CTLA-4 immunotherapy to treat cancer metastasis. A recent study used carbon nanotubes to perform PTT with 808 nm NIR laser irradiation [116]. The combined therapy resulted in significantly reduced lung metastasis. In addition to carbon nanotubes, poly(lactic-co-glycolic) acid (PLGA) nanoparticles with indocyanine green (ICG) and imiquimod have been developed and used for photothermal treatment. The generated cancer vaccine effect has been found to be able to attack residue tumor cells as well as prevent tumor recurrence [199].

18.8 Theranostic Plasmonic Nanoparticles

Nanotheranostics, combining diagnostic and therapeutic functions into one nanoplatform, has attracted great attention due to its promise as a powerful tool for personalized therapy. Theranostic nanoprobes can potentially be used for image-guided cancer therapy with improved targeting specificity and therapeutic efficacy [200]. Imaging-guided therapy is significant in solving the challenges of cancer heterogeneity and adaptation. The knowledge of the exact location of the tumor is essential to provide therapeutic treatments that are more specific to individuals and therefore enhance the therapeutic efficiency and reduce side effects to nearby organs.

Our group has developed multifunctional theranostic GNS nanoprobe for combined *in vivo* cancer imaging (SERS, CT, TPL) and PTT on mice with primary sarcomas [54]. Radiolabeling as well as CT and optical imaging were performed to investigate the GNS probe's biodistribution and intratumoral uptake at both macroscopic and microscopic scales. GNS were tagged with Raman active pMBA for *in vivo* SERS imaging under 785 nm excitation. Shortly following IV injection, GNS are mainly located inside the blood vessels; however, after 24 h, the GNS nanoprobe was found to accumulate in the tumor periphery. Three days after GNS intravenous administration, unique SERS peaks of pMBA at 1,067 and 1,588 cm^{-1} were detected within the tumor but not in the normal muscle. Due to the easily tunable plasmonic peak and tip-enhanced plasmonics of GNS, superior PTT under NIR excitation is a major advantage of these nanoprobes for image-guided PTT. Under NIR laser irradiation (980 nm), the GNS nanoprobe's high photon-to-heat conversion efficiency allowed for concentrated heating at a performance level greater than gold nanoshells, a widely used PTT agent. Effective tumor ablation after photothermal treatment was demonstrated *in vivo* for xenograft sarcoma mouse model. It was found that under the same laser irradiation, tumors with GNS showed much higher temperatures (more than 50°C) as compared to those without GNS (less than 40°C). The tumors were ablated with temperatures higher than 50°C and a decrease in tumor size for mice with GNS injection was observed, while the tumor size continued to increase for control mice. The multifunctional GNS nanoprobe demonstrated potential for future use as an agent in intraoperative tumor margin delineation by SERS and TPL and image-guided PTT.

Seo et al. developed methylene blue-loaded mesoporous silica-coated GNRs (MB-GNR@SiO$_2$) for combined cancer imaging and photothermal/photodynamic dual therapy [201]. It was found that encapsulated MB molecules had both monomer and dimer forms, resulting in an increase in the photosensitizing effect, and the embedded MB molecules also showed NIR light-induced SERS performance with a Raman enhancement factor of 3.0×10^{10}, which is enough for the detection of a single cancer cell. Furthermore, the MB-GNR@SiO$_2$ nanoparticles exhibited a synergistic effect of PDT and PTT under single-wavelength NIR laser irradiation (780 nm). More recently, Li and colleagues reported the use of gold nanoclusters for confocal laser endomicroscopy-guided dual PTT/PDT therapy for pancreatic ductal adenocarcinoma (PDAC) [202]. The nanocluster platform consisted of four components: the gold nanocluster PTT agent, an active targeting ligand U11 peptide, a Cathepsin E (CTSE)-sensitive PDT therapy prodrug and a CTSE-sensitive imaging agent (cyanine dye Cy5.5). In this system, they observed that CTSE-sensitive nanoclusters with targeting U11 peptide can significantly increase the uptake and apoptosis of pancreatic cancer cells, compared with that of the nontargeted nanocluster AuS-PEG and the insensitive nanocluster AuC-PEG. They reported that enzyme-triggered drug release of 5-ALA with tumor targeting in nanoclusters AuS-U11 could achieve optimal therapeutic efficacy with endomicroscopy-guided PTT/PDT, with reduced side effects.

Our group has also developed a folate receptor (FR)-targeted theranostic nanoparticle for SERS-based detection and PDT [203]. The diagnostic aspect of the construct comes from a silver-embedded gold nanostar tagged with a Raman active label (GNS-pMBA@Ag) that acts as a SERS tag for Raman imaging [204]. Protoporphyrin IX (PpIX) PS molecules are loaded onto the SERS tag by encapsulating them in a silica shell for PDT. The selective detection of FR-positive cells was demonstrated using Raman imaging, showing a high SERS signal from FR-positive cells and little-to-no signal from FR-negative cells. The selective therapeutic effect was demonstrated by performing an *in vitro* PDT study demonstrating that only the FR-positive cells were affected by the treatment. Use of other small-molecule Raman dyes can provide the particles with unique SERS spectra for the application in multiplex detection when combined with other specific targeting agents, such as antibodies.

In 2017, Huang et al. coated FA functionalized gold nanoparticles with bovine serum albumin for pH-sensitive, FR-targeted drug delivery through the encapsulation of DOX [205]. These Au–BSA–DOX–FA nanocomposites proved to be efficient and durable *in vivo* CT contrast agents for targeted CT imaging of the FR overexpressed in cancer tissues. *In vivo* experiments demonstrated that Au–BSA–DOX–FA nanocomposites had selective antitumor activity effects on FR-overexpressing tumors and no adverse effects on normal tissues and organs. The Au–BSA–DOX–FA nanocomposite exhibited selective targeting activity, X-ray attenuation activity and pH-sensitive drug release activity.

Magnetic nanoparticles are well known as MRI imaging agents. In 2013, Yang et al. developed non-cytotoxic magnetic polymer-modified gold nanorods (MPGNRs) to act as dual MRI and PA imaging contrast agents [16]. MPGNRs significantly enhanced the NIR-laser-induced photothermal effect due to their increased thermal stability. MPGNRs thus provide a promising new theranostic platform for cancer diagnosis and treatment by combining dual MR/PA imaging with highly effective targeted PTT. Li et al. constructed Au-coated iron oxide (Fe_3O_4@Au) nanoroses to integrate five distinct functions, including aptamer-based targeting, MR and optical imaging, PTT and chemotherapy, into one single probe [206]. The inner Fe_3O_4 core functioned as an MRI agent, while the gold shell allows for PTT and also facilitates the release of DOX carried by the nanoroses. In addition, aptamers were immobilized on the nanorose surface, enabling efficient and selective drug delivery, imaging and PTT with high specificity. More recently, Hu et al. developed multifunctional FA-targeted Fe_3O_4@Au core/shell nanostars for targeted multimodal MR/CT/PA imaging and PTT of tumors [207]. Fe_3O_4@Ag were used as seeds and exposed to Au growth solution to induce the formation of Fe_3O_4@Au core/shell nanostars. Due to co-existence of Fe_3O_4 core and star-shaped Au shell, the NSs can be used for MR and CT imaging of tumors, respectively. In addition, the NIR plasmonic absorption of these nanoconstructs enables their use for PA imaging and PTT of tumors.

18.9 Conclusion

Gold nanoparticles have become a promising platform for cancer imaging and diagnostic applications as well as direct therapeutic applications, such as plasmonic PTT or PDT. Recent biomedical uses of AuNPs have increasingly combined several of these properties into single, multimodal platforms that have demonstrated great potential for theranostic applications. The increasing selectivity and efficacy of such platforms hold significant promise for the translation into clinical applications. Specifically, our group has developed a surfactant-free synthesis of plasmonics–active GNSs with capabilities spanning multimodality sensing (SERS, TPL CT, MRI, PET) and therapy (PTT, PDT and immunotherapy). Due to their desirable features, including the tunable plasmonic bands in the NIR region, lack of photobleaching and photodegradation, narrow spectral fingerprints and capability to allow multiplexed bioanalysis, GNS are promising theranostic candidates for *in vivo* applications. Furthermore, recent studies combining checkpoint blockade immunotherapy with GNS-mediated PTT show promise in addressing one of the most challenging problems in the treatment of metastatic cancer and ultimately induce effective long-lasting immunity against cancer. The potential of the plasmonic properties of nanoparticles in improving sensitivities of bioassays and cancer therapy efficiency suggests that the field of the plasmonic nanoparticles for biomedical applications remains promising.

Abbreviations

AuNP:	gold nanoparticles
CT:	computed tomography
DAMPs:	damage-associated molecular pattern molecules
DOX:	doxorubicin
EGFR:	epidermal growth factor receptor
EM:	electromagnetic
EPR:	enhanced permeability and retention
FA:	folic acid
GNR:	gold nanorods
GNS:	gold nanostars
HT:	hyperthermia
iMS:	inverse molecular sentinel
MPM:	multiphoton microscopy
MS:	molecular sentinel
MRI:	magnetic resonance imaging
NIR:	near infrared
PCI:	plasmon coupling interference
PD-L1:	programmed death ligand 1
PDT:	photodynamic therapy
PEG:	polyethylene glycol
PEI:	polyethyleneimine
PpIX:	protoporphyrin IX
PTT:	photothermal therapy
ROS:	reactive oxygen species
SERRS:	surface-enhanced resonance Raman scattering
SERS:	surface-enhanced Raman scattering
SPR:	surface plasmon resonance
SYMPHONY:	synergistic immune photothermal nanotherapy
TPL:	two-photon luminescence

Acknowledgments

This work was sponsored by the National Institute of Health (R21Ca196426), the Department of Defense # W81XWH-17-1-0567 and the Duke Faculty Exploratory Research Fund. This material is based upon work of BMC supported by the National Science Foundation Graduate Research Fellowship (No. 1106401).

References

1. Institute, N.N.C. *Cancer Statistics*. Understanding Cancer 2018; Available from: https://www.cancer.gov/about-cancer/understanding/statistics.
2. Jeanmaire, D.L. and R.P. Van Duyne, Surface Raman spectroelectrochemistry: Part I. Heterocyclic, aromatic, and aliphatic amines adsorbed on the anodized silver electrode. *Journal of the*

Electroanalytical Chemistry and Interfacial Electrochemistry, 1977. **84**(1): pp. 1–20.

3. Albrecht, M.G. and J.A. Creighton, Anomalously intense Raman spectra of pyridine at a silver electrode. *Journal of the American Chemical Society*, 1977. **99**(15): pp. 5215–5217.

4. Vo-Dinh, T. et al., Surface-enhanced Raman spectrometry for trace organic analysis. *Analytical Chemistry*, 1984. **56**(9): pp. 1667–1670.

5. Sokolov, K. et al., Real-time vital optical imaging of precancer using anti-epidermal growth factor receptor antibodies conjugated to gold nanoparticles. *Cancer Research*, 2003. **63**(9): pp. 1999–2004.

6. El-Sayed, I.H., X. Huang, and M.A. El-Sayed, Surface plasmon resonance scattering and absorption of anti-EGFR antibody conjugated gold nanoparticles in cancer diagnostics: Applications in oral cancer. *Nano Letters*, 2005. **5**(5): pp. 829–834.

7. Huang, X. et al., Cancer cell imaging and photothermal therapy in the near-infrared region by using gold nanorods. *Journal of the American Chemical Society*, 2006. **128**(6): pp. 2115–2120.

8. Loo, C. et al., Immunotargeted nanoshells for integrated cancer imaging and therapy. *Nano Lett*, 2005. **5**(4): pp. 709–711.

9. Aioub, M. et al., Biological targeting of plasmonic nanoparticles improves cellular imaging via the enhanced scattering in the aggregates formed. *Journal of Physical Chemistry Letters*, 2014. **5**(15): pp. 2555–2561.

10. Aioub, M., L.A. Austin, and M.A. El-Sayed, Determining drug efficacy using plasmonically enhanced imaging of the morphological changes of cells upon death. *Journal of Physical Chemistry Letters*, 2014. **5**(20): pp. 3514–3518.

11. Link, S. and M.A. El-Sayed, Spectral properties and relaxation dynamics of surface plasmon electronic oscillations in gold and silver nanodots and nanorods. *Journal of Physical Chemistry B*, 1999, **103**: pp. 8410–8426.

12. Li, W. and X. Chen, Gold nanoparticles for photoacoustic imaging. *Nanomedicine (Lond)*, 2015. **10**(2): pp. 299–320.

13. Copland, J.A. et al., Bioconjugated gold nanoparticles as a molecular based contrast agent: Implications for imaging of deep tumors using optoacoustic tomography. *Molecular Imaging & Biology*, 2004. **6**(5): pp. 341–349.

14. Comenge, J. et al., Preventing plasmon coupling between gold nanorods improves the sensitivity of photoacoustic detection of labeled stem cells in vivo. *ACS Nano*, 2016. **10**(7): pp. 7106–7116.

15. Jokerst, J.V. et al., Nanoparticle PEGylation for imaging and therapy. *Nanomedicine*, 2011. **6**(4): pp. 715–728.

16. Yang, H.-W. et al., Magnetic gold-nanorod/ PNIPAAmMA nanoparticles for dual magnetic resonance and photoacoustic imaging and targeted photothermal therapy. *Biomaterials*, 2013. **34**(22): pp. 5651–5660.

17. Wang, Y. et al., Photoacoustic tomography of a nanoshell contrast agent in the in vivo rat brain. *Nano Letters*, 2004. **4**(9): pp. 1689–1692.

18. Yang, X. et al., Photoacoustic tomography of a rat cerebral cortex in vivo with au nanocages as an optical contrast agent. *Nano letters*, 2007. **7**(12): pp. 3798–3802.

19. Raghavan, V. et al. Dual plasmonic gold nanoparticles for multispectral photoacoustic imaging application. In A.A. Oraevsky and L. Wang (eds.), *Photons Plus Ultrasound: Imaging and Sensing 2014*, SPIE BiOS, 2014, San Francisco, CA. 2014. International Society for Optics and Photonics.

20. Yuan, H. et al., Plasmonics-enhanced and optically modulated delivery of gold nanostars into brain tumor. *Nanoscale*, 2014. **6**(8): pp. 4078–4082.

21. Kim, C. et al., In vivo photoacoustic mapping of lymphatic systems with plasmon-resonant nanostars. *Journal of the Materials Chemistry*, 2011. **21**(9): pp. 2841–2844.

22. Xia, J. et al., Whole-body ring-shaped confocal photoacoustic computed tomography of small animals in vivo. *Journal of Biomedical Optics*, 2012. **17**(5): p. 050506.

23. Lu, W. et al., Photoacoustic imaging of living mouse brain vasculature using hollow gold nanospheres. *Biomaterials*, 2010. **31**(9): pp. 2617–2626.

24. Manivasagan, P. et al., Paclitaxel-loaded chitosan oligosaccharide-stabilized gold nanoparticles as novel agents for drug delivery and photoacoustic imaging of cancer cells. *International Journal of Pharmaceutics*, 2016. **511**(1): pp. 367–379.

25. Truby, R.L., S.Y. Emelianov, and K.A. Homan, Ligand-mediated self-assembly of hybrid plasmonic and superparamagnetic nanostructures. *Langmuir*, 2013. **29**(8): pp. 2465–2470.

26. Song, J. et al., "Smart" gold nanoparticles for photoacoustic imaging: An imaging contrast agent responsive to the cancer microenvironment and signal amplification via pH-induced aggregation. *Chemical Communications*, 2016. **52**(53): pp. 8287–8290.

27. Carr, D.H. et al., Gadolinium-DTPA as a contrast agent in MRI: Initial clinical experience in 20 patients. *AJR American Journal of Roentgenology*, 1984. **143**(2): pp. 215–224.

28. Runge, V.M. et al., MR imaging of rat brain glioma: Gd-DTPA versus Gd-DOTA. *Radiology*, 1988. **166**(3): pp. 835–838.

29. Debouttière, P.J. et al., Design of gold nanoparticles for magnetic resonance imaging. *Advanced Functional Materials*, 2006. **16**(18): pp. 2330–2339.

30. Moriggi, L.C. et al., Gold nanoparticles functionalized with gadolinium chelates as high-relaxivity MRI

contrast agents. *Journal of the American Chemical Society*, 2009. **131**(31): pp. 10828–10829.

31. Holbrook, R.J. et al., Gd(III)-dithiolane gold nanoparticles for T1-weighted magnetic resonance imaging of the pancreas. *Nano Letters*, 2016. **16**(5): pp. 3202–3209.

32. Alivov, Y. et al., Optimization of the K-edge imaging for vulnerable plaques using gold nanoparticles and energy-resolved photon counting detectors: A simulation study. *Physics in Medicine and Biology*, 2014. **59**(1): pp. 135–152.

33. Reuveni, T. et al., Targeted gold nanoparticles enable molecular CT imaging of cancer: An in vivo study. *International Journal of Nanomedicine*, 2011. **6**: pp. 2859–2864.

34. Peng, C. et al., Facile formation of dendrimer-stabilized gold nanoparticles modified with diatrizoic acid for enhanced computed tomography imaging applications. *Nanoscale*, 2012. **4**(21): pp. 6768–6778.

35. Zhou, B. et al., Synthesis and characterization of PEGylated polyethylenimine-entrapped gold nanoparticles for blood pool and tumor CT imaging. *ACS Applied Materials Interfaces*, 2014. **6**(19): pp. 17190–17199.

36. Misawa M., Takahashi J., Generation of reactive oxygen species induced by gold nanoparticles under x-ray and UV Irradiations. *Nanomedicine* (2011), **7**: pp. 604–614.

37. Zhang, Y. et al., Ultrastable polyethyleneimine-stabilized gold nanoparticles modified with polyethylene glycol for blood pool, lymph node and tumor CT imaging. *Nanoscale*, 2016. **8**(10): pp. 5567–5577.

38. Sun, X. et al., Chelator-free 64Cu-integrated gold nanomaterials for positron emission tomography imaging guided photothermal cancer therapy. *ACS Nano*, 2014. **8**(8): pp. 8438–8446.

39. Frellsen, A.F. et al., Mouse positron emission tomography study of the biodistribution of gold nanoparticles with different surface coatings using embedded copper-64. *ACS Nano*, 2016. **10**(11): pp. 9887–9898.

40. Liu, Y. et al., Plasmonic gold nanostars for multimodality sensing and diagnostics. *Sensors*, 2015. **15**(2): pp. 3706–3720.

41. Jimenez de Aberasturi, D. et al., Surface enhanced raman scattering encoded gold nanostars for multiplexed cell discrimination. *Chemistry of Materials*, 2016. **28**(18): pp. 6779–6790.

42. Matschulat, A., D. Drescher, and J. Kneipp, Surface-enhanced Raman scattering hybrid nanoprobe multiplexing and imaging in biological systems. *ACS Nano*, 2010. **4**(6): pp. 3259–3269.

43. Zavaleta, C.L. et al., Multiplexed imaging of surface enhanced Raman scattering nanotags in living mice using noninvasive Raman spectroscopy. *Proceedings of the National Academy of Sciences of the United States of America*, 2009. **106**(32): pp. 13511–13516.

44. Zhang, Y. et al., Molecular imaging with SERS-active nanoparticles. *Small*, 2011. **7**(23): pp. 3261–3269.

45. Park, H. et al., SERS imaging of HER2-overexpressed MCF7 cells using antibody-conjugated gold nanorods. *Physical Chemistry Chemical Physics*, 2009. **11**(34): pp. 7444–7449.

46. Lee, S. et al., Surface-enhanced Raman scattering imaging of HER2 cancer markers overexpressed in single MCF7 cells using antibody conjugated hollow gold nanospheres. *Biosensors and Bioelectronics*, 2009. **24**(7): pp. 2260–2263.

47. Lee, S. et al., Rapid and sensitive phenotypic marker detection on breast cancer cells using surface-enhanced Raman scattering (SERS) imaging. *Biosensors and Bioelectronics*, 2014. **51**: pp. 238–243.

48. Khoury, C.G. and T. Vo-Dinh, Gold nanostars for surface-enhanced Raman scattering: Synthesis, characterization and optimization. *Journal of Physical Chemistry. C, Nanomaterials and Interfaces*, 2008. **2008**(112): pp. 18849–18859.

49. Yuan, H. et al., Gold nanostars: Surfactant-free synthesis, 3D modelling, and two-photon photoluminescence imaging. *Nanotechnology*, 2012. **23**(7): p. 075102.

50. Yuan, H. et al., Spectral characterization and intracellular detection of Surface-Enhanced Raman Scattering (SERS)-encoded plasmonic gold nanostars. *Journal of Raman Spectroscopy*, 2013. **44**(2): pp. 234–239.

51. Liu, Y. et al., pH–sensing nanostar probe using surface-enhanced Raman scattering (SERS): Theoretical and experimental studies. *Journal of Raman Spectroscopy*, 2013. **44**(7): pp. 980–986.

52. Schutz, M. et al., Hydrophilically stabilized gold nanostars as SERS labels for tissue imaging of the tumor suppressor p63 by immuno-SERS microscopy. *Chemical Communications* (Camb), 2011. **47**(14): pp. 4216–4218.

53. Webb, J.A. et al., Theranostic gold nanoantennas for simultaneous multiplexed Raman imaging of immunomarkers and photothermal therapy. *ACS Omega*, 2017. **2**(7): pp. 3583–3594.

54. Liu, Y. et al., A plasmonic gold nanostar theranostic probe for in vivo tumor imaging and photothermal therapy. *Theranostics*, 2015. **5**(9): pp. 946.

55. Register, J.K. et al., In vivo detection of SERS-encoded plasmonic nanostars in human skin grafts and live animal models. *Analytical and Bioanalytical Chemistry*, 2015. **407**(27): pp. 8215–8224.

56. Harmsen, S. et al., Surface-enhanced resonance Raman scattering nanostars for high-precision cancer imaging. *Science Translational Medicine*, 2015. **7**(271): pp. 271ra7

57. Wang, H. et al., In vitro and in vivo two-photon luminescence imaging of single gold nanorods. *Proceedings of the National Academy of Sciences of the United States of America*, 2005. **102**(44): pp. 15752–15756.

58. Gao, N. et al., Shape-dependent two-photon photoluminescence of single gold nanoparticles. *The Journal of Physical Chemistry C*, 2014. **118**(25): pp. 13904–13911.

59. Yuan, H. et al., In vivo particle tracking and photothermal ablation using plasmon-resonant gold nanostars. *Nanomedicine: Nanotechnology, Biology and Medicine, 2012.* **8**(8): pp. 1355–1363.

60. Liu, Y. et al., Quintuple-modality (SERS-MRI-CT-TPL-PTT) plasmonic nanoprobe for theranostics. *Nanoscale*, 2013. **5**(24): pp. 12126–12131.

61. Crawford, B.M. et al., Photothermal ablation of inflammatory breast cancer tumor emboli using plasmonic gold nanostars. *International Journal of Nanomedicine*, 2017. **12**: p. 6259–6272.

62. Shammas, R.L. et al., Human adipose-derived stem cells labeled with plasmonic gold nanostars for cellular tracking and photothermal cancer cell ablation. *Plastic and Reconstructive Surgery*, 2017. **139**(4): pp. 900e–910e.

63. Yuan, H., A.M. Fales, and T. Vo-Dinh, TAT peptide-functionalized gold nanostars: Enhanced intracellular delivery and efficient NIR photothermal therapy using ultralow irradiance. *Journal of the American Chemical Society*, 2012. **134**(28): pp. 11358–11361.

64. Arifin, D.R. et al., Trimodal gadolinium-gold microcapsules containing pancreatic islet cells restore normoglycemia in diabetic mice and can be tracked by using US, CT, and positive-contrast MR imaging. *Radiology*, 2011. **260**(3): pp. 790–798.

65. Kircher, M.F. et al., A brain tumor molecular imaging strategy using a new triple-modality MRI-photoacoustic-Raman nanoparticle. *Nature Medicine*, 2012. **18**(5): pp. 829–834.

66. Hirsch, L.R. et al., A whole blood immunoassay using gold nanoshells. *Analytical Chemistry*, 2003. **75**(10): pp. 2377–2381.

67. Sonnichsen, C. et al., A molecular ruler based on plasmon coupling of single gold and silver nanoparticles. *Nature Biotechnology*, 2005. **23**(6): pp. 741–745.

68. Prasad, P.N., *Nanophotonics*. 2004, Hoboken, NJ: Wiley-Interscience.

69. Malicka, J., I. Gryczynski, and J.R. Lakowicz, DNA hybridization assays using metal-enhanced fluorescence. *Biochemical and Biophysical Research Communications*, 2003. **306**(1): pp. 213–218.

70. Zheng, D. et al., Aptamer nano-flares for molecular detection in living cells. *Nano Letters*, 2009. **9**(9): pp. 3258–3261.

71. Rana, S. et al., A multichannel nanosensor for instantaneous readout of cancer drug mechanisms. *Nature Nanotechnology*, 2015. **10**(1): pp. 65–69.

72. You, C.C. et al., Detection and identification of proteins using nanoparticle-fluorescent polymer 'chemical nose' sensors. *Nature Nanotechnology*, 2007. **2**(5): pp. 318–323.

73. Kneipp, K. et al., Single molecule detection using surface-enhanced Raman scattering (SERS). *Physical Review Letters*, 1997. **78**(9): p. 1667.

74. Nie, S. and S.R. Emory, Probing single molecules and single nanoparticles by surface-enhanced Raman scattering. *Science*, 1997. **275**(5303): pp. 1102–1106.

75. Zrimsek, A.B., A.-I. Henry, and R.P. Van Duyne, Single molecule surface-enhanced Raman spectroscopy without nanogaps. *The Journal of Physical Chemistry Letters*, 2013. **4**(19): pp. 3206–3210.

76. Darby, B.L., P.G. Etchegoin, and E.C. Le Ru, Single-molecule surface-enhanced Raman spectroscopy with nanowatt excitation. *Physical Chemistry Chemical Physics*, 2014. **16**(43): pp. 23895–23899.

77. Faulds, K. et al., SERRS as a more sensitive technique for the detection of labelled oligonucleotides compared to fluorescence. *Analyst*, 2004. **129**(7): pp. 567–568.

78. von Maltzahn, G. et al., SERS-coded gold nanorods as a multifunctional platform for densely multiplexed near-infrared imaging and photothermal heating. *Advanced Materials*, 2009. **21**(31): pp. 3175–3180.

79. Chao, J. et al., Nanostructure-based surface-enhanced Raman scattering biosensors for nucleic acids and proteins. *Journal of Materials Chemistry B*, 2016. **4**(10): pp. 1757–1769.

80. Laing, S., K. Gracie, and K. Faulds, Multiplex in vitro detection using SERS. *Chemical Society Reviews*, 2016. **45**(7): pp. 1901–1918.

81. Cao, Y.C., R. Jin, and C.A. Mirkin, Nanoparticles with Raman spectroscopic fingerprints for DNA and RNA detection. *Science*, 2002. **297**(5586): pp. 1536–1540.

82. Li, M. et al., Three-dimensional hierarchical plasmonic nano-architecture enhanced surface-enhanced Raman scattering immunosensor for cancer biomarker detection in blood plasma. *ACS Nano*, 2013. **7**(6): pp. 4967–4976.

83. Qian, X.M. and S.M. Nie, Single-molecule and single-nanoparticle SERS: From fundamental mechanisms to biomedical applications. *Chemical Society Reviews*, 2008. **37**(5): pp. 912–920.

84. Dinish, U.S. et al., Actively targeted in vivo multiplex detection of intrinsic cancer biomarkers using biocompatible SERS nanotags. *Scientific Reports*, 2014. **4**: p. 4075.

85. Wang, H.N. and T. Vo-Dinh, Plasmonic coupling interference (PCI) nanoprobes for nucleic acid detection. *Small*, 2011. **7**(21): pp. 3067–3074.

86. Vo-Dinh, T., K. Houck, and D. Stokes, Surface-enhanced Raman gene probes. *Analytical Chemistry*, 1994. **66**(20): pp. 3379–3383.

87. Vo-Dinh, T. et al., SERS nanosensors and nanoreporters: Golden opportunities in biomedical applications. Wiley Interdisciplinary Reviews. *Nanomedicine and Nanobiotechnology*, 2015. **7**(1): pp. 17–33.

88. Wabuyele, M.B. and T. Vo-Dinh, Detection of human immunodeficiency virus type 1 DNA sequence using plasmonics nanoprobes. *Analytical Chemistry*, 2005. **77**(23): pp. 7810–7815.

89. Wang, H.N. and T. Vo-Dinh, Multiplex detection of breast cancer biomarkers using plasmonic molecular sentinel nanoprobes. *Nanotechnology*, 2009. **20**(6): p. 065101.

90. Tyagi, S. and F.R. Kramer, Molecular beacons: Probes that fluoresce upon hybridization. *Nature Biotechnology*, 1996. **14**(3): pp. 303–308.

91. Wang, H.N., A.M. Fales, and T. Vo-Dinh, Plasmonics-based SERS nanobiosensor for homogeneous nucleic acid detection. *Nanomedicine*, 2015. **11**(4): pp. 811–814.

92. Wang, H.-N. et al., Mutiplexed detection of MicroRNA biomarkers using SERS-based inverse molecular sentinel (iMS) Nanoprobes. *The Journal of Physical Chemistry C*, 2016. **120**(37): pp. 21047–21055.

93. Wang, H.-N., B.M. Crawford, and T. Vo-Dinh, Molecular SERS nanoprobes for medical diagnostics. In T. Vo-Dinh (ed.), *Nanotechnology in Biology and Medicine: Methods, Devices, and Applications.* 2017, Boca Raton: CRC Press, p. 289.

94. Wang, H.-N. et al., Surface-enhanced Raman scattering nanosensors for in vivo detection of nucleic acid targets in a large animal model. *Nano Research*, 2018. **11**: pp. 1–12.

95. Khoury, C.G. and T. Vo-Dinh, Plasmonic nanowave substrates for SERS: Fabrication and numerical analysis. *The Journal of Physical Chemistry C*, 2012. **116**(13): pp. 7534–7545.

96. Ngo, H.T. et al., Label-free DNA biosensor based on SERS molecular sentinel on nanowave chip. *Analytical Chemistry*, 2013. **85**(13): pp. 6378–6383.

97. Ngo, H.T. et al., Multiplex detection of disease biomarkers using SERS molecular sentinel-on-chip. *Analytical and Bioanalytical Chemistry*, 2014. **406**(14): pp. 3335–3344.

98. Ngo, H.T. et al., DNA bioassay-on-chip using SERS detection for dengue diagnosis. *Analyst*, 2014. **139**(22): pp. 5655–5659.

99. Ngo, H.T. et al., Sensitive DNA detection and SNP discrimination using ultrabright SERS nanorattles and magnetic beads for malaria diagnostics. Biosensors and Bioelectronics, 2016. **81**: pp. 8–14.

100. Zhang, H. et al., Surface-enhanced Raman scattering detection of DNA derived from the west nile virus genome using magnetic capture of Raman-active gold nanoparticles. *Analytical Chemistry*, 2011. **83**(1): p. 254-260.

101. Donnelly, T. et al., Silver and magnetic nanoparticles for sensitive DNA detection by SERS, 2014. **50**(85): pp. 12907–12910.

102. Vohra, P. et al., Squamous cell carcinoma DNA detection using ultrabright SERS nanorattles and magnetic beads for head and neck cancer molecular diagnostics. *Analytical Methods*, 2017. **9**(37): p. 5550–5556.

103. Bertram, J.S., The molecular biology of cancer. *Molecular Aspects of Medicine*, 2000. **21**(6): pp. 167–223.

104. Sun, L., C. Yu, and J. Irudayaraj, Raman multiplexers for alternative gene splicing. *Analytical Chemistry*, 2008. **80**(9): pp. 3342–3349.

105. Wang, X. et al., Detection of circulating tumor cells in human peripheral blood using surface-enhanced Raman scattering nanoparticles. *Cancer Research*, 2011. **71**(5): pp. 1526–1532.

106. Wu, X. et al., Improved SERS nanoparticles for direct detection of circulating tumor cells in the blood. *ACS Applied Materials & Interfaces*, 2015. **7**(18): pp. 9965–9971.

107. Ye, S. et al., Asymmetric signal amplification for simultaneous SERS detection of multiple cancer markers with significantly different levels. *Analytical Chemistry*, 2015. **87**(16): pp. 8242–8249.

108. Falk, M. and R. Issels, Hyperthermia in oncology. *International Journal of Hyperthermia*, 2001. **17**(1): pp. 1–18.

109. Frey, B. et al., Old and new facts about hyperthermia-induced modulations of the immune system. *International Journal of Hyperthermia*, 2012. **28**(6): pp. 528–542.

110. Hildebrandt, B. et al., The cellular and molecular basis of hyperthermia. *Critical Reviews in Oncology/Hematology*, 2002. **43**(1): pp. 33–56.

111. Eifler, A.C. and C.S. Thaxton, Nanoparticle therapeutics: FDA approval, clinical trials, regulatory pathways, and case study. In S.J. Hurst (ed.), *Biomedical Nanotechnology. Methods in Molecular Biology (Methods and Protocols)*, 2011, Vol. 726. Humana Press, Totowa, NJ. pp. 325–338.

112. Loo, C. et al., Nanoshell-enabled photonics-based imaging and therapy of cancer. *Technology in Cancer Research & Treatment*, 2004. **3**(1): pp. 33–40.

113. Pandita, T.K., S. Pandita, and S.R. Bhaumik, Molecular parameters of hyperthermia for radiosensitization. *Critical Reviews in Eukaryotic Gene Expression*, 2009. **19**(3): pp. 235–251.

114. Takada, T. et al., Growth inhibition of re-challenge B16 melanoma transplant by conjugates of melanogenesis substrate and magnetite nanoparticles as the basis for developing melanoma-targeted chemo-thermo-immunotherapy. *Journal of Biomedicine and Biotechnology*, 2009. **2009**: p. 457936.

115. Koning, G.A. et al., Hyperthermia and thermosensitive liposomes for improved delivery of chemotherapeutic drugs to solid tumors. *Pharmaceutical Research*, 2010. **27**(8): p. 1750–1754.

116. Wang, C. et al., Immunological responses triggered by photothermal therapy with carbon nanotubes in combination with anti-CTLA-4 therapy to inhibit cancer metastasis. *Advanced Materials*, 2014. **26**(48): pp. 8154–8162.

117. Oldenburg, S. et al., Nanoengineering of optical resonances. *Chemical Physics Letters*, 1998. **288**(2–4): pp. 243–247.

118. Prodan, E. et al., A hybridization model for the plasmon response of complex nanostructures. *Science*, 2003. **302**(5644): pp. 419–422.

119. Hirsch, L.R. et al., Metal nanoshells. *Annals of Biomedical Engineering*, 2006. **34**(1): pp. 15–22.

120. Jain, P.K. and M.A. El-Sayed, Universal scaling of plasmon coupling in metal nanostructures: Extension from particle pairs to nanoshells. *Nano Letters*, 2007. **7**(9): pp. 2854–2858.

121. Hirsch, L.R. et al., Nanoshell-mediated near-infrared thermal therapy of tumors under magnetic resonance guidance. *Proceedings of the National Academy of Sciences*, 2003. **100**(23): pp. 13549–13554.

122. O'Neal, D.P. et al., Photo-thermal tumor ablation in mice using near infrared-absorbing nanoparticles. *Cancer Letters*, 2004. **209**(2): pp. 171–176.

123. Loo, C. et al., Gold nanoshell bioconjugates for molecular imaging in living cells. *Optics Letters*, 2005. **30**(9): pp. 1012–1014.

124. Lowery, A.R. et al., Immunonanoshells for targeted photothermal ablation of tumor cells. *International Journal of Nanomedicine*, 2006. **1**(2): pp. 149–154.

125. Gobin, A.M., J.J. Moon, and J.L. West, EphrinAl-targeted nanoshells for photothermal ablation of prostate cancer cells. *International Journal of Nanomedicine*, 2008. **3**(3): p. 351.

126. Stern, J.M. et al., Efficacy of laser-activated gold nanoshells in ablating prostate cancer cells in vitro. *Journal of Endourology*, 2007. **21**(8): pp. 939–943.

127. Bernardi, R.J. et al., Immunonanoshells for targeted photothermal ablation in medulloblastoma and glioma: An in vitro evaluation using human cell lines. *Journal of Neuro-Oncology*, 2008. **86**(2): pp. 165–172.

128. Liu, S.-Y. et al., In vitro photothermal study of gold nanoshells functionalized with small targeting peptides to liver cancer cells. *Journal of Materials Science: Materials in Medicine*, 2010. **21**(2): pp. 665–674.

129. Schwartz, J.A. et al., Feasibility study of particle-assisted laser ablation of brain tumors in orthotopic canine model. *Cancer Research*, 2009. **69**(4): pp. 1659–1667.

130. Choi, M.-R. et al., A cellular Trojan Horse for delivery of therapeutic nanoparticles into tumors. *Nano Letters*, 2007. **7**(12): pp. 3759–3765.

131. Pitsillides, C.M. et al., Selective cell targeting with light-absorbing microparticles and nanoparticles. *Biophysical Journal*, 2003. **84**(6): pp. 4023–4032.

132. Zharov, V.P., V. Galitovsky, and M. Viegas, Photothermal detection of local thermal effects during selective nanophotothermolysis. *Applied Physics Letters*, 2003. **83**(24): pp. 4897–4899.

133. Zharov, V.P., E. Galitovskaya, and M. Viegas. Photothermal guidance for selective photothermolysis with nanoparticles. In *Laser Interaction with Tissue and Cells Xv*. 2004, Vol. 5319. San Jose, CA: International Society for Optics and Photonics.

134. Zharov, V.P. et al., Synergistic enhancement of selective nanophotothermolysis with gold nanoclusters: Potential for cancer therapy. *Lasers in Surgery and Medicine*, 2005. **37**(3): pp. 219–226.

135. Zharov, V.P. et al., Self-assembling nanoclusters in living systems: Application for integrated photothermal nanodiagnostics and nanotherapy. *Nanomedicine: Nanotechnology, Biology and Medicine*, 2005. **1**(4): pp. 326–345.

136. Hleb, E.Y. et al., LANTCET: Elimination of solid tumor cells with photothermal bubbles generated around clusters of gold nanoparticles. *Nanomedicine*, 2008. **3**: pp. 647–667.

137. Huang, X. et al., The potential use of the enhanced nonlinear properties of gold nanospheres in photothermal cancer therapy. *Lasers in Surgery and Medicine*, 2007. **39**(9): pp. 747–753.

138. Lapotko, D. et al., Method of laser activated nano-thermolysis for elimination of tumor cells. *Cancer Letters*, 2006. **239**(1): pp. 36–45.

139. El-Sayed, I.H., X. Huang, and M.A. El-Sayed, Selective laser photo-thermal therapy of epithelial carcinoma using anti-EGFR antibody conjugated gold nanoparticles. *Cancer Letters*, 2006. **239**(1): pp. 129–135.

140. Huang, X. et al., Comparative study of photothermolysis of cancer cells with nuclear-targeted or cytoplasm-targeted gold nanospheres: Continuous wave or pulsed lasers. *Journal of Biomedical Optics*, 2010. **15**(5): p. 058002.

141. Kennedy, L.C. et al., T cells enhance gold nanoparticle delivery to tumors in vivo. *Nanoscale Research Letters*, 2011. **6**(1): p. 283.

142. Liu, M. and P. Guyot-Sionnest, Mechanism of silver (I)-assisted growth of gold nanorods and bipyramids. *The Journal of Physical Chemistry B*, 2005. **109**(47): pp. 22192–22200.

143. Nikoobakht, B. and M.A. El-Sayed, Preparation and growth mechanism of gold nanorods (NRs) using seed-mediated growth method. *Chemistry of Materials*, 2003. **15**(10): pp. 1957–1962.

144. Murphy, C.J. et al., Anisotropic metal nanoparticles: Synthesis, assembly, and optical applications. *J. Phys. Chem. B*, 2005. **109**: pp. 13857–13870.

145. Jain, P.K. et al., Calculated absorption and scattering properties of gold nanoparticles of different size, shape, and composition: Applications in biological imaging and biomedicine. *The Journal of Physical Chemistry B*, 2006. **110**(14): pp. 7238–7248.

146. Takahashi, H. et al., Gold nanorod-sensitized cell death: Microscopic observation of single living cells irradiated by pulsed near-infrared laser light in the presence of gold nanorods. *Chemistry Letters*, 2006. **35**(5): pp. 500–501.

147. Takahashi, H. et al., Photothermal reshaping of gold nanorods prevents further cell death. *Nanotechnology*, 2006. **17**(17): p. 4431.

148. Dickerson, E.B. et al., Gold nanorod assisted near-infrared plasmonic photothermal therapy (PPTT) of squamous cell carcinoma in mice. *Cancer Letters*, 2008. **269**(1): pp. 57–66.

149. Goodrich, G.P. et al., Photothermal therapy in a murine colon cancer model using near-infrared absorbing gold nanorods. *Journal of Biomedical Optics*, 2010. **15**(1): p. 018001.

150. Von Maltzahn, G. et al., Computationally guided photothermal tumor therapy using long-circulating gold nanorod antennas. *Cancer Research*, 2009. **69**(9): pp. 3892–3900.

151. Park, J.H. et al., Magnetic iron oxide nanoworms for tumor targeting and imaging. *Advanced Materials*, 2008. **20**(9): pp. 1630–1635.

152. Schwartzberg, A.M. et al., Synthesis, characterization, and tunable optical properties of hollow gold nanospheres. *The Journal of Physical Chemistry B*, 2006. **110**(40): pp. 19935–19944.

153. Melancon, M.P. et al., In vitro and in vivo targeting of hollow gold nanoshells directed at epidermal growth factor receptor for photothermal ablation therapy. *Molecular Cancer Therapeutics*, 2008. **7**(6): pp. 1730–1739.

154. Lu, W. et al., Targeted photothermal ablation of murine melanomas with melanocyte-stimulating hormone analog–conjugated hollow gold nanospheres. *Clinical Cancer Research*, 2009. **15**(3): pp. 876–886.

155. Kumar, R. et al., Hollow gold nanoparticles encapsulating horseradish peroxidase. *Biomaterials*, 2005. **26**(33): pp. 6743–6753.

156. Ke, H. et al., Gold nanoshelled liquid perfluorocarbon nanocapsules for combined dual modal ultrasound/CT imaging and photothermal therapy of cancer. *Small*, 2014. **10**(6): pp. 1220–1227.

157. Sun, Y., B.T. Mayers, and Y. Xia, Template-engaged replacement reaction: A one-step approach to the large-scale synthesis of metal nanostructures with hollow interiors. *Nano Letters*, 2002. **2**(5): pp. 481–485.

158. Chen, J. et al., Immuno gold nanocages with tailored optical properties for targeted photothermal destruction of cancer cells. *Nano Letters*, 2007. **7**(5): pp. 1318–1322.

159. Au, L. et al., A quantitative study on the photothermal effect of immuno gold nanocages targeted to breast cancer cells. *ACS Nano*, 2008. **2**(8): pp. 1645–1652.

160. Chen, J. et al., Gold nanocages as photothermal transducers for cancer treatment. *Small*, 2010. **6**(7): pp. 811–817.

161. Cai, W. et al., Applications of gold nanoparticles in cancer nanotechnology. *Nanotechnology, Science and Applications*, 2008. **1**: p. 17.

162. Mendes, R., A. Fernandes, and P. Baptista, Gold nanoparticle approach to the selective delivery of gene silencing in cancer—The case for combined delivery? *Genes*, 2017. **8**(3): p. 94.

163. Timko, B.P. et al., Near-infrared–actuated devices for remotely controlled drug delivery. *Proceedings of the National Academy of Sciences*, 2014. **111**(4): pp. 1349–1354.

164. Shi, P. et al., Near–infrared light–encoded orthogonally triggered and logical intracellular release using gold nanocage@ smart polymer shell. *Advanced Functional Materials*, 2014. **24**(6): pp. 826–834.

165. Schmaljohann, D., Thermo-and pH-responsive polymers in drug delivery. *Advanced Drug Delivery Reviews*, 2006. **58**(15): pp. 1655–1670.

166. Yavuz, M.S. et al., Gold nanocages covered by smart polymers for controlled release with near-infrared light. *Nature Materials*, 2009. **8**(12): p. 935.

167. You, J. et al., Photothermal-chemotherapy with doxorubicin-loaded hollow gold nanospheres: A platform for near-infrared light-trigged drug release. *Journal of Controlled Release*, 2012. **158**(2): pp. 319–328.

168. Casares, N. et al., Caspase-dependent immunogenicity of doxorubicin-induced tumor cell death. *Journal of Experimental Medicine*, 2005. **202**(12): pp. 1691–1701.

169. Srivastava, A. et al., Nanosomes carrying doxorubicin exhibit potent anticancer activity against human lung cancer cells. *Scientific Reports*, 2016. **6**: p. 38541.

170. Fire, A. et al., Potent and specific genetic interference by double-stranded RNA in *C. elegans*. *Nature*, 1998. **391**: pp. 806–811.

171. Choung S., et al., Chemical modification of siRNAs to improve serum stability without loss of efficacy. *Biochemical and Biophysical Research Communications*, 2006. **342**(3): pp. 919–927.

172. Sardo, C. et al., Gold nanostar–polymer hybrids for siRNA delivery: Polymer design towards colloidal stability and in vitro studies on breast cancer cells. *International Journal of Pharmaceutics*, 2017. **519**(1–2): pp. 113–124.

173. Wu, G. et al., Remotely triggered liposome release by near-infrared light absorption via hollow gold nanoshells. *Journal of the American Chemical Society*, 2008. **130**(26): pp. 8175–8177.

174. Huang, X. et al., Modular plasmonic nanocarriers for efficient and targeted delivery of cancer-therapeutic siRNA. *Nano Letters*, 2014. **14**(4): pp. 2046–2051.

175. Braun, G.B. et al., Laser-activated gene silencing via gold nanoshell-siRNA conjugates. *ACS Nano*, 2009. **3**(7): pp. 2007–2015.

176. Zharov, V., R. Letfullin, and E. Galitovskaya, Microbubbles-overlapping mode for laser killing of cancer cells with absorbing nanoparticle clusters. *Journal of Physics D: Applied Physics*, 2005. **38**(15): p. 2571.

177. Lapotko, D.O., E. Lukianova, and A.A. Oraevsky, Selective laser nano-thermolysis of human leukemia cells with microbubbles generated around clusters of gold nanoparticles. *Lasers in Surgery and Medicine*, 2006. **38**(6): pp. 631–642.

178. Barhoumi, A. et al., Light-induced release of DNA from plasmon-resonant nanoparticles: Towards light-controlled gene therapy. *Chemical Physics Letters*, 2009. **482**(4–6): pp. 171–179.

179. Huschka, R. et al., Gene silencing by gold nanoshell-mediated delivery and laser-triggered release of antisense oligonucleotide and siRNA. *ACS Nano*, 2012. **6**(9): pp. 7681–7691.

180. Huschka, R. et al., Visualizing light-triggered release of molecules inside living cells. *Nano Letters*, 2010. **10**(10): pp. 4117–4122.

181. Castano, A.P., T.N. Demidova, and M.R. Hamblin, Mechanisms in photodynamic therapy: Part one-photosensitizers, photochemistry and cellular localization. *Photodiagnosis and Photodynamic Therapy*, 2004. **1**(4): pp. 279–293.

182. Rossi, L.M. et al., Protoporphyrin IX nanoparticle carrier: Preparation, optical properties, and singlet oxygen generation. *Langmuir*, 2008. **24**(21): pp. 12534–12538.

183. Hone, D.C. et al., Generation of cytotoxic singlet oxygen via phthalocyanine-stabilized gold nanoparticles: A potential delivery vehicle for photodynamic therapy. *Langmuir*, 2002. **18**(8): pp. 2985–2987.

184. Camerin, M. et al., The in vivo efficacy of phthalocyanine–nanoparticle conjugates for the photodynamic therapy of amelanotic melanoma. *European Journal of Cancer*, 2010. **46**(10): pp. 1910–1918.

185. Stuchinskaya, T. et al., Targeted photodynamic therapy of breast cancer cells using antibody–phthalocyanine–gold nanoparticle conjugates. *Photochemical & Photobiological Sciences*, 2011. **10**(5): pp. 822–831.

186. Fales, A.M., H. Yuan, and T. Vo-Dinh, Cell-penetrating peptide enhanced intracellular Raman imaging and photodynamic therapy. *Molecular Pharmaceutics*, 2013. **10**(6): pp. 2291–2298.

187. Shi, Z. et al., Stability enhanced polyelectrolyte-coated gold nanorod-photosensitizer complexes for high/low power density photodynamic therapy. *Biomaterials*, 2014. **35**(25): pp. 7058–7067.

188. Raviraj, V. et al., First demonstration of gold nanorods-mediated photodynamic therapeutic destruction of tumors via near infra-red light activation. *Small*, 2014. **10**(8): pp. 1612–1622.

189. Kumar, A. et al., Oxidative nanopeeling chemistry-based synthesis and photodynamic and photothermal therapeutic applications of plasmonic core-petal nanostructures. *Journal of the American Chemical Society*, 2014. **136**(46): pp. 16317–16325.

190. Jang, B. et al., Gold nanorod-photosensitizer complex for near-infrared fluorescence imaging and photodynamic/photothermal therapy in vivo. *ACS Nano*, 2011. **5**(2): pp. 1086–1094.

191. Khlebtsov, B. et al., Nanocomposites containing silica-coated gold–silver nanocages and Yb–2, 4-Dimethoxyhematoporphyrin: Multifunctional capability of IR-luminescence detection, photosensitization, and photothermolysis. *ACS Nano*, 2011. **5**(9): pp. 7077–7089.

192. Gao, L. et al., Hypocrellin-loaded gold nanocages with high two-photon efficiency for photothermal/photodynamic cancer therapy in vitro. *ACS Nano*, 2012. **6**(9): pp. 8030–8040.

193. Wang, S. et al., Single continuous wave laser induced photodynamic/plasmonic photothermal therapy using photosensitizer-functionalized gold nanostars. *Advanced Materials*, 2013. **25**(22): pp. 3055–3061.

194. Zhou, F.F. et al., Antitumor immunologically modified carbon nanotubes for photothermal therapy. *Biomaterials*, 2012. **33**(11): pp. 3235–3242.

195. Liu, Y. et al., Synergistic Immuno Photothermal Nanotherapy (SYMPHONY) for the treatment of unresectable and metastatic cancers. *Scientific Reports*, 2017. **7**(1): p. 8606.

196. Vo-Dinh, T. and B.A. Inman, What potential does plasmonics-amplified synergistic immuno photothermal nanotherapy have for treatment of cancer? *Nanomedicine*, 2018. **13**: pp. 139–144

197. Chowdhury, F. et al., PD-L1 and CD8+ PD1+ lymphocytes exist as targets in the pediatric tumor microenvironment for immunomodulatory therapy. *Oncoimmunology*, 2015. **4**(10): p. e1029701.

198. Bertucci, F. and A. Gonalves, Immunotherapy in breast cancer: The emerging role of PD-1 and PD-L1. *Current Oncology Reports*, 2017. **19**(10): p. 64.

199. Chen, Q. et al., Photothermal therapy with immune-adjuvant nanoparticles together with checkpoint blockade for effective cancer immunotherapy. *Nature Communications*, 2016. **7**: p. 13193.

200. Palekar-Shanbhag, P. et al., Theranostics for cancer therapy. *Current Drug Delivery*, 2013. **10**(3): pp. 357–362.

201. Seo, S.-H. et al., NIR-light-induced surface-enhanced Raman scattering for detection and photothermal/photodynamic therapy of cancer cells using methylene blue-embedded gold nanorod@ SiO$_2$ nanocomposites. *Biomaterials*, 2014. **35**(10): pp. 3309–3318.

202. Li, H. et al., Combination of active targeting, enzyme-triggered release and fluorescent dye into gold nanoclusters for endomicroscopy-guided photothermal/photodynamic therapy to pancreatic ductal adenocarcinoma. *Biomaterials*, 2017. **139**: pp. 30–38.

203. Fales, A.M., B.M. Crawford, and T. Vo-Dinh, Folate receptor-targeted theranostic nanoconstruct for surface-enhanced Raman scattering imaging and photodynamic therapy. *ACS Omega*, 2016. **1**(4): pp. 730–735.

204. Fales, A.M. and T. Vo-Dinh, Silver embedded nanostars for SERS with internal reference (SENSIR). *Journal of Materials Chemistry C*, 2015. **3**(28): pp. 7319–7324.

205. Huang, H. et al., pH-sensitive Au–BSA–DOX–FA nanocomposites for combined CT imaging and targeted drug delivery. *International Journal of Nanomedicine*, 2017. **12**: p. 2829.

206. Li, C. et al., Gold-coated Fe$_3$O$_4$ nanoroses with five unique functions for cancer cell targeting, imaging and therapy. *Advanced Functional Materials*, 2014. **24**(12): pp. 1772–1780.

207. Hu, Y. et al., Multifunctional Fe$_3$O$_4$@ Au core/shell nanostars: A unique platform for multimode imaging and photothermal therapy of tumors. *Scientific Reports*, 2016. **6**: p. 28325.

Nanotechnology in Cell Delivery Systems

Ali Golchin
Urmia University of Medical Sciences

Parisa Kangari
Shiraz University of Medical Sciences

Sepideh Mousazadehe
Iran University of Medical Sciences

Faeza Moradi
Tarbiat Modares University

Simzar Hosseinzadeh
Shahid Beheshti University of Medical Sciences

19.1 Introduction

Cell-based therapy is one of the regenerative medicine branches that consists of using living cells as biological medicine for therapeutic purposes. A wide range of cells including stem cells, somatic cells, engineered cells and their production like exosomes (Golchin, Farahany, et al. 2019). from different sources can be used for managing incurable diseases (Golchin, Farahany, et al. 2018). Recently, cell-based therapy has focused on stem cell therapy as a strong tool for repair of damaged tissues. However, one of the main limitations of cell-based therapeutic approaches is the constraint of cell delivery systems (CDSs). As mentioned, cell therapy based on stem cells is an emerging area of research in regenerative medicine science, and significant advances have been reported for many patients living with serious and currently incurable diseases, including cancer, Alzheimer, diabetes, cardiac disease, bone and wound repair (Golchin et al. 2017; C. Zhao et al. 2013; Gao et al. 2013). Therefore, stem cell-based therapy creates new approaches in regeneration of tissue/organs. Totally, there are three kinds of stem cell sources: (i) pluripotent embryonic stem (ES) cells, (ii) Adult stem cells and (iii) induced pluripotent stem (iPS) cells (Kögler et al. 2004; International Stem Cell Initiative et al. 2007; Wei et al. 2013). The generation of iPSCs is a novel personalized-regenerative medicine technology, which can transform autologous somatic cells into pluripotent stem cells and differentiate into specific-lineage cells (Chen et al. 2013; Shah et al. 2013). Thus, these cells have enormous potential in the field of regenerative medicine. Due to the limitation of using ES and iPSCs in the clinic, great interest has developed in Adult stem cells including mesenchymal stem cells (MSCs), hematopoietic stem cells (HSCs) and other sources (International Stem Cell Initiative et al. 2007). To regenerate new tissues, stem cells should be delivered in a three-dimensional (3D) biodegradable porous matrices, named "scaffold", to organize the cells into a higher ordered assembly so as to achieve selected tissue function (C. Zhao et al. 2013). Presently, ~6,800 clinical trials around the world involve some form of cell therapy, including therapies for cancer, cardiac disease, stroke, diabetes, and bone repair (https://clinicaltrials.gov). However, cell-based therapy has several challenges such as inadequate cell sources, immunogenicity and teratogenicity. Therefore, studies have focused on improvement of cell proliferation, differentiation, cell imaging, tracking and delivery in vitro or in vivo along mentioned limitations.

Nanotechnology is a multidisciplinary field that provides a suitable situation for medical researchers to manipulate biomaterials, tissues, cells, drugs and nucleic acids for use in different medical applications. Therefore, nanomaterials, with a size range of 1–1,000 nm (or smaller than one-tenth of a micrometer), can be used for many applications in cell-based therapy especially CDSs (Golchin et al. 2017). Nanostructures or nanomaterials in biomedicine are defined as nanostructured materials that are applied to construct a designed system to interface with cells or tissue components in the efficacy of their function. These materials include any nanoscaled materials or devices such as nanoparticles (NPs), nanofibers, nanocrystals, nanotubes and nanowires. Nanomaterials can be used for many applications in cell-based therapy, in particular, CDSs; cellular machinery to facilitate the delivery of drugs; DNA; small molecular, signaling factors; functionalizing cells, targeting, imaging and monitoring injected cells, controlling the release of drugs and growth factors (Wimpenny et al. 2012; Golchin et al. 2017).

This chapter, in light of the above introduction, describes the use of nanotechnology and nanomaterials in cell-based therapy and CDSs in particular.

19.2 Nanoscaled Materials

Nanostructures are nano-microscopic materials that have one dimension less than 100 nm at least and are confined to the nanoscale in all three dimensions (Laurent et al. 2008). These nanostructure materials (NSMs) classify based on dimension as zero-dimensional (0D), one-dimensional (1D), two-dimensional (2D) and three-dimensional (3D) (Tiwari et al. 2012). The most commonly used nanomaterials in stem cell-based research are organic and inorganic NPs, liposomes, polyplexes and carbon nanotubes (CNTs) (Mocan et al. 2012; Martino et al. 2012). However, different biomaterials can be utilized as nanomaterials in cellular therapy (CT) projects and especially CDSs include natural materials such as collagen, chitosan and synthetic materials such as polylactide-co-glycolide (PLGA), poly-caprolactone (PCL) (Bamrungsap et al. 2012; Golchin et al. 2017), Poly(METH)acrylates and polyacrylamides, poly(N-isopropyl acrylamide), oligo(poly PEG fumarate), polyamido amine (PAMAM), polyvinyl alcohol (PVA), PLGA, poly(l-lysine) (PLL), and PEG poly(ethylene glycol) (Golchin et al. 2017).

NPs are divided into various categories depending on their morphology, size and chemical properties including carbon-based NPs such as fullerenes and CNTs, metal NPs, ceramics NPs, semiconductor NPs, polymeric NPs and lipid-based NPs. NPs have been used for various applications because of different physical and chemical properties such as large surface area, mechanically strong, optically active and chemically reactive (Khan et al. 2017). One of the most applicable fields of NPs is CbT field. These nanoscale particles are used for the delivery of chemical and biological molecules such as drugs, proteins, enzymes and antibodies to injured sites in to the body for different disease treatment (McMillan et al. 2011). According to the type of cells and aim considered, options of selected nanomaterials and NPs can vary. For instance, human neural stem cells (hNSCs) are exposed to five types of NPs in vitro, containing three silica particles with a diameter of 30 nm, 70 nm and < 44 μm (SP30, SP70 and SPM) and two titanium oxide particles 80 nm and <44 μm in diameter (TP80 and TPM) and subsequently show especial behavior (Zhang et al. 2018). Moreover, significant advancements are done for obtaining new carriers based on NPs in CDSs in recent years. NPs' incorporation within CDSs provides suitable spatial for cell biological function and controlled release of growth factors and cytokines from cells that help cell tissue-based therapy fields (Huang & Fu 2010).

19.3 Nanomaterials and Cell Delivery Systems

That's where combined nanotechnology with cell-based therapy can promote some of the limitations significantly. Nanomaterials that regulate cell adhesion, proliferation and differentiation can be used for targeting transplanted cells' migration and delivery to localized target injured sites (Zhang et al. 2018). Nanomaterials as a defined media have several advantages of facilitating cell viability and differentiation into specific lineages during the process of cell-based therapy approaches. For instance, both silica and titanium NPs successfully promote NSC differentiation in human and mouse NSC lines (Fujioka et al. 2014; Liu et al. 2010). Moreover, to better mimic the natural extre-cellular matrix (ECM) nanostructure, scaffolds produce nanomaterials such as nanofibers, nanotubes, NPs and hydrogel to investigate cells' fate manipulation (C. Zhao et al. 2013) that resemble the ECM and efficiently replace defective tissues (Gong et al. 2015).

Nanomaterials have better functional, mechanical, physical and customizable features along superior biocompatibility compared to macro-scaled materials. Hence, nanomaterials have attracted more attention to cell-based therapy and particular CDSs because they can intimately mimic natural biological environments. So these materials can be applied to construct nanostructures and other properties of native tissues. Based on nanomaterial properties, using them in CDSs offers many advantages; however, there are a number of disadvantages too, which should not be ignored. Nanomaterial properties that make them suitable candidate in this regard are included in Golchin et al. 2017:

- Nanotechnological advancements of nanostructure fabrication including surface modifications and controlling their sizes lead to their feasible applications in desired characterizations. Nanomaterials can provide larger functional areas for cells in CDSs by reducing the waste physical interaction.

- Nanomaterials are suitable conductors for the delivery of bioactive molecules and different cells spontaneously which is an important factor for the simultaneous interaction between them.

- Nanomaterials are capable of the precise release of therapeutic agents to the different cells in a defined period that is vital for the tissue engineering approaches.

- Several nanomaterials have become standardized products that improve imaging for monitoring of treatment processes and validation of cell therapeutic efficacy.

- Nanomaterials can induce the combination therapy to improve comprehensive therapeutic approach, such as gene therapy with stem cell therapy.

- Nanomaterial-based scaffold is capable of extinguishing free radicals within the lesion tissue, thereby decreasing the oxidative stress and increasing survival and engraftment of cells in the high oxidative stress environments.
- Nanostructures can improve the stability of cells and protect them from various natural destructive factors of damaged tissue like enzymatic degradation.

The aim of a delivery system is to be able to transplant the right amount of cells to the desired site and attain the maximal cell. For instance, the fabrication of the 3D polymeric scaffolds coated with heparin/PLL NP provides an ideal matrix for cell adhesion and growth, which causes cell delivery improvement. The unique properties of NPs, including their biocompatibility, uniform distribution on a substrate and chemical or biological action in cells, may help to solve some of the problems thus far associated with tissue regeneration (Kun Na et al. 2007).

Polycation–DNA complexes and CNTs have been employed for such applications. The polycations are typically polymers such as polyethyleneimine (PEI) and chitosan, but functionalized CNTs have also been employed recently. PEI as poly cation forms a complex with DNA and facilitates the entry of DNA into the host cells for transfection. Also, when cells were delivered by a matrix containing nanoparticulate DNA vector due to transfection with appropriate growth factors, the growth factor can be produced in a more prolonged and sustained manner (Wan & Ying 2010).

Nanoporous and nanofibrous polymer matrices can be constructed by electrospinning, phase separation, particulate leaching, chemical etching and 3D printing methods (Gong et al. 2015). Nanomaterials have unique advantages in controlling cell function and CDSs that provide informative microenvironments mimicking a physiological niche (Mocan et al. 2012; Martino et al. 2012). Biomaterials, based on their composition and structures, transfer specific signals to cells that decode into biochemical signals. Hence, topography, chemistry and physical characteristics of nanomaterials are critical factors for directing stem cell differentiation (Mocan et al. 2012). In some instances, these interactions lead to toxicity, which can be of serious concern. Particularly, toxic responses have been released by the degradation of injected nanomaterials, due to the wear debris with nanofeatures and heavy metals (e.g. iron, nickel, and cobalt catalysts) (Gong et al. 2015).

19.4 Nanomaterial-Cell Delivery Systems for Therapeutic Application

Bone and joint injuries are one of the important candidates for using nanoscience in cell tissue-based therapeutic approaches. Therefore, researchers are emerging cells and nanomaterials as one of the best studied applications in this field. The proposed approach is isolation and expandation of stem cells from the patients and their seeding on suitable scaffolds (Mocan et al. 2012). Engineered tissues are implanted into the injured area to regenerate the defective bone/cartilage as the scaffold degrades. One of the most important stages of engineering is to find a suitable material to produce an applicable scaffold for the selected tissue. Nanophase ceramics, especially nanohydroxyapatite (nanoHA), are popular bone substitutes due to their ability to promote mineralization (Gong et al. 2015). Nanoceramic scaffolds rise osteoblast functions, such as adhesion, proliferation and differentiation. These properties are achieved because of nanometer grain sizes and high surface fraction of grain boundaries in nanoceramics. Similar capacity has been shown for other nanoceramics including alumina, zinc oxide and titania (Gong et al. 2015). To stimulate stem cells to the osteogenic/chondrogenic lineage differentiation, they are usually exposed to signaling molecules (e.g. growth factors and other osteoinductive/chondroinductive molecules) during the in vitro culture period (Qi et al. 2013). Therefore, osteogenic or chondrogenic differentiation of stem cells can be achieved by intracellular delivery of stimulating factors mediated by nanomaterials (Mocan et al. 2012). Another silico-phosphate chain nanomaterials are bioactive glasses and they are used in the therapy of periodontal bone injuries. In addition, electrospun PCL nanofiber meshes, electrospun PCL and PLGA nanofiber scaffolds have been established to support MSCs' growth into chondrogenic differentiation in vitro (Qi et al. 2013; C. Zhao et al. 2013). The Food and Drug Administration (FDA) approved all of these nanomaterials.

Neural tissue defects are another candidate for CbT and CDSs (C. Zhao et al. 2013; Martino et al. 2012). Therefore, cell-based therapy could represent a promise for various neurological disorders (Tukmachev et al. 2015). Recently, nanomaterials have been applied in the neurological field for tracking and nervous system diseases' treatments with a wide improvement (Tukmachev et al. 2015). Some synthetic and natural biopolymers can generate aligned support, such as poly (L-lactide) (PLLA), PLGA, PCL, collagen and chitosan, to produce a neural guidance conduit for neuronal repair and regeneration in neural injuries (C. Zhao et al. 2013; Martino et al. 2012). The combination of stem cells and nanomaterials has huge effects on clinical diagnosis and regeneration of multiple neurological disorders (Zhang et al. 2018). Additionally, nanomaterials can be used for gene/drug delivery in cell-based therapeutic applications for numerous central nervous system diseases, such as spinal cord injury (SCI), multiple sclerosis (MS) and Parkinson's and Alzheimer's diseases (Zhang et al. 2018). Currently, as the best medicine, only prevention is identified to be restorative therapy of SCI (Tukmachev et al. 2015). The magnetic system was used to add stem cells labeled with superparamagnetic iron oxide nanoparticles (SPION) and the loading of stem cells that guarantees sufficient attractive magnetic forces was completed to SCI therapies (Tukmachev et al. 2015). Further, the magnetic system rapidly directed the SPION-labeled cells exactly to the injury site (Tukmachev et al. 2015). Definitely, magnetic forces have already been effectively approved in directing SPION-labeled cells to a vascular graft, the femoral artery, bone, cartilage and the retina (Tukmachev et al. 2015). However, NPs have potential cytotoxic effects on nerve cells (Zhang et al. 2018).

Besides the enormous development detailed above, the combination of stem cells and nanomaterials also brings powerful tools to regenerate other defected tissues such as skin, eye, muscles, cardiovascular system, bladder and tendon. A study has shown that hMSCs grow up significantly on collagen/PLLCL nanofibers than on PLLCL scaffolds using proliferation assay (Jin et al. 2011). Several recent studies have been developed in the ophthalmic regeneration of tissue based on stem cells and nanomaterials and provide new strategies. One of them showed that murine retinal progenitor cells (RPCs) grown on chitosan graft PCL/plain PCL (20/80) scaffolds have more similarity to differentiate into retinal neurons than those grown on PCL scaffolds (Gu et al. 2011). McKeon-Fischer et al. created three composite scaffolds made of PLLA and electrospun gold (Au) NPs; they were 7% Au-PLLA, 13% Au-PLLA and 21% Au-PLLA (Gong et al. 2015). Others are trying to develop novel nano-biomaterials for muscle repair using electrospun chitosan microfibers (Gong et al. 2015). So far, chitosan microfibers have been shown to support the attachment and viability of rat muscle-derived SCs and their subcutaneous implantation (Gong et al. 2015). For bladder regeneration, a highly porous 3D PLLA scaffold was fabricated to provide a framework to regenerate smooth muscle cells with contractile function (C. Zhao et al. 2013). Another demonstrated platform was an aligned electrospun PLLA nanofibrous scaffold to provide an informative microenvironment for human tendon stem cells to differentiate into adult tendon from the corroboration of good expression of tendon specific markers (C. Zhao et al. 2013). Moreover, cardiovascular diseases are the biggest cause of mortality in developed countries and nanotechnology helps in its tissue engineering studies.

19.5 Nanomaterials Structure and Cell Delivery

Since many of cell interactions and its regulation signaling can be adjusted at a natural nanoscale level, using nanomaterials in CDSs can provide conditions similar to the natural environment of cells. In addition, self-assembling networks that can degrade within a controlled time scale have great potential for cell delivery with maximal functions in cells and minimal invasion which nanomaterials showed suitable behavior in this regard (Y.-L. Wang et al. 2016; Barros et al. 2015). There are several methods to activate the surface of nanostructure post-fabrication such as grafting cationic polymers onto electrospun nanofibers,

chemical etching, plasma treatment and ultraviolet (UV) irradiation (SULLENGER et al. 2015; Park et al. 2005; Zhang et al. 2012). As mentioned, nanomaterials (NSMs) are classified in five groups based on dimensions including zero-dimensional (0D), one-dimensional (1D), two-dimensional (2D) and three-dimensional (3D) (25) that among these 2D and 3D structures are conventionally used in the cell tissue-based therapy field. Nanomaterials in 3D structure can improve cell viability, proliferation, differentiation and maintenance of different cells in CDSs (Vasita & Katti 2006) which due to this advantage 3D nanotechnological fabrication methods has attracted more attention. Generally, the structural properties of nanomaterials forcefully depend on their atomic arrangements. However, 2D nanomembranes are different compared to 3D nanostructures because one of their dimensions is formed from only a few atomic layers thick (Nicolosi et al. 2013). Moreover, fabrication of nanoscale structures could be a suitable approach to reach a self-organized new tissue by preparing cellular bridge formation (Golchin et al. 2017). In summary, 3D nanostructures have protective support, functioning in the biological condition, maintenance of tissue shape, forming a suitable defensive shield to protect vital organs (Shin et al. 2006). Thereby, constructs that fabricated via nanotechnological methods and nanomaterials can be functionalized before application and improve their biological behaviors. In the following, introduce and discuss the different structures of nanomaterials that are used in CDSs.

19.6 Nanomembrane Structure

Cellular membranes are a good pattern for CDSs that these biomimetic CDSs whose morphology, surface properties and size imitate natural tissue possess special functions to enhance delivery and cell homing in target tissues (Zhang et al. 2017). However, investigation on 2D nanomaterials field is still in its infancy and only a few types of 2D nanomaterials have been evaluated for cyto- and biocompatibility, but recent rapid progressions in 2D nanomaterials have shown enhanced physicochemical and biological functionality of 2D nanomembranes along their uniform shapes, surface charge, and high surface-to-volume ratios (Chimene et al. 2015). Two-dimensional nanomembrane could potentially lead to 3D mesostructures with excellent properties (Chimene et al. 2015). Fabrication of nanomembrane is exploited by different nanomaterials. These nanomaterials are classified into two groups which include nanomaterials producing single monolayer such as graphene, silicene, germanene, hexagonal boron nitride (hBN) and graphitic carbon nitride (C3N4), and nanomaterials such as 2D clays, LDHs, TMOs and TMDs that are composed of stable, single crystal units and produce two or multilayer structures (Chimene et al. 2015). For fabricating high-quality and ultrathin nanosheets showing their own merits and demerits in preparing 2D nanomaterials, several techniques have been developed such as micromechanical exfoliation,

Electrospun Nanofiber Scaffold

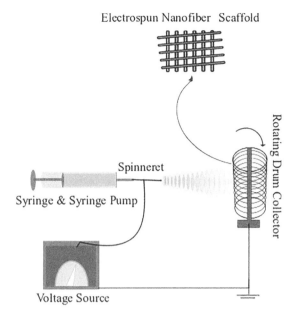

FIGURE 19.1 Schematic representation of electrospinning technique for fabrication of nanofibrous structures.

ultrasonic exfoliation, solution-based molecular assembly techniques, hydrothermal method, topochemical transformation and the chemical vapor deposition method (Ma & Sasaki 2015; J. Wang et al. 2016). However, among different methods, electrospinning has attracted attention for cell delivery applications (see Figure 19.1) (Moradi et al. 2018). Electrospinning is known as a common method for fabricating both 2D and 3D nanostructures. Hence, nanofibrous membranes have drawn a great deal of attention due to their significant advantages. For instance, PCL membrane has shown suitable features for design of CDSs (He et al. 2013). Membranes are well manipulated for different types of investigations, including the antibacterial usage (Yao et al. 2008) and cell fate effects (Yoo & Kim 2009). In summary, the biomedical applications of nanomembrane are a rapidly advancing and promising development in the regenerative medicine field particularly CDSs for applications in tissue engineering, stem cell research, and wound dressings.

19.7 Nanoporous Structure

Porous biomaterials have been widely used as scaffolds in cell-based therapies of several diseases. The porous scaffolds can be passive and active. Passive porous scaffolds, as 3D structures, deliver cells mostly through mechanisms including molecular diffusion, material degradation and cell migration, so that they do not allow for dynamic external regulations. In contrast, active porous scaffolds possess the ability to deliver biological agents under the controls of external motivation, which can be achieved by appropriate arrangement of porous biomaterials (Zhao et al. 2011).

The porous scaffolds due to a high surface-to-volume ratio provide interested structure for improving cell adhesion,

cell viability and cell engraftment. Scaffold structure affects cell behavior. These scaffolds facilitate cell movement into desired site, transporting nutrients and waste (Smith et al. 2009). A porous surface facilitates connection between scaffold and surrounding tissues. Porous scaffolds typically have pore sizes between 50 and 100 μm or larger which allows for cell growth and vascular network development (Hatton 2013). The pore size varies for different types of cells and affects cell function and migration. In addition, the porosity and pore size affect mechanical properties, so that the large pore size and high pore density lead to a reduction in the mechanical strength of the scaffold. Therefore, pore size, porosity, pore density, and interconnectivity are important factors for the ideal porous scaffolding development that needs to be considered (Smith et al. 2009).

Nanoporous materials as a subset of 3D nanostructures with pore diameter between 1 and 100 nm possess specially a high surface-to-volume ratio. These materials have unique properties including high surface area, large porosity and very ordered and uniform structure (Lu & Zhao 2004). Nanoporous scaffolds can be made from organic (polymers), inorganic (e.g. ceramic and metal) and composite materials. Anodic alumina is an inorganic nanoporous membrane having well-ordered pores. Some of the typical polymeric nanoporous materials are polycarbonate (PC), polyethylene terephthalate (PET) and polysulfide. There are many techniques available to fabricate nanoporous scaffolds including electrochemical leaching, lithography, focused ion beam (FIB) etching, ion-track etching, phase separation technique, and sol-gel process (Adiga et al. 2009). The basic technology for the nanopores themselves was based on the use of a sacrificial oxide layer, sandwiched between two structural layers, for reaching a desired pore size (Desai et al. 1998; Desai et al. 2000). Nanoporous membranes can be classified based on the pore size as microporous (<2 nm), mesoporous (2–50 nm) and microporous (>50 nm). Nanoporous scaffolds due to some unique properties are interested in biomedical as one of the most applicable devices. They have suitable pore size and distribution that can exactly control molecular transport. Also, these materials have low membrane thickness and high porosity that is ideal property for high flux. On the other hand, adequate mechanical strength, sufficient thermal and chemical stability under a wide range of biological environment and finally biocompatibility and resistance to biofouling are important characteristics for nanoporous scaffolds which makes them a suitable device for in vivo applications. It is noteworthy that these characteristics prevent immune responses. Nanoporous structures are applicable in separation and sorting macromolecules such as DNA, RNA and proteins for single-molecule analysis (Adiga et al. 2009). Nanopore size membranes play an important role in the formation of collagen fibers and ECM, also these membranes provide organized smooth surface that improve cell attachment and cell functioning (Bružauskaitė et al. 2016). The studies showed that the pore size of nanoporous membranes allows small molecules such as oxygen, glucose and insulin to pass but prevents the passage of much larger immune system molecules such as immunoglobulin. Accordingly, these membranes act as immune-isolation shield, which protects transplanted cells from the body's immune system (Nyitray et al. 2015). Zhanga et al. demonstrated that a dual nanoporous capsule can provide a suitable 3D microenvironment for cell growth, immune-protection, fibrosis prevention and controllable release of secreted product in a biodegradable device (Zhang et al. 2008). Immune-protection of transplanted cells by semipermeable scaffolds is an attractive approach for cell-based therapy to treat patients with diseases such as Alzheimer's, diabetes mellitus, hepatic disease, amyotrophic lateral sclerosis, hemophilia, hypothyroidism and Parkinson's disease (Colton 1995; Desai et al. 2000; Lanza et al. 2000; Aebischer et al. 1988). Chang et al. investigated a polymeric nanoporous immune-protective cell encapsulation device (NID) through its ability to immunoprotect and sustain the function and viability of human stem cell-derived β-cells both in vitro and in vivo. They reported that this scaffold is capable of excluding immune molecules while allowing exchange of oxygen and nutrients necessary for in vitro and in vivo stem cell viability and function. On other hand, the device promotes neovascular formation and cells' immune-protection effects versus immune system in vivo. This study indicated engraftment, viability and function of cells encapsulated in nanoporous scaffolds after several months (Chang et al. 2017). Nanoporous scaffolds are due to some effects on cell growth and development. There are various examples showing the effects of nanoscale surface topologies on cell growth, particularly in the field of bone implant surface engineering and functional self-assembling peptide nanofiber hydrogel for peripheral nerve regeneration (Hatton 2013).

Nanoporous scaffolds affect cell-surface interaction, cell attachment, cell growth and proliferation, and cell functions as well. Altogether, pore size is one of the most critical issues for the successful scaffold application in cell therapy and tissue regeneration.

Nowadays, production of nanoscaffolds can be realized via numerous applications such as:

Salt leaching: In this method, one must place porogen or salt crystals (e.g. sodium chloride) on a mold and a polymer must then be augmented to fill up what is left of the gaps. The polymer is then hardened and the salt is extracted via dissolution in a solvent. Once all the salt is withdrawn, a porous polymer is what remains (Loh & Choong 2013). The conspicuous benefit of this method is its economy and low amounts of polymer used. However, the inter-pore openings and pore shape of scaffolds that are produced afterwards are not so easily controlled (Ma 2004).

Phase separation: A polymer should first be dissolved in a reliable solvent, and subsequently, it should be relocated in a solvent until it is freezes. The solvent is drawn out through freeze-drying, leaving nothing but polymer pores. There is no requisite for an extra leaching step, but adding organic solvents like ethanol or methanol restricts

the amalgamation of biologically active molecules or cells in the midst of scaffold generation.

3D printing: Three-dimensional printing involves the laying down of successive layers of material to form 3D models, thus enabling better control of pore sizes, pore morphology, and porosity of matrix as compared with other fabrication methods. This method is done by accumulating consecutive layers of material to form 3D models, which will in turn allow more controlled pores and pore sizes. In general, the binder solution is added onto the powder bed from an "inkjet" print head. A 2D layer will be printed and a new layer of powder will be laid down subsequently. The process repeats and the layers merge when fresh binder solution is deposited each time. The finished component will be removed and any unbound powder will be left behind.

Stereolithography: It involves the use of UV laser to polymerize liquid UV curable photopolymer resin layer by layer (LbL), which then solidifies to form a 3D model (Draget et al. 1997). The excess resin will be drained after the complete model is raised out of the vat for curing in an oven. The resolution of each layer is dependent on the resolution of the elevator layer and the spot size of the laser (Zein et al. 2002). However, due to the application of additional curing step to improve the model's property, the final resolution is comprised of shrinkage that typically occurs in this post-processing.

Electrospinning: During solution electrospinning, a voltage is applied to generate a potential difference between a polymeric solution and a collecting target (Lim & Mao 2009; Cipitria et al. 2011). A polymeric jet of fluid will be ejected from a spinneret or tip of a capillary when the surface tension of the fluid is surmounted by the electrical charges (Figure 19.1). The macromolecular entanglements present in the polymeric solution will result in the fluid jet being drawn toward the collector as it does not go through Rayleigh instabilities, that is, formation of droplets due to the fluid jet dismantling. As the fluid jet nears the collector, twisting occurs due to the presence of higher surface charge density. Consequently, a mesh of electrospun fibers will be collected. The diameter of each fiber can be controlled by modifying the concentration and flow rate of the polymer solution and varying the distance between the needle and collector (Moradi et al. 2018; Lim & Mao 2009). On the other hand, using computer-aided design (CAD) modeling (Liu Tsang & Bhatia 2004) requires good fabrication skills to sustain uniformity in scaffold architecture. Although rapid prototyping techniques may have several advantages, some limitations and challenges still remain such as the limited number of biomaterials that can be processed by rapid prototyping compared to conventional techniques. Other approaches that ape the way that tissues are formed in the body and their nanofibrous structure can be found in modular assembly methods and electrospinning methods (Yang et al. 2001). Surrogate applications to prefabrication of scaffolds prior to cell seeding include cell encapsulation.

19.8 Encapsulation

Encapsulation is defined as a membrane with a pore structure to entrap cells or tissue in a biocompatible material that has permeability for nutrients, oxygen and hormones while preventing entry of immune cells into the cell-containing matrix. Cell encapsulation represents a strategy for regulated delivery of protein and morphogen, drug delivery, cell delivery and 3D culture in stem cell research field. Encapsulated cell-based therapies are under study for a variety of diseases such as bone and cartilage defects, treatment of chronic anemia, myocardial regeneration and neurological diseases (Orive et al. 2014).

One of the most common nanotechnology applications is using self-assembly systems through LbL technique for producing polyelectrolyte (PE) capsules (Decher 1997). In other words, PE capsules are assembling of oppositely charged synthetic polymer layers around a charged circular core that this core can be removed after layers' formation to obtain a hollow capsule (Gil et al. 2008).

The polymeric layers are interconnected with each other through electrostatic interactions on a charged template and form LbL (Yang et al. 2015; Wilson et al. 2011). The electrostatic attraction is the most common force in the coupling of opposing structures to create PE multipliers (Iler 1966). The LbL multilayers are made by other interactions between layers that include hydrogen bonding (Eugenia Kharlampieva et al. 2005), van der Waals forces (Masahito Sano et al. 2002), hydrophobic interactions and covalent bonding (Teramura & Iwata 2010).

These nanoscale capsules usually have a wall thickness in nanometers and a diameter ranging from tens of nanometers to several micrometers depending on the original core size. The reduction in the thickness of the layers provides permeability in these capsules. These capsules are not affected by pH-wide changes and organic solvents (Antipov et al. 2002; Sukhorukov et al. 2000). Hua Ai and coworkers used the electrostatic LbL self-assembly technique for platelets in 2002 (Ai et al. n.d.). Krol et al. investigated yeast cell encapsulation in LbL shells and reported that PE nanocapsules possess the capability for living cell encapsulation and provide conditions necessary for the growth and proliferation of cells (Krol et al. 2004). Hachim et al. showed that the arrangement of polymers in layers can affect the function and survival of human adipocyte-derived stem cells (Hachim et al. 2013). Veerabadran et al. demonstrated that individual MSCs maintain their morphology and viability long term in a shell consisting of nano-layer-shells of hyaluronic acid (HA) and PLL using the LbL technique (Veerabadran et al. 2007). Studies have shown that encapsulated living cells into nano-capsules possess metabolic activity (Svaldo-Lanero et al. 2007; Alberto Diaspro et al. 2002). Different types of mammalian including stem cells, endothelial cells, fibroblast, and erythrocytes have been coated by nano-shells (Fakhrullin et al. 2012).

Nano-shells, as bio-functional particles, have selective permeability than other types of ions and molecules, so that this feature makes them suitable 3D matrices for cell assembling (Park et al. 2014). Recent researches based on encapsulation of cells with nano-shells have indicated cytoprotective effects of nano encapsulation. This encapsulation strategy protects cells against physical, chemical, environmental and biological stresses (Matsuzawa et al. 2013; Hong et al. 2013) such as heat (Ko et al. 2013), osmotic pressure (Yang et al. 2009) and lytic enzymes (Magrassi et al. 2010; Yang et al. 2011). The nano-capsules affect cell growth rate by controlling cell division timing (Yang et al. 2009; Yang et al. 2011). Yang and coworkers by assembling of inner layer and cross-linked outer layer make a new mimicking ECM polymer nano-capsule simple that prevents the penetration of cytotoxic foreign aggressions and preserves the cell in long term. Thickness of the encapsulation layers provides hydrated condition and nutrition diffusion. They reported that this nano-capsule is suitable for different mammalian cells such as stem cells and due to biocompatibility can be used in vivo cell delivery (Yang et al. 2017).

Therefore, in recent years, cell encapsulation by nano coating is a novel strategy that provides facilitating cell implantation and high viability protection against external physical, chemical and biological stresses, preserving cell functionality and cell delivering to predefined organs. With regard to the abovementioned, nanoencapsulation as a suitable strategy for cell delivery possesses applicable potential in cell therapy and tissue engineering.

19.9 Hydrogel

Hydrogels as most commonly used biomaterials in regenerative medicine, cell encapsulation and tissue engineering are hydrophilic polymeric networks that retain a lot of water. These have cross-links between polymeric chains and do not dissolve in water (Vedadghavami et al. 2017; Akhtar et al. 2016). The polymers used for procuring hydrogels involve natural polymers such as fibrin, collagen, HA, chitosan and alginate, as well as synthetic polymers such as PEG, poly(acrylamide) (PAAM) and PVA (Verhulsel et al. 2014). Natural polymers increase cell viability and lead cellular functionality. Though synthetic polymers are stable under different environmental conditions and favorable mechanical behavior, they should be modified to be bioactive (Vedadghavami et al. 2017; Akhtar et al. 2016).

Hydrogels have some specific properties such as biodegradability, biocompatibility, three-dimensional (3D) structure, elasticity and flexibility, soft nature, less irritability effect to the neighboring tissue, water loving surface that can stick to cells surfaces, similarity to natural tissue and ability to transfer effective body. Furthermore, hydrogels are favorable candidates in tissue engineering, cell delivery, drug delivery, wound healing and bone and cartilage repair (Vedadghavami et al. 2017; Akhtar et al. 2016).

Recently nano-hydrogels have emerged as NPs in the nanometer scale from tens to hundreds of nanometers that show huge loading capacity of water-soluble compounds and play important role in drug delivery for tissue regeneration. Some properties of nano-hydrogels include high stability in physiological media, distinct responsiveness to environmental factors such as pH and temperature and high cellular uptake due to the endocytosis mechanisms. Nano-hydrogels are made by various self-assembly NP technologies such as solvent emulsion, diffusion and precipitation methods (Qian et al. 2013).

Peptide amphiphiles, as self-assemble peptide, are one of the main nanomaterials in hydrogel formation. These amphipathic peptides can form nanofibers that establish a water-entrapped network as hydrogel (Wan & Ying 2010). Self-assembling peptide nanofiber scaffolds (SAPNS) have three-dimensional (3D) structure with a diameter of about 10 nm and pore diameter of 5–200 nm that retain high content of water. These hydrogels are stable in temperature and pH changes and do not degrade by proteolysis in vitro (Zhang et al. 1993). However, SAPNS are biodegradable by a variety of proteases in the body and their bioproducts have high biocompatibility (Yoshimi et al. 2009). A number of studies have reported some specific properties such as biomimetic tissue environment (Yoshimi et al. 2009), cell attachment culture (Zhang et al. 1995), proliferation and cell survival maintenance (Kisiday et al. 2002), promoting cell differentiation and migration (Horii et al. 2007) and injectable (Davis et al. 2005). Nanogel, like a PE network, possesses the ability to bind to oppositely charged particles. On the other hand, these reveal hydrophilic nonionic network properties and can be stabled in aqueous solution without sedimentation (Vinogradov et al. 2002). In designing a scaffold for a tissue, it is important to identify a material that can maintain a normal rate of proliferation of cells and a high rate of cell synthesis of specific ECM macromolecules and mechanical properties (Yoshimi et al. 2009). Nanofiber-gel layers possess the ability to maintain hepatocyte phenotype in long term to be used for toxicology (Wu et al. 2010), and with biologically active motifs, peptide amphiphiles have been investigated for various applications. An investigation on ischemic tissue has demonstrated that bone marrow-derived stem cells encapsulated within nanopetide gel have enhancing viability in vivo, as well as in vitro study showed proliferation and adhesion of associated cells (Webber et al. 2010). So, these nanogels are usable as scaffold in cell-based therapy for cardiovascular disease.

Moore et al. injected nanofibrous peptide hydrogel without any small molecule subcutaneously in rat and observed blood vessel formation as strong angiogenic response and neural fascicles within hydrogel. Therefore, nanopeptide-based hydrogels can be used in bio-regeneration.

Nanocomposite hydrogels are hybrid hydrogels that are either nanoscale copolymer network or cross-linked with NPs or nanostructures. Nanocomposite hydrogels can be based on NPs such as carbon-based nanomaterial (e.g. CNTs), polymeric NPs (e.g. dendrimers), inorganic/ceramic NPs (hydroxyapatite) and metal/metal-oxide NPs (gold) (Gaharwar et al. 2014). Nanocomposite hydrogels due to high surface interactions between the NPs and the polymer chains are applicable in biomedical area. In these hydrogels, due to the combination of multiple phases and different properties, incorporation obtained suitable matrix mimicking native ECM (Wang et al. 2010).

Engineered nanocomposite hydrogels due to better mimicking can be recapitulated to the native tissue microenvironments. Nano-hydrogels can be highly porous structure and have high mechanical strength so that they are able to control cell behavior and delivery (Memic et al. 2015).

Nano-hydrogels as next-generation polymeric nanoparticulate system are used for a wide variety of drug-delivery applications including cells for various disease conditions. They are injectable and precise drug delivery. Accordingly, nano-hydrogels are favorable tools or can deliver cells to the desired site and are applicable in cell therapy as regeneration medicine.

19.10 Nanomaterials and Biofactors Delivery

An ideal and effective drug-delivery system can be qualified to their small size, reduced drug toxicity, controlled time of release and modification of drug pharmacokinetics and biological distribution (Golchin et al. 2017). NPs containing encapsulated, dispersed, absorbed or conjugated drugs have unique properties to suppress several types of cancers (Golchin et al. 2017). Cancer stem cell (CSC) biomarkers, both specific and universal, could be considered in the design of innovative and effective drug-delivery systems for cancer therapy (Mokhtarzadeh et al. 2017). Targeting CSCs could lead to the incident reduction in drug resistance, CSCs suppression and consequent cancer treatment (Lou & Dean 2007). In the cancer therapy based on CSCs, small chemical ligands, lipids, peptides and polysaccharides with selective affinity to CSCs biomarkers could be attached to the therapeutic nanomaterials or drug nanocarriers (Mokhtarzadeh et al. 2017). Several anti-cancer drugs including paclitaxel, doxorubicin, 5-fluorouracil and dexamethasone have been efficiently made using nanomaterials (DUGGAL 2009). Dexamethasone, a glucocorticoid with an intracellular site of action, are encapsulated by PLGA and PLA based on NPs (DUGGAL 2009). Dexamethasone is a chemotherapeutic agent that has anti-proliferative and anti-inflammatory effects (DUGGAL 2009). According to the hopeful consequence, the excellent aim of cancer suppression may be achieved by targeting CSCs biomarkers with nanomaterials in cancer diagnosis and therapy in the near future.

To achieve the effective delivery of genetic material systems into target stem cells to specifically direct differentiation is severely hampered in the clinic of stem cell-based therapies (Shah et al. 2013). Several recent studies have shown that magnetic nanoparticles (MNPs) have numerous advantages that are especially attractive for stem cell research applications (Shah et al. 2013). Typically, MNPs are composed of a magnetic core that can consist of metals or metal oxides, metal alloys, and, more recently, doped metals (Shah et al. 2013). MNPs are multifunctional because of its inherent magnetism and can be used to magnetically facilitate delivery, cell targeting, and magnetic resonance imaging (MRI) contrast agents (Shah et al. 2013). Also, MNPs are formulated to delivery of genetic materials and short interfering RNA (siRNA) into cells in a highly efficient, spatiotemporally controlled, and biocompatible method (Li et al. 2013). Another gene delivery system is mesoporous silica nanoparticles (MSNs) to quickly produce iPSC-derived specific-lineage cells (Chen et al. 2013). FITC-conjugated MSNs (FMSNs) were effectively used for labeling of iPSCs and utilized the multifunctional properties of FMSNs for a suitable carrier for biomolecule delivery (Chen et al. 2013). FMSNs have multiple advantages not only to serve as an ideal vector for stem cell labeling but also for safe genetic material delivery to induce the production of special cells from iPSCs (Chen et al. 2013). However, stem cell-based therapy has remarkable limitations so far, while transplanted stem cells are largely invasive at predetermined time points after transplant (Gao et al. 2013). Consequently, non-invasive imaging systems are extremely needed to screen transplanted stem cells qualitatively and quantitatively. This will facilitate the prediction of treatment efficacy and reveal optimal transplantation conditions, allowing the cell dosage, delivery route, and timing of transplantations to be determined (Gao et al. 2013). Therefore, being able to monitor the in vivo behavior of transplanted stem cells and understand the fate of these cells is essential to achieve full safety and efficient therapeutic potentials of stem cells (Qi et al. 2013). To address this requires an effective, non-invasive and non-toxic technique for cell tracking based on NPs (Qi et al. 2013). NPs are contrast agents of fluorescence, magnetic resonance and photoacoustic imaging, as three frequently used imaging modalities, for stem cell labeling and monitoring (Wang et al. 2013). MRI is a very sensitive, repetitive and non-invasive in vivo monitor of magnetically labeled stem cells (Andreas et al. 2012). Nanomaterials can be used as MRI, optical imaging, and photoacoustic imaging contrast agents (Janowski et al. 2012; Fernandez-Fernandez et al. 2011). MRIs are used to follow SPIONs-labeled SCs and have been anticipated as a golden standard for monitoring the in vivo migration of transplanted SPION-labeled SCs (Andreas et al. 2012). Labeling with commercial SPIONs is still problematic because of low labeling efficiencies and the need of potentially toxic transfection agents (Li et al. 2013; Andreas et al. 2012). But citrate SPIONs are effective intracellular magnetic labels and in vivo tracer of SCs without transfecting agents by MRI. PEI-upconversion nanoparticles (UCNPs) also are another promising factors for SCs labeling and tracking (L. Zhao et al. 2013). PEI-UCNP labeled MSCs in normal early proliferation stage (L. Zhao et al. 2013). Finally, PEI-UCNP labeled MSCs were able to undergo osteogenic and adipogenic differentiation upon in vitro stimulation, although the osteogenesis of labeled MSCs seemed to be fewer potent than that of the unlabeled MSCs (L. Zhao et al. 2013). UCNP is another sensitive and reliable exogenous contrast mediator for SC labeling and in vivo tracking (Wang et al. 2012).

Overall, the combination of stem cells and nanomaterial scaffold is a useful strategy in tissue engineering and covered a wide range of applications in tissue repairment or replacement. More application areas and development into close level of clinical applications can be explored in the not-too distant future. There are another potential applications of NPs in stem cell research including non-invasive tracking of stem cells transplanted in vivo; intracellular delivery of DNA, RNA interference molecules, proteins, peptides, genes, and small drugs for stem cell differentiation or survival and as biosensors to monitor intracellular levels of relevant biomolecules/enzymes in real time. Thereupon, using nanotechnology, quicker and much cheaper treatments can be developed.

19.11 Emerging Trends and Future Outlook

With great progress being made in the synthesis of nanostructures, there are interesting new opportunities for regenerative medicine, in particular, cell therapy and CDSs. Research indicated that the nanofeatures in substrate cause increasing interaction between cells and substrate, also nanofeatures cause leading improvement cell adhesion, cell growth and cytoskeletal organization in regenerative medicine applications. Therefore, nanomaterials have opened a new viewpoint in cell-based therapy, particularly CDSs.

Nanotechnology uses various nanofabrication methods to develop ideal CDSs with nanofeature that mimics the structure and function of the natural biological compounds. Continuity of researches can be focused on the preparation of a library of different nanomaterials that are appropriate for CDSs. Moreover, most physicochemical evaluations to date have been part of the nanomaterial research, and the long-term effects and the fate of nanomaterials in vitro and in vivo are clearly not known. Currently, nanomaterials play a central role in cell tissue-based therapy and regenerative medicine. The structural similarities to natural tissue in nanoscale and their smart behavior create a suitable cellular environment for their applications in CDSs. However, used with proper fabrication techniques, proper nano-biomaterials, and in combination with appropriate cells, the potential of nanomaterials can be unlimited from biofactor delivery to CDS applications.

References

Adiga, S.P. et al., 2009. Nanoporous membranes for medical and biological applications. *Wiley Interdisciplinary Reviews: Nanomedicine and Nanobiotechnology,*

1(5), pp. 568–581. Available at: www.ncbi.nlm.nih.gov/pubmed/20049818 [Accessed November 22, 2018].

Aebischer, P., Winn, S.R. & Galletti, P.M., 1988. Transplantation of neural tissue in polymer capsules. *Brain Research*, 448(2), pp. 364–368. Available at: www.sciencedirect.com/science/article/pii/0006899388912784 [Accessed November 22, 2018].

Akhtar, M.F., Hanif, M. & Ranjha, N.M., 2016. Methods of synthesis of hydrogels … A review. *Saudi Pharmaceutical Journal*, 24(5), pp. 554–559. Available at: www.sciencedirect.com/science/article/pii/S1319016415000857 [Accessed November 22, 2018].

Ai, H. et al., 2002. Applications of the electrostatic layer-by-layer self-assembly technique in biomedical engineering. In *Proceedings of the Second Joint 24th Annual Conference and the Annual Fall Meeting of the Biomedical Engineering Society Engineering in Medicine and Biology*. IEEE, Houston, TX, pp. 502–503. Available at: http://ieeexplore.ieee.org/document/1136916/ [Accessed November 22, 2018].

Andreas, K. et al., 2012. Highly efficient magnetic stem cell labeling with citrate-coated superparamagnetic iron oxide nanoparticles for MRI tracking. *Biomaterials*, 33(18), pp. 4515–4525. Available at: www.ncbi.nlm.nih.gov/pubmed/22445482 [Accessed November 20, 2018].

Antipov, A.A. et al., 2002. *Colloids and Surfaces. A, Physicochemical and Engineering Aspects*. Elsevier Science (Firm), Elsevier Science Pub. Co. Available at: www.tib.eu/en/search/id/BLSE%3ARN109256626/Polyelectrolyte-multilayer-capsule-permeability/ [Accessed November 22, 2018].

Bamrungsap, S. et al., 2012. Nanotechnology in therapeutics: a focus on nanoparticles as a drug delivery system. *Nanomedicine*, 7(8), pp. 1253–1271. Available at: www.ncbi.nlm.nih.gov/pubmed/22931450 [Accessed November 19, 2018].

Barros, D., Amaral, I.F. & Pêgo, A.P., 2015. Biomimetic synthetic self-assembled hydrogels for cell transplantation. *Current Topics in Medicinal Chemistry*, 15(13), pp. 1209–1226. Available at: www.ncbi.nlm.nih.gov/pubmed/25858129 [Accessed November 20, 2018].

Bružauskaitė, I. et al., 2016. Scaffolds and cells for tissue regeneration: different scaffold pore sizes—different cell effects. *Cytotechnology*, 68(3), pp. 355–369. Available at: www.ncbi.nlm.nih.gov/pubmed/26091616 [Accessed November 22, 2018].

Chang, R. et al., 2017. Nanoporous immunoprotective device for stem-cell-derived β-cell replacement therapy. *ACS Nano*, 11(8), pp. 7747–7757. Available at: www.ncbi.nlm.nih.gov/pubmed/28763191 [Accessed November 22, 2018].

Chen, W. et al., 2013. Nonviral cell labeling and differentiation agent for induced pluripotent stem cells based on mesoporous silica nanoparticles. *ACS Nano*, 7(10), pp. 8423–8440. Available at: www.ncbi.nlm.nih.gov/pubmed/24063246 [Accessed November 19, 2018].

Chimene, D., Alge, D.L. & Gaharwar, A.K., 2015. Two-dimensional nanomaterials for biomedical applications: emerging trends and future prospects. *Advanced Materials*, 27(45), pp. 7261–7284. Available at: http://doi.wiley.com/10.1002/adma.201502422 [Accessed November 26, 2018].

Cipitria, A. et al., 2011. Design, fabrication and characterization of PCL electrospun scaffolds—a review. *Journal of Materials Chemistry*, 21(26), p. 9419. Available at: http://xlink.rsc.org/?DOI=c0jm04502k [Accessed November 22, 2018].

Colton, C.K., 1995. Implantable biohybrid artificial organs. *Cell transplantation*, 4(4), pp. 415–436. Available at: www.ncbi.nlm.nih.gov/pubmed/7582573 [Accessed November 22, 2018].

Davis, M.E. et al., 2005. Injectable self-assembling peptide nanofibers create intramyocardial microenvironments for endothelial cells. *Circulation*, 111(4), pp. 442–450. Available at: www.ncbi.nlm.nih.gov/pubmed/15687132 [Accessed April 14, 2017].

Decher, G., 1997. Fuzzy nanoassemblies: toward layered polymeric multicomposites. *Science*, 277(5330), pp. 1232–1237. Available at: www.sciencemag.org/cgi/doi/10.1126/science.277.5330.1232 [Accessed November 22, 2018].

Desai, T.A. et al., 1998. Microfabricated immunoisolating biocapsules. *Biotechnology and Bioengineering*, 57(1), pp. 118–120. Available at: http://doi.wiley.com/10.1002/%28SICI%291097-0290%2819980105%2957%3A1%3C118%3A%3AAID-BIT14%3E3.0.CO%3B2-G [Accessed November 22, 2018].

Desai, T.A., Hansford, D.J. & Ferrari, M., 2000. Micromachined interfaces: new approaches in cell immunoisolation and biomolecular separation. *Biomolecular Engineering*, 17(1), pp. 23–36. Available at: www.ncbi.nlm.nih.gov/pubmed/11042474 [Accessed November 22, 2018].

Diaspro, A. et al., 2002. Single living cell encapsulation in nano-organized polyelectrolyte shells. Available at: https://pubs.acs.org/doi/10.1021/la025646e [Accessed November 22, 2018].

Draget, K.I., Skjåk-Braek, G. & Smidsrød, O., 1997. Alginate based new materials. *International Journal of Biological Macromolecules*, 21(1–2), pp. 47–55. Available at: www.ncbi.nlm.nih.gov/pubmed/9283015 [Accessed November 22, 2018].

Duggal, D., 2009. Role of nanotechnology in new drug delivery systems. *International Journal of Drug Development & Research*. Available at: www.ijddr.in/drug-development/role-of-nanotechnology-in-new-drug-delivery-systems.php?aid=5666 [Accessed November 20, 2018].

Fakhrullin, R.F. et al., 2012. Cyborg cells: functionalisation of living cells with polymers and nanomaterials. *Chemical Society Reviews*, 41(11), p. 4189. Available at:

www.ncbi.nlm.nih.gov/pubmed/22509497 [Accessed November 22, 2018].

Fernandez-Fernandez, A., Manchanda, R. & McGoron, A.J., 2011. Theranostic applications of nanomaterials in cancer: drug delivery, image-guided therapy, and multifunctional platforms. *Applied Biochemistry and Biotechnology*, 165(7–8), pp. 1628–1651. Available at: www.ncbi.nlm.nih.gov/pubmed/21947761 [Accessed November 20, 2018].

Fujioka, K. et al., 2014. Effects of silica and titanium oxide particles on a human neural stem cell line: morphology, mitochondrial activity, and gene expression of differentiation markers. *International Journal of Molecular Sciences*, 15(7), pp. 11742–11759. Available at: www.ncbi.nlm.nih.gov/pubmed/24992594 [Accessed November 25, 2018].

Gaharwar, A.K., Peppas, N.A. & Khademhosseini, A., 2014. Nanocomposite hydrogels for biomedical applications. *Biotechnology and Bioengineering*, 111(3), pp. 441–453. Available at: www.ncbi.nlm.nih.gov/pubmed/24264728 [Accessed November 22, 2018].

Gao, Y. et al., 2013. Stem cell tracking with optically active nanoparticles. *American Journal of Nuclear Medicine and Molecular Imaging*, 3(3), pp. 232–246. Available at: www.ncbi.nlm.nih.gov/pubmed/23638335 [Accessed November 19, 2018].

Gil, P.R. et al., 2008. Nanoparticle-modified polyelectrolyte capsules. *Nano Today*, 3(3–4), pp. 12–21. Available at: http://linkinghub.elsevier.com/retrieve/pii/S1748013208700409 [Accessed November 22, 2018].

Golchin, A., Farahany, T.Z., et al., 2018. The clinical trials of mesenchymal stem cell therapy in skin diseases: an update and concise review. *Current Stem Cell Research & Therapy*. Available at: www.ncbi.nlm.nih.gov/pubmed/30210006 [Accessed October 15, 2018].

Golchin, A., Farahany, T.Z., et al., 2019. Biological products: cellular therapy and FDA approved products. *Stem Cell Reviews and Reports*, 15 (2), 166–175. Available at: https://link.springer.com/article/10.1007/s12015-018-9866-1 [Accessed March 28, 2018].

Golchin, A., Hosseinzadeh, S. & Roshangar, L., 2017. The role of nanomaterials in cell delivery systems. *Medical Molecular Morphology*, 51(1), pp. 1–12. Available at: www.ncbi.nlm.nih.gov/pubmed/29170827 [Accessed January 17, 2018].

Gong, T. et al., 2015. Nanomaterials and bone regeneration. *Bone Research*, 3(1), p. 15029. Available at: www.nature.com/articles/boneres201529 [Accessed November 19, 2018].

Gu, J. et al., 2011. Electrospun chitosan-graft-poly (ε-caprolactone)/poly (ε-caprolactone) nanofibrous scaffolds for retinal tissue engineering. *International Journal of Nanomedicine*, 6, p. 453. Available at: www.ncbi.nlm.nih.gov/pubmed/21499434 [Accessed November 19, 2018].

Hachim, D., Melendez, J. & Ebensperger, R., 2013. Nanoencapsulation of human adipose mesenchymal stem cells: experimental factors role to successfully preserve viability and functionality of cells. *Journal of Encapsulation and Adsorption Sciences*, 03(01), pp. 1–12. Available at: www.scirp.org/journal/doi.aspx?DOI=10.4236/jeas.2013.31001 [Accessed November 22, 2018].

Hatton, B.D., 2013. Engineering nanoporous biomaterials. *Nanomaterials in Tissue Engineering*, pp. 64–88. Available at: www.sciencedirect.com/science/article/pii/B9780857095961500031 [Accessed November 22, 2018].

He, H. et al., 2013. A naonoporous cell-therapy device with controllable biodegradation for long-term drug release. *Journal of Controlled Release*, 165(3), pp. 226–233. Available at: www.ncbi.nlm.nih.gov/pubmed/23228849 [Accessed November 20, 2018].

Hong, D. et al., 2013. Artificial spores: cytoprotective nanoencapsulation of living cells. *Trends in Biotechnology*, 31(8), pp. 442–447. Available at: www.ncbi.nlm.nih.gov/pubmed/23791238 [Accessed November 22, 2018].

Horii, A. et al., 2007. Biological designer self-assembling peptide nanofiber scaffolds significantly enhance osteoblast proliferation, differentiation and 3-D migration. *PLoS ONE*, 2(2), p. e190, M. Isalan, ed. Available at: http://dx.plos.org/10.1371/journal.pone.0000190 [Accessed November 22, 2018].

Huang, S. & Fu, X., 2010. Naturally derived materials-based cell and drug delivery systems in skin regeneration. *Journal of Controlled Release*, 142(2), pp. 149–159. Available at: www.ncbi.nlm.nih.gov/pubmed/19850093 [Accessed May 14, 2017].

Iler, R.K., 1966. Multilayers of colloidal particles. *Journal of Colloid and Interface Science*, 21(6), pp. 569–594. Available at: www.sciencedirect.com/science/article/pii/0095852266900183 [Accessed November 22, 2018].

International Stem Cell Initiative et al., 2007. Characterization of human embryonic stem cell lines by the International Stem Cell Initiative. *Nature Biotechnology*, 25(7), pp. 803–816. Available at: www.ncbi.nlm.nih.gov/pubmed/17572666 [Accessed November 19, 2018].

Janowski, M., Bulte, J.W.M. & Walczak, P., 2012. Personalized nanomedicine advancements for stem cell tracking. *Advanced Drug Delivery Reviews*, 64(13), pp. 1488–1507. Available at: www.ncbi.nlm.nih.gov/pubmed/22820528 [Accessed November 20, 2018].

Jin, G., Prabhakaran, M.P. & Ramakrishna, S., 2011. Stem cell differentiation to epidermal lineages on electrospun nanofibrous substrates for skin tissue engineering. *Acta Biomaterialia*, 7(8), pp. 3113–3122. Available at: www.ncbi.nlm.nih.gov/pubmed/21550425 [Accessed November 19, 2018].

Khan, I., Saeed, K. & Khan, I., 2017. Nanoparticles: properties, applications and toxicities. *Arabian Journal of Chemistry*. Available at: www.sciencedirect.com/science/article/pii/S1878535217300990 [Accessed November 22, 2018].

Kharlampieva, E. et al., 2005. Hydrogen-bonded multilayers of thermoresponsive polymers. *Macromolecules*. Available at: https://pubs.acs.org/doi/10.1021/ma0516891 [Accessed November 22, 2018].

Kisiday, J. et al., 2002. Self-assembling peptide hydrogel fosters chondrocyte extracellular matrix production and cell division: implications for cartilage tissue repair. *Proceedings of the National Academy of Sciences of the United States of America*, 99(15), pp. 9996–10001. Available at: www.ncbi.nlm.nih.gov/pubmed/12119393 [Accessed November 22, 2018].

Ko, E.H. et al., 2013. Bioinspired, cytocompatible mineralization of silica-titania composites: thermoprotective nanoshell formation for individual *chlorella* cells. *Angewandte Chemie International Edition*, 52(47), pp. 12279–12282. Available at: http://doi.wiley.com/10.1002/anie.201305081 [Accessed November 22, 2018].

Kögler, G. et al., 2004. A new human somatic stem cell from placental cord blood with intrinsic pluripotent differentiation potential. *The Journal of Experimental Medicine*, 200(2), pp. 123–135. Available at: www.ncbi.nlm.nih.gov/pubmed/15263023 [Accessed November 19, 2018].

Krol, S. et al., 2004. Nanocapsules: coating for living cells. *IEEE Transactions on Nanobioscience*, 3(1), pp. 32–38. Available at: http://ieeexplore.ieee.org/document/1273505/ [Accessed November 22, 2018].

Kun Na, et al., 2007. Heparin/Poly(l-lysine) nanoparticle-coated polymeric microspheres for stem-cell therapy. Available at: https://pubs.acs.org/doi/10.1021/ja067707r [Accessed November 22, 2018].

Lanza, R.P., Kuhtreiber, W.M. & Chick, W.L., 2000. Microcapsules and composite microreactors for immunoisolation of cells. Available at: www.freepatentsonline.com/6126936.html [Accessed November 22, 2018].

Laurent, S. et al., 2008. Magnetic iron oxide nanoparticles: synthesis, stabilization, vectorization, physicochemical characterizations, and biological applications. *Chemical Reviews*, 108(6), pp. 2064–2110. Available at: http://pubs.acs.org/doi/abs/10.1021/cr068445e [Accessed November 22, 2018].

Li, L. et al., 2013. Superparamagnetic iron oxide nanoparticles as MRI contrast agents for non-invasive stem cell labeling and tracking. *Theranostics*, 3(8), pp. 595–615. Available at: www.ncbi.nlm.nih.gov/pubmed/23946825 [Accessed November 20, 2018].

Lim, S.H. & Mao, H.-Q., 2009. Electrospun scaffolds for stem cell engineering. *Advanced Drug Delivery Reviews*, 61(12), pp. 1084–1096. Available at: www.ncbi.nlm.nih.gov/pubmed/19647024 [Accessed November 22, 2018].

Liu, X. et al., 2010. A protein interaction network for the analysis of the neuronal differentiation of neural stem cells in response to titanium dioxide nanoparticles. *Biomaterials*, 31(11), pp. 3063–3070. Available at: www.ncbi.nlm.nih.gov/pubmed/20071024 [Accessed November 25, 2018].

Liu Tsang, V. & Bhatia, S.N., 2004. Three-dimensional tissue fabrication. *Advanced Drug Delivery Reviews*, 56(11), pp. 1635–1647. Available at: www.ncbi.nlm.nih.gov/pubmed/15350293 [Accessed November 22, 2018].

Loh, Q.L., & Choong, C., 2013. Three-dimensional scaffolds for tissue engineering applications: role of porosity and pore size. *Tissue Engineering. Part B, Reviews*, 19(6), pp. 485–502. Available at: www.ncbi.nlm.nih.gov/pubmed/23672709 [Accessed November 22, 2018].

Lou, H. & Dean, M., 2007. Targeted therapy for cancer stem cells: the patched pathway and ABC transporters. *Oncogene*, 26(9), pp. 1357–1360. Available at: www.ncbi.nlm.nih.gov/pubmed/17322922 [Accessed November 20, 2018].

Lu, G.Q. & Zhao, X.S., 2004. *Nanoporous Materials: Science and Engineering*, Published by Imperial College Press and Distributed by World Scientific Publishing Co. Available at: www.worldscientific.com/worldscibooks/10.1142/p181 [Accessed November 22, 2018].

Ma, P.X., 2004. Scaffolds for tissue fabrication. *Materials Today*, 7(5), pp. 30–40. Available at: www.sciencedirect.com/science/article/pii/S1369702104002330 [Accessed November 22, 2018].

Ma, R. & Sasaki, T., 2015. Two-dimensional oxide and hydroxide nanosheets: controllable high-quality exfoliation, molecular assembly, and exploration of functionality. *Accounts of Chemical Research*, 48(1), pp. 136–143. Available at: http://pubs.acs.org/doi/10.1021/ar500311w [Accessed November 27, 2018].

Magrassi, R. et al., 2010. Protection capabilities of nanostructured shells toward cell encapsulation: a Saccharomyces/Paramecium model. *Microscopy Research and Technique*, 73(10), pp. 931–936. Available at: www.ncbi.nlm.nih.gov/pubmed/20872735 [Accessed November 22, 2018].

Martino, S. et al., 2012. Stem cell-biomaterial interactions for regenerative medicine. *Biotechnology Advances*, 30(1), pp. 338–351. Available at: www.ncbi.nlm.nih.gov/pubmed/21740963 [Accessed November 19, 2018].

Matsuzawa, A., Matsusaki, M. & Akashi, M., 2013. Effectiveness of nanometer-sized extracellular matrix layer-by-layer assembled films for a cell membrane coating protecting cells from physical stress. *Langmuir*, 29(24), pp. 7362–7368. Available at: www.ncbi.nlm.nih.gov/pubmed/23092370 [Accessed November 22, 2018].

McMillan, J., Batrakova, E. & Gendelman, H.E., 2011. Cell delivery of therapeutic nanoparticles. *Progress in Molecular Biology and Translational Science*. pp. 563–601. Available at: www.ncbi.nlm.nih.gov/pubmed/22093229 [Accessed November 22, 2018].

Memic, A. et al., 2015. Hydrogels 2.0: improved properties with nanomaterial composites for biomedical applications. *Biomedical Materials*, 11(1), p. 014104. Available at: www.ncbi.nlm.nih.gov/pubmed/26694229 [Accessed November 22, 2018].

Mocan, L. et al., 2012. Influence of nanomaterials on stem cell differentiation: designing an appropriate nanobiointerface. *International Journal of Nanomedicine*, 7, p. 2211. Available at: www.ncbi.nlm.nih.gov/pubmed/22619557 [Accessed November 19, 2018].

Mokhtarzadeh, A. et al., 2017. Nano-delivery system targeting to cancer stem cell cluster of differentiation biomarkers. *Journal of Controlled Release*, 266, pp. 166–186. Available at: www.ncbi.nlm.nih.gov/pubmed/28941992 [Accessed November 20, 2018].

Moradi, S.L. et al., 2018. Bone tissue engineering: adult stem cells in combination with electrospun nanofibrous scaffolds. *Journal of Cellular Physiology*, 233(10), pp. 6509–6522. Available at: http://doi.wiley.com/10.1002/jcp.26606 [Accessed August 7, 2018].

Nicolosi, V. et al., 2013. Liquid exfoliation of layered materials. *Science*, 340(6139). Available at: https://srg.mit.edu/liquid-exfoliation-layered-materials [Accessed November 26, 2018].

Nyitray, C.E. et al., 2015. Polycaprolactone thin-film micro- and nanoporous cell-encapsulation devices. *ACS Nano*, 9(6), pp. 5675–5682. Available at: www.ncbi.nlm.nih.gov/pubmed/25950860 [Accessed November 22, 2018].

Orive, G. et al., 2014. Application of cell encapsulation for controlled delivery of biological therapeutics. *Advanced Drug Delivery Reviews*, 67–68, pp. 3–14. Available at: www.ncbi.nlm.nih.gov/pubmed/23886766 [Accessed November 22, 2018].

Park, G.E. et al., 2005. Accelerated chondrocyte functions on NaOH-treated PLGA scaffolds. *Biomaterials*, 26(16), pp. 3075–3082. Available at: www.sciencedirect.com/science/article/pii/S0142961204007100 [Accessed April 10, 2017].

Park, J.H. et al., 2014. Nanocoating of single cells: from maintenance of cell viability to manipulation of cellular activities. *Advanced Materials*, 26(13), pp. 2001–2010. Available at: www.ncbi.nlm.nih.gov/pubmed/24452932 [Accessed November 22, 2018].

Qi, Y. et al., 2013. The application of super paramagnetic iron oxide-labeled mesenchymal stem cells in cell-based therapy. *Molecular Biology Reports*, 40(3), pp. 2733–2740. Available at: www.ncbi.nlm.nih.gov/pubmed/23269616 [Accessed November 19, 2018].

Qian, Z.-Y., Fu, S.-Z. & Feng, S.-S., 2013. Nanohydrogels as a prospective member of the nanomedicine family. *Nanomedicine*, 8(2), pp. 161–164. Available at: www.ncbi.nlm.nih.gov/pubmed/23394150 [Accessed November 22, 2018].

Sano, M. et al., 2002. Noncovalent self-assembly of carbon nanotubes for construction of "Cages." Available at: https://pubs.acs.org/doi/abs/10.1021/nl025525z [Accessed November 22, 2018].

Shah, B. et al., 2013. Multimodal magnetic core-shell nanoparticles for effective stem-cell differentiation and imaging. *Angewandte Chemie International Edition*, 52(24), pp. 6190–6195. Available at: www.ncbi.nlm.nih.gov/pubmed/23650180 [Accessed November 19, 2018].

Shin, H.J. et al., 2006. Electrospun PLGA nanofiber scaffolds for articular cartilage reconstruction: mechanical stability, degradation and cellular responses under mechanical stimulation in vitro. *Journal of Biomaterials Science. Polymer Edition*, 17(1–2), pp. 103–19. Available at: www.ncbi.nlm.nih.gov/pubmed/16411602 [Accessed November 25, 2018].

Smith, I.O. et al., 2009. Nanostructured polymer scaffolds for tissue engineering and regenerative medicine. *Wiley Interdisciplinary Reviews. Nanomedicine and Nanobiotechnology*, 1(2), pp. 226–36. Available at: www.ncbi.nlm.nih.gov/pubmed/20049793 [Accessed November 22, 2018].

Sukhorukov, G.B. et al., 2000. Microencapsulation by means of step-wise adsorption of polyelectrolytes. *Journal of Microencapsulation*, 17(2), pp. 177–185. Available at: www.ncbi.nlm.nih.gov/pubmed/10738693 [Accessed November 22, 2018].

Sullenger, B.A. et al., 2015. Electrospun cationic nanofibers and methods of making and using the same. Available at: https://patentscope.wipo.int/search/en/detail.jsf?docId=WO2015161094 [Accessed November 20, 2018].

Svaldo-Lanero, T. et al., 2007. Morphology, mechanical properties and viability of encapsulated cells. *Ultramicroscopy*, 107(10–11), pp. 913–921. Available at: www.ncbi.nlm.nih.gov/pubmed/17555876 [Accessed November 22, 2018].

Teramura, Y. & Iwata, H., 2010. Cell surface modification with polymers for biomedical studies. *Soft Matter*, 6(6), p. 1081. Available at: http://xlink.rsc.org/?DOI=b913621e [Accessed November 22, 2018].

Tiwari, J.N., Tiwari, R.N. & Kim, K.S., 2012. Zero-dimensional, one-dimensional, two-dimensional and three-dimensional nanostructured materials for advanced electrochemical energy devices. *Progress in Materials Science*, 57(4), pp. 724–803. Available at: www.sciencedirect.com/science/article/pii/S0079642511001034 [Accessed November 22, 2018].

Tukmachev, D. et al., 2015. An effective strategy of magnetic stem cell delivery for spinal cord injury therapy. *Nanoscale*, 7(9), pp. 3954–3958. Available at: www.ncbi.nlm.nih.gov/pubmed/25652717 [Accessed November 19, 2018].

Vasita, R. & Katti, D.S., 2006. Nanofibers and their applications in tissue engineering. *International Journal of Nanomedicine*, 1(1), pp. 15–30. Available at: www.ncbi.nlm.nih.gov/pubmed/17722259 [Accessed November 25, 2018].

Vedadghavami, A. et al., 2017. Manufacturing of hydrogel biomaterials with controlled mechanical properties for tissue engineering applications. *Acta Biomaterialia*, 62, pp. 42–63. Available at: www.sciencedirect.com/science/article/pii/S1742706117304658 [Accessed November 22, 2018].

Veerabadran, N.G. et al., 2007. Nanoencapsulation of stem cells within polyelectrolyte multilayer shells.

Macromolecular Bioscience, 7(7), pp. 877–882. Available at: http://doi.wiley.com/10.1002/mabi.200700061 [Accessed November 22, 2018].

Verhulsel, M. et al., 2014. A review of microfabrication and hydrogel engineering for micro-organs on chips. *Biomaterials*, 35(6), pp. 1816–1832. Available at: www.sciencedirect.com/science/article/pii/S0142961213013744 [Accessed November 22, 2018].

Vinogradov, S. V, Bronich, T.K. & Kabanov, A. V, 2002. Nanosized cationic hydrogels for drug delivery: preparation, properties and interactions with cells. *Advanced Drug Delivery Reviews*, 54(1), pp. 135–147. Available at: www.sciencedirect.com/science/article/pii/S0169409X01002459 [Accessed November 22, 2018].

Wan, A.C.A. & Ying, J.Y., 2010. Nanomaterials for in situ cell delivery and tissue regeneration. *Advanced Drug Delivery Reviews*, 62(7–8), pp. 731–740. Available at: www.ncbi.nlm.nih.gov/pubmed/20156499 [Accessed November 22, 2018].

Wang, C. et al., 2012. Towards whole-body imaging at the single cell level using ultra-sensitive stem cell labeling with oligo-arginine modified upconversion nanoparticles. *Biomaterials*, 33(19), pp. 4872–4881. Available at: www.ncbi.nlm.nih.gov/pubmed/22483011 [Accessed November 20, 2018].

Wang, J., Li, G. & Li, L., 2016. Synthesis strategies about 2D materials. In *Two-dimensional Materials – Synthesis, Characterization and Potential Applications*. InTech. Available at: www.intechopen.com/books/two-dimensional-materials-synthesis-characterization-and-potential-applications/synthesis-strategies-about-2d-materials [Accessed November 27, 2018].

Wang, Q. et al., 2010. High-water-content mouldable hydrogels by mixing clay and a dendritic molecular binder. *Nature*, 463(7279), pp. 339–343. Available at: www.nature.com/articles/nature08693 [Accessed November 22, 2018].

Wang, Y.-L. et al., 2016. Self-assembled peptide-based hydrogels as scaffolds for proliferation and multi-differentiation of mesenchymal stem cells. *Macromolecular Bioscience*, 17(4), pp. 1–13. Available at: http://doi.wiley.com/10.1002/mabi.201600192 [Accessed November 20, 2018].

Wang, Y., Xu, C. & Ow, H., 2013. Commercial nanoparticles for stem cell labeling and tracking. *Theranostics*, 3(8), pp. 544–560. Available at: www.ncbi.nlm.nih.gov/pubmed/23946821 [Accessed November 20, 2018].

Webber, M.J. et al., 2010. Development of bioactive peptide amphiphiles for therapeutic cell delivery. *Acta Biomaterialia*, 6(1), pp. 3–11. Available at: www.ncbi.nlm.nih.gov/pubmed/19635599 [Accessed November 22, 2018].

Wei, X. et al., 2013. Mesenchymal stem cells: a new trend for cell therapy. *Acta Pharmacologica Sinica*, 34(6), pp. 747–754. Available at: www.nature.com/doifinder/10.1038/aps.2013.50 [Accessed November 2, 2017].

Wilson, J.T. et al., 2011. Cell surface engineering with polyelectrolyte multilayer thin films. *Journal of the American Chemical Society*, 133(18), pp. 7054–7064. Available at: http://pubs.acs.org/doi/abs/10.1021/ja110926s [Accessed November 22, 2018].

Wimpenny, I., Markides, H. & El Haj, A.J., 2012. Orthopaedic applications of nanoparticle-based stem cell therapies. *Stem Cell Research & Therapy*, 3(2), p. 13. Available at: www.ncbi.nlm.nih.gov/pubmed/22520594 [Accessed November 19, 2018].

Wu, J. et al., 2010. Nanometric self-assembling peptide layers maintain adult hepatocyte phenotype in sandwich cultures. *Journal of Nanobiotechnology*, 8(1), p. 29. Available at: http://jnanobiotechnology.biomedcentral.com/articles/10.1186/1477-3155-8-29 [Accessed November 22, 2018].

Yang, J. et al., 2017. Nanoencapsulation of individual mammalian cells with cytoprotective polymer shell. *Biomaterials*, 133, pp. 253–262. Available at: www.ncbi.nlm.nih.gov/pubmed/28445804 [Accessed November 22, 2018].

Yang, S. et al., 2001. The design of scaffolds for use in tissue engineering. Part I. Traditional factors. *Tissue Engineering*, 7(6), pp. 679–689. Available at: www.ncbi.nlm.nih.gov/pubmed/11749726 [Accessed November 22, 2018].

Yang, S.H. et al., 2009. Biomimetic encapsulation of individual cells with silica. *Angewandte Chemie International Edition*, 48(48), pp. 9160–9163. Available at: http://doi.wiley.com/10.1002/anie.200903010 [Accessed November 22, 2018].

Yang, S.H. et al., 2015. Cytocompatible in situ cross-linking of degradable LbL films based on thiol–exchange reaction. *Chemical Science*, 6(8), pp. 4698–4703. Available at: http://xlink.rsc.org/?DOI=C5SC01225B [Accessed November 22, 2018].

Yang, S.H. et al., 2011. Mussel-inspired encapsulation and functionalization of individual yeast cells. *Journal of the American Chemical Society*, 133(9), pp. 2795–2797. Available at: www.ncbi.nlm.nih.gov/pubmed/21265522 [Accessed November 22, 2018].

Yao, C. et al., 2008. Surface modification and antibacterial activity of electrospun polyurethane fibrous membranes with quaternary ammonium moieties. *Journal of Membrane Science*, 320, pp. 259–267. Available at: www.sciencedirect.com/science/article/pii/S0376738808003189 [Accessed May 14, 2017].

Yoo, H.S. & Kim, T.G., 2009. Surface-functionalized electrospun nanofibers for tissue engineering and drug delivery. *Advanced Drug Delivery Reviews*, 61(12), pp. 1033–1042. Available at: www.sciencedirect.com/science/article/pii/S0169409X09002324 [Accessed May 14, 2017].

Yoshimi, R. et al., 2009. Self-assembling peptide nanofiber scaffolds, platelet-rich plasma, and mesenchymal stem cells for injectable bone regeneration with tissue engineering. *Journal of Craniofacial Surgery*, 20(5), pp. 1523–1530. Available at: www.ncbi.nlm.nih.gov/pubmed/19816290 [Accessed November 22, 2018].

Zein, I. et al., 2002. Fused deposition modeling of novel scaffold architectures for tissue engineering applications. *Biomaterials*, 23(4), pp. 1169–1185. Available at: www.sciencedirect.com/science/article/pii/S0142961201002320 [Accessed November 22, 2018].

Zhang, G. et al., 2018. The application of nanomaterials in stem cell therapy for some neurological diseases. *Current Drug Targets*, 19(3), pp. 279–298. Available at: www.ncbi.nlm.nih.gov/pubmed/28356028 [Accessed November 19, 2018].

Zhang, P., Liu, G. & Chen, X., 2017. Nanobiotechnology: cell membrane-based delivery systems. *Nano Today*, 13, pp. 7–9. Available at: www.sciencedirect.com/science/article/pii/S1748013215301444 [Accessed November 20, 2018].

Zhang, S. et al., 1995. Self-complementary oligopeptide matrices support mammalian cell attachment. *Biomaterials*, 16(18), pp. 1385–1393. Available at: www.ncbi.nlm.nih.gov/pubmed/8590765 [Accessed November 22, 2018].

Zhang, S. et al., 1993. Spontaneous assembly of a self-complementary oligopeptide to form a stable macroscopic membrane. *Proceedings of the National Academy of Sciences of the United States of America*, 90(8), pp. 3334–3338. Available at: www.ncbi.nlm.nih.gov/pubmed/7682699 [Accessed November 22, 2018].

Zhang, X. et al., 2008. A biodegradable, immunoprotective, dual nanoporous capsule for cell-based therapies. *Biomaterials*, 29(31), pp. 4253–4259. Available at: www.ncbi.nlm.nih.gov/pubmed/18694595 [Accessed November 22, 2018].

Zhang, Z., Hu, J. & Ma, P.X., 2012. Nanofiber-based delivery of bioactive agents and stem cells to bone sites. *Advanced Drug Delivery Reviews*, 64(12), pp. 1129–1141. Available at: www.ncbi.nlm.nih.gov/pubmed/22579758 [Accessed October 27, 2017].

Zhao, C. et al., 2013. Nanomaterial scaffolds for stem cell proliferation and differentiation in tissue engineering. *Biotechnology Advances*, 31(5), pp.654–668. Available at: www.ncbi.nlm.nih.gov/pubmed/22902273 [Accessed November 19, 2018].

Zhao, L. et al., 2013. Stem cell labeling using polyethylenimine conjugated (α-NaYbF$_4$:Tm^{3+})/CaF$_2$ upconversion nanoparticles. *Theranostics*, 3(4), pp. 249–257. Available at: www.ncbi.nlm.nih.gov/pubmed/23606911 [Accessed November 20, 2018].

Zhao, X. et al., 2011. Active scaffolds for on-demand drug and cell delivery. *Proceedings of the National Academy of Sciences of the United States of America*, 108(1), pp. 67–72. Available at: www.ncbi.nlm.nih.gov/pubmed/21149682 [Accessed May 14, 2017].

Bio-Nanoparticles: Nanoscale Probes for Nanoscale Pathogens

Mohamed S. Draz
Brigham and Women's Hospital
Harvard University
Tanta University

Yiwei Tang
Agricultural University of Hebei

Pengfei Zhang
Tongji University School of Medicine
Shanghai Skin Disease Hospital

20.1 Introduction

Manipulating materials at the nanoscale size allows the creation of new nanomaterials that come with interesting and unique properties. However, this is not usually the case in biology, where size has a very different connotation—small microbes are difficult to treat or control. Viruses are among the smallest (mostly in the nanoscale) microbial structures and are able to infect all the known cellular forms of life, including humans, animals, plants, and even other microorganisms such as bacteria, yeast, and fungi [1]. The concept of applying nanoparticles modified with biomolecules is considered to be a promising approach for controlling and eliminating viral infections. When nanoparticles are applied as biosensing probes, the detection limit reaches zeptomolar concentrations, thereby opening up the possibilities for ultrasensitive detections [2,3]. The ability to detect marker biomolecules and analytes at this high sensitivity would not only allow accurate diagnostics but also enable efficient monitoring and treatment of infections.

The rapid, accurate, and early detection of viruses is crucial to initiate proper control procedures—particularly for highly contagious and deadly infections such as influenza [4], acquired immunodeficiency syndrome (AIDS) [5], hepatitis [6], dengue hemorrhagic fever and/or dengue shock syndrome (DHF/DSS) [7], and severe acute respiratory syndrome (SARS) [8]. The current gold standard for the detection of viral infections is the enzyme-linked immunosorbent assay (ELISA) technique, which detects the corresponding serological markers using enzyme-conjugated antibodies. Molecular biology methods—mainly based on deoxyribonucleic acid (DNA) fingerprinting using polymerase chain reaction (PCR) techniques such as multiplex PCR, gene chips, quantitative PCR, and real-time fluorescence quantitative PCR techniques—have also been widely described for sensitive and specific testing of viruses in both clinical and laboratory settings. However, most of these techniques still suffer from a number of potential limitations in accuracy and repeatability. Molecular assays are particularly expensive, are time-consuming, and typically require special instruments and skilled personnel [9,10]. Bio-nanoparticle probes for virus detection have been successfully prepared and tested through various modalities and formats [2,3,11–19]. Here, we discuss how bio-nanoparticle probes can be applied to overcome the limitations of standard techniques for virus detection (i.e., ELISA and PCR). The major types of nanoparticle systems used in virus detection (e.g., metal nanoparticles, carbon nanotubes, quantum dots [QDs], upconversion nanoparticles [UCNPs], polymer nanoparticles) are discussed along with the developed detection formats for different viruses.

20.2 Nanoprobes: Small, but Smart

The term "nanoparticle" widely encompasses a large number of nanostructures (in the range from 1 to 100 nm) that usually have size-dependent physicochemical properties, such as a large fraction of surface atoms, a high surface energy, spatial confinement effects, and reduced imperfections [20,21]. Nanoparticles are now easily produced in various shapes (e.g., spheres, tubes, wires, rods, cubes, fibers, films, porous, and other concentric core–shell structures, including hollow capsules or alloys) and size ranges through different chemical and physical synthesis approaches [21]. Metal nanoparticles, carbon nanotubes, semiconductor nanoparticles, UCNPs,

and polymer nanoparticles are widely reported for virus detection (Table 20.1). Coupled with biomolecules such as DNA, ribonucleic acid (RNA), antibody, antigen, peptide, streptavidin, and doxorubicin, nanoprobe systems were developed for different DNA and RNA viruses (Table 20.1). Single nanoparticle systems that include one type of nanoparticle to generate the biosensing signal are reported along with multi-nanoparticle systems that benefit from the properties of more than one type of nanoparticles. For example, magnetic nanoparticles (MNPs) serve to enrich and separate the target virus, while the sensing signals are generated by other nanoparticles such as gold nanoparticles or QDs. Although the biosensing activity of nanoprobes relies mostly on the physicochemical properties of the core nanoparticles, the size similarity of nanoprobes to viruses allows nanoprobes to serve as "smart systems" for virus detection. The nanoscale size of the detection probes permits (i) high reactivity due to the increased surface area available for biomolecule conjugation, (ii) rapid inward configuration of surface biomolecules, (iii) efficient dynamic probe–target interactions that significantly increase the specificity and sensitivity of detection, and (iv) reduced time and material costs per reaction. In addition to the advantages of using nanoprobes in virus sensing, different types of nanoparticles possess additional advantages for biosensing, and the superiority of nanoparticle-based systems over traditional approaches (i.e., ELISA and PCR) in terms of size,

TABLE 20.1   Bio-Nanoparticle Probes Used in Virus Detection

Bio-Nanoparticle System				
Nanoparticle		Biomolecule	Target Viruses[a]	References
Metal nanoparticles	Gold nanoparticles	DNA	DENV, EOBV, HAV, HBV, HCV, HCMV, KSHV, HEV, HIV, IV, SARS	[22–45]
		Antibody	HBV, HIV, HPV, IV, HSV, HTNV, RVFV	[46–59]
		Protein A	IV, HBV, HCV	[49,60]
		Peptide	HBV	[61]
		Streptavidin	HBV, HIV, IV	[33,62–64]
	Silver nanoparticles	DNA	EBOV, HBV, EBV, HCMV, HIV, HSV	[29,65–69]
		Antibody	HBV, HCV, HIV, RSV	[70–73]
	Iron oxide nanoparticles	DNA	HBV, HCV, HIV	[74–76]
		Antibody	HBV, IV, HSV	[77–83]
		Streptavidin	HBV	[78]
Carbon nanotubes		DNA	HCV, HBV, IV	[84–88]
		Antibody	IV, DENV, HPV, HBV, HIV	[89–93]
Graphene oxide		DNA	HBV, IV, EBOV, DENV, SV40	[94–98]
		Antibody	RV, DENV, IV, HBV, RSV	[99–104]
Semiconductor nanoparticles	QDs	DNA	HBV, HCV, IV, DENV, HPV, HIV, ZKIV	[105–110,111,112]
		Antibody	HBV, HCMV, HSV, IV, RSV	[61,113–117]
UCNPs		DNA	HPV, EBOV, HBV	[118–120]
		Antigen	Adenovirus, HIV	[121,122]
Polymer nanoparticles	Polystyrene nanoparticles	DNA	HSV	[123]
		Antibody	HIV	[124]
	Polystyrene/polyacrylic acid nanoparticles	DNA	HBV	[125]
		Antibody	IV	[126]

[a]DENV, dengue virus; EBOV, Ebola virus; EBV, Epstein–Barr virus; HAV, hepatitis A virus; HBV, hepatitis B virus; HCMV, human cytomegalovirus; HCV, hepatitis C virus; HEV, hepatitis E virus; HIV, human immunodeficiency virus; HPV, human papillomavirus; HSV, herpes simplex virus; HTNV, Hantaan virus; IV, influenza virus; KSHV, Kaposi's sarcoma-associated herpesvirus; RSV, respiratory syncytial virus; RV, rotavirus; RVFV, Rift Valley fever virus; SARS, severe acute respiratory syndrome; SV40, simian virus 40; WNV, West Nile virus; ZIKV, Zika virus.

performance, specificity, signal sensitivity, and stability was widely reported. The major biosensing-related advantages and disadvantages of different types of nanoparticles are summarized in Table 20.2.

20.3 Novel Biosensing Functions in Virus Detection: Exclusive to Nanoparticles

The unique properties and various size-based phenomena (e.g., large surface area-to-volume ratio, quantum confinement, superparamagnetism, and nanoplasmonics) of nanoparticles are now well established as interesting biosensing functions that can be particularly useful in overcoming major hurdles in virus detection—mainly arising from their nanoscale size and high genetic variability [139]. In this section, we discuss the nanoparticle-related phenomena which guided the emergence of sensing concepts and functions recently applied in virus detection.

20.3.1 Nano-Bioimmobilization and Encapsulation of Signaling Molecules

Reducing the particle size to the nanoscale results in an increased surface area-to-volume ratio, which is one of the core concepts for using nanoparticles in biosensing applications. The increase in the surface area improves the performance of nanoprobes by increasing the amount of surface available for (i) immobilization of biomolecules, (ii) interaction with the target analyte, and (iii) encapsulating signal molecules.

Immobilization of biomolecules on the surface of nanoparticles (nano-bioimmobilization) is a crucial step in the preparation of nanoprobes, and the successful immobilization of biomolecules directly affects the performance of the biosensing platform. There are two major strategies for the immobilization of biomolecules: (i) physical adsorption and (ii) chemical coupling (covalent or non-covalent bonding by amine (NH_2), thiol (SH_2), silanol (SiOH), and

TABLE 20.2 Summary of the Advantages and Disadvantages of Nanoparticles in Virus Detection

Nanoparticle	Potential in Virus Detection		References
	Advantages	Disadvantages	
Gold nanoparticles	Easy synthesis and modification Low cost and easy use Size and shape tenability High surface-to-volume ratio Efficient optical, electric, catalytic properties Chemical stability and biocompatibility	High sensitivity for the presence of electrolytes and pH changes Particle stickiness and its related instability and nonspecific surface interactions Limited fluorescence yields and broad emission bands	[127,128]
Silver nanoparticles	Size and shape tenability Small size/high surface-to-volume ratio Efficient optical, electric, and catalytic properties Biocompatibility	Difficult surface modification Chemical degradation and instability	[129]
Iron oxide nanoparticles	Simple and inexpensive synthesis and modification Small size/high surface-to-volume ratio Lower mass transfer resistance Stable and strong magnetism	Limited biocompatibility Thermal instability Wide size and moment distribution Uncontrolled clustering of its bioconjugates in biological systems	[130,131]
Carbon nanotubes	High surface-to-volume ratio Useful electronic, catalytic, and mechanical properties Semiconductor behavior makes them ideal for nanoscale field-effect transistors (FETs)	High manufacture and purification cost Difficult processing and manipulation Lack of solubility and susceptibility for clumping Toxicity	[132,133]
Graphene oxide	Facile surface modifications Large flat/smooth effective surface area Unique optical, electrical, catalytic, and high mechanical properties Chemical stability Water solubility and biocompatibility Acceptable processability	Lower electrical conductivity and poor thermal stability Low-yield production	[134]
QDs	Large surface-to-volume ratio Flexible solution processing Broad excitation with narrow and tunable emission spectra Large molar extinction coefficients and high quantum yields Photo and chemical stability	Heterogeneity in size Insolubility Difficult surface functionalization Rapidly decaying fluorescence gives rise to autofluorescence Blinking Toxicity and biologically incompatible	[135,136]
UCNPs	Infrared-to-visible upconversion behavior High quantum efficiency Narrow emission spectra and long lifetimes No autofluorescence Photochemical stability No blinking and limited photobleaching Reduced phototoxicity Improved biocompatibility High-resolution multiplexing capabilities	Complex preparation and surface modification Limited solubility in water	[137]
Polymer nanoparticles	Easy preparation and modification Transparency High loading and encapsulation capacities Chemical stability Water solubility and biocompatibility Lyophilization and redispersion possibilities	Nonspecific surface interactions Tendency to give a high background signal	[138]

aldehyde (CHO) groups) [140–142]. By following any of these approaches, the increased surface area of nanoparticles allows easy surface modification and high-yield immobilization of single or multiple biomolecules, thus enhancing the activity, stability, and regioselectivity of nanoprobes. Also, nanoparticles act as three-dimensional (3D) immobilization supports with a considerably low diffusional resistance that allows the immobilized molecules to behave freely as in the soluble state.

The ability to entrap thousands of functional molecules into one nanoparticle (i.e., encapsulation) offers a new concept for preparing biosensing probes—mainly by providing effective transport and protection for signaling molecules such as fluorescent dyes, enzymes, and electrolytes. In addition, the ability to use a large number of signaling molecules (e.g., dyes and electrolytes) coupled with an increased number of targeting biomolecules results in high signal amplification and detection sensitivity using nanoprobes. Silica nanoparticles, polymer nanoparticles, and liposomes are one of the most common nanoparticle encapsulation systems reported for virus sensing: silica and polystyrene nanoparticles provide a transparent matrix for the protection of organic fluorophores (i.e., from photobleaching and chemical instability) without reducing the fluorescence signal [143,144], while liposomes have been described to allow for a controlled release of the biosensing signal (e.g., electrical, optical, mechanical) that can be correlated with the presence and absence of a target virus [145,146].

20.3.2 Light Confinement and Quantized Conductance

Quantum confinement is a nanoscale phenomenon that refers to the change in the electronic and optical properties of nanoparticles due to the spatial localization of electrons and holes to dimensions that approach the so-called exciton Bohr radius. At this size, the principle of conservation of momentum breaks down, and new energy levels appear, which gives rise to new enhanced optical and conduction properties, such as (i) light confinement in semiconductor nanoparticles and (ii) quantized conductance [147,148].

Semiconductor quantum confinement into nanoscale dimensions approaching the exciton Bohr radius, typically ranging from 2 to 60 nm depending on the type of semiconductor, results in an increase in bandgap with decreasing size. The larger the bandgap is, the greater the energy needed for excitation and concurrently, the greater the resulting optical emission. As a corollary, it is possible to control the emission color resulting from semiconductor nanoparticles by altering their size with high precision. This size-dependent tuning of color is a key concept in the use of nanoscale semiconductors in several optical-based biosensing applications [149,150]. QDs, which are confined in all three dimensions, are widely studied and exploited in fluorescence virus-sensing applications. In simple terms,

their outstanding size-tunable, high-yield emission enables various highly flexible sensing applications, including high-resolution fluorescent tagging, multiplexed fluorometry, and nanoparticle-based fluorescence resonance energy transfer (FRET) assays [150].

Quantized conductance is one of a variety of unusual electron transport properties that originate from the quantum confinement effect at nanoscale dimensions. In nanoparticle systems such as nanowires and nanotubes, the size is extremely confined in one dimension, and the electronic mean free path λ of the particle becomes larger than its length (L), allowing the particles to behave similarly to electron waveguides. In this case, electron transport is ballistic, and each waveguide mode or conduction channel contributes equally to the total conductance of the wire; thus, the conductance becomes quantized and is given by NG_0, where $G_0 = 2e^2/h$ is the conductance quantum, N is the number of modes, e is the electron charge, and h is Plank's constant [151,152]. The conductance quantization in these one-dimensional (1D) structures is highly sensitive to the surface adsorption of molecules, which results in scattering of the conducting electrons on the particle surface and directly disrupts the number of conductance channels, causing variations in the conductance. This discrete electrical transduction of surface immobilization events offers a powerful biosensing concept similar to that of FETs. Nanotubes are entirely composed of surface atoms, and nanowires have a very high surface area; these properties endow them better modulation of electrical properties along with more efficient detection signal amplification than other nanostructures. These nano-based structural and electrical properties enabled these nanoparticles to be easily configured for ultrahigh-sensitive FET biosensing and potentially integrated into simple, real-time, and label-free virus-sensing systems [153].

20.3.3 Superparamagnetism: The Concept of External Target Control in Biosensing

Superparamagnetism is a nanoscale phenomenon that describes the state of single-domain particles of magnetic materials. This phenomenon takes place when the particle dimensions are critically reduced to a few nanometers and subdomains merge into one domain creating what are called single-domain particles. These nanoscale particles, as a single domain, are known to display uniform magnetization behavior with no domain-like magnetic interactions in the well-dispersed state. Hence, an object consisting of highly uniform and dispersed particles would exhibit magnetic properties under an applied magnetic field, but permanent magnetization would not persist when the applied magnetic field is removed [77,154]. This concept of a sharp magnetic–nonmagnetic transition that is characteristic of superparamagnetic nanoparticles provides several novel biosensing functionalities and applications. The "external control" concept, where superparamagnetic nanoparticles are simply used to induce controlled movements within the sensing system using an external magnetic field, is one of the most promising functionalities. This advantage offers many exciting and versatile bioconcentration and bioseparation functions in virus-sensing platforms. It is also ideal for enabling an efficient enrichment of nanoscale virus pathogens and undoubtedly increases and enhances the detection sensitivity in large-volume samples. Superparamagnetic nanomaterials modified with biomolecules such as DNA, antibodies, and streptavidin are commonly used to capture and concentrate the detection targets in various virus-sensing platforms [155].

20.3.4 Nanoplasmonics

Nanoplasmonics is a multitude of nanoscale photonic phenomena associated with the existence of surface plasmons (SPs) on metal nanoparticles. SPs—also known as surface plasmon polaritons (SPPs)—are electromagnetic waves that propagate and travel along the surfaces of metal nanostructures [20]. Excitation of SPs with light can cause free surface electrons to oscillate away from a metal nanoparticle in one direction, creating a dipole that can switch direction with a change in the electrical field of the incident light. When the frequency of the dipole plasmon is approximately the same as that of the incident light, a resonance condition is reached, leading to constructive interference and the strongest signal for the plasmon. This condition is referred to as localized surface plasmon resonance (LSPR) [156]. LSPR was observed to cause nanoscale localization of optical energy creating intense light absorption, scattering, enhancement, and even quenching effects that are characteristic of metal nanoparticles. These optical localization effects at the nanoscale make plasmonic nanostructures promising for various biosensing concepts and applications [156,157]. The high sensitivity of LSPR to changes in the structure, size, shape, and surrounding environment of metal nanoparticles was utilized in different plasmonic-sensing functionalities. Based on the specific changes in these parameters in response to the presence and absence of the target virus, highly sensitive assays were validated by measuring the absorption, scattering, enhancement and quenching, and optical properties of nanoparticles [158]. These specific properties have driven the ongoing intense interest in localized surface plasmons (LSPs) and have prompted the construction of many LSP-based virus detection probes using the resonant properties of plasmonic gold and silver nanoparticles [139,159].

20.4 Application of Bio-Nanoparticle Systems for Virus Detection

The concept of applying nanoparticles in virus sensing was first introduced in the late 1990s for human papillomavirus

using gold nanoparticles. Currently, there is a diverse range of nanomaterials, including metal nanoparticles, carbon nanotubes, silica nanoparticles, QDs, UCNPs, and polymer nanoparticles, that are being comprehensively investigated for virus detection [77,154,156]. One of the most common approaches for exploiting these nanostructures in virus detection is the development of bio-nanoparticle hybrid systems that contain one or more biomolecules derived from viruses, such as DNA, RNA, antibodies, antigens, or peptides (Table 20.1). The applications of bio-nanoparticle systems usually hinge on leveraging the labeling and signal transduction functions of nanoparticles and the specific activity of the conjugated biomolecules [77,154,157]. Such virus-specific nanoprobes have been widely reported in various optical, fluorometric, electrochemical, and electrical assays for the detection and monitoring of human viruses. The results of most of these studies demonstrate the potential of these nanoprobes, along with numerous advantages over traditional approaches, in terms of size, performance, specificity, signal sensitivity, and stability.

20.4.1 Metal Nanoparticles (Gold Nanoparticles, Silver Nanoparticles, Magnetic Nanoparticles)

Two classes of metal nanoparticles have been demonstrated in biosensing applications: noble metal nanoparticles (i.e., Au, Ag, Pt, and Pd) and transition metal nanoparticles (i.e., Fe, Ni, and Cu). Metal nanoparticles are easy to synthesize and be modified with biomolecules. In addition to their many attractive optical, electronic, catalytic, and magnetic properties, metal nanoparticles have great potential to build versatile biosensing systems. In particular, nanoparticles prepared from gold, silver, and iron are the most common nanoparticle-sensing probes applied in virus detection. The plasmon resonance properties of gold and silver nanoparticles are strongly sensitive to the size and shape of the nanoparticles and the dielectric properties of the surrounding medium and can thus be readily tuned for various optical and electrical sensing modalities. For example, spherical- and rod-shaped gold nanoparticles were applied for the colorimetric, gravimetric, fluorometric, and electrochemical detection of viruses [22–44,46–58,60–63]. Likewise, spherical silver nanoparticles with a size range of 5–50 nm were used in colorimetric, fluorometric, and electrochemical detection [29,65–72]. On the other hand, iron oxide nanoparticles have attracted significant attention in virus-sensing applications because of their interesting superparamagnetic properties. MNPs were primarily integrated with PCR, immunological, and electrochemical assays to concentrate and enrich the target viral nucleic acid and protein, allowing further enhancements in the sensitivity and specificity of detection [74–83]. For example, MNPs functionalized with antibodies were applied as magnetic relaxation switches for the detection of HBV, IV, and human simplex virus (HSV) [77].

20.4.2 Graphene and Carbon Nanotubes

Graphene and carbon nanotubes are two allotropic forms of carbon. Graphene is a two-dimensional (2D) carbon nanostructure consisting of one-atom-thick planar sheets in which sp^2-hybridized carbon atoms are bonded together in a hexagonal network. Carbon nanotubes are 3D cylindrical carbon nanostructures composed of rolled graphene sheets and can exist as single-walled or multiwalled tubes. A single-walled carbon nanotube (SWCNT) is made of a single graphene sheet rolled in a form of cylinder. A multiwalled carbon nanotube (MWCNT) comprises concentrically nested cylinders of graphene sheets with an interlayer spacing of 0.34–0.36 nm [160]. Their unique physical, chemical, and mechanical characteristics, such as high surface-to-volume ratio, high catalytic activity, and efficient electron transfer capabilities, make them ideal materials for the development of advanced and reliable virus biosensing systems [160,161]. Graphene oxide, which is oxygenated graphene, modified with DNA and whole antibodies was applied for the electrochemical detection of HBV and rotavirus, respectively [94,99]. The recently observed plasmonic quenching properties of graphene oxide were used in the fluorometric detection of SV40 [95]. Both SWCNTs and MWCNTs were tailored for the electrochemical monitoring and detection of HBV DNA [84,85]. The further quantized electronic conductance characteristics of carbon nanotubes allow their use as nanotube FETs for the electrical sensing of HCV and IV [86–88].

20.4.3 Semiconductor Nanoparticles (Quantum Dots)

QDs are highly crystalline fluorescent semiconductor (e.g., II–VI) nanocrystals made of semiconductors with strong and tunable spectral emission. The electrons within QDs are confined to discrete energy levels, which are governed by the QD size in a phenomenon known as quantum confinement; therefore, the emission bands are tunable by changing the size of the QDs [162,163]. Typical QDs are composed of three layers: a crystal core layer of binary alloys, such as cadmium selenide (CdSe), cadmium sulfide (CdS), indium arsenide (InAr), or indium phosphide (InP) coated with a layer of ZnS, which has a larger bandgap and efficiently confines excitation to the core to improve the optical properties, and a surface layer of polymer shell responsible for solubility and functionalization [163]. As a class of unique fluorescent nanomaterials, QDs have been described by several reports as effective alternatives to traditional organic dyes due to their size- and structure-dependent spectral emission, broad excitation spectra, narrow emission spectra, and outstanding photostability. These intrinsic nanophotonic properties have expanded their applications in biosensing, and currently, a wide variety of biofunctionalized QDs (i.e., those modified with DNA, RNA, and antibodies) are applied as fluorescent probes for the detection of human viruses, such as

DENV, HBV, HCV, HCMV, HIV, HSV, IV, RSV, and ZIKV [61,105–112].

20.4.4 Upconversion Nanoparticles (UCNPs)

UCNPs are a class of nanomaterials able to emit high-energy visible light upon excitation with low-energy near-infrared (NIR) light through a process referred to as upconversion luminescence (UCL) [164–166]. The most common types of UCNPs are hybrid structure materials of rare-earth lanthanide ions, such as Pr, Eu, Ho, Er, Tm, and Yb, doped in a crystalline host and generally known as lanthanide chelates. The electron configuration of these ions is characterized by the presence of 4f electrons that are unusually shielded by the filled $5s^2$ and $5p^6$ subshells. This unusual electronic structure results in very interesting luminescent characteristics in lanthanide-doped UCNPs, such as sharp line-like emission bands, large anti-Stokes shifts (up to 500 nm) that separate discrete emission peaks from the infrared excitation light, long-lived emission on a micro- to millisecond timescale, and high photostability [166]. Because of the remarkable light penetration depth and the absence of autofluorescence in biological specimens of infrared excitation, these UCNPs are ideal luminescent probes for highly sensitive biosensing applications [164,167]. Direct, homogeneous, and time-resolved assays are among the most intriguing applications recently facilitated by UCNP probes. UCNPs are also expected to obtain properties common at the nanoscale, including high surface-to-volume ratio and quantum confinement effects that result in additional novel applications in virus sensing and detection [118–121]. Recently, $NaYF_4$: Yb^{3+}, Er^{3+} UCNPs of 55 nm in size showing UCL in the 520–560 nm (green emission) and in the 650–700 nm (red emission) regions were synthesized and employed to detect a selection of ten common adenovirus genotypes causing human infections [122]. In another adenovirus detection assay, antibody-modified europium (Eu^{3+}) UCNPs were applied as highly fluorescent probes for viral antigen detection. Eu^{3+} nanoparticles coated with streptavidin and HIV-derived antigens were also applied in HIV detection assays [121].

20.4.5 Polymer Nanoparticles (Polystyrene Nanoparticles, Polyacrylic Acid Nanoparticles)

Because polymers can be readily made in large quantities with precise control over their structural and functional properties, they are uniquely versatile precursors for nanoparticle synthesis. Polymer nanoparticles with well-defined colloidal and surface characteristics have attracted considerable recent interest as superior building blocks for nanostructures [168,169]. In the biosensing field, these nanoparticles are frequently used as nanoscale detection platforms. In particular, polymer nanoparticle systems rely on immunological recognition reactions between an antigen and an antibody or on the hybridization of two complementary DNA single strands. The high surface-to-volume ratio and good structural and morphological controllability of polymer nanoparticles make them ideal uniform platforms for virus sensing with enhanced immobilization and bioconjugation capacity [170]. The remarkable encapsulation capability of polymer materials is another prominent feature of polymer nanoparticles that is of great interest in biosensing applications. Polymer nanoparticle encapsulants are currently used to enhance the photonic and chemical stability of different fluorescent dyes and to enhance signal amplification by increasing the number of impregnated dye molecules [144,171]. A variety of polymer nanoparticles for detection purposes have been prepared. Among them, styrene/acrylic acid copolymer nanoparticles that were impregnated with ArcaDia BF 530 dye and that carried influenza B virus antibodies were used to improve influenza B virus detection performance in a fluorescent nanoparticle tracer-based approach [126]. Fluorescent carboxylated polystyrene nanoparticles (mean diameter, ∼92 nm; Dragon Green dye, excitation 480 nm, emission 520 nm) were also conjugated with a gp120 monoclonal antibody specific for HIV [124].

20.5 Gold Nanoparticle for the Molecular and Immunological Detection of Viruses

20.5.1 Colorimetric Assays

Colorimetric assays are among the most popular biosensing methods used in virus detection. These assays have drawn considerable attention as simple and inexpensive tests in which the molecular detection events can be transformed into color changes that are easily observed by only the naked eye without the need for sophisticated instruments [172]. Generally, solutions of well-dispersed gold nanoparticles display a ruby red color, while solutions of aggregated particles exhibit a purple or blue color. This color change of the nanoparticle solution is associated with interparticle plasmon coupling that generates a significant absorption band shift in the visible region of the electromagnetic spectrum (Figure 20.1) [173]. Therefore, several sensing platforms based on the characteristic red color of gold nanoparticles or their color change from red to purple upon aggregation were developed (Table 20.3). In these platforms, two main approaches for the molecular detection of the viral nucleic acid were reported. The first approach relies on the simple use of gold nanoparticles as a labeling probe

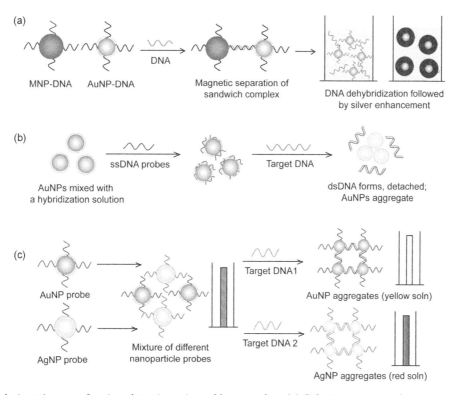

FIGURE 20.1 Colorimetric assays for virus detection using gold nanoprobes. (a) Colorimetric assay of viruses using gold nanoparticles (AuNPs) and silver metal enhancement in solution [24]. (b) Colorimetric detection of viruses based on the change of the color of AuNP's solution caused by the aggregation of AuNPs [27,31]. (c) Multiplex colorimetric one-pot nanoparticle-based assay. This assay includes the use of a mixture of silver nanoparticle (AgNP)–DNA and AuNP–DNA conjugates, and upon the addition of sample DNA, complementary DNA hybridization triggers the aggregation of bare nanoparticle conjugates and the subsequent decrease or disappearance of its corresponding color [36].

TABLE 20.3 Gold Nanoparticle-Based Assays for the Molecular and Immunological Detection of Viruses

Assay	Virus[a]	Bio-Nanoparticle System[b] — Gold Nanoparticle (nm)	Biomolecule	Detection Limit	Detection Range	Detection Platform	References
Colorimetric	HAV	Spherical 13	DNA	20 fM	At 20 fM	Chip	[22]
	HBV	Spherical 15 to 8	DNA	0.36 pM	0.3 µM to 0.3 pM	Chip	[23]
	HCV	Spherical 8–15	DNA	0.36 pM	0.3 µM to 0.3 pM	Chip	[23]
		40	DNA	2 fmol	30 to 2 fmol	Stripe	[27]
	HEV	Spherical 14	DNA	100 fM	1×10^4 to 50 fM	Chip	[174]
	HIV	Spherical 60	DNA	~11 $\log_{10}$ copies/mL	11 to 12.5 $\log_{10}$ copies	Stripe	[38]
		Spherical 13	DNA	10 fM	10 nM to 1 fM	Microtiter plate	[24]
	HSV	Spherical ND	Antibody	ND	ND	Strip	[59]
	KSHV	Spherical 15	DNA	~1 nM	1 mM to 10 pM	Tube	[36]
	IV	Rod $L \leq 30$, $D \leq 10$	None	1 pg	0.1 ng to 0.01 pg	Tube	[175]
		Spherical 32	Antibody	40 to 0.1 ng	100 to 0.1 ng	Stripe	[176]
		Spherical 10	Pentabody	10 ng/mL	10^{-1} to 10^{-4} µg/mL	Tube	[177]
	SARS	Spherical 13	DNA	60 fmol	ND	Tube	[31]
Electrical, electrochemical	HBV	Spherical 16	Antibody	0.1 ng/mL	650 to 0.5 ng/mL	Electrodes	[46]
		~10	Antibody	2.3 pg/mL	1.0 to 0.01 ng/mL	Electrodes	[53]
		16	Antibody	87 pg/mL	1,500 to 0.1 ng/mL	Electrochemical cell	[54]
		5	Avidin	0.7 ng/mL	1.47 to 0.7 ng/mL	Electrochemical cell	[33]
	HCMV	Spherical 20	DNA	5 pM	5,000 to 5 pM	Electrodes	[25]
	HCV	Spherical 10–15	DNA	~1 pM	2.0 to 0.01 nM	Electrochemical cell	[35]
		12	DNA antibody	1 pg/µL	10 ng/µL to 1 pg/µL	Electrodes	[178]
	HIV	Spherical 3	DNA	0.34 fM	1.0 µM to 0.1 pM	Electrodes	[39]
	IV	Spherical ND	Avidin	577 pM	3 to 0.001 pmol	Electrodes	[62]
		ND	Daunorubicin	0.4 pM	1.0 nM to 5.0 pM	Electrodes	[41]
		3	DNA	10 pM	10 pM to 100 nM	Chip	[42]
	SARS	Spherical 70	DNA	2.5 pM	50 to 2.5 pM	Electrodes	[179]
Immuno-PCR (IPCR)	HIV	Spherical 30	Antibody/DNA	1 pg/mL	10,000 to 1 pg/mL	Microtiter plate	[55]
	HTNV	Spherical 30	Antibody/DNA	10 fg/mL	10^5 to 1 fg/mL	Microtiter plate	[49]
	HCV	Spherical 8	DNA	300 fM	550 to 15 fM	ND	[32]
Fluorometric	HBV	Rod $L = 46$, $D = 13$	Antibody	8.3 ng/mL	Up to 264 ng/mL	ND	[180]
		ND, $D = 13$	DNA	15 pM	6.0 to 0.045 nM	ND	[26]
	IV	Spherical 5.5	Peptide	0.1 pg/mL	0.1 to 0.0001 ng/mL	Microtiter plate	[61]
		Spherical 25	Protein A	13.9 pg/mL	800 to 12.5 ng/mL	ND	[181]
		ND	DNA	25 nM	50 to 25 nM	ND	[182]
		22	Antibody	0.09 ng/mL	12 to 0.27 ng/mL	Strip	[57]
Light scattering	HPV	Spherical 16	DNA	1 fM	100 to 0.001 pM	Microbead array	[183]
	EBOV	Spherical 13	DNA	20 fM	20 fM	Chip	[29]
	HAV	Spherical 13	DNA	20 fM	At 20 fM	Chip	[29]
	HBV	Spherical 10, 50, and 100	Antibody	0.005 IU/mL	1–0.005 IU/mL	Tube	[51]

(Continued)

TABLE 20.3 (Continued) Gold Nanoparticle–Based Assays for the Molecular and Immunological Detection of Viruses

Assay	Virus[a]	Bio-Nanoparticle System[b]		Biomolecule	Detection Limit	Detection Range	Detection Platform	References
		Gold Nanoparticle	(nm)					
	HCV	Spherical	8–15	DNA	0.36 pM	0.3 µM to 0.3 pM	Chip	[23]
	HIV	Spherical	15	Avidin/DNA	0.2 pM	31.4 pM to 1.6 fM	Chip	[64]
			13	DNA	10 fM	10 nM to 1 fM	Microtiter plate	[24]
	IV	Spherical	20	Antibody	<100 $TCID_{50}$/mL	1.4×10^6 to 5.5×10^3 $TCID_{50}$/mL; 4.8×10^5 to 7.2×10^1 $TCID_{50}$/mL	Tube	[58]
Quartz crystal microbalance (QCM)	DENV	Spherical	13	DNA	2 PFU/mL	2×10^6 to 2 PFU/mL	Chip	[28]
Atomic force microscopy (AFM)	HIV	Spherical	20	Antibody	0.025 pg/mL	3,500 to 50 copies/mL	Chip	[184]
Surface-enhanced Raman scattering (SERS)	IV	Spherical	ND	DNA	25 nM	50 to 25 nM	ND	[182]
	RVFV	Spherical	60	Antibody	5 fg/mL	50 to 0.005 pg/mL	Glass slide	[48]
	WNV	Spherical	60	Antibody	5 fg/mL	50 to 0.005 pg/mL	Tube	[48]
Inductively coupled plasma mass spectrometry (ICP-MS)	DENV	Spherical	ND	DNA	1.6 fM	5 pM to 5 fM	Tube	[30]

[a]DENV, dengue virus; EBOV, Ebola virus; HAV, hepatitis A virus; HBV, hepatitis B virus; HCMV, human cytomegalovirus; HCV, hepatitis C virus; HEV, hepatitis E virus; HIV, human immunodeficiency virus; HPV, human papillomavirus; HSV, herpes simplex virus; HTNV, Hantaan virus; IV, influenza virus; KSHV, Kaposi's sarcoma-associated herpesvirus; RVFV, Rift Valley fever virus; SARS, severe acute respiratory syndrome; WNV, West Nile virus.

[b]ND, not determined; D, diameter; L, length.

followed by a color enhancement step using silver or gold metal ions and is generally called the "amplification coloring technique" [22,23]. Using this approach, a colorimetric assay for HIV detection was developed. In this assay, MNP-DNA probes capture the target HIV sequences, which are then magnetically collected and labeled with gold nanoparticle–DNA probes. After repetitive washing, the gold nanoparticles are released into solution by a dehybridization step and subjected to silver metal enhancement, resulting in particle growth and a change in the color of the solution (Figure 20.1a) [24]. The second approach called the "gold aggregation technique" depends on the color change of gold nanoparticle solutions from red to purple in response to particle aggregation. This gold nanoparticle aggregation could be either non-crosslinking aggregation caused by the target-triggered removal of stabilizing ligands from surfaces of gold nanoparticles or interparticle-crosslinking aggregation caused by the binding of ligands modified onto gold nanoparticles with target analytes (Figure 20.1b,c) [144,148,149,151–153]. A simple colorimetric hybridization assay based on a non-crosslinking gold aggregation was developed for SARS detection. In this assay, the difference in electrostatic properties between single-stranded DNA (ssDNA) and double-stranded DNA (dsDNA) was used to sense the presence of the target DNA. The working principle of this assay is that the ssDNA molecules can readily stick to the surface of gold nanoparticles and stabilize the nanoparticle solution. Upon hybridization with the target viral nucleic acid, dsDNA is formed, and the gold nanoparticles aggregate, as dsDNA does not tend to be adsorbed on the surface of nanoparticles, causing a visible color change to blue (gray in print) (Figure 20.1b) [31]. Using the target-triggered aggregation of gold nanoparticles, a colorimetric one-pot nanoparticle-based assay for KSHV was developed to differentiate KSHV infection from infection by a frequently confounding Bartonella species. With the addition of sample DNA to a mixture of silver nanoparticle–DNA and gold nanoparticle–DNA conjugates specific for KSHV and Bartonella DNA, respectively, complementary DNA hybridization triggers the aggregation of bare nanoparticle conjugates, causing a significant change or decrease in the color intensity of the corresponding nanoparticles (Figure 20.1c) [32].

20.5.2 Electrical and Electrochemical Assays

Electrical and electrochemical assays are usually presented as elegant methods for low cost, easy-to-use, reliable, and robust biosensing platforms that have excellent detection limits and that use small sample volumes [185,186]. A typical electrochemical sensing platform comprises a sensing electrode (or a working electrode) and a counter electrode separated by a thin layer of electrolyte. The integration of the excellent conductive and catalytic properties of gold nanoparticles into electrochemical sensing platforms resulted in the development of various new electrochemical detection techniques in which the nanoparticles are used to either label the capture target or modify the detection electrodes, allowing enhanced selectivity and sensitivity of the detection [185]. Recently, novel strategies for developing multiple electrochemical assays for virus detection with high sensitivity using gold nanoparticles have been proposed (Table 20.3). Modification of electrode surfaces with gold nanoparticles is commonly performed to enhance the direct electron transfer to the electrode surface through their conducting channels and thus allow sensitive and rapid detection of the target analyte captured on the surface of the electrodes [179]. Using this approach, a gold nanoparticle-screen-printed carbon electrode was used for the electrochemical detection of SARS infection. The gold nanoparticle-modified electrode was modified with capture DNA probes allowed to hybridize with biotinylated targets. Streptavidin–alkaline phosphatase (S-AP) was then applied for the indirect reduction of silver ions in the solution into a metallic deposit. The formed silver metal deposits were then electrochemically measured to determine the target virus DNA concentration (Figure 20.2a) [179]. Gold nanoparticle–antibody conjugates labeled with HRP were used for the electrochemical detection of HBV. The target HBV surface antigen was captured and labeled with gold nanoparticle conjugates on the surface of the gold electrode. Then, the detection signal was measured using differential pulse voltammetry (DPV) to assess the reduction of H_2O_2 catalyzed by the bound HRP (Figure 20.2b) [46]. Similar electrochemical assays that combine gold nanoparticles with DNA hybridization were developed for the molecular detection of viral nucleic acid. These strategies include (i) the electrochemical detection of gold nanoparticle labels by detecting the silver ions or gold ions released after acidic dissolution. Specifically, HCMV was electrochemically detected by the acidic dissolution of gold nanoparticles. HCMV DNA was directly labeled with specific gold nanoparticle modified with DNA probes. After the acidic dissolution of the gold nanoparticles, Au(III) ions were measured by anodic stripping voltammetry (ASA) (Figure 20.2c) [25]. (ii) Gold nanoparticles were used as carriers of other nanoprobes of gold nanoparticles. Specifically, a bio-barcode amplification (BCA)-enhanced electrochemical immunoassay using gold nanoparticles dually modified with bio-barcode DNA and antibodies was reported. Gold nanoparticle probes are used to label the target HCV antigen captured by MNP probes. The bio-barcode ssDNA is released and employed to form gold nanoparticle multilayers on the electrodes to increase the produced electric current (Figure 20.2d) [178].

20.5.3 Immuno-PCR Assays

Gold nanoparticle-based PCR assays are fundamentally exploiting the prominent amplification capabilities of the PCR technique together with the large surface area-to-volume ratio of nanoparticles to enable the highly sensitive

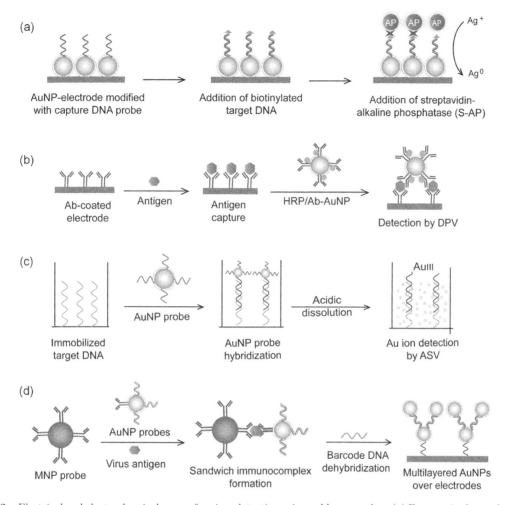

FIGURE 20.2 Electrical and electrochemical assays for virus detection using gold nanoprobes. (a) Enzymatic electrochemical detection of viruses by ASA using a gold nanoparticle (AuNP)-screen-printed carbon electrode [179]. (b) An enzyme-catalyzed electrochemical immunoassay of the virus using AuNP-Ab/HRP, and Ab-modified Au electrode [46]. (c) Electrochemical detection of viral nucleic acid by the acidic dissolution of AuNPs and ASA [25]. (d) BCA-enhanced electrochemical immunoassay using AuNPs dually modified with bio-barcode DNA and antibodies and MNP probes [187].

molecular detection of viruses. In gold nanoparticle-based reverse transcription PCR (RT-PCR) and real-time PCR detection schemes, nanoparticles were found to increase the amplification yield of the PCR product and shorten the PCR time compared to those for the conventional reaction protocol [188,189]. Additionally, IPCR was developed using gold nanoparticle. IPCR is a hybrid technology that combines the primary protocols of both PCR and ELISA by replacing the antibody–enzyme conjugates used in ELISA with antibody–DNA conjugates that are amplified by the PCR technique. In IPCR, the large surface area-to-volume ratio of nanoparticles increases the ratio of DNA conjugated to the detection antibody compared to that obtained with traditional antibody–DNA coupling chemistry and consequently improves the sensitivity and specificity of target detection. IPCR was applied to greatly enhance the immunological detection of Hantaan virus and HIV using gold nanoparticle-based antibody–DNA conjugates (Figure 20.3 and Table 20.3) [49,55].

20.5.4 Fluorometric Assays

Fluorescence assays that rely on the quenching efficiency of gold nanoparticles were efficiently applied for the detection of both viral nucleic acids [32] and antigens [190] (Table 20.3). Due to the ability of gold nanoparticles to increase the nonradiative rate and to reduce the radiative rate of fluorophores in their proximity, an intrinsic quenching efficiency of gold nanoparticles has been reported and is known to extend over much larger distances (up to 21 nm) than those associated with molecular quenchers in the Förster mechanism [26,27,50,61,182]. In these assays, gold nanoparticles act as efficient fluorescent dye quenchers, leading to further increases in the efficacy of biosensing systems. FRET-induced quenching of fluorophores on the surface of gold nanoparticles was applied for the detection of HBV using several protocols. Spherical- and rod-shaped gold nanoparticles were applied to quench the fluorescence of fluorophores, such as QDs and fluorescein amidite (FAM)

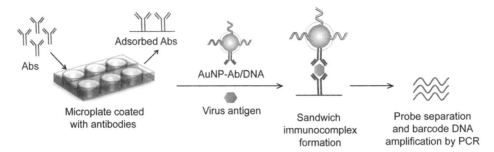

FIGURE 20.3 Immuno-PCR assay for virus detection. Gold nanoparticle (AuNP) probes dually functionalized with antibody (Ab) and dsDNA are used to label virus antigens pre-captured on a microplate. The probes are carrying dsDNA that includes barcode ssDNA for signal amplification. The barcode DNA is separated, amplified, and detected by gel electrophoresis [49].

(Figure 20.4a–c) [26,50,61]. A similar gold nanoparticle FRET-based scheme for the detection of IV has also been reported. In this assay, ssDNA probes modified with a fluorescent dye are attracted to the surface of gold nanoparticles, resulting in quenched emission. The presence of target DNA sequences results in the formation of dsDNA in a suspension of citrate-reduced gold nanoparticles and increases the fluorescence recovery intensities due to the desorption of the dsDNA–dye conjugates from the surface of the gold nanoparticles (Figure 20.4d) [27]. In addition, fluorometric detection of HCV using Cy3-tagged ssRNA that is electrostatically adsorbed or covalently coupled through

thiol (SH) chemistry to the surface of gold nanoparticles was reported. The coupling of gold nanoparticles and ssRNA probes brings Cy3 in the vicinity of the surface of the gold nanoparticles and eventually results in emission quenching; upon hybridization with the target RNA, the quenched emission is released (Figure 20.4e) [182].

20.5.5 Light Scattering Assays

Light scattering-based detection involves the measurement of the amount of light scattered by a solution using light spectroscopy techniques. Due to their nanoscale size and

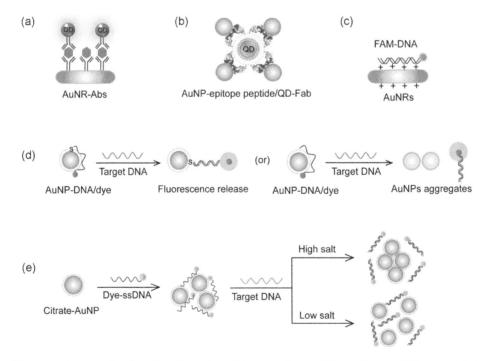

FIGURE 20.4 Fluorometric assays for virus detection using gold nanoprobes and fluorophores. (a) Fluorometric immunoassay by the FRET-induced quenching of QDs using gold nanorods (AuNRs) [180]. (b) Fluorometric immunoassay by the FRET-induced quenching of QDs using spherical gold nanoparticles (AuNPs) [61]. (c) Fluorometric assay for virus nucleic acid detection by the FRET-induced quenching of FAM using gold nanorods (AuNRs) [26]. (d) Fluorometric detection of the nucleic acid of viruses using Cy3-tagged ssRNA that is electrostatically adsorbed or covalently coupled through thiol (SH) chemistry to the surface of AuNPs [27]. (e) Fluorometric detection by the FRET-induced quenching of fluorophores by AuNPs and based on the difference of binding affinity of ssDNA and dsDNA to the surface of AuNPs [182].

solution stability, gold nanoparticles have very strong light scattering properties, which offer the possibility for the simple, rapid, and sensitive detection of viral nucleic acids. Additional deposition of silver metal on the surface of gold nanoparticles increases the nanoparticle size and improves the detection sensitivity of the technique by amplifying the scattering signal [24]. According to this silver enhancement approach, a light scattering-based assays for the nucleic acid detection of EBOV, HAV, HCV, and HIV were developed, and similar schemes for the immunological detection of HBV and IV were reported (Figure 20.5 and Table 20.3) [24,42,60].

20.5.6 Quartz Crystal Microbalance Assays

QCM is a simple, cost-effective, high-resolution mass-sensing technique. The core part of the QCM is the piezoelectric AT-cut quartz crystal sandwiched between a pair of electrodes. When the electrodes are connected to an oscillator and an AC voltage is applied over the electrodes, the quartz crystal starts to oscillate at its resonance frequency due to the piezoelectric effect [28,191,192]. QCM sensors modified with gold nanoparticle–DNA conjugates were applied for the detection of DENV type 1 (DENV-1) (Table 20.3). Gold nanoparticles have a definite mass and

relatively large surface area for conjugation and molecular interaction, and when integrated with QCM, gold nanoparticles act as effective mass amplifiers to enhance the detection sensitivity. In this system, the QCM works as a highly sensitive transducer for the mass increase of the DNA-modified quartz crystal upon DNA hybridization, and gold nanoparticles were applied as a signal amplifier (Figure 20.6) [28].

20.5.7 Atomic Force Microscopy Assays

AFM is one of the techniques most widely used to study surface features. By allowing the visualization and quantification of surface-immobilized biomolecules in a specified sensing region, AFM has drawn great attention to its biosensing applications. Using surface immobilization techniques for biomolecules combined with coherent surface characterization using AFM led to the development of several biosensing assays and platforms [193,194]. Gold nanoparticles were utilized to label the captured biomolecules, causing significant topography changes that directly amplify the detection signal. This approach has typically been applied for the immunological detection of viruses such as HIV through the immobilization of virus-specific antibodies on the surface of a chip followed by the binding of the target virus protein and labeling with gold

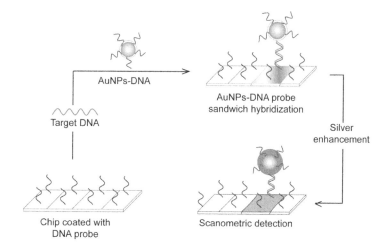

FIGURE 20.5 Light scattering-based assay for virus detection on-chip using gold nanoprobes and silver staining. The target DNA is captured on chip modified with DNA capturing probes and then directly labeled by gold nanoprobes, followed by silver enhancement to allow for a highly sensitive scanometric detection based on light scattering. This detection scheme has been used to detect viruses, such as EBOV, HAV, HCV, and HIV [24].

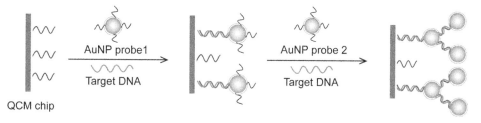

FIGURE 20.6 QCM-based assay for virus detection using DNA–gold nanoprobes. The presence of the target DNA induces the assembly of AuNP multilayers on a QCM chip and amplifies the detection signal [28].

nanoparticles. The nanoparticle immunocomplexes formed on the surface of the chip are readily detected by AFM, allowing the very sensitive and specific detection of the virus [47] (Figure 20.7 and Table 20.3).

20.5.8 Surface-Enhanced Raman Spectroscopy Assays

SERS is a Raman spectroscopic (RS) technique that provides Raman scattering enhanced by molecules adsorbed on rough metal surfaces (i.e., Ag, Au, and Cu). The intensity of the Raman signal can be enhanced up to 10^8 to 10^{14} times, which means that a single molecule can be detected using this technique [195,196]. Similar to RS, the main protocol of this technique involves shining a laser onto the sample, which is placed close to a metallic surface. Of the light absorbed by the sample, most of it will be scattered back at the same laser wavelength; however, a small portion of it will be inelastically scattered at a series of different wavelengths that are indicative of the vibrational transitions in the molecules forming the tested sample. Because different molecules have different vibrational modes, the obtained spectrum of the scattered light can be considered a molecular fingerprint and used to uniquely identify the interrogated molecules [197]. The discovery of SERS-based testing of molecules conjugated or positioned near the surface of metal nanostructures has renewed interest in plasmon resonances of gold nanoparticles. The excited SPs of gold nanoparticles were found to enhance the electromagnetic energy in the vicinity of the target molecule and significantly increase the intensity of the resulting Raman signal. This signal enhancement is commonly explained by a combination of an electromagnetic mechanism describing the surface electron movement and a chemical mechanism related to a charge transfer (CT) process [195,197,198]. High molecular fingerprinting ability with enhanced detection signal and relatively narrow spectral bandwidths enables ultrasensitive and multiplexing applications in virus detection (Table 20.3). A multiplexed SERS assay for nucleic acids has been developed for the detection of RVFV and WNV using Raman reporter dye-coated gold nanoparticles and MNPs. Gold nanoparticles and MNPs are conjugated with a polyclonal antibody specific for the target virus antigen, forming gold nanoparticle/virus antigen/magnetic complexes that, upon excitation with a laser, yield signature spectral peaks of the Raman dye with an intensity proportional to the virus antigen concentration (Figure 20.8) [48]. In addition, a chip-based SERS assay was developed for EBOV detection using gold nanoparticles modified with Cy3 dye. The target DNA oligonucleotides are captured on the surface of the chip and labeled with Raman-active gold nanoparticle–DNA probes followed by silver enhancement. The deposition of silver metal enhances the Raman signal of Cy3 present on the surface of the gold nanoparticles, allowing highly sensitive detection of the virus. This assay has also been reported for the detection of a wide range of viral nucleic acids, including those of HAV, HBV, and HCV, which belong to different virus groups [29].

20.5.9 Inductively Coupled Plasma Mass Spectrometry Assays

ICP-MS is a mass spectrometry-based elemental analysis technique. ICP-MS analysis relies on employing inductively

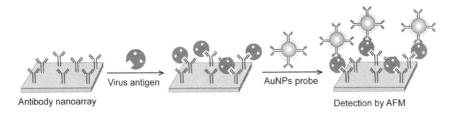

FIGURE 20.7 AFM-based assay for virus detection using gold nanoprobes and antibody nanoarray.

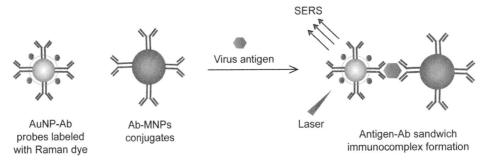

FIGURE 20.8 SERS-based assay for virus detection using gold nanoparticles. Raman reporter dye-coated gold nanoparticles (AuNPs) and MNPs were used to sandwich virus antigens forming AuNP/virus antigen/MNP complexes. Then, a 785 nm laser excites the magnetically concentrated AuNP/virus antigen/MNP complexes, and the presence of Raman dye results in specific SERS signature peaks [48].

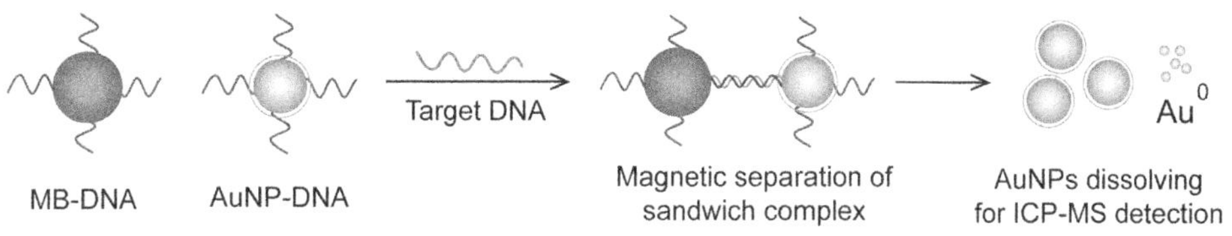

FIGURE 20.9 ICP-MS-based assay for virus detection using gold nanoprobes and magnetic separation technique. MB–DNA and AuNP–DNA sandwich the target DNA sequence, forming three-component hybridization complexes, which are magnetically separated and analyzed by ICP-MS after acidic dissolution [30].

coupled plasma as an ionization source and a mass spectrometer (MS), usually with a quadrupole mass filter, to separate and detect the ions produced. The ICP-MS technique is capable of performing qualitative, semiquantitative, and quantitative analyses with an excellent sensitivity limit down to the sub-nanogram per liter or parts per trillion level [199,200]. Based on the unprecedented sensitivity of the ICP-MS technique in target detection and analysis, elemental tags were applied to develop biosensing and detection systems. In these systems, ICP-MS provides advantages over other techniques in terms of its high sensitivity, large dynamic range, high specificity, high throughput, low matrix effects, low background interference, great potential for multiplexed detection, and superior spectral resolution and sensitivity [30,200]. Gold nanoparticles, due to their uniform size and well-defined atomic composition and atom number per bioconjugate, are attractive biosensing elemental tags. In addition, the active surface properties of gold nanoparticles allow signal amplification via further metal enhancement steps to achieve highly sensitive detection. An ICP-MS-based assay using magnetic separation and labeling with gold nanoparticle–DNA probes for the molecular detection of DENV-1 was developed. Magnetic bead (MB)-DNA and gold nanoparticle–DNA complexes sandwich the target DNA sequence, forming three-component hybridization complexes, which are separated by magnet and then analyzed by ICP-MS (Figure 20.9 and Table 20.3) [30].

20.6 Summary and Concluding Remarks

The applications of bio-nanoparticle systems in virus detection have expanded to encompass most of the clinically relevant groups of human viruses. The synthesis and preparation protocols of most types of nanoparticles, such as metal nanoparticles, carbon nanoparticles, semiconductor nanoparticles, UCNPs, and polymer nanoparticles, are widely available, and procedures for their conjugation with different biomolecules are well addressed in the literature. A wide variety of bio-nanoparticle hybrid systems can be readily prepared and specifically designed in a way that provides smarter detection systems to match the clinical needs of each virus. Nanoparticles have proven to be powerful sensing tools with major advantages over the techniques used for virus testing and control.

Nanoparticles enable simple, rapid, and sensitive detection with an unprecedented multiplexing capability. In addition, bio-nanoparticles have matured from tools that are added to improve traditional methods, such as PCR and ELISA, to the basis of completely novel techniques for virus detection developed and introduced for the first time, such as SERS-, ICP-MS-, AFM-, and QCM-based assays. Interestingly, an unlimited number of detection assays with optimal detection simplicity, speed, and cost and with sensitivity levels that reach zeptomolar concentrations can be developed using nanoparticles. Nanoparticles are rapidly moving to become the main components in different biomedical industries, and they will likely occupy the interest of scientists from all disciplines, whether fundamental, environmental, clinical, or industrial, in the future.

References

1. Prangishvili D, Forterre P, Garrett RA: Viruses of the Archaea: A unifying view. *Nature Reviews Microbiology* 2006, **4**(11):837.
2. Kumar CS: *Nanomaterials for Biosensors*. New York: John Wiley & Sons; 2007.
3. Jain KK: Nanodiagnostics: Application of nanotechnology in molecular diagnostics. *Expert Review of Molecular Diagnostics* 2003, **3**(2):153–161.
4. Lu PS: Early diagnosis of avian influenza. *Science* 2006, **312**(5772):337.
5. Drew O, Patel R: HIV–early diagnosis: Key to improved prognosis and the reduction of onward transmission. *Clinical Medicine* 2011, **11**(4): 358–360.
6. Schnuriger A, Dominguez S, Valantin M-A, Tubiana R, Duvivier C, Ghosn J, Simon A, Katlama C, Thibault V: Early detection of hepatitis C virus infection by use of a new combined antigen-antibody detection assay: Potential use for high-risk individuals. *Journal of Clinical Microbiology* 2006, **44**(4):1561–1563.
7. Chunhakan S, Butthep P, Yoksan S, Tangnararatchakit K, Chuansumrit A: Early diagnosis of dengue virus infection by detection of dengue viral antigen in peripheral blood mononuclear cell. *The Pediatric Infectious Disease Journal* 2009, **28**(12):1085–1088.
8. Lo Y: SARS diagnosis, monitoring and prognostication by SARS-coronavirus RNA detection. *Hong Kong*

Medical Journal Xianggang yi xue za zhi 2009, **15**:11–14.

9. Wang E, Paessler S, Aguilar PV, Carrara A-S, Ni H, Greene IP, Weaver SC: Reverse transcription-PCR-enzyme-linked immunosorbent assay for rapid detection and differentiation of alphavirus infections. *Journal of Clinical Microbiology* 2006, **44**(11):4000–4008.

10. Yanik AA, Huang M, Kamohara O, Artar A, Geisbert TW, Connor JH, Altug H: An optofluidic nanoplasmonic biosensor for direct detection of live viruses from biological media. *Nano Letters* 2010, **10**(12):4962–4969.

11. Wang X, Liu L-H, Ramstroem O, Yan M: Engineering nanomaterial surfaces for biomedical applications. *Experimental Biology and Medicine* 2009, **234**(10):1128–1139.

12. Kumar CS: *Nanomaterials for Medical Diagnosis and Therapy*. Weinheim: Wiley-VCH; 2007.

13. Rosi NL, Mirkin CA: Nanostructures in biodiagnostics. *Chemical Reviews* 2005, **105**(4):1547–1562.

14. Alivisatos P: The use of nanocrystals in biological detection. *Nature Biotechnology* 2004, **22**(1):47.

15. McNeil SE: Nanotechnology for the biologist. *Journal of Leukocyte Biology* 2005, **78**(3):585–594.

16. Lewin M, Carlesso N, Tung C-H, Tang X-W, Cory D, Scadden DT, Weissleder R: Tat peptide-derivatized magnetic nanoparticles allow in vivo tracking and recovery of progenitor cells. *Nature Biotechnology* 2000, **18**(4):410.

17. Hirsch LR, Stafford RJ, Bankson J, Sershen SR, Rivera B, Price R, Hazle JD, Halas NJ, West JL: Nanoshell-mediated near-infrared thermal therapy of tumors under magnetic resonance guidance. *Proceedings of the National Academy of Sciences* 2003, **100**(23):13549–13554.

18. Kim S, Lim YT, Soltesz EG, De Grand AM, Lee J, Nakayama A, Parker JA, Mihaljevic T, Laurence RG, Dor DM: Near-infrared fluorescent type II quantum dots for sentinel lymph node mapping. *Nature Biotechnology* 2004, **22**(1):93.

19. Liu J, Liu J, Yang L, Chen X, Zhang M, Meng F, Luo T, Li M: Nanomaterial-assisted signal enhancement of hybridization for DNA biosensors: A review. *Sensors* 2009, **9**(9):7343–7364.

20. Wang H, Brandl DW, Nordlander P, Halas NJ: Plasmonic nanostructures: Artificial molecules. *Accounts of Chemical Research* 2007, **40**(1): 53–62.

21. Sajanlal PR, Sreeprasad TS, Samal AK, Pradeep T: Anisotropic nanomaterials: Structure, growth, assembly, and functions. *Nano Reviews* 2011, **2**(1):5883.

22. Wan Z, Wang Y, Li SS-c, Duan L, Zhai J: Development of array-based technology for detection of HAV using gold-DNA probes. *BMB Reports* 2005, **38**(4):399–406.

23. Wang YF, Pang DW, Zhang ZL, Zheng HZ, Cao JP, Shen JT: Visual gene diagnosis of HBV and HCV based on nanoparticle probe amplification and silver staining enhancement. *Journal of Medical Virology* 2003, **70**(2):205–211.

24. Xu X, Georganopoulou DG, Hill HD, Mirkin CA: Homogeneous detection of nucleic acids based upon the light scattering properties of silver-coated nanoparticle probes. *Analytical Chemistry* 2007, **79**(17):6650–6654.

25. Authier L, Grossiord C, Brossier P, Limoges B: Gold nanoparticle-based quantitative electrochemical detection of amplified human cytomegalovirus DNA using disposable microband electrodes. *Analytical Chemistry* 2001, **73**(18):4450–4456.

26. Lu X, Dong X, Zhang K, Han X, Fang X, Zhang Y: A gold nanorods-based fluorescent biosensor for the detection of hepatitis B virus DNA based on fluorescence resonance energy transfer. *Analyst* 2013, **138**(2):642–650.

27. Glynou K, Ioannou PC, Christopoulos TK, Syriopoulou V: Oligonucleotide-functionalized gold nanoparticles as probes in a dry-reagent strip biosensor for DNA analysis by hybridization. *Analytical Chemistry* 2003, **75**(16):4155–4160.

28. Chen S-H, Chuang Y-C, Lu Y-C, Lin H-C, Yang Y-L, Lin C-S: A method of layer-by-layer gold nanoparticle hybridization in a quartz crystal microbalance DNA sensing system used to detect dengue virus. *Nanotechnology* 2009, **20**(21):215501.

29. Cao YC, Jin R, Mirkin CA: Nanoparticles with Raman spectroscopic fingerprints for DNA and RNA detection. *Science* 2002, **297**(5586):1536–1540.

30. Hsu I-H, Chen W-H, Wu T-K, Sun Y-C: Gold nanoparticle-based inductively coupled plasma mass spectrometry amplification and magnetic separation for the sensitive detection of a virus-specific RNA sequence. *Journal of Chromatography A* 2011, **1218**(14):1795–1801.

31. Li H, Rothberg L: Colorimetric detection of DNA sequences based on electrostatic interactions with unmodified gold nanoparticles. *Proceedings of the National Academy of Sciences* 2004, **101**(39):14036–14039.

32. Griffin J, Singh AK, Senapati D, Rhodes P, Mitchell K, Robinson B, Yu E, Ray PC: Size- and distance-dependent nanoparticle surface-energy transfer (NSET) method for selective sensing of hepatitis C virus RNA. *Chemistry* 2009, **15**(2):342–351.

33. Hanaee H, Ghourchian H, Ziaee A: Nanoparticle-based electrochemical detection of hepatitis B virus using stripping chronopotentiometry. *Analytical Biochemistry* 2007, **370**(2):195–200.

34. Shawky SM, Bald D, Azzazy HM: Direct detection of unamplified hepatitis C virus RNA using unmodified gold nanoparticles. *Clinical Biochemistry* 2010, **43**(13–14):1163–1168.

35. Li W, Wu P, Zhang H, Cai C: Catalytic signal amplification of gold nanoparticles combining with conformation-switched hairpin DNA probe for hepatitis C virus quantification. *Chemical Communications* 2012, **48**(63):7877–7879.

36. Mancuso M, Jiang L, Cesarman E, Erickson D: Multiplexed colorimetric detection of Kaposi's sarcoma associated herpesvirus and Bartonella DNA using gold and silver nanoparticles. *Nanoscale* 2013, **5**(4):1678–1686.

37. Liu H-H, Cao X, Yang Y, Liu M-G, Wang Y-F: Array-based nano-amplification technique was applied in detection of hepatitis E virus. *BMB Reports* 2006, **39**(3):247–252.

38. Rohrman BA, Leautaud V, Molyneux E, Richards-Kortum RR: A lateral flow assay for quantitative detection of amplified HIV-1 RNA. *PLoS One* 2012, **7**(9):e45611.

39. Hu Y, Hua S, Li F, Jiang Y, Bai X, Li D, Niu L: Green-synthesized gold nanoparticles decorated graphene sheets for label-free electrochemical impedance DNA hybridization biosensing. *Biosensors and Bioelectronics* 2011, **26**(11):4355–4361.

40. Wu J-C, Chen C-H, Fu J-W, Yang H-C: Electrophoresis-enhanced detection of deoxyribonucleic acids on a membrane-based lateral flow strip using avian influenza H5 genetic sequence as the model. *Sensors* 2014, **14**(3):4399–4415.

41. Liu X, Cheng Z, Fan H, Ai S, Han R: Electrochemical detection of avian influenza virus H5N1 gene sequence using a DNA aptamer immobilized onto a hybrid nanomaterial-modified electrode. *Electrochimica Acta* 2011, **56**(18):6266–6270.

42. Cho H, Jung J, Chung BH: Scanometric analysis of DNA microarrays using DNA intercalator-conjugated gold nanoparticles. *Chemical Communications* 2012, **48**(61):7601–7603.

43. Zhao J, Tang S, Storhoff J, Marla S, Bao YP, Wang X, Wong EY, Ragupathy V, Ye Z, Hewlett IK: Multiplexed, rapid detection of H5N1 using a PCR-free nanoparticle-based genomic microarray assay. *BMC Biotechnology* 2010, **10**(1):74.

44. Kim SK, Cho H, Jeong J, Kwon JN, Jung Y, Chung BH: Label-free and naked eye detection of PNA/DNA hybridization using enhancement of gold nanoparticles. *Chemical Communications* 2010, **46**(19):3315–3317.

45. Karakoc AE, Berkem R, Beyaz E: Detection of hepatitis B virus deoxyribonucleic acid using real-time polymerase chain reaction method in pooled-plasma samples of Turkish blood donors. *Clinical Laboratory* 2009, **55**(5–6):229–234.

46. Wu S, Zhong Z, Wang D, Li M, Qing Y, Dai N, Li Z: Gold nanoparticle-labeled detection antibodies for use in an enhanced electrochemical immunoassay of hepatitis B surface antigen in human serum. *Microchimica Acta* 2009, **166**(3–4):269–275.

47. Lee K-B, Kim E-Y, Mirkin CA, Wolinsky SM: The use of nanoarrays for highly sensitive and selective detection of human immunodeficiency virus type 1 in plasma. *Nano Letters* 2004, **4**(10):1869–1872.

48. Neng J, Harpster MH, Wilson WC, Johnson PA: Surface-enhanced Raman scattering (SERS) detection of multiple viral antigens using magnetic capture of SERS-active nanoparticles. *Biosensors and Bioelectronics* 2013, **41**:316–321.

49. Chen L, Wei H, Guo Y, Cui Z, Zhang Z, Zhang X-E: Gold nanoparticle enhanced immuno-PCR for ultrasensitive detection of Hantaan virus nucleocapsid protein. *Journal of Immunological Methods* 2009, **346**(1–2):64–70.

50. Zeng Q, Zhang Y, Liu X, Tu L, Kong X, Zhang H: Multiple homogeneous immunoassays based on a quantum dots–gold nanorods FRET nanoplatform. *Chemical Communications* 2012, **48**(12): 1781–1783.

51. Wang X, Li Y, Quan D, Wang J, Zhang Y, Du J, Peng J, Fu Q, Zhou Y, Jia S et al.: Detection of hepatitis B surface antigen by target-induced aggregation monitored by dynamic light scattering. *Analytical Biochemistry* 2012, **428**(2):119–125.

52. Wang X, Li Y, Wang H, Fu Q, Peng J, Wang Y, Du J, Zhou Y, Zhan L: Gold nanorod-based localized surface plasmon resonance biosensor for sensitive detection of hepatitis B virus in buffer, blood serum and plasma. *Biosensors and Bioelectronics* 2010, **26**(2):404–410.

53. Ding C, Li H, Hu K, Lin J-M: Electrochemical immunoassay of hepatitis B surface antigen by the amplification of gold nanoparticles based on the nanoporous gold electrode. *Talanta* 2010, **80**(3):1385–1391.

54. Shen G, Zhang Y: Highly sensitive electrochemical stripping detection of hepatitis B surface antigen based on copper-enhanced gold nanoparticle tags and magnetic nanoparticles. *Analytica Chimica Acta* 2010, **674**(1):27–31.

55. Dong H, Liu J, Zhu H, Ou C-Y, Xing W, Qiu M, Zhang G, Xiao Y, Yao J, Pan P et al.: Two types of nanoparticle-based bio-barcode amplification assays to detect HIV-1 p24 antigen. *Virology Journal* 2012, **9**(1):180.

56. Baek TJ, Park PY, Han KN, Kwon HT, Seong GH: Development of a photodiode array biochip using a bipolar semiconductor and its application to detection of human papilloma virus. *Analytical and Bioanalytical Chemistry* 2008, **390**(5): 1373–1378.

57. Li X, Lu D, Sheng Z, Chen K, Guo X, Jin M, Han H: A fast and sensitive immunoassay of avian influenza virus based on label-free quantum dot probe and lateral flow test strip. *Talanta* 2012, **100**:1–6.

58. Driskell JD, Jones CA, Tompkins SM, Tripp RA: One-step assay for detecting influenza virus using

dynamic light scattering and gold nanoparticles. *Analyst* 2011, **136**(15):3083–3090.

59. Laderman EI, Whitworth E, Dumaual E, Jones M, Hudak A, Hogrefe W, Carney J, Groen J: Rapid, sensitive, and specific lateral-flow immunochromatographic point-of-care device for detection of herpes simplex virus type 2-specific immunoglobulin G antibodies in serum and whole blood. *Clinical and Vaccine Immunology* 2008, **15**(1):159–163.

60. Duan L, Wang Y, Li SS-c, Wan Z, Zhai J: Rapid and simultaneous detection of human hepatitis B virus and hepatitis C virus antibodies based on a protein chip assay using nano-gold immunological amplification and silver staining method. *BMC Infectious Diseases* 2005, **5**(1):53.

61. Draz MS, Fang BA, Li L, Chen Z, Wang Y, Xu Y, Yang J, Killeen K, Chen FF: Hybrid nanocluster plasmonic resonator for immunological detection of hepatitis B virus. *ACS Nano* 2012, **6**(9):7634–7643.

62. Bonanni A, Pividori M, Del Valle M: Impedimetric detection of influenza A (H1N1) DNA sequence using carbon nanotubes platform and gold nanoparticles amplification. *Analyst* 2010, **135**(7):1765–1772.

63. Tang S, Hewlett I: Nanoparticle-based immunoassays for sensitive and early detection of HIV-1 capsid (p24) antigen. *Journal of Infectious Diseases* 2010, **201**(Supplement 1):S59–S64.

64. Kim E-Y, Stanton J, Korber BT, Krebs K, Bogdan D, Kunstman K, Wu S, Phair JP, Mirkin CA, Wolinsky SM: Detection of HIV-1 p24 Gag in plasma by a nanoparticle-based bio-barcode-amplification method. *Nanomedicine (London, England)* 2008, **3**:293–303.

65. Yen C-W, de Puig H, Tam JO, Gómez-Márquez J, Bosch I, Hamad-Schifferli K, Gehrke L: Multicolored silver nanoparticles for multiplexed disease diagnostics: Distinguishing dengue, yellow fever, and Ebola viruses. *Lab on a Chip* 2015, **15**(7):1638–1641.

66. Li X, Scida K, Crooks RM: Detection of hepatitis B virus DNA with a paper electrochemical sensor. *Analytical Chemistry* 2015, **87**(17):9009–9015.

67. Li H, Sun Z, Zhong W, Hao N, Xu D, Chen H-Y: Ultra-sensitive electrochemical detection for DNA arrays based on silver nanoparticle aggregates. *Analytical Chemistry* 2010, **82**(13):5477–5483.

68. Wabuyele MB, Vo-Dinh T: Detection of human immunodeficiency virus type 1 DNA sequence using plasmonics nanoprobes. *Analytical Chemistry* 2005, **77**(23):7810–7815.

69. Liu Y, Huang CZ: One-step conjugation chemistry of DNA with highly scattered silver nanoparticles for sandwich detection of DNA. *Analyst* 2012, **137**(15):3434–3436.

70. Zhao L-J, Yu R-J, Ma W, Han H-X, Tian H, Qian R-C, Long Y-T: Sensitive detection of protein biomarkers using silver nanoparticles enhanced immunofluorescence assay. *Theranostics* 2017, **7**(4):876.

71. Valipour A, Roushani M: Using silver nanoparticle and thiol graphene quantum dots nanocomposite as a substratum to load antibody for detection of hepatitis C virus core antigen: Electrochemical oxidation of riboflavin was used as redox probe. *Biosensors and Bioelectronics* 2017, **89**:946–951.

72. Valdez J, Bawage S, Gomez I, Singh SR: Facile and rapid detection of respiratory syncytial virus using metallic nanoparticles. *Journal of Nanobiotechnology* 2016, **14**(1):13.

73. Kurdekar AD, Chunduri LA, Chelli SM, Haleyurgirisetty MK, Bulagonda EP, Zheng J, Hewlett IK, Kamisetti V: Fluorescent silver nanoparticle based highly sensitive immunoassay for early detection of HIV infection. *RSC Advances* 2017, **7**(32):19863–19877.

74. He N, Wang F, Ma C, Li C, Zeng X, Deng Y, Zhang L, Li Z: Chemiluminescence analysis for HBV-DNA hybridization detection with magnetic nanoparticles based DNA extraction from positive whole blood samples. *Journal of Biomedical Nanotechnology* 2013, **9**(2):267–273.

75. Delaviz N, Gill P, Ajami A, Aarabi M: Aptamer-conjugated magnetic nanoparticles for the efficient removal of HCV particles from human plasma samples. *RSC Advances* 2015, **5**(97):79433–79439.

76. Hassen WM, Chaix C, Abdelghani A, Bessueille F, Leonard D, Jaffrezic-Renault N: An impedimetric DNA sensor based on functionalized magnetic nanoparticles for HIV and HBV detection. *Sensors and Actuators B: Chemical* 2008, **134**(2):755–760.

77. Perez JM, Simeone FJ, Saeki Y, Josephson L, Weissleder R: Viral-induced self-assembly of magnetic nanoparticles allows the detection of viral particles in biological media. *Journal of the American Chemical Society* 2003, **125**(34):10192–10193.

78. Nourani S, Ghourchian H, Boutorabi SM: Magnetic nanoparticle-based immunosensor for electrochemical detection of hepatitis B surface antigen. *Analytical Biochemistry* 2013, **441**(1):1–7.

79. Xi Z, Gong Q, Wang C, Zheng B: Highly sensitive chemiluminescent aptasensor for detecting HBV infection based on rapid magnetic separation and double-functionalized gold nanoparticles. *Scientific Reports* 2018, **8**(1):9444.

80. Chou T-C, Hsu W, Wang C-H, Chen Y-J, Fang J-M: Rapid and specific influenza virus detection by functionalized magnetic nanoparticles and mass spectrometry. *Journal of Nanobiotechnology* 2011, **9**(1):52.

81. Kamikawa TL, Mikolajczyk MG, Kennedy M, Zhong L, Zhang P, Setterington EB, Scott DE, Alocilja EC: Pandemic influenza detection by electrically active magnetic nanoparticles and surface Plasmon

resonance. *IEEE Transactions on Nanotechnology* 2012, **11**(1):88–96.

82. Oh S, Kim J, Tran VT, Lee DK, Ahmed SR, Hong JC, Lee J, Park EY, Lee J: Magnetic nanozyme-linked immunosorbent assay for ultrasensitive influenza A virus detection. *ACS Applied Materials and Interfaces* 2018, **10**(15):12534–12543.

83. Li X, Zhang Q, Hou P, Chen M, Hui W, Vermorken A, Luo Z, Li H, Li Q, Cui Y: Gold magnetic nanoparticle conjugate-based lateral flow assay for the detection of IgM class antibodies related to TORCH infections. *International Journal of Molecular Medicine* 2015, **36**(5):1319–1326.

84. Narang J, Singhal C, Malhotra N, Narang S, Gupta R, Kansal R, Pundir C: Impedimetric genosensor for ultratrace detection of hepatitis B virus DNA in patient samples assisted by zeolites and MWCNT nano-composites. *Biosensors and Bioelectronics* 2016, **86**:566–574.

85. Ariksoysal DO, Karadeniz H, Erdem A, Sengonul A, Sayiner AA, Ozsoz M: Label-free electrochemical hybridization genosensor for the detection of hepatitis B virus genotype on the development of lamivudine resistance. *Analytical Chemistry* 2005, **77**(15):4908–4917.

86. Dastagir T, Forzani ES, Zhang R, Amlani I, Nagahara LA, Tsui R, Tao N: Electrical detection of hepatitis C virus RNA on single wall carbon nanotube-field effect transistors. *Analyst* 2007, **132**(8):738–740.

87. Gui EL, Li L-J, Zhang K, Xu Y, Dong X, Ho X, Lee PS, Kasim J, Shen Z, Rogers JA: DNA sensing by field-effect transistors based on networks of carbon nanotubes. *Journal of the American Chemical Society* 2007, **129**(46):14427–14432.

88. Tran TL, Nguyen TT, Tran TTH, Chu VT, Tran QT, Mai AT: Detection of influenza A virus using carbon nanotubes field effect transistor based DNA sensor. *Physica E: Low-Dimensional Systems and Nanostructures* 2017, **93**:83–86.

89. Lee D, Chander Y, Goyal SM, Cui T: Carbon nanotube electric immunoassay for the detection of swine influenza virus H1N1. *Biosensors and Bioelectronics* 2011, **26**(8):3482–3487.

90. Wasik D, Mulchandani A, Yates M: Salivary detection of dengue virus NS1 protein with a label-free immunosensor for early dengue diagnosis. *Sensors* 2018, **18**(8):2641.

91. Gopinath P, Anitha V, Mastani SA: Design of biosensor array with current boost and signal conditioning circuits for HPV detection. *Alexandria Engineering Journal* 2018, **57**(2):671–681.

92. Gao Z, Li Y, Zhang X, Feng J, Kong L, Wang P, Chen Z, Dong Y, Wei Q: Ultrasensitive electrochemical immunosensor for quantitative detection of HBeAg using Au@ Pd/MoS2@ MWCNTs nanocomposite as enzyme-mimetic labels. *Biosensors and Bioelectronics* 2018, **102**:189–195.

93. Fang Y-S, Huang X-J, Wang L-S, Wang J-F: An enhanced sensitive electrochemical immunosensor based on efficient encapsulation of enzyme in silica matrix for the detection of human immunodeficiency virus p24. *Biosensors and Bioelectronics* 2015, **64**:324–332.

94. Wu M, Kempaiah R, Huang P-JJ, Maheshwari V, Liu J: Adsorption and desorption of DNA on graphene oxide studied by fluorescently labeled oligonucleotides. *Langmuir* 2011, **27**(6):2731–2738.

95. Wu C, Zhou Y, Miao X, Ling L: A novel fluorescent biosensor for sequence-specific recognition of double-stranded DNA with the platform of graphene oxide. *Analyst* 2011, **136**(10):2106–2110.

96. Jeong S, Kim D-M, An S-Y, Kim DH, Kim D-E: Fluorometric detection of influenza viral RNA using graphene oxide. *Analytical Biochemistry* 2018, **561**:66–69.

97. Wen J, Li W, Li J, Tao B, Xu Y, Li H, Lu A, Sun S: Study on rolling circle amplification of Ebola virus and fluorescence detection based on graphene oxide. *Sensors and Actuators B: Chemical* 2016, **227**: 655–659.

98. Jin S-A, Poudyal S, Marinero EE, Kuhn RJ, Stanciu LA: Impedimetric dengue biosensor based on functionalized graphene oxide wrapped silica particles. *Electrochimica Acta* 2016, **194**:422–430.

99. Jung JH, Liu F, Seo TS: Graphene oxide-based immunobiosensor for ultrasensitive pathogen detection. *10th IEEE International Conference on Nanotechnology*, Seoul, South Korea, IEEE; 2010: 692–695.

100. Kamil YM, Bakar MHA, Yaacob MH, Syahir A, Lim HN, Mahdi MA: Dengue E protein detection using a graphene oxide integrated tapered optical fiber sensor. *IEEE Journal of Selected Topics in Quantum Electronics* 2019, **25**(1):1–8.

101. Singh R, Hong S, Jang J: Label-free detection of influenza viruses using a reduced graphene oxide-based electrochemical immunosensor integrated with a microfluidic platform. *Scientific Reports* 2017, **7**:42771.

102. Zhao F, Bai Y, Zeng R, Cao L, Zhu J, Han G, Chen Z: An electrochemical immunosensor with graphene-oxide-ferrocene-based nanocomposites for hepatitis B surface antigen detection. *Electroanalysis* 2018.

103. Liu M, Zheng C, Cui M, Zhang X, Yang D-P, Wang X, Cui D: Graphene oxide wrapped with gold nanorods as a tag in a SERS based immunoassay for the hepatitis B surface antigen. *Microchimica Acta* 2018, **185**(10):458.

104. Zhan L, Li CM, Wu WB, Huang CZ: A colorimetric immunoassay for respiratory syncytial virus detection based on gold nanoparticles–graphene oxide hybrids with mercury-enhanced peroxidase-like activity. *Chemical Communications* 2014, **50**(78):11526–11528.

105. Dutta Chowdhury A, Ganganboina AB, Nasrin F, Takemura K, Doong R-A, Utomo DIS, Lee J, Khoris IM, Park EY: Femtomolar detection of dengue virus DNA with serotype identification ability. *Analytical Chemistry* 2018, **90**(21): 12464–12474.

106. Jimenez AMJ, Moulick A, Richtera L, Krejcova L, Kalina L, Datta R, Svobodova M, Hynek D, Masarik M, Heger Z: Dual-color quantum dots-based simultaneous detection of HPV-HIV co-infection. *Sensors and Actuators B: Chemical* 2018, **258**:295–303.

107. Lee N, Wang C, Park J: User-friendly point-of-care detection of influenza A (H1N1) virus using light guide in three-dimensional photonic crystal. *RSC Advances* 2018, **8**(41):22991–22997.

108. Adegoke O, Morita M, Kato T, Ito M, Suzuki T, Park EY: Localized surface plasmon resonance-mediated fluorescence signals in plasmonic nanoparticle-quantum dot hybrids for ultrasensitive Zika virus RNA detection via hairpin hybridization assays. *Biosensors and Bioelectronics* 2017, **94**:513–522.

109. Jimenez AMJ, Rodrigo MAM, Milosavljevic V, Krizkova S, Kopel P, Heger Z, Adam V: Gold nanoparticles-modified nanomaghemite and quantum dots-based hybridization assay for detection of HPV. *Sensors and Actuators B: Chemical* 2017, **240**:503–510.

110. Liu L, Wang X, Ma Q, Lin Z, Chen S, Li Y, Lu L, Qu H, Su X: Multiplex electrochemiluminescence DNA sensor for determination of hepatitis B virus and hepatitis C virus based on multicolor quantum dots and Au nanoparticles. *Analytica Chimica Acta* 2016, **916**:92–101.

111. Ribeiro JF, Pereira MI, Assis LG, Cabral Filho PE, Santos BS, Pereira GA, Chaves CR, Campos GS, Sardi SI, Pereira G: Quantum dots-based fluoroimmunoassay for anti-Zika virus IgG antibodies detection. *Journal of Photochemistry and Photobiology B: Biology* 2019, **194**:135–139.

112. Kim Y-G, Moon S, Kuritzkes DR, Demirci U: Quantum dot-based HIV capture and imaging in a microfluidic channel. *Biosensors and Bioelectronics* 2009, **25**(1):253–258.

113. Wu F, Yuan H, Zhou C, Mao M, Liu Q, Shen H, Cen Y, Qin Z, Ma L, Li LS: Multiplexed detection of influenza A virus subtype H5 and H9 via quantum dot-based immunoassay. *Biosensors and Bioelectronics* 2016, **77**:464–470.

114. Lei Y-M, Zhou J, Chai Y-Q, Zhuo Y, Yuan R: SnS$_2$ quantum dots as new emitters with strong electrochemiluminescence for ultrasensitive antibody detection. *Analytical Chemistry* 2018, **90**(20):12270–12277.

115. Wang X, Wang G, Li W, Zhao B, Xing B, Leng Y, Dou H, Sun K, Shen L, Yuan X: NIR-emitting quantum dot-encoded microbeads through membrane emulsification for multiplexed immunoassays. *Small* 2013, **9**(19):3327–3335.

116. Zhang X, Li D, Wang C, Zhi X, Zhang C, Wang K, Cui D: A CCD-based reader combined quantum dots-labeled lateral flow strips for ultrasensitive quantitative detection of anti-HBs antibody. *Journal of Biomedical Nanotechnology* 2012, **8**(3): 372–379.

117. Yang H, Guo Q, He R, Li D, Zhang X, Bao C, Hu H, Cui D: A quick and parallel analytical method based on quantum dots labeling for ToRCH-related antibodies. *Nanoscale Research Letters* 2009, **4**(12):1469.

118. Zhou L, Fan Y, Wang R, Li X, Fan L, Zhang F: High-capacity upconversion wavelength and lifetime binary encoding for multiplexed biodetection. *Angewandte Chemie International Edition* 2018, **57**(39):12824–12829.

119. Tsang M-K, Ye W, Wang G, Li J, Yang M, Hao J: Ultrasensitive detection of Ebola virus oligonucleotide based on upconversion nanoprobe/nanoporous membrane system. *ACS Nano* 2016, **10**(1):598–605.

120. Zhu H, Lu F, Wu X-C, Zhu J-J: An upconversion fluorescent resonant energy transfer biosensor for hepatitis B virus (HBV) DNA hybridization detection. *Analyst* 2015, **140**(22):7622–7628.

121. Wu Y-M, Cen Y, Huang L-J, Yu R-Q, Chu X: Upconversion fluorescence resonance energy transfer biosensor for sensitive detection of human immunodeficiency virus antibodies in human serum. *Chemical Communications* 2014, **50**(36): 4759–4762.

122. Ylihärsilä M, Harju E, Arppe R, Hattara L, Hölsä J, Saviranta P, Soukka T, Waris M: Genotyping of clinically relevant human adenoviruses by array-in-well hybridization assay. *Clinical Microbiology and Infection* 2013, **19**(6):551–557.

123. Thomson DA, Dimitrov K, Cooper MA: Amplification free detection of herpes simplex virus DNA. *Analyst* 2011, **136**(8):1599–1607.

124. Kim BC, Ju MK, Dan-Chin-Yu A, Sommer P: Quantitative detection of HIV-1 particles using HIV-1 neutralizing antibody-conjugated beads. *Analytical Chemistry* 2009, **81**(6):2388–2393.

125. Ding L, Xiang C, Zhou G: Silica nanoparticles coated by poly (acrylic acid) brushes via host-guest interactions for detecting DNA sequence of Hepatitis B virus. *Talanta* 2018, **181**:65–72.

126. Koskinen JO, Vainionpää R, Meltola NJ, Soukka J, Hänninen PE, Soini AE: Rapid method for detection of influenza A and B virus antigens by use of a two-photon excitation assay technique and dry-chemistry reagents. *Journal of Clinical Microbiology* 2007, **45**(11):3581–3588.

127. Boisselier E, Astruc D: Gold nanoparticles in nanomedicine: Preparations, imaging, diagnostics,

therapies and toxicity. *Chemical Society Reviews* 2009, **38**(6):1759–1782.

128. Draz MS, Fang BA, Zhang P, Hu Z, Gu S, Weng KC, Gray JW, Chen FF: Nanoparticle-mediated systemic delivery of siRNA for treatment of cancers and viral infections. *Theranostics* 2014, **4**(9):872.

129. Lee SH, Jun B-H: Silver nanoparticles: Synthesis and application for nanomedicine. *International Journal of Molecular Sciences* 2019, **20**(4):865.

130. Schrand AM, Rahman MF, Hussain SM, Schlager JJ, Smith DA, Syed AF: Metal-based nanoparticles and their toxicity assessment. *Wiley Interdisciplinary Reviews: Nanomedicine and Nanobiotechnology* 2010, **2**(5):544–568.

131. Park JW, Bae KH, Kim C, Park TG: Clustered magnetite nanocrystals cross-linked with PEI for efficient siRNA delivery. *Biomacromolecules* 2010, **12**(2):457–465.

132. Liu Z, Tabakman SM, Chen Z, Dai H: Preparation of carbon nanotube bioconjugates for biomedical applications. *Nature Protocols* 2009, **4**(9):1372.

133. Kolosnjaj J, Szwarc H, Moussa F: Toxicity studies of carbon nanotubes. In: Warren CW (Ed.), *Bio-Applications of Nanoparticles*. New York: Springer; 2007: 181–204.

134. Mukhopadhyay P, Gupta RK: *Graphite, Graphene, and their Polymer Nanocomposites*. Boca Raton, FL: CRC Press; 2012.

135. Hardman R: A toxicologic review of quantum dots: Toxicity depends on physicochemical and environmental factors. *Environmental Health Perspectives* 2005, **114**(2):165–172.

136. Banerjee A, Pons T, Lequeux N, Dubertret B: Quantum dots–DNA bioconjugates: Synthesis to applications. *Interface Focus* 2016, **6**(6):20160064.

137. Muhr V, Wilhelm S, Hirsch T, Wolfbeis OS: Upconversion nanoparticles: From hydrophobic to hydrophilic surfaces. *Accounts of Chemical Research* 2014, **47**(12):3481–3493.

138. Li S, Ge Y, Li H: *Smart Nanomaterials for Sensor Application*. Beijing, China: Bentham Science Publishers; 2012.

139. Draz MS, Shafiee H: Applications of gold nanoparticles in virus detection. *Theranostics* 2018, **8**(7):1985.

140. Wang H, Shen G, Yu R. Aspects of recent development of immunosensors. In: (ed.) Yu R. Electrochemical Sensors, Biosensors and their Biomedical Applications. Amsterdam, The Netherlands: Elsevier. 2008:237–260.

141. Zeng S, Yong K-T, Roy I, Dinh X-Q, Yu X, Luan F: A review on functionalized gold nanoparticles for biosensing applications. *Plasmonics* 2011, **6**(3):491–506.

142. Alex S, Tiwari A: Functionalized gold nanoparticles: Synthesis, properties and applications: A review. *Journal of Nanoscience and Nanotechnology* 2015, **15**(3):1869–1894.

143. Wolfbeis OS: An overview of nanoparticles commonly used in fluorescent bioimaging. *Chemical Society Reviews* 2015, **44**(14):4743–4768.

144. Reisch A, Klymchenko AS: Fluorescent polymer nanoparticles based on dyes: Seeking brighter tools for bioimaging. *Small* 2016, **12**(15):1968–1992.

145. Yigit MV, Mishra A, Tong R, Cheng J, Wong GC, Lu Y: Inorganic mercury detection and controlled release of chelating agents from ion-responsive liposomes. *Chemistry and Biology* 2009, **16**(9):937–942.

146. Mufamadi MS, Pillay V, Choonara YE, Du Toit LC, Modi G, Naidoo D, Ndesendo VM: A review on composite liposomal technologies for specialized drug delivery. *Journal of Drug Delivery* 2011, **2011**:1–19.

147. Prasad PN: *Nanophotonics*. Hoboken, NJ: John Wiley & Sons; 2004.

148. Nalwa HS: *Nanostructured Materials and Nanotechnology: Concise Edition*. Houston, TX: Gulf Professional Publishing; 2001.

149. Maksimenko S, Slepyan GY, Ledentsov N, Kalosha V, Hoffmann A, Bimberg D: Light confinement in a quantum dot. *Semiconductor Science and Technology* 2000, **15**(6):491.

150. Shirasaki Y, Supran GJ, Bawendi MG, Bulović V: Emergence of colloidal quantum-dot light-emitting technologies. *Nature Photonics* 2013, **7**(1):13.

151. Wharam D, Thornton TJ, Newbury R, Pepper M, Ahmed H, Frost J, Hasko D, Peacock D, Ritchie D, Jones G: One-dimensional transport and the quantisation of the ballistic resistance. *Journal of Physics C: Solid State Physics* 1988, **21**(8):L209.

152. Van Wees B, Van Houten H, Beenakker C, Williamson JG, Kouwenhoven L, Van der Marel D, Foxon C: Quantized conductance of point contacts in a two-dimensional electron gas. *Physical Review Letters* 1988, **60**(9):848.

153. Li J., To S., You L., Sun Y. (2016) Nanostructure Field Effect Transistor Biosensors. In: Bhushan B. (eds) *Encyclopedia of Nanotechnology*. Springer, Dordrecht. 2713–2725.

154. Kolhatkar A, Jamison A, Litvinov D, Willson R, Lee T: Tuning the magnetic properties of nanoparticles. *International Journal of Molecular Sciences* 2013, **14**(8):15977–16009.

155. Haun JB, Yoon TJ, Lee H, Weissleder R: Magnetic nanoparticle biosensors. *Wiley Interdisciplinary Reviews: Nanomedicine and Nanobiotechnology* 2010, **2**(3):291–304.

156. Willets KA, Van Duyne RP: Localized surface plasmon resonance spectroscopy and sensing. *Annual Review of Physical Chemistry* 2007, **58**:267–297.

157. Mayer KM, Hafner JH: Localized surface plasmon resonance sensors. *Chemical Reviews* 2011, **111**(6):3828–3857.

158. Li Y, Schluesener HJ, Xu S: Gold nanoparticle-based biosensors. *Gold Bulletin* 2010, **43**(1):29–41.

159. Doria G, Conde J, Veigas B, Giestas L, Almeida C, Assunção M, Rosa J, Baptista PV: Noble metal nanoparticles for biosensing applications. *Sensors* 2012, **12**(2):1657–1687.

160. Jariwala D, Sangwan VK, Lauhon LJ, Marks TJ, Hersam MC: Carbon nanomaterials for electronics, optoelectronics, photovoltaics, and sensing. *Chemical Society Reviews* 2013, **42**(7):2824–2860.

161. Tîlmaciu C-M, Morris MC: Carbon nanotube biosensors. *Frontiers in Chemistry* 2015, **3**:59.

162. Smith AM, Nie S: Semiconductor nanocrystals: Structure, properties, and band gap engineering. *Accounts of Chemical Research* 2009, **43**(2):190–200.

163. Alivisatos AP, Gu W, Larabell C: Quantum dots as cellular probes. *Annual Review of Biomedical Engineering* 2005, **7**:55–76.

164. Wang M, Abbineni G, Clevenger A, Mao C, Xu S: Upconversion nanoparticles: Synthesis, surface modification and biological applications. *Nanomedicine: Nanotechnology, Biology and Medicine* 2011, **7**(6):710–729.

165. Auzel F: Upconversion and anti-stokes processes with f and d ions in solids. *Chemical Reviews* 2004, **104**(1):139–174.

166. Wang F, Liu X: Recent advances in the chemistry of lanthanide-doped upconversion nanocrystals. *Chemical Society Reviews* 2009, **38**(4):976–989.

167. Wang F, Banerjee D, Liu Y, Chen X, Liu X: Upconversion nanoparticles in biological labeling, imaging, and therapy. *Analyst* 2010, **135**(8):1839–1854.

168. Rao JP, Geckeler KE: Polymer nanoparticles: Preparation techniques and size-control parameters. *Progress in Polymer Science* 2011, **36**(7):887–913.

169. Hanemann T, Szabó DV: Polymer-nanoparticle composites: From synthesis to modern applications. *Materials* 2010, **3**(6):3468–3517.

170. Lu X-Y, Wu D-C, Li Z-J, Chen G-Q: Polymer nanoparticles. In: Teplow D (Ed.), *Progress in Molecular Biology and Translational Science,* vol. 104. Burlington, VT: Elsevier; 2011: 299–323.

171. Li K, Liu B: Polymer-encapsulated organic nanoparticles for fluorescence and photoacoustic imaging. *Chemical Society Reviews* 2014, **43**(18):6570–6597.

172. Liu D, Wang Z, Jiang X: Gold nanoparticles for the colorimetric and fluorescent detection of ions and small organic molecules. *Nanoscale* 2011, **3**(4):1421–1433.

173. Ghosh SK, Pal T: Interparticle coupling effect on the surface plasmon resonance of gold nanoparticles: From theory to applications. *Chemical Reviews* 2007, **107**(11):4797–4862.

174. Liu HH, Cao X, Yang Y, Liu MG, Wang YF: Array-based nano-amplification technique was applied in detection of hepatitis E virus. *Journal of Biochemistry and Molecular Biology* 2006, **39**(3):247–252.

175. Hassan Nikbakht PG, Tabarraei A, Niazi A: Nanomolecular detection of human influenza virus type A using reverse transcription loop-mediated isothermal amplification assisted with rod-shaped gold nanoparticles. *RSC Advances* 2014, **4**: 13575–13580.

176. Wu JC, Chen CH, Fu JW, Yang HC: Electrophoresis-enhanced detection of deoxyribonucleic acids on a membrane-based lateral flow strip using avian influenza H5 genetic sequence as the model. *Sensors (Basel)* 2014, **14**(3):4399–4415.

177. Mu B, Huang X, Bu P, Zhuang J, Cheng Z, Feng J, Yang D, Dong C, Zhang J, Yan X: Influenza virus detection with pentabody-activated nanoparticles. *Journal of Virological Methods* 2010, **169**(2): 282–289.

178. Chang T-L, Tsai C-Y, Sun C-C, Chen C-C, Kuo L-S, Chen P-H: Ultrasensitive electrical detection of protein using nanogap electrodes and nanoparticle-based DNA amplification. *Biosensors and Bioelectronics* 2007, **22**(12):3139–3145.

179. Martínez-Paredes G, González-García MB, Costa-García A: Genosensor for SARS virus detection based on gold nanostructured screen-printed carbon electrodes. *Electroanalysis: An International Journal Devoted to Fundamental and Practical Aspects of Electroanalysis* 2009, **21**(3–5):379–385.

180. Zeng Q, Zhang Y, Liu X, Tu L, Kong X, Zhang H: Multiple homogeneous immunoassays based on a quantum dots-gold nanorods FRET nanoplatform. *Chemical Communications* 2012, **48**(12): 1781–1783.

181. Chang Y-F, Wang S-F, Huang JC, Su L-C, Yao L, Li Y-C, Wu S-C, Chen Y-MA, Hsieh J-P, Chou C: Detection of swine-origin influenza A (H1N1) viruses using a localized surface plasmon coupled fluorescence fiber-optic biosensor. *Biosensors and Bioelectronics* 2010, **26**(3):1068–1073.

182. Ganbold E-O, Kang T, Lee K, Lee SY, Joo S-W: Aggregation effects of gold nanoparticles for single-base mismatch detection in influenza A (H1N1) DNA sequences using fluorescence and Raman measurements. *Colloids and Surfaces B: Biointerfaces* 2012, **93**:148–153.

183. Zhang H, Liu L, Li CW, Fu H, Chen Y, Yang M: Multienzyme-nanoparticles amplification for sensitive virus genotyping in microfluidic microbeads array using Au nanoparticle probes and quantum dots as labels. *Biosens Bioelectron* 2011, **29**(1): 89–96.

184. Lee KB, Kim EY, Mirkin CA, Wolinsky SM: The use of nanoarrays for highly sensitive and selective detection of human immunodeficiency virus type 1 in plasma. *Nano Letters* 2004, **4**(10): 1869–1872.

185. Guo S, Wang E: Synthesis and electrochemical applications of gold nanoparticles. *Analytica Chimica Acta* 2007, **598**(2):181–192.

186. Welch CM, Compton RG: The use of nanoparticles in electroanalysis: A review. *Analytical and Bioanalytical Chemistry* 2006, **384**(3):601–619.

187. Martinez-Paredes G, Gonzalez-Garcia MB, Costa-Garcia A: Genosensor for SARS virus detection based on gold nanostructured screen-printed carbon electrodes. *Electroanalysis* 2009, **21**(3–5):379–385.

188. Kim E-Y, Stanton J, Vega RA, Kunstman KJ, Mirkin CA, Wolinsky SM: A real-time PCR-based method for determining the surface coverage of thiol-capped oligonucleotides bound onto gold nanoparticles. *Nucleic Acids Research* 2006, **34**(7):e54–e54.

189. Huang S-H, Yang T-C, Tsai M-H, Tsai I-S, Lu H-C, Chuang P-H, Wan L, Lin Y-J, Lai C-H, Lin C-W: Gold nanoparticle-based RT-PCR and real-time quantitative RT-PCR assays for detection of Japanese encephalitis virus. *Nanotechnology* 2008, **19**(40):405101.

190. Stringer RC, Schommer S, Hoehn D, Grant SA: Development of an optical biosensor using gold nanoparticles and quantum dots for the detection of porcine reproductive and respiratory syndrome virus. *Sensors and Actuators B: Chemical* 2008, **134**(2):427–431.

191. Steinem C, Janshoff A: *Piezoelectric Sensors*, vol. 5. Berlin Heidelberg: Springer Science & Business Media; 2007.

192. Wu T-Z, Su C-C, Chen L-K, Yang H-H, Tai D-F, Peng K-C: Piezoelectric immunochip for the detection of dengue fever in viremia phase. *Biosensors and Bioelectronics* 2005, **21**(5):689–695.

193. Zhang P, Tan W: Atomic force microscopy for the characterization of immobilized enzyme molecules on biosensor surfaces. *Fresenius' Journal of Analytical Chemistry* 2001, **369**(3–4):302–307.

194. Ray S, Mehta G, Srivastava S: Label-free detection techniques for protein microarrays: Prospects, merits and challenges. *Proteomics* 2010, **10**(4):731–748.

195. Kneipp K, Kneipp H, Itzkan I, Dasari RR, Feld MS: Ultrasensitive chemical analysis by Raman spectroscopy. *Chemical Reviews* 1999, **99**(10):2957–2976.

196. Le Ru EC, Meyer M, Etchegoin PG: Proof of single-molecule sensitivity in surface enhanced Raman scattering (SERS) by means of a two-analyte technique. *The Journal of Physical Chemistry B* 2006, **110**(4):1944–1948.

197. Huh YS, Chung AJ, Erickson D: Surface enhanced Raman spectroscopy and its application to molecular and cellular analysis. *Microfluidics and Nanofluidics* 2009, **6**(3):285.

198. Ko H, Singamaneni S, Tsukruk VV: Nanostructured surfaces and assemblies as SERS media. *Small* 2008, **4**(10):1576–1599.

199. Taylor HE, Taylor HM: *Inductively Coupled Plasma-Mass Spectrometry: Practices and Techniques*. Amsterdam: Academic Press; 2001.

200. Merkoçi A. *Biosensing Using Nanomaterials*. Hoboken, NJ: John Wiley & Sons; 2009.

Index

Printed and bound by CPI Group (UK) Ltd, Croydon, CR0 4YY

17/10/2024

01775672-0014